INHIBITION OF MATRIX METALLOPROTEINASES: THERAPEUTIC POTENTIAL

ANNALS OF THE NEW YORK ACADEMY OF SCIENCES
Volume 732

INHIBITION OF MATRIX METALLOPROTEINASES: THERAPEUTIC POTENTIAL

Edited by Robert A. Greenwald and Lorne M. Golub

The New York Academy of Sciences
New York, New York
1994

⊗ The paper used in this publication meets the minimum requirements of American National Standard for Information Sciences—Permanence of Paper for Printed Library Materials, ANSI Z39.48-1984.

Cover illustration: Artistic representation of a prototypical MMP, modeled after human fibroblast collagenase, showing a generic inhibitor occupying the catalytic cleft and interacting with the catalytic Zn^{2+}. Sincerest thanks to Brett Lovejoy of Glaxo Research Institute for designing the front cover illustration using Insight II software (Biosym).

Library of Congress Cataloging-in-Publication Data

Inhibition of matrix metalloproteinases: therapeutic potential /
edited by Robert A. Greenwald and Lorne M. Golub.
p. cm. — (Annals of the New York Academy of Sciences, ISSN
0077-8923; v. 732)
Includes bibliographical references and indexes.
ISBN 0-89766-899-5 (cloth: alk. paper. — ISBN 0-89766-900-2
(paper: alk. paper)
1. Metalloproteinases—Inhibitors—Therapeutic use—Congresses.
2. Extracellular enzymes—Congresses. I. Greenwald, Robert A.,
1943- . II. Golub, Lorne M. III. New York Academy of Sciences.
IV. Series.
[DNLM: 1. Metalloproteinases—antagonists & inhibitors—
congresses. 2, Metalloproteinases—metabolism—congresses.
3. Extracellular Matrix—enzymology—congresses. W1 AN626YL v. 732
1994 / QU 136 I55 1994]
Q11.N5 vol. 732
[RM666.M512]
500 s—dc20
[612'01516]
DNLM/DLC
for Library of Congress 94-28727
CIP

SP
Printed in the United States of America
ISBN 0-89766-899-5 (cloth)
ISBN 0-89766-900-2 (paper)
ISSN 0077-8923

ANNALS OF THE NEW YORK ACADEMY OF SCIENCES

Volume 732
September 6, 1994

INHIBITION OF MATRIX METALLOPROTEINASES: THERAPEUTIC POTENTIAL[a]

Editors and Conference Organizers
ROBERT A. GREENWALD AND LORNE M. GOLUB

CONTENTS

[a]This volume contains papers presented at a conference entitled Inhibition of Matrix Metalloproteinases: Therapeutic Potential, which was sponsored by the New York Academy of Sciences and the Long Island Jewish Medical Center and held in Tampa, Florida on January 19–22, 1994.

Part III. Tetracyclines—Dental

Part IV. Tetracyclines—Medical

Part V. MMPs and Inhibitors in Cancer/Metastasis

Part VI. Other Inhibitors

Poster Papers

Financial assistance was received from:

Major Funders

- LONG ISLAND JEWISH MEDICAL CENTER
- NATIONAL INSTITUTES OF HEALTH
  - NATIONAL CANCER INSTITUTE
  - NATIONAL INSTITUTE OF ARTHRITIS AND MUSCULOSKELETAL AND SKIN DISEASES
  - NATIONAL INSTITUTE OF DENTAL RESEARCH

Supporters

- COLLAGENEX, INC.
- PFIZER INC.
- R. J. REYNOLDS TOBACCO COMPANY

Contributors

- ABBOTT LABORATORIES
- BLOCK DRUG COMPANY, INC.
- BOEHRINGER INGELHEIM
- THE COUNCIL FOR TOBACCO RESEARCH—U.S.A., INC.
- GLYCOMED, INC.
- HOFFMANN-LA ROCHE INC.
- MERCK RESEARCH LABORATORIES
- MILES INC.
- OSTEOARTHRITIS SCIENCES, INC.
- PARKE-DAVIS PHARMACEUTICAL RESEARCH
- THE PROCTER & GAMBLE COMPANY
- SMITH KLINE BEECHAM PHARMACEUTICALS
- SYNTEX (U.S.A.) INC.
- THE UPJOHN COMPANY
- WARNER-LAMBERT COMPANY
- WYETH-AYERST RESEARCH

Preface

ROBERT A. GREENWALD

Division of Rheumatology
Long Island Jewish Medical Center
New Hyde Park, New York 11042

Two hundred and six scientists, clinicians, and clinician/scientists from 14 countries registered in Tampa, Florida in January, 1994 for a four-day conference entitled Inhibition of Matrix Metalloproteinases: Therapeutic Potential. Although matrix metalloproteinases (MMPs) have been frequent foci for discussion at many other meetings, and despite the fact that there had been a Gordon Conference six months prior, enthusiasm and interest ran high and was well sustained throughout the entire meeting, which was jointly sponsored by the New York Academy of Sciences and my home institution, the Long Island Jewish Medical Center, the Long Island campus for the Albert Einstein College of Medicine.

The discussants and attendees covered a broad array of disciplines, including rheumatology, periodontal disease, oncology, ocular disease, otolaryngology, obstetrics, surgery, dermatology, biochemistry, pharmacology, enzymology, and physiology. Many facets of the physiologic and/or pathologic role of the MMPs and approaches that might be taken to suppress the pathologically excessive activity which characterizes many disease processes were explored in the 31 podium presentations and 50 posters.

In my opening remarks, I suggested several themes that had played a part in the planning of the meeting. First, it has been apparent for some time that a disparity exists between the *in vitro* and *in vivo* actions of various inhibitors. Some agents show only limited activity by standard enzymatic parameters *in vitro,* especially when tested against low molecular weight substrates. Yet these same agents can be shown, often at very low doses, to have marked suppressive effects on MMP activity when given to animals or humans; this can be proven directly, by assessment of tissue enzyme activity, or indirectly, by measures of MMP-related events such as bone loss or release of matrix degradation products. On the other hand, many synthetic, low molecular weight inhibitors, whose configuration has been carefully refined to produce optimal enzyme inhibitory characteristics *in vitro,* have disappointingly little activity when administered to animals.

Second, the "power" of an inhibitor often varies with the assay system. Test tube systems using natural substrates give different results when compared to low molecular weight surrogate substrates, and neither result may correlate with the observations made in an intact organism. Variability from one laboratory to another is also not uncommon. Third, many MMP inhibitors have been "on the shelf" in the pharmaceutical industry for years, if not decades, but none has been brought to market, and only recently have any even entered Phase III trials. I suggested that perhaps this might be true because model systems for evaluating such agents have not been fully developed, and that a major contribution which the conference might engender would be agreement on standards for such studies. The clinical round table, which was transcribed and summarized as a manuscript in this volume, was designed to address that question specifically.

As the third day wound down, everyone agreed that the blend of basic and clinical science had been "just what the inhibitor-seeker had ordered." For the younger workers, especially those caught up in the complexities of cell biology and macromolecular ultrastructure, the opportunity to appreciate the overall clinical perspective was invaluable. The papers in this volume, including the transcription of the clinical round table, will provide a broad viewpoint on the mechanisms, clinical indications, and plans for implementation of MMP inhibition for human disease.

Acknowledgments: Lorne (Larry) Golub, my close friend and collaborator for 20 years, is Professor of Oral Biology and Medicine at the School of Dental Medicine, State University of New York at Stony Brook. Larry's interest in collagenase started some 28 years ago when he spent three years at the Harvard Dental School in the company of the pioneers in the field, including Paul Goldhaber and Mel Glimcher (Harvard Medical School). Few scientists ever get to make a non-obvious discovery (*viz,* tetracycline inhibition of collagenase) as did Larry; the fact that it may also prove eminently useful to mankind is an additional blessing. As he relates in his paper elsewhere in this volume and in other writings, it was careful science that led to the discovery of this phenomenon, which in turn eventually led to my involvement, which engendered our fruitful collaborations, which begot this conference. For his consistent support, faithful attention to the planning (and especially the funding) of this meeting, and solid friendship, I am much indebted.

Susan Moak in my laboratory and Rama (Dr. N. S. Ramamurthy) and Hsi-Ming Lee in Larry's laboratory have been unflinching lieutenants for both of us for years; without them, most of our scientific work would have gone undone, and we would both still be buried in unmounted posters, unanswered faxes, and unpublished papers. Drs. Seymour Cohen, Philip Lanzkowsky, and David Dantzker at the Long Island Jewish Medical Center were instrumental in garnering the support needed to run the meeting; we are also indebted to the many commercial supporters and to the National Institutes of Health/National Institute of Dental Research for their financial contributions. Sherryl Greenberg from the New York Academy of Sciences was wonderful, as were its other staff, especially Lynn Serra, Geraldine Busacco, Bill Boland, Sheila Kane, and Angela Fink. Stan Zucker and Sandy Simon helped us with the abstract reviews. Finally, to my wife, Elaine, and to Larry's wife, Bonny, I want to express my sincerest thanks for having steadfastly endured the exigencies of life in academia, thereby allowing us to have reached this milestone.

Clinical Importance of Metalloproteinases and Their Inhibitors[a]

STEPHEN M. KRANE

Department of Medicine
Harvard Medical School and the Medical Services
(Arthritis Unit), Massachusetts General Hospital
Boston, Massachusetts 02114

The integrity of connective tissues is determined by the balance of resorption and repair of components of their extracellular matrix (ECM). The activity of proteolytic enzymes is rate-limiting for the degradation and therefore the resorption of the collagen and other macromolecular constituents of the ECM.[1–6] There is considerable evidence that, among potential proteinases, the matrix metalloproteinases (MMPs) have a major role in physiological resorption of collagen and other macromolecules in development and postnatal remodeling and in pathological resorption associated, for example, with local invasiveness of malignant tumors, resorption of the periodontal structures in periodontal disease, and the destruction of joints in rheumatoid arthritis.[7] It is thus of considerable interest that the MMP genes are among the most abundant of those expressed by cells in these inflammatory and malignant lesions.

The matrix metalloproteinases (MMPs) or matrixins are members of a large subfamily of proteinases that contain tightly bound zinc. Other related subfamily members of an even larger superfamily[8] include thermolysin, astacin, and the serratia and snake venom metalloproteinases. There is a catalytic zinc-binding domain in all of these enzymes that includes a sequence motif HEXXH in which the Glu (E) acts as a catalytic base. The matrixin subfamily of MMPs comprises at least twelve members, each of which is the product of a different gene.[3,4,9] Included in the gene subfamily are those that encode at least two interstitial collagenases, three stromelysins, and several gelatinases. The matrixins have several structural features in common that include a propeptide domain that contains the "cysteine switch,"[10,11] the catalytic zinc-binding domain with the sequence HEXGHXXGXXHS, and a hemopexin-like domain.[6,9]

COLLAGENASES

Collagenase (the fibroblast enzyme) was the first of the animal metalloproteinases described in 1962 and during the following two decades received the greatest attention.[1–6] Most of my comments relate to collagenase (MMP-1), although some are applicable to understanding the biology of other metalloproteinases that function in other systems and have activity towards substrates other than interstitial collagens. Despite the considerable amino acid sequence identity and the conservation of amino acid sequence motifs among the MMPs and the capacity of several MMPs to cleave gelatins and type IV collagen, only the "fibroblast" collagenase (MMP-1) and the "neutrophil" collagenase (MMP-8) can cleave native, undena-

[a]This work was supported by National Institutes of Health grants AR-03564 and AR-07258.

tured, "interstitial" collagens (types I, II, III, X) at neutral pH. The structural basis for this remarkable proteolytic specificity is still not determined. The specific sequence requirements for collagenase cleavage have been identified by measuring the rates of hydrolysis of synthetic oligopeptides that cover the P_4 through P_5 subsites of the natural collagenous substrates.[12–14] The subsite preferences identified for the human MMP-1 are similar to those of several noncollagenous substrates, for example, gelatins and α_2-macroglobulins, that are also cleaved by this enzyme. Nevertheless, the specificities revealed in these studies do not account for the failure of either the fibroblast or neutrophil[15,16] collagenases to hydrolyze other potentially cleavable sites in collagens or explain why the native helical collagens are better substrates than the denatured collagens (gelatins). A high percentage of sequence identity exists among the members of the matrixin family. For example, human MMP-1[17,18] has 60% identity to human MMP-8, and human MMP-3 (stromelysin 1) has 56% identity with human MMP-1.[19] The deduced amino acid sequences also do not explain the unique capacity of the collagenases to cleave helical collagens whereas the stromelysins do not have this capacity. Analysis of chimeras of collagenase and stromelysin suggests that the C-terminal domain of collagenase *is* a determinant of substrate specificity, derived from its capacity to specifically bind to the collagens.[20]

The MMPs are synthesized as inactive zymogens. *In vivo,* it is likely that the procollagenases are activated by proteolytic cleavage catalyzed by other proteinases such as plasmin; the activity of stromelysin is probably required for complete activation[21–23] (FIG. 1). Once activated, the neutral collagenases, so far identified, cleave the helical molecules across the three chains of the homotrimeric (types II and III) or heterotrimeric (type I) collagen molecules at specific sites. In the type I collagen molecule, cleavage occurs between helical residues Gly_{775} and Ile_{776} in the $\alpha 1$(I) chain and between the corresponding Gly and Leu in the $\alpha 2$(I) chain to yield a $\sim 3/4^{(A)}$ and $\sim 1/4^{(B)}$ helical, trimeric fragment.[1,6,12,21] Similar sequences are present in the collagenase-cleavage domains in types II and III collagen. As discussed above, the basis for this remarkable specificity is not understood. Our approach has been to use site-directed mutagenesis to alter the sequences around the cleavage site in the

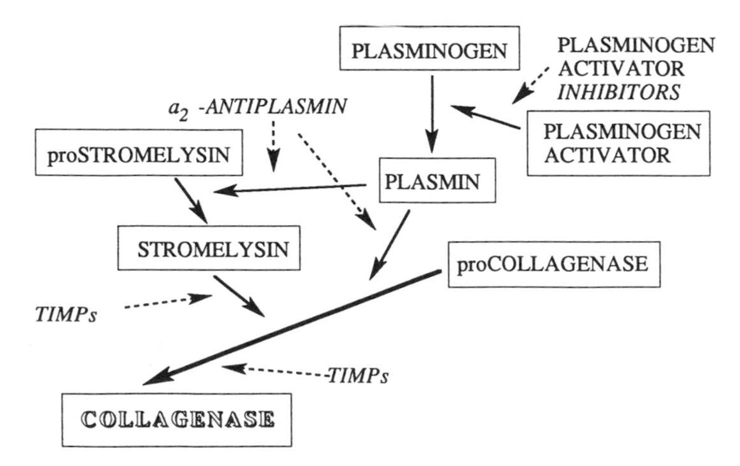

FIGURE 1. Activation/inhibition cascade. A schematic representation of the pathways for activation and inhibition of two metalloproteinases, collagenase and stromelysin (stromelysin 1).

TABLE 1. Evidence for the Role of a Metalloproteinase in Pathological Matrix Remodeling

- Presence of metalloproteinase (MMP) mRNA in lesional cells
- Presence of MMP protein and enzymatic activity in tissue extracts; presence of MMP protein in appropriate locations in the lesion by immunohistochemistry
- Production of MMP by cells cultured from the lesions
- Presence of specific reaction products in lesions

α1(I) chain to probe enzymatic mechanisms and dissect the role of collagenases in connective tissue remodeling.[24–26]

ROLES OF METALLOPROTEINASES *IN VIVO*

Evidence for the role of any particular metalloproteinase in a pathological process is provided by findings such as those listed in TABLE 1. Such evidence suggests that collagenase could have an important role in degradation of the ECM of articular and periarticular structures in a disease such as rheumatoid arthritis.[7,27–30] For example, several groups have succeeded in using *in situ* hybridization with specific riboprobes to demonstrate the presence of mRNA of MMP-1 and MMP-3 in fibroblastic cells in sections of rheumatoid synovial tissues.[31–33] Collagenase has been detected in synovial fluid, and collagen fragments derived from the specific cleavage of collagen by MMP-1 or MMP-8 have been localized in synovial tissues. Abundant active collagenase is secreted by fragments of rheumatoid synovium and in some instances by cells dispersed from these fragments.[27,28,34] The results of studies of cartilage from rheumatoid joints using an antibody that reacts with epitopes present only on "unwound" or fragmented type II collagen are also consistent with action of collagenase *in vivo.*[35,36] These observations suggest that collagenase is a reasonable target for the design of various types of inhibitors that could be active *in vivo* to diminish the abnormal degradation of joint structures, which is a major cause of disability in many patients with this form of arthritis, even though activation of metalloproteinase production is "downstream" from many of the morbid events. Other forms of joint disease such as osteoarthritis could also be approached in this manner inasmuch as cartilage degradation is also a major pathological feature.

CONTROLS OF METALLOPROTEINASE PRODUCTION

Several groups of investigators have been interested in identifying controls of collagenase production by the component cells of inflammatory lesions. Our group has made use of cell culture models comprising dispersed adherent human synovial cells as well as other fibroblasts and chondrocytes[27,34] cocultured with inflammatory cells in order to understand some of the features of the cellular cascade responsible for activation of the target cells that produce the enzyme (FIG. 2). Synovial fibroblasts and possibly chondrocytes produce most of the collagenase in the rheumatoid synovium. Coculture of monocyte/macrophages or their conditioned medium with the mesenchymal cells results in stimulation of production (synthesis and release) of the metalloproteinases. Abundant evidence now exists that interleukin-1 (particularly IL-1β) or tumor necrosis factor-α (TNF-α) are the predominant cytokines responsible for this stimulation, although other factors (components of the ECM,

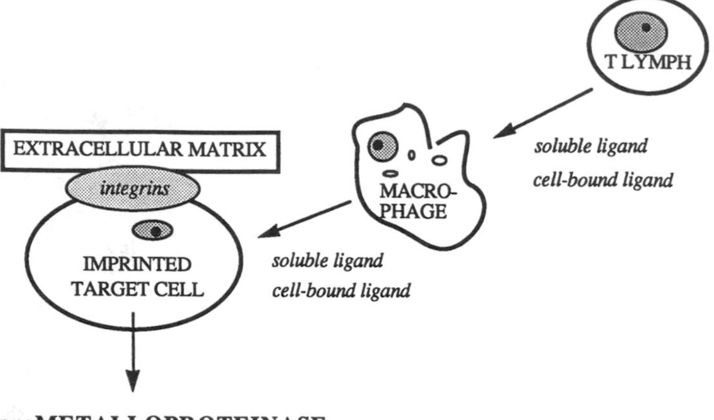

FIGURE 2. Cellular cascade. A schematic representation of some of the cellular interactions in inflammation that lead to production of metalloproteinases.

neuropeptides) also participate in this process.[7] These ligands interact specifically with their cellular receptors and induce a set of signals that result in activation of transcription of the MMP genes through a complex "signal transduction/synthesis cascade" depicted in the schema in FIGURE 3. Induction of these genes by phorbol ester (PMA)[37] has been demonstrated to involve the PMA-responsive protein complex AP-1, a transcriptional transactivator comprising Fos and Jun proteins linked through a leucine zipper dimerization domain that binds to specific sequences (TRE) in the 5′ regulatory regions of the gene.[38–47] There is some evidence that IL-1

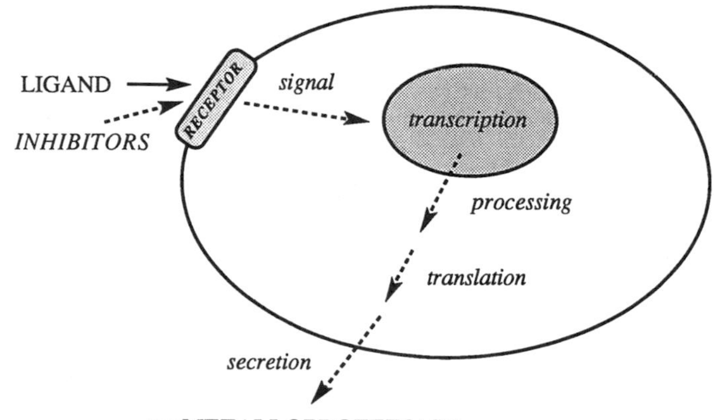

FIGURE 3. Metalloproteinase signal transduction/synthesis cascade. A schematic representation of the events that follow binding of a ligand to its receptor that lead ultimately to increased production of the metalloproteinases. Examples of the ligands that could function in this capacity in the inflammation of rheumatoid arthritis are interleukin-1β and tumor necrosis factor-α.

also induces procollagenase through AP-1 binding to sites in the 5′ regulatory region of its gene and IL-1 induces transient increases in the levels of *fos* and *jun* mRNAs.[48,49] Our group has made use of a mutated IL-1β (IL-1$\beta_{R\rightarrow G}$ mutein) that binds to high-affinity IL-1 receptors but does not stimulate T cell proliferation.[50] In fibroblasts, this IL-1$\beta_{R\rightarrow G}$ mutein elicits early transduction signals such as transcription of *fos* and *jun*, but does not induce transcription of the procollagenase or the prostromelysin genes.[51] Thus, *fos* and *jun* transcription is not sufficient for induction of transcription of the procollagenase gene; other mechanisms must be operative. Sites in the 5′-regulatory region of these metalloproteinase genes other than the TRE, must be important for IL-1 action or, alternatively, components of the AP-1 complex require critical posttranslational modifications for biological activity.[47] The AP-1 components must dimerize before they bind to DNA. Although Jun proteins can bind as homodimers or Jun heterodimers, action of Fos requires that it heterodimerizes with a Jun protein. Furthermore, phosphorylation of Jun can alter its function.[47] Phosphorylation of specific residues in the N-terminal domain increases ability to stimulate transcription whereas phosphorylation in the C-terminal domain is inhibitory. Thus, many events determine the effects of a given ligand. Understanding the processes whereby potent cytokines such as IL-1β increase production of MMPs is critical to understand the pathogenesis of the destructive lesions of a disorder such as rheumatoid arthritis.

TABLE 2. Proof for the Role of a Metalloproteinase in Pathological Matrix Remodeling

- Alter remodeling with a specific drug or antibody to the metalloproteinase (MMP)
- Reproduce the alteration by overexpression of the MMP gene in transgenic animals
- Abolish the (experimental) lesion by deleting the MMP gene
- Identify spontaneous mutations in the MMP gene and characterize the phenotypes
- Identify spontaneous mutations in the target substrate gene or induce mutations in the gene that reproduce the lesion

GENETIC MODELS FOR THE STUDY OF METALLOPROTEINASES

Although considerable evidence suggests a role for collagenase in matrix resorption in inflammatory joint disease, it is difficult to obtain rigorous proof for this role (TABLE 2). Even though drugs may be designed as highly specific inhibitors of collagenase, it does not necessarily follow that these drugs would have no other additional effects that might account for their *in vivo* action. In order to define how collagenases (MMP-1 and MMP-8) work enzymatically with such exquisite specificity for the helical collagen substrates, it would be of great value to identify spontaneous mutations in the enzyme protein that are associated with either loss or gain of function. The existence of mutations that lead to loss of enzymatic activity could establish whether or not collagenase is essential for physiological or pathological connective tissue remodeling. An alternative approach to define the role that collagenase plays in remodeling of the ECM and/or to clarify how the enzyme functions is to study mutations in the substrate (collagen) rather than the enzyme. Although many mutations in *human* COL1A1 and COL1A2 genes have been identified,[52,53] none so far has directly involved the collagenase-cleavage domain. We have attempted to alter sequences around the collagenase-cleavage site in *murine* α1(I) chains by site-directed mutagenesis of the murine Col1a1 gene[24–26] in order to obtain information to interpret human pathological situations. We chose to study the mouse because (1) full-length murine Col1a1 genomic clones are available; (2) the mutated collagen genes can be expressed and their products analyzed in homozygous Mov-13 fibroblasts in which the endogenous Col1a1 gene is inactive; (3) techniques are available for insertion of mutations in murine genes by homologous recombination in embryonic stem (ES) cells; and (4) the ES cells containing the mutated gene can be transplanted into blastocysts of pseudopregnant mice and expressed, and the phenotype can be determined.[54,55] We have begun to obtain interesting information using this murine model.[56]

Only the collagenases (MMP-1 and MMP-8) derived from all of the species tested so far appear to have the capacity to catalyze the hydrolysis of the peptide bond in native collagens that follows the Gly_{775} residue of the sequences (P_1'-P_2') [Ile or Leu]-[Ala or Leu] as follows (subsite designation [P] is used according to the designation of Schechter and Berger[57]):

Residue:	772	773	774	775	776	777	778	779	780	781	782	783
Amino Acid:	Gly	Pro	Gln	Gly	Ile	Ala	Gly	Gln	Arg	Gly	Val	Val
Subsite:	P_4	P_3	P_2	P_1	P_1'	P_2'	P_3'	P_4'	P_5'	P_6'	P_7'	P_8'

We therefore devised the strategy first to engineer mutations in this region of the Col1a1 gene and express these mutated genes in homozygous Mov-13 fibroblasts.[24] Type I collagen molecules were produced which comprise heterotrimers with "rescued" α2(I) chains and, in some clones, α1(I) homotrimers. These endogenously labeled mutated collagens were then tested for susceptibility to collagenase cleavage; analysis of reaction products utilized SDS-PAGE. We found that a nonconservative substitution of Pro for Ile_{776} (P_1') resulted in trimers that were not cleaved by human or rabbit fibroblast collagenase. Nonconservative double substitutions of Pro for Gln_{774} and Ala_{777} at the P_2 and P_2' subsites leaving P_1' intact were also collagenase-resistant. Double substitution of Ala for $Val_{782,783}$ ($P_7'P_8'$) had been designed in order to generate a "cassette" that would facilitate introduction of oligonucleotides containing the various point mutations surrounding the cleavage site into Col1a1. Contrary to our predictions, however, collagenase cleavage of type I collagen which contained these substitutions was slower than that of wild-type (wt) collagens. Examination of sequences in the data base for types I–III collagen from several species shows that one or two hydrophobic amino acids are regularly found in this position; their presence therefore could be important for maximal rate of cleavage. It should also be emphasized that in those molecules that contain a collagenase-resistant α1(I) chain, the wt α2(I) chain is also not cleaved.

We have been successful in expressing one of these mutations that was inserted into the endogenous Col1a1 gene by homologous recombination in the germ line in mice and are currently analyzing the phenotype as the animals age to detect spontaneous changes in collagen content of bone, dermis, and internal organs as well as responses to resorptive stimuli.[56] Eventually it should be possible to obtain definitive information using these mice to establish specific roles for collagenase in bone and soft tissue remodeling in embryonic development and growth, healing wounds or in the pathological remodeling of destructive arthritis. Changes in selected organs would provide further information about specific regulation of remodeling of the extracellular matrix in these organs. A full-length mouse fibroblast collagenase cDNA has recently been cloned[19] using the rat uterine fibroblast cDNA[58] as a probe. The rat and mouse DNAs have 97% sequence identity, but, surprisingly, they differ significantly from the DNAs that encode collagenases from other species (52–55% identical to human, rabbit, pig, and cow).[19] It might be expected that these structural differences would be reflected in functional differences. It should now be possible to examine the substrate specificity of the rodent enzymes using the murine substrates that are prepared by site-directed mutagenesis.

It is possible that there are spontaneous mutations in type I collagen genes that mimic those we produced by site-directed mutagenesis and result in abnormal resorption of collagenous structures. It has been postulated, for example, that osteoblasts normally synthesize a collagenase that degrades the surface layer of hypomineralized collagen in bone to permit osteoclasts to bind to and resorb mineralized bone.[59–62] Administration of tetracycline antibiotics, which have been shown to have collagenase inhibitory activity *in vitro,* can restore bone mass in experimental animal models of osteopenia.[63] The presence of collagenase-resistant collagen might then produce an osteopetrosis phenotype. Several rodent models of osteopetrosis have been described,[64] but none of these has been shown to result from mutations around the collagenase-cleavage site in type I collagen, the predominant collagen of bone matrix. A phenotype of osteopetrosis, dermal or other organ fibrosis might also result from mutations in the MMP-1 gene associated with decreased functional activity of collagenase or, alternatively, with mutations in the type I collagen substrate.

At this conference several presentations will describe different attempts to alter the production and/or action of metalloproteinases. Most of the attempts are directed at inhibiting the activity of metalloproteinases with specific drugs. Nevertheless, it is important to appreciate the potential for other ways to target the metalloproteinases by interruption at several steps in the activation/inhibition, cellular and signal transduction/synthesis cascades depicted in FIGURES 1–3.

REFERENCES

1. GROSS, J. 1974. Collagen biology: Structure, degradation and disease. Harvey Lect. **68:** 351–432.
2. BIRKEDAL-HANSEN, H. 1987. Catabolism and turnover of collagens: Collagenases. Methods Enzymol. **144:** 140–171.
3. WOESSNER, J. F., JR. 1991. Matrix metalloproteinases and their inhibitors in connective tissue remodeling. FASEB J. **5:** 2145–2154.
4. BIRKEDAL-HANSEN, H., Z. WERB, H. G. WELGUS & H. E. VAN WART, Eds. 1992. Matrix Metalloproteinases and Inhibitors. Gustav Fischer Verlag. Stuttgart.
5. MURPHY, G. & A. J. P. DOCHERTY. 1992. The matrix metalloproteinases and their inhibitors. Am. J. Respir. Cell Mol. Biol. **7:** 120–125.
6. BIRKEDAL-HANSEN, H., W. G. I. MOORE, M. K. BODDEN, L. J. WINDSOR, B. BIRKEDAL-HANSEN, A. DECARLO & J. A. ENGLER. 1993. Matrix metalloproteinases: A review. Crit. Rev. Oral Biol. Med. **4:** 197–250.
7. KRANE, S. M. 1993. Mechanisms of tissue destruction in rheumatoid arthritis. *In* Arthritis and Allied Conditions. A Textbook of Rheumatology. 12th edit. D. J. McCarty & W. J. Koopman, Eds.: 763–779. Lea & Febiger. Philadelphia, PA.
8. JIANG, W. & J. S. BOND. 1992. Families of metalloendopeptidases and their relationships. FEBS Lett. **312:** 100–114.
9. DOCHERTY, A. J. P. & G. MURPHY. 1990. The tissue metalloproteinase family and the inhibitor TIMP: A study using DNAs and recombinant proteins. Ann. Rheum. Dis. **49:** 469–479.
10. VAN WART, H. E. & H. BIRKEDAL-HANSEN. 1990. The cysteine switch: A principle of regulation of metalloproteinase activity with potential applicability to the entire matrix metalloproteinase gene family. Proc. Natl. Acad. Sci. USA **87:** 5578–5582.
11. SPRINGMAN, E. B., E. L. ANGLETON, H. BIRKEDAL-HANSEN & H. E. VAN WART. 1990. Multiple modes of activation of latent human fibroblast collagenase: Evidence for the

role of Cys[73] active-site zinc complex in latency and a "cysteine switch" mechanism for activation. Proc. Natl. Acad. Sci. USA **87:** 364–368.
12. FIELDS, G. B., H. E. VAN WART & H. BIRKEDAL-HANSEN. 1987. Sequence specificity of human skin fibroblast collagenase. J. Biol. Chem. **262:** 6221–6226.
13. FIELDS, G. B., S. J. NETZEL-ARNETT, L. J. WINDSOR, J. A. ENGLER, H. BIRKEDAL-HANSEN & H. E. VAN WART. 1990. Proteolytic activities of human fibroblast collagenase: Hydrolysis of a broad range of substrates at a single active site. Biochemistry **29:** 6670–6677.
14. NETZEL-ARNETT, S., G. FIELDS, H. BIRKEDAL-HANSEN & H. E. VAN WART. 1991. Sequence specificities of human fibroblast and neutrophil collagenases. J. Biol. Chem. **266:** 6747–6755.
15. HASTY, K. A., J. J. JEFFREY, M. S. HIBBS & H. G. WELGUS. 1987. The collagen substrate specificity of human neutrophil collagenase. J. Biol. Chem. **262:** 10048–10052.
16. HASTY, K. A., T. F. POURMOTABBED, G. I. GOLDBERG, J. P. THOMPSON, D. G. SPINELLA, R. M. STEVENS & C. L. MAINARDI. 1990. Human neutrophil collagenase. A distinct gene product with homology to other matrix metalloproteinases. J. Biol. Chem. **265:** 11421–11424.
17. GOLDBERG, G. I., S. M. WILHELM, A. KRONBERGER, E. A. BAUER, G. A. GRANT & A. Z. EISEN. 1986. Human fibroblast collagenase. Complete primary structure and homology to an oncogene transformation-induced rat protein. J. Biol. Chem. **261:** 6600–6605.
18. WHITHAM, S. E., G. MURPHY, P. ANGEL, H.-J. RAHMSDORF, B. J. SMITH, A. LYONS, T. J. R. HARRIS, J. J. REYNOLDS, P. HERRLICH & A. J. P. DOCHERTY. 1986. Comparison of human stromelysin and collagenase by cloning and sequence analysis. Biochem. J. **240:** 913–916.
19. HENRIET, P., G. G. ROUSSEAU & Y. EECKHOUT. 1992. Cloning and sequencing of mouse collagenase cDNA. FEBS Lett. **310:** 175–178.
20. MURPHY, G., J. A. ALLAN, F. WILLENBROCK, M. I. COCKETT, J. P. O'CONNELL & A. J. P. DOCHERTY. 1992. The role of the C-terminal domain in collagenase and stromelysin specificity. J. Biol. Chem. **267:** 9612–9618.
21. HARRIS, E. D., JR. & C. A. VATER. 1982. Vertebrate collagenases. Methods Enzymol. **82:** 423–452.
22. GRANT, G. A., A. Z. EISEN, B. L. MARMER, W. T. ROSWIT & G. I. GOLDBERG. 1987. The activation of human skin fibroblast collagenase. Sequence identification of the major conversion products. J. Biol. Chem. **262:** 5886–5889.
23. GOLDBERG, G. I., S. M. FRISCH, C. HE, S. M. WILHELM, R. REICH & I. E. COLLIER. 1990. Secreted proteases. Regulation of their activity and their possible role in metastases. Ann. N.Y. Acad. Sci. **580:** 375–384.
24. WU, H., M. H. BYRNE, A. STACEY, M. B. GOLDRING, J. R. BIRKHEAD, R. JAENISCH & S. M. KRANE. 1990. Generation of collagenase-resistant collagen by site-directed mutagenesis of murine proα1(I) collagen gene. Proc. Natl. Acad. Sci. USA **87:** 5888–5892.
25. KRANE, S. M. & R. JAENISCH. 1992. Site-directed mutagenesis of type I collagen: Effect on susceptibility to collagenase. Matrix **1(Suppl):** 64–67.
26. HASTY, K. A., H. WU, M. BYRNE, M. B. GOLDRING, J. M. SEYER, R. JAENISCH, S. M. KRANE & C. L. MAINARDI. 1993. Susceptibility of type I collagen containing mutated α1(I) chains to cleavage by human neutrophil collagenase. Matrix **13:** 181–186.
27. HARRIS, E. D., JR. & S. M. KRANE. 1974. Collagenases. N. Engl. J. Med. **291:** 557–563, 605–609, 652–661.
28. HARRIS, E. D., JR., H. G. WELGUS & S. M. KRANE. 1984. Regulation of the mammalian collagenases. Collagen Relat. Res. **4:** 493–512.
29. WOOLLEY, D. E. 1984. Mammalian collagenase. *In* Extracellular Matrix Biochemistry. A. Piez & A. H. Reddi, Eds.: 119–157. Elsevier. Amsterdam.
30. KRANE, S. M., W. CONCA, M. L. STEPHENSON, E. P. AMENTO & M. B. GOLDRING. 1990. Mechanisms of matrix degradation in rheumatoid arthritis. Ann. N.Y. Acad. Sci. **580:** 340–354.
31. GRAVALLESE, E. M., J. M. DARLING, A. L. LADD, J. N. KATZ & L. H. GLIMCHER. 1991. In situ hybridization studies of stromelysin and collagenase messenger RNA expression in rheumatoid synovium. Arthritis Rheum. **34:** 1076–1084.

32. McCachren, S. S. 1991. Expression of metalloproteinases and metalloproteinase inhibitors in human arthritic synovium. Arthritis Rheum. **34:** 1085–1093.
33. Firestein, G. S., M. M. Paine & B. H. Littmann. 1991. Gene expression (collagenase, tissue inhibitor of metalloproteinases, complement, and HLA-DR) in rheumatoid arthritis and osteoarthritis synovium. Arthritis Rheum. **34:** 1094–1105.
34. Dayer, J.-M. & S. M. Krane. 1978. The interaction of immunocompetent cells and chronic inflammation as exemplified by rheumatoid arthritis. Clin. Rheum. Dis. **4:** 517–537.
35. Dodge, G. R. & A. R. Poole. 1989. Immunohistochemical detection and immunochemical analysis of type II collagen degradation in human normal, rheumatoid and osteoarthritic articular cartilages and in explants of bovine articular cartilage cultured with interleukin 1. J. Clin. Invest. **83:** 647–661.
36. Dodge, G. R., I. Pidoux & A. R. Poole. 1991. The degradation of type II collagen in rheumatoid arthritis: An immunoelectron microscopic study. Matrix **11:** 330–338.
37. Brinckerhoff, C. E., R. M. McMillan, J. V. Fahey & E. D. Harris, Jr. 1979. Collagenase production by synovial fibroblasts treated with phorbol myristate acetate. Arthritis Rheum. **22:** 1109–1116.
38. Angel, P., A. Poting, U. Mallick, H. J. Rahmsdorf, M. Schorpp & P. Herrlich. 1986. Induction of metallothionein and other mRNA species by carcinogens and tumor promoters in primary human skin fibroblasts. Mol. Cell. Biol. **6:** 1760–1766.
39. Angel, P., I. Baumann, B. Stein, H. Delius, H. J. Rahmsdorf & P. Herrlich. 1987. 12-*O*-tetradecanoyl-phorbol-13-acetate induction of the human collagenase gene is mediated by an inducible enhancer element located in the 5′-flanking region. Mol. Cell. Biol. **7:** 2256–2266.
40. Angel, P., M. Imagawa, R. Chiu, B. Stein, R. J. Imbra, H. J. Ramhsdorf, C. Jonat, P. Herrlich & M. Karin. 1987. Phorbol ester-inducible genes contain a common *cis* element recognized by a TPA-modulated *trans*-acting factor. Cell **49:** 729–739.
41. Lee, W., P. Mitchell & R. Tjian. 1987. Purified transcription factor AP-1 interacts with TPA-inducible enhancer elements. Cell **49:** 741–752.
42. Curran, T. & B. R. Franza, Jr. 1988. Fos and Jun: The AP-1 connection. Cell **55:** 395–397.
43. Cohen, D. R., P. C. P. Ferreira, R. Gentz, B. R. Franza, Jr. & T. Curran. 1989. The product of a *fos*-related gene, *fra*-1, binds cooperatively to the AP-1 site with Jun: transcription factor AP-1 is comprised of multiple protein complexes. Genes Dev. **3:** 173–184.
44. Turner, R. & R. Tjian. 1989. Leucine repeats and an adjacent DNA binding domain mediate the formation of functional cFos-cJun heterodimers. Science **243:** 1689–1694.
45. Gentz, R., F. J. Rauscher III, C. Abate & T. Curran. 1989. Parallel association of fos and jun leucine zippers juxtaposes DNA binding domains. Science **243:** 1695–1699.
46. Vinson, C. R., P. B. Sigler & S. L. McKnight. 1989. Scissors-grip model for DNA recognition by a family of leucine zipper proteins. Science **246:** 911–916.
47. Karin, M. & T. Smeal. 1992. Control of transcription factors by signal transduction pathways: The beginning of the end. TIBS **17:** 418–422.
48. Conca, W., P. B. Kaplan & S. M. Krane. 1989. Increases in levels of procollagenase messenger RNA in cultured fibroblasts induced by human recombinant interleukin 1β or serum follow c-*jun* expression and are dependent on new protein synthesis. J. Clin. Invest. **83:** 1753–1757.
49. Conca, W., P. B. Kaplan & S. M. Krane. 1989. Increases in levels of procollagenase mRNA in human fibroblasts induced by interleukin-1, tumor necrosis factor-α, or serum follow c-*jun* expression and are dependent on new protein synthesis. Trans. Assoc. Am. Physicians **102:** 195–203.
50. Gehrke, L., S. A. Jobling, L. S. K. Paik, B. McDonald, L. J. Rosenwasser & P. E. Auron. 1990. A point mutation uncouples human interleukin-1β biological activity and receptor binding. J. Biol. Chem. **265:** 5922–5925.
51. Conca, W., P. E. Auron, M. Aoun-Wathne, N. Bennett, P. Seckinger, H. G. Welgus, S. R. Goldring, S. P. Eisenberg, J.-M. Dayer, S. M. Krane & L. Gehrke. 1991. An

interleukin 1β point mutant demonstrates that *jun/fos* expression is not sufficient for fibroblast metalloproteinase expression. J. Biol. Chem. **266:** 16265–16267.
52. KUIVANIEMI, H., G. TROMP & D. J. PROCKOP. 1991. Mutations in collagen genes: Causes of rare and some common diseases in humans. FASEB J. **5:** 2052–2060.
53. ROYCE, P. M. & B. STEINMANN, Eds. 1992. Connective Tissue and Its Heritable Disorders: Molecular, Genetic and Medical Aspects. Wiley-Liss. New York.
54. HOGAN, B. & K. LYONS. 1988. Gene targeting. Getting nearer the mark. Nature **336:** 304–305.
55. MANSOUR, S. L., K. R. THOMAS & M. R. CAPECCHI. 1988. Disruption of the proto-oncogene *int-2* in mouse embryo-derived stem cells: A general strategy for targeting mutations to non-selectable genes. Nature **336:** 348–352.
56. WU, H., X. LIU, M. BYRNE, S. KRANE & R. JAENISCH. 1993. Germ-line transmission of a collagenase-resistant Col1a1 gene produced by homologous recombination in embryonic stem cells. J. Bone Miner. Res. **8(Suppl):** S146.
57. SCHECHTER, I. & A. BERGER. 1967. On the size of the active site in proteases. I. Papain. Biochem. Biophys. Res. Commun. **27:** 157–162.
58. QUINN, C. O., D. K. SCOTT, C. E. BRINCKERHOFF, L. M. MATRISIAN, J. J. JEFFREY & N. C. PARTRIDGE. 1990. Rat collagenase. Cloning, amino acid sequence comparison, and parathyroid hormone regulation in osteoblastic cells. J. Biol. Chem. **265:** 22342–22347.
59. CHAMBERS, T. J., J. A. DARBY & K. FULLER. 1985. Mammalian collagenase predisposes bone surfaces to osteoclastic resorption. Cell Tissue Res. **241:** 671–675.
60. PARTRIDGE, N. C., J. J. JEFFREY, L. S. EHLICH, S. L. TEITELBAUM, C. FLISZAR, H. G. WELGUS & A. J. KAHN. 1987. Hormonal regulation of the production of collagenase and a collagenase inhibitor activity by rat osteogenic sarcoma cells. Endocrinology **120:** 1956–1962.
61. SHEN, V., G. KOHLER, J. J. JEFFREY & W. A. PECK. 1988. Bone-resorbing agents promote and interferon-γ inhibits bone cell collagenase production. J. Bone Miner. Res. **3:** 657–666.
62. VAES, G., J.-M. DELAISSE & Y. EECKHOUT. 1992. Relative roles of collagenase and lysosomal cysteine-proteinases in bone resorption. Matrix **1(Suppl):** 383–388.
63. GOLUB, L. M., N. S. RAMAMURTHY, T. F. MCNAMARA, R. A. GREENWALD & B. R. RIFKIN. 1991. Tetracyclines inhibit connective tissue breakdown: New therapeutic implications for an old family of drugs. Crit. Rev. Oral Biol. Med. **2:** 297–322.
64. MARKS, S. C. J. 1987. Osteopetrosis: Multiple pathways for the interruption of osteoclast function. Appl. Pathol. **5:** 172–183.

The Family of Matrix Metalloproteinases[a]

J. FREDERICK WOESSNER, JR.

Departments of Biochemistry & Molecular Biology and Medicine
University of Miami School of Medicine
Miami, Florida 33101

WHAT IS THE MATRIXIN FAMILY AND WHEN DOES AN ENZYME BELONG?

The matrix metalloproteinases (MMPs) are now recognized as members of a broader family that are all related to interstitial collagenase. The family name "matrixin" was suggested and seven members were defined, together with suggested MMP numbers and Enzyme Commission names and numbers at the Destin Beach Matrix Metalloproteinase meeting in 1989.[1] Since then, several additional members have been added for a total of nine, as listed in TABLE 1. Twenty-four species have been detailed at the cDNA level, the gene structure is known for eight species and the chromosomal localization for seven. Earlier work on these enzymes has recently been extensively reviewed by Matrisian[2] and Birkedal-Hansen *et al.*[3] Since my earlier review[4] and comprehensive bibliography[5] appeared, more than 1400 papers have been published in this area, so only a few advances can be described here.

Suggestions have been made to divide the family into three subclasses: collagenases (MMP-1 and MMP-8), gelatinases (MMP-2 and MMP-9), and stromelysins (everything else). This scheme, based largely on substrate specificities, is probably satisfactory for the collagenases, in view of their unique ability to digest interstitial collagens of type I, II, and III. It is less satisfactory for the gelatinases, sometimes referred to as type IV collagenases, because stromelysin digests type IV collagen better than the gelatinases. Further, each of the remaining MMPs (3, 7, 10, 11, 12) has distinctive properties and substrate specificities that are not fully understood, and their action on a large body of matrix proteins has not yet been examined. Stromelysin-3 is evolutionarily far removed from stromelysin-1.[6] Although these subclasses may be useful for mnemonic purposes, they may not be so for codifying enzyme relationships.

Membership in the family requires that at least four criteria be met: the enzyme should display proteolytic activity and should function outside the cell, and its cDNA should code for a protein sequence for the cysteine switch mechanism (PRCGxPD) and for the binding of the catalytic zinc (HExGHxxGxxHS/T). The presence of Ser/Thr after the third His residue distinguishes the matrixins from three other groups of metalloproteinases.[7] By these criteria, stromelysin-3 (MMP-11)[8] and macrophage elastase (MMP-12)[9] are family members. In the absence of cDNA data, one would look for latency and action on matrix components that is blocked by 1,10-phenanthroline and tissue inhibitor of metalloproteinases (TIMP).

The cDNA criteria lead one to consider enzymes from life forms lower than the tadpole. The sea urchin hatching enzyme (envelysin) fits satisfactorily and has about 40% sequence identity to mammalian interstitial collagenase.[10] A proteinase from

[a]This work was supported by National Institutes of Health grants AM-16940 and HD-06773.

TABLE 1. Known Sequences and Gene Structures of Matrixins and TIMPS

Protein	MMP	Human	Rabbit	Rat	Mouse	Bovine	Other
Collagenase	MMP-1	G 11q22	G	■	■	■	■ Pig
Gelatinase A	MMP-2	G 16q21		■	■		
Stromelysin-1	MMP-3	■ 11q22	■	G	■		■ Dog
Matrilysin	MMP-7	G		G			
PMN collagenase	MMP-8	■					
Gelatinase B	MMP-9	G 16q					
Stromelysin-2	MMP-10	■ 11q22		G			
Stromelysin-3	MMP-11	■ 22q11			■		
Macrophage elastase	MMP-12	■			■ 9		
TIMP-1		G Xp11	■		■ Xp21	■	■Pig
TIMP-2		■ 17q25		■	■ 11	■	
TIMP-3		■ 22					■ Chick

Symbols: ■, DNA sequence; G, genomic sequence; 11q, chromosome location.

the leaf of the soybean plant also appears to belong to this family; it has the zinc-binding sequence HEIGHLLGLGHS and is inhibited by TIMP. However, a proenzyme form with cysteine-switch region has not yet been identified. However, because this enzyme has about 40% identity to collagenase and is inhibited by TIMP-1, it should be included.[11] Finally, the gamete lytic enzyme of *Chlamydomonas* contains the zinc-binding domain HExxHxxGxxHA and the cysteine-switch region PRCxVPx; it is secreted in proenzyme form which is then activated by proteolysis.[12] There is little further homology with collagenase beyond these two regions, so assignment to the matrixin family is less certain.

WHERE DID THE MATRIXINS COME FROM?

The presence of collagenase-like features in plant and algal metalloproteinases suggests that the precursor of the matrixins arose before the evolutionary branching of plants and animals. Jiang and Bond[13] suggest that the broad class of metalloproteinases be split into those with an HExxH zinc-binding motif and those with an HxxEH motif (such as the insulin-degrading protease of *Drosophila*, see FIG. 1). The HExxH group can then be divided into those with Glu as the third ligand (thermolysin, neprilysin) and those with His as the third ligand. Here four subgroups are suggested: astacin, serralysin, snake venom, and matrixins. These ideas can now be put on a much firmer basis with crystallographic data for the first three groups in hand (structures obtained for matrixins have not yet been published). On the basis of structural similarity as well as sequence, Bode *et al.*[7] recommend that the four groups be combined into one common family of metzincins. These all share the motif HExxHxxGxxH and a Met-turn that forms the base of the active center. The four groups differ in the residue immediately following the last His: astacins have Glu, serralysins have Pro, adamalysins have Asp, and the matrixins have Ser (Thr). These authors would place the thermolysin group earliest in evolution, with astacins and serralysins/matrixins splitting somewhat later. The serralysins and matrixins then split from one another (FIG. 1). Murphy *et al.*[6] have analyzed the evolutionary relationship among the matrixins (FIG. 1).

None of the phylogenetic trees presented to date take into account the urchin and soy enzymes, nor *Chlamydomonas*. However, the general picture that emerges is that the matrilysin lineage dates back about 3×10^9 years—including the features of extracellular action and proenzyme forms.

DO THE MATRIXINS HAVE A ROLE IN LIVING ANIMALS?

TABLE 2 lists examples of connective tissue remodeling in which matrixins are believed to play an important role; this includes both normal physiological processes and various disease states. The word "believed" is used advisedly because in almost every instance the evidence for such involvement is circumstantial: the enzyme activity is found in the tissue, mRNA is found by Northern blots or *in situ* hybridization, enzyme is seen by immunohistochemistry, etc. However, the presence of the enzyme does not prove it is digesting something or that it is digesting what the investigator suspects. There are very few cases *in vivo* in which specific inhibition of a matrixin or knock-out of its gene has blocked a particular biological remodeling process. A few studies have been done in which hydroxamate inhibitors have been used to block ovulation,[14] metastatic tumor growth,[15] and corneal ulceration.[16] Transfection of metastatic rat cells with TIMP-2 reduces metastasis when the cells are injected into nude mice.[17] On the other hand, antisense to TIMP-1 was introduced into Swiss 3T3 cells of mice and conferred metastatic properties on these cells.[18] These preliminary lines of evidence are, of course, buttressed by extensive studies of the role of the matrixins in cell and tissue explant cultures.

In lieu of discussion of these multiple systems in which matrixins and their inhibitors appear to be of critical importance, some recent reviews will be cited. Jeffrey[19] discusses processes of pregnancy and parturition; cervical dilatation is

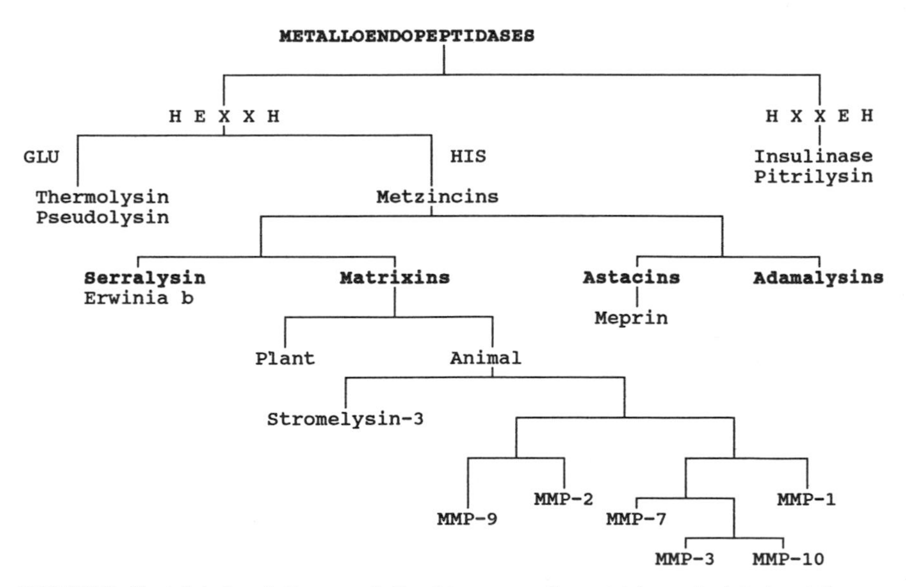

FIGURE 1. Postulated evolutionary relationships among the matrixins and related metalloproteinases (based on refs. 6, 7, and 13). The vertical scale is not proportional to time. The first division is into proteases with HEXXH and HXXEH as two ligands to the zinc atom. The first group then splits again according to the third ligand to zinc—Glu or His.

TABLE 2. Matrixin Involvement in Tissue Resorption/Degradation

Normal Processes	Pathological Processes
Ovulation	Cancer invasion
Endometrial cycling	Tumor metastasis
Blastocyst implantation	Rheumatoid arthritis
Embryogenesis	Osteoarthritic cartilage
Salivary gland morphogenesis	Periodontal disease
Mammary development/Involution	Wound/Fracture healing
Cervical dilatation	Fibrotic lung disease
Fetal membrane rupture	Liver cirrhosis
Uterine involution	Corneal ulceration
Bone growth plate	Gastric ulcer
Bone remodeling	Dilated cardiomyopathy
Angiogenesis	Aortic aneurysm
Tooth eruption	Atherosclerosis
Hair follicle cycle	Otosclerosis
Macrophage function	Epidermolysis bullosa
Neutrophil function	

briefly outlined by Rechberger and Woessner.[20] Other normal processes include bone resorption,[21] blastocyst implantation,[22] and mammary involution.[23] Recent reviews of disease processes include inflammation,[24] osteoarthritis,[25] periodontal disease,[26] wound healing,[27] and invasion and metastasis.[28] In summary, the metalloproteinases are indispensable for the breakdown of the extracellular matrix, but their regulation by TIMPs and other inhibitors is critically important. Any imbalance can lead to fibrotic processes (inhibitor > enzyme) or to excessive tissue destruction or invasion (enzyme > inhibitor).

WHAT ROLE DO THE VARIOUS DOMAINS PLAY IN MATRIXIN FUNCTION?

The domain structure of a typical matrixin is illustrated in FIGURE 2. The secreted zymogen loses it signal peptide. Upon activation, the propeptide domain is cleaved off in several steps. The catalytic domain includes the zinc-binding domain, which is considered separately because it arises from another exon and because 175-residue inserts are found in gelatinase A and B immediately ahead of this region. These inserts are triple repeats of the gelatin-binding domain of fibronectin and are believed to assist in binding of gelatinases to their substrates. Between the zinc-binding domain and the hemopexin-like domain is a variable hinge region, rich in proline, that may permit the C-terminal domain to fold back on the catalytic domain. Gelatinase B has a further insert in this hinge region of a type V collagen domain of 53 residues which is of unknown function. The enzyme illustrated has M_r 57,000, as do the stromelysins and macrophage elastase. Additional domains increase this to M_r 72,000 and 92,000 for gelatinase A and B; matrilysin lacks the hemopexin domain and has M_r 28,000.

Propeptide Domain

The function of the propeptide is to maintain latency of the matrixins until a signal for activation is given. Mounting evidence shows that the propeptide functions by virtue of the cysteine switch mechanism, involving the coordination of Cys^{73} in the conserved sequence PRCGVPDG (FIG. 2) to the active center zinc atom. The switch mechanisms and the steps in activation by mercurials and proteases have been extensively reviewed.[3,4] Confirmation is given by mutagenesis studies showing the Arg and Cys are critical for maintenance of latency,[29] by extended X-ray absorption fine structure (EXAFS) data showing that the Cys contacts zinc,[30] and by the finding that a synthetic dodecapeptide spanning the conserved Cys is highly inhibitory to the enzyme.[31] Cysteine by itself does not explain latency, other regions of the propeptide binding to the enzyme help to hold the Cys in place. Thus, Chen *et al.*[32] have shown that Cys can be blocked with iodoacetamide and the enzyme still retain its latency. Surprisingly, activation can then be achieved with aminophenylmercuric acetate, which is obviously acting on residues other than cysteine to destabilize the propeptide structure or interaction with the enzyme.

Activation of the gelatinases presents some special problems. Gelatinase A is commonly found with one mole of TIMP-2 bound to it, and gelatinase B with TIMP-1. This binding is in the hemopexin domain, but it is close enough to the active center to interfere with the autoactivation step and with the activated form of the enzymes. Typically, gelatinase A displays about 10% of its maximum activity if TIMP-2 is present during the activation step with mercurials; if stromelysin-1 is also present the activity obtained is eightfold higher.[33] (Is this due to proteolysis by stromelysin or to transfer of TIMP-2 to the active center of stromelysin?) Activation of gelatinase A seems to be accomplished *in vivo* by a membrane-bound enzyme.[34] This enzyme is metal-dependent and its action is followed by autolytic cleavage to the final active form.[35] Monsky *et al.*[36] have shown that gelatinase A binds to "invadopodia" of invasive tumor cells, leading directly to activation and to concentration of enzyme at the leading edge of invasion directly on the substrate. It is possible that binding of gelatinase A to cell receptors may alter its conformation, permitting autolytic activation.

Catalytic Domain

It has long been recognized that the sequence HExGH must be the zinc-binding site of matrixins. However, it is only with the elucidation of the closely related astacin

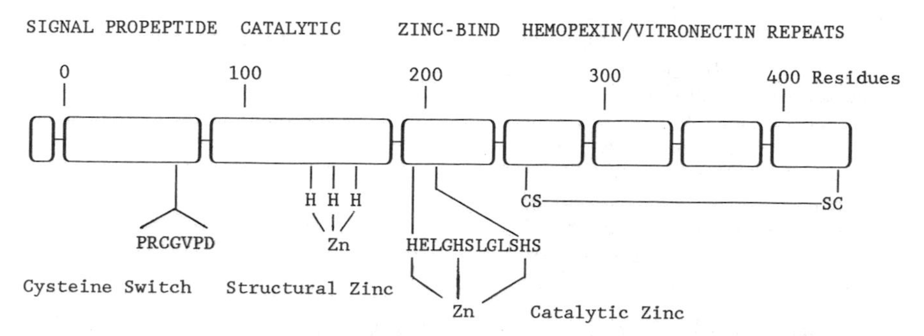

FIGURE 2. The domain structure of human fibroblast collagenase. A scale of residue numbers is indicated. The following positions are noted: Cys-72 for the cysteine switch; His-149, -164, and -179 for the structural zinc; His-199, -203, and -210 for the catalytic zinc; and Cys-259 and -466 for the disulfide bridge. The zinc ligands are placed by analogy to stromelysin-1.

and adamalysin X-ray structures that it has become accepted that the third zinc ligand is also His in the sequence HExGHxxGxxHS.[7] X-ray structures, presented at meetings but not yet published, indicate that the matrixins will be quite similar to the astacins in their structure. A further surprise has been the discovery that the matrixins all contain a second, structural zinc atom. Three His ligands have been identified by NMR in human stromelysin-1: residues 151, 166, and 179.[37] These correspond to His residues 149, 164, and 177 in interstitial collagenase (FIG. 2). Similarly, the catalytic zinc would be bound to His residues 199, 203, and 210.

Extensive specificity studies of five matrixins on 60 synthetic peptides have been tabulated by Netzel-Arnett *et al.*[38] However, we still possess a very incomplete picture of the natural substrates of the matrixins and the specificity of cleavage of such substrates. The matrixins attack peptides and natural substrates optimally at pH 7–8. However, in the case of human stromelysin-1, the pH optimum on such substrates falls in the range pH 5.5–6.5.[39,40] This unusual optimum might be due to the unique His^{225} found only in this enzyme and brought into the active site by the Met fold. A more detailed discussion of enzyme structure and function can be found in the next article by Van Wart and colleagues.

Hemopexin / Vitronectin Domain

There are four repeats in this region that have weak homology to hemopexin and vitronectin.[41] The amino end contains a hinge region and a disulfide bridge connects the extreme ends of this domain. The function of the domain is not completely clear, and, indeed, matrilysin and macrophage elastase function perfectly well without this domain. Interstitial collagenase (MMP-1) has good catalytic activity without this domain (enzymatically or mutagenically removed), but it loses its ability to digest the triple helix of collagen. This is also true for neutrophil collagenase (MMP-8).[42] Various chimeras have been produced by manipulation of the cDNA whereby the C-terminal domains of various matrixins are interchanged. The C-domain of interstitial collagenase can cause the catalytic domain of stromelysin-1 to bind to collagen, but does not confer the ability to digest collagen.[43] Collagenase with the stromelysin C-domain also fails to digest collagen. However, collagenase activity depends not only on the C-domain but also on specific sequences within the zinc-binding domain.[44]

The C-domain also modulates the binding of TIMPs to the active centers of the matrixins.[43] Further, it was mentioned above that TIMP-2 binds to the C-domain of progelatinase A and TIMP-1 to progelatinase B. From this position the TIMP is able to interfere with the active center of the gelatinases when they are activated. It is suggested that TIMP-2 can shift its binding, occupying the site in the C-domain of latent enzyme and then rearranging to maintain this link while also binding to the active center.[45] This is readily conceivable because there are two distinct domains in TIMP; the C-domain binds the C-domain of gelatinase[46] so that the N-domain is free to interact with other matrixins or with the active center of the same molecule when it becomes active. Folding at the hinge region may result in the N-domain of the TIMPs being in close proximity to the active center of the gelatinases.

HOW ARE ACTIVE METALLOPROTEINASES CONTROLLED?

The matrixins are regulated at many points. Regulation at the gene level has been reviewed by several of the speakers in this volume and elsewhere.[47,48] Regulation at the level of secretion has recently been shown to be important in epithelial cells.[49] Regulation at the level of zymogen activation has been touched upon briefly above. Once the zymogen is activated, the major point of control lies in the proteinase inhibitors of the extracellular matrix. These are chiefly α_2-macroglobulin and members of the TIMP family. TABLE 1 lists 11 species of TIMP falling into three groups. All three types of TIMP contain two domains: the N-domain which reacts with the active centers of most of the matrixins and the C-domain, involved in binding to other components such as the C-domain of the gelatinases (see above).[46] Truncated TIMP-1 and TIMP-2 both function perfectly well as inhibitors.[50] It should be kept in mind that the TIMPs are also active factors in growth-promotion;[51,52] perhaps this function lies in the C-domain.

The mechanism by which TIMP binds so strongly to the active center of matrixins is poorly understood, particularly in the absence of three-dimensional structures. Mutagenesis shows that residues 3–13 are important in binding,[53] but this may not be the whole story. As far as is known the various types of TIMP can inhibit all of the matrixins; however, there are clear differences in the binding affinity of each type of TIMP for the various enzymes.[54] It is only recently that the third type of TIMP has been isolated from chickens (ChIMP-3).[55] This has unusual solubility properties and appears to be firmly anchored in the matrix. TIMP-3 has the additional properties of stimulating cell growth and promoting the detachment of transformed cells from the matrix.[56] This latter function is not understood, but perhaps relates to interaction with an enzyme such as gelatinase A held at the cell surface by receptors (see above).

WHAT ARE THE INTERRELATIONSHIPS BETWEEN EXTRACELLULAR MATRIX AND MATRIXINS?

The effect of TIMP-3 on cell attachment is but one aspect of a broader problem. It is increasingly recognized that the extracellular matrix (ECM) is extremely important in governing the behavior of cells and that changes in the nature of the cell substratum can produce radical changes in the behavior of cells grown thereon.[57] Signal transduction from the ECM is beginning to be understood, particularly as mediated by integrins.[58] However, thus far little information exists on how the matrix governs the production of matrixins. The cells must have ways of determining the amounts of enzymes and inhibitors present in their environment. Furthermore, little attention has been paid to the fact that almost all of the matrixins are firmly anchored in the ECM and can be extracted only with great difficulty. We noted this for the collagenase of the rat uterus as early as 1976,[59] and speculated that the enzyme might be anchored to collagen. However, latent interstitial collagenase has little affinity for collagen, so other anchors must be sought.

A number of examples have been found recently of matrix components stimulating cells to secrete matrixins. Thus, benign tumor cells are made invasive and produce gelatinase A upon stimulation by a peptide from the laminin A chain.[60] There is evidence that melanoma cell synthesis of gelatinase A is stimulated by contact between cell vitronectin receptor (integrin) and vitronectin, but not fibronectin.[61] On the other hand, antibodies against the keratinocyte integrin for fibronectin stimulate the production of gelatinase B, and this response may be related to the signaling which tells the cells to separate from their ECM.[62] This keratinocyte enzyme also responds to type I collagen matrix but is decreased by type IV collagen.[63] Mechanical interactions between cell and matrix are also important; it has been known for many years that cell shape is an important determinant of matrixin expression. This idea is extended by Lambert *et al.*,[64] who show that mechanical forces applied to fibroblast cultures regulate expression of interstitial collagenase. This promises to be an active and important area for future research.

REFERENCES

1. NAGASE, H., A. J. BARRETT & J. F. WOESSNER, JR. 1992. Nomenclature and glossary of the matrix metalloproteinases. Matrix (**Suppl. 1**): 421–424.
2. MATRISIAN, L. M. 1992. The matrix-degrading metalloproteinases. BioEssays **14:** 455–463.
3. BIRKEDAL-HANSEN, H., W. G. I. MOORE, M. K. BODDEN, L. J. WINDSOR, B. BIRKEDAL-HANSEN, A. DECARLO & J. A. ENGLER. 1993. Matrix metalloproteinases: A review. Crit. Rev. Oral Biol. Med. **4:** 197–250.
4. WOESSNER, J. F., JR. 1991. Matrix metalloproteinases and their inhibitors in connective tissue remodeling. FASEB J. **5:** 2145–2154.
5. WOESSNER, J. F., JR. 1992. Literature on vertebrate matrix metalloproteinases and their tissue inhibitors. Matrix (**Suppl. 1**): 425–501.
6. MURPHY, G. J. P., G. MURPHY & J. J. REYNOLDS. 1991. The origin of matrix metalloproteinases and their familial relationships. FEBS Lett. **289:** 4–7.
7. BODE, W., F-X. GOMIS-RÜTH & W. STÖCKER. 1993. Astacins, serralysins, snake venom and matrix metalloproteinases exhibit identical zinc-binding environments (HEXX-HXXGXXH and Met-turn) and topologies and should be grouped into a common family, the "metzincins." FEBS Lett. **331:** 134–140.
8. MURPHY, G., J-P. SEGAIN, M. O'SHEA, M. COCKETT, C. IOANNOU, O. LEFEBVRE, P. CHAMBON & P. BASSET. 1993. The 28-kDa N-terminal domain of mouse stromelysin-3 has the general properties of a weak metalloproteinase. J. Biol. Chem. **268:** 15435–15441.
9. SHAPIRO, S. D., D. K. KOBAYASHI & T. J. LEY. 1993. Cloning and characterization of a unique elastolytic metalloproteinase produced by human alveolar macrophages. J. Biol. Chem. **268:** 23824–23829.
10. LEPAGE, T. & C. GACHE. 1990. Early expression of a collagenase-like hatching enzyme gene in the sea urchin embryo. EMBO J. **9:** 3003–3012.
11. MCGEEHAN, G., W. BURKHART, R. ANDEREGG, J. D. BECHERER, J. W. GILLIKIN & J. S. GRAHAM. 1992. Sequencing and characterization of the soybean leaf metalloproteinase. Structural and functional similarity to the matrix metalloproteinase family. Plant Physiol. **99:** 1179–1183.
12. KINOSHITA, T., H. FUKUZAWA, T. SHIMADA, T. SAITO & Y. MATSUDA. 1992. Primary structure and expression of a gamete lytic enzyme in *Chlamydomonas reinhardtii:* Similarity of functional domains to matrix metalloproteases. Proc. Natl. Acad. Sci. USA **89:** 4693–4697.
13. WEIPING, J. & J. S. BOND. 1992. Families of metalloendopeptidases and their relationships. FEBS LETT. **312:** 110–114.
14. BUTLER, T. A., C. ZHU, R. A. MUELLER, G. C. FULLER, W. J. LEMAIRE & J. F. WOESSNER, JR. 1991. Inhibition of ovulation in the perfused rat ovary by the synthetic collagenase inhibitor SC-44463. Biol. Reprod. **44:** 1183–1188.
15. AXELROD, J. H., R. REICH & R. MISKIN. 1989. Expression of human recombinant plasminogen activators enhances invasion and experimental metastasis of H-*ras*-transformed NIH 3T3 cells. Mol. Cell. Biol. **9:** 2133–2141.
16. SCHULTZ, G. S., S. STRELOW, G. A. STERN, N. CHEGINI, M. B. GRANT, R. E. GALARDY, D. GROBELNY, J. J. ROWSEY, K. STONECIPHER, V. PARMLEY & P. T. KHAW. 1992. Treatment of alkali-injured rabbit corneas with a synthetic inhibitor of matrix metalloproteinases. Invest. Ophthalmol. Visual Sci. **33:** 3325–3331.
17. DECLERCK, Y. A., N. PEREZ, H. SHIMADA, T. C. BOONE, K. E. LANGLEY & S. M. TAYLOR. 1992. Inhibition of invasion and metastasis in cells transfected with an inhibitor of metalloproteinases. Cancer Res. **52:** 701–708.
18. KHOKHA, R., P. WATERHOUSE, S. YAGEL, P. K. LALA, C. M. OVERALL, G. NORTON & D. T. DENHARDT. 1989. Antisense RNA-induced reduction in murine TIMP levels confers oncogenicity on Swiss 3T3 cells. Science **243:** 947–950.

19. JEFFREY, J. J. 1991. Collagen and collagenase: Pregnancy and parturition. Semin. Perinatol. **15:** 118–126.
20. RECHBERGER, T. & J. F. WOESSNER, JR. 1993. Collagenase, its inhibitors, and decorin in the lower uterine segment in pregnant women. Am. J. Obstet. Gynecol. **168:** 1598–1603.
21. DELAISSÉ, J-M. & G. VAES. 1992. Mechanism of mineral solubilization and matrix degradation in osteoclastic bone resorption. *In* Biology and Physiology of the Osteoclast. B. R. Rifkin & C. V. Gay, Eds.: 290–314. CRC Press. Boca Raton, FL.
22. BEHRENDTSEN, O., C. M. ALEXANDER & Z. WERB. 1992. Metalloproteinases mediate extracellular matrix degradation by cells from mouse blastocyst outgrowths. Development **114:** 447–456.
23. TALHOUK, R. S., M. J. BISSELL & Z. WERB. 1992. Coordinated expression of extracellular matrix-degrading proteinases and their inhibitors regulates mammary epithelial function during involution. J. Cell Biol. **118:** 1271–1282.
24. WOESSNER, J. F., JR. 1992. Role of cellular proteinases and their protein inhibitors in inflammation. *In* Immunology and Medicine, Biochemistry of Inflammation. J. T. Whicher & S. W. Evans, Eds.: 57–89. Kluwer Academic Publishers. Dordrecht.
25. NAGASE, H. & J. F. WOESSNER, JR. 1993. Role of endogenous proteinases in the degradation of cartilage matrix. *In* Joint Cartilage Degradation: Basic and Clinical Aspects. J. F. Woessner, Jr. & D. S. Howell, Eds.: 159–185. Marcel Dekker, Inc. New York.
26. BIRKEDAL-HANSEN, H. 1993. Role of matrix metalloproteinases in human periodontal diseases. J. Periodontol. **64:** 474–484.
27. JEFFREY, J. J. 1992. Collagen degradation. *In* Wound Healing: Biochemical and Clinical Aspects. R. F. Diegelmann & W. J. Lindblad, Eds.: 177–194. W. B. Saunders Company. Philadelphia, PA.
28. STETLER-STEVENSON, W. G., L. A. LIOTTA & D. E. KLEINER, JR. 1993. Extracellular matrix 6: Role of matrix metalloproteinases in tumor invasion and metastasis. FASEB J. **7:** 1434–1441.
29. PARK, A. J., L. M. MATRISIAN, A. F. KELLS, R. PEARSON, Z. G. YUAN & M. NAVRE. 1991. Mutational analysis of the transin (rat stromelysin) autoinhibitor region demonstrates a role for residues surrounding the cysteine switch. J. Biol. Chem. **266:** 1584–1590.
30. HOLZ, R. C., S. P. SALOWE, C. K. SMITH, G. C. CUCA & L. QUE, JR. 1992. EXAFS evidence for a "cysteine switch" in the activation of prostromelysin. J. Am. Chem. Soc. **114:** 9611–9614.
31. STETLER-STEVENSON, W. G., J. A. TALENO, M. E. GALLAGHER, H. C. KRUTZSCH & L. A. LIOTTA. 1991. Inhibition of human type IV collagenase by a highly conserved peptide sequence derived from its prosegment. Am. J. Med. Sci. **302:** 163–170.
32. CHEN, L. C., M. E. NOELKEN & H. NAGASE. 1993. Disruption of the cysteine-75 and zinc ion coordination is not sufficient to activate the precursor of human matrix metalloproteinase 3 (stromelysin 1). Biochemistry **32:** 10289–10295.
33. MIYAZAKI, K., F. UMENISHI, K. FUNAHASHI, N. KOSHIKAWA, H. YASUMITSU & N. UMEDA. 1992. Activation of TIMP-2/progelatinase A complex by stromelysin. Biochem. Biophys. Res. Commun. **185:** 852–859.
34. WARD, R. V., S. J. ATKINSON, P. M. SLOCOMBE, A. J. P. DOCHERTY, J. J. REYNOLDS & G. MURPHY. 1991. Tissue inhibitor of metalloproteinases-2 inhibits the activation of 72 kDa progelatinase by fibroblast membranes. Biochim. Biophys. Acta **1079:** 242–246.
35. BROWN, P. D., D. E. KLEINER, E. J. UNSWORTH & W. G. STETLER-STEVENSON. 1993. Cellular activation of the 72 kDa type IV procollagenase/TIMP-2 complex. Kidney Int. **43:** 163–170.
36. MONSKY, W. L., T. KELLY, C-Y. LIN, Y. YEH, W. G. STETLER-STEVENSON, S. C. MUELLER & W-T. CHEN. 1993. Binding and localization of M_r 72,000 matrix metalloproteinases at cell surface invadopodia. Cancer Res. **53:** 3159–3164.
37. GOOLEY, P. R., B. A. JOHNSON, A. I. MARCY, G. C. CUCA, S. P. SALOWE, W. K. HAGMANN, C. K. ESSER & J. P. SPRINGER. 1993. Secondary structure and zinc ligation of human recombinant short-form stromelysin by multidimensional heteronuclear NMR. Biochemistry **32:** 13098–13108.
38. NETZEL-ARNETT, S., Q. X. SANG, W. G. I. MOORE, M. NAVRE, H. BIRKEDAL-HANSEN &

H. E. VAN WART. 1993. Comparative sequence specificities of human 72-kDa and 92-kDa gelatinases (type-IV collagenases) and PUMP (matrilysin). Biochemistry **32:** 6427–6432.

39. WILHELM, S. M., Z. H. SHAO, T. J. HOUSLEY, P. K. SEPERACK, A. P. BAUMANN, Z. GUNJA-SMITH & J. F. WOESSNER, JR. 1993. Matrix metalloproteinase-3 (stromelysin-1): Identification as the cartilage acid metalloprotease and effect of pH on catalytic properties and calcium affinity. J. Biol. Chem. **268:** 21906–21913.
40. HARRISON, R. K., B. CHANG, L. NIEDZWIECKI & R. L. STEIN. 1992. Mechanistic studies on the human matrix metalloproteinase stromelysin. Biochemistry **31:** 10757–10762.
41. JENNE, D. & K. K. STANLEY. 1987. Nucleotide sequence and organization of the human S-protein gene: Repeating peptide motifs in the "pexin" family and a model for their evolution. Biochemistry **26:** 6735–6742.
42. KNÄUPER, V., A. OSTHUES, Y. A. DECLERCK, K. E. LANGLEY, J. BLÄSER & H. TSCHESCHE. 1993. Fragmentation of human polymorphonuclear-leucocyte collagenase. Biochem. J. **291:** 847–854.
43. MURPHY, G., J. A. ALLAN, F. WILLENBROCK, M. E. COCKETT, J. P. O'CONNELL & A. J. P. DOCHERTY. 1992. The role of the C-terminal domain in collagenase and stromelysin specificity. J. Biol. Chem. **267:** 9612–9618.
44. SANCHEZ-LOPEZ, R., C. M. ALEXANDER, O. BEHRENDTSEN, R. BREATHNACH & Z. WERB. 1993. Role of zinc-binding– and hemopexin domain–encoded sequences in the substrate specificity of collagenase and stromelysin-2 as revealed by chimeric proteins. J. Biol. Chem. **268:** 7238–7247.
45. KLEINER, D. E., JR., E. J. UNSWORTH, H. C. KRUTZSCH & W. G. STETLER-STEVENSON. 1992. Higher-order complex formation between the 72-kilodalton type IV collagenase and tissue inhibitor of metalloproteinases-2. Biochemistry **31:** 1665–1672.
46. WILLENBROCK, F., T. CRABBE, P. M. SLOCOMBE, C. W. SUTTON, A. J. P. DOCHERTY, M. I. COCKETT, M. O'SHEA, K. BROCKLEHURST, I. R. PHILLIPS & G. MURPHY. 1993. The activity of the tissue inhibitors of metalloproteinases is regulated by C-terminal domain interactions: A kinetic analysis of the inhibition of gelatinase A. Biochemistry **32:** 4330–4337.
47. BRINCKERHOFF, C. E. 1992. Regulation of metalloproteinase gene expression: Implications for osteoarthritis. Crit. Rev. Eukaryotic Gene Expression **2:** 145–164.
48. MATRISIAN, L. M., G. L. GANSER, L. D. KERR, R. W. PELTON & L. D. WOOD. 1992. Negative regulation of gene expression by TGF-β. Mol. Reprod. Dev. **32:** 111–120.
49. WHITELOCK, J. M., M. L. PAINE, J. R. GIBBINS, R. F. KEFFORD & R. L. O'GRADY. 1993. Multiple levels of post-transcriptional regulation of collagenase (matrix metalloproteinase-1) in an epithelial cell line. Immunol. Cell Biol. **71:** 39–47.
50. DECLERCK, Y. A., T-D. YEAN, Y. LEE, J. M. TOMICH & K. E. LANGLEY. 1993. Characterization of the functional domain of tissue inhibitor of metalloproteinases-2 (TIMP-2). Biochem. J. **289:** 65–69.
51. HAYAKAWA, T., K. YAMASHITA, K. TANZAWA, E. UCHIJIMA & K. IWATA. 1992. Growth-promoting activity of tissue inhibitor of metalloproteinases-1 (TIMP-1) for a wide range of cells: A possible new growth factor in serum. FEBS Lett. **298:** 29–32.
52. NEMETH, J. A. & C. L. GOOLSBY. 1993. TIMP-2, a growth-stimulatory protein from SV40-transformed human fibroblasts. Exp. Cell Res. **207:** 376–382.
53. O'SHEA, M., F. WILLENBROCK, R. A. WILLIAMSON, M. I. COCKETT, R. B. FREEDMAN, J. J. REYNOLDS, A. J. P. DOCHERTY & G. MURPHY. 1992. Site-directed mutations that alter the inhibitory activity of the tissue inhibitor of metalloproteinases-1: Importance of the N-terminal region between cysteine 3 and cysteine 13. Biochemistry **31:** 10146–10152.
54. WARD, R. V., R. M. HEMBRY, J. J. REYNOLDS & G. MURPHY. 1991. The purification of tissue inhibitor of metalloproteinases-2 form its 72 kDa progelatinase complex: Demonstration of the biochemical similarities of tissue inhibitor of metalloproteinases-2 and of tissue inhibitor of metalloproteinases-1. Biochem. J. **278:** 179–187.
55. PAVLOFF, N., P. W. STASKUS, N. S. KISHNANI & S. P. HAWKES. 1992. A new inhibitor of metalloproteinases from chicken: ChIMP-3. J. Biol. Chem. **267:** 17321–17326.
56. YANG, T-T. & S. P. HAWKES. 1992. Role of the 21-kDa protein TIMP-3 in oncogenic

transformation of cultured chicken embryo fibroblasts. Proc. Natl. Acad. Sci. USA **89:** 10676–10680.
57. LIN, C. Q. & M. J. BISSELL. 1993. Multi-faceted regulation of cell differentiation by extracellular matrix. FASEB J. **7:** 737–743.
58. JULIANO, R. L. & S. HASKILL. 1993. Signal transduction from the extracellular matrix. J. Cell Biol. **120:** 577–585.
59. WEEKS, J. G., J. HALME & J. F. WOESSNER, JR. 1976. Extraction of collagenase from the involuting rat uterus. Biochem. Biophys. Acta **445:** 205–214.
60. ROYCE, L. S., G. R. MARTIN & H. K. KLEINMAN. 1992. Induction of an invasive phenotype in benign tumor cells with a laminin A-chain synthetic peptide. Invasion Metastasis **12:** 149–155.
61. SEFTOR, R. E. B., E. A. SEFTOR, K. R. GEHLSEN, W. G. STETLER-STEVENSON, P. D. BROWN, E. RUOSLAHTI & M. J. C. HENDRIX. 1992. Role of the alpha$_v$beta$_3$ integrin in human melanoma cell invasion. Proc. Natl. Acad. Sci. USA **89:** 1557–1561.
62. LARJAVA, H., J. G. LYONS, T. SALO, M. MAKELA, L. KOIVISTO, H. BIRKEDAL-HANSEN, S. K. AKIYAMA, K. M. YAMADA & J. HEINO. 1993. Anti-integrin antibodies induce type-IV collagenase expression in keratinocytes. J. Cell. Physiol. **157:** 190–200.
63. SARRETT, Y., D. T. WOODLEY, G. S. GOLDBERG, A. KRONBERGER & K. C. WYNN. 1992. Constitutive synthesis of a 92-kDa keratinocyte-derived type IV collagenase is enhanced by type I collagen and decreased by type IV collagen matrices. J. Invest. Dermatol. **99:** 836–841.
64. LAMBERT, C. A., E. P. SOUDANT, B. V. NUSGENS & C. M. LAPIÈRE. 1992. Pretranslational regulation of extracellular matrix macromolecules and collagenase expression in fibroblasts by mechanical forces. Lab. Invest. **66:** 444–451.

Evidence for a Triple Helix Recognition Site in the Hemopexin-like Domains of Human Fibroblast and Neutrophil Interstitial Collagenases[a]

SARAH NETZEL-ARNETT, AMID SALARI, UMESH B. GOLI,[b] AND HAROLD E. VAN WART[c]

Institute of Biochemistry and Cell Biology
Syntex Discovery Research, S3-1
Palo Alto, California 94303

[b]*Department of Chemistry*
Florida State University
Tallahassee, Florida 32306

The metabolic turnover of the major protein constituents of the extracellular matrix is catalyzed by a family of enzymes called the matrix metalloproteinases (MMP).[1] Two members of this gene family, human fibroblast collagenase (HFC) and human neutrophil collagenase (HNC), are capable of hydrolyzing the fibrillar type I, II, and III collagens at an appreciable rate under physiological conditions. They do so in a highly specific and characteristic manner by making a single scission across all three alpha (α) chains of the triple helical tropocollagen (TC) monomers at a specific, sensitive locus approximately three-quarters from the N-terminus to produce TC^A and TC^B fragments.[2,3] This reaction is the committed step in collagen turnover. At body temperature, the TC^A and TC^B fragments subsequently dissociate from the collagen fibrils, denature into unordered gelatin chains, and are hydrolyzed into smaller peptides by synergistic proteinases.

The unique mode of cleavage of the interstitial collagens by human collagenases is one of the most specific and remarkable events in enzymology. The interstitial collagens are the most abundant proteins in humans, and HFC and HNC are the only two enzymes that are able to catalyze this specific cleavage. Conversely, only the interstitial collagen types I, II, and III are susceptible to this extremely specific attack. The underlying structural basis for this recognition is not well understood, but must be related to the local conformation of the collagen substrates at the cleavage site. In an earlier study, we presented a hypothesis to explain the cleavage site specificity of tissue collagenases.[4] According to this model, the cleavage site is in an apolar region of these collagens located at a boundary between an imino acid-rich, tightly triple helical segment and a loosely wound, imino acid-deficient section. The loose region on the carboxyl side of the scissile peptide bond allows the individual α chains to be unwound and bind at the active site of the collagenases in order to allow the catalytic, bond-breaking step to proceed. The tight triple helical segment, on the amino terminal side, is believed to be a recognition site that facilitates the binding of

[a]This work was supported by National Institutes of Health grant GM27939.
[c]Corresponding author.

this region to the collagenases, accounting for the low value of K_m for the collagens ($\sim 1\ \mu M$) compared to that for the hydrolysis of short peptide substrates with the identical collagen cleavage site sequences (~ 1 mM).[5,6]

Based on this view of the interstitial collagenases, their specificities toward collagen substrates may be conceptually divided into their "sequence specificities" toward short peptide cleavage sequences based on substrate interactions with the active site located in the catalytic domain, plus an interaction between a triple helical part of the substrate with an "exosite" portion of the enzyme located elsewhere. It is known that the hemopexin-like domain modulates the protein substrate specificity of the collagenases. When they lose this domain autolytically, they lose collagenase activity but retain activity against gelatin, casein, and peptide substrates.[7–9] Thus, the triple helix recognition site referred to above is probably located on the hemopexin-like domain of these collagenases. In this study, we have prepared synthetic triple helical peptides that are shown to compete with collagen in tissue collagenase assays. This supports the existence of the triple helix binding site in these collagenases.

MATERIALS AND METHODS

Materials. Pro-HNC, pro-HFC, human fibroblast prostromelysin (pro-HFS), and human promatrilysin (pro-MAT) were purified to homogeneity and activated prior to the inhibition experiments. Pro-HNC was isolated from human buffy coats as described earlier.[10] Pro-HFC and pro-HFS were purified from the harvest media of human gingival fibroblasts.[11] Pro-MAT was isolated from Chinese hamster ovary cells carrying an amplified pro-MAT cDNA.[12] [^{3}H]acetylated substrates were prepared according to Mallya *et al.*[13] Pro-OBzl · TSA and Fmoc-6-aminohexanoic acid (AHA) were prepared according to Bodanszky and Bodanszky.[14] DNP-Pro-Leu-Ala-Leu-Trp-Ala-Arg (DNP-PLALWAR) was synthesized and purified as described previously.[15]

Chemicals were purchased from the following sources: Boc-Gly, Hyp-OBzl · HCl, Boc-Gly-PAM resin, Fmoc-Lys (ϵ-Boc) from Bachem; acetic anhydride, dichloromethane (DCM), N,N-dimethylformamide (DMF), N-methylpyrrolidinone (NMP), dicyclohexylcarbodiimide (DCC), trifuoromethanesulfonic acid (TFMSA), isopropanol, ethanol, piperidine, triethylamine (TEA), AHA, toluene sulfonic acid (TSA), acetonitrile, dioxane, ether, methanol, trichloroacetic acid and trifluoroacetic acid (TFA) from Fisher; N-hydroxybenzotriazole and O-benzotriazol-1-yl-N,N,N′,N′-tetramethyluronium hexafluorophosphate (HBTU) from Applied Biosystems Inc; fluorescamine and Hepes from Sigma; Ready Safe scintillation cocktail from Beckman; and 1,10-phenanthroline from Lancaster synthesis.

Peptide synthesis. The triple helical peptides were synthesized using both solution- and solid-phase methods.[14,16] Boc-Gly-Pro-Hyp was prepared by solution-phase synthesis by first coupling Boc-Gly to Pro-OBzl using DCC followed by removal of the Bzl group by hydrogenolysis (H_2 with 10% palladium-on-charcoal). Hyp-OBzl was then coupled to Boc-Gly-Pro by the same procedure and deprotected to give Boc-Gly-Pro-Hyp. Intermediates and products were purified by silica chromatography. Next a three-forked "bridge" was prepared by two rounds of coupling of Fmoc-Lys (ϵ-Boc) to Gly-PAM resin using DCC followed by addition of Fmoc-AHA. The Fmoc and Boc blocking groups were removed in 20% piperidine/DMF and 25% TFA/DCM, respectively. The product, referred to as the "bridge," contains three sites (one α- and two ϵ-NH_2 groups of the Lys residues) that can be used for extension of the peptide. Last, the Boc-Gly-Pro-Hyp triplets were coupled to the bridge by segment synthesis. The first two sets of triplets were coupled overnight with DCC in 20% DMF/DCM. The coupling of subsequent triplets was carried out with the reagent HBTU in the presence of N-hydroxybenzotriazole in 25% NMP/DMF. Resin samples were removed after the addition of 6, 7, 8, and 10 triplets. Peptides were cleaved from the resin with 10% TFMSA/TFA, ether precipitated and purified by HPLC.

Circular dichroism spectroscopy. Circular dichroism (CD) spectra were acquired with a Jasco J720 spectropolarimeter equipped with a Brinkmann Lauda RM6 recirculating bath. Collagen and peptide samples were dissolved in 10 mM Hepes, 0.2 M NaCl, 5 mM $CaCl_2$, pH 7.5 and examined in a 1-cm pathlength cell in the range of 250–190 nm at 0.5 nm intervals with a 4-s integration time. All spectra were background corrected. For the melting curves, changes in ellipticity at 225 nm, θ_{225}, were monitored as a function of temperature in the 5–70 °C range.

Collagenase assays. Collagenolytic activity was measured using soluble [^{3}H]acetylated rat tail tendon type I collagen as the substrate.[13] For the inhibition measurements, the collagenases were incubated with the triple helical peptides (0.025–1.5 mM) for 25 min at 23 °C in assay buffer (50 mM Hepes, 0.2 M NaCl, 10 mM $CaCl_2$, 50 μM $ZnSO_4$, pH 7.5). The reaction was initiated by addition of 50 μg of [^{3}H]acetylated collagen to give a final concentration of 0.3 μM. Aliquots were removed at selected times and quenched with 7 mM 1,10-phenanthroline. The unhydrolyzed collagen was precipitated by addition of dioxane (final concentration 50% v/v) at 4 °C and centrifuged. The ^{3}H concentration of the supernatants was determined using a Beckman LS 1800 scintillation counter.

Casein assays. HFS and MAT activities were quantitated by monitoring the hydrolysis of [^{3}H]acetylated casein at a substrate concentration of 0.1 mg/mL as described earlier.[17] The undigested casein was precipitated with 7.5% trichloroacetic acid.

DNP-PLALWAR assay. Peptidase activity was measured by monitoring the hydrolysis of DNP-PLALWAR.[15] The DNP-PLALWAR concentration was determined spectrophotometrically (ϵ_{365} = 17,300 M^{-1} cm^{-1}). For the inhibition measurements, the collagenases were incubated with the triple helical peptides (0.025–1.5 mM) for 25 min at 23 °C in assay buffer. The reactions were initiated by addition of DNP-PLALWAR to a final concentration of 15 and 40 μM for HNC and HFC, respectively. The fluorescence changes were recorded continuously with excitation at 280 nm and emission at 360 nm using a Perkin-Elmer Model LS-50B fluorometer.

RESULTS AND DISCUSSION

In order to provide evidence for the existence of a triple helix recognition site on the hemopexin-like domains of the tissue collagenases, a series of synthetic triple helical peptides was prepared. If the tissue collagenases utilize such an interaction to bind their collagen substrates, then these synthetic triple helical peptides should compete with collagen for binding to these enzymes in collagenase assays. This competition should be manifested as a dose-dependent inhibition. The rationale behind using synthetic triple helical peptides is that short, naturally occurring portions of triple helical collagens are difficult to prepare and are not generally thermally stable. However, it has been proven possible to prepare such molecules by solid-phase peptide synthesis.[18–21]

The synthetic approach taken to prepare the triple helical peptides for this study follows that of Heideman and associates[18–20] and was used earlier by us[21] in which a three-stranded bridge consisting of a Lys-Lys-Gly unit is first synthesized on a PAM resin. This allows the preparation of a peptide that consists of three parallel strands with the same sequence. This trimeric linkage helps to overcome the unfavorable entropy of nucleating the triple helix in the resultant peptide and greatly reduces the possibility of individual chains associating out of register. Next, AHA moieties are added to the single α- and two ϵ-amino groups of the Lys-Lys-Gly unit. These groups act as "flexible arms" that allow the three chains the freedom to align in register. Inasmuch as Gly-Pro-Hyp triplets provide the maximum stabilization of the triple helix,[22] a series of these has been consecutively added to the bridge, as illustrated in FIGURE 1. The degree of triple helicity of such peptides is expected to be dependent on the number of triplets. Thus, four different peptides, denoted $[(\text{Gly-Pro-Hyp})_n]_3$-bridge, have been prepared, where n = 6, 7, 8, and 10. A sample of the bridge alone has also been prepared.

CD spectroscopy has been used to investigate the conformation of the peptides prepared. Triple helical collagens are known to exhibit distinctive CD spectra

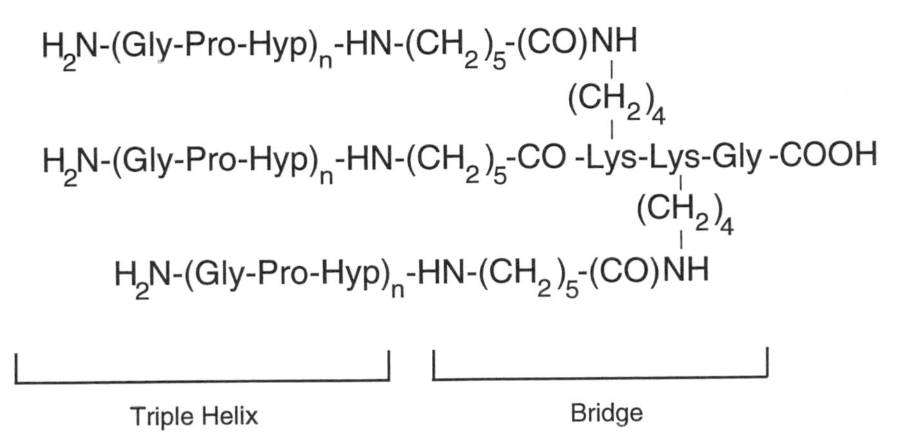

FIGURE 1. Schematic of the peptides synthesized in this study. These peptides are denoted $[(\text{Gly-Pro-Hyp})_n]_3$-bridge, where n = 6, 7, 8, 10, and consist of an N-terminal portion with a variable propensity to be triple helical and a C-terminal three-stranded bridge.

characterized by a peak near 225 nm.[23] This is illustrated in FIGURE 2A for acetylated rat tendon type I collagen. As the temperature is increased and the triple helix melts, this band decreases in intensity until there is a weak trough at this wavelength. A plot of the ellipticity at 225 nm, θ_{225}, as a function of temperature is shown in FIGURE 3. This curve shows that collagen denatures in a highly cooperative manner with a transition temperature of approximately 38 °C.

Similarly, CD spectra have been recorded for the peptides described above. The spectra of $[(\text{Gly-Pro-Hyp})_{10}]_3$-bridge at several different temperatures are shown in FIGURE 2B. The spectrum recorded at 5 °C shows a prominent peak at 225 nm, indicating that this peptide is substantially triple helical at this temperature. As with collagen, this band decreases in intensity as the solution is heated until it is replaced by a trough at 75 °C. Similar trends are found for the n = 6, 7, and 8 analogs, except that the ellipticities of the 225 nm band are respectively lower for the peptides with the smaller number of Gly-Pro-Hyp triplets. Melting curves for three of these

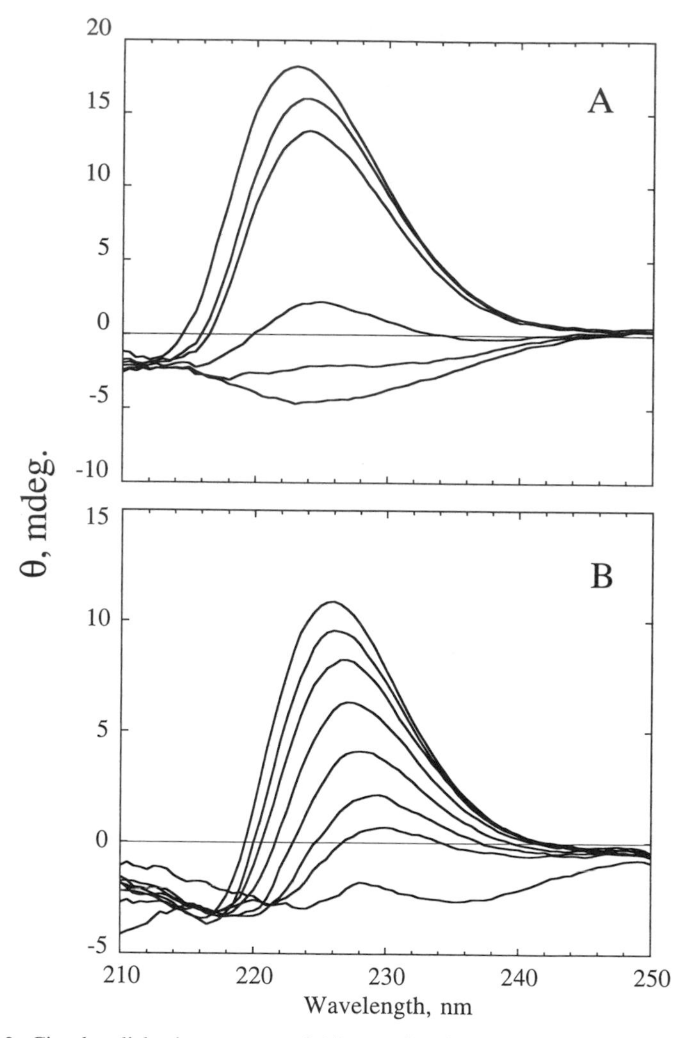

FIGURE 2. Circular dichroism spectra of **(A)** acetylated rat tendon type I collagen (0.075 mg/mL) at 15, 30, 33, 35, 38, and 55 °C and **(B)** $[(Gly\text{-}Pro\text{-}Hyp)_{10}]_3$-bridge (0.05 mg/mL) at 5, 25, 35, 45, 55, 65, and 75 °C. All samples were dissolved in 10 mM Hepes, 0.2 M NaCl, 5 mM $CaCl_2$, pH 7.5.

peptides are also plotted in FIGURE 3. These curves and that for collagen have been normalized for solutions of the same concentration (0.05 mg/mL) so that the θ_{225} value can be used as an approximate measure of percent triple helicity. The results show that $[(Gly\text{-}Pro\text{-}Hyp)_{10}]_3$-bridge is almost fully triple helical at 5 °C and undergoes a broad triple helix-to-coil transition centered at approximately 42 °C. $[(Gly\text{-}Pro\text{-}Hyp)_6]_3$-bridge and $[(Gly\text{-}Pro\text{-}Hyp)_8]_3$-bridge are correspondingly less triple helical at each temperature and have lower melting points.

A series of experiments has been carried out to assess whether the triple helical peptides prepared here compete with different MMP for their substrates to prevent hydrolysis. The first experiments involved competition studies with the collagenases HFC and HNC toward both type I collagen and the peptide substrate DNP-PLALWAR carried out at 23 °C. All of these assays were carried out at the same ratio of $[S_0]$ to K_m (0.3 K_m) in order to allow direct comparison of the results. Thus, a collagen concentration of 0.3 μM was used for both HFC and HNC, because the K_m values for hydrolysis of type I collagen by both collagenases are ~1 μM.[5] Because the K_m values for the hydrolysis of DNP-PLALWAR by HFC and HNC are 130 and 40 μM, respectively, these assays were carried out at substrate concentrations of 40 and 15 μM, respectively.[15] The effect of increasing concentrations of $[(\text{Gly-Pro-Hyp})_{10}]_3$-bridge on the hydrolysis of both substrates by HFC is shown in FIGURE 4. This peptide has no effect on the hydrolysis of DNP-PLALWAR, nor do any of the other peptides or the bridge itself. However, $[(\text{Gly-Pro-Hyp})_{10}]_3$-bridge inhibits the hydrolysis of type I collagen in a dose-dependent manner with an IC_{50} value of 0.1 mM. $[(\text{Gly-Pro-Hyp})_7]_3$-bridge also inhibits, but with the weaker IC_{50} value of 0.8 mM (TABLE 1). There is no inhibition by the bridge itself. Very similar results are found with HNC where both $[(\text{Gly-Pro-Hyp})_{10}]_3$-bridge and $[(\text{Gly-Pro-Hyp})_7]_3$-bridge are slightly better inhibitors of collagen hydrolysis than for HFC. Once again, there is no inhibition of collagen hydrolysis by the bridge, and no inhibition of DNP-PLALWAR hydrolysis by any of the peptides or the bridge.

To investigate whether the inhibition observed above for HFC and HNC was unique to the collagenases, two other members of the MMP family have been

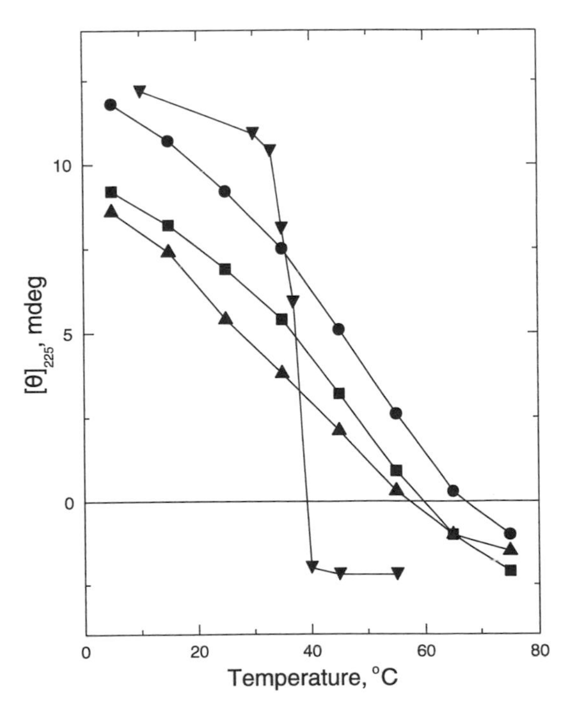

FIGURE 3. Melting curves for (▼) acetylated rat tendon type I collagen, (●) $[(\text{Gly-Pro-Hyp})_{10}]_3$-bridge, (■) $[(\text{Gly-Pro-Hyp})_8]_3$-bridge, and (▲) $[(\text{Gly-Pro-Hyp})_6]_3$-bridge obtained from CD spectroscopy. All samples were dissolved in 10 mM Hepes, 0.2 M NaCl, 5 mM $CaCl_2$, pH 7.5 at a concentration of 0.05 mg/mL.

studied. Matrilysin (MAT) has been studied because it is the smallest member of the MMP family and does not have a hemopexin-like domain. In contrast, human fibroblast stromelysin (HFS) is of interest because it does have a hemopexin-like domain, but does not have collagenase activity. The effect of $[(\text{Gly-Pro-Hyp})_{10}]_3$-bridge on the hydrolysis of the protein substrate casein and the peptide DNP-PLALWAR by these two MMP has been measured. No inhibition was found in any of the four sets of assays carried out (TABLE 1).

The results of the competition assays described above are consistent with the presence of a triple helix recognition site in the hemopexin-like domains of the interstitial collagenases HFC and HNC. The ability of the peptides prepared in this study to compete with collagen, but not DNP-PLALWAR, in assays carried out at a temperature where they are partially triple helical, indicates that this site is spatially removed from the active site and can be appropriately considered an exosite as hypothesized by us earlier.[4] The fact that HFC and HNC lose their ability to hydrolyze triple helical collagens when they autolytically lose their hemopexin-like domains[7–9] strongly suggests that this exosite is located in this domain. In support of this concept, the triple helical peptides have no effect on MAT or HFS. Although HFS has this domain, it does not have collagenase activity. Thus, it seems likely that only MMP that hydrolyze interstitial collagens possess this triple helix binding site.

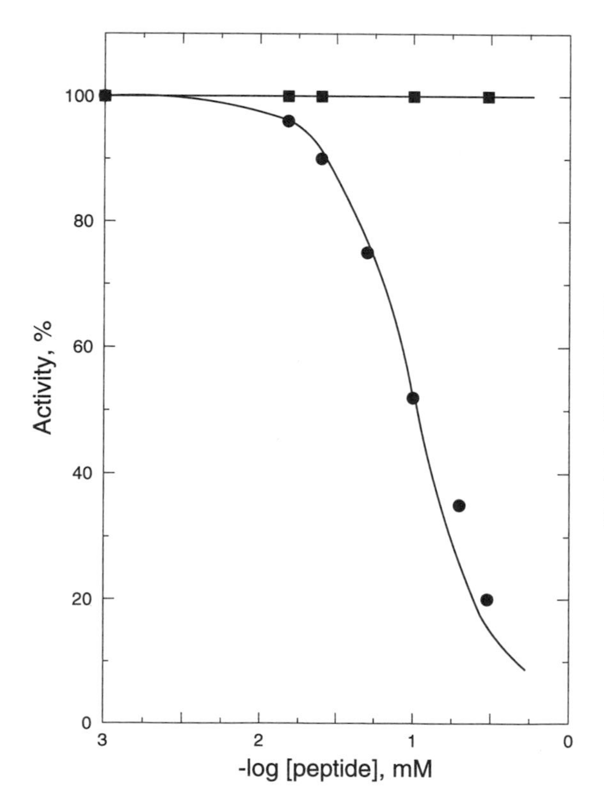

FIGURE 4. Effect of $[(\text{Gly-Pro-Hyp})_{10}]_3$-bridge on the hydrolysis of (●) acetylated rat tendon type I collagen (0.30 μM) and (■) DNP-Pro-Leu-Ala-Leu-Trp-Ala-Arg (40 μM) by HFC. Assays were carried out in 50 mM Hepes, 0.2 M NaCl, 10 mM $CaCl_2$, 50 μM $ZnSO_4$, pH 7.5 at 23 °C.

TABLE 1. Effect of Triple Helical Peptides on the Activities of Various Matrix Metalloproteinases

MMP	Substrate	Peptide	IC_{50} (mM)
HFC	Type I collagen	Bridge	> 10
		$[(GPHy)_7]_3$-bridge	0.8
		$[(GPHy)_{10}]_3$-bridge	0.1
	DNP-peptide[a]	Bridge	> 10
		$[(GPHy)_7]_3$-bridge	> 5
		$[(GPHy)_{10}]_3$-bridge	> 5
HNC	Type I collagen	Bridge	> 10
		$[(GPHy)_7]_3$-bridge	0.6
		$[(GPHy)_{10}]_3$-bridge	0.08
	DNP-peptide[a]	Bridge	> 10
		$[(GPHy)_7]_3$-bridge	> 5
		$[(GPHy)_{10}]_3$-bridge	> 5
MAT	Casein	$[(GPHy)_{10}]_3$-bridge	> 5
	DNP-peptide[a]	$[(GPHy)_{10}]_3$-bridge	> 5
HFS	Casein	$[(GPHy)_{10}]_3$-bridge	> 5
	DNP-peptide[a]	$[(GPHy)_{10}]_3$-bridge	> 5

[a]DNP-Pro-Leu-Ala-Leu-Trp-Ala-Arg.

Abbreviations: HFC, human fibroblast collagenase; HNC, human neutrophil collagenase; MAT, matrilysin; HFS, human fibroblast stromelysin.

REFERENCES

1. BIRKEDAL-HANSEN, H., W. G. I. MOORE, M. K. BODDEN, L. J. WINDSOR, B. BIRKEDAL-HANSEN, A. DECARLO & J. A. ENGLER. 1993. Matrix metalloproteinases: A review. Crit. Rev. Oral Biol. Med. **4:** 197–250.
2. GROSS, J. & C. M. LAPIERE. 1962. Collagenolytic activity in amphibian tissues: A tissue culture assay. Proc. Natl. Acad. Sci. USA **48:** 1014–1022.
3. GROSS, J. & Y. NAGAI. 1965. Specific degradation of the collagen molecule by tadpole collagenolytic enzyme. Proc. Natl. Acad. Sci. USA **54:** 1197–1204.
4. FIELDS, G. B. & H. E. VAN WART. 1992. Unique features of the tissue collagenase cleavage site in interstitial collagens. Matrix (**Suppl. 1**): 68–70.
5. MALLYA, S. K., K. A. MOOKHTIAR, Y. GAO, K. BREW, M. DIOSZEGI, H. BIRKEDAL-HANSEN & H. E. VAN WART. 1990. Characterization of 58 kDa human neutrophil collagenase: Comparison with human fibroblast collagenase. Biochemistry **29:** 10628–10634.
6. NETZEL-ARNETT, S., G. FIELDS, H. BIRKEDAL-HANSEN & H. E. VAN WART. 1991. Sequence specificities of human fibroblast and neutrophil collagenases. J. Biol. Chem. **266:** 6747–6755.
7. BIRKEDAL-HANSEN, B., W. G. I. MOORE, R. E. TAYLOR, A. S. BHOWN & H. BIRKEDAL-HANSEN. 1988. Monoclonal antibodies to human fibroblast procollagenase. Inhibition of enzymatic activity, affinity purification of the enzyme, and evidence for clustering of epitopes in the NH_2-terminal end of the activated enzyme. Biochemistry **27:** 6751–6758.
8. KNAUPER, V., A. OSTHUES, Y. A. DECLERCK, K. E. LANGLEY, J. BLASER & H. TSCHESCHE. 1993. Fragmentation of human polymorphonuclear-leucocyte collagenase. Biochem. J. **291:** 847–854.
9. CLARK, I. M. & T. E. CAWSTON. 1989. Fragments of human fibroblast collagenase: Purification and characterization. Biochem. J. **263:** 201–206.
10. MOOKHTIAR, K. A. & H. E. VAN WART. 1990. Purification to homogeneity of latent and active 58 kDa forms of human neutrophil collagenase. Biochemistry **29:** 10620–10627.

11. BIRKEDAL-HANSEN, H. 1987. *In* Methods in Enzymology. L. W. Cunningham, Ed.: 140–171. Academic Press. San Diego, CA.
12. BARNETT, J., K. STRAUB, B. NGUYEN, J. CHOW, R. SUTTMAN, K. THOMPSON, S. TSING, P. BENTON, R. SCHATZMAN, M. CHEN & H. CHAN. 1994. Production, purification, and characterization of human matrilysin (PUMP) from recombinant Chinese hamster ovary cells. Protein Express. Purif. **5**: 27–36.
13. MALLYA, S. K., K. A. MOOKHTIAR & H. E. VAN WART. 1986. Accurate quantitative assays for the hydrolysis of soluble type I, II and III [^{3}H]-acetylated collagens by bacterial and tissue collagenases. Anal. Biochem. **158:** 334–345.
14. BODANSZKY, M. & A. BODANSZKY. 1984. The Practice of Peptide Synthesis. Springer-Verlag. Berlin.
15. NETZEL-ARNETT, S., S. K. MALLYA, H. NAGASE, H. BIRKEDAL-HANSEN & H. E. VAN WART. 1991. Continuously recording fluorescent assays optimized for five human matrix metalloproteinases. Anal. Biochem. **195:** 86–92.
16. STEWART, S. M. & S. D. YOUNG. 1984. Solid Phase Peptide Synthesis. Pierce Chemical Co. Pierce, IL.
17. BOND, M. D. & H. E. VAN WART. 1984. Purification and separation of individual collagenases of *Clostridium histolyticum* using red dye ligand chromatography. Biochemistry **23:** 3077–3085.
18. ROTH, W. & E. HEIDEMANN. 1980. Triple helix-coil transition of covalently bridged collagenlike peptides. Biopolymers **19:** 1909–1917.
19. GERMANN, H.-P. & E. HEIDEMANN. 1988. A synthetic model of collagen: An experimental investigation of the triple-helix stability. Biopolymers **27:** 157–163.
20. THAKUR, S., D. VADOLAS, H.-P. GERMANN & E. HEIDEMANN. 1986. Influence of different tripeptides on the stability of the collagen triple helix. II. An experimental approach with appropriate variations of a trimer model oligotripeptide. Biopolymers **25:** 1081–1086.
21. GOLI, U. B., G. B. FIELDS & H. E. VAN WART. 1992. Synthetic triple helical models for the collagen cleavage site in interstitial collagens. Matrix (**Suppl. 1**): 71–72.
22. PIEZ, K. A. 1984. Molecular and aggregate structures of the collagens. *In* Extracellular Matrix Biochemistry. K. A. Piez & A. H. Reddi, Eds.: 1–39. Elsevier Science Publishing Co. New York.
23. HAYASHI, T., S. CURRAN-PATEL & D. J. PROCKOP. 1979. Thermal stability of the triple helix of type I procollagen and collagen. Precautions for minimizing ultraviolet damage to proteins during circular dichroism studies. Biochemistry **18:** 4182–4187.

Regulation of Matrix Metalloproteinase Activity[a]

GILLIAN MURPHY,[b,c] FRANCES WILLENBROCK,[d]
THOMAS CRABBE,[e] MARK O'SHEA,[b] ROBIN WARD,[b]
SUSAN ATKINSON,[b] JAMES O'CONNELL,[e]
AND ANDREW DOCHERTY[e]

[b]*Strangeways Research Laboratory*
Worts Causeway
Cambridge CB1 4RN, United Kingdom

[d]*Department of Biochemistry*
Queen Mary and Westfield College
London E1 4NS, United Kingdom

[e]*Celltech Research*
Slough SL1 4EN, United Kingdom

Matrix metalloproteinases (MMPs) are thought to initiate the degradation of the extracellular matrix during the remodeling of connective tissues. A clear understanding of the mechanisms governing the regulation of their activity during normal physiological processes should give further insights into the uncontrolled remodeling occurring in degradative pathologies. Regulation of the MMPs occurs at the level of gene expression, with precise spatial and temporal compartmentalization of both synthesis and secretion by resident cells as well as by those cells invading the tissue. Extracellularly, MMPs are further regulated by the extent of processing of the proform to an active enzyme and by the relative production of the specific inhibitors of MMPs, the tissue inhibitors of metalloproteinases (TIMPs). Furthermore, the potential for association of MMPs with the cell surface or extracellular matrix components further constrains their relationship with substrates, activators and inhibitors, acting as a further regulator of MMP activity.

We initiated a program of study of the MMPs and TIMPs to ascertain the relation between their structure and their function, with particular emphasis on the mechanisms of biological regulation. Recombinant wild-type proteins and specific mutants, including deletion and site mutations, have been prepared using a mammalian expression system.[1] Aspects of our findings, which illustrate the fundamental importance of the domain structure of MMPs and TIMPs in their biology, are presented here.

ACTIVATION OF MATRIX METALLOPROTEINASES

All the MMPs have an N-terminal propeptide of 77 to 87 amino acids that determines the latency of the proenzyme form. This is thought to be due in part to

[a]This work was supported by the Arthritis and Rheumatism Council (G.M. and R.W.), the Wellcome Trust (M.O'S.), and the Medical Research Council, UK (F.W. and S.A.).
[c]Corresponding author.

the presence of a conserved sequence PRCGV/NPD, in which the cysteine interacts with the Zn(II) of the adjoining catalytic domain, displacing the H_2O molecule required for catalytic activity.[2] Further weak interactions between the propeptide and the catalytic domain further stabilize this binding. Activation of most MMPs involves sequential exogenous or endogenous cleavages of the propeptide which destabilize the cysteine-Zn(II) interaction, modify the enzyme conformation, and permit further exogenous or autocatalytic processing to the final active form. Potential physiological activators of the MMPs have been described[3] of which plasmin is thought to be the most significant activity *in vivo.* In the human, MMPs including prostromelysins 1 and 2 and fibroblast and neutrophil procollagenases have identifiable propeptide motifs cleavable by plasmin. This concept has been

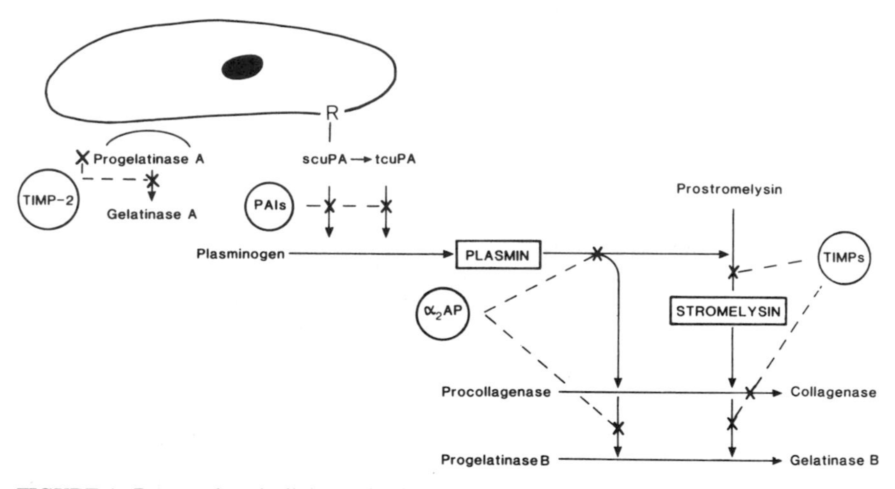

FIGURE 1. Proposed pericellular activation cascades for the matrix metalloproteinases. The majority of MMPs may be activated by the action of plasmin generated at the cell surface by the juxtaposition of receptor-bound urokinase-like plasminogen activator (scuPA, tcuPA) and membrane-bound plasminogen. The plasmin-mediated activation of stromelysin is central to the cascade and is able to productively cleave plasmin-processed collagenase and progelatinase B to yield active forms. Other MMP interactions may also occur, leading to processing events. Progelatinase A follows a different pathway but this is apparently cell membrane associated, involving specific binding and proteolytic processing that may be autocatalytic.

indirectly supported by a number of studies using cell model systems.[4–7] The generation of plasmin from plasminogen by the action of plasminogen activators occurs largely at the cell surface, where both plasminogen and urokinase-like plasminogen activator can be specifically localized.[8] It is likely that in localized pericellular regions, the action of α2 anti-plasmin is limited. Prostromelysin-1, which can be sequestered on the collagenous matrix,[9] is susceptible to plasmin activation and once activated can potentiate collagenase activity and act as a progelatinase B activator (FIG. 1). TIMPs may regulate the activation process to some extent, because it has been shown that they can slow down or prevent processing of MMPs.[10,11]

Progelatinase A has no apparent plasmin-susceptible propeptide cleavage site and plasmin-independent activation can be demonstrated in human and rabbit cell systems.[5,12] *In vitro* progelatinase A can be induced to self-process to remove the propeptide and generate an active form.[3,13,14] We have mutated progelatinase A such that the proposed active site glutamic acid residue, Glu375, was replaced by Ala (Glu375Ala; FIG. 2). In the presence of the MMP activator 4-aminophenylmercuric acetate, the Glu375Ala mutant of progelatinase A was unable to undertake autocatalytic processing of the propeptide in the same way as the wild-type enzyme (1 h at 23 °C gave complete processing of the wild type). The mutant underwent limited self-processing in the presence of the organomercurial over 168 h at 37 °C (FIG. 3). However, in the presence of an active truncated ($\Delta_{416-613}$) form of gelatinase A (see below and FIG. 2) much more rapid processing of the mutant occurred (FIG. 3), indicating that intermolecular cleavages of gelatinase A molecules can effect processing of the propeptide. The molecular mass of the processed Glu375Ala gelatinase A mutant was 66 kDa, and the N-terminal sequence was shown to be Tyr-Asn-Phe-Phe-Pro, identical to the active wild-type gelatinase A. However, the processed mutant displayed negligible gelatinolytic activity (less than 0.01% of wild type).[15]

A number of independent studies have described a cell-membrane mediated activation process for progelatinase A. The membrane "activator," which is specific for gelatinase A, can be induced in a number of cell types, such as fibroblasts, osteoblasts, chondrocytes, endothelial cells, and tumor cells, by effectors such as concanavalin A, cytochalasin D, phorbol esters, transforming growth factor β1[11,16–18] or by exposure to collagen.[18,19] Activation by membrane preparations from such cells is prevented by metalloproteinase inhibitors, including 1,10-phenanthroline, peptide inhibitors, and the TIMPs.[11] TIMP-2 is much more efficient in the inhibition of progelatinase A activation than TIMP-1, which is thought to be due to the ability of TIMP-2 to bind to progelatinase A through the C-terminal domain of the enzyme.[11,20,21] Consequently, we prepared a recombinant truncated form of progelatinase A consisting of the catalytic N-terminal domain and fibronectin-like domain but lacking the complete C-terminal domain ($\Delta_{416-613}$; FIG. 2).[6] This mutant could be activated with the same kinetics as wild-type gelatinase A by exposure to organomercurials and demonstrated similar proteolytic activity towards a number of substrates. However, it could not be activated by concanavalin-A–stimulated fibroblasts or membrane preparations derived from them, which efficiently convert full-length gelatinase A into its active form.[6] This implied a role for the C-terminal domain of progelatinase A in its binding to stimulated cell membranes as a prerequisite to the activation process. Using both immunolocalization and radiolabeling techniques we showed that progelatinase A binds to concanavalin A-stimulated fibroblasts in a saturable manner (FIG. 4). The binding is largely inhibited by either isolated gelatinase A C-terminal domain or TIMP-2, but not by the truncated mutant lacking the C-terminal domain. Normal fibroblasts bound progelatinase A to a much lower extent; this binding was not affected by the C-terminal domain, but was prevented by the addition of the N-terminal domain mutant (FIG. 4).

We concluded that some binding of progelatinase A through the N-terminal domain occurs, possibly due to the association of the fibronectin-like region with cell-bound collagenous matrix.[22] This mode of binding does not appear to initiate the progelatinase A activation process. It was found that a mutant of progelatinase A lacking the fibronectin-like region ($\Delta_{191-364}$; FIG. 2) was efficiently processed by exposure to stimulated cells or their membranes.[23] Strongin *et al.*[24] similarly demonstrated that TIMP-2 and a 26-kDa peptide derived from the C-terminal domain prevented cell membrane mediated progelatinase A activation. They showed that membrane-dependent activation of progelatinase A results in an initial propeptide cleavage at Asn37-Leu38, followed by the Asn80-Phe81 cleavage which occurs during autoproteolytic cleavage in the presence of organomercurial.[13] It has not yet

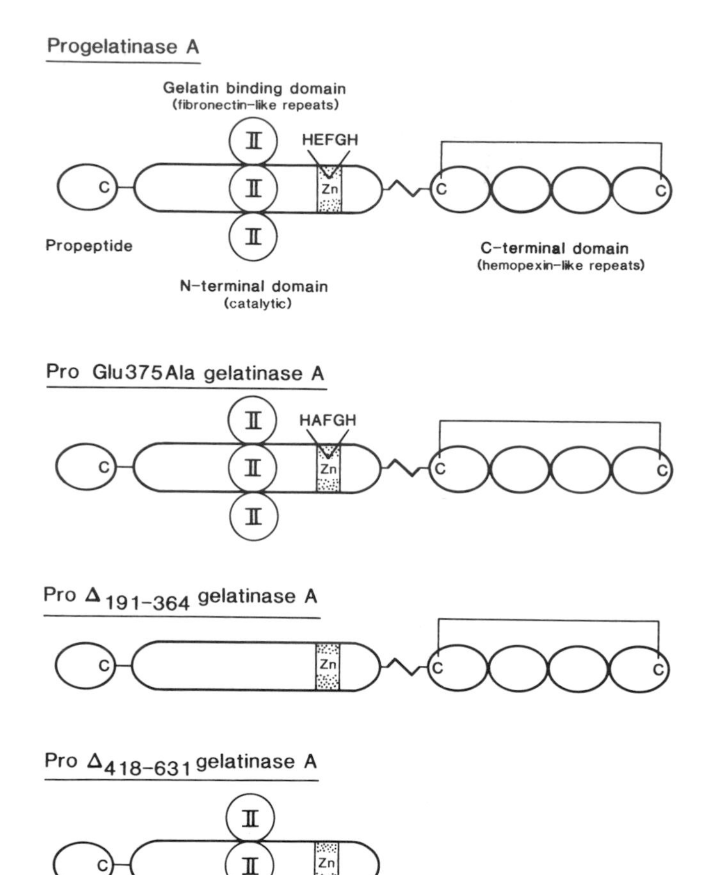

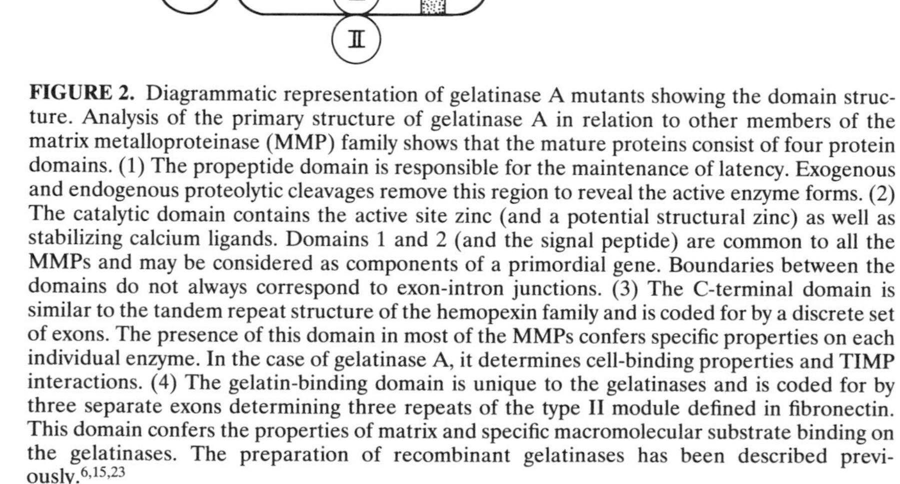

FIGURE 2. Diagrammatic representation of gelatinase A mutants showing the domain structure. Analysis of the primary structure of gelatinase A in relation to other members of the matrix metalloproteinase (MMP) family shows that the mature proteins consist of four protein domains. (1) The propeptide domain is responsible for the maintenance of latency. Exogenous and endogenous proteolytic cleavages remove this region to reveal the active enzyme forms. (2) The catalytic domain contains the active site zinc (and a potential structural zinc) as well as stabilizing calcium ligands. Domains 1 and 2 (and the signal peptide) are common to all the MMPs and may be considered as components of a primordial gene. Boundaries between the domains do not always correspond to exon-intron junctions. (3) The C-terminal domain is similar to the tandem repeat structure of the hemopexin family and is coded for by a discrete set of exons. The presence of this domain in most of the MMPs confers specific properties on each individual enzyme. In the case of gelatinase A, it determines cell-binding properties and TIMP interactions. (4) The gelatin-binding domain is unique to the gelatinases and is coded for by three separate exons determining three repeats of the type II module defined in fibronectin. This domain confers the properties of matrix and specific macromolecular substrate binding on the gelatinases. The preparation of recombinant gelatinases has been described previously.[6,15,23]

been clearly established whether the membrane activation process involves a separate membrane-bound proteinase or if the interaction of gelatinase A with the cell-membrane initiates autocatalytic cleavages. It is known that progelatinase A can self-activate at high concentrations, with demonstrable cleavage of the Asn80-Phe81 bond, and that the binding of the enzyme to heparin or dextran sulphate can potentiate gelatinase A self-cleavage at much lower concentrations.[14,25] Whatever the mechanism, we can conclude that progelatinase A activation may well occur at the cell surface *in vivo* (FIG. 1). The precise role of TIMP-2 in the regulation of binding as well as enzyme activity is not clear, but it is likely to represent a significant level of control of gelatinase activity. A major question concerns the stage at which progelatinase A–TIMP-2 complexes form, intra- or extracellularly, and how this process is regulated. Binding of progelatinase A to some tumor cells without concomitant activation has also been observed.[26,27] It will be interesting to analyze the nature of this binding relative to the C-terminal domain mediated process

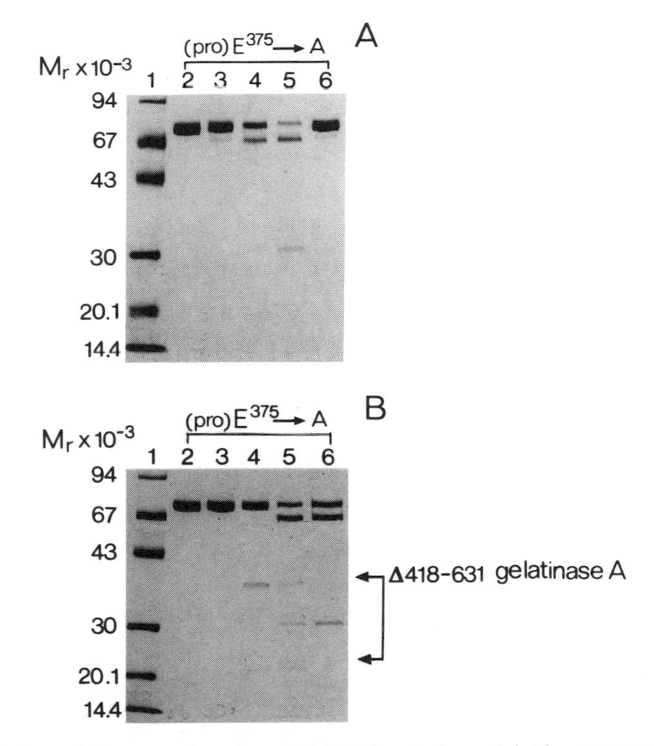

FIGURE 3. Proteolytic processing of pro Glu375Ala gelatinase A in the presence of 4-aminophenylmercuric acetate or $\Delta_{418-631}$ gelatinase A. **(A)** Pro Glu375Ala gelatinase A (7.1 μM) was incubated with 1 mM 4-aminophenylmercuric acetate at 37 °C and aliquots removed at the indicated time points for analysis on SDS polyacrylamide gels (Coomassie blue staining). **Lane 1,** marker proteins; **lane 2,** 0 h; **lane 3,** 24 h; **lane 4,** 96 h; **lane 5,** 168 h. **Lane 6** indicates the stability of the protein to incubation alone at 37 °C for 168 h. **(B)** Pro Glu375Ala gelatinase (5.5 μM) was incubated alone or with 0.9 μM active $\Delta_{418-631}$ gelatinase at 37 °C for the times indicated and then purified as described by Crabbe *et al.*[15] **Lane 1,** marker proteins; **lane 2,** 0 h alone; **lane 3,** 16 h alone; **lane 4,** 0 h plus $\Delta_{418-631}$ gelatinase A; **lane 5,** 16 h plus $\Delta_{418-631}$ gelatinase A; **lane 6,** as lane 5 after removal of $\Delta_{418-631}$ gelatinase A.

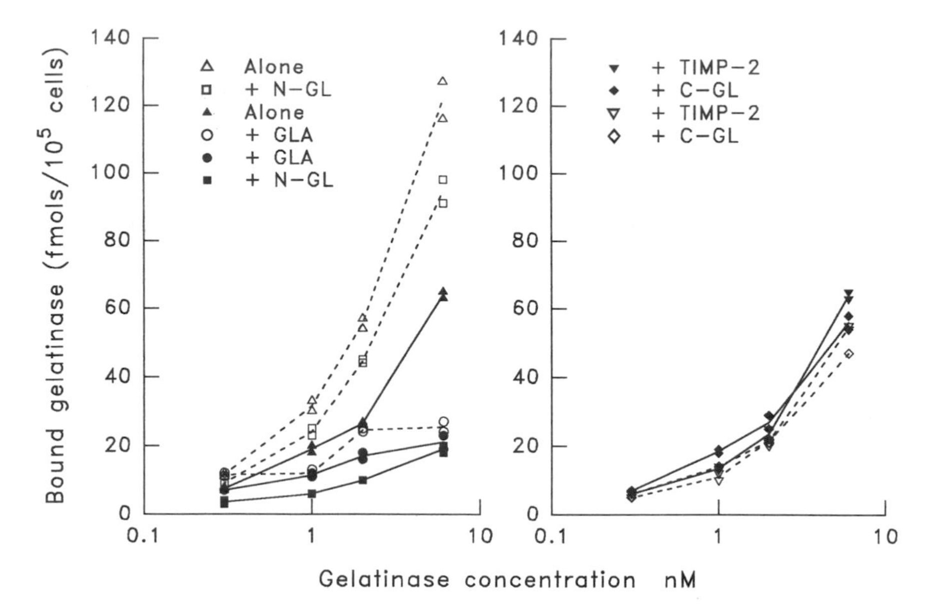

FIGURE 4. The effect of TIMP-2, C-terminal (C-GL), N-terminal (N-GL), and full-length progelatinase A (GLA) on the binding of [^{125}I]progelatinase A to concanavalin A stimulated and nonstimulated human skin fibroblasts. The binding of [^{125}I]progelatinase to fibroblasts was studied as described by Ward *et al.*[22] The effect of various domains of progelatinase A or TIMP-2 added in 25M excess is documented. C-GL is $\triangle_{1-414}$ gelatinase A; N-GL is $\triangle_{418-631}$ gelatinase A. Concanavalin A stimulated (*dashed line*) and nonstimulated (*solid line*) human skin fibroblasts.

initiating the progelatinase A activation mechanism. The physiological relevance of these binding processes is not yet clear, although gelatinase A bound to tumor cell surfaces has frequently been reported in immunohistochemical studies of tumor tissue.[28] However, the mRNA for the enzyme is largely associated with adjacent stromal cells.[29]

INHIBITION OF MATRIX METALLOPROTEINASES

The tissue inhibitors of the MMP (TIMP) family appear to be the major local inhibitors of MMPs, although the ubiquitous proteinase inhibitor α2-macroglobulin undoubtedly plays a regulatory role in certain situations. Three forms have been unequivocally identified by cloning and sequencing from a number of species. TIMP-1, a 28.5-kDa glycoprotein,[30] and TIMP-2, a 21-kDa unglycosylated protein,[31] are produced by many cell types and can be demonstrated in body fluids and tissue extracts. Their regulation at the gene level appears to be very different, TIMP-1 expression being very tightly controlled by many cytokines and growth factors whereas little has yet been found that regulates TIMP-2. TIMP-3, a 21-kDa protein which may by glycosylated, is less well understood, although it has been cloned from chicken, mouse, and human.[32-34] The chicken protein, Chimp-3, was initially reported to be a component of the extracellular matrix of embryo fibroblasts during the early stages of transformation.[35] The human and mouse mRNAs for TIMP-3 have been shown to be very widely expressed in connective tissues, including during development.[33,34]

Although the individual TIMP forms show as little as 40% sequence identity, they share considerably higher structural similarity. This is notably due to the conservation of 12 cysteine residues that have been shown to form disulphide bonds in TIMP-1, giving a 6-loop structure.[36] They all form high-affinity, noncovalent, and essentially irreversible complexes with the active forms of MMPs with a 1:1 stoichiometry. It has been shown that active, unmodified TIMP-1 can be recovered from enzyme complexes. TIMP-1 can also bind, albeit rather less tightly, to progelatinase B[37] and TIMP-2 to progelatinase A,[21] properties that are thought to allow tight regulation of the activation of these enzymes.

Mutagenesis of both TIMP-1 and TIMP-2 to remove sections of each protein has demonstrated that they consist of two structurally distinct domains; the N-terminal domain consists of 3 loops and the C-terminal domain of the remaining, smaller loops. The N-terminal domain can fold independently of the C-terminal domain to give a functional MMP inhibitor, which interacts with the enzyme catalytic domain. Because the assay systems that are currently available for the MMPs[38] are too insensitive, it has not been possible to compare K_i values for the TIMPs and their truncated mutants. However, we have shown that the rate of inhibition of active MMPs by TIMP-1 and TIMP-2 is significantly decreased by the removal of the inhibitor C-terminal domain (TABLE 1). In the case of gelatinase A, using deletion mutagenesis we were able to demonstrate that it is interactions between the C-terminal domains of TIMP-1 and TIMP-2 with the enzyme C-terminal domain that play a role in regulating the rate of complex formation (TABLE 1). In the case of TIMP-2, ionic interactions, including those with the exposed C-terminal negatively charged "tail" of the inhibitor, are particularly important in determining a very fast rate of binding.[39] A scheme depicting the potential domain interactions between gelatinase A and TIMPs based on these data is shown in FIGURE 5. Even if TIMP-2 is already associated with the proform of gelatinase A through C-terminal domain interactions, subsequent activation of the progelatinase can occur, and binding of the N-terminal domains of TIMP-2 and enzyme appears to be considerably slower. The

TABLE 1. Association Rate Constants for the Binding of TIMP-1 and TIMP-2 with Gelatinase A and B and Stromelysin-1. Effect of Domain Deletions[a]

	TIMP-1	N-Terminal Domain of TIMP-1 ($\Delta_{127-184}$)	TIMP-2	N-Terminal Domain of TIMP-2 ($\Delta_{128-194}$)
	$10^{-4} \times k_{on}(M^{-1}s^{-1})$			
Stromelysin-1	190	117	30	5
N-terminal domain of stromelysin-1 ($\Delta_{248-460}$)	190	120	47	5
Gelatinase A	470	9	3800	36
N-terminal domain of gelatinase A ($\Delta_{418-631}$)	1.4	1.4	30	30
Gelatinase B	1110	8	26	26
N-terminal domain of gelatinase B ($\Delta_{426-688}$)[b]	2	13	37	21
Matrilysin	30	40	nd	nd

[a]Values for the association constants for the binding of MMPs and TIMPs and their domain deletion mutants were determined using a quenched fluorescent substrate.[23,38,39,46,47] Studies were conducted at 37 °C, or [b]at 25 °C.

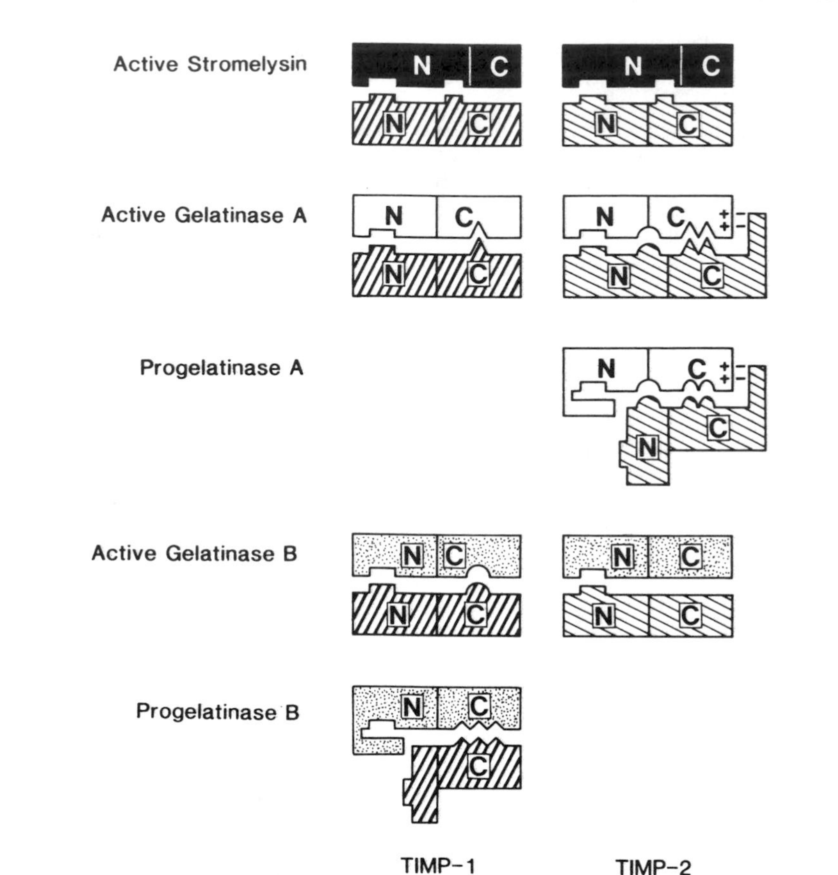

FIGURE 5. Domain interactions between TIMP-1 and TIMP-2 and active stromelysin-1, gelatinase A, progelatinase A, gelatinase B, and progelatinase B during complex formation. Both the inhibitors and the enzymes are shown as an N-terminal domain (N) and a C-terminal domain (C). The N-terminal domain of both TIMP-1 and TIMP-2 contain the inhibitory site that interacts with the catalytic N-terminal domain of each of the active MMPs. The C-terminal domain interactions appear to involve different sites for different MMPs. In the case of stromelysin-1, which binds TIMP-1 more rapidly than TIMP-2, the TIMP C-terminal domain interactions are with the catalytic domain of the enzyme. For gelatinase A, binding of TIMP-2 is more rapid than the binding of TIMP-1. This appears to be due in part to the ability of the nine amino acid negatively charged C terminus to bind specifically within the gelatinase A C-terminal domain. Gelatinase B binds TIMP-1 more rapidly and, again, the C-terminal domain interactions of both inhibitor and enzyme determine this property. The interactions between the TIMPs and gelatinase C-terminal domains defined using the active enzymes may be those that determine the specific proenzyme binding patterns, but this has not been studied, and the binding sites are therefore represented differently in the figure.

nature of these latter interactions is not yet well understood. However, we found that neither deletion of the fibronectin-like region of the gelatinase A N-terminal domain, nor modification of the active site Glu375 of the zinc box of the enzyme active site to Ala invokes any changes in the rate of TIMP binding, because mutants with these features had comparable k_{on} values to the wild-type gelatinase A.[15,23]

By contrast, similar mutagenesis studies with stromelysin-1 have shown that TIMP-1 binds more quickly than TIMP-2 to the active form of this MMP (TABLE 1). The rate of inhibition of stromelysin is affected to a much lesser extent by the C-terminal domain of either TIMP than is gelatinase A, and ionic interactions cannot be demonstrated. Removal of the C-terminal domain of stromelysin also does not result in any significant changes in the association rate constants, and we conclude that any TIMP C-terminal domain interactions are with the catalytic domain of stromelysin only (FIG. 5).

TIMP-1 binds more rapidly to gelatinase B than TIMP-2 (TABLE 1). As for the binding of TIMP-2 to gelatinase A, it seems likely that this is due to more efficient interaction between the C-terminal domains of the two proteins (FIG. 5). The fact that TIMP-1 can bind to progelatinase B through binding of these domains may similarly modify its ability to inhibit the exposed catalytic domain of gelatinase B as activation occurs.

As already discussed, TIMP inhibition of the active MMPs largely occurs as a result of binding the catalytic (N-terminal) domains of the enzymes and, hence, the active site. Competition between TIMPs and low molecular weight peptide inhibitors (based on substrate sequences) has been observed.[40,41] The amino acid residues in the TIMPs that are responsible for binding at or near the enzyme active site have not yet been defined. Two sequences in TIMP-1 (Asp16-Lys22 and Glu82-Glu87) could fulfill the MMP specificity requirements.[42] However, mutations within these regions did not alter the affinity of TIMP-1 for matrilysin,[43] or the rate of association of TIMP-1 and gelatinase A,[44] suggesting that these sequences are not important for TIMP activity. The data from a systematic site-directed mutagenesis study of TIMP-1 showed that residues His7 and Gln9 had some role because mutations at these positions decreased the affinity for matrilysin.[43] Further studies have shown that other residues in this region (Thr10-Phe12) have no demonstrable involvement in enzyme binding and that His7 and Gln9 mutations simply result in a form of TIMP-1 that has the kinetic characteristics of the N-terminal domain mutant when assessed using different MMPs (TABLE 2).[45] It is therefore possible that these residues are responsible in some way for maintaining the overall conformation of full-length TIMP so that the N- and C-terminal domains are correctly orientated.

We can conclude that the mechanism of TIMP action is complex, involving numerous points of interaction with the MMPs. The TIMP C-terminal domain has a number of enzyme binding sites which differ according to the MMP, and act to increase the rate of inhibition. The mechanism for this rate enhancement is by an increase of the probability of interaction of the two N-terminal domains. Further analysis of this binding should be facilitated by structural studies of enzyme-inhibitor complexes.

TABLE 2. Kinetic Analysis of TIMP-1 Mutants

	Wild Type	H7A	Q9A
		$10^{-4} \times k_{on}$ ($M^{-1}s^{-1}$)	
Matrilysin	29	42	nd
(K_inM	0.37	2.25	1.20)
Gelatinase A	420	30	20
N-gelatinase A	1.5	5.0	nd
Stromelysin	200	190	140
N-stromelysin	190	140	200

ACKNOWLEDGMENTS

We thank Mark Cockett and Mary Harrison for their invaluable contributions to this work and John Reynolds for discussions.

REFERENCES

1. BEBBINGTON, C. R., G. RENNER, S. THOMSON, D. KING, D. ABRAMS & G. T. YARRANTON. 1992. Bio/Technology **10:** 169–175.
2. SPRINGMAN, E. B., E. L. ANGLETON, H. BIRKEDAL-HANSEN & H. E. VAN WART. 1990. Proc. Natl. Acad. Sci. USA **87:** 364–368.
3. NAGASE, H., Y. OGATA, K. SUZUKI, J. J. ENGHILD & G. SALVESEN. 1991. Biochem. Soc. Trans. **19:** 715–718.
4. GAVRILOVIC, J., R. M. HEMBRY, J. J. REYNOLDS & G. MURPHY. 1987. J. Cell Sci. **87:** 357–362.
5. MOSCATELLI, D. & D. B. RIFKIN. 1988. Biochim. Biophys. Acta **948:** 67–85.
6. MURPHY, G., F. WILLENBROCK, R. V. WARD, M. I. COCKETT, D. EATON & A. J. P. DOCHERTY. 1992. Biochem. J. **283:** 637–641.
7. MURPHY, G., J. A. ALLAN, F. WILLENBROCK, M. I. COCKETT, J. P. O'CONNELL & A. J. P. DOCHERTY. 1992. J. Biol. Chem. **267:** 9612–9618.
8. VASSALLI, J.-D., A.-P. SAPPINO & D. BELIN. 1991. J. Clin. Invest. **88:** 1067–1072.
9. ALLAN, J. A., R. M. HEMBRY, S. ANGAL, J. J. REYNOLDS & G. MURPHY. 1991. J. Cell Sci. **99:** 789–795.
10. WARD, R. V., R. M. HEMBRY, J. J. REYNOLDS & G. MURPHY. 1991. Biochem. J. **278:** 179–187.
11. WARD, R. V., S. J. ATKINSON, P. M. SLOCOMBE, A. J. P. DOCHERTY, J. J. REYNOLDS & G. MURPHY. 1991. Biochim. Biophys. Acta **1079:** 242–246.
12. ATKINSON, S. J., R. V. WARD, J. J. REYNOLDS & G. MURPHY. 1992. Biochem. J. **288:** 605–611.
13. STETLER-STEVENSON, W. G., H. C. KRUTZSCH, M. P. WACHER, I. M. K. MARGULIES & L. A. LIOTTA. 1989. J. Biol. Chem. **264:** 1353–1356.
14. CRABBE, T., C. IOANNOU & A. J. P. DOCHERTY. 1993. Eur. J. Biochem. **218**: 431–438.
15. CRABBE, T., S. ZUCKER, M. I. COCKETT, F. WILLENBROCK, S. TICKLE, J. P. O'CONNELL, J. M. SCOTHERN, G. MURPHY & A. J. P. DOCHERTY. 1994. Biochemistry. In press.
16. BROWN, P. D., A. T. LEVY, I. M. K. MARGULIES, L. A. LIOTTA & W. G. STETLER-STEVENSON. 1990. Cancer Res. **50:** 6184–6191.
17. OVERALL, C. M. & J. SODEK. 1990. J. Biol. Chem. **265:** 21141–21151.
18. ATKINSON, S. Unpublished data.
19. AZZAM, H. S. & E. W. THOMPSON. 1992. Cancer Res. **52:** 4540–4544.
20. HOWARD, E. W. & M. J. BANDA. 1991. J. Biol. Chem. **266:** 17972–17977.
21. FRIDMAN, R., R. R. FUERST, R. E. BIRD, M. HOYHTYA, M. OELKUCT, S. KRAUS, D. KOMAREK, L. A. LIOTTA, M. L. BERMAN & W. G. STETLER-STEVENSON. 1992. J. Biol. Chem. **267:** 15398–15405.
22. WARD, R. V., S. J. ATKINSON, J. J. REYNOLDS & G. MURPHY. 1994. Biochem. J. In press.
23. MURPHY, G., Q. NGUYEN, M. I. COCKETT, S. J. ATKINSON, J. A. ALLAN, C. G. KNIGHT, F. WILLENBROCK & A. J. P. DOCHERTY. 1994. J. Biol. Chem.
24. STRONGIN, A. Y., I. E. COLLIER, P. A. KRASNOV, L. T. GENRICH, B. L. MARMER & G. I. GOLDBERG. 1993. Kidney Int. **43:** 158–162.
25. MURPHY, G. Unpublished data.
26. EMONARD, H. P., A. G. REMACLE, A. C. NOEL, J.-A. GRIMAUD, W. G. STETLER-STEVENSON & J.-M. FOIDART. 1992. Cancer Res. **52:** 5845–5848.
27. MURPHY, G. & R. V. WARD. Unpublished data.
28. POULSOM, R., A. M. HANBY, M. PIGNATELLI, R. E. JEFFERY, J. M. LONGCROFT, L. ROGERS & G. W. H. STAMP. 1993. J. Clin. Pathol. **46:** 429–436.
29. PYKE, C., E. RALFKIAER, K. TRYGGVASON & K. DANO. 1993. Am. J. Pathol. **142:** 359–365.
30. DOCHERTY, A. J. P., A. LYONS, B. J. SMITH, E. M. WRIGHT, P. E. STEPHENS, T. J. R.

HARRIS, G. MURPHY & J. J. REYNOLDS. 1985. Nature **318:** 66–69.
31. BOONE, T. C., M. J. JOHNSON, Y. A. DECLERCK & K. E. LANGLEY. 1990. Proc. Natl. Acad. Sci. USA **87:** 2800–2804.
32. PAVLOFF, N., P. W. STASKUS, N. S. KISHNANI & S. P. HAWKES. 1992. J. Biol. Chem. **267:** 17321–17326.
33. APTE, S. S., M.-G. MATTEI & B. R. OLSEN. 1994. Genomics **19:** 86–90.
34. APTE, S. S., K. HAYASNI, M. F. SELDIN, M.-G. MATTEI, M. HAYASHI & B. R. OLSEN. 1994. Development. In press.
35. STASKUS, P. W., F. R. MASIARZ, L. J. PALLANCK & S. P. HAWKES. 1991. J. Biol. Chem. **266:** 449–454.
36. WILLIAMSON, R. A., F. A. O. MARSTON, S. ANGAL, P. KOKLITIS, M. PANICO, H. R. MORRIS, A. F. CARNE, B. J. SMITH, T. J. R. HARRIS & R. B. FREEDMAN. 1990. Biochem. J. **268:** 267–274.
37. GOLDBERG, G. I., A. STRONGIN, I. E. COLLIER, L. T. GENRICH & B. L. MARMER. 1992. J. Biol. Chem. **267:** 4583–4591.
38. KNIGHT, C. G., F. WILLENBROCK & G. MURPHY. 1992. FEBS Lett. **296:** 263–266.
39. WILLENBROCK, F., T. CRABBE, P. M. SLOCOMBE, C. W. SUTTON, A. J. P. DOCHERTY, M. I. COCKETT, M. O'SHEA, K. BROCKLEHURST, I. R. PHILLIPS & G. MURPHY. 1993. Biochemistry **32:** 4330–4337.
40. LELIEVRE, Y., R. BOUBOUTOU, J. BOIZIAU, D. FAUCHER, D. ACHARD & T. CARTWRIGHT. 1990. Matrix **10:** 292–299.
41. MURPHY, G. Unpublished data.
42. WOESSNER, J. F. 1991. FASEB J. **5:** 2145–2154.
43. O'SHEA, M., F. WILLENBROCK, R. A. WILLIAMSON, M. I. COCKETT, R. B. FREEDMAN, J. J. REYNOLDS, A. J. P. DOCHERTY & G. MURPHY. 1992. Biochemistry **31:** 10146–10152.
44. WILLENBROCK, F. Unpublished data.
45. WILLENBROCK, F., H. M. MORRIS & G. MURPHY. Unpublished data.
46. NGUYEN, Q., F. WILLENBROCK, M. I. COCKETT, M. O'SHEA, A. J. P. DOCHERTY & G. MURPHY. 1994. Biochemistry. In press.
47. O'CONNELL, J. P., F. WILLENBROCK, A. J. P. DOCHERTY, D. EATON & G. MURPHY. 1994. J. Biol. Chem. In press.

Matrix Metalloproteinase Gene Expression[a]

LYNN M. MATRISIAN

Department of Cell Biology
Vanderbilt University
Nashville, Tennessee 37232

The dramatic overexpression of members of the matrix-degrading metalloproteinase (MMP) family in pathological conditions characterized by connective tissue destruction, including diseases such as arthritis, periodontitis, and cancer, suggests that tight regulation of MMP genes is critical for normal tissue homeostasis. An understanding of the molecular mechanisms controlling normal MMP gene expression and of the misregulation that occurs in disease processes may therefore provide leads for eventual rational therapeutic interventions.

NORMAL MMP GENE EXPRESSION

The regulation of MMP genes in normal tissues has not been thoroughly examined at this point, but initial studies point to complex and highly individualized patterns of expression for the various members of the MMP family. Examples of cell type- and tissue-specific regulation, inducible and constitutive expression, and discrepancies between *in vitro* and *in vivo* patterns of expression add to the complexity. Although the field is still evolving, certain generalities are emerging from available data.

Of the MMPs examined thus far, the 72-kDa gelatinase A appears to be the most widely expressed. In developing mouse embryos, gelatinase A mRNA is expressed in most mesenchymal-derived tissues and highly expressed in newborn lung, heart, and kidney.[1] The enzyme is frequently detected in normal human adult tissues and is elevated in cancerous lesions of these tissues.[2,3]

Although the 72-kDa gelatinase A is expressed in many normal tissues, the mRNA is generally restricted to connective tissue cells and rarely observed in epithelial tissues *in vivo.* Stromelysin-3 mRNA is also expressed in connective tissue, but the expression is much more restricted and has been observed in specific areas such as in developing digits[4] and involuting mammary gland.[5] In contrast, the mRNA for matrilysin has been observed in glandular epithelial cells of normal mouse small intestine (unpublished data) and the cycling human endometrium,[6] and is not seen in stromal cells of the same tissues.

Hematopoietic cells also show distinct and highly specific patterns of expression of MMPs. Neutrophils contain a collagenase distinct from interstitial collagenase.[7] The expression of MMPs in cells of the monocyte/macrophage lineage demonstrates stage-specific expression patterns. Promonocytes produce matrilysin mRNA,[8] whereas activated macrophages produce stromelysin-1 and interstitial collagenase mRNA.[9,10]

The 92-kDa gelatinase B is also expressed in hematopoietic cells.[11] Interestingly, other cell types, such as fibroblast and epithelial cells, have been shown to produce

[a]This work is supported by grants from the National Institutes of Health (CA46843 and CA60867) and Syntex Research.

gelatinase B in culture.[11-13] *In situ* hybridization studies of developing mouse embryos, however, suggest that the mRNA for this enzyme is restricted to osteoclasts and bone marrow cells *in vivo.*[14] Such studies suggest that culture conditions may alter the normal expression patterns of MMPs.

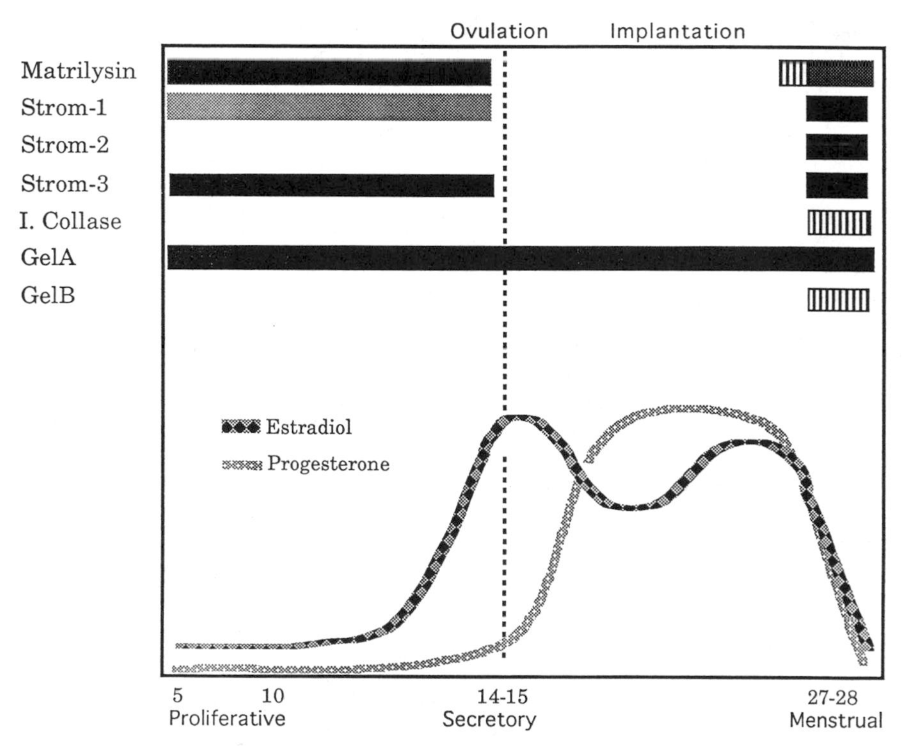

FIGURE 1. Summary of localization of metalloproteinase mRNA in the cycling human endometrium. Biopsy and hysterectomy samples were analyzed for the expression of the indicated metalloproteinases by *in situ* hybridization.[6,15] Bars indicate where samples were positive for MMP mRNA. The lighter grey bar indicates that stromelysin-1 was not expressed in all proliferative phase samples examined. The alternating black and white bars indicate focal expression of the MMP. The relative levels of circulating estradiol and progesterone through the stages of the cycle are indicated. The numbers indicate the day of the cycle. *Abbreviations:* Strom, stromelysin; I. Collase, interstitial collagenase; Gel, gelatinase.

MMP EXPRESSION IN THE CYCLING HUMAN ENDOMETRIUM

Perhaps the most dramatic expression of MMPs in a normal tissue is seen in the cycling human endometrium. An examination of MMP mRNA expression in this tissue clearly demonstrates the differences in the regulation of each individual MMP family member (summarized in FIG. 1).

The mRNA for the MMPs interstitial collagenase, stromelysin-1, -2, and -3, matrilysin, gelatinase A, and gelatinase B are all expressed during the menstrual phase in the human endometrium, but display differences in cellular distributions.[6,15]

Matrilysin is expressed in the surface epithelium and glandular epithelial cells throughout the tissue. Interstitial collagenase and gelatinase B are in stromal cells, but concentrated in the luminal region of the tissue. The transcripts for stromelysin-3 and gelatinase A are localized throughout the stromal component of the tissue. The widespread expression of MMPs in all compartments of this tissue suggests that the concerted effort of several MMP family members may play an important role in the breakdown and release of endometrial tissue during menstruation.

The mRNA for the 72-kDa gelatinase A is the only MMP transcript observed throughout the entire menstrual cycle, although there is some variation in the intensity of the signal.[15] The level of gelatinase A mRNA is somewhat reduced during the secretory phase of the cycle when progesterone levels are elevated. The mRNA for all other MMPs is absent during this phase of the cycle, suggesting an inhibitory effect of progesterone on the expression of each of these genes. Inasmuch as the tissue is preparing for implantation at this stage of the cycle, it is possible that repression of MMP expression may be critical for maintaining an appropriate endometrial environment for embryo implantation and development.

In addition to the expression of gelatinase A during the proliferative phase of the cycle, matrilysin[6] and stromelysin-3[15] transcripts are detected in proliferative epithelial and stromal cells, respectively. This pattern suggests that distinct mechanisms are regulating the expression of the genes for these MMPs. The expression of specific MMP proteins during the proliferative phase of the cycle may contribute to the growth and restructuring of the tissue that occur at this time.

MMP EXPRESSION IN CANCER AND ARTHRITIS

For many years, MMPs have been described as being overexpressed in cancer. Recent studies using *in situ* hybridization to localize MMP transcripts to specific tumor components have produced surprising and interesting results. Elevated levels of 72-kDa gelatinase are present in many human tumors, including breast and colon.[2,3] Interestingly, the mRNA is localized to stromal cells surrounding the tumor cells,[16,17] whereas the protein is detected in tumor cells themselves.[18] Other MMPs are also expressed in the stroma surrounding the tumor cells in breast and colon adenocarcinomas, including stromelysin-1 and stromelysin-3.[19] In contrast, matrilysin is expressed in the malignant epithelium of breast,[19] gastric,[20] prostate,[21] and colon[22] adenocarcinomas. The tissue-specificity of the expression may be related to the tissue specificity observed in normal tissues, that is, the restriction of matrilysin to epithelial cells and stromelysin-1 and stromelysin-3 to connective tissue cells (see above).

An examination of MMP expression in cartilage from patients with rheumatoid arthritis demonstrated elevated expression of stromelysin and collagenase mRNA and protein in the synovial lining, hyperproliferative pannus cells overlaying destroyed cartilage, and in the chondrocytes themselves.[23–28] McCracken[26] has determined that the mRNA for stromelysin-1 is localized to cells positive for a monocyte/macrophage marker as well as cells of fibroblast lineage. The expression of stromelysin and interstitial collagenase in fibroblasts and chondrocytes is consistent with the concept that these enzymes are expressed primarily in tissues of mesenchymal origin.

INDUCERS AND INHIBITORS OF MMP GENE EXPRESSION

MMP expression in normal tissues appears to be restricted to specific tissues at specific times, with the possible exception of the 72-kDa gelatinase, which in general appears to be more widespread. In several cases, for example in the cycling human endometrium (see above), involuting uterus,[29] and involuting mammary gland,[5,30] the pattern of expression suggests regulation by reproductive hormones. *In vitro,* estrogen and progesterone have been shown to regulate collagenase and stromelysin expression.[31,32] It is not clear that the regulation is always direct, and evidence for paracrine mediators of hormonal regulation of metalloproteinases has been obtained in cultures of isolated endometrial stromal and epithelial cells.[33] Growth factors such as EGF or TGFβ are candidate regulators of MMP expression, because there are numerous examples of MMP regulation by growth factors in tissue culture cells.[34]

The presence of high levels of MMP mRNA in specific disease states appears to be the result of a specific inductive mechanism that alters MMP gene expression. TNFα and IL-1 are believed to be mediators of stromelysin and collagenase induction in rheumatoid arthritis.[23,27,28] The genetic changes that activate oncogenes in the carcinogenic process may also result in the induction of MMP genes. Indeed, several oncogenes have been shown to induce the genes for stromelysin and collagenase in tissue culture systems.[35,36]

MOLECULAR MECHANISMS REGULATING MMP GENE EXPRESSION

The mechanism by which MMP genes are induced is generally believed to be mediated at the transcriptional level, although evidence also exists for regulation of mRNA stability of stromelysin, collagenase, and gelatinase A.[37,38] Interestingly, an analysis of the promoter sequence of several MMP genes gives insights into the similarities and differences observed in the expression patterns of the different MMP family members.

The human gelatinase A promoter has several of the characteristics of a housekeeping, or constitutive promoter[39] (FIG. 2). This observation may help to explain the widespread expression of gelatinase A. Overall *et al.*[40] suggested that gelatinase A may play a role in removing unfolded or abnormal collagen that is not incorporated into collagen fibrils, and thus contributes to the formation of a functional connective tissue.

The promoters for gelatinase B, interstitial collagenase, stromelysin-1, stromelysin-2, and matrilysin have some intriguing similarities (FIG. 2). These genes have a classical TATA element at position -30 ± 2, and have an AP-1 element at position -70 ± 4. AP-1 elements bind the transcription factor AP-1, consisting of a complex of proteins including members of the c-fos and c-jun family. The AP-1 element is required for significant transcription from the collagenase[41] and stromelysin[42] genes, and c-fos and c-jun have been shown to be required for growth factor[43] and oncogene[36] regulation of stromelysin and collagenase gene expression. These promoters also have either one or two PEA-3 transcription elements, recognizing transcription factors in the c-ets family,[44] located at a variable distance from the AP-1 site but within the first 300 bases of each of these promoters. The combination of AP-1 and PEA-3 sites has been referred to as a tumor promoter and oncogene-response element, and is responsible for growth factor and oncogene induction of collagenase[45] and stromelysin.[46] The similarities in these elements between the different MMPs may help to explain why several members of the MMP family are induced in similar situations, that is, during menstruation and in advanced-stage cancers.

The molecular mechanism controlling inhibition of MMP gene expression has been examined for several cases: dexamethasone inhibition of interstitial collagenase,[47,48] retinoic acid repression of rat stromelysin,[49] and TGFβ repression of rat

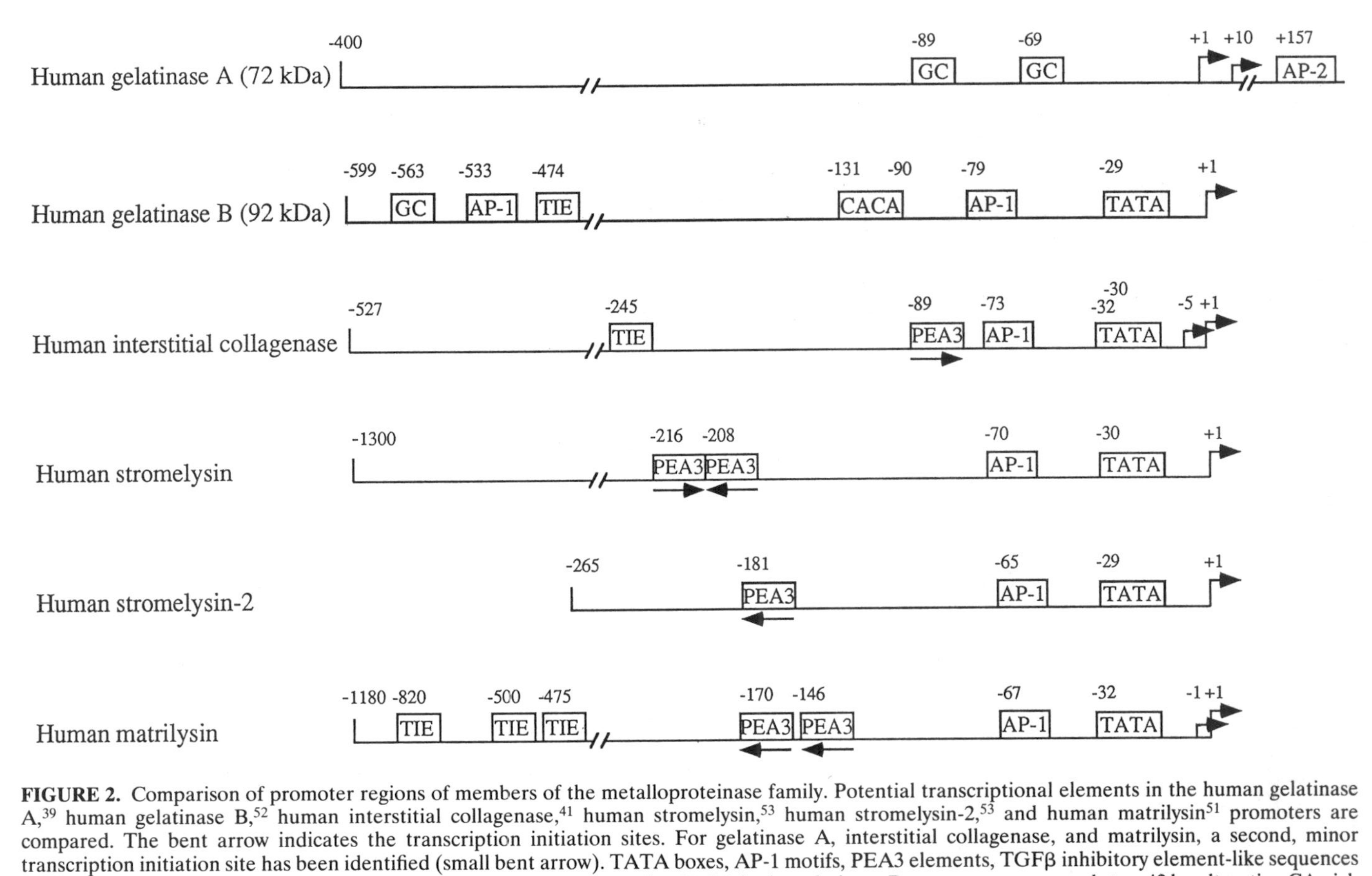

FIGURE 2. Comparison of promoter regions of members of the metalloproteinase family. Potential transcriptional elements in the human gelatinase A,[39] human gelatinase B,[52] human interstitial collagenase,[41] human stromelysin,[53] human stromelysin-2,[53] and human matrilysin[51] promoters are compared. The bent arrow indicates the transcription initiation sites. For gelatinase A, interstitial collagenase, and matrilysin, a second, minor transcription initiation site has been identified (small bent arrow). TATA boxes, AP-1 motifs, PEA3 elements, TGFβ inhibitory element-like sequences (TIE), AP-2 sites, and GC box/SP-1 binding sites (GC) are indicated. The CACA in the gelatinase B promoter corresponds to a 42 bp alterating CA-rich sequence between positions −131 and −90 in this promoter.[52] The orientation of the PEA3 elements according to the consensus sequence is indicated by an arrow. The nucleotide position of the identified elements relative to the transcriptional start site is shown above the element. Diagram is not to scale.

stromelysin.[50] In the case of dexamethasone and retinoic acid, a complex between the steroid and its receptor appears to directly interfere with AP-1 mediated expression, repressing transcription of the gene.[47–49] TGFβ appears to act differently, and a specific TGFβ response element, referred to as the TGFβ inhibitory element (TIE) in the rat stromelysin gene, is required for the repressive effect of TGFβ.[50] Interestingly, this repression may also involve the transcription factor c-fos, further complicating the understanding of positive and negative regulation of MMP gene expression.

The elements regulating the tissue specificity of MMP genes have not yet been examined. Initial studies comparing growth factor and tumor promoter induction of the promoter for human matrilysin and rat stromelysin demonstrated that, in HeLa cells which do not express the endogenous gene for either MMP, constructs containing the AP-1 or AP-1 plus PEA-3 elements have roughly equivalent activity.[51] It will be of interest to determine the specific elements that regulate the expression of matrilysin in epithelial cells and stromelysin-1 in mesenchymal-derived cells.

THERAPEUTIC IMPLICATIONS

The expression patterns of MMPs have interesting implications for the use of metalloproteinase inhibitors as therapeutic agents. Insights might be gained as to the preference for a general MMP inhibitor as opposed to an inhibitor designed to be specific for certain MMP family members as it relates to a defined disease state, and may give clues to potential side effects. For example, the expression of matrilysin, stromelysin-1, and stromelysin-3 in late-stage colon cancers suggests that the most efficacious inhibitor may be one designed to limit the activity of all of these enzymes. However, limiting the activity of the 72-kDa gelatinase B may produce side effects related to the housekeeping functions of this enzyme. Inhibitors of matrilysin and stromelysin-3 may have dramatic effects on the female reproductive cycle because of the apparent functions of these enzymes in proliferative endometrium. Although clearly the real answers to these questions can only be determined empirically, further examination of the expression patterns of the MMP genes in normal and pathological conditions can contribute significantly to the rational design of anti-MMP therapies.

Efforts to inhibit specific MMPs in disease states might also involve taking advantage of the intricate specificity built into the regulation of each MMP gene. An understanding of the molecular mechanism regulating the induction and repression of a specific MMP, as compared to other family members, may provide valuable leads to therapeutic agents with specificities not achievable by more conventional approaches.

REFERENCES

1. Reponen, P., C. Sahlberg, P. Huhtala, T. Hurskainen, I. Thesleff & K. Tryggvason. 1992. Molecular cloning of murine 72-kDa type IV collagenase and its expression during mouse development. J. Biol. Chem. **267:** 7856–7862.
2. Stetler-Stevenson, W. G. 1990. Type IV collagenases in tumor invasion and metastasis. Cancer Metastasis Rev. **9:** 289–303.
3. Tryggvason, K., M. Höyhtyä & C. Pyke. 1993. Type IV collagenases in invasive tumors. Breast Cancer Res. Treat. **24:** 209–218.
4. Basset, P., J. P. Bellocq, C. Wolf, I. Stoll, P. Hutin, J. M. Limacher, O. L.

PODHAJCER, M. P. CHENARD, M. C. RIO & P. CHAMBON. 1990. A novel metalloproteinase gene specifically expressed in stromal cells of breast carcinomas. Nature **348:** 699–704.

5. LEFEBVRE, O., C. WOLF, J.-M. LIMACHER, P. HUTIN, C. WENDLING, M. LEMEUR, P. BASSET & M.-C. RIO. 1992. The breast cancer-associated stromelysin-3 gene is expressed during mouse mammary gland apoptosis. J. Cell Biol. **119:** 997–1002.
6. RODGERS, W. H., K. G. OSTEEN, L. M. MATRISIAN, M. NAVRE, L. C. GIUDICE & F. GORSTEIN. 1993. Expression and localization of matrilysin, a matrix metalloproteinase, in human endometrium during the reproductive cycle. Am. J. Obstet. Gynecol. **168:** 253–260.
7. HASTY, K. A., T. F. POURMOTABBED, G. I. GOLDBERG, J. P. THOMPSON, D. G. SPINELLA, R. M. STEVENS & C. L. MAINARDI. 1990. Human neutrophil collagenase. A distinct gene product with homology to other matrix metalloproteinases. J. Biol. Chem. **265:** 11421–11424.
8. BUSIEK, D. F., F. P. ROSS, S. MCDONNELL, G. MURPHY, L. M. MATRISIAN & H. G. WELGUS. 1992. The matrix metalloproteinase matrilysin is expressed in developing human mononuclear phagocytes. J. Biol. Chem. **267:** 9087–9092.
9. CAMPBELL, E. J., J. D. CURY, S. D. SHAPIRO, G. I. GOLDBERG & H. G. WELGUS. 1991. Neutral proteinases of human mononuclear phagocytes. Cellular differentiation markedly alters cell phenotype for serine proteinases, metalloproteinases, and tissue inhibitor of metalloproteinases. J. Immunol. **146:** 1286–1293.
10. SHAPIRO, S. D., E. J. CAMPBELL, R. M. SENIOR & H. G. WELGUS. 1991. Proteinases secreted by human mononuclear phagocytes. J. Rheumatol. **18:** 95–98.
11. HIBBS, M. S. 1992. Expression of 92kDa phagocyte gelatinase by inflammatory and connective tissue cells. *In* Matrix Metalloproteinases and Inhibitors. H. Birkedal-Hansen, Z. Werb, H. G. Welgus & H. E. Van Wart, Eds.: 51–57. Gustav Fischer. Stuttgart.
12. SALO, T., J. G. LYONS, F. RAHEMTULLA, H. BIRKEDAL-HANSEN & H. LARJAVA. 1991. Transforming growth factor-β1 up-regulates type IV collagenase expression in cultured human keratinocytes. J. Biol. Chem. **266:** 11436–11441.
13. WATANABE, K., S. KINOSHITA & H. NAKAGAWA. 1990. Gelatinase secretion by glomerular epithelial cells. Nephron **56:** 405–409.
14. REPONEN, P., C. SAHLBERG, C. MUNAUT, I. THESLEFF & K. TRYGGVASON. 1994. High expression of 92-kDa type IV collagenase (gelatinase) in the osteoclast lineage during mouse development. N.Y. Acad. Sci. This volume.
15. RODGERS, W. H., L. M. MATRISIAN, L. C. GIUDICE, B. DSUPIN, P. CANNON, C. SVITEK, F. GORSTEIN & K. G. OSTEEN. 1994. Patterns of matrix metalloproteinase expression in cycling endometrium imply differential functions and regulation by steroid hormones. J. Clin. Invest. In press.
16. PYKE, C., E. RALFKLAER, K. TRYGGVASON & K. DANO. 1993. Messenger RNA for two type IV collagenases is located in stromal cells in human colon cancer. Am. J. Pathol. **142:** 359–365.
17. POULSOM, R., M. PIGNATELLI, W. G. STETLER-STEVENSON, L. A. LIOTTA, P. A. WRIGHT, R. E. JEFFERY, J. M. LONGCROFT, L. ROGERS & G. W. H. STAMP. 1992. Stromal expression of 72 kDa type IV collagenase (MMP-2) and TIMP-2 mRNAs in colorectal neoplasia. Am. J. Pathol. **141:** 389–396.
18. LEVY, A. T., V. CIOCE, M. E. SOBEL, S. GARBISA, W. F. GRIGIONI, L. A. LIOTTA & W. G. STETLER-STEVENSON. 1991. Increased expression of the M_r 72,000 type IV collagenase in human colonic adenocarcinoma. Cancer Res. **51:** 439–444.
19. WOLF, C., N. ROUYER, Y. LUTZ, C. ADIDA, M. LORIOT, J.-P. BELLOCQ, P. CHAMBON & P. BASSET. 1993. Stromelysin 3 belongs to a subgroup of proteinases expressed in breast carcinoma fibroblastic cells and possibly implicated in tumor progression. Proc. Natl. Acad. Sci. USA **90:** 1843–1847.
20. MCDONNELL, S., M. NAVRE, R. J. COFFEY & L. M. MATRISIAN. 1991. Expression and localization of the matrix metalloproteinase pump-1 (MMP-7) in human gastric and colon carcinomas. Mol. Carcinog. **4:** 527–533.
21. PAJOUH, M. S., R. B. NAGLE, R. BREATHNACH, J. S. FINCH, M. K. BRAWER & G. T.

Bowden. 1991. Expression of metalloproteinase genes in human prostate cancer. J. Cancer Res. Clin. Oncol. **117:** 144–150.

22. Newell, K., J. Witty, W. H. Rodgers & L. M. Matrisian. 1994. Expression and localization of matrix-degrading metalloproteinases during colorectal tumorigenesis. Mol. Carcinog. In press.
23. Nguyen, Q., J. S. Mort & P. J. Roughley. 1992. Preferential mRNA expression of prostromelysin relative to procollagenase and in situ localization in human articular cartilage. J. Clin. Invest. **89:** 1189–1197.
24. Case, J. P., H. Sano, R. Lafyatis, E. F. Remmers, G. K. Kumkumian & R. L. Wilder. 1989. Transin/stromelysin expression in the synovium of rats with experimental erosive arthritis. In situ localization and kinetics of expression of the transformation-associated metalloproteinase in euthymic and athymic Lewis rats. J. Clin. Invest. **24:** 1731–1740.
25. Firestein, G. S., M. M. Paine & B. H. Littman. 1991. Gene expression (collagenase, tissue inhibitor of metalloproteinases, complement, and HLA-DR) in rheumatoid arthritis and osteoarthritis synovium: Quantitative analysis and effect of intraarticular corticosteroids. Arthritis Rheum. **34:** 1094–1105.
26. McCachren, S. S. 1991. Expression of metalloproteinases and metalloproteinase inhibitor in human arthritic synovium. Arthritis Rheum. **34:** 1085–1093.
27. Hasty, K., R. A. Reife, A. H. Kang & J. M. Stuart. 1990. The role of stromelysin in the cartilage destruction that accompanies inflammatory arthritis. Arthritis Rheum. **33:** 388–397.
28. Okada, Y., M. Shinmei, O. Tanaka, K. Naka, A. Kimura, I. Nakanishi, M. T. Bayliss, K. Iwata & H. Nagase. 1992. Localization of matrix metalloproteinase 3 (stromelysin) in osteoarthritic cartilage and synovium. Lab. Invest. **66:** 680–690.
29. Woessner, J. F. & C. Taplin. 1988. Purification and properties of a small latent matrix metalloproteinase of the rat uterus. J. Biol. Chem. **263:** 16918–16925.
30. Talhouk, R. S., M. J. Bissell & Z. Werb. 1992. Coordinated expression of extracellular matrix-degrading proteinases and their inhibitors regulates mammary epithelial function during involution. J. Cell Biol. **118:** 1271–1282.
31. Marbaix, E., J. Donnez, P. J. Courtoy & Y. Eeckhout. 1992. Progesterone regulates the activity of collagenase and related gelatinases A and B in human endometrial explants. Proc. Natl. Acad. Sci. USA **89:** 11789–11793.
32. Sato, T., A. Ito, Y. Mori, K. Yamashita, T. Hayakawa & H. Nagase. 1991. Hormonal regulation of collagenolysis in uterine cervical fibroblasts. Modulation of synthesis of procollagenase, prostromelysin and tissue inhibitor of metalloproteinases (TIMP) by progesterone and oestradiol-17β. Biochem. J. **275:** 645–650.
33. Osteen, K. G., W. H. Rodgers, M. Gaire, J. T. Hargrove, F. Gorstein & L. M. Matrisian. 1994. Stromal-epithelial interactions regulate steroidal metalloproteinase expression in the human endometrium. Submitted.
34. Matrisian, L. M. & B. L. M. Hogan. 1990. Growth factor-regulated proteases and extracellular matrix remodeling during mammalian development. Curr. Top Dev. Biol. **24:** 219–259.
35. Matrisian, L. M., N. Glaichenhaus, M. C. Gesnel & R. Breathnach. 1985. Epidermal growth factor and oncogenes induce transcription of the same cellular mRNA in rat fibroblasts. EMBO J. **4:** 1435–1440.
36. Schonthal, A., P. Herrlich, H. J. Rahmsdorf & H. Ponta. 1988. Requirement for fos gene expression in the transcriptional activation of collagenase by other oncogenes and phorbol esters. Cell **35:** 325–334.
37. Delany, A. M. & C. E. Brinckerhoff. 1992. Post-transcriptional regulation of collagenase and stromelysin gene expression by epidermal growth factor and dexamethasone in cultured human fibroblasts. J. Cell. Biochem. **50:** 400–410.
38. Overall, C. M., J. L. Wrana & J. Sodek. 1991. Transcriptional and post-transcriptional regulation of 72-kDa gelatinase/type IV collagenase by transforming growth factor-beta1 in human fibroblasts: Comparisons with collagenase and tissue inhibitor of matrix metalloproteinase gene expression. J. Biol. Chem. **266:** 14064–14071.
39. Huhtala, P., L. T. Chow & K. Tryggvason. 1990. Structure of the human type IV collagenase gene. J. Biol. Chem. **265:** 11077–11082.

40. OVERALL, C. M., J. L. WRANA & J. SODEK. 1991. Independent regulation of collagenase, 72-kDa progelatinase, and metalloendoproteinase inhibitor expression in human fibroblasts by transforming growth factor-beta. J. Biol. Chem. **264:** 1860–1869.
41. ANGEL, P., I. BAUMANN, B. STEIN, H. DELIUS, H. J. RAHMSDORF & P. HERRLICH. 1987. 12-*O*-tetradecanoyl-phorbol-13-acetate induction of the human collagenase gene is mediated by an inducible enhancer element located in the 5′-flanking region. Mol. Cell. Biol. **7:** 2256–2266.
42. MATRISIAN, L. M., S. MCDONNELL, D. B. MILLER, M. NAVRE, E. A. SEFTOR & M. J. C. HENDRIX. 1991. The role of the matrix metalloproteinase stromelysin in the progression of squamous cell carcinomas. Am. J. Med. Sci. **302:** 157–162.
43. MCDONNELL, S. E., L. D. KERR & L. M. MATRISIAN. 1990. Epidermal growth factor stimulation of stromelysin mRNA in rat fibroblasts requires induction of proto-oncogenes c-*fos* and c-*jun* and activation of protein kinase C. Mol. Cell. Biol. **10:** 4284–4293.
44. WASYLYK, C., P. FLORES, A. GUTMAN & B. WASYLYK. 1989. PEA3 is a nuclear target for transcription activation by non-nuclear oncogenes. EMBO J. **8:** 3371–3378.
45. GUTMANN, A. & B. WASYLYK. 1990. The collagenase gene promoter contains a TPA and oncogene-responsive unit encompassing the PEA3 and AP-1 binding sites. EMBO J. **9:** 2241–2246.
46. WASYLYK, C., A. GUTMAN, R. NICHOLSON & B. WASYLYK. 1991. The c-Ets oncoprotein activates the stromelysin promoter through the same elements as several non-nuclear oncoproteins. EMBO J. **10:** 1127–1134.
47. JONAT, C., H. J. RAHMSDORF, K-K. PARK, A. C. B. CATO, S. GEBEL, H. PONTA & P. HERRLICH. 1990. Anti-tumor promotion and anti-inflammation: Down-modulation of AP-1 (Fos/Jun) activity by glucocorticoid hormone. Cell **62:** 1189–1204.
48. YANG-YEN, H.-F., J.-C. CHAMBARD, Y.-L. SUN, T. SMEAL, T. J. SCHMIDT, J. DROUIN & M. KARIN. 1990. Transcriptional interference between c-Jun and the glucocorticoid receptor: Mutual inhibition of DNA binding due to direct protein-protein interaction. Cell **62:** 1205–1215.
49. NICHOLSON, R. C., S. MADER, S. NAGPAL, M. LEID, C. ROCHETTE-EGLY & P. CHAMBON. 1990. Negative regulation of the rat stromelysin gene promoter by retinoic acid is mediated by an AP1 binding site. EMBO J. **9:** 4443–4454.
50. KERR, L. D., D. B. MILLER & L. M. MATRISIAN. 1990. TGF-beta1 inhibition of transin-stromelysin gene expression is mediated through a fos binding sequence. Cell **61:** 267–278.
51. GAIRE, M., Z. MAGBANUA, S. MCDONNELL, L. MCNEIL, D. H. LOVETT & L. M. MATRISIAN. 1994. Structure and expression of the human gene for the matrix metalloproteinase matrilysin. J. Biol. Chem. **269:** 2032–2040.
52. HUHTALA, P., A. TUUTTILA, L. T. CHOW, J. LOHI, J. KESKI-OJA & K. TRYGGVASON. 1991. Complete structure of the human gene for 92-kDa type IV collagenase. Divergent regulation of expression for the 92- and 72-kilodalton enzyme genes in HT-1080 cells. J. Biol. Chem. **266:** 16485–16490.
53. SIRUM, K. L. & C. E. BRINCKERHOFF. 1989. Cloning of the genes for human stromelysin and stromelysin-2: Differential expression in rheumatoid synovial fibroblasts. Biochemistry **28:** 8691–8698.

Regulation of Tissue Inhibitor of Matrix Metalloproteinase Expression[a]

CHRISTOPHER M. OVERALL

Faculty of Dentistry
University of British Columbia
2199 Wesbrook Mall
Vancouver, B.C. V6T 1Z3, Canada
(email: overall @ unixg.ubc.ca)

Changes in the balance between the synthesis and degradation of connective tissue matrix molecules can have profound effects on the composition and organization of the extracellular matrix. At the tissue level, this affects tissue morphology and function, while at the cellular level, changes in the composition of the extracellular matrix influence cell adhesion, migration, differentiation, and activity. Therefore, connective tissue formation and degradation are processes of critical importance in embryogenesis and tissue development, in normal tissue maintenance and remodeling, in the progression of many diseases, and in tissue restitution following wounding and pathology. Indeed, many morphogens and growth factors involved in cellular differentiation and tissue morphogenesis such as *all-trans*-retinoic acid and transforming growth factor-β1 (TGF-β1) have been shown to regulate the expression of extracellular matrix molecules, matrix metalloproteinases (MMPs), and the tissue inhibitors of matrix metalloproteinases (TIMPs).[1]

The interaction of cells with cell-attachment proteins influences cell phenotype and gene expression. Accordingly, in addition to direct effects, growth factors may contribute indirectly to the regulation of cell function by changes in the extracellular matrix composition. Thus, control of connective tissue degradation is of critical importance, not only with respect to the direct effects on the tissue, but also for the indirect consequences of altered cell phenotype and gene expression resulting from alterations in tissue architecture and loss of cell attachment. Therefore, considering the potentially destructive effects of uncontrolled tissue degradation, the balance between matrix synthesis and degradation must be precisely regulated to maintain the structural integrity and function of tissues.

Extracellular control of MMP activity can be achieved through the inhibition of activated MMPs by the specific inhibitor TIMP-1[2] and the related inhibitor TIMP-2.[3,4] TIMP is secreted by many cells[3,5] and is found in serum, amniotic fluid,[5,6] and saliva.[7] It forms high-affinity ($K_i < 10^{-10}$ M) complexes with active collagenase, 72-kDa gelatinase, and stromelysin.[3] TIMP-2 binds and inhibits active 72-kDa gelatinase and also complexes with latent 72-kDa gelatinase.[8,9] Thus, remodeling of connective tissue matrices can be controlled through the precise modulation of MMP and TIMP expression.

Collagenase, stromelysin-1, and TIMP-1 are regulated coordinately by a variety of stimuli,[10–12,18] prompting early workers to postulate that this was a general feature of MMP and TIMP regulation. However, despite these examples of "coordinate regulation," it is now clear that MMP and TIMP expression can be independently or even reciprocally regulated. For example, TGF-β1, a "formative" mediator that

[a] This work was supported by a grant from the Medical Research Council of Canada. C.M.O. was supported by a Medical Research Council of Canada Dental Clinician Scientist Award.

induces connective tissue deposition by fibroblasts[13] and bone cell populations,[14] suppresses collagenase synthesis while concomitantly increasing TIMP-1 and 72-kDa gelatinase expression.[15,16] TGF-β1 also reverses the increase in collagenase and amplifies the induction of TIMP-1 mediated by epidermal growth factor and fibroblast growth factor.[17] In contrast, other cytokines, such as interleukin-1 (IL-1),[18] epidermal growth factor, and fibroblast growth factor,[17] and the tumor promoter 12-*O*-tetradecanoylphorbol-13-acetate (TPA)[10,18,19,27] increase collagenase, stromelysin-1, and TIMP-1 expression coordinately, but the increase in TIMP-1 appears to negate the promotion of a "resorptive" cell phenotype.

Inasmuch as an understanding of both the regulation of MMP and TIMP expression and the extracellular activation of MMPs is of considerable importance in delineating the role of MMPs and TIMPs in pathological and physiological processes, we have investigated the mechanistic aspects of the regulation of MMPs and TIMPs by the morphogen *all-trans*-retinoic acid (retinoic acid), TGF-β1, and the lectin concanavalin A (ConA) in cells from both soft and mineralized connective tissues.

BONE CELL CULTURE

In an attempt to determine the effect of TGF-β1 and osseotropic hormones on TIMP-1 and MMP expression in normal bone cells, primary cultures of bone cell populations were prepared from 21-day old fetal rat calvaria by sequential enzymic digestion as described previously.[20] By use of this procedure distinct bone cell populations are separated. The cells recovered after the first enzymic digestion of the minced calvaria, rat calvarial population I (RCI), have been extensively characterized previously[14,20] and, in addition to osteoblasts, have been shown to contain cells having a more fibroblastic character. In contrast, the cells eluting in the last enzymic digest (RCIV), although still a mixed population, are predominantly osteoblastic in nature. In particular, these cells synthesize bone proteins and form bone nodules in culture.[21]

Primary cultures of RCI and RCIV cells were prepared by plating 1.3×10^6 cells in 100-mm tissue culture dishes and were then grown to confluence ($5–6 \times 10^6$) in α-minimum essential medium, pH 7.6, supplemented with 15% (v/v) heat-treated (56 °C, 20 min) fetal bovine serum and antibiotics at 37 °C in a humidified atmospheric mix of air/CO_2 (19:1 v/v). For experiments, confluent cultures were rendered quiescent after incubation in serum-depleted medium (0.5% v/v serum) for 24 h. The quiescent cultures were then incubated in the presence of either TGF-β1 (1.0 ng/mL), *all-trans*-retinoic acid (10^{-6} M), parathyroid hormone (PTH) (1.0 U/mL), IL-1 (10 U/mL), or vehicle for 24–72 h, the medium being collected and replenished each day as appropriate. Cell numbers were determined by use of a Coulter counter and protein synthesis determined after labeling cultures with 4 μCi/mL [^{35}S]methionine and precipitation with 10% (w/v) trichloroacetic acid as described previously.[15] RNA was prepared and analyzed by Northern blots and slot blots using cDNA probes labeled with [^{32}P]dCTP.[15] High-sensitivity zymography was performed according to Overall and Limeback.[22]

TIMP-1 AND MMP REGULATION IN BONE CELLS

The effects of retinoic acid, TGF-β1, PTH, and IL-1 on TIMP-1 mRNA expression were investigated by Northern and slot blot analyses. TIMP-1 mRNA, identified in the total cellular RNA isolated from RCIV cells at days 1, 2, and 3 after treatment, was elevated ~3.6-fold by TGF-β1 and by ~2-fold with PTH and IL-1 (FIG. 1). Contrasting these results, retinoic acid significantly reduced the levels of TIMP-1 mRNA by ~60%. In the more fibroblastic bone cell population (RCI), the effects were similar to those found in the RCIV cells, but some differences were noted. In particular, TGF-β1 and IL-1, although found to stimulate TIMP-1 mRNA levels (FIG. 2), did not increase the already high levels of TIMP-1 message to the same extent as that found with the RCIV cells. These results indicated that TIMP-1 expression in RCI cells was already at or near maximal and could not be further stimulated. That these cells were responsive to TGF-β1 has been previously shown by several criteria including a characteristic increase in the synthesis of specific connective tissue matrix components after TGF-β1 treatment.[14,23]

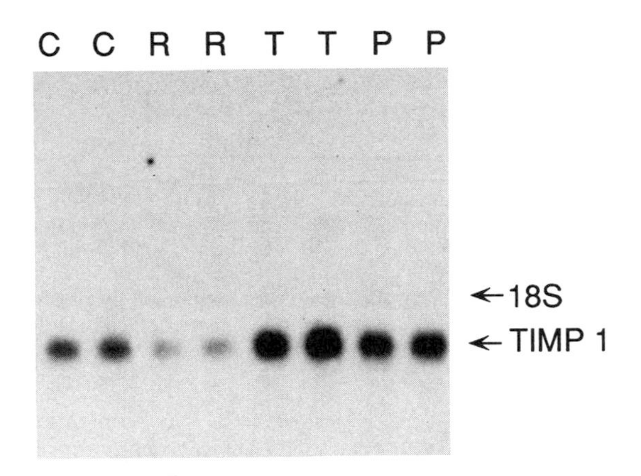

FIGURE 1. Northern hybridization analysis of TIMP-1 gene expression in rat bone cell populations. Quiescent confluent cultures of normal RCIV rat bone cells were treated in quadruplicate with either 1×10^{-6} M retinoic acid (R), 1.0 ng/mL TGF-β1 (T), 1.0 U/mL PTH (P), or vehicle (C) for 24 h. Cellular RNA was prepared and 20 μg RNA aliquots fractionated on a 1.2% agarose gel. The RNA blots were hybridized with [^{32}P]dCTP-labeled murine TIMP-1 cDNA (0.1 μg, specific activity 4.8×10^8 cpm/μg), washed finally in 0.2% SSC at 45 °C, and exposed to SB-5 X-ray film for 48 h. TIMP-1 mRNA and the migration position of the 18S ribosomal RNA are indicated.

Of particular interest in the experiments reported here were the observations that in addition to reducing TIMP-1 expression retinoic acid elevated stromelysin-1 mRNA levels in RCIV cells (FIG. 3). Zymography was also used to analyze gelatinase levels in the conditioned medium. 72-kDa gelatinase protein levels in the medium were unaffected by retinoic acid treatment (FIG. 4), but TGF-β1 elevated the synthesis of 72-kDa gelatinase in RCIV cells as found before with human gingival fibroblasts.[15,24]

MECHANISTIC ASPECTS OF TIMP-1 REGULATION IN BONE CELLS BY *ALL-TRANS*-RETINOIC ACID AND TGF-β1

The mechanistic aspects of the retinoic acid–induced decrease in TIMP-1 mRNA levels and the TGF-β1–induced increase were studied in further detail and com-

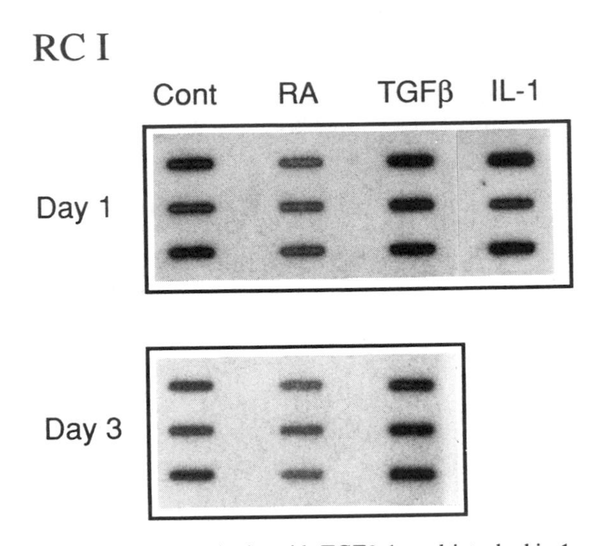

FIGURE 2. Influence of *all-trans*-retinoic acid, TGFβ-1, and interleukin-1 on TIMP-1 gene expression in RCI bone cell populations. Confluent, quiescent cultures of normal RCI rat bone cells were treated in triplicate with either 1×10^{-6} M retinoic acid (RA), 1.0 ng/mL TGF-β1 (TGFβ), 10 U/mL interleukin-1 (IL-1), or vehicle (Cont) for 24 or 72 h. Cellular RNA was prepared and 5 μg of RNA immobilized on BioTrans membrane by slot blotting. The RNA blots were hybridized with [^{32}P]dCTP-labeled murine TIMP-1 cDNA (50 ng, specific activity 4.8×10^8 cpm/μg), washed in 0.2% SSC at 50 °C, and exposed to X-ray film.

pared. A dose response analysis using retinoic acid at concentrations ranging from 10^{-6} to 10^{-11} M revealed that TIMP-1 mRNA levels were repressed in a dose-dependent manner (FIG. 5) with a 60% reduction in mRNA levels occurring at 10^{-6} M retinoic acid. This concentration was then used throughout the rest of the investigation.

To investigate whether the retinoic acid and TGF-β1–induced changes in mRNA levels occurred posttranscriptionally through alterations in the stability of the TIMP-1 message, the mRNA half-lives of TIMP-1 were determined after treatment of RCIV cells with either retinoic acid or TGF-β1 or vehicle for 24 h. 5,6-Dichloro-1-β-D-ribofuranosylbenzimidazole, a specific RNA polymerase II inhibitor, was then added to cultures after the 24-h treatments to follow the decrease in specific mRNA levels. Over the subsequent 24 h, RNA was collected at each time point from quadruplicate cultures and analyzed by slot blot analysis using stringency conditions determined from prior Northern blot analyses (e.g., FIG. 1). Compared with control cultures, the turnover of TIMP-1 mRNA was unaltered by either retinoic acid or TGF-β1 treatment with the TIMP-1 mRNA having a $t_{1/2}$ calculated to be ~14 h (Overall, manuscript submitted). Therefore, these data indicated that both the reduction in TIMP-1 gene expression by retinoic acid and the elevation of TIMP-1 mRNA levels by TGF-β1 occurred transcriptionally without posttranscriptional regulation of mRNA stability.

To determine the requirement for newly synthesized protein in the mediation of the retinoic acid and TGF-β1 specific responses, cycloheximide was added to cultures at a concentration (20 μg/mL) which was shown to reduce protein synthesis in these cells from 70–90%. However, after the addition of cycloheximide, mRNA levels specific for TIMP-1 were induced >10-fold regardless of whether the cells were also treated with retinoic acid or TGF-β1 or vehicle (manuscript submitted). The superinduction of TIMP-1 mRNA by cycloheximide therefore precluded the analysis of the requirement for newly synthesized protein to effect the observed increase (for TGF-β1) or decrease (for retinoic acid) in TIMP-1 mRNA levels in response to these agents. However, unlike TIMP-1, the increase in stromelysin-1 mRNA levels in response to 10^{-6} M retinoic acid was found to be cycloheximide-sensitive 12 h after treatment (FIG. 3). Thus, these data indicate that the induction of stromelysin-1 mRNA by retinoic acid requires the synthesis of new protein.

COMPARISONS WITH TIMP-1 REGULATION IN HUMAN FIBROBLASTS BY TGF-β1

The experiments reported here demonstrate that the effects of TGF-β1 on TIMP-1 gene expression in rat bone cells are similar to those we found previously using human fibroblasts. In these earlier studies, TGF-β1 increased the steady-state levels of TIMP-1 mRNA ~2.5-fold[15] through the direct effects of TGF-β1 on TIMP-1 gene transcriptional activity. However, the induction of TIMP-1 gene transcription was delayed compared to that of the extracellular matrix components α1(I) procollagen, fibronectin, and 72-kDa gelatinase genes which exhibited a high level of gene transcription 7 h after TGF-β1 treatment.[24] Further, and as found here, TGF-β1 did not alter the stability of the TIMP-1 mRNA.[24] This was also unlike the

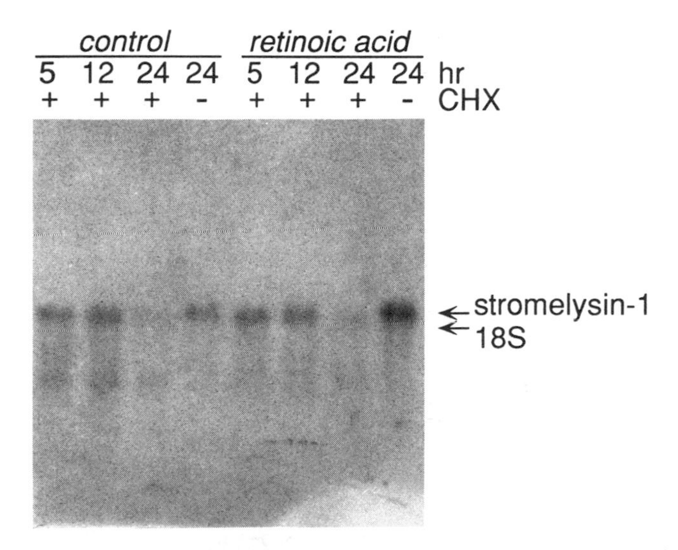

FIGURE 3. Northern hybridization analysis of the cycloheximide sensitivity of the *all-trans*-retinoic acid induced increase in stromelysin-1 mRNA in rat bone cell populations. Quiescent confluent cultures of normal RCIV rat bone cells were treated in the presence (+) or absence (−) of 20 μg/mL cycloheximide (CHX) with 1×10^{-6} M retinoic acid or vehicle for up to 24 h. Cellular RNA was prepared at the times indicated and 10 μg RNA aliquots were fractionated on a 1.2% agarose gel. The RNA blots were hybridized with [^{32}P]dCTP-labeled stromelysin-1 cDNA (0.1 μg, specific activity = 9×10^8 cpm/μg), final washes were in 0.2% SSC at 50 °C, and the blot exposed to X-Omat X-ray film for 24 h. Stromelysin-1 mRNA and the migration position of the 18S ribosomal RNA are indicated.

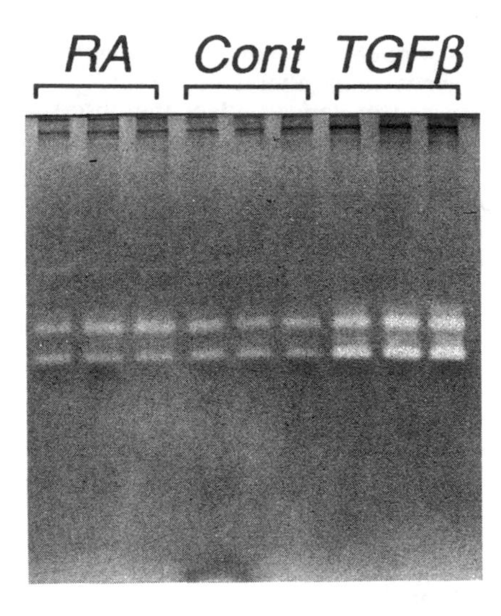

FIGURE 4. Zymogram analysis of conditioned culture medium from normal RCIV rat bone cell populations. Confluent quiescent cultures of RCIV cells were incubated either in the presence of 1×10^{-6} M retinoic acid (RA), 1.0 ng/mL TGF-β1 (TGFβ), or vehicle (Cont) for 48–72 h, the media collected and 2-μL aliquots electrophoresed in the presence of SDS on 10% polyacrylamide gels containing gelatin (40 μg/mL) as substrate. Following electrophoresis the gels were processed for enzymography, incubated in assay buffer for 120 min at 37 °C, and stained in Coomassie blue G-250.

effect of TGF-β1 on α1(I) procollagen, fibronectin, and 72-kDa gelatinase expression in which TGF-β1 was found to increase the half-life of the mRNAs from these genes.[24,25] Thus, it is clear that multiple pathways are utilized by TGF-β1 to modulate TIMP, MMP, and extracellular matrix protein gene expression. That the steady-state levels of TIMP-1 mRNA increased by TGF-β1 were translated into a corresponding amount of secreted, functional protein was confirmed by functional assays, performed after separation of TIMP-1 from MMPs by use of mini-affinity columns and by immunoprecipitation analyses.[15]

In these earlier studies, fibroblasts (Gin-1) from adult human gingival tissue were used to investigate the effects of TGF-β1 on MMP and TIMP-1 expression. In the present investigation fetal bone cells were used. However, Edwards *et al.*[17] found different effects of TGF-β1 on TIMP-1 and collagenase expression using human fetal lung (MRC 5) fibroblasts. In their report, TGF-β1 alone had *no* effect on TIMP-1, collagenase, collagen, or fibronectin expression. However, when TGF-β1 was added in combination with epidermal growth factor or fibroblast growth factor, TGF-β1 further increased the stimulatory effects of these growth factors on TIMP-1 expression and reduced the increase in collagenase mRNA levels. Unlike the study in which very late-passage (24–34) MRC 5 fetal lung fibroblasts were used,[17] in my study primary cultures of fetal rat bone cells were used. In addition, the human gingival fibroblasts used previously[15,24] were early-passage adult cells. Cells at high-passage number show transcriptional repression of c-*fos* induction,[26] which may form part of the explanation of the discrepancy between these results.

TIMP-1 GENE MODULATION BY CONCANAVALIN A AND PHORBOL ESTER

TIMP expression was markedly reduced by ConA[27] as revealed by functional assays of TIMP-1 in conditioned medium fractionated[15] to separate TIMP-1 from MMPs. This was confirmed by immunoprecipitation of TIMP-1 after metabolic labeling using an anti-TIMP-1 antibody that did not recognize TIMP-2. FIGURE 6 shows one such immunoprecipitation analysis in which the levels of TIMP-1 were found to be reduced by ~20% on day 1 and by ~50% on day 2 after ConA treatment. These data contrast the effects of TPA which caused a 1.4-fold *increase* in TIMP-1 protein levels on day 1 and a 1.9-fold increase on day 2 (FIG. 6) with a corresponding increase in TIMP mRNA levels at 24 h. The figure also shows that in addition to TIMP-1, active 72-kDa gelatinase [M_r of 59 kDa − dithiothreitol (DTT); 61 kDa + DTT)] complexed with TIMP-1 was also immunoprecipitated from the medium of ConA-treated cultures but not from either control or TPA-treated cells.

ConA also reduced TIMP-1 transcript levels from control levels at 24 h to barely detectable levels by 72 h.[27] However, the levels of immunoprecipitable and functional TIMP free in the medium were already decreased after 24 h of ConA treatment (FIG. 6), probably due to complexing of TIMP with the endogenously activated collagenase and 72-kDa gelatinase. That the transcriptional activity of the *timp-1* gene was decreased by ConA was confirmed by nuclear run-on analyses where it was found that *timp-1* transcription was reduced by ~80% after 24 h. Cycloheximide blocked both *timp-1* gene transcription and the reduction in TIMP-1 mRNA levels, indicating that these events were mediated, in part, by proteins with relatively short half-lives.[27]

MMP GENE MODULATION BY CONCANAVALIN A AND PHORBOL ESTER

The level of 72-kDa gelatinase in the medium was increased ~3-fold by ConA and, unlike the 66-kDa (−DTT) latent form synthesized by control and TPA-treated cells, the 72-kDa gelatinase was present as an activated 59-kDa (−DTT) form as analyzed by zymography (FIG. 6, lower panel). This is significant because, although some active collagenase and 72-kDa gelatinase are released from organ cultures,[28,29] active collagenase and active 72-kDa gelatinase have rarely been reported in short-term cell cultures. The endogenous activation of 72-kDa gelatinase to produce a 61-kDa form (+DTT) after ConA treatment of cells occurred in a TIMP-2–depleted environment, after TIMP-2 down-regulation by ConA treatment. This supports the possibility that TIMP-2 stabilizes the 66-kDa (+DTT) activation

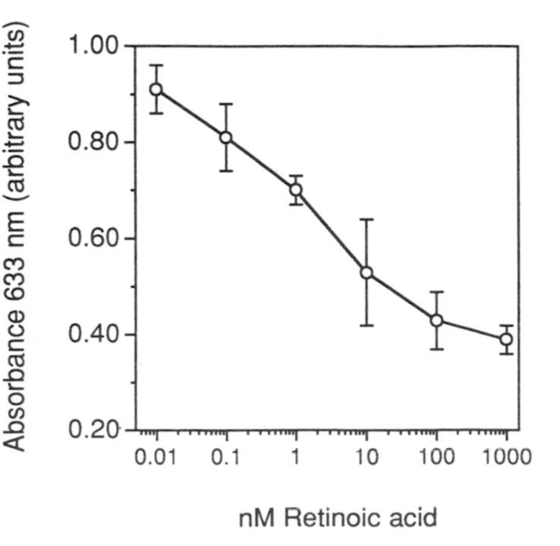

FIGURE 5. Dosiometric analysis of TIMP-1 mRNA levels in response to *all-trans*-retinoic acid. Total cellular RNA was prepared 24 h after stimulating confluent quiescent cultures of RCIV cells with retinoic acid at the concentrations indicated in quadruplicate. Ten-microgram RNA samples were then immobilized on Bio-Trans membrane with a slot blotting apparatus. As determined from prior Northern blot analyses, the slot blots were prepared using hybridization and washing conditions that did not result in nonspecific hybridization. After exposure of the X-ray film in the linear range of the film's response, the blots were densitometrically scanned and the mean ± S.D. plotted.

intermediate of 72-kDa gelatinase and prevents the final activation cleavage to convert the 66-kDa form to the fully active 61-kDa (+DTT, 59 kDa − DTT) species or to the high specific activity 43-kDa "mini"-N-gelatinase degradation product.[27] Thus, modulation of TIMP-2 expression may control the activation of 72-kDa

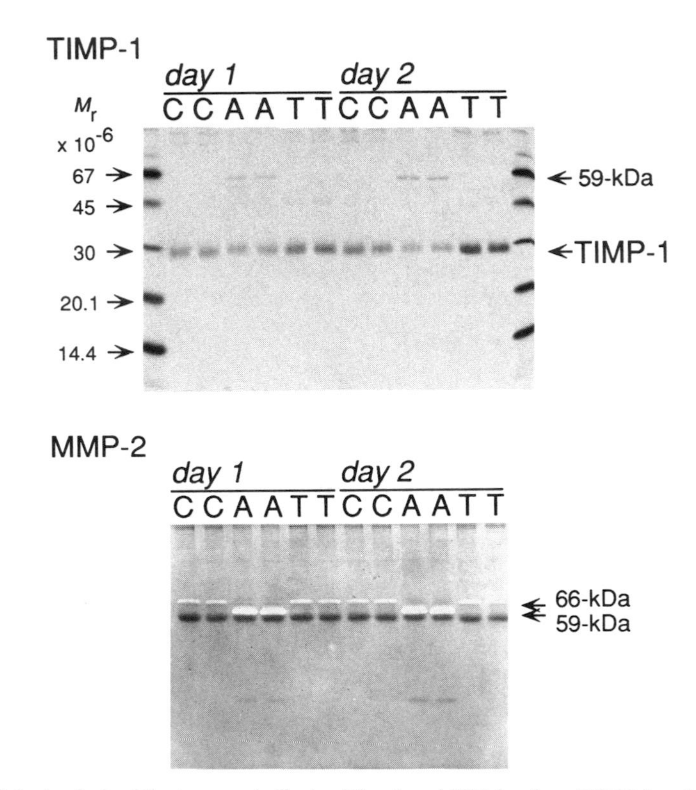

FIGURE 6. Analysis of the temporal effects of ConA and TPA levels on TIMP-1 and gelatinase activity in human fibroblasts. Using rabbit anti-human TIMP antibody, immunoprecipitates were prepared from [^{35}S]methionine-labeled conditioned culture medium collected from confluent cultures after treatment with either 20 μg/mL ConA (A), 50 ng/mL TPA (T), or vehicle (C). The immunoprecipitates from media, collected at the time points indicated, were then electrophoresed under reducing conditions on 15% SDS-PAGE gels and processed for fluorography. The positions of TIMP-1 and the 59-kDa band, identified as the activated form of the 72-kDa gelatinase that was complexed with TIMP in the medium and immunoprecipitated with the TIMP-antibody complex, are indicated. Molecular weight markers are labeled as appropriate. In the **lower panel,** a zymogram shows the gelatinolytic activity present in the conditioned culture medium at 24 or 48 h collected from human gingival fibroblasts treated, as described in the **upper panel.** Of note, essentially all the 72-kDa gelatinase was activated after ConA treatment and converted to the 59-kDa form from the latent 66-kDa form.

gelatinase to either the 61-kDa or the 43-kDa high specific activity forms. Since the first report on the endogenous activation of MMP-2 by ConA,[27] more recent work by Murphy's group in the United Kingdom and Stetler-Stevenson's group at the National Institutes of Health, Washington, D.C. have elucidated further details of the activation of 72-kDa gelatinase after ConA stimulation which now appears to be EDTA-sensitive and to occur on the cell membrane at specific 72-kDa gelatinase or TIMP-2 receptors.

A substantial increase ($\sim$16-fold) in collagenase mRNA was also observed after treatment of cells with ConA for 24–72 h. Although matrilysin (pump-1) mRNA was never detected in control cells, ConA treatment induced the expression of the matrilysin transcript.[27] In contrast, the 72-kDa gelatinase transcript was increased only slightly ($<$2-fold) by ConA. Thus, the reduced expression of TIMP-1 and TIMP-2 by ConA, the comparatively low-level increases in 72-kDa gelatinase expression compared with collagenase, and the lack of any stromelysin-1 induction by ConA, provide additional evidence for the independent regulation of MMPs and TIMPs.

The actions of ConA differed markedly from the effects of TPA, an established stimulator of collagenase and TIMP-1 expression.[18,19,30,31] ConA treatment induced considerable cell rounding and arborization that was not seen following TPA treatment.[27] Although the degree of TPA stimulation of collagenase protein ($>$30-fold at 24 h) and mRNA was greater than that for ConA ($\sim$16–20-fold), TPA did not alter the levels of either 72-kDa gelatinase protein in the medium (FIG. 6) or mRNA and did not generate activated forms of MMPs.

DISCUSSION

These studies have helped establish a regulatory role for TGF-β1 and *all-trans*-retinoic acid in the expression of fibroblast and osteoblast collagenolytic activity. TGF-β1 was found to suppress collagenase activity directly, by decreasing collagenase transcription, and indirectly, by increasing TIMP-1 transcription. TGF-β1 increased TIMP-1 protein and mRNA in human fibroblasts[15] as a result of altered transcriptional activities, through pathways that required protein synthesis, and without changes in mRNA degradation rates.[24] In rat bone cells TGF-β1 also markedly elevated TIMP-1 mRNA levels without alterations in mRNA half-life ($\sim$14 h). The general reduction in proteolytic activity by TGF-β1 would augment the synthesis of extracellular matrix, a process stimulated by TGF-β1, inducing a formative cellular phenotype.[32] Notably, 72-kDa gelatinase expression is stimulated by TGF-β1,[15,16,24] indicating that the production of 72-kDa gelatinase in latent form may be an integral part of connective tissue formation as well as resorption.

The actions of TGF-β1 also reveal a reciprocity with the induction of a resorptive cellular phenotype by ConA that was characterized by both an increased expression and activation of MMPs and a 10-fold reduction in TIMP-1 and TIMP-2 expression that occurred through transcriptional pathways requiring protein synthesis and without changes in mRNA stability.[27] The reciprocal regulation of MMPs and TIMPs would achieve a greater degradative activity than that obtained by increases in MMP expression alone. This appears to be a unique feature of the ConA stimulation of fibroblasts that has not been previously demonstrated with other growth factors or hormones.

Because ConA interacts with many types of cell surface receptors and cell membrane proteins, ConA may mimic the "multiple hits" that a cell responds to in the tissue milieu. Thus, although ConA is not a physiological cell activator, the receptor stimulation and intracellular signals induced by ConA *in vitro* may be more physiological than the "single hit" produced by a growth factor or hormone added alone to cell cultures.

The dose-dependent reduction in TIMP-1 mRNA levels following administration of retinoic acid was found in both the RCI and the RCIV bone cell populations.

Few reports have investigated the retinoic acid regulation of TIMP-1 in any system. Notably, my findings contrast those of both Clark *et al.*,[33] who reported that TIMP-1 expression was elevated ~2-fold by retinoic acid in human fibroblasts, and Wright *et al.*,[34] who found significant increases in TIMP-1 levels in two of three human synovial cell lines after retinoic acid treatment. The data I report here, coupled with the retinoic acid stimulation of collagenase[35] and stromelysin-1 in rat bone cells, reveal that retinoic acid can modulate MMP and TIMP-1 expression in a reciprocal manner to achieve a resorptive cell phenotype in fetal rat bone cells.

Although it is unlikely that the changes in the net levels of MMPs and TIMP-1 induced by retinoic acid are solely responsible for the major tissue changes that occur during retinoic acid–induced cell differentiation and morphogenesis in the developing embryo, the increased proteolytic activity of bone cells coupled with a reduced TIMP-1 expression would be expected to favor bone matrix resorption and remodeling. Moreover, osseous deformities associated with *in utero* exposure to retinoic acid may be the result of inappropriate modulation of MMP and TIMP-1 gene expression.

The reduced levels of TIMP-1 and the elevated collagenase and stromelysin-1 expression in bone cells following retinoic acid treatment are particularly interesting and appear to be tissue (or species) specific given the well-characterized reduction in fibroblast collagenase levels induced by retinoic acid.[36–40] Although the nuclear pathways involved with TIMP-1 regulation by retinoic acid have yet to be elucidated, considerable work has shown the involvement of retinoic acid receptor-γ 1 in reducing the levels of basal collagenase expression[41] and retinoic receptors α1, β1, and γ1 in reducing phorbol ester–stimulated collagenase expression.[41] These effects seem to be due, at least in part, to interaction of the retinoic acid receptors with the AP-1 proteins, *fos* and *jun*,[41,42] or through repression of c-*fos* induction.[43] Thus, the tissue-specific induction of collagenase and the down-regulation of TIMP-1 by retinoic acid in bone cells indicates that important differences occur either in the retinoic acid receptor repertoire utilized by these cells or in the involvement of other *trans*-acting factors such as *cEts1*(*p68*),[44] which is also involved in the retinoic acid and AP-1 response of stromelysin-1 and collagenase.[45]

As has been shown by many authors, with agents such as platelet-derived growth factor, epidermal growth factor, TPA, and IL-1 the expression of both MMPs and TIMPs is often up-regulated together. However, after the cloning of the genomic DNAs of the MMPs and TIMPs it became clear from the analysis of the 5′ regulatory regions that these controlling elements, although often similar, differ between the MMPs and TIMPs. Moreover, tissue-specific influences, such as those described here for TIMP-1 regulation by retinoic acid, have the potential to further modulate MMP and TIMP-1 gene expression. Thus, although coordinated regulation can occur, it is not necessarily due to identical regulation pathways, but rather from coincidental similar net effects after *trans* activation or repression of the MMP and TIMP genes. Indeed, the differential regulation of collagenase, stromelysin-1, 72-kDa gelatinase and matrilysin, and the reciprocal regulation of TIMP-1 by TGF-β1, ConA, and retinoic acid reviewed here clearly demonstrates that MMPs can be independently regulated.

A MODEL FOR MMP AND TIMP REGULATION

The data presented and cited show that precise, independent control of MMP and TIMP expression can occur by cells and that different MMPs may participate in different physiological and pathological processes at different times and at different tissue sites. As proposed in FIGURE 7, gradients of growth factors or morphogens (biological response modifiers) across sites of tissue remodeling may therefore differentially modulate MMP, TIMP, and extracellular matrix protein expression both spatially and temporally. Overall, normal connective tissue is characterized by a low constitutive level of extracellular matrix protein and TIMP expression and, other than 72-kDa gelatinase, an absence of MMP expression. On the other hand, at a remodeling connective tissue site a net resorptive cell phenotype may be expected at the focus of degradation, characterized on a per cell basis by increased MMP and reduced TIMP expression. The surrounding normal tissue is likely protected from MMP activity by a "green belt" comprising cells that exhibit an active formative phenotype induced by "formative factors" such as TGF-β, and which is marked by an up-regulation of TIMP, 72-kDa gelatinase, and matrix protein synthesis to both restitute local tissue and to buffer adjacent normal tissue from inappropriate MMP activity diffusing from the focus of degradation. In addition to a spatial distribution of growth factors or other regulatory influences, temporal gradients of these biologi-

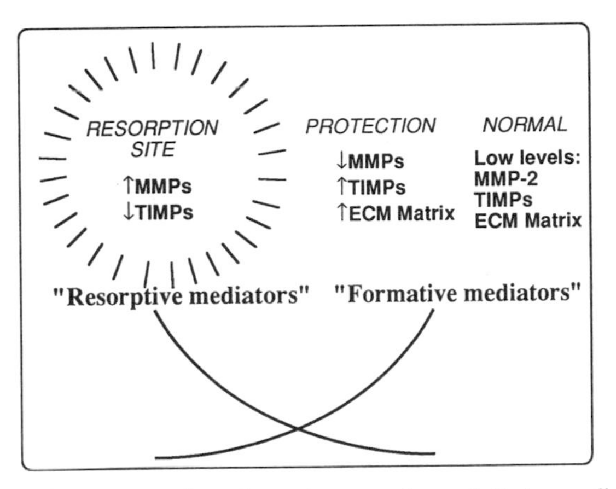

FIGURE 7. Schematic representation of morphogen and growth factor gradients across a site of tissue remodeling and their possible influence on connective tissue matrix degradation and formation according to location.

cal response modifiers would also occur as a particular site came under the influence of alterations in concentrations of the "biological response modifiers" and changes in connective tissue matrix integrity during the normal progression of the tissue process. In particular, the very important "turning off" of a resorptive cell phenotype, whether by a passive rundown of the inducing signal as it decays in the local environment or by an induced active shutdown of MMP expression, such as can occur by growth factor–induced autocrine and paracrine expression of TGF-β1, must also occur with time at a particular tissue site. In this model, the spatial and temporal distribution of biological response modifiers would also be reflected by spatial and temporal differences in the expression of the different MMPs and TIMPs, often in a tissue-specific manner, to meet the particular needs of the tissue and biological process.

It is important to emphasize that although the reciprocal and independent regulation of MMP and TIMP expression at the single-cell level would influence the local environment in a concerted manner, these changes may not be necessarily reflected at the tissue level. Here the tissue MMP and TIMP levels are the net result of cell responses at many sites in the tissue, often from mixed cell populations, in response to the different concentrations in growth factors occurring in the tissue. There is also considerable experimental and histological evidence that not all cells in a population will respond in an identical manner to the same combination and concentration of growth factors. Tissue- or differentiation-specific imprinting will alter the conditions under which responder cells will be influenced to regulate MMP and TIMP expression. Thus, MMP and TIMP levels in tissue extracts or inflammatory exudates do not necessarily reflect the response of individual cells, but rather of the cell population as a whole.

In summary, short-term phenotypic changes in tissue-cell populations, induced by temporal and spatial differences in concentrations of growth factors and morphogens across remodeling tissue, are required to modulate the changes in extracellular matrix formation and remodeling seen during physiological and pathological processes. The independent and opposite regulation of MMP and TIMP-1 expression induced by ConA, TGF-β1, and retinoic acid illustrates that gene expression in mesenchymal cells can be regulated in a concerted manner to achieve either a net degradative or a net formative cellular phenotype. The challenge is to determine whether these phenotypic changes—a switching back and forth between formative and resorptive phenotypes, can occur in the same cell over time *in vivo,* or whether they occur at the tissue level through differential responses of different cell populations or cell strains. Equally important is to identify the biological response modifier "teams" of the relevant individual growth factors and morphogen "players" that modulate MMP and TIMP expression *in vivo* to effect the physiological and pathological processes in different tissues. This will then provide potential targets for future therapeutic intervention to modulate connective tissue degradation, formation, and remodeling during disease and tissue healing.

REFERENCES

1. OVERALL, C. M. 1991. Recent advances in matrix metalloproteinase research. Trends Glycosci. Glycotechnol. **3:** 384–400.
2. CAWSTON, T. E., W. A. GALLOWAY, E. MERCER, G. MURPHY & J. J. REYNOLDS. 1981. Purification of rabbit bone inhibitor of collagenase. Biochem. J. **195:** 159–165.
3. STETLER-STEVENSON, W. G., H. C. KRUTZSCH & L. A. LIOTTA. 1989. Tissue inhibitor of metalloproteinase (TIMP-2): A new member of the metalloproteinase inhibitor family. J. Biol. Chem. **264:** 17374–17378.
4. GOLDBERG, G. I., B. L. MARMER, G. A. GRANT, A. Z. EISEN, S. WILHELM & C. HE. 1989. Human 72K type IV collagenase forms a complex with a tissue inhibitor of metalloproteases designated TIMP-2. Proc. Natl. Acad. Sci. USA **86:** 8207–8211.
5. WELGUS, H. G. & G. P. STRICKLIN. 1983. Human skin fibroblast collagenase inhibitor. Comparative studies in human connective tissues, serum and amniotic fluid. J. Biol. Chem. **258:** 12259–12264.
6. MURPHY, G., T. E. CAWSTON & J. J. REYNOLDS. 1981. An inhibitor of collagenase from human amniotic fluid. Purification, characterization and action on metalloproteinases. Biochem. J. **195:** 167–170.
7. DROUIN, L., C. M. OVERALL & J. SODEK. 1988. Identification of matrix metalloendoproteinase inhibitor (TIMP) in human parotid and submandibular saliva: Partial purification and characterization. J. Periodontal Res. **23:** 370–377.
8. STETLER-STEVENSON, W. G., H. C. KRUTZSCH & L. A. LIOTTA. 1992. TIMP-2: Identification and characterization of a new member of the metalloproteinase inhibitor family. *In* Matrix Metalloproteinases and Inhibitors. H. Birkedal-Hansen, Ed.: 299–306. Gustav Fischer. Stuttgart.
9. GOLDBERG, G. I., I. E. COLLIER, A. Z. EISEN, G. A. GRANT, B. L. MARMER & S. M.

WILHELM. 1992. Mosaic structure of the secreted ECM metalloproteases and interaction of the type IV collagenases with inhibitors. *In* Matrix Metalloproteinases and Inhibitors. H. Birkedal-Hansen, Ed.: 25–30. Gustav Fischer. Stuttgart.
10. BRINCKERHOFF, C. E., R. M. MCMILLAN, J. V. FAHEY & E. D. HARRIS, JR. 1979. Collagenase production by synovial fibroblasts treated with phorbol myristate acetate. Arthritis Rheum. **22:** 1109–1116.
11. WERB, Z., P. M. TREMBLE, O. BEHRENDTSEN, E. CROWLEY & C. H. DAMSKY. 1989. Signal transduction through the fibronectin receptor induces collagenase and stromelysin gene expression. J. Cell Biol. **109:** 877–889.
12. WERB, Z. & J. J. REYNOLDS. 1974. Stimulation by endocytosis of secretion of collagenase and neutral proteinase from rabbit synovial fibroblasts. J. Exp. Med. **140:** 1482–1497.
13. WRANA, J. L., J. SODEK, R. BER & C.G. BELLOWS. 1986. The effects of platelet derived transforming growth factor-β on normal human diploid gingival fibroblasts. Eur. J. Biochem. **159:** 69–76.
14. WRANA, J. L., M. MAENO, B. HAWERLYSHYN, K.-L. YAO, C. DOMENICUCCI & J. SODEK. 1988. Differential effects of transforming growth factor-β on the synthesis of extracellular matrix proteins by normal rat bone cell populations. J. Cell Biol. **106:** 915–924.
15. OVERALL, C. M., J. L. WRANA & J. SODEK. 1989. Independent regulation of collagenase, 72kDa progelatinase and metalloendoproteinase inhibitor expression in human fibroblasts by transforming growth factor beta. J. Biol. Chem. **264:** 1860–1869.
16. OVERALL, C. M., J. L. WRANA & J. SODEK. 1989. Transforming growth factor-beta regulation of collagenase, 72 kDa progelatinase, TIMP and PAI-1 expression in rat bone cell populations and human fibroblasts. Connect. Tissue Res. **20:** 289–294.
17. EDWARDS, D. R., G. MURPHY, J. J. REYNOLDS, S. E. WHITHAM, A. J. P. DOCHERTY & P. ANGEL. 1987. Transforming growth factor beta modulates the expression of collagenase and metalloproteinase inhibitor. EMBO J. **6:** 1899–1904.
18. MURPHY, G., J. J. REYNOLDS & Z. WERB. 1985. Biosynthesis of tissue inhibitor of metalloproteinases by human fibroblasts in culture: Stimulation by 12-O-tetradecanoylphorbol 13-acetate and interleukin 1 in parallel with collagenase. J. Biol. Chem. **260:** 3079–3083.
19. CLARK, S. D., S. M. WILHELM, G. P. STRICKLIN & H. G. WELGUS. 1985. Coregulation of collagenase and collagenase inhibitor production by phorbol myristate acetate in human skin fibroblasts. Arch. Biochem. Biophys. **241:** 36–49.
20. SODEK, J. & F. A. BERKMAN. 1987. Bone cell cultures. Methods Enzymol. **145:** 303–324.
21. BELLOWS, C. G., J. E. AUBIN, J. N. N. HEERSCHE & M. E. ANTOSZ. 1986. Mineralized bone nodules formed in vitro from enzymatically released rat calvaria cell populations. Calcif. Tissue Int. **38:** 143–154.
22. OVERALL, C. M. & H. LIMEBACK. 1988. Identification and characterization of enamel proteinases isolated from developing enamel. Amelogeninolytic serine proteinases are associated with enamel maturation. Biochem. J. **256:** 965–972.
23. KASUGAI, S., Q. ZHANG, C. M. OVERALL, J. WRANA, W. T. BUTLER & J. SODEK. 1991. Differential regulation of the 55 kDa and the 44 kDa forms of secreted phosphoprotein 1 (SPP-1, oeteopontin) in normal and transformed rat bone cells by osteotropic hormones, growth factors and a tumor promoter. Bone Miner. Res. **13:** 235–250.
24. OVERALL, C. M., J. L. WRANA & J. SODEK. 1991. Transcriptional and posttranscriptional regulation of 72 kDa gelatinase/type IV collagenase by transforming growth factor-beta-1 in human fibroblasts. Comparisons with collagenase and tissue inhibitor of matrix metalloproteinase. J. Biol. Chem. **266:** 14064–14071.
25. WRANA, J. L., C. M. OVERALL & J. SODEK. 1991. Regulation of SPARC and extracellular matrix gene expression in human fibroblasts by transforming growth factor β. Eur. J. Biochem. **197:** 519–528.
26. SESHADRI, T. & J. CAMPISI. 1990. Repression of c-fos transcription and an altered genetic program in senescent human fibroblasts. Science **247:** 205–209.
27. OVERALL, C. M. & J. SODEK. 1990. Concanavalin-A produces a matrix-degradative phenotype in human fibroblasts. Induction/endogenous activation of collagenase, 72 kDa gelatinase, & Pump-1 is accompanied by suppression of tissue inhibitor of matrix metalloproteinases. J. Biol. Chem. **265:** 21141–21151.

28. STRICKLIN, G. P., E. A. BAUER, J. J. JEFFREY & A. Z. EISEN. 1977. Human skin collagenase: Isolation of precursor and active forms from both fibroblast and organ cultures. Biochemistry **16:** 1607–1615.
29. SELTZER, J. L., S. A. ADAMS, G. A. GRANT & A. Z. EISEN. 1981. Purification and properties of a gelatin-specific neutral protease from human skin. J. Biol. Chem. **256:** 4662–4668.
30. BRINCKERHOFF, C. E., I. M. PLUCINSKA, L. A. SHELDON & G. T. O'CONNOR. 1986. Half-life of synovial cell collagenase mRNA is modulated by phorbol myristate acetate but not by all-trans-retinoic acid or dexamethasone. Biochemistry **25:** 6378–6384.
31. ANGEL, P., I. BAUMANN, B. STEIN, H. DELIUS, H. J. RAHMSDORF & P. HERRLICH. 1987. 12-*O*-Tetradecanoyl-phorbol-13-acetate induction of the human collagenase gene is mediated by an inducible enhancer element located in the 5′-flanking region. Mol. Cell. Biol. **7:** 2256–2266.
32. OVERALL, C. M., J. L. WRANA & J. SODEK. 1991. Induction of formative and resorptive cellular phenotypes in human gingival fibroblasts by TGF-beta-1 and concanavalin-A. Regulation of matrix metalloproteinases and TIMP. J. Periodontal Res. **26:** 279–282.
33. CLARK, S. D., D. K. KOBAYASHI & H. G. WELGUS. 1987. Regulation of the expression of tissue inhibitor of metalloproteinases and collagenase by retinoids and glucocorticoids in human fibroblasts. J. Clin. Invest. **80:** 1280–1288.
34. WRIGHT, J. K., I. M. CLARK, T. E. CAWSTON & B. L. HAZLEMAN. 1991. The secretion of the tissue inhibitor of metalloproteinases (TIMP) by human synovial fibroblasts is modulated by all-trans-retinoic acid. Biochim. Biophys. Acta **1133:** 25–30.
35. SODEK, J. & C. M. OVERALL. 1992. Matrix metalloproteinases in periodontal tissue remodelling. *In* Matrix Metalloproteinases and Inhibitors. H. Birkedal-Hansen, Ed.: 352–362. Gustav Fischer. Stuttgart.
36. BRINCKERHOFF, C. E., R. M. MCMILLAN, J.-M. DAYER & E. D. HARRIS, JR. 1980. Inhibition by retinoic acid of collagenase production in rheumatoid synovial cells. N. Engl. J. Med. **303:** 432–436.
37. BRINCKERHOFF, C. E. & E. D. HARRIS, JR. 1981. Modulation by retinoic acid and corticosteroid of collagenase production by rabbit synovial fibroblasts treated with phorbol myristate acetate or poly(ethylene glycol). Biochim. Biophys. Acta **677:** 424–432.
38. BAUER, E. A., J. L. SELTZER & A. Z. EISEN. 1983. Retinoic acid inhibition of collagenase and gelatinase expression in human skin fibroblast cultures. Evidence for a dual mechanism. J. Invest. Dermatol. **81:** 162–168.
39. TEMPLETON, N. S., P. D. BROWN, A. T. LEVY, I. M. K. MARGULIES, L. A. LIOTTA & W. G. STETLER-STEVENSON. 1990. Cloning and characterization of human tumor cell interstitial collagenase. Cancer Res. **50:** 5431–5437.
40. BAILLY, C., S. DREZE, D. ASSELINEAU, B. NUSGENS, C. M. LAPERE & M. DARMON. 1990. Retinoic acid inhibits production of collagenase by human epidermal keratinocytes. J. Invest. Dermatol. **94:** 47–51.
41. PAN, L., S. H. CHAMBERLAIN & C. E. BRINKERHOFF. 1992. Differential Regulation of collagenase gene expression by retinoic acid receptors alpha, bete, and gamma. Nucl. Acids Res. **20:** 3105–3111.
42. YANG-YEN, H. F., X. K. ZHANG, G. GRAUPNER, M. TZUKERMAN, B. SAKAMOTO, M. KARIN & M. PFAHL. 1991. Antagonism between retinoic acid receptors and AP-1: Implications for tumor promotion and inflammation. New Biologist **3:** 1206–1219.
43. LAFYATIS, R., S. J. KIM, P. ANGEL, A. B. ROBERTS, M. B. SPORN, M. KARIN & R. L. WILDER. 1990. Interleukin-1 stimulates and all-trans-retinoic acid inhibits collagenase gene expression through its 5′ activator protein-1-binding site. Mol. Endocrinol. **4:** 973–980.
44. EDWARDS, D. R., H. ROCHELEAU, R. R. SHARMA, A. J. WILLS, A. COWIE, J. A. HASSELL & J. K. HEATH. 1992. Involvement of AP1 and PEA3 binding sites in the regulation of murine tissue inhibitor of metalloproteinases-1 (TIMP-1) transcription. Biochim. Biophys. Acta **1171:** 41–55.
45. WASYLYK, C. & B. WASYLYK. 1992. Oncogenic conversion alters the transcriptional properties of *ets.* Cell Growth Differen. **3:** 617–625.

Structure-Function Studies of Mouse Tissue Inhibitor of Metalloproteinases-1[a]

DAVID T. DENHARDT, SHYAMALA RAJAN,
AND SUSAN E. WALTHER

Department of Biological Sciences
Rutgers University
Piscataway, New Jersey 08855

The three members of the family of collagenase inhibitors known as tissue inhibitor of metalloproteinases are TIMP-1, TIMP-2, and TIMP-3. They are extracellular proteins that are effective inhibitors of all members of the family of metalloproteinases known as matrix metalloproteinases, which include interstitial and neutrophil collagenases, gelatinases A and B (also known respectively as the 72- and 92-kD type IV collagenases), and the stromelysins (for reviews see refs. 1 and 2). TIMP-1 and TIMP-2 form specific complexes with progelatinase B and A, respectively, at a site, distinct from the active site, that involves the carboxyl-terminal domain of the enzyme; formation of the TIMP-prometalloproteinase complex modulates the activation of the proenzyme.[3–6] TIMP-1 and TIMP-2 have also been shown to promote the proliferation of erythroid precursors and certain other cell types. The basis for the growth factor activity and its relation to the metalloproteinase activity is unclear. TIMP-3 has an affinity for components of the extracellular matrix and as a consequence is largely sequestered there; it also appears to enhance the expression of the transformed phenotype of chick embryo fibroblasts.[7,8] The ability of TIMP-1 and -2 to inhibit invasion and metastasis has been documented in a number of experimental paradigms.[9–13]

STRUCTURE

FIGURE 1 shows a representation of the TIMP molecule. Amino acids that are conserved in the 10 TIMPs whose sequences have been reported are in large, underlined letters; those that are conserved only in the TIMP-1 and TIMP-2 family are in large, bold type. Inasmuch as the cysteines are conserved, it is generally believed that the disulfide links identified by Williamson *et al.*[14] for TIMP-1 exist also for TIMP-2[15] and TIMP-3. This comparison of the TIMP sequences suggests that certain regions may be important for the function of the protein, and has prompted us to mutate his7, asp16, and his95 to arg, tyr, and arg, respectively. In the next section we describe our initial studies on the ability of these mutant TIMPs to influence cell migration.

O'Shea *et al.*[16] made a more extensive set of TIMP-1 mutants and characterized the mutant proteins in terms of their metalloproteinase inhibitory activity. Despite the highly or absolutely conserved character of the altered amino acids, the mutant proteins studied (his7ala, his7gln, gln9ala, arg20ala, lys22ala, glu28asp, tyr38val,

[a]This research was supported by National Institutes of Health grant CA50183 and the Charles and Johanna Busch endowment.

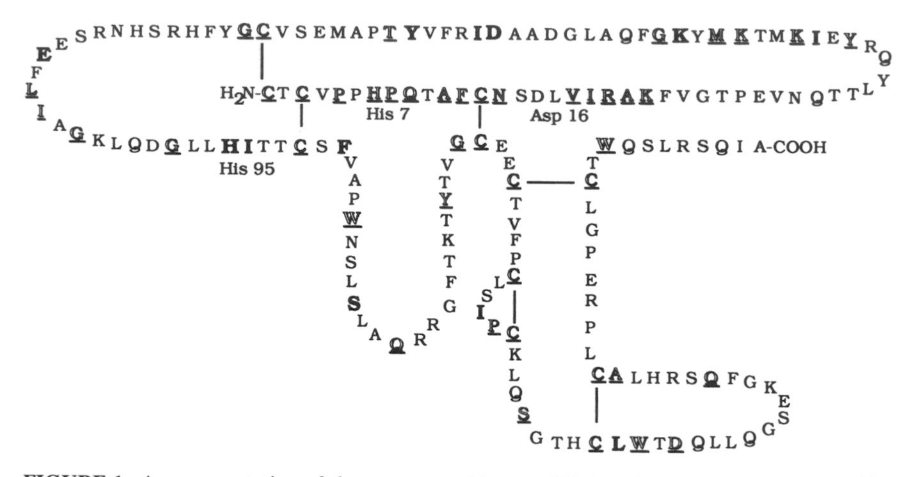

FIGURE 1. A representation of the structure of human TIMP-1 (adapted from Woessner[35]). The cysteines that are joined in disulfide linkages are indicated in bold underlined letters together with the other amino acids that are conserved in the 10 TIMP (-1, -2, -3) proteins whose sequences have been reported so far.[2,36,37] Residues conserved only in the TIMP-1 and TIMP-2 species are in large, bold letters.

lys41ala, his95ala, his95gln, trp105ala, and the double mutants his74ala/his77ala and glu81ala/glu82/ala) were functional metalloproteinase inhibitors. Mutations in his7 or gln9 yielded proteins that retained 15–41% of wild-type activity depending upon the amino acid substitution. These mutant proteins appeared to fold correctly, though perhaps with different inherent stabilities, and were in some cases not expressed as well as wild-type protein. The impaired activity of the his7 and gln9 mutants implicates the constellation of amino acids in this region as important in metalloproteinase inhibition. Interestingly, inactivation of TIMP with diethylpyrocarbonate suggested that his95 was important in TIMP activity[17]—a result not in agreement with the mutagenesis studies. Modification of his144 and/or his164 by diethylpyrocarbonate was also sufficient to inactivate the protein. These data do not lead to a simple picture of the interaction of TIMP with the metalloproteinase.

We initiated structural studies on a truncated version of mouse TIMP-1 (FIG. 2). The region between cys124 and cys127 was mutated in a TIMP-1 cDNA expression vector so that restriction with *Hpa*I would separate the regions encoding the N-terminal 2/3 and the C-terminal 1/3 of TIMP-1. The N-terminal 2/3 segment, known to possess metalloproteinase inhibitor activity,[18] was expressed in *Escherichia coli* using the pET3c expression vector. We called the N-terminal fragment of TIMP-1, and the expression construct making it, NF-TIMP and pNF-TIMP, respectively. The truncated but active NF-TIMP protein is expressed at high levels in the cytosol, from which it was purified by anion and cation exchange chromatography followed by reverse phase HPLC. FIGURE 3 shows a composite of SDS gel analyses of the protein at various stages of purification. We estimate that the protein in the peak HPLC fractions is more than 98% pure with a yield of about 0.5 mg/L culture and possesses a specific activity suggesting that most of the protein is active.

Circular dichroism studies of the purified protein (FIG. 4) indicate substantial α-helical structure (42%) at pH 8.3, with some random coil (35%), but little beta sheet or turns. A theoretical analysis of the secondary structure of NF-TIMP using the Gascuel and Gomlard method[19] gives an estimate of ~38% α-helical and ~48%

random coil, and ~14% in the extended state. Denaturation at higher (>pH 9.4) or lower (<pH 5.5) pH values produced polypeptide chains with mostly random structure. This nonglycosylated 14,720-Da protein is currently the subject of a detailed NMR analysis. If the protein can be crystallized, an X-ray diffraction analysis will also be undertaken.

FUNCTION

When TIMP-1 production by mouse 3T3 cells is down-regulated by the endogenous expression of an antisense TIMP-1 RNA, the cells acquire the ability to form tumors.[20] Intrigued by this property of TIMP-1, we have been searching for an explanation for how it could behave as a tumor suppressor in 3T3 cells. Tumorigenicity was assessed in the mouse by the formation of a tumor where a bolus of cells

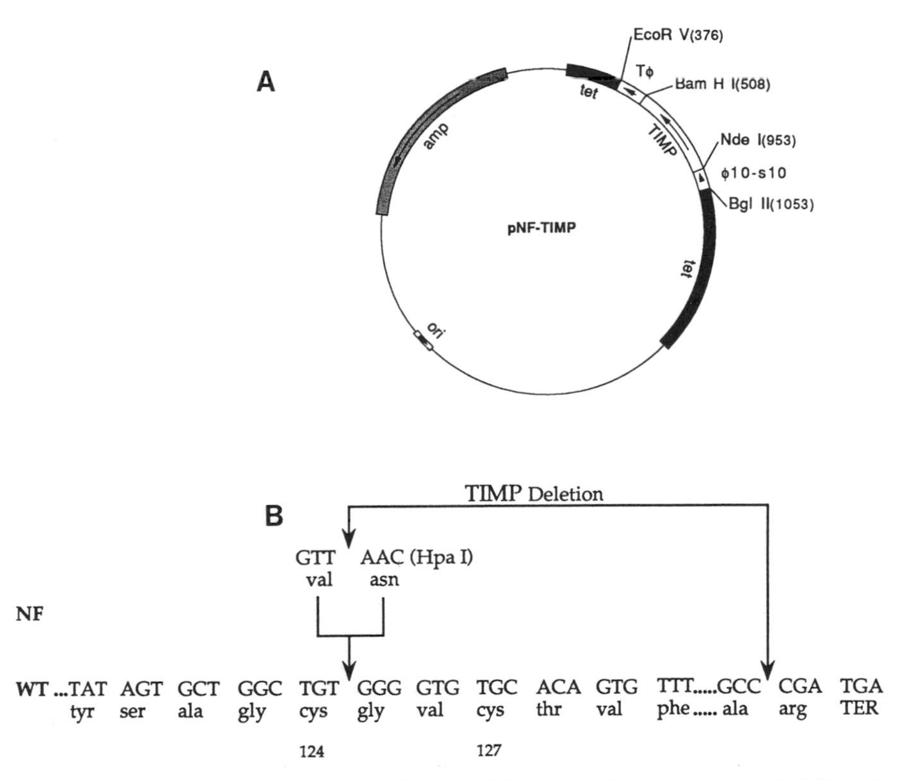

FIGURE 2. Expression of recombinant TIMP-1. (**A**) Structure of the plasmid pNF-TIMP used to produce recombinant TIMP-1 in *E. coli.* The T7 gene 10 promoter is used to drive transcription of the inserted TIMP-1 coding sequence. (**B**) Sequence in the region of the mouse TIMP-1 cDNA that was altered so that the minigene could be cleaved between cys124 and cys127 by *Hpa*I. The two codons encoding val and asn were inserted as indicated using PCR mutagenesis. Cleavage with *Hpa*I and *Apa*I removed the C-terminal domain of TIMP-1, and religation of the residual plasmid produced a vector capable of making the N-terminal 2/3 domain of TIMP-1 terminating in a val-arg sequence that did not seem to impair function.

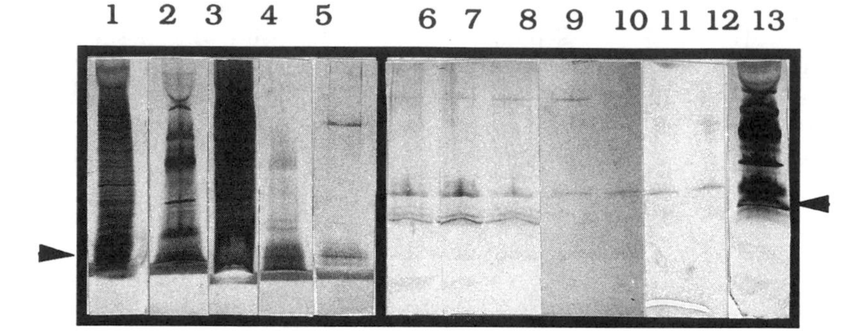

FIGURE 3. Purification of recombinant truncated mouse TIMP-1. Shown is an SDS-polyacrylamide Phast gel of the protein preparation at various stages of purification. **Lane 1,** cytosol from induced cells; **lanes 2 and 13,** molecular weight markers; **lane 3,** flow-through from DEAE-Sepharose; **lanes 4 and 5,** peak fractions off BioRex-70; **lanes 6–8,** peak fractions from the heparin-Sepharose chromatography; **lanes 9–12,** peak fractions from the octadecylsilane (ODS) column. The proteins were eluted from the BioRex-70 and heparin-Sepharose columns with linear gradients of 0–1 M NaCl in 50 mM Tris HCl, pH 8.3. Proteins were eluted from the reverse phase ODS column with a linear gradient of 100% H_2O to 100% CH_3CN. The left five lanes were electrophoresed under conditions such that NF-TIMP is at the position indicated by the left arrow. The right eight lanes were electrophoresed such that NF-TIMP is at the position indicated by the arrow on the right.

down-regulated for TIMP-1 expression had been injected subcutaneously. In the complementary experiment, expression of TIMP-1[10,11] or TIMP-2[12] in B16F10 melanoma cells or *ras*-transformed rat embryo fibroblasts, respectively, caused a reduction in the invasive and tumorigenic properties of the cells. Inactivation of the single TIMP-1 gene (on the X-chromosome) in (male) embryonic stem (ES) cells by homologous recombination has been shown to enhance the invasiveness of mesenchymal cells derived from the ES cells.[21] In seeming contradiction with the concept that TIMP inhibits tumor development, there are a number of reports in the literature where *increased* TIMP (1 or 2) expression in tumor tissue appeared to correlate with tumor formation.[21A,22] In at least some human lymphomas the apparent increase in TIMP activity in a tumor may represent expression from stromal cells rather than from the tumor cells themselves.[23]

TIMP-1 (and TIMP-2) has two known functions: one is to inhibit matrix metalloproteinases and the second is to promote the proliferation of receptive cells. A secreted protein, TIMP will reduce the activity of metalloproteinases in the extracellular environment. Thus any reduction in TIMP expression is likely to lead to a higher level of metalloproteinase activity in the vicinity of an injected bolus of cells.[24] Increased metalloproteinase activity could facilitate tumor development by breaking down host barriers and interfering with host defenses. This could in turn promote the survival of the 3T3 cells in the host animal for a period of time sufficient for variants to arise that are capable of proliferating and forming a tumor despite host defenses. In support of this possibility is the observation[25] that cells derived from tumors that did arise from the TIMP down-modulated cells showed altered patterns of gene expression and, in particular, the augmented expression of several other proteins frequently found to be up-regulated in tumor cells (e.g. transin, cathepsin L, calcyclin, and osteopontin).

TIMP-1 (and TIMP-2) also has growth factor activity on several cell types. As erythroid potentiating activity (EPA), it promotes the proliferation of erythroid precursors.[26,27] It can also stimulate the multiplication of a variety of cell types including human erythroleukemia cells, skin keratinocytes, gingival fibroblasts, and normal human foreskin fibroblasts.[28–32] Since it seems paradoxical to believe that *reduced* expression of this growth-promoting activity would enhance the tumor-forming capacity of 3T3 cells, we do not think the erythroid potentiating activity is responsible for the tumor-suppressing activity of TIMP. However, in a counter argument, if receptors are present on the cell surface that recognize TIMP and mediate the EPA, they could also mediate a tumor-suppressing activity.

Our experimental approach to the problem of understanding how TIMP-1 could be a tumor suppressor has been to make site-directed mutations in conserved regions of the molecule and to ask if the mutant protein retains the ability of wild-type TIMP to suppress the tumorigenicity of 3T3 cells. The mutations we chose to make altered three conserved amino acids, his7 to arg, asp16 to tyr, and his95 to arg. (Recent sequence data have revealed that the chicken TIMP-3 protein has tyrosine in place of his95 and the ovine TIMP-1 has a glutamic residue in place of asp16.) Although we have not yet succeeded in expressing these mutant proteins in sufficient quantity to assess their metalloproteinase inhibitory activity, we have initiated studies on how they affect the migratory phenotype of B16F10 cells.

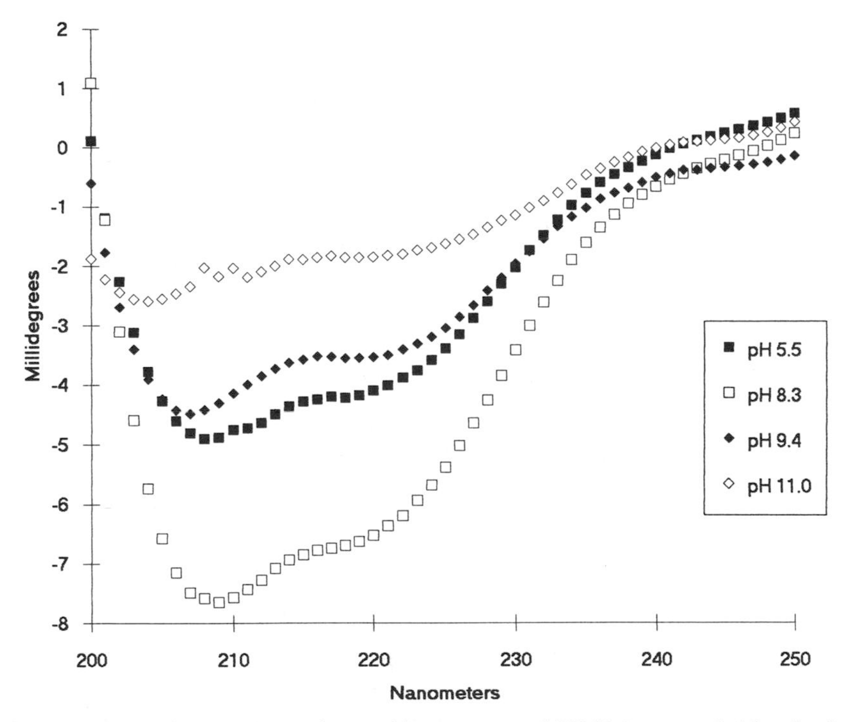

FIGURE 4. Secondary structure of recombinant truncated TIMP-1 as revealed by circular dichroism. The UV-CD spectra of NF-TIMP at different pHs were determined using an AVIV model 60DS CD spectrometer. The protein was dissolved at 25–50 μg/mL in 100 mM NaCl and 50 mM MES (2-[N-morpholino]ethanesulfonic acid) (pH 5.5), or Tris HCl (pH 8.3), or CHES (2-[N-cyclohexylamino]ethanesulfonic acid) (pH 9.4) or CAPS (3-[cyclohexylamino]-1-propane sulfonic acid) (pH 11.0). Data were analyzed as described.[38]

B16F10 cells, which are tumorigenic and metastatic, do not make detectable levels of endogenous TIMP. However, when they are engineered to express wild-type TIMP-1, their metastatic capability is reduced.[11] We observed that the way the cells migrate on a plastic surface when a confluent monolayer is wounded so as to alleviate contact inhibition is also modulated by TIMP expression. FIGURE 5 shows a typical pattern of migration of B16F10 cells and B16F10 cells expressing wild-type TIMP-1. Some of the clones expressing mutant versions of TIMP-1 also show suppressed migration, suggesting that alteration of any one of the three targeted amino acids did not substantially impair the ability of the mutant protein to inhibit cell migration.

Research currently in progress is focused on characterizing these cell lines in more detail. In particular, we want to relate the invasiveness and tumorigenicity of B16F10 cells expressing the different mutant TIMPs to the extent of expression of the protein. These and other data should reveal whether any one of these three amino acids is essential for TIMP-1 to inhibit invasion, tumor formation or growth promotion.

DISCUSSION

TIMP appears to be a compact, highly folded structure that interacts with the active site of the metalloproteinase by making contacts with both the zinc ion and key amino acids. Although inhibition appears not to be critically dependent upon any one amino acid, modification of certain residues (ethylation of his95, his144 or his164) can significantly reduce activity, presumably by blocking access of the inhibitor to the active site.[17] If this view is correct, then it raises the question as to why certain amino acids have been so rigorously conserved. The answer to this question may lie in the multiple and key functions the TIMPs play in various metabolic processes. Tissue remodeling is accomplished in part by metalloproteinases, whose activity must be exquisitely controlled. Even minor deviations in the efficiency with which TIMP can inhibit metalloproteinases may have significant deleterious effects on the host organism.

An alternative answer to the question of amino acid conservation is that his7, for example, is essential for a critical activity of the protein *in vivo,* an activity that is not easily related to the ability of the protein to inhibit metalloproteinases *in vitro.* What might this activity be? The only other known activity of TIMP is its EPA, its ability to promote the proliferation of specific cell types. Avalos *et al.*[28] have obtained evidence for the presence of a receptor on K562 human erythroleukemia cells that may be required for TIMP to promote proliferation of these cells. A requirement for a particular TIMP to enable an essential cell type to survive in the organism could explain the conservation of sequence in a region (separate from that involved in metalloproteinase inhibition) that interacted with the receptor. For this hypothesis to attain a degree of credibility, it is necessary (1) to identify a cell type dependent upon TIMP for survival in culture, and (2) to establish that the conserved amino acid in question is essential for that activity.

To our knowledge, this is the first report documenting a role for TIMP-1 in cell migration. A variety of reports has implicated TIMP as an inhibitor of invasion. Monsky *et al.*[33] have reported that gelatinase B associates with the invadopodia of transformed chick embryo fibroblasts. In order for cells to pass through an extracellular matrix, degradation of the macromolecular components of that barrier is presumed to occur by an enzymatic process often involving matrix metalloproteinases. Thus it is not surprising that TIMP might inhibit that process. What is less evident is how TIMP might influence cell migration in circumstances where there is

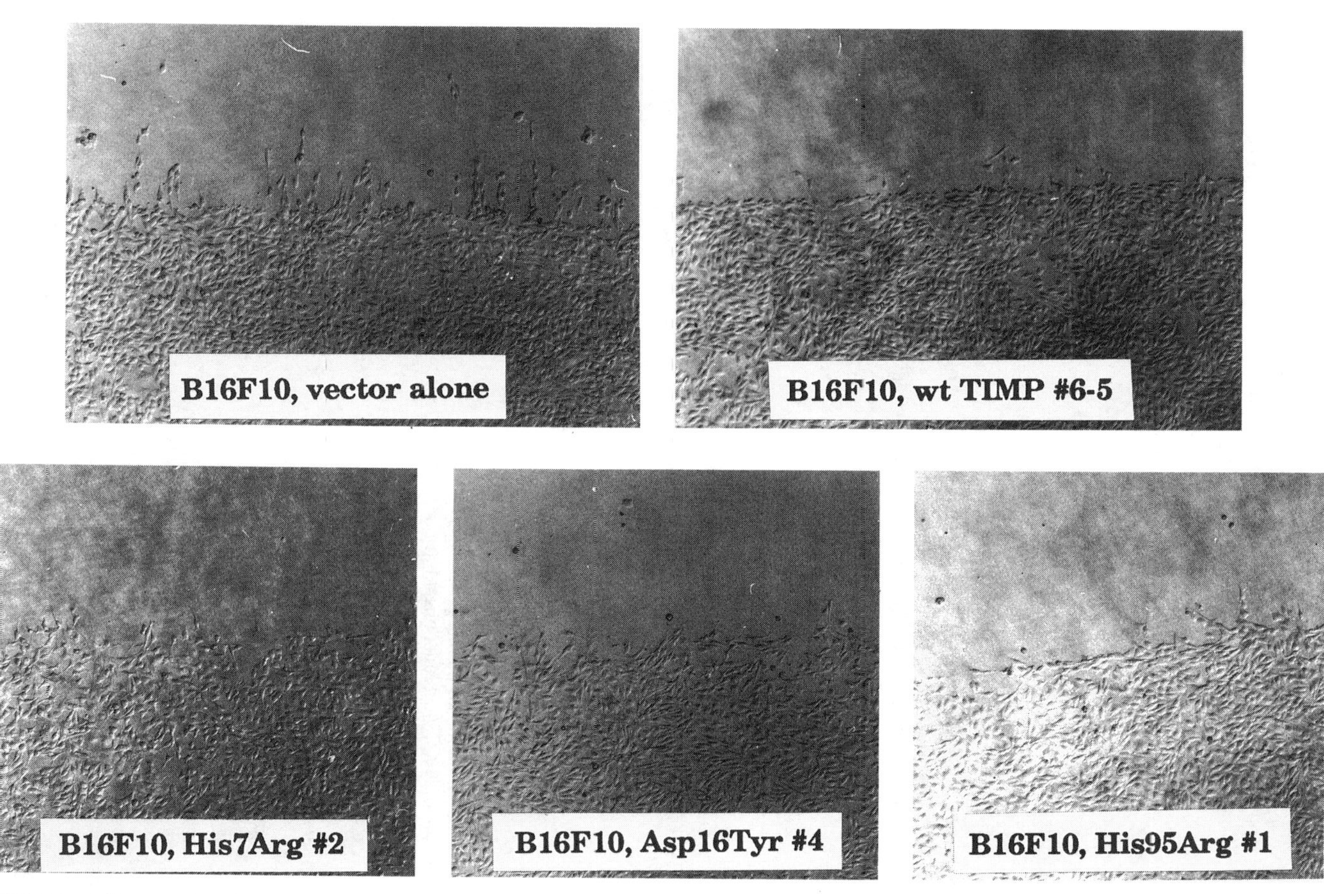

FIGURE 5. Influence of endogenously produced wild-type and mutant TIMP-1 on the migration of B16F10 cells. Migration was assessed 18 h after wounding a ~80% confluent monolayer by scraping away one-half of the monolayer in a 6-well plate.[39] The medium on the cells was replaced with serum-free DMEM for the 18-h period to minimize the proliferation of the cells. The cells in each monolayer are indicated on the figure.

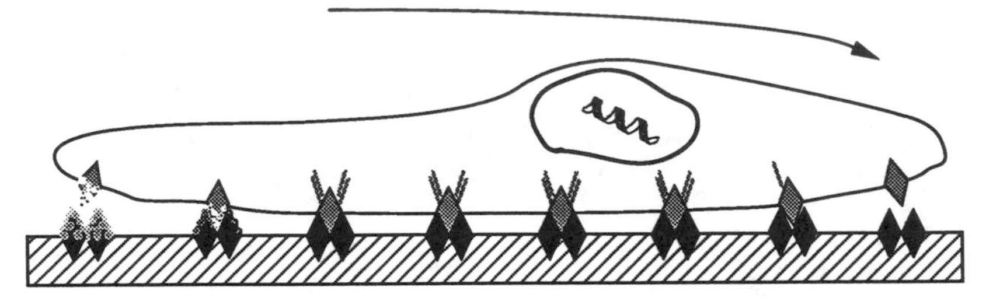

FIGURE 6. Model for cell migration (adapted from Stossel[34]).

no barrier to penetrate. Our interpretation of these intriguing observations is diagrammed in FIGURE 6, adapted from Stossel.[34] During cell migration, cells are known to make and break connections with the substratum; however, the enzymology of this process has not been characterized. Our results suggest that in order for a cell to break adhesive contacts efficiently with the substratum a metalloproteinase is essential, at least for B16F10 migration, and that TIMP-1 can inhibit the process.

ACKNOWLEDGMENTS

The authors thank K. Curtis for help with the figures, K. Callanan for the fermentor scale-up of NF-TIMP production, and Drs. J. Elliott and N. Greenfield for help with CD spectroscopy.

REFERENCES

1. BIRKEDAL-HANSEN, H., Z. WERB, H. WELGUS & H. VAN WART, EDS. 1992. Matrix Metalloproteinases and Inhibitors. Gustav Fischer Verlag. New York.
2. DENHARDT, D. T., B. FENG, D. R. EDWARDS, E. T. COCUZZI & U. M. MALYANKAR. 1993. Tissue inhibitor of metalloproteinases (TIMP, aka TPA): Structure, control of expression and biological functions. Pharmacol. Ther. **59:** 329–341.
3. GOLDBERG, G. I., A. STRONGIN, I. E. COLLIER, L. T. GENRICH & B. L. MARMER. 1992. Interaction of 92-kDa Type IV collagenase with the tissue inhibitor of metalloproteinases prevents dimerization, complex formation with interstitial collagenase, and activation of the proenzyme with stromelysin. J. Biol. Chem. **267:** 4583–4591.
4. MURPHY, G., F. WILLENBROCK, R. V. WARD, M. I. COCKETT, D. EATON & A. J. P. DOCHERTY. 1992. The C-terminal domain of 72-kDa gelatinase A is not required for catalysis, but is essential for membrane activation and modulates interactions with tissue inhibitors of metalloproteinases. Biochem. J. **283:** 637–641.
5. KLEINER, D. E., JR., A. TUUTTILA, K. TRYGGVASON & W. G. STETLER-STEVENSON. 1993. Stability analysis of latent and active 72-kDa type IV collagenase: The role of tissue inhibitor of metalloproteinases-2 (TIMP-2). Biochemistry **32:** 1583–1592.
6. STRONGIN, A. Y., B. L. MARMER, G. A. GRANT & G. I. GOLDBERG. 1993. Plasma membrane-dependent activation of the 72-kDa Type IV collagenase is prevented by complex formation by TIMP-2. J. Biol. Chem. **268:** 14033–14039.
7. PAVLOFF, N., P. W. STASKUS, N. S. KISHANI & S. P. HAWKES. 1992. A new inhibitor of metalloproteinases from chicken: ChIMP-3. J. Biol. Chem. **267:** 17321–17326.
8. YANG, T. T. & S. P. HAWKES. 1992. Role of the 21-kDa protein TIMP-3 in oncogenic transformation of cultured chicken embryo fibroblasts. Proc. Natl. Acad. Sci. USA **89:** 10676–10680.

9. SCHULTZ, R. M., S. SILBERMAN, B. PERSKY, A. S. BAJKOWSKI & D. F. CARMICHAEL. 1988. Inhibition by human recombinant tissue inhibitor of metalloproteinases of human amnion invasion and lung colonization by murine B16-F10 melanoma cells. Cancer Res. **48:** 5539–5545.
10. KHOKHA, R., M. J. ZIMMER, C. H. GRAHAM, P. K. LALA & P. WATERHOUSE. 1992. Suppression of invasion by inducible expression of tissue inhibitor of metalloproteinase-1 (TIMP-1) in B16F10 melanoma cells. J. Natl. Cancer Inst. **84:** 1017–1022.
11. KHOKHA, R., M. J. ZIMMER, S. M. WILSON & A. F. CHAMBERS. 1992. Up-regulation of TIMP-1 expression in B16-F10 melanoma cells suppresses their metastatic ability in chick embryo. Clin. Exp. Metastasis **10:** 365–370.
12. DECLERCK, Y. A., N. PEREZ, H. SHIMADA, T. C. BOONE, K. E. LANGLEY & S. M. TAYLOR. 1992. Inhibition of invasion and metastasis in cells transfected with an inhibitor of metalloproteinases. Cancer Res. **52:** 701–708.
13. TSUCHIYA, Y., H. SATO, Y. ENDO, Y. OKADO, M. MAI, T. SASAKI & M. SEIKI. 1993. Tissue inhibitor of metalloproteinase 1 is a negative regulator of the metastatic ability of human gastric cancer cell line, KKLS, in the chick embryo. Cancer Res. **53:** 1397–1402.
14. WILLIAMSON, R. A., F. A. O. MARSTON, S. ANGAL, P. KOKLITIS, M. PANICO, H. R. MORRIS, A. F. CARNE, B. J. SMITH, T. J. R. HARRIS & R. F. FREEDMAN. 1990. Disulphide bond assignment in human tissue inhibitor of metalloproteinases (TIMP). Biochem. J. **268:** 267–274.
15. DECLERCK, Y. A., T.-D. YEAN, Y. LEE, J. M. TOMICH & K. E. LANGLEY. 1993. Characterization of the functional domain of tissue inhibitor of metalloproteinases-2 (TIMP-2). Biochem. J. **289:** 65–69.
16. O'SHEA, M., F. WILLENBROCK, R. A. WILLIAMSON, M. I. COCKETT, R. B. FREEDMAN, J. J. REYNOLDS, A. J. P. DOCKERTY & G. MURPHY. 1992. Site-directed mutations that alter the inhibitory activity of the tissue inhibitor of metalloproteinases-1: Importance of the N-terminal region between cysteine 3 and cysteine 13. Biochemistry **31:** 10146–10152.
17. WILLIAMSON, R. A., B. J. SMITH, S. ANGAL & R. B. FREEDMAN. 1993. Chemical modification of tissue inhibitor of metalloproteinases-1 and its inactivation by diethylpyrocarbonate. Biochim. Biophys. Acta **1203:** 147–154.
18. MURPHY, G., A. HOUBRECHTS, M. I. COCKETT, R. A. WILLIAMSON, M. O'SHEA & A. J. P. DOCHERTY. 1991. The N-terminal domain of tissue inhibitor of metalloproteinases retains metalloproteinase inhibitory activity. Biochemistry **30:** 8097–8102.
19. GASCUEL, O. & J. L. GOLMARD. 1988. A simple method for predicting the secondary structure of globular proteins: Implications and accuracy. Comput. Appl. Biosci. **4:** 357–365.
20. KHOKHA, R., P. WATERHOUSE, S. YAGEL, P. K. LALA, C. M. OVERALL, G. NORTON & D. T. DENHARDT. 1989. Antisense RNA-induced reduction in murine TIMP levels confers oncogenicity on Swiss 3T3 cells. Science **243:** 947–950.
21. ALEXANDER, C. M. & Z. WERB. 1992. Targeted disruption of the tissue inhibitor of metalloproteinases gene increases the invasive behavior of primitive mesenchymal cells derived from embryonic stem cells in vitro. J. Cell Biol. **118:** 727–739.
21A. LU, X., M. LEVY, I. B. WEINSTEIN & R. M. SANTELLA. 1991. Immunological quantitation of levels of tissue inhibitor of metalloproteinase-1 in human colon cancer. Cancer Res. **51:** 6231–6235.
22. URBANSKI, S. J., D. R. EDWARDS, N. HERSHFIELD, S. A. HUCHCROFT, E. SHAFFER, L. SUTHERLAND & A. E. KOSSAKOWSKA. 1993. Expression pattern of metalloproteinases and their inhibitors changes with the progression of human sporadic colorectal neoplasia. Diagn. Mol. Pathol. **2:** 81–89.
23. KOSSAKOWSKA, A. E., S. J. URBANSKI, A. WATSON, L. J. HAYDEN & D. R. EDWARDS. 1993. Patterns of expression of metalloproteinases and their inhibitors in human malignant lymphomas. Oncology Res. **5:** 19–28.
24. PONTON, A., B. COULOMBE & D. SKUP. 1991. Decreased expression of tissue inhibitor of metalloproteinases in metastatic tumor cells leading to increased levels of collagenase activity. Cancer Res. **51:** 2138–2143.
25. KHOKHA, R., P. WATERHOUSE, P. LALA, M. ZIMMER & D. T. DENHARDT. 1991. Increased

proteinase expression during tumor progression of cell lines down-modulated for TIMP levels: A new transformation paradigm? J. Cancer Res. Clin. Oncol. **117:** 333–338.

26. Gasson, J. C., D. W. Golde, S. E. Kaufman, C. A. Westbrook, R. M. Hewick, R. J. Kaufman, G. G. Wong, P. A. Temple, A. C. Leary, E. L. Grown, E. C. Orr & S. S. Clark. 1985. Molecular characterization and expression of the gene encoding human erythroid-potentiating activity. Nature **315:** 768–771.
27. Stetler-Stevenson, W. G., N. Bersch & D. E. Golde. 1992. Tissue inhibitor of metalloproteinase-2 (TIMP-2) has erythroid potentiating activity. FEBS Lett. **296:** 231–234.
28. Avalos, B. R., S. E. Kaufman, M. Tomonaga, R. E. Williams, D. W. Golde & J. C. Gasson. 1988. K562 cells produce and respond to human erythroid-potentiating activity. Blood **71:** 1720–1725.
29. Bertaux, B., W. Hornebeck, A. Z. Eisen & L. Dubertret. 1991. Growth stimulation of human keratinocytes by tissue inhibitor of metalloproteinases. J. Invest. Dermatol. **97:** 679–685.
30. Hayakawa, T., K. Yamashita, K. Tanzawa, E. Uchijima & K. Iwata. 1992. Growth-promoting activity of tissue inhibitor of metalloproteinases-1 (TIMP-1) for a wide range of cells. FEBS Lett. **298:** 29–32.
31. Murate, T., K. Yamashita, H. Ohashi, Y. Kagami, K. Tsushita, T. Kinoshita, T. Hotta, H. Saito, S. Yoshida, K. J. Mori & T. Hayakawa. 1993. Erythroid potentiating activity of tissue inhibitor of metalloproteinases on the differentiation of erythropoietin-responsive mouse erythroleukemia cell line, ELM-I-1-3, is closely related to its cell growth potentiating activity. Exp. Hematol. **21:** 169–176.
32. Nemeth, J. A. & C. L. Goolsby. 1993. TIMP-2, a growth-stimulatory protein from SV40-transformed human fibroblasts. Exp. Cell Res. **207:** 376–382.
33. Monsky, W. L., T. Kelly, C.-Y. Lin, Y. Yeh, W. G. Stetler-Stevenson, S. C. Mueller & W.-T. Chen. 1993. Binding and localization of Mr 72,000 matrix metalloproteinase at cell surface invadopodia. Cancer Res. **53:** 3159–3164.
34. Stossel, T. P. 1993. On the crawling of animal cells. Science **260:** 1086–1094.
35. Woessner, J. F., Jr. 1991. Matrix metalloproteinases and their inhibitors in connective tissue remodeling. FASEB J. **5:** 2145–2154.
36. Smith, G. W., T. L. Goetz, R. V. Anthony & M. F. Smith. 1994. Molecular cloning of an ovine tissue inhibitor of metalloproteinases: Ontogeny of messenger ribonucleic acid expression and *in situ* localization within preovulatory follicles and luteal tissue. Endocrinology **134:** 344–352.
37. Leco, K.-J., R. Khokha, N. Pavloff, S. P. Hawkes & D. R. Edwards. 1994. Tissue inhibitor of metalloproteinase-3 (TIMP-3) is an extracellular matrix associated protein with a distinctive pattern of expression in mouse cells and tissues. J. Biol. Chem. **269:** 9352–9360.
38. Brahms, S. & J. Brahms. 1980. Determination of protein structure in solution by vacuum ultraviolet circular dichroism. J. Mol. Biol. **138:** 149–178.
39. Murugesan, G., G. M. Chisolm & P. L. Fox. 1993. Oxidized low density lipoprotein inhibits the migration of aortic endothelial cells in vitro. J. Cell Biol. **120:** 1011–1019.

Role of TIMP and MMP Inhibition in Preventing Connective Tissue Breakdown[a]

TIM CAWSTON, TRACY PLUMPTON, VALERIE CURRY, ALISON ELLIS, AND LIZ POWELL

Rheumatology Research Unit, Box 194
Addenbrookes Hospital
Hill's Road
Cambridge CB2 2QQ, United Kingdom

The matrix metalloproteinases (MMPs) are a unique family of enzymes that in concert can degrade all the components of the extracellular matrix.[1] These potent enzymes are made in a proenzyme form, and activation occurs after leaving the cell. Extracellular activity is also controlled by specific inhibitor called tissue inhibitor of metalloproteinases (TIMPs)[2] that bind to the active forms of the enzyme forming 1:1 complexes and block their activity.

The inflammatory cytokine interleukin-1 (IL-1) can induce resorption of cartilage *in vitro*[3] and *in vivo* when injected into rabbit joints[4] and is also found in joint fluids from patients with rheumatoid arthritis.[5,6] IL-1 is also known to stimulate the production of the MMPs collagenase and stromelysin from human synovial cells and chondrocytes.[7] Raised levels of these enzymes have been localized at both the mRNA and protein level within diseased cartilage from different rheumatic diseases.[8] Consequently, the MMPs are a valid therapeutic target for inhibition in the rheumatic diseases.[9]

A number of studies have shown that IL-1–stimulated cartilage model systems can be used to test the effectiveness of proteinase inhibitors. The release of glycosaminoglycan (GAG) fragments can be prevented by the addition of low molecular weight synthetic inhibitors,[10–14] but not by the addition of TIMP.[15–16] However, although the release of GAG fragments from cartilage appears to be rapid *in vivo* in response to IL-1[4,17] it is also quickly resynthesized by chondrocytes.[18] In contrast the release of collagen from cartilage is much slower, is less reproducible, and appears to be irreversible as resynthesis is difficult to achieve.[19]

In this study we have examined the release of both collagen and proteoglycan fragments from resorbing cartilage to determine if the naturally occurring MMP inhibitors TIMP and TIMP-2 and low molecular weight synthetic inhibitors can block both proteoglycan and collagen degradation in IL-1 stimulated cartilage.

MATERIALS AND METHODS

Cartilage Degradation Assay

Control culture medium was Dulbecco's modification of Eagle's medium containing 25 mM HEPES (Gibco) supplemented with sodium bicarbonate (0.5 g/L),

[a]This work was supported by the Arthritis and Rheumatism Council.

glutamine (2 mM), streptomycin (100 μg/mL), penicillin (100 U/mL), and amphotericin (2.5 μg/mL). Cartilage slices were dissected from porcine joints and incubated as described previously[16] except that the incubations were continued in some experiments for 28 days to follow the release of collagen fragments. Cartilage slices were dissected from bovine nasal septum cartilage. Discs were cut from the slices (2 × 2 mm) and washed twice in HBSS (Gibco). The discs were then incubated at 37 °C in groups of three in control medium (600 μL) in a 24-well plate for 24 h for stabilization. Control medium (600 μL) with or without inhibitors and rhIL 1α (generous gifts from Glaxo Group Research plc, UK or Roche Products Ltd, UK) was added and the plate incubated at 37 °C for 7 days. The supernates were harvested and replaced with fresh medium containing identical test reagents to day 1. The experiment was continued for a further 7 days and day 7 and day 14 supernates were stored at −20 °C until assay.

In order to determine the total GAG and hydroxyproline (OHPro) content of the cartilage fragments, the remaining cartilage was digested with papain (2.5 μg/mL; Sigma) in 0.1 M phosphate buffer, pH 6.5, containing 5 mM EDTA and 5 mM cysteine hydrochloride, incubating at 65 °C until digestion was complete (16 h).

Inhibitors

TIMP-1 and TIMP-2 were prepared from WI-38 culture medium. TIMP-1 was purified by a monoclonal-based affinity column[20] and TIMP-2 as previously described.[21] Synthetic low molecular inhibitors (British Biotechnology, Oxford, UK; and SB Pharmaceuticals, Harlow, UK) were dissolved in ethanol and diluted in control medium to appropriate concentrations.

Proteoglycan Degradation

Media samples and papain digests were assayed for sulphated glycosaminoglycans (as a measure of proteoglycan release) using a modification of the 1,9-dimethylmcthylene blue dye binding assay.[22] Sample or standard (40 μL) was mixed with dye reagent (250 μL), prepared as described, in the well of a microtiter plate, and the absorbance at 525 nm determined immediately. Chondroitin sulphate from shark fin (5–40 μg/mL) was used as a standard. The complex formed with 1,9-dimethylmethylene blue results in a decrease in absorbance at 525 nm. Hyaluronate causes no absorbance change.[23]

Collagen Degradation

Hydroxyproline release was assayed (as a measure of collagen degradation) using a microtiter plate modification of the method of Bergman and Loxley.[24] Chloramine T (7% w/v) was diluted 1:4 in acetate-citrate buffer (57 g sodium acetate, 37.5 g tri-sodium citrate, 5.5 g citrate acid, 385 mL propan-2-ol per liter water). P-dimethylaminobenzaldehyde (DAB; 20 g in 30 mL 60% perchloric acid) was diluted 1:3 in propan-2-ol. Specimens were hydrolyzed in 6M HCl for 20 h at 105 °C and the hydrolysate neutralized by drying over NaOH *in vacuo*. The residue was dissolved in water and 40 μL sample or standard (hydroxyproline; 5–30 μg/mL) added to microtiter plate together with chloramine-T reagent (25 μL) and then DAB reagent (150 μL) after 4 min. The plate was covered and heated to 60 °C for 35 min, cooled for 5 min to room temperature, and the absorbance at 560 nm determined.

RESULTS

Early experiments were conducted with pig articular cartilage stimulated to resorb with IL-1. FIGURE 1 illustrates a dose-response curve for the release of GAG fragments after three days in culture and treatment with increasing concentrations of IL-1. Our previous studies have shown that addition of TIMP to this system does not

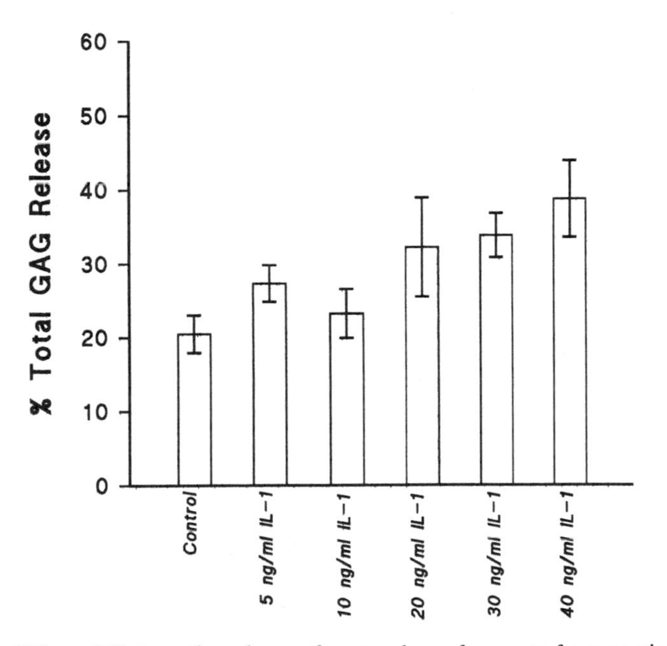

FIGURE 1. Effect of IL-1 on the release of proteoglycan fragments from porcine articular cartilage. Porcine articular cartilage was treated with increasing concentrations of IL-1α for 72 h and the level of GAG fragments measured in the culture medium and expressed as a percentage of the total GAG.

prevent the release of GAG fragments, but we have shown that TIMP is unable to penetrate the cartilage matrix.[16] However both TGF-β and IFN-γ can prevent the release of proteoglycan fragments. TGF-β is likely to up-regulate TIMP and thus prevent cartilage breakdown; IFN-γ down-regulates the production of metalloproteinases leading to a reduction in resorption.[25,26]

These previous studies were all completed following the release of GAG fragments. In this study the culture period was extended to follow the release of collagen fragments from pig articular cartilage. FIGURE 2 illustrates that no effect was seen on the release of GAG when a synthetic inhibitor was added to the cartilage at 10^{-5} M. FIGURE 3 illustrates the time course of the release of collagen from pig articular cartilage which often occurred during the later stages of culture. A low molecular

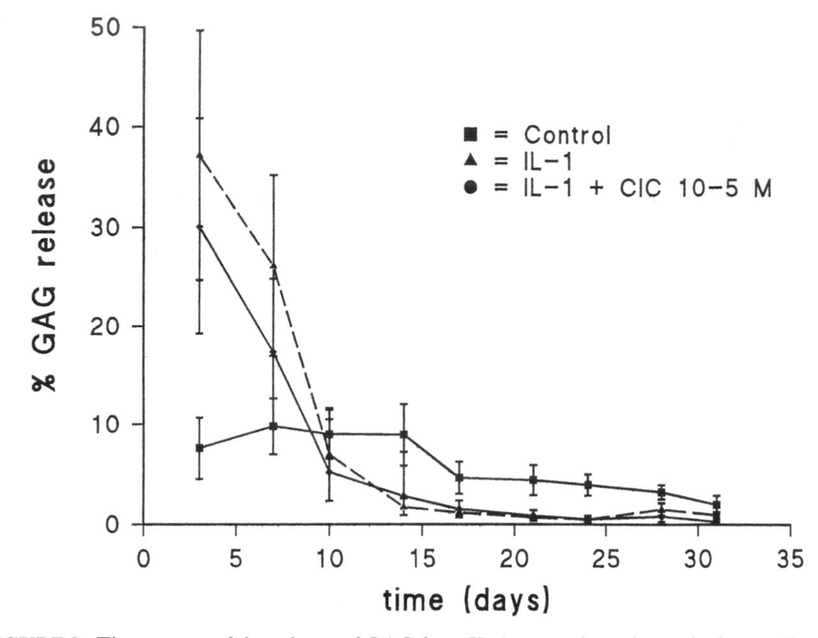

FIGURE 2. Time course of the release of GAG from IL-1–treated porcine articular cartilage in the presence of a synthetic MMP inhibitor. Porcine articular cartilage was cultured in the presence and absence of IL-1 for a total of 32 days and the medium removed approximately every 3 days. The level of GAG was measured at each time point and expressed as a percentage of the total. Inclusion of a synthetic MMP inhibitor[16] at 10^{-5} M had no significant effect on the release of GAG fragments.

weight synthetic inhibitor used in previous studies[16] was able to prevent this release of collagen. FIGURE 4 illustrates the effect of adding a different synthetic inhibitor BB87 to pig articular cartilage. At 10^{-5} M BB87 is unable to prevent the release of proteoglycan fragments but completely prevents the release of collagen. This figure illustrates the variable time course of collagen release from pig cartilage; it occurred in this experiment at day 8–9. In some experiments no collagen release was observed in the first 28 days of culture, and so the use of bovine nasal cartilage was investigated as an alternative and found to be a much more reproducible system.

Bovine nasal cartilage was cultured with high concentrations of IL-1 essentially as described by Nixon *et al.*[11] Media was removed at day 7 and assayed for GAG and also at day 14 and assayed for collagen. A high dose of human recombinant IL-1 (1000 units/mL) was used to stimulate the release of matrix components from bovine nasal cartilage discs placed in culture medium. After 7 days the medium was removed and assayed for GAG fragments. The medium was replenished and the cultures continued for a further 7 days. At the end of 14 days the medium was removed and assayed for both GAG and collagen fragments. The cartilage remaining was then digested with papain and the GAG and collagen remaining then calculated. Most of the GAG was released by day 7 with very little collagen, and up to 70% of the collagen was released by day 14. Addition of TIMP and TIMP-2 (100 units/mL) had no effect on the release of GAG fragments into the medium at day 7 (data not shown) as found previously for a pig articular cartilage model system.[16] However, both addition of TIMP and TIMP-2, at the same concentration as for the GAG release, completely blocked the release of collagen fragments from the tissue (data not shown).

Additional experiments were completed with this same model system but with the addition of low molecular weight, synthetic inhibitor of collagenase. Addition of BB87 (10^{-5}–10^{-8} M) was unable to prevent the release of GAG fragments from the cartilage. The same concentrations of BB87 were able to prevent release of collagen components in a dose-dependent fashion. FIGURE 5 illustrates that approximately 70% of the cartilage collagen is released into the medium after the addition of IL-1 and subsequent culture for 14 days. In the presence of 10^{-5} and 10^{-6} M BB87, complete inhibition of collagen release was observed with increasing release at 10^{-7} and 10^{-8} M BB87.

DISCUSSION

Much previous work has been published on the prevention of cartilage damage as a rational therapeutic target in the rheumatic diseases. Numerous animal models and *in vitro* cartilage models have been used. Many of these have focused almost exclusively on monitoring the release of GAG fragments from the tissue as a measure

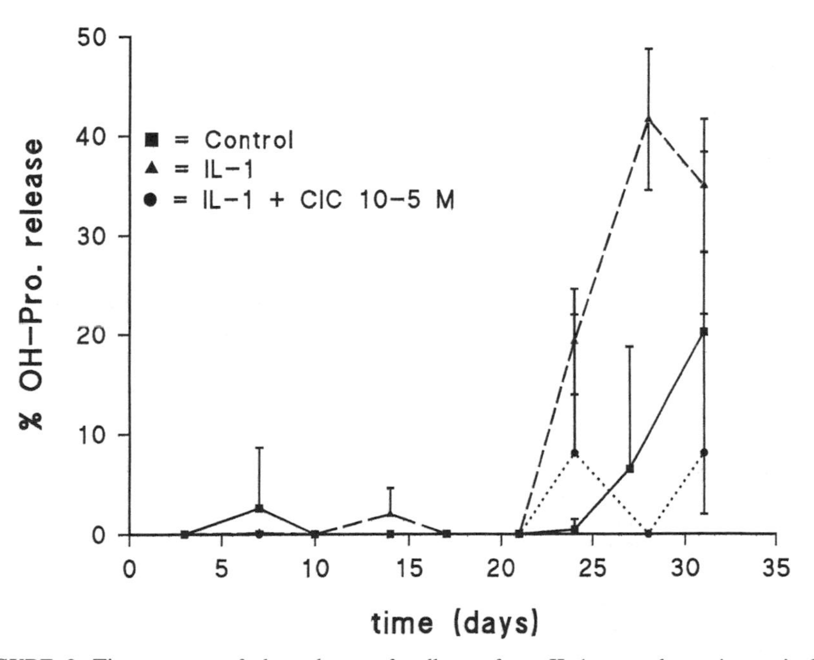

FIGURE 3. Time course of the release of collagen from IL-1–treated porcine articular cartilage in the presence of a synthetic MMP inhibitor. Porcine articular cartilage was cultured in the presence and absence of IL-1 for a total of 32 days and the medium removed approximately every 3 days. The level of OH-proline was measured at each time point and expressed as a percentage of the total. Inclusion of a synthetic MMP inhibitor[16] prevented the release of OH-proline on day 27 onward.

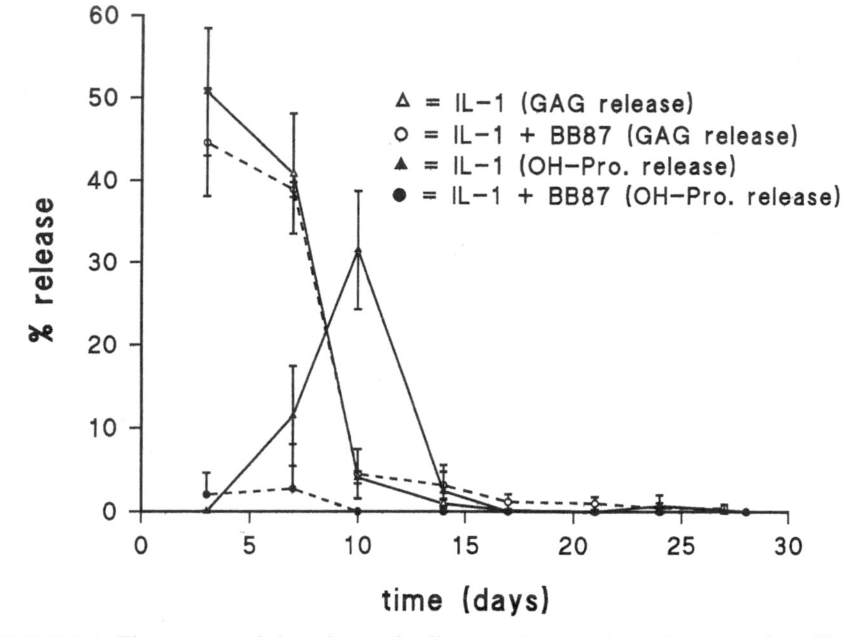

FIGURE 4. Time course of the release of collagen and proteoglycan fragments from IL-1–treated porcine articular cartilage: Prevention of collagen release with a synthetic MMP inhibitor, BB87. Porcine articular cartilage was treated as in FIGURES 2 and 3. The synthetic inhibitor BB87 at 10^{-5} M did not significantly affect the release of GAG from the cartilage, but did prevent the release of OH-proline from the IL-1–stimulated cartilage, which in this experiment was released at day 10.

of damage to the articular cartilage because the release of GAG is rapid and easy to measure. However, recent results would suggest[17] that cartilage can tolerate quite high levels of GAG turnover before permanent damage ensues and GAG can also be rapidly replaced by the chondrocytes.[18] A more appropriate measure of cartilage degradation is to follow the release of collagen fragments because cleavage of the collagen fibers is usually followed by irreversible damage.[19] However, release of collagen from cartilage is not always reproducible after treatment with cytokines, and is always delayed. In the present study we have shown that release of collagen fragments occurs when bovine nasal cartilage is stimulated with high concentrations of IL-1, but that pig articular cartilage is less reliable as a model system. The reason for this difference is unclear. Treatment of bovine nasal cartilage with high concentrations of IL-1 released collagen fragments after day 7 but before day 14. TIMP and TIMP-2 had no effect on GAG release from bovine nasal cartilage confirming our previously published results using pig articular cartilage.[16] However, both inhibitors completely prevented the release of collagen fragments. Although it is possible that the release of GAG fragments is initiated by a proteinase not inhibited by TIMP, this seems unlikely because low molecular weight inhibitors were able to block release.[10–16] All members of the MMP family purified to date are inhibited by the TIMPs. It is more likely that both TIMPs are unable to gain access to the cartilage matrix in the presence of high concentrations of the highly charged GAG fragments. In our previous study with pig articular cartilage[16] we were able to show that although

TIMP was bound to the cartilage it did not penetrate the matrix to any extent and so was unable to block the release of GAG. Removal of GAG fragments from the cartilage on day 7 in this study allows the inhibitors to penetrate the cartilage inasmuch as over 75% of the GAG has been released at this time. TIMP and TIMP-2 are thus able to reach the sites of cartilage collagen destruction surrounding the cells and prevent collagenase from acting on collagen.

The results with the low molecular weight inhibitors are also interesting. A marked difference in concentration is required to block GAG release (relatively high concentrations required) compared to collagen release (lower concentrations required). This could be explained in the same way as results obtained with addition of TIMPs. The highly charged GAGs at high concentration possibly perturb the interaction of proteinase with inhibitor by competing as the substrate, and therefore a much higher concentration of inhibitor is required than would be expected from the inhibition constants obtained with isolated enzymes. However, high concentrations of collagen are also present as substrate, and the same effect would also be expected with the concentration of inhibitor required to inhibit collagenase. Alternatively, the inhibitors are targeted to known members of the MMP family, particularly collagenase and stromelysin. If release of the PG fragments is effected by a proteinase other than stromelysin or collagenase, then the currently available inhibitors would not necessarily be expected to inhibit GAG release at the same range of concentrations as found effective against stromelysin. Recent studies by

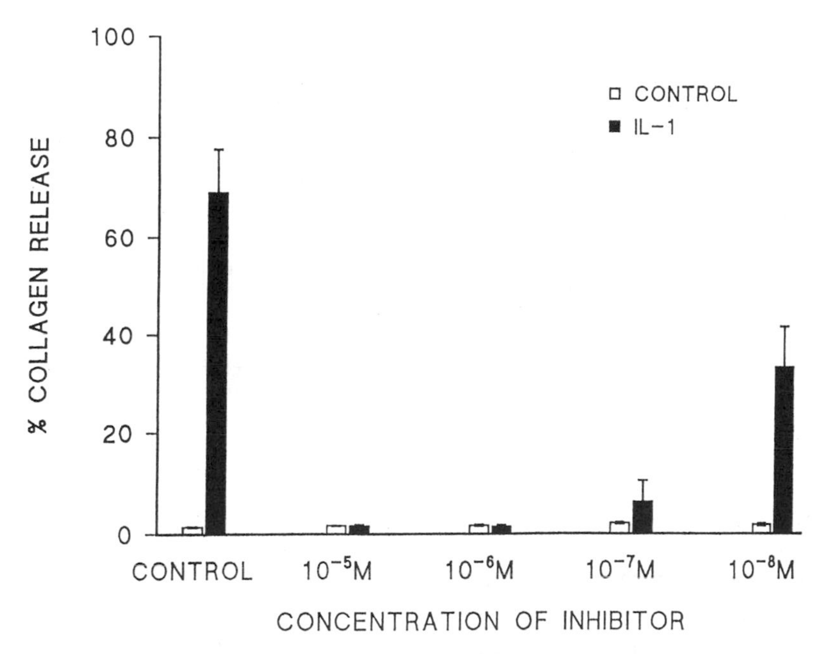

FIGURE 5. Effect of a low molecular weight synthetic inhibitor, BB87, on collagen release from IL-1–stimulated bovine nasal cartilage. IL-1 (1000 units/mL) and BB87 were added to the cartilage on day 1. Collagen release was measured on day 7 and day 14 using the hydroxyproline assay. BB87 gave almost total inhibition of collagen release at the higher concentrations of 10^{-7} M and 10^{-6} M. No inhibition of GAG release was observed at these concentrations of inhibitor in medium on day 7.

Sandy *et al.*[27] have proposed that stromelysin is not the proteinase responsible because the fragments released after cartilage resorption *in vitro* and *in vivo* are produced by cleavage of the aggrecan molecule at a position not cleaved by stromelysin.

CONCLUSION

IL-1-stimulated porcine articular and bovine nasal cartilage fragments were cultured in the presence and absence of various metalloproteinase inhibitors. Tissue inhibitor of metalloproteinases (TIMP) and tissue inhibitor of metalloproteinases-2 (TIMP-2) were unable to prevent the release of proteoglycan from the cartilage, but completely blocked the release of collagen. Incubation of cartilage with low molecular weight synthetic inhibitors of metalloproteinases inhibited the release of both collagen and proteoglycan in a dose-dependent fashion. However, significantly greater levels of inhibitor were required to block proteoglycan release than those required to block collagen release. The results show that these synthetic metalloproteinase inhibitors will be important to prevent the destruction of cartilage collagen in conditions such as rheumatoid arthritis.

ACKNOWLEDGMENTS

The authors thank British Biotechnology and SB Pharmaceuticals for the generous supply of the synthetic MMP inhibitors.

REFERENCES

1. Woessner, J. F., Jr. 1991. Matrix metalloproteinases and their inhibitors in connective tissue remodeling. FASEB J. **5:** 2145–2154.
2. Cawston, T. E., W. A. Galloway, E. Mercer, G. Murphy & J. J. Reynolds. 1981. Purification of rabbit bone inhibitor of collagenase. Biochem. J. **195:** 159–165.
3. Saklatvala, J., L. M. C. Pilsworth, S. J. Sarsfield, J. Gavrilovic & J. K. Heath. 1984. Pig catabolin is a form of interleukin 1. Biochem. J. **224:** 461–466.
4. Pettipher, E. R., G. A. Higgs & B. Henderson. 1986. Interleukin 1 induces leukocyte infiltration and cartilage proteoglycan degradation in the synovial joint. Proc. Natl. Acad. Sci. USA **85:** 8749–8753.
5. Wood, D. D., E. J. Ihrie, C. A. Dinarello & P. L. Cohen. 1983. Isolation of an interleukin-1-like factor from human joint effusions. Arthritis Rheum. **26:** 975–983.
6. Di Giovine, F., D. Sitanayake & G. W. Duff. 1987. Radioimmunoassay (RIA) of interleukin 1 beta (IL1) in acute and chronic rheumatic diseases. Br. J. Rheumatol. **26(Suppl. 1):** 34.
7. Bunning, R. A. D., H. J. Richardson, A. Crawford, H. Skjodt, D. Hughes, D. B. Evans, M. Gowen, P. R. M. Dobson, B. L. Brown & R. G. G. Russell. 1986. The effect of IL1 on connective tissue metabolism and its relevance to arthritis. Agents Actions Suppl. **18:** 131–152.
8. Brinckerhoff, C. E. 1991. Joint destruction in arthritis: Metalloproteinases in the spotlight. Arthritis Rheum. **34:** 1073–1075.
9. Cawston, T. 1993. Blocking cartilage destruction with metalloproteinase inhibitors: A valid therapeutic target. Ann. Rheum. Dis. **52:** 769–770.
10. Buttle, D. J., J. Saklatvala, M. Tamai & A. J. Barrett. 1992. Inhibition of interleukin 1-stimulated cartilage proteoglycan degradation by a lipophilic inactivator of cysteine endopeptidases. Biochem. J. **281:** 175–177.

11. NIXON, J. S., K. M. K. BOTTOMLEY, M. J. BROADHURST, P. A. BROWN, W. H. JOHNSON, G. LAWTON, J. MARLEY, A. D. SEDGWICK & S. E. WILKINSON. 1991. Potent collagenase inhibitors prevent interleukin-1-induced cartilage degradation in vitro. Int. J. Tissue React. **13:** 237–243.
12. CAPUTO, C. B., L. A. SYGAWSKI, D. J. WOLANIN, S. P. POTTON, R. G. CACCESE, A. SHAW, R. A. ROBERTS & G. DIPASQUALE. 1987. Effect of synthetic metalloproteinase inhibitors on cartilage autolysis in vitro. J. Pharmacol. Exp. Ther. **240:** 460–465.
13. DIPASQUALE, G., R. G. CACCESE, R. PASTERNAK, J. CONATY, S. HUBBS & K. PERRY. 1986. Proteoglycan- and collagen-degrading enzymes from human interleukin 1-stimulated chondrocytes from several species: Proteoglycanase and collagenase inhibitors as potentially new disease-modifying antiarthritis agents (42416). Proc. Soc. Exp. Biol. Med. **183:** 262–267.
14. SEED, M. P., S. ISMAIEL, C. Y. CHEUNG, T. A. THOMSON, C. R. GARDNER, R. M. ATKINS & C. J. ELSON. 1993. Inhibition of interleukin 1β induced rat and human cartilage degradation in vitro by the metalloproteinase inhibitor U27391. Ann. Rheum. Dis. **52:** 37–43.
15. SAKLATVALA, J. & T. BIRD. 1987. Interleukin 1, tumour necrosis factor and cartilage degradation. *In* The Control of Tissue Damage. The Arthritis and Rheumatism Council Conference Proceedings No. 2. P. Maddison, Ed.: 6. ARC. Chesterfield, UK.
16. ANDREWS, H. J., T. A. PLUMPTON, G. P. HARPER & T. E. CAWSTON. 1992. A synthetic peptide metalloproteinase inhibitor, but not TIMP, prevents the breakdown of proteoglycan within articular cartilage *in vitro.* Agents Actions **37:** 147–154.
17. DINGLE, J. T., D. P. PAGE THOMAS, B. KING & D. R. BARD. 1987. In vivo studies of articular tissue damage mediated by catabolin/interleukin 1. Ann. Rheum. Dis. **46:** 527–533.
18. THOMAS, D. P. P., B. KING, T. STEPHENS & J. T. DINGLE. 1991. In vivo studies of cartilage regeneration after damage induced by catabolin/interleukin-1. Ann. Rheum. Dis. **50:** 75–80.
19. FELL, H. B. & R. W. JUBB. 1980. The breakdown of collagen by chondrocytes. J. Pathol. **130:** 159–167.
20. HODGES, D. J., D. REID & T. E. CAWSTON. 1994. Purification and secondary structural analysis of tissue inhibitor of metalloproteinases-1. Biochem. Biophys. Acta. In press.
21. CURRY, V. A., I. M. CLARK, H. BIGG & T. E. CAWSTON. 1992. Large inhibitor of metalloproteinases (LIMP) contains tissue inhibitor of metalloproteinases (TIMP)-2 bound to 72000-M_r progelatinase. Biochem. J. **285:** 143–147.
22. FARNDALE, R. W., D. J. BUTTLE & A. J. BARRETT. 1986. Improved quantitation and discrimination of sulphated glycosaminoglycans by use of dimethylmethylene blue. Biochem. Biophys. Acta **883:** 173–177.
23. RATCLIFFE, A., M. DOCHERTY, R. N. MAINI & T. E. HARDINGHAM. 1988. Increased concentrations of proteoglycan components in the synovial fluids of patients with acute but not chronic joint disease. Ann. Rheum. Dis. **46:** 826–832.
24. BERGMAN, I. & R. LOXLEY. 1963. Two improved and simplified methods for the spectrophotometric determination of hydroxyproline. Anal. Biochem. **35:** 1961–1965.
25. ANDREWS, H. J., T. A. EDWARDS, T. E. CAWSTON & B. L. HAZLEMAN. 1989. Transforming growth factor-beta causes partial inhibition of interleukin 1-stimulated cartilage degradation in vitro. Biochem. Biophys. Res. Commun. **162:** 144–150.
26. ANDREWS, H. J., R. A. D. BUNNING, T. A. PLUMPTON, I. M. CLARK, R. G. G. RUSSELL & T. CAWSTON. 1990. Inhibition of interleukin-1-induced collagenase production in human articular chondrocytes in vitro by recombinant human interferon-gamma. Arthritis Rheum. **33:** 1733–1738.
27. SANDY, J. D., C. R. FLANNERY, P. J. NEAME & L. S. LOHMANDER. 1992. The structure of aggrecan fragments in human synovial fluid. Evidence for the involvement in osteoarthritis of a novel proteinase which cleaves the Glu 373-Ala 374 bond of the interglobular domain. J. Clin. Invest. **89:** 1512–1516.

Analysis of the TIMP-1/FIB-CL Complex[a]

M. KIRBY BODDEN,[b,c] L. JACK WINDSOR,[b,d]
NANCY C. M. CATERINA,[d] AUDRA YERMOVSKY,[d]
BENTE BIRKEDAL-HANSEN,[e] GRAZYNA GALAZKA,[d]
JEFFREY A. ENGLER,[b,d]
AND HENNING BIRKEDAL-HANSEN[b,f]

[b]*Department of Oral Biology and Research Center in Oral Biology*
[e]*Department of Diagnostic Sciences*
University of Alabama School of Dentistry

Departments of Pathology[c] *and of Biochemistry and Molecular Genetics*[d]
University of Alabama at Birmingham
Birmingham, Alabama 35294

THE FIBROBLAST-TYPE COLLAGENASE MOLECULE

The secreted form of fibroblast-type collagenase (FIB-CL) shares with seven of eight other matrix metalloproteinases (MMPs) a characteristic four-domain structure with the central M_r 18,000 ($\approx$150 residue) catalytic domain that contains the highly conserved active-site zinc (Zn)-binding sequence.[1,2] The catalytic domain is preceded by a propeptide which endows the enzyme with catalytic latency and is followed by a short hinge region and a large 200-residue COOH-terminal domain of four pexin-like repeats. The only disulfide bond of the molecule links the COOH-terminus to the end of the hinge region. In the nascent zymogen a single unpaired Cys residue located at the end of the propeptide (Cys73) forms a coordination bond with the active-site Zn-atom. The domain arrangement is shown in FIGURE 1.

THE TIMP MOLECULE

Six disulfide bonds apportion the 184-residue human tissue inhibitor of metalloproteinase-1 (TIMP-1) molecule into six distinct loop structures as shown in FIGURE 2. The six disulfide bonds are arranged in two to three knot-like structures, the first composed of disulfide bonds C1–C70 and C3–C99; the second of disulfide bonds C13–C124 and C127–174.[3] The remaining two disulfide bonds are C132–C137 and C145–C166 in the COOH-terminal domain of the molecule. The large 70-residue loop 1 connects in two directions the two disulfide knots, through a short sequence composed of residues 3–13 and through a long sequence of residues 13–70. Throughout the molecule, sequences that are highly conserved are interrupted by more variable domains. The most highly conserved sequence is the first 20 residues of the NH_2-terminal region (FIG. 3).

[a]This work was supported by U.S. Public Health grants DE08228, DE10631, and DE 00283.
[f]Address correspondence to Dr. Henning Birkedal-Hansen, National Institute of Dental Research, National Institutes of Health, Building 30, Room 132, Bethesda, Maryland 20892.

COMPLEX FORMATION

MMP complexes with TIMPs, unlike those formed with α_2-M,[4,5] are not dependent upon prior proteolytic cleavage of the inhibitor and do not entail covalent trapping of the enzyme. As a result, the inhibitor and the enzyme can again be released and recovered from the complex in intact form.[6,7] TIMP-1 and TIMP-2 also form complexes with the catalytically inactive zymogens of M_r 72K gelatinase (TIMP-2) and M_r 92K gelatinase (TIMP-1), but the relationship of these complexes to complexes formed with the active forms of these enzymes, or with other activated MMP, is not clear. The objective of this review is to summarize recent findings in our laboratory that have begun to define in functional and topographic terms the complex formed between activated human FIB-CL and human TIMP-1.

TIMP-1 inhibits FIB-CL effectively and forms a noncovalent, bimolecular complex with K_d is in the range 4.1×10^{-9}–10^{-11}.[8–10] Complex formation can be demonstrated by SDS-PAGE analysis by the method of DeClerck *et al.*[11] based on the observation that exposure to low concentrations of SDS (0.1%) does not dissociate the complexes. Intact complexes can therefore be resolved from the free

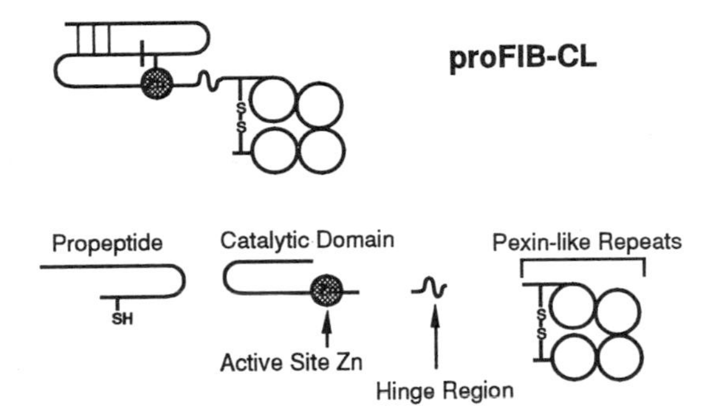

FIGURE 1. Domain structure of the zymogen of human fibroblast-type collagenase.

components by SDS-PAGE and subsequently identified either by autoradiography or Western analysis. Activated forms of FIB-CL, as well as SL-1 and SL-2, form SDS-resistant complexes with both TIMP-1 and TIMP-2 whereas the full-length latent proenzymes do not. Partial processing of the propeptide by cleavage near the trypsin-sensitive region, as frequently occurs when recombinant human FIB-CL is expressed in *Escherichia coli,* however, yields a M_r 45,000 intermediate, which spontaneously forms a small amount of SDS-resistant complex.[12] If the enzyme is activated by APMA before addition of TIMP-1, a single M_r 59,000 complex with the processed M_r 42,000 enzyme is formed (data not shown). If APMA is added at the same time as TIMP, two complexes are generated, one (M_r 59,000) with the processed form and another (M_r 71,000) with the apparently unprocessed, full-length M_r 52,000 FIB-CL (FIG. 4A). We interpret these findings to suggest that interaction with APMA generates a "switch open" intermediate[13,14] prior to autolytic processing which is also captured by TIMP. A significant fraction of the switch-open form, however, is processed autolytically and is captured after excision of the propeptide. We observed that seemingly minor mutations of the FIB-CL sequence (FIG. 4B) dictate whether capture occurs before or after propeptide processing. Mutation of a

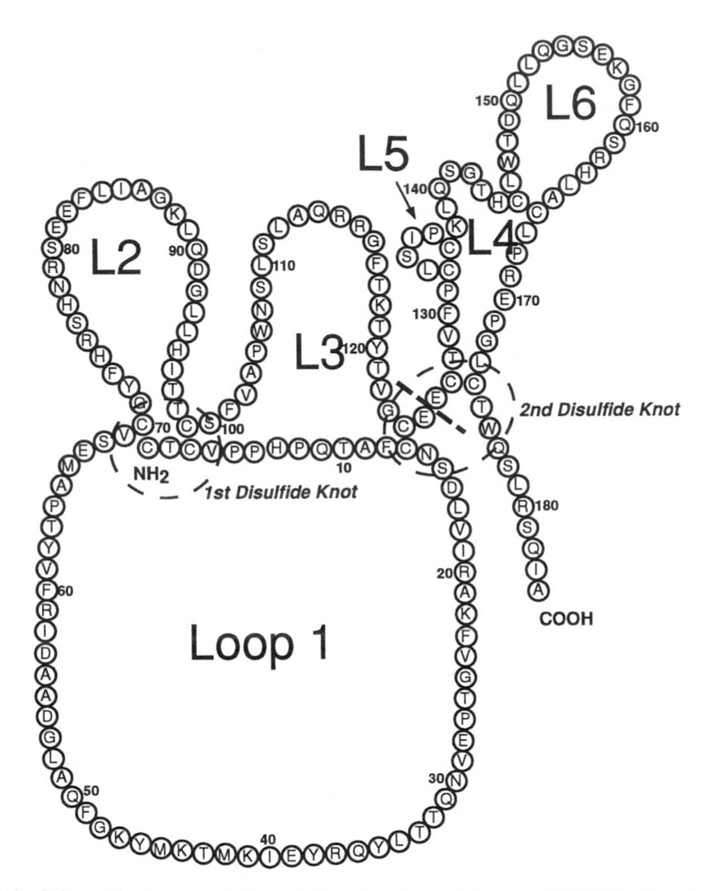

FIGURE 2. Schematic representation of the structure of human TIMP-1. (Bodden *et al.*[10] Reproduced with permission from the *Journal of Biological Chemistry.*)

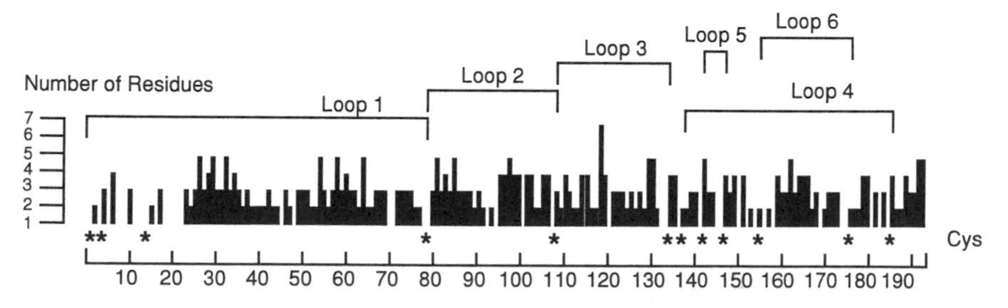

FIGURE 3. Homology of tissue inhibitor of metalloproteinases (TIMPs). Complete or partial sequences of 11 TIMPs were aligned to yield maximal homology, and the degree of divergence (number of residues that occupy each position) was plotted as a function of residue location from the NH_2-terminus (residue 1) to the COOH-terminus of human TIMP-1 (residue 194). Because the alignment creates gaps in some sequences, the numbering is slightly different from that used in FIGURE 2. The invariant position of Cys residues (*asterisk*) and the loop structures that result from disulfide pairing are also shown. Sequences used were human TIMP-1,[21] bovine TIMP-1,[20] rabbit TIMP-1,[20] mouse TIMP-1,[19] rat TIMP-1,[23] human TIMP-2,[24] bovine TIMP-2,[24] mouse TIMP-2,[25] rat TIMP-2,[23] chicken TIMP-2,[26] and chicken TIMP-3.[27]

poorly conserved sequence, HR, immediately preceding the active-site Zn-binding sequence of FIB-CL, significantly alters whether the processed or unprocessed form is captured (FIG. 4C). In other MMPs, HR is replaced with YR, FI, FL, and the two residues are not even conserved among FIB-CLs from different species. Wild-type and H194S and H194F mutants form both the M_r 59,000 and 71,000 complexes, mutant H194R forms exclusively the M_r 71,000 complex, and double mutant HR-FL only forms a complex with the fully processed form (M_r 59,000) (FIG. 4C). The variation in degree of autolytic processing before capture with TIMP-1 most likely is a result of changes in the association rate of the complex, the rate of autolytic processing, or both. The pexin-like COOH-terminal domain is not an absolute requirement for complex formation with TIMP-1. Deletion of the four pexin-like repeats in FIB-CL (mini-FIB-CL) yields a mutant that still forms SDS-resistant complex with TIMP-1 (data not shown).

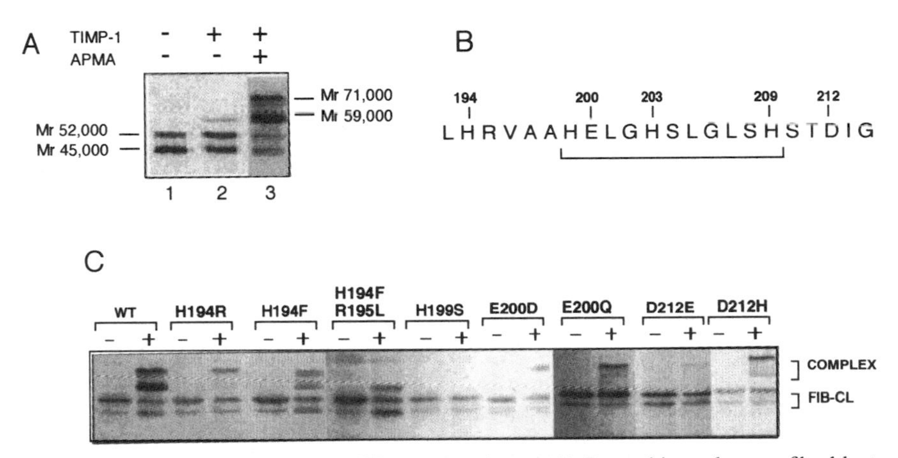

FIGURE 4. Formation of TIMP-1/FIB-CL complexes. **(A)** Recombinant human fibroblast-type collagenase expressed in *E. coli*[12,28] (*lane 1*) was incubated with TIMP-1 (*lane 2*) or with TIMP-1 plus 1.0 mM APMA. The samples were resolved by SDS-PAGE by the method of DeClerck[11] and stained with mAb to FIB-CL. Complexes formed are M_r 71,000 and M_r 59,000. Data from ref. 12. **(B)** Mutation sites in the sequence surrounding the active site Zn-binding domain of human FIB-CL. **(C)** TIMP-1 capture by wild type (wt) and mutants of human FIB-CL using the technique described in **panel A**. Samples were incubated with TIMP-1 plus 1.0 mM APMA (+) or with buffer alone (−). Mutants examined were H194F, H194R, H194S (not shown), H194F/R195L, H199S, H199C (not shown), H203S (not shown), H209S (not shown), E200D, E200Q, D212E, and D212H. (Data from Windsor *et al.*[12])

ROLE OF THE ACTIVE SITE IN COMPLEX FORMATION

Mutations in the active-site Zn-binding ligands His199, His203, and His209, yield proteins devoid of catalytic activity that are rapidly and completely degraded by low concentrations of trypsin and therefore appear to be improperly folded. None of these mutants is captured by TIMP-1[12] (FIG. 4C). These findings suggest that an intact, properly folded Zn-binding site is a requirement for complex formation with TIMP-1. The active site, however, does not have to be catalytically functional. A catalytically incompetent, but properly folded, mutant of the active site Glu200 residue readily forms a complex with TIMP-1. Because this mutant is incapable of autolytic processing, only the high-M_r 71,000 complex is formed. Essentially similar results were obtained with another catalytically incompetent mutant, D212E. D212, which is located immediately downstream of the Zn-binding sequence, is occupied by D in all collagenases, in matrilysin, and in M_r 72K GL, but is replaced by A, N or V in other MMPs. Although this mutant showed reduced ability to form a complex with TIMP-1, it formed a complex with TIMP-2 fully as well as wild type. Taken together these findings suggest that a properly folded but not necessarily catalytically competent active site is required for formation of a complex with TIMP-1. Based on these findings, we propose the model shown in FIGURE 5 for the activation/TIMP-1 capture mechanism. Opening of the switch during activation uncovers a cryptic binding site for TIMP-1 and leads to capture of TIMP-1 either before or after autolytic processing of the propeptide. It is still unresolved whether organomercurials merely react with the free thiol group of the "switch-open" form as it exists in equilibrium with the "switch-closed" form or whether the "switch" remains closed in the absence of APMA but is forced open by interaction of the organomercurial with

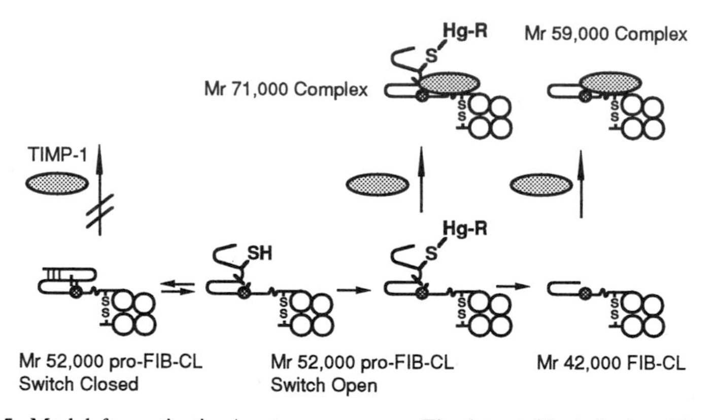

FIGURE 5. Model for activation/capture sequence. The latent ("switch-closed") zymogen exists in equilibrium with a "switch-open" intermediate. Reaction with organomercurials locks the enzyme in the open position and exposes the TIMP-1 binding site. This form is subsequently processed autolytically to yield the M_r 42,000 fully processed enzyme. The switch-closed zymogen does not bind TIMP-1, but the M_r 52,000 switch-open form and the processed M_r 42,000 molecule form complexes (M_r 71,000 and M_r 59,000).

the propeptide. Exposure of the latent proenzyme to SDS rapidly (within 30 min at 37 °C) leads to complete labeling with maleimide compounds of the otherwise cryptic Cys residue in the propeptide, but in the absence of SDS very little spontaneous labeling of this residue (1–5%) occurs for up to 24 h, suggesting that $t_{1/2}$ is $\geq$500 h. These findings suggest that the switch essentially stays closed in the natural state and lead us to speculate that organomercurials, like SDS, induce conformational changes that force the switch open or that activated molecules process other molecules while the switch is still closed in a manner similar to trypsin.

MUTATIONAL ANALYSIS OF TIMP-1

In an attempt to identify domains of TIMP-1 that are important to complex formation with FIB-CL we conducted a series of mutations of the human TIMP-1 molecule. In designing these experiments, we were interested primarily in identifying mutants that render TIMP-1 nonfunctional (knock-out mutations) rather than mutants with modestly increased K_i that still form complexes with FIB-CL. Replacement of the two glycosylation sites singly or in combination did not alter the activity of TIMP-1 against FIB-CL and confirmed that the carbohydrate attachments are probably not involved in complex formation.[15] Replacement of highly conserved residues in the NH_2-terminal sequence, P5A and P9A, also had only minor effects and shifted IC_{50} upward by less than twofold. That was the case also with a mutant in the highly conserved VIRAK sequence (VIRAK to VIAAA) and with a COOH-terminal domain deletion mutant that ends at residue E126 after the first disulfide bridge of the second disulfide knot. The latter mutant is identical to that previously constructed by Murphy and coworkers,[16] and the results are similar. Replacement of Cys1 or Cys3 with Ser which prevents formation of the disulfide bonds Cys1—Cys70 or Cys3—Cys99, however, completely abolished inhibitory activity and rendered the molecule inactive. These findings show that drastic mutations are required to inactivate TIMP-1 consistent with the conclusions reached by Murphy and coworkers.[16,17] Although these findings might suggest that the residues targeted in the mutation experiments conducted by us and by others[16,17] are of minor importance to complex formation, an equally plausible explanation might be that TIMP-1 maintains an extended contact surface with FIB-CL so that elimination of a single contact site only moderately affects K_i. We therefore pursued alternate strategies to identify functionally important domains in TIMP-1.

BLOCKING ANTIBODIES

Although it might be argued that immunoglobulin (Ig) molecules are at least five times larger than the TIMP-1 molecule and that virtually any antibody that binds strongly enough would be expected to block complex formation, a survey of a collection of monoclonal (mAb) and polyclonal (pAb) antibodies undertaken in our laboratory suggested that many antibodies fail to block TIMP-1 activity. We succeeded in identifying two blocking antibodies, one mAb and one pAb, for which the epitopes could be reasonably well defined. Binding studies based on ELISA plates coated with a series of long peptides (20–34 residues) modeled after the human TIMP-1 sequence showed that the epitope for the blocking mAb 7-23G9 (developed by Hayakawa and coworkers[18] and supplied to us by Dr. K. Iwata, Fuji, Japan) was located in loop 3 (peptide T99-125). By use of a series of shorter peptides, it was observed that peptides T121–129 (TVGCEECTV), T121–127 (TVGCEEC), and T123–127 (GCEEC), but not T121–124 (TVGC), competed for binding of the mAb to TIMP-1, suggesting that the epitope was located near the second disulfide knot. Inasmuch as this mAb was raised against bovine TIMP-1,[18] we examined the murine, bovine, and human sequences to determine what the mouse immune system might have "seen" in bovine TIMP-1 that is also present in human TIMP-1.

```
m-TIMP-1 Q Q R V F S K K N Y S A G C G V C T V 19
b-TIMP-1 Q R R G F T K T – Y A A G C E E C T V 20
                  *  *     *        *  *

                          121
h-TIMP-1 Q R R G F T K T – Y T V G C E E C T V 21
                                *  *
```

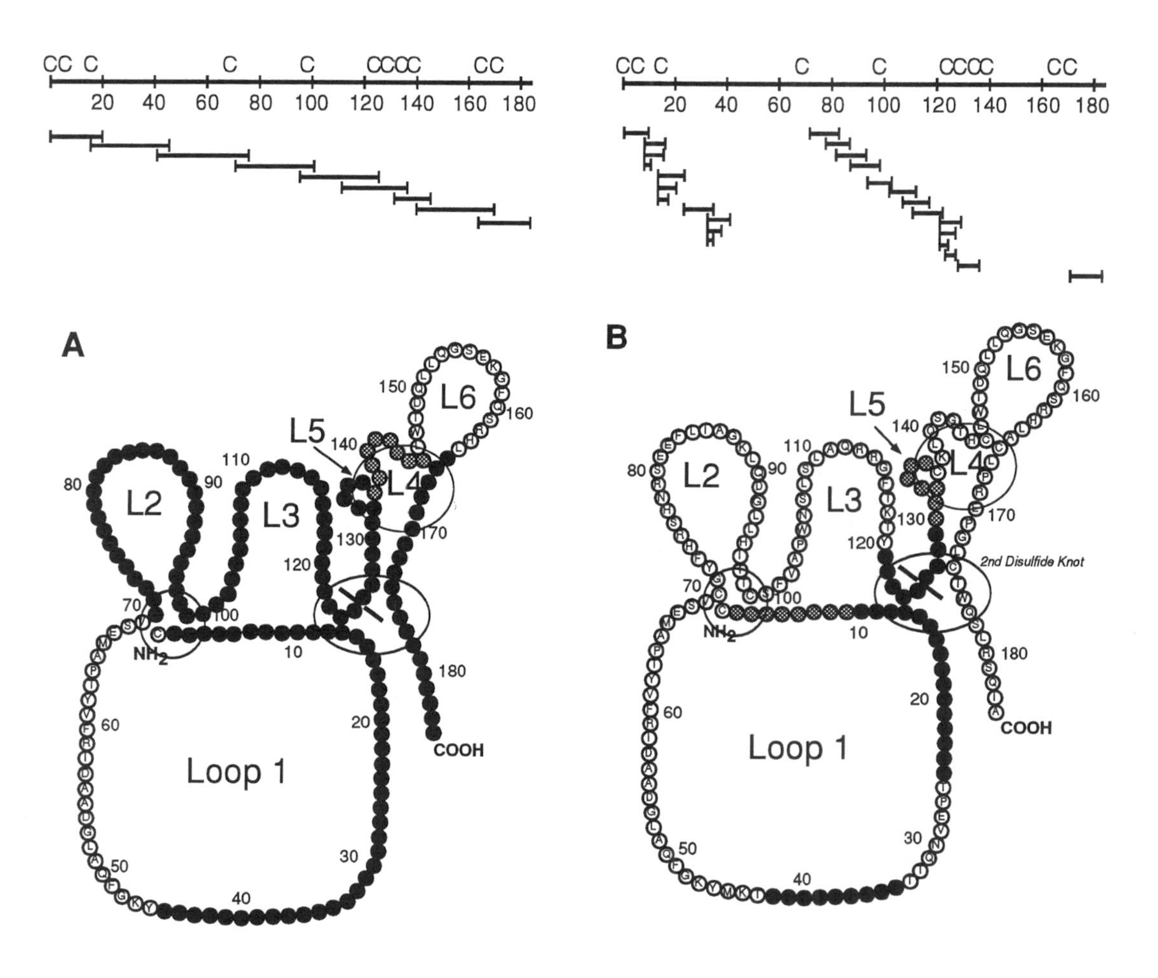
A
B
Loop 1
L2
L3
L4
L5
L6
NH2
COOH
2nd Disulfide Knot
CC C
C
C
CCCC
CC
20 40 60 80 100 120 140 160 180

Within the T121–129 sequence, bovine and human TIMP-1 contain the residues EE which are replaced in mouse with GV. These residues are probably important for binding although they do not necessarily constitute the entire epitope. Using similar techniques we determined that a nonblocking mAb which bound with equal strength to human TIMP-1 ($K_d = 3.9 \times 10^{-9}$ M versus 1.2×10^{-9}) recognized an epitope located in a nine-residue peptide in loop 6 (CLPREPGLC) in the COOH-terminal domain of TIMP-1. The human sequence in this region differs in two positions from that of mouse:

m-TIMP-1 C L P R N L G L C
h-TIMP-1 C L P R E P G L C
* *

The blocking pAb (raised against recombinant human TIMP-1 refolded from *E. coli* extracts—a generous gift by Dr. Carmichael, Synergen) recognized predominantly loop 3 and the last 20 residues of the molecule (T165–184). By contrast, a nonblocking pAb raised in rabbits against natural TIMP-1 did not recognize loop 3. The predominant epitope(s) were located in loop 1, and additional immunoreactivity was found in peptide T141–170 and in the COOH-terminal tail. These findings when held together suggest, but do not prove, that loop 3 and the transition region to loop 4 may be important to inhibitory function.

PEPTIDE COMPETITION STUDIES

To further identify domains in TIMP-1 that are involved in contact with FIB-CL we employed a peptide competition approach. To this end, we examined the ability of synthetic peptides modeled after the human TIMP-1 sequence to complete for binding to FIB-CL, and to neutralize the inhibitory effect of TIMP-1 against this enzyme. We initially chose relatively long peptides ($n = 20$–34) because they are more likely than shorter ones to adopt secondary structure in solution. A first round of competition experiments showed that peptides together covering more than two-thirds of the TIMP-1 sequence neutralized the inhibitory activity of TIMP-1 against FIB-CL (FIG. 6). Peptides that competed for binding to FIB-CL neutralized TIMP-1 in a dose-dependent manner in a narrow concentration range from 0.1–1.0 mM. None of these peptides directly inhibited the enzyme at concentrations up to 1 mM. Peptides that neutralized TIMP-1 activity included the first 45 residues of loop 1, but not the last 30 residues, which contained the predominant epitopes for the nonblocking pAb. In addition, peptides covering loops 2 and 3 as well as the

←

FIGURE 6. Competition of synthetic peptides for binding of TIMP-1 to FIB-CL. (**A**) First round of competition with long peptides (20–34 residues) modeled after the human TIMP-1 sequence. Peptide alignment is shown schematically in **top panel**. Increasing concentrations of synthetic peptides (0.01–1.0 mM) were incubated with mixtures of FIB-CL and TIMP-1 adjusted to yield 80–90% inhibition of the enzyme. The degree of neutralization of TIMP-1 activity was quantified by the reemergence of FIB-CL activity. Color markings indicate location of peptides that completely (*black*) or partially (*gray*) neutralized TIMP-1 at 1.0 mM (*gray*). Peptides that did not measurably neutralize TIMP-1 at 1.0 mM are shown in white. (**B**)Second round of competition experiments using shorter peptides from selected regions of human TIMP-1 shown schematically in **top panel.** The methods were as described in panel **A**. (Data in **A** and **B** are from Bodden *et al.*[10]; figure adapted from the *Journal of Biological Chemistry.*)

beginning portion of loop 4 and the COOH-terminal tail were also effective. Although only a few residues in each peptide may be directly involved in contact with FIB-CL, these findings suggested that important contact sites are distributed throughout the molecule and are not confined to a single region.

To further refine the map of neutralizing peptides, a series of shorter peptides modeled after selected segments identified in the first round was analyzed. The result is shown in FIGURE 6. These studies drew particular attention to the region in and around the second disulfide knot (Cys13–Cys124 and Cys127–Cys174). In loop 1, each of three shorter peptides from the NH_2-terminal region (T2–11, T10–18, and T15–25) neutralized TIMP activity with T15–25 being the most effective and T2–11 the least. IC_{50} values for these peptides were approximately 1.0, 0.5, and 0.2 mM. Peptide T2–11 contains several of the residues previously mutated by O'Shea and coworkers,[17] as well as residues mutated in our laboratory (N. Caterina *et al.,* in preparation) as described above. It is interesting to note that none of the mutations in the two studies entirely abolished inhibitory activity but caused rather modest changes of K_i. Peptide T15–25 (SDLVIRAKFVG) contains the VIRAK sequence which is absolutely conserved in all TIMPs. Further shortening of this peptide to SDLVIRAK and SDLVI abolished neutralizing activity. Further downstream, T34–42 (LYQRYEIKM) also neutralized TIMP-1, whereas the intervening peptide which carries one of the two N-glycosylation sites did not. T34–42 contains a sequence, YEIK, that is conserved in all TIMPs-1 and TIMPs-2. In loop 2, none of four 8–10 residue, overlapping peptides modeled after T70–99 was able to neutralize TIMP-1 activity and precluded further refinement. In loop 3, only peptide T121–129 (TVGCEECTV) which contains the transition region to loop 4 (CEEC), previously implicated by the mAb binding studies, was able to compete with TIMP-1 and neutralize its activity. Shortened versions of this peptide, TVGCEEC, GCEEC, and TVGC all were unable to compete with TIMP-1, whereas an overlapping peptide, TVFPCLSIP, partially neutralized TIMP-1 activity. Beyond this peptide, two shorter peptides in the COOH-terminal tail were tested in an effort to more narrowly define important domains in the COOH-terminal region, but neither T166–174 nor T171–184 was able to compete effectively at ≤ 1.0 mM concentrations and therefore precluded further refinement.

The findings summarized above drew our attention to the second disulfide knot because peptides of both the "upper" and "lower" strand of this structure in the depiction used in FIGURE 2 neutralized TIMP-1 activity. Moreover, out of 34 different peptides covering the entire TIMP-1 molecule, only peptides T121–129 and T10–18, which constitute the upper and lower strand of the second disulfide knot, directly inhibited FIB-CL activity in the absence of TIMP-1. IC_{50} values for inhibition of FIB-CL were ≈ 0.8 mM. Under similar conditions, free cysteine and cysteamine yielded IC_{50} values 15–20-fold higher. Since both of the inhibiting peptides contain a free thiol group, we suspected that the direct inhibitory effect against FIB-CL might in part be attributed to the Zn-coordinating capabilities of the thiol groups. The suspicion was verified because alkylation with iodoacetamide abrogated the inhibitory effect against FIB-CL but, somewhat surprisingly, did not abolish the neutralizing effect against TIMP-1. The FIB-CL inhibitory activity therefore could be dissociated from the TIMP-1 neutralizing effect. A survey of all 34 peptides, as well as 5 additional peptides synthesized with an extra terminal Cys residue for linkage to carrier matrices, showed that 23 of the peptides contained at least one Cys residue and 15 contained two or more. Yet only peptides TAFCNSDLV and TVGCEECTV directly inhibited FIB-CL at concentrations ≤ 1 mM. These findings show that although the free thiol groups of peptides T10–18 and T121–129 are involved in the inhibition of FIB-CL, the presence of one or more Cys residues in itself is not sufficient to endow the peptides with inhibitory activity against FIB-CL at concentrations ≤1 mM. Inasmuch as the two inhibiting peptides also neutralized TIMP-1, we speculate that the peptide sequence that binds to FIB-CL contributes to the 15–20-fold lower K_i values for these thiol peptides compared to those of cysteine and cysteamine. We propose as a working hypothesis that the peptides of the second disulfide knot bind near enough to the active-site Zn to permit formation of coordination bonds with the Zn atom and that this bonding is responsible for the inhibitory effect. After blocking of the free thiol group by alkylation, the peptide still

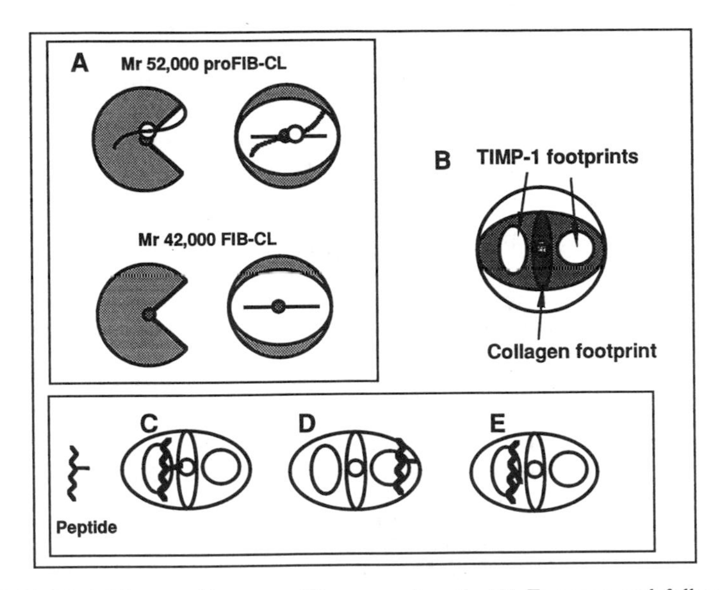

FIGURE 7. Model for peptide competition experiments. **(A)** Zymogen and fully activated forms of human FIB-CL. In the M_r 52,000 zymogen form, the propeptide blocks access to the active site Zn by forming a coordination bond with the Zn-atom (Zn-Cys73). Cleavage of the propeptide in the fully processed M_r 42,000 enzyme opens access to the active site. **(B)** Schematic representation of model proposing different binding sites ("footprints") for the collagen triple helix and TIMP-1 on the exposed FIB-CL surface. **(C–E)** Schematic representation of model proposing placement of peptides from the second disulfide knot near the FIB-CL active site Zn to explain the dissociation of the TIMP-1-neutralizing from the collagenase-inhibitory effect. In panel **C**, the peptide neutralizes TIMP-1 by binding in the TIMP-1 footprint and inhibits FIB-CL by coordinately binding to the active site Zn-atom. Peptides that contain Cys residue(s) but bind at distant sites in the TIMP-1 footprint neutralize TIMP-1 but do not inhibit FIB-CL (panel **D**). In panel **E**, the collagenase-inhibitory activity of the peptide is abolished by alkylation of the free thiol group. The peptide, however, still neutralizes TIMP-1.

binds well enough to compete with TIMP-1 but does not interfere with collagen binding. Because the K_d for collagen binding (0.5×10^{-6} M[22]) is at least 100-fold higher than for TIMP-1 binding, it is unlikely that TIMP-1 is displaced from FIB-CL more readily than collagen. We therefore propose that collagen and TIMP-1 engage different binding sites, that is, the two molecules have different "footprints" on the surface of the FIB-CL molcculc (FIG. 7).

It cannot be precluded, however, that the FIB-CL inhibitory effect and the TIMP-1 neutralizing effect of peptides T10–18 and T121–129 are caused by binding at two entirely different sites, that is, FIB-CL inhibition through binding to the active-site Zn solely through the free thiol group, and neutralization of TIMP-1 through binding to a different site via the peptide sequence. This explanation appears less likely, however, because the peptides are 15–20-fold more effective inhibitors than free cysteine or cysteamine alone, and because other Cys-containing peptides fail to inhibit FIB-CL. The observation that alkylation of the free thiol in these peptides abolishes FIB-CL inhibitory activity but not TIMP-1 neutralizing activity, moreover, suggests as highly likely that the active-site Zn is not directly involved in TIMP-1 binding. It does appear likely, however, that the second disulfide knot is positioned within a few angstroms of the active-site Zn atom. It is somewhat surprising, however, that the second disulfide knot also appears to play a major role in the formation of the zymogen-inhibitor complex with proM_r 92K GL. This observation raises the question whether the second disulfide knot is bound near the active-site Zn atom also in the proenzyme-TIMP-1 complex, but this question can probably only be resolved by X-ray crystallography.

REFERENCES

1. Birkedal-Hansen, H., W. G. I. Moore, M. K. Bodden, L. J. Windsor, B. Birkedal-Hansen, A. DeCarlo & J. A. Engler. 1993. Crit. Rev. Oral Biol. Med. **4:** 197–250.
2. Woessner, J. F., Jr. 1991. FASEB J. **5:** 2145–2154.
3. Williamson, R. A., F. A. O. Marston, S. Angal, P. Koklitis, M. Panico, H. R. Morris, A. F. Carne, B. J. Smith, T. J. R. Harris & R. B. Freedman. 1990. Biochem. J. **268:** 267–274.
4. Sottrup-Jensen, L. & H. Birkedal-Hansen. 1989. J. Biol. Chem. **264:** 393–401.
5. Enghild, J. J., G. Salvesen, K. Brew & H. Nagase. 1989. J. Biol. Chem. **264:** 8779–8785.
6. Murphy, G., P. Koklitis & A. F. Carne. 1989. Biochem. J. **261:** 1031–1034.
7. Okada, Y., Y. Gonoji, K. Naka, K. Tomita, I. Nakanishi, K. Iwata, K. Yamashita & T. Hayakawa. 1992. J. Biol. Chem. **267:** 21712–21719.
8. Cawston, T. E., G. Murphy, E. Mercer, W. A. Galloway, B. L. Hazleman & J. J. Reynolds. 1983. Biochem. J. **211:** 313–318.
9. Stricklin, G. P. & H. G. Welgus. 1983. J. Biol. Chem. **258:** 12252–12258.
10. Bodden, M. K., G. J. Harber, B. Birkedal-Hansen, L. J. Windsor, N. C. M. Caterina, J. A. Engler & H. Birkedal-Hansen. J. Biol. Chem. In press.
11. DeClerck, Y. A., T. D. Yean, H. S. Lu, J. Ting & K. E. Langley. 1991. J. Biol. Chem. **266:** 3893–3899.
12. Windsor, L. J., M. K. Bodden, B. Birkedal-Hansen, J. A. Engler & H. Birkedal-Hansen. Submitted.
13. Springman, E. B., E. L. Angleton, H. Birkedal-Hansen & H. E. Van Wart. 1990. Proc. Natl. Acad. Sci. USA **87:** 364–368.
14. Van Wart, H. E. & H. Birkedal-Hansen. 1990. Proc. Natl. Acad. Sci. USA **87:** 5578–5582.
15. Stricklin, G. P. 1986. Collagen Relat. Res. **6:** 219–228.
16. Murphy, G., A. Houbrechts, M. I. Cockett, R. A. Williamson, M. O'Shea & A. J. P. Docherty. 1991. Biochemistry **30:** 8097–8102.
17. O'Shea, M., F. Willenbrock, R. A. Williamson, M. I. Cockett, R. B. Freedman, J. J. Reynolds, A. J. P. Docherty & G. Murphy. 1992. Biochemistry **31:** 10146–10152.
18. Kodama, S., J. Kishi, K. Obata, K. Iwata & T. Hayakawa. 1987. Matrix **7:** 341–350.
19. Gewert, D. R., B. Coulombe, M. Castelino, D. Skup & B. R. G. Williams. 1987. EMBO J. **6:** 651–657.
20. Freudenstein, J., S. Wagner, R. M. Luck, R. Einspanier & K. H. Scheit. 1990. Biochem. Biophys. Res. Commun. **171:** 250–256.
21. Docherty, A. J. P., A. Lyons, B. J. Smith, E. M. Wright, P. E. Stephens, T. J. R. Harris, G. Murphy & J. J. Reynolds. 1985. Nature **318:** 66–69.

22. WELGUS, H. G., J. J. JEFFREY, G. P. STRICKLIN, W. T. ROSWIT & A. Z. EISEN. 1980. J. Biol. Chem. **255:** 6806–6813.
23. ROSWIT, W. T., D. W. MCCOURT, N. C. PARTRIDGE & J. J. JEFFREY. 1992. Arch. Biochem. Biophys. **292:** 402–410.
24. BOONE, T. C., M. J. JOHNSON, Y. A. DECLERCK & K. E. LANGLEY. 1990. Proc. Natl. Acad. Sci. USA **87:** 2800–2804.
25. SHIMIZU, S., K. MALIK, H. SEJIMA, J. KISHI, T. HAYAKAWA & O. KOIWAI. 1992. Gene **114:** 291–292.
26. STASKUS, P. W., F. R. MASIARZ, L. J. PALLANCK & S. P. HAWKES. 1991. J. Biol. Chem. **266:** 449–464.
27. PAVLOFF, N., P. W. STASKUS, N. S. KISHNANI & S. P. HAWKES. 1992. J. Biol. Chem. **267:** 17321–17326.
28. WINDSOR, L. J., H. BIRKEDAL-HANSEN, B. BIRKEDAL-HANSEN & J. A. ENGLER. 1991. Biochemistry **30:** 641–647.

A Non-Antimicrobial Tetracycline Inhibits Gingival Matrix Metalloproteinases and Bone Loss in *Porphyromonas gingivalis*-induced Periodontitis in Rats[a]

LORNE M. GOLUB,[b] R. T. EVANS,[c] T. F. McNAMARA,[b]
H. M. LEE,[b] AND NUNGAVARUM S. RAMAMURTHY[b]

[b]*Department of Oral Biology & Pathology*
School of Dental Medicine
State University of New York at Stony Brook
Stony Brook, New York 11794-8702

[c]*Department of Oral Biology*
School of Dentistry
State University of New York at Buffalo
Buffalo, New York 14214

A number of studies suggest that oral microorganisms can induce matrix metalloproteinase (MMP; e.g., collagenase, gelatinase) production and/or activation in the host tissues as a key mechanism in the pathogenesis of periodontal disease,[1–3] one of the most common diseases worldwide. Of direct relevance to this symposium, the pharmacologic inhibition of MMP activity has recently been recognized as a desirable goal of periodontal therapy. This approach, as well as the inhibition of prostaglandin synthesis by nonsteroidal anti-inflammatory drugs, was one of the two strategies highlighted in a recent position paper by the American Academy of Periodontology on the therapeutic modulation of host response.[4] As reviewed in recent years,[5–9] the tetracycline (TC) antibiotics, which have long been used as adjuncts in periodontal therapy based on their antimicrobial effectiveness against a variety of suspected periodontopathic microorganisms, are now known to *also* inhibit pathologically excessive host-derived MMP activity during periodontal and other diseases, although the mechanisms of action are not yet well defined. In fact, the discovery of the anticollagenolytic properties of the TCs was made using an animal model of both pathologically excessive collagenase activity in gingival tissues and periodontal breakdown.[10–13]

Over the past decade, *Porphyromonas* (previously called *Bacteroides*) *gingivalis*, an anaerobic gram-negative oral microorganism, has been considered a major, but not the only, bacterial etiologic factor in human periodontal disease.[14–16] A number of observations have linked this organism to periodontal breakdown in humans, including its relatively large numbers in subgingival plaque of active periodontal lesions, and absence or reduced numbers in healthy sites;[14] the positive relationship between elevated serum antibodies against this organism and periodontal break-

[a]This research was supported by grants from the National Institute of Dental Research (RE37 DE-03987) and from Collagenex, Inc., Wayne, Pennsylvania.

down;[17] and its ability to promote a number of pathogenic responses in the host tissues including, but not limited to, the release of its lipopolysaccharide which can suppress gingival fibroblast proliferation and induce production of cytokines, prostaglandins, and tissue-destructive enzymes (see Klausen *et al.*[1] and Page[2] for reviews). Moreover, inoculation of this human periodontopathogen, *P. gingivalis,* into the oral cavity of experimental animals (rats, monkeys) has been found to induce periodontal bone loss[18,19] and to increase tissue-destructive proteinase (collagenase, gelatinase, and cathepsin B and L) activity in gingiva.[1,20,21] Of extreme interest is the finding that when the infected rats were immunized with heat-killed *P. gingivalis* cells, or with partially purified fimbriae from these microorganisms, periodontal breakdown was reduced to levels seen in the noninfected control rats.[1]

In a recent preliminary study, we observed that treating germ-free and pathogen-reduced rats, inoculated with *P. gingivalis,* by the oral administration of either antimicrobial or chemically modified non-antimicrobial TCs significantly inhibited periodontal bone loss.[22] In the current study, mechanisms of MMP inhibition by TCs were examined in this animal model of disease. The ultimate goal is to determine whether the anti-MMP properties of these drugs and their analogs can provide a therapeutic strategy for periodontitis in humans and other MMP-mediated diseases, including (but not limited to) various forms of arthritides; tumor invasion, metastasis, and angiogenesis; sterile corneal ulcers; and bullous and ulcerating skin lesions.[5,6,9]

MATERIALS AND METHODS

The details of this animal model and procedures described below, including inoculation with *P. gingivalis,* preparation of gingival extracts, and measurement of both alveolar bone loss and activities of MMPs, have been described in detail by us previously[1,20–22] and, therefore, are only briefly described herein.

Experimental Protocol; Preparation of Animals

In a typical experiment, 22 barrier-raised adult male Sprague-Dawley rats (Taconic Farms, Germantown, NY), body weight approximately 350 g, were maintained under germ-free conditions in inflatable vinyl isolators (Standard Safety Equipment, Palatine, IL) and fed sterile food and water *ad libitum* over the entire 7-week experimental protocol. The rats were distributed into four experimental groups (n = 4–6 rats per group) including noninfected controls (NIC group) and three other groups that were inoculated with *Porphyromonas gingivalis* (strain 381), obtained from a human with adult periodontitis (supplied by Drs. Evans and Genco, State University of New York at Buffalo). One of the three inoculated groups was treated daily with vehicle alone (2% carboxymethylcellulose; CMC) (the *P. gingivalis*-infected [PgI] group), whereas the other two groups were treated once daily by oral gavage with either 5 mg doxycycline (PgI + DOXY) or the same dose of 4-de-dimethylamino-tetracycline (PgI + CMT-1), both drugs suspended in the vehicle. Doxycycline is a commercially available antimicrobial TC whereas CMT-1 is a TC that has been chemically modified to eliminate its antimicrobial effect, while retaining its anticollagenase activity;[5,6] details of the synthesis and characterization of CMT-1 have been described by us previously.[13,23]

Prior to inoculation with the periodontopathogen, all rats were pretreated once daily for 3 days with a mixture of 20 mg each of ampicillin and kanamycin to suppress the already minimal oral microflora, thus promoting infection of the gingival sulci with the inoculum; infection with *P. gingivalis* was accomplished by suspension of the organism in 5% CMC and topically applying the suspension around the gingival margins throughout the oral cavity, followed by colorectal application once daily for 3 days. Note that inoculation was initiated 3 days after the last dose of ampicillin/kanamycin to allow a "wash-out" period of the antibiotic mixture, and the gingival sulcular infection was promoted by the coprophagous nature of the rats. The success of this protocol has been demonstrated in our laboratory by elevated serum antibody levels against *P. gingivalis* and by microbiologic examination of subgingival plaque samples.[1,21,22]

Preparation of Gingival Extracts and Assessment of Alveolar Bone Loss

As described by Klausen *et al.*,[1] the buccal gingiva were dissected from the upper and lower jaws (about 20 mg wet weight/rat) at the end of the 7-week protocol and pooled by group; insufficient tissue was available for individual analysis. The pooled gingival tissues were weighed, minced (all procedures at 4 °C) and extracted first with Tris-NaCl-$CaCl_2$ buffer containing sucrose followed by buffer containing 5 M urea. The extracts were exhaustively dialyzed, partially purified by precipitation with ammonium sulfate added to 60% saturation and concentrated with an Amicon P-10 filter.[10,12,24] Aliquots of the gingival extracts were measured for protein using the Bio-Rad protein assay kit (Bio-Rad, Richmond, CA) prior to measurement of neutral proteinase activities (see below).

As described by us previously,[1,21,22] after defleshing the jaws, alveolar bone loss was measured at 17 different sites in all molar regions in one half-maxilla per rat (the other half-maxilla was processed for standard histology; findings to be presented elsewhere) under a microscope fitted with a micrometer eye piece (each unit = 0.1 mm) by recording the distance from the cemento-enamel junction to the alveolar bone crest. It should be noted that this morphometric analysis of periodontal bone loss has been found to yield similar conclusions as the quantitative analysis of radiographs.[22]

Collagenase, Gelatinase, and Elastase Assays: In Vitro *Activation and Inhibition*

Collagenase activity was assessed in aliquots of gingival extracts using rat skin type I collagen labeled *in vitro* with [^{3}H]formaldehyde. The collagen degradation fragments (α^A and α^B), generated by the gingival extracts after a 24-h incubation at 22 °C with the [^{3}H]collagen substrate, were identified by a combination of SDS-PAGE and autoradiography; collagenase activity was measured by scanning the autoradiograms using an LKB laser densitometer (for details, see refs. 1 and 10). In addition, aliquots of gingival extract from the different groups of rats were incubated with [^{3}H-methyl] collagen at 27 °C for 18 h, the reaction stopped by the addition of 1,10-phenanthroline and the undigested collagen precipitated by the addition of methylated nonradioactive collagen, as carrier, and 1,4 dioxane. After centrifugation, 50 μL aliquots of supernatant containing [^{3}H-methyl] collagen degradation fragments were counted in a liquid scintillation spectrometer. The details of *in vitro* MMP activation (with either aminophenylmercuric acetate, APMA, or hypochlorous acid, HOCl) and inhibition are described in the legends to the FIGURES.

Gelatinase activity was measured using a modification of the method of McCroskery *et al.*[25] In brief, [^{3}H-methyl] type I collagen was denatured by heating at 60 °C for 20 min to serve as the gelatin substrate. Aliquots of the gingival extract were preincubated with APMA and soybean trypsin inhibitor added at a final concentration of 1.4 mM and 200 μg/mL, respectively, for 1 h at 22 °C; then the [^{3}H-methyl] gelatin was added to the incubation mixture which was incubated for an additional 4 h at 37 °C. The reaction was terminated by addition of nonradioactive methylated gelatin (20 mg/mL) and trichloroacetic acid at a final concentration of 45%, the mixture was cooled to 4 °C, centrifuged (13,000 × *g*), and the release of [^{3}H]labeled gelatin degradation products in the supernatant was counted in a liquid scintillation counter. In addition, the gelatinase activity was demonstrated and characterized by zymography in 10% SDS polyacrylamide gels co-polymerized with 0.8 mg/mL denatured type I collagen (gelatin) as detailed by Brown *et al.*[26] Elastase activity was measured as described by Ramamurthy and Golub[24] by spectrophotometrically measuring the degradation of the specific synthetic peptide substrate, succinyl-(L-alanyl)$_3$-*p*-nitroanilide.

RESULTS

As described in our earlier studies on this animal model,[1,21] microbiologic examination of oral and fecal samples exhibited no evidence of infection with *P. gingivalis* in the NIC group, whereas the other three groups showed evidence of infection. In recent preliminary experiments, we demonstrated that therapy with DOXY reduced, but did not eliminate, this organism in the oral cavity (and gut) whereas the CMT-1–treated rats showed a similar level of infection as the sham-treated PgI rats.[22] Based on morphometric analysis of the defleshed rat jaws, the PgI group showed significantly greater alveolar bone loss ($p < 0.05$) at all 17 different sites in each half-maxilla than the other groups and, as described previously,[22] this bone loss was greatest at site 7, which corresponds to the interproximal area between the 1st and 2nd molar teeth (FIGS. 1 and 2). Treating the *P. gingivalis*-infected rats with either doxycycline (PgI + DOXY group) or CMT-1 (PgI + CMT-1 group) generally "normalized" the level of alveolar bone except at sites 7, 8, and 9. However, even at these three sites, both TC compounds significantly reduced the severe pathologic bone loss (FIG. 2).

The oral administration of both TC compounds was found to reduce by >90% the pathologically elevated latent collagenase activity in the gingival extracts of the *P. gingivalis*-infected rats (FIGS. 3 and 4). Assessing the APMA-activated collagenase either at 27 °C and measuring the solubilized degradation products in a liquid scintillation spectrometer (FIG. 3), or at 22 °C and detecting the degradation of the α_1 and α_2 collagen components to the specific breakdown products, α_1^A and α_2^A, by SDS-PAGE/autoradiography (FIG. 4) demonstrated: (1) no detectable active or latent collagenase in the gingiva of the NIC rats and a dramatic increase in collagenase activity when the rats were infected with *P. gingivalis* (complete enzyme inhibition *in vitro* by CMT-1 and by the chelators, EDTA and 1,10-phenanthroline, and the lack of effect of leupeptin and PMSF—the latter two inhibit neutral proteinases other than MMPs—provides additional evidence that collagenase activity is being measured in these gingival extracts), and (2) that DOXY and CMT-1 administration *in vivo* reduced this pathologically elevated latent collagenase activity in the gingiva to near-normal, that is, undetectable, levels. When the APMA-activated collagenase in the gingiva of the sham-treated *P. gingivalis* infected rats was incubated *in vitro* with final concentrations of CMT-1 ranging from 2–100 μM, all of

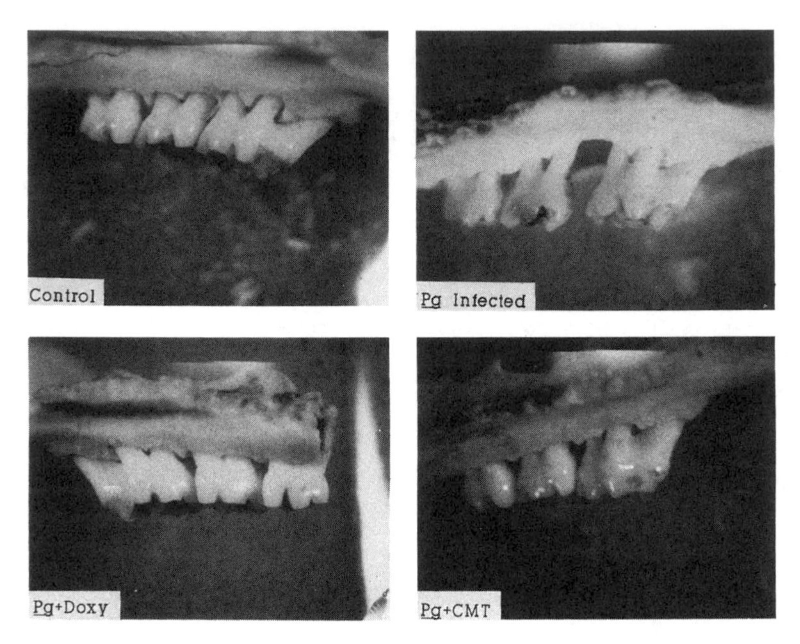

FIGURE 1. Alveolar bone loss in *P. gingivalis*-infected (PgI) rats: Inhibition by *in vivo* administration of either DOXY or CMT-1. Defleshed half-maxillas from representative animals in each experimental group are shown. Note the severe bone loss particularly in the interproximal areas, as well as the drifting of and open contacts between the molar teeth, in the untreated PgI rats; these effects were not seen in the jaws of the DOXY and CMT-1 treated animals.

these levels of drug inhibited the MMP activity (FIG. 4). Only at the 2 μM concentration of CMT-1 could trace amounts of $\alpha_1{}^A$ and $\alpha_2{}^A$ be detected; however, even this minimal level of collagen degradation was much less than that generated by the sham-treated PgI rat gingiva (FIG. 4).

As shown in FIGURE 5A, using the combination of SDS-PAGE and autoradiography also demonstrated that the collagenase in the *P. gingivalis* rat gingiva at the end of the 7-week protocol was largely present as the inactive or latent proenzyme form because when APMA was deleted from the incubation mixture no collagen degradation was seen (compare lanes 6 and 7). When the reactive oxygen metabolite, HOCl, was added *in vitro* to the latent procollagenase in the *P. gingivalis* rat gingiva, collagenase activity could be detected (compare lanes 3 and 6). However, after *in vivo* treatment of the Pg-infected rats with either DOXY or CMT-1, the subsequent addition of HOCl did not generate collagenase activity *in vitro* (lanes 4 and 5). As shown in FIGURE 5B, the addition of 50 mM $CaCl_2$ to the incubation mixture did not prevent the CMT-1 from blocking HOCl activation of latent gingival collagenase when these agents were added at the beginning of the preincubation (see lanes 8 and 9). However, delaying the addition of CMT-1 plus $CaCl_2$ by 1 h did eliminate the ability of CMT-1 to inhibit gingival collagenase even though the addition of CMT-1 by itself, after a 1-h delay, did inhibit collagenase activity (lanes 10 and 11).

Gelatin zymography (FIG. 6) indicated that the pathologically elevated gelatinase activity in the gingiva of the PgI rats (lanes 4 and 5): (*a*) was likely of PMN leukocyte origin (compare to lanes 1 and 9) and (*b*) was completely undetectable in the gingiva of the PgI rats treated *in vivo* by the oral administration of either doxycycline or CMT-1 (lanes 6–8).

Additional data (not shown) generated by the current study demonstrated the following:

1. Incubation of aliquots of the different gingival extracts with [^{3}H-methyl] gelatin, then measuring the release of solubilized degradation products by liquid scintillation spectrometry, showed a similar pattern of change, that is, a dramatic increase in gingival gelatinase activity in the sham-treated PgI rats and a reduction to essentially normal levels (like that exhibited by the NIC group) when the rats were orally administered either DOXY or CMT-1.

2. Elastase was also measured in extracts of gingiva (and skin); was characterized as a serine-proteinase, not a metalloproteinase, based on the use of several different proteinase inhibitors (PMSF, EDTA, 1,10-phenanthroline, CMT-1); and exhibited the same pattern of change in the different groups of rats as previously described, that is, a reduction to normal levels of the pathologically excessive elastase activity in the gingiva of the PgI rats by *in vivo* CMT-1 administration. Moreover, when the gingival extracts from the different groups of rats were incubated with [^{3}H-methyl] α_1-antitrypsin (α_1-AT; the major endogenous inhibitor of neutrophil elastase, a serine-proteinase) in the presence of APMA, and the reaction mixture examined by SDS-PAGE and autoradiography, only the gingiva from the sham-treated PgI rats

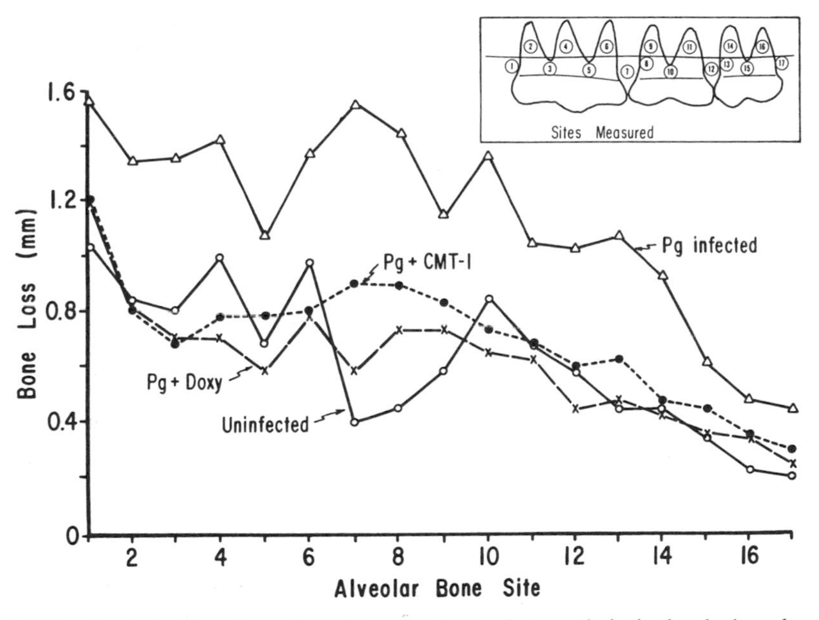

FIGURE 2. Oral administration of DOXY and CMT-1 reduces pathologic alveolar bone loss to near-normal levels. Bone loss was measured morphometrically at 17 different sites in each half-maxilla of the different groups of rats, and each value represents the mean of 4–6 measurements. (One half-maxilla per rat was analyzed morphometrically and the other was prepared for and examined by standard histology; data to be presented elsewhere). Note the increased bone loss at all sites around the molar teeth in the untreated *P. gingivalis*-infected rats compared to that seen in the other three groups.

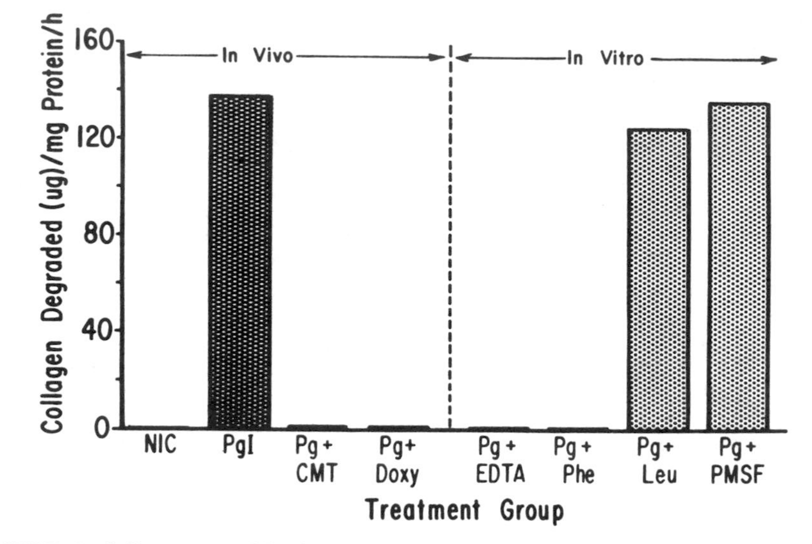

FIGURE 3. Collagenase activity in gingiva of *P. gingivalis*-infected (PgI) rats: Effect of doxycycline and CMT-1 therapy. The gingival tissues (~20 mg per rat) were pooled by group and extracted first in Tris-NaCl-$CaCl_2$ buffer (pH 7.5) containing 0.4 M sucrose, then in the same buffer with 5 M urea. The extracts were exhaustively dialyzed, and the collagenase was partially purified by precipitation with $(NH_4)_2SO_4$ added to produce 60% saturation. Aliquots of the gingival extracts were incubated with [^{3}H-methyl] collagen in solution (27 °C, 18 h) with APMA added in a final concentration of 1.2 mM and the solubilized [^{3}H] collagen breakdown products counted in a liquid scintillation spectrometer. To characterize the enzyme activity, aliquots of gingival extract from the sham-treated PgI rats were incubated *in vitro* with each of the following proteinase inhibitors: EDTA (25 mM), 1,10-phenanthroline (Phe; 1 mM), leupeptin (Leu; 1 μM) or PMSF (1 mM).

converted the 52-kDa α_1-AT to the inactive 48- and 4-kDa degradation fragments; the gingiva from the DOXY- and CMT-1–treated PgI rats and the NIC, did not show evidence of α_1-AT degradation.

3. Infecting the rats with *P. gingivalis* significantly increased the activity of collagenase, gelatinase, and elastase in extracts of gingiva, but *not* in skin, indicating that these alterations in proteinase activity reflected a local interaction between the subgingival bacteria (or their metabolic products) and the adjacent gingival tissues, rather than a systemic effect on a distal connective tissue such as skin.

DISCUSSION

A current view on the pathogenesis of human periodontal disease can be summarized as follows. The accumulation of various microorganisms in the subgingival plaque—*Porphyromonas gingivalis* being just one, albeit a prominent and relevant member of this ecosystem—initiates a cascade of reactions in the adjacent gingival tissues resulting in enhanced MMP production by resident (epithelial cells, fibroblasts) or infiltrating (PMNLs, macrophages) cells, followed by connective tissue destruction.[2,27] One example of such microbially induced host responses is the ability of products of gram-positive and gram-negative plaque bacteria to stimulate production of cytokines (e.g., IL-1β, TNFα) by mononuclear inflammatory cells; another is the ability of lipopolysaccharide/endotoxin from gram-negative microorganisms to stimulate production of arachidonic acid metabolites such as prostaglandin E_2 by gingival cells. Ultimately, and for reasons not yet understood, similar host responses presumably occur in the deeper periodontal tissues, that is, the periodontal ligament and alveolar bone, and result in the connective tissue breakdown that characterizes progressive destructive periodontitis.[28-31]

Similar patterns of periodontal breakdown have been observed in animal models of disease. For example, Holt *et al.*[19] reported that oral inoculation of *P. gingivalis* induced a progressive periodontitis in monkeys, and Klausen *et al.*[1] and Chang *et al.*[22] reviewed the studies showing similarly initiated disease in rats. In the current study, which addressed the therapeutic potential of MMP inhibitors in periodontitis, the latter model was selected, in part because of relevant earlier studies in our laboratory that showed (1) infection of germ-free rats with each of several different anaerobic gram-negative periodontopathogens including *P. gingivalis, Actinobacillus actinomycetemconitans,* and *Capnocytophaga spp.* significantly enhanced host-derived collagenase activity in gingival tissue[20] and (2) the increased activity of several different tissue-destructive proteinases (collagenase, gelatinase, cathepsins B and L) and enhanced alveolar bone loss, both induced by infection with *P. gingivalis,* were all prevented when the rats were immunized against this microorganism.[1]

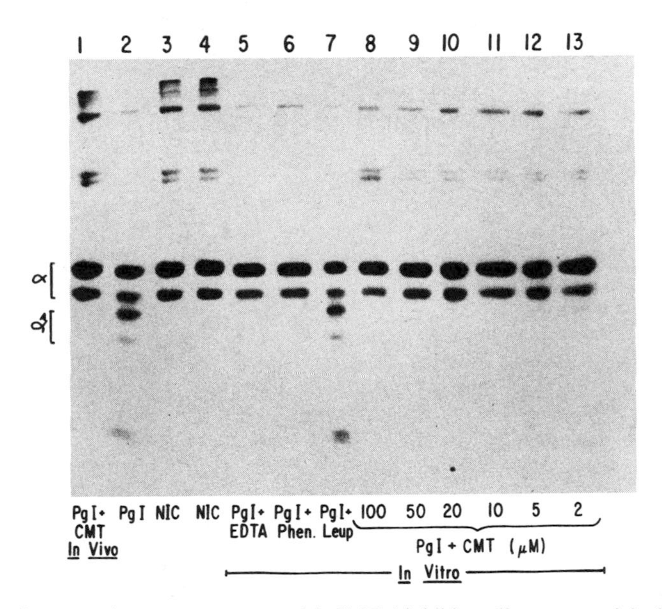

FIGURE 4. *In vivo* and *in vitro* treatment with CMT-1 inhibits collagenase activity in gingiva of rats infected with *P. gingivalis.* Aliquots of gingival extract from noninfected control (NIC; *lanes 3 and 4*), and CMT-1 treated (*lane 1*) and untreated (*lane 2*) *P. gingivalis*-infected (PgI) rats were incubated for 24 h (22 °C) with [^{3}H-methyl] collagen, and the breakdown of α collagen components to α^A three-quarter fragments was assessed by SDS-PAGE/fluorography. All incubations (*lanes 1–13*) were carried out with APMA added in a final concentration of 1.2 mM. In addition, aliquots of gingival extract from the PgI group of rats (sham treated) were incubated with the proteinase inhibitors, EDTA (25 mM), 1,10-phenanthroline (1 mM) and leupeptin (1 μM) (*lanes 5–7*), or with CMT-1 added *in vitro* at a final concentration of 2–100 μM (*lanes 8–13*).

In fact, it was a rat model of pathologically excessive gingival collagenase activity that led to the discovery of the anti-MMP properties of TC antibiotics[5,6,10–12] and the finding that a chemically modified analog of TC (4-de-dimethylaminotetracycline), lacking antimicrobial activity, could still retain its anticollagenase properties.[5,13,23,32] Since then, both commercially available antimicrobial TCs such as minocycline, DOXY, and TC itself, plus a series of newly developed CMTs (CMT 1–10, except for

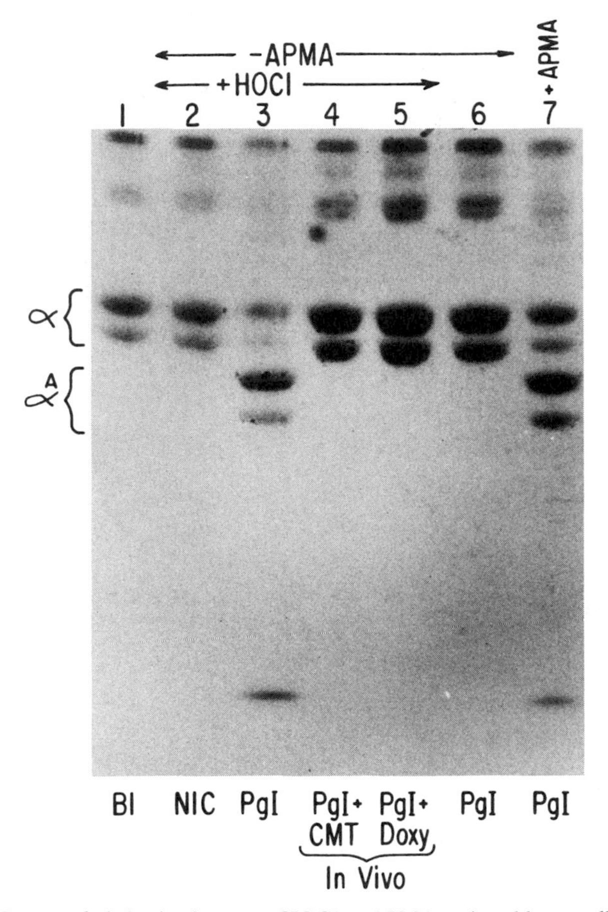

FIGURE 5A. *P. gingivalis* infection increases HOC1 or APMA-activatable procollagenase in rat gingiva: Inhibition by *in vivo* therapy with non-antimicrobial (CMT-1) and antimicrobial (doxycycline) tetracyclines. Aliquots of gingival extract from the different groups of rats were incubated (22 °C, 24 h) with [^{3}H-methyl] collagen and substrate degradation assessed by SDS-PAGE/fluorography. **Lane 1:** [^{3}H]collagen itself. **Lanes 2–5:** Gingival extract from noninfected control (NIC) and sham-treated (PgI), CMT-1–treated (PgI + CMT), and doxycycline-treated (PgI + DOXY) PgI rats were incubated for 1 h (22 °C) with HOC1 added in a final concentration of 5 μM. After quenching the HOC1 with 50 μM methionine, the extracts were incubated at 22 °C (24 h) with [^{3}H]collagen. **Lanes 6 and 7:** Gingival extract from the PgI (sham-treated) group was incubated with [^{3}H]collagen in the absence (*lane 6*) or presence (*lane 7*) of 1.2 mM APMA; no HOC1 was added to these incubations.

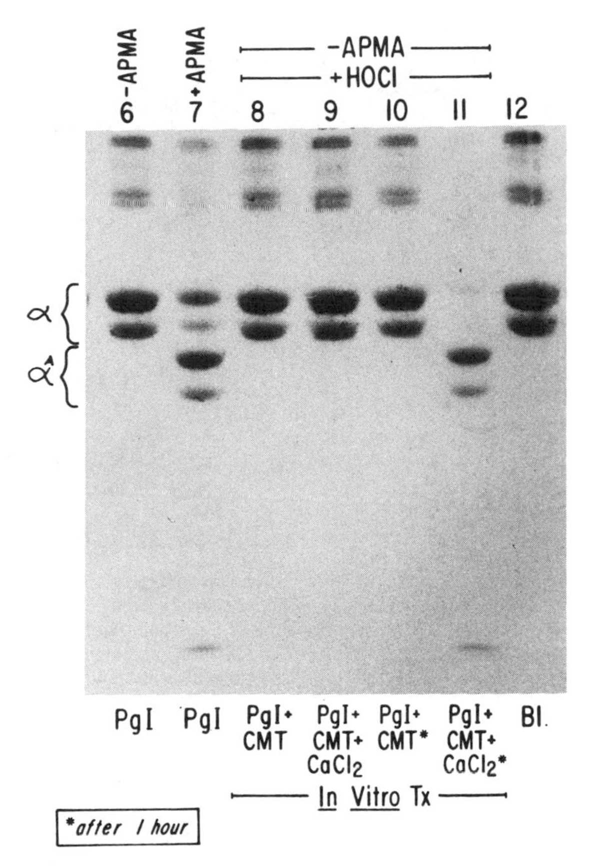

FIGURE 5B. CMT inhibits collagenase in gingiva of *P. gingivalis*-infected rats by (at least) two mechanisms: (1) by preventing oxidative (HOC1) activation of latent procollagenase, and (2) by blocking already-active collagenase. Aliquots of gingival extract from the PgI-infected (sham-treated) rats were incubated with [^{3}H]collagen and its degradation assessed by SDS-PAGE/fluorography, as described in FIGURE 5A except for the modifications now described. **Lanes 6 and 7:** For description, see same lanes in FIGURE 5A. **Lanes 8 and 9:** Prior to incubation of PgI gingival extract with [^{3}H]collagen, aliquots of the extract were preincubated (1 h, 22 °C) with either 5 μM HOC1 and 50 μM CMT-1 simultaneously added to the mixture (*lane 8*) or 50 mM $CaCl_2$ was also added to the HOC1/CMT-1 preincubation (*lane 9*). **Lanes 10 and 11:** PgI gingival extract was preincubated with HOC1 for 1 h (22 °C) *prior* to the addition of either CMT-1 alone (*lane 10*) or CMT-1 plus $CaCl_2$ (*lane 11*). **Lane 12:** [^{3}H]collagen.

CMT-5), have been found to inhibit collagenases and gelatinases (as well as macrophage elastase)—all Ca^{2+}- and Zn^{2+}-dependent MMPs—from a variety of human and animal sources.[5,9,33] To date, several mechanisms of action of these drugs have been identified including their ability to block already-active MMPs, presumably by binding Zn^{2+} in the enzyme molecule; in this regard, the mechanism appears to be noncompetitive,[34,35] and the addition of micromolar concentrations of Zn^{2+} (or millimolar concentrations of Ca^{2+}) to the enzyme-substrate-TC reaction mixture has been found to prevent TC inhibition of the MMP.[36,37] During the symposium, one

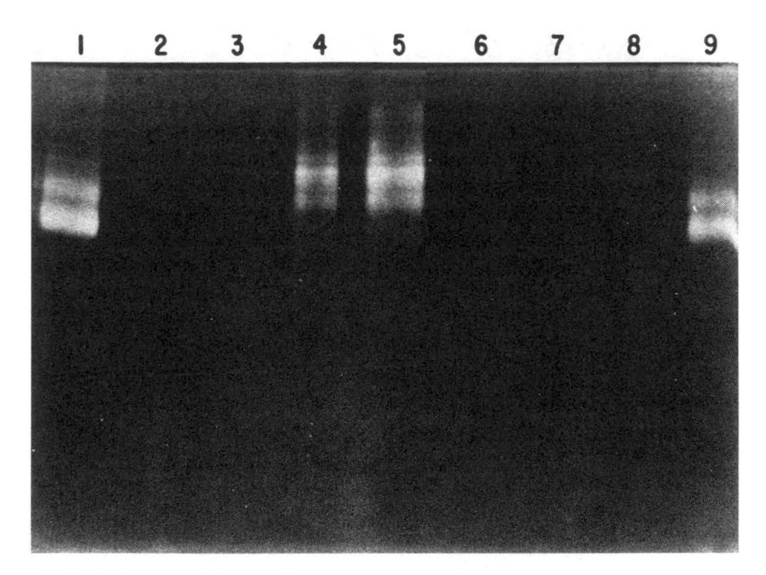

FIGURE 6. Gelatinase activity in gingiva of *P. gingivalis*-infected (PgI) rats: Effect of *in vivo* doxycycline and CMT-1 therapy. Extracts of gingiva from the different groups of rats were examined by gelatin zymography (10% polyacrylamide co-polymerized with 0.8 mg/mL denatured type I collagen; electrophoresis carried out under nonreducing conditions, 5 mA/lane). **Lanes 1 and 9:** Partially purified extract of PMN leukocytes; **lanes 2 and 3:** Extract (2 and 10 μL, respectively) from NIC rats; **lanes 4 and 5:** Gingival extract (2 and 10 μL, respectively) from sham-treated PgI rats, respectively; **lanes 6 and 7:** Extract (2 and 10 μL, respectively) from doxycycline-treated PgI rats; **lane 8:** Gingival extract (10 μL) from CMT-1 treated PgI rats.

hypothesis that emerged was that TCs may bind to a *secondary zinc* (or Ca^{2+}), not at the active site, in the enzyme resulting in a conformational change and loss of catalytic activity of the MMP.[38] Moreover, TCs can prevent the oxidative activation (at least by hypochlorous acid) of latent procollagenase and progelatinase secreted by several cell types (PMN leukocytes, osteoblasts) found in the periodontium.[39,40] Based on a paper presented at this meeting by Smith *et al.*,[41] the following mechanism has also been invoked: When PMN procollagenase is activated with trypsin or APMA *in vitro,* it is converted to active enzyme. However, in the presence of doxycycline, the proenzyme may, instead, be converted to smaller *inactive* fragments, possibly reflecting a drug-induced altered conformation of the MMP zymogen rendering it more susceptible to proteolysis. This mechanism was challenged, however, because Sorsa *et al.*[35] were unable to detect degradation products of Pro-MMP-8 or MMP-9 during the inhibition of these proteinases by TCs.

Current studies are exploring additional mechanisms by which TCs and CMTs can inhibit connective tissue breakdown (see FIG. 7 for an overview of proposed direct and indirect mechanisms of action). These include (1) the decreased synthesis of Pro-MMPs.[9,42] Uitto *et al.*[42] observed decreased MMP expression by skin keratinocytes exposed to doxycycline and several CMTs in cell culture. (2) Protection of endogenous MMP inhibitors (TIMP-1)[34] and other neutral proteinase inhibitors (α_1-antitrypsin)[43] from degradation and inactivation. The ability of TCs/CMTs to protect α_1-AT from inactivation by collagenase and gelatinase may contribute to reduced elastase activity *in vivo* (see data described above) albeit by an indirect mechanism; moreover, inasmuch as elastase has recently been observed to degrade TIMPs,[44] and because TCs might also prevent the oxidative inactivation of TIMPs by reactive oxygen species such as HOCl,[45,46] the TCs could also contribute indirectly to inhibition of additional MMPs, i.e., stromelysin which, unlike collagenase and gelatinase, may not be directly inhibited by these drugs. (3) Enhanced *production* of collagen[47,48] and bone[49] when these anabolic processes are suppressed during disease such as diabetes-induced skin and periodontal ligament atrophy and osteopenia. In this regard, Yu *et al.*[50] recently reported that the pathologically suppressed steady-state levels of type I procollagen mRNA in the atrophic skin of diabetic rats were increased to near-normal levels when these rats were orally administered CMT-1, even though the severity of hyperglycemia was unaffected.

In fact, several of the above therapeutic mechanisms of action were observed in the current study on experimental periodontitis in rats. Infection with *P. gingivalis* increased the level of latent procollagenase in extracts of rat gingival tissue. However, already-active collagenase was not detectable in the gingiva of these rats (autocatalytic loss of this form of the enzyme during processing of the tissue extracts is one possibility; others are discussed below), and the other groups (NIC, PgI + DOXY, PgI + CMT-1) showed no evidence of latent or active collagenase. Thus, the pathologically excessive collagenase activity in the gingiva of the PgI rats was only observed after treatment with APMA (an organomercurial, thiol-active compound) or hypochlorous acid (a reactive oxygen metabolite), both of which presumably "turned-off" the cysteine switch in the procollagenase molecule, thus activating the enzyme and allowing the Zn^{2+} in its catalytic domain to participate in the hydrolysis of collagen substrate.[27] Sorsa *et al.*[51] recently demonstrated that although a variety of periodontopathic microorganisms such as *Actinobacillus actinomycetemcomitans, Fusobacterium nucleatum,* and *Prevotella intermedia* were unable to activate latent fibroblast-type or PMN-type procollagenase (the latter is likely predominant in the PgI gingiva [see below] although both types are likely present),

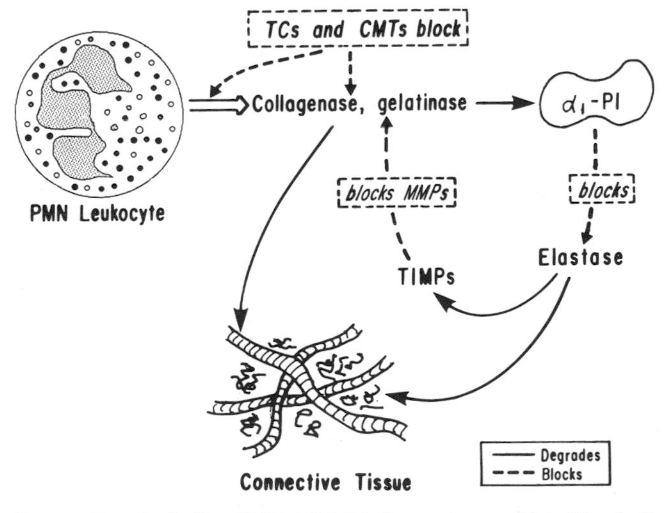

FIGURE 7. Tetracyclines including CMTs inhibit inflammatory-mediated (and other pathways of) connective tissue breakdown by *directly* blocking collagenase and gelatinase and *indirectly* blocking other proteinases (e.g., elastase and stromelysin): A working hypothesis.

an 80-kDa trypsin-like proteinase from *P. gingivalis* was able to activate both types of latent enzyme. In view of these findings, the following hypothesis may explain the predominantly latent, rather than active, collagenase observed in the gingiva of the rats infected with *P. gingivalis*: First, it is important to note that gingival collagenase (and gelatinase) activity was only measured at the termination of the 7-week experimental protocol and has not yet been measured at earlier time periods. It is entirely possible that the intensity of infection with this organism actually *waned* 7 weeks after inoculation compared to earlier times, for example at 2–3 weeks. If so, then sufficient infection and elevated bacterial (e.g., 80-kDa trypsin-like) proteinase may have been present at these earlier time periods to increase the synthesis and activation of host-derived MMPs,[51,52] thus promoting tissue breakdown including bone resorption. However, with this hypothesis in mind, diminished infection at the end of the 7-week protocol (compared with earlier time periods) could result in decreased *P. gingivalis*-derived proteinase (also decreased bacterial endotoxin), decreased PMN leukocyte response (including decreased generation of reactive oxygen metabolites such as HOCl), and perhaps also decreased breakdown of TIMPs (or increased TIMP production), all of which could result in pathologically elevated latent but not active MMP levels. The bone loss actively generated during heightened host response (and enhanced osteoclastic activity) at earlier time periods might not recover at 7 weeks, thus leaving a "scar" of past active disease. Clearly, future studies are required to monitor these changes at different time periods and to map out the temporal sequence of events, after initial infection with *P. gingivalis* (or other periodonto-pathogens), that lead to enhanced MMP activity and bone loss; these are currently under way.

Treatment of the *P. gingivalis*-infected rats with doxycycline, a commercially available antimicrobial TC, or with CMT-1, which has been chemically modified to eliminate the antimicrobial activity of the drug, both reduced gingival MMP activity to the undetectable levels seen in the noninfected control rats. These results are consistent with previous studies demonstrating that TCs can inhibit host-derived MMPs by mechanisms unrelated to the antimicrobial activity of these drugs. Moreover, in the current study, *in vitro* CMT concentrations as low as 2 μM were able to inhibit the collagenase activity in the PgI rat gingiva. These levels of CMT are readily achieved *in vivo* in the serum of rats that are orally administered this drug.[23] Thus, one mechanism by which *in vivo* treatment with CMT-1 could reduce MMP levels is a direct block of already-active enzyme in the extracellular matrix of the gingiva. However, as discussed in a recent review article,[9] other mechanisms may also be involved including scavenging the reactive oxygen species, produced by PMN leukocytes during the inflammatory response, thus preventing the conversion of latent procollagenase to active collagenase and, perhaps, even CMT inhibition of pro-MMP synthesis.[42] These effects could also account for the reduction of elastase activity in the PgI gingiva even though CMT did not directly inhibit this serine-proteinase based on *in vitro* studies (ref. 5 and current study); CMT treatment, by decreasing MMP activity in the gingiva, could prevent the degradation and loss of α_1-antitrypsin (ref. 43; also see RESULTS) and, thus, indirectly inhibit gingival elastase *in vivo* (see FIG. 7 for summary).

Further insight into mechanisms is addressed by the following data: (1) CMT therapy inhibited gelatinase (as well as collagenase) activity in the gingiva of the PgI rats. Based on its molecular weight characteristics assessed by zymography, plus its sensitivity to CMT inhibition *in vitro,* this MMP activity appears to originate primarily from PMN leukocytes which would be expected to infiltrate the gingival tissue in response to infection. (2) CMT inhibited both already-active collagenase and the conversion of procollagenase to the active enzyme by oxidative mechanisms. In fact, these two anti-MMP functions of CMT-1 may reside in different parts of the drug molecule based on the following observations: When the CMT-1 and Ca^{2+} were added to the reaction mixture *together* with HOCl, CMT-1 was still able to prevent HOCl activation of the gingival procollagenase. However, when the procollagenase was activated with HOCl *prior* to the addition of the CMT, the addition of Ca^{2+} then prevented the drug from blocking collagenase activity; these data are consistent with our previous studies using different systems showing (1) that the same concentration (50 mM) of Ca^{2+} can prevent TC from inhibiting already-activated PMN leukocyte collagenase[10] and (2) that the addition of excess Ca^{2+} (50 mM) or Zn^{2+} (50 μM) does not prevent TCs from inhibiting the oxidative activation of procollagenase secreted by osteoblasts in culture.[40] These experiments also suggest that CMT-1 requires its metal-ion binding site to block already-active collagenase (consistent with previous reports[5,10,36,37]), but does not require this chelating property of TC to prevent oxidative activation of the latent proenzyme. Future studies will be carried out using the pyrazole derivative of TC (CMT-5) which lacks anticollagenase activity due to the loss of the Ca^{2+}/Zn^{2+} binding site at the carbon-11 and -12 positions.[5] If CMT-5 is found to prevent procollagenase activation by HOCl, it would confirm that this property of TC (unlike its ability to block already-active collagenase) is independent of its metal-ion binding capabilities (at least at physiologic pH).

Concerning bone loss, Rifkin *et al.*[9] recently reviewed the mechanisms of its inhibition (at least *in vitro*) by TCs and their chemically modified analogs. TCs inhibit osteoblast collagenase and gelatinase both directly and by preventing oxidative activation of the pro-MMPs[40] (and perhaps by other mechanisms); of importance is that MMP production has now been observed in human as well as rat osteoblasts,[53] which addresses the relevance of this cellular source of MMPs to bone resorption in human disease such as periodontitis.[27] Based on current understanding of bone resorption mechanisms, this action of the drug(s) would be expected to inhibit osteoid breakdown, thus decreasing access of the underlying calcified bone to osteoclast attack.[5] Although the mechanisms of this drug effect are not yet clear, TCs can also inhibit osteoclast-mediated bone resorption directly.[9] TC inhibition of collagenase and gelatinase (Ramamurthy *et al.*, unpublished data) produced by these cells.[54,55] in addition to other effects of these drugs including attenuation of the osteoclast ruffled border, decreased acid production, and effects on the Ca^{2+} receptor of the cell, have also been invoked.[9] Although the current study clearly demonstrates *in vivo* inhibition of pathologic bone loss by a non-antimicrobial (as well as antimicrobial) TC, it is not yet known whether this therapeutic effect reflects a direct inhibition of bone resorption during the entire 7-week protocol of the current study, or inhibition of episodes of bone resorption early on but not later in the study—or additional pro-anabolic effects on bone formation. These issues, as well as others such as the temporal relationship of MMP induction and bone loss, will be addressed in future studies on CMTs and this rat model of periodontal disease.

ACKNOWLEDGMENTS

The authors acknowledge the excellent technical assistance of Xia Zhang and typing of Gloria Haas and Susan Choudhari. The authors also thank Dr. Sanford Simon for helpful discussions on the manuscript.

REFERENCES

1. Klausen, B., R. T. Evans, N. S. Ramamurthy, L. M. Golub, C. Sfintescu, J. Y. Lee, G. L. Bedi, J. J. Zambon & R. J. Genco. 1991. Oral Microbiol. Immunol. **6:** 193–201.
2. Page, R. C. 1991. J. Periodontal Res. **26:** 230–242.
3. Uitto, V. J. 1986. Proteinases. *In* Inflammation and Tumor Invasion. H. Tschesche, Ed.: 211–223. Walter de Gruyter. New York.
4. Howell, T. H. & R. C. Williams. 1992. Pharmacologic Blocking of Host Response as an Adjunct in the Management of Periodontal Diseases: A Research Update. American Academy of Periodontolology, Research, Science & Therapy Committee: 1–8.
5. Golub, L. M., N. S. Ramamurthy, T. F. McNamara, R. A. Greenwald & B. R. Rifkin. 1991. Crit. Rev. Oral Biol. Med. **2:** 297–322.
6. Golub, L. M., K. Suomalainen & T. Sorsa. 1992. Curr. Opin. Dent. **2:** 80–90.
7. Golub, L. M., M. Wolff, S. Roberts, H. M. Lee, M. Leung & G. Payonk. 1994. JADA. **125:** 163–169.
8. Ingman, T., T. Sorsa, K. Suomalainen, S. Haliknen, O. Lindy, A. Lauhio, H. Saari, Y. T. Konttinen & L. M. Golub. 1993. J. Periodontol. **64:** 82–88.
9. Rifkin, B. R., A. T. Vernillo & L. M. Golub. 1993. J. Periodontol. **64 (Suppl.):** 819–827.
10. Golub, L. M., H. M. Lee, G. Lehrer, A. Nemiroff, T. F. McNamara, R. Kaplan & N. S. Ramamurthy. 1983. J. Periodontal Res. **18:** 516–526.
11. Golub, L. M., N. S. Ramamurthy, T. F. McNamara, B. Gomes, M. Wolff, S. Ciancio & H. Perry. 1984. J. Periodontal Res. **19:** 651–655.
12. Golub, L. M., M. S. Wolff, H. M. Lee, T. F. McNamara, N. S. Ramamurthy, J. Zambon & S. G. Ciancio. 1985. J. Periodontal Res. **20:** 12–23.
13. Golub, L. M., T. F. McNamara, G. D'Angelo, R. A. Greenwald & N. S. Ramamurthy. 1987. J. Dent. Res. **66:** 1310–1314.
14. Slots, J. 1982. Importance of black-pigmented *Bacteroides* in human periodontal disease. *In* Host-Parasite Interactions in Periodontal Diseases. R. J. Genco & S. E. Mergenhagen, Eds.: 27–45. American Society for Microbiology. Washington, D.C.
15. Moore, W. E. C. 1987. J. Periodontal Res. **22:** 335–341.
16. Genco, R. J., J. J. Zambon & L. A. Christersson. 1988. Adv. Dent. Res. **2:** 245–259.
17. Gmur, R., K. Hrodek, U. P. Saen & B. Guggenheim. 1986. Infect. Immun. **52:** 768–776.
18. Wyss, C. & B. Guggenheim. 1984. J. Periodontal Res. **19:** 574–577.
19. Holt, S. C., J. Ebersole, J. Felton, M. Brunsvold & K. S. Kornman. 1988. Science **239:** 55–57.
20. Chang, K. M., N. S. Ramamurthy, T. F. McNamara, R. J. Genco & L. M. Golub. 1988. J. Periodontal Res. **23:** 239–244.
21. Evans, R. T., B. Klausen, N. S. Ramamurthy, L. M. Golub, C. Sfintescu & R. J. Genco. 1992. Arch. Oral Biol. **37:** 813–819.
22. Chang, K. M., N. S. Ramamurthy, T. F. McNamara, R. T. Evans, B. Klausen, P. A. Murray & L. M. Golub. 1994. J. Periodontal Res. In press.
23. Yu, A., M. K. Leung, N. S. Ramamurthy, T. F. McNamara & L. M. Golub. 1992. Biochem. Med. Metabol. Biol. **47:** 10–20.
24. Ramamurthy, N. S. & L. M. Golub. 1983. J. Periodontal Res. **18:** 23–30.
25. McCroskery, P. A., J. F. Richards & E. Harris, Jr. 1975. Biochem. J. **152:** 131–142.
26. Brown, P. D., A. T. Leung, I. M. K. Marguiles, L. A. Liotta & W. G. Stettler-Stevenson. 1990. Cancer Res. **50:** 6184–6191.
27. Birkedal-Hansen, H., W. G. I. Moore, H. K. Bodden, L. M. Windsor, B. Birkedal-Hansen, A. DeCarlo & J. A. Engler. 1993. Crit. Rev. Oral Biol. Med. **4:** 197–250.
28. Heath, J. K., S. J. Atkinson, R. M. Hembry, J. J. Reynolds & M. C. Meikle. 1987. Infect. Immun. **55:** 2148–2154.
29. Meikle, M. C., J. K. Heath & J. J. Reynolds. 1986. J. Oral Pathol. **15:** 239–250.
30. Wahl, L. M. & S. E. Mergenhagen. 1988. J. Oral Pathol. **17:** 452–455.
31. Howell, T. H. & R. C. Williams. 1993. Crit. Rev. Oral Biol. Med. **4:** 177–196.
32. McNamara, T. F., L. M. Golub, G. D'Angelo & N. S. Ramamurthy. 1986. J. Dent. Res. **65 (Spec. Issue):** AADR Abstr. 515.

33. GOLUB, L. M., R. GREENWALD, N. S. RAMAMURTHY, S. ZUCKER, L. RAMSAMMY & T. MCNAMARA. 1992. Matrix (Suppl. 1): 315–316.
34. SORSA, T., O. TERONEN, T. SALO, T. INGMAN, K. SUOMALAINEN, O. LINDY, Y. T. KONTTINEN & L. M. GOLUB. 1993. Molecular basis for pathogenesis and molecular targeting in periodontal diseases: A centennial symposium. School of Dental Medicine, State University of New York at Buffalo. Program and Abstracts: 84–85.
35. SORSA, T., Y. DING, T. SALO, A. LAUHIO, O. TERONEN, T. INGMAN & Y. T. KONTTINEN. 1994. Ann. N. Y. Acad. Sci. This volume.
36. LEE, H. M., J. E. HALL, T. SORSA, S. SIMON & L. M. GOLUB. 1992. J. Dent. Res. **71 (Spec. Issue):** AADR Abstr. 1120.
37. YU, L. P., G. N. SMITH, K. A. HASTY & K. D. BRANDT. 1991. J. Rheumatol. **18:** 1450–1452.
38. LOVEJOY, B., A. CLEASBY, A. M. HASSELL, M. A. LUTHER, D. WEIGL, G. MCGEEHAN & S. R. JORDAN. 1994. Ann. N. Y. Acad. Sci. This volume.
39. LAUHIO, A., T. SORSA, O. LINDY, K. SUOMALAINEN, H. SAARI, L. M. GOLUB & Y. T. KONTTINEN. 1992. Arthritis Rheum. **35:** 195–198.
40. RAMAMURTHY, N. S., A. T. VERNILLO, R. A. GREENWALD, H. M. LEE, T. SORSA, L. M. GOLUB & B. R. RIFKIN. 1993. J. Bone Miner. Res. **8:** 1247–1253.
41. SMITH, G. N., JR., K. D. BRANDT & K. A. HASTY. 1994. Ann. N. Y. Acad. Sci. This volume.
42. UITTO, V. J., J. D. FIRTH, Y. M. PAN, L. NIP & L. GOLUB. 1994. Ann. N. Y. Acad. Sci. This volume.
43. SORSA, T., O. LINDY, Y. T. KONTTINEN, K. SUOMALAINEN, T. INGMAN, H. SAARI, H. HALINEN, H. M. LEE, L. M. GOLUB, J. HALL & S. SIMON. 1993. Antimicrob. Agents Chemother. **37:** 592–594.
44. OKADA, Y., S. WATANABE, I. NAKANISHI, *et al.* 1988. FEBS Lett. **229:** 157–160.
45. LAUHIO, A., T. SORSA, O. LINDY, K. SUOMALAINEN, H. SAARI, L. M. GOLUB & Y. T. KONTTINEN. 1992. Arthritis Rheum. **35:** 195–198.
46. STRICKLIN, G. P. & J. R. HOIDAL. 1992. *In* Matrix Metalloproteinases and Inhibitors. H. Birkedal-Hansen, Z. Werb, H. Welgus & H. V. Van Wart, Eds.: 325. Gustav Fischer Verlag. New York.
47. SCHNEIR, M., N. S. RAMAMURTHY & L. M. GOLUB. 1990. Matrix **10:** 112–123.
48. SASAKI, T., N. S. RAMAMURTHY, Z. YU & L. M. GOLUB. 1992. J. Periodontal Res. **27:** 631–639.
49. SASAKI, T., N. S. RAMAMURTHY & L. M. GOLUB. 1992. Calcif. Tissue Int. **50:** 411–419.
50. YU, Z., R. CRAIG, L. XU, R. BARRA, N. S. RAMAMURTHY & L. M. GOLUB. 1993. J. Dent. Res. **72 (Spec. Issue):** IADR Abstr. 1520.
51. SORSA, T., T. INGMAN, K. SUOMALAINEN, M. HAAPASALO, Y. T. KONTTINEN, O. LINDY, H. SAARI & V. J. UITTO. 1992. Infect. Immun. **60:** 4491–4495.
52. UITTO, V. J., H. LARJAVA, J. HEINO & T. SORSA. 1988. Infect. Immun. **57:** 213–218.
53. MEIKLE, M. C., S. BORD, R. M. HEMBRY, J. COMPSTON, P. I. CROUCHER & J. J. REYNOLDS. 1992. J. Cell Sci. **103:** 1093–1099.
54. EVERTS, V., J. M. DELAISSE, W. KORPER, A. NIEHOF, G. VAEST & W. BEERTSEN. 1992. J. Cell Physiol. **150:** 221–231.
55. BARON, R., Y. EECKHOUT & L. NEFF. 1990. J. Bone Miner. Res. **5 (Suppl.):** 2.

Effects of Tetracyclines on Neutrophil, Gingival, and Salivary Collagenases

A Functional and Western-Blot Assessment with Special Reference to Their Cellular Sources in Periodontal Diseases[a]

TIMO SORSA,[b-d] YANLI DING,[b] TUULA SALO,[e]
ANNELI LAUHIO,[c] OLLI TERONEN,[f] TUULA INGMAN,[b,g]
HARUO OHTANI,[h] NORIAKI ANDOH,[h]
SATOSHI TAKEHA,[h] AND YRJÖ T. KONTTINEN[g]

Departments of Periodontology,[b] Anatomy,[g] Medical Chemistry,[c] and Oral and Maxillofacial Surgery[f]
University of Helsinki
Helsinki, Finland

[e]Department of Oral Surgery and Pathology
University of Oulu
Oulu, Finland

[h]Department of Pathology
Tohoku University School of Medicine
Sendai, Japan

The periodontium is an organ that functions in tooth support. It comprises both hard and soft connective tissues. The cementum, which covers the tooth root surface, and the alveolar bone, which forms the tooth socket, both represent mineralized periodontal connective tissues. The periodontal ligament, which anchors the tooth into the socket, and the lamina propria of the gingiva represent periodontal soft connective tissue. The connective tissues of periodontium, with the exception of cementum, remodel rather rapidly when compared to other mature connective tissues.[1] Inflammation of the periodontium leads to connective tissue degradation and eventual tooth loss.[1] Extracellular matrix degradation during the course of periodontal diseases is thought to be mediated by a complex cascade(s) involving both host- and microbial-derived proteinases.[2-7] In this regard, the host-derived matrix metalloproteinases (MMPs) are thought to play key roles, and the characterization as well as regulation of MMPs have been studied to determine their specific roles in periodontal connective tissue destruction.[1-7] In fact, the interaction of several types of plaque bacteria with resident gingival cells (fibroblasts, epithelial cells) as well as infiltrating

[a]This work was supported by grants from the Academy of Finland, the Finnish Medical Society Duodecim, the Finnish Dental Society, the Finnish Rheumatism Foundation, the ITI Straumann Research Foundation, the Pharmacal Research Foundation, and the Yrjö Jahnsson Foundation. Timo Sorsa is a Young Research Fellow and Tuula Ingman is a Research Assistant of the Academy of Finland.

[d]Address correspondence to Dr. Timo Sorsa, Department of Periodontology, Institute of Dentistry, University of Helsinki, P.O. Box 41 (Mannerheimintie 172) FIN-00014 Helsinki, Finland.

inflammatory cells (polymorphonuclear leukocytes and monocyte/macrophages) has increased synthesis and release of the tissue-destructive proteinases by such triggered cells.[1–5] Mechanisms include the ability of microbial products, such as lipopolysaccharide, proteinases, and lectins to induce MMP expression at the transcriptional level.[1–5] Alternatively, the microbial products may indirectly stimulate MMP expression by host cells through the action of lymphocyte and monocyte/macrophage-derived proinflammatory agents such as prostaglandins, cytokines (i.e., IL-1β, TNF-α) and growth factors.[1–4] Regardless of mechanisms, elevated levels of MMPs and other host-derived proteinases (cathepsins, elastase, tryptases/trypsin-like proteinases) with capacities to degrade the various constituents of the extracellular matrix (ECM) have been detected in inflamed human gingiva, gingival crevicular fluid (GCF), and in saliva and/or mouthrinse samples with different forms of periodontal diseases, as well as in GCF of patients with osteointegrated dental implants.[1–15] Moreover, periodontal therapy has been found to decrease the levels and activities of these proteinases.[8,10,12]

Recent studies have addressed not only the roles of interstitial collagenases (MMP-1, MMP-8) and other MMPs such as gelatinase/type IV collagenases (MMP-2, MMP-9) in the pathogenesis of periodontal connective tissue breakdown,[1–16] but also their response to the therapeutic use of tetracyclines.[8,14,15] The rationale has been that tetracyclines and their chemically modified nonantimicrobial analogs (CMT molecules) can inhibit pathologically elevated MMP activities (and bone resorption at least in organ and cell cultures) as well as prevent the oxidative activation of procollagenases (proMMP-1 and proMMP-8).[8,14,15–17] Thus, determination of the MMP release mechanisms and cellular sources of MMPs associated with different forms of periodontal diseases is not only important to clarify the pathogenic mechanisms, but may also direct, at least in part, the therapeutic strategy to reduce the pathologically excessive MMP levels associated with different forms of periodontal diseases.[8,14] In this regard, Golub *et al.*[18] recently administered to humans with adult periodonitis (AP) a 2-week regimen of specifically formulated low-dose doxycycline (LDD) capsules and observed a reduction in collagenase activities in both GCF and gingival tissue of these patients. However, it was not clear whether this *in vivo* effect was direct or indirect. These questions, plus others such as the cellular source(s) of these gingival MMPs and their susceptibility and molecular interactions associated with doxycycline (DOXY) inhibition, were addressed in the current study. Also the presence and molecular forms of tissue inhibitors of matrix metalloproteinases (TIMP-1 and TIMP-2) present in dental plaque, GCF, and saliva were studied.

MATERIALS AND METHODS

Doxycycline (DOXY), aminoparaphenyl mercuric acetetate (APMA), gold thioglucose (GTG), phospholipase C (PLC; type XI, proteinase free), phorbol myristic acetate (PMA) and *p*-nitrophenylphosphorylcholine (NPPC) were purchased from Sigma Chemicals (St. Louis, MO). Phenylmercuric chloride (PMC) was obtained from EGA-Chemie (Steinham/Albuch, Germany). Sodium hypochlorite (NaOCl) was purchased from the Pharmacy of the University of Helsinki, Helsinki. Soluble native [^{3}H-methyl] type I collagen was prepared as described previously by Golub *et al.*[18] It was extracted from bovine tendon and further purified by selective salt precipitation at acidic and neutral pH. The purity of collagen was examined by cyanogen bromide analysis.[9] Soluble type I gelatin was prepared from the purified nonlabeled type I collagen by heating at 100 °C for 20 min. All other chemicals were highest available purity and, if not otherwise stated, commercially purchased from Sigma Chemicals. Human polymorphonuclear neutrophilic leukocyte collagenase (MMP-8) and 92-kDa type IV collagenase/gelatinase (MMP-9) were purified to apparent homogeneity as described previously.[13,19]

Isolation and Treatment of Peripheral Blood Polymorphonuclear Neutrophilic Leukocytes with Phorbol Myristate Acetate and Phospholipase C

Heparinized peripheral blood (PB) was centrifugated for 10 min at 200 *g*. Supernatant was removed and the cells were washed twice with Hank's balanced salt solution (HBSS, Orion Diagnostica D-27, Helsinki). PB polymorhonuclear neutrophilic leukocytes (PMNs) were prepared by dextran sedimentation followed by hypotonic lysis of contaminating red cells in 50 mM Tris, 6 mM NH_4Cl, pH 7.2 buffer before Lymphoprep (specific activity 1.078 g/mL, Nyegaard, Oslo) density gradient isolation. The proportion of contaminating mononuclear cells was less than 5%. The isolated PB PMNs were diluted in HBSS in a density of 1×10^7 cells/mL. The PMNs were degranulated by incubation with buffer, phorbol myristate acetate (PMA; 50 ng/mL) and phospholipase C (PLC; 1 mU/mL) for indicated time periods in humified 5% CO_2-in-air at 37 °C.[20,21] After incubation the cells were centrifugated for 10 min at 500 *g* and the supernatants were used for the measurement of total (PMC+) and endogenously active (PMC−) collagenase activities (see below for details).[6–12]

Collection and Preparation of Gingival Tissue, Dental Plaque Extracts, Crevicular Fluid, and Salivary Samples

Inflamed gingival tissue extracts, dental plaque extracts, GCF and salivary samples were collected from AP and localized juvenile periodontitis (LJP) patients and further processed for functional collagenase and gelatinase activity measurements and Western blotting (see below) as described previously.[8–13,22] We had previously characterized the clinical characteristics of the AP and LJP periodontitis patients included in these studies.[9,10,12,13] Study protocols were carried out according to the Declaration of Helsinki, and with approval of the ethical committee of the Institute of Dentistry, University of Helsinki. All patients included in the studies gave informed consent.

Collagenase and Gelatinase Assay

MMP-8 and MMP-1 as well as gingival tissue extracts, dental plaque extracts, GCF, and salivary samples from periodontitis patients were incubated at 22 °C for indicated time periods with native nonlabeled or [^{3}H-methyl]-labeled soluble 1.5 μM type I collagen in the presence and absence of 1.0–1.2 mM APMA, and collagenase activity was measured in a combination of SDS/PAGE/fluorography as described previously.[18,23,24] DOXY was added to the reaction mixtures in final concentrations 0–1000 μM. The SDS-PAGE gels and fluorograms were scanned in a LKB Ultroscan XL Laser Densitometry to assess the conversion of the intact α-chain to αA (3/4 α)-collagenase degradation products, and the IC_{50} (that is, the concentration of the drug to inhibit 50% of the enzyme activity) was determined as described previously.[23,24] Gelatinase activity was assayed by type I gelatin zymography according to Ingman *et al.*[11,13] In brief, type I gelatin at a final concentration of 1 mg/mL was mixed with 10% polyacrylamide and the slab gel was cast.[11,13] Aliquots of MMP-2 and MMP-9 and gingival tissue, crevicular fluid, and salivary samples were electrophoresed (30 mA) for 2 h at 4 °C. The slab gels were then incubated for 2 h with 2.5% Triton X-100 (30 min) and then washed with water. The appropriate lanes were then dissected and incubated at 37 °C in Tris-NaCl-$CaCl_2$ (TNC) buffer with DOXY added to final concentration ranging from 0 to 500 μM. The gels were stained with Coomassie brilliant blue, destained with 20% methanol/5% acetic acid (22 °C), and the molecular weight of the gelatinolytic zones were compared to prestained molecular weight protein standards.

Determination of Apparent K_m of Human MMP-8 for Type I Collagen in the Absence and Presence of 75 μM DOXY

The cleavage rate of type I collagen as function of substrate concentration (0.1–3 μM) by 1 mM APMA-treated MMP-8 was studied.[25] Substrate-enzyme incubations (15 min at 22 °C) were carried out in the absence and presence of 75 μM DOXY; this amount of DOXY inhibits efficiently but not completely MMP-8.[23,24] Data was analyzed by the double-reciprocal plots.[25]

Western Blotting

The molecular forms and cellular sources of collagenases (MMP-1 and MMP-8) and gelatinases/type IV collagenases (MMP-2 and MMP-9) present in inflamed gingival tissue extracts, dental plaque extracts, gingival crevicular fluid, and salivary samples were characterized by Western blotting using specific antisera. The specific antisera against human polymorphonuclear neutrophilic collagenase (MMP-8) were kindly provided to us by Dr. Jurgen Michaelis, Department of Pathology, Christchurch Medical School, Christchurch, New Zealand.[26] The polyclonal rabbit antisera against human polymorphonuclear leukocyte gelatinase (MMP-9) and human 72 kDa gelatinase/type IV collagenase (MMP-2) were kindly provided to us by Drs. Lars Kjeldsen and Nils Borregaard, Granulocyte Research Laboratory, Rigshospitalet, Copenhagen, Denmark and Dr. Taina Turpeenniemi-Hujanen, Department of Oncology, Central Hospital, University of Oulu, Oulu, Finland. The polyclonal rabbit antisera against fibroblast-type collagenase (MMP-1) and TIMP-1 were kindly provided to us by Dr. Henning Birkedal-Hansen, Department of Oral Biology, University School of Dentistry at Alabama, University of Alabama, Alabama. The polyclonal rabbit antisera against TIMP-2 were kindly provided by Dr. William G. Stettler-Stevenson, National Cancer Institute, Bethesda, Maryland. The assayed PMN extracts, gingival tissue, dental plaque, GCF, and salivary samples contained 5–20 μg protein as indicated in each experiment. The samples were treated with Laemli-buffer, pH 6.8 containing 5 mM dithiothreitol (DTT) and heated for 5 min at 100 °C. As molecular weight markers, high and low range prestained SDS-PAGE standards (Bio-Rad, Richmond, Canada) were used. The PMN extracts, dental plaque extracts, gingival tissue extracts, GCF, and salivary samples as well as purified antigens were separated on 8–10% SDS-polyacrylamide (SDS-PAGE) cross-linked gels at 200 V for 45 min and then electrophoretically transferred to nitrocellulose membrane at 100 V for 45 min (Bio-Rad Laboratories). Gelatin (3%) in 10 mM Tris-HCl, pH 8.0–0.5% Triton X-100 22 mM NaCl (TST) was used to block nonspecific binding sites on the nitrocellulose membrane. After washes with TST (3 × 15 min), the membrane was incubated with anti-MMP-8 antibody (1:500

dilution in TST), with anti-MMP-9 antibody (1:1000 dilution in TST), with anti-MMP-2 antibody (1:1000 dilution in TST), with anti-MMP-1 antibody (1:1000 dilution in TST), with anti-TIMP-1 and anti-TIMP-2 (both 1:1000 dilutions in TST) for 10 h. After washes with TST (3 × 15 min), the membrane was incubated with goat anti-rabbit antibody-alkaline phosphatase conjugate (1:1000 dilution in TST, Sigma, St. Louis, MO) for 1 h. After washings with TST (1 × 15 min) and with 10 mM Tris-HCl (1 × 15 min), pH 8.0–22 mM NaCl, the immunoblots, by addition of Nitro blue tetrazolium (NTB) and 5-bromo-chloro-3-indolyl-phosphate (BCIP), were diluted to N-N-dimethylformide (Sigma) in 100 mM Tris-HCl, 5 mM $MgCl_2$, 100 mM NaCl, pH 9.5. All incubations were carried out at 22 °C.[27] The secondary antibody did not react with the bands detected by Western blotting.

Measurement of Gingival Crevicular Fluid Phospholipase C Activity

The assay for PLC was carried out according to Kurioka and Matsuda.[28] It is based on hydrolysis *p*-nitrophenylphosphocholine (NPPC) that results in the release of chromogen, *p*-nitrophenol (NP), which was monitored spectrophotometrically at 405 nm.[28] GCF samples were collected from periodontally healthy controls and patients with adult periodontitis (AP), as described previously.[8–11,13]

Immunohistochemical Analysis of MMP-8 and MMP-9 in Periodontitis

Immunohistochemical analysis was performed using formalin-fixed, paraffin-embedded sections in four cases of untreated adult periodontitis gingival tissue specimens obtained by excision of inflamed human gingiva around the periodontally affected teeth before extraction of the teeth. After deparaffinization, sections were pretreated in distilled water either by heating by autoclave (120 °C, 10 min) or microwave oven (95 °C, 10 min)[29,30] for the purpose of antigen retrieval. Specificity of this procedure was confirmed by the fact that results with this procedure were essentially the same as those of frozen sections (Ohtani *et al.,* unpublished results). Indirect immunoperoxidase methods were employed.[31] Both anti-MMP-8 and anti-MMP-9 antibodies were diluted to 1:1000 and applied overnight at 4 °C. The specimens were then treated in methanol containing 0.03% hydrogen peroxide. Horseradish peroxidase-conjugated F(ab)2 fragments of anti-rabbit Ig (Amersham, Poole, UK) were diluted at 1:100 and applied for 8–16 h at 4 °C. Diaminobenzidine tetrahydrochloride (Dojin, Kumamoto, Japan) was used as chromogen.

RESULTS

IC_{50} values of DOXY inhibition for MMP-1, MMP-8, MMP-9, GCF collagenase, dental plaque collagenase, as well as gingival tissue collagenase and gelatinase assayed in the presence and absence of 1 mM APMA against their natural substrates type I collagen and gelatin, were measured as described.[22–24,32] Values of 10–30 μM were obtained for MMP-8, gingival tissue collagenase, GCF collagenase, and dental plaque collagenase without and with the presence of 1 mM APMA. IC_{50} values of DOXY inhibition ranging from 30–50 μM for MMP-9, gingival tissue, and GCF gelatinases were obtained as assessed by type I gelatin zymography.[32] Also collagenase from *Porphyromonas* (previously called *Bacteroides*) *gingivalis*, a potent periodontopathogen, was inhibited by DOXY with an IC_{50} value of 20 μM, a value similar to those seen with MMP-8, GCF, gingival tissue and dental plaque collagenases.[22,32] IC_{50} value of 280–300 μM was obtained for MMP-1 in agreement with our previous findings.[22–24]

FIGURE 1 shows the Western-blot analysis of GCF and salivary collagenases and gelatinases. GCF from periodontally and systemically healthy adults did not show the presence of MMP-8, but slight immunoreactivity for MMP-9 was detected (FIG. 1A and C, lanes 1–4 and 3 and 4, respectively). Clear immunoreactivity was detected for both latent 75-kDa proMMP-8 and 65-kD active MMP-8 in GCF of AP patients (FIG. 1A, lanes 5–9) and saliva (not shown). However, in GCF of patients with LJP immunoreactivity for MMP-8 (mainly latent 75-kDa form) was detected only in GCF samples from periodontitis sites that showed clear clinical signs of periodontal inflammation; LJP GCF from periodontitis samples without clear clinical signs of periodontal inflammation did not show immunoreactivity for MMP-8 as assessed by Western blotting (FIG. 1A, lanes 10–12); however, in these samples slight immunoreactivity for MMP-1 and MMP-2 were detected (FIG. 1B, lanes 4–8). In general, in salivary samples clear immunoreactivity for MMP-8 and MMP-9 as well as slight immunoreactivity for MMP-1 and MMP-2 were detected (not shown). In dental plaque extracts, both supra- and subgingival, 58-kDa immunoreactivity for MMP-8 and low molecular weight enzyme species (20–50 kDa) for MMP-2 and MMP-9 were detected (FIG. 1D). In gingival tissue extracts latent 75-kDa and active 65-kDa MMP-8 and 90-kD MMP-9 species were detected (FIG. 1A and C, lanes 14 and 15; 8 and 9, respectively). Slight, if any, immunoreactivities were detected for MMP-1 and MMP-2 in extracts of inflamed human gingiva from AP patients (not shown). The presence and molecular forms of TIMP-1 and TIMP-2 in untreated AP and LJP GCF as well as dental plaque extracts of AP patients were also analyzed by Western blotting using specific antisera (FIGS. 2 and 3). The Western blots show that the TIMP-1 antibody recognized 28-kDa TIMP-1 produced by cultured human gingival fibroblasts and oral keratinocytes (FIG. 2A and B, lanes 1 and lanes marked HGK [human gingival keratinocytes] and HGF [human gingival fibroblasts], respectively). However, extracts of human peripheral blood PMNs, concentrated GCF samples from healthy, AP, and LJP patients did not show any immunoreactivity for TIMP-1 antibody (FIG. 2A, lanes 2–4, 7, and 9–11). On the other hand, in dental plaque extracts of untreated AP patients, a strong 28-kDa band was detected by anti-human TIMP-1 antibody (FIG. 2A, lane 6). We also found in saliva of untreated AP patients strong TIMP-1 immunoreactivity that existed in multiple molecular weight forms (28–120 kDa) probably representing TIMP-1 polymers and/or TIMP-1 bound to different native and/or fragmented MMPs (FIG. 2B, lanes 4–7). TIMP-2 (21-kDa) was detected in saliva and dental plaque extracts of AP and LJP patients, but not in GCF. Furthermore, in PMN extracts high molecular weight 92-kDa TIMP-2 immunoreactivity was detected (FIG. 3, lanes 1–6).

The cleavage rate of type I collagen as a function of substrate concentration was studied and analysis of the data by double-reciprocal plot indicated an apparent K_m value of 0.3–0.6 μM for APMA-activated human MMP-8. DOXY (75 μM) inhibited the MMP-8 in a manner that did not result in changes in apparent K_m but did prevent the initial degradation from reaching the V_{max} (FIG. 4).

FIGURE 5 shows the Western-blot analysis of effects 0–1000 μM DOXY treatment on molecular forms on purified PMN MMP-8 and MMP-8 present in GCF as well as purified PMN MMP-9. No changes and/or further fragmentations of these MMPs that resulted from DOXY treatment were detected (FIG. 5A–C). FIGURE 5D and E shows the results of interaction of the MMP-activating oxidant, 10 μM NaOCl

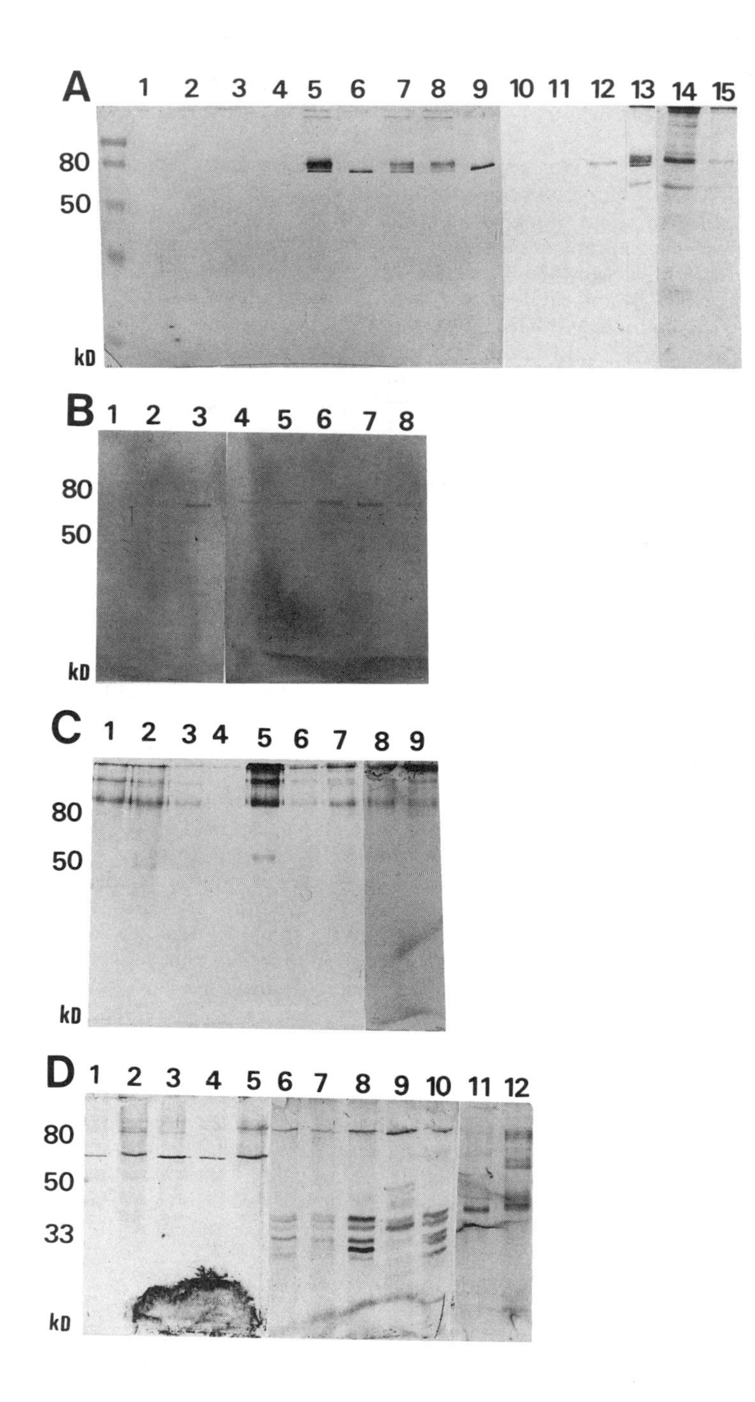
A
1 2 3 4 5 6 7 8 9 10 11 12 13 14 15
80
50
kD
B
1 2 3 4 5 6 7 8
80
50
kD
C
1 2 3 4 5 6 7 8 9
80
50
kD
D
1 2 3 4 5 6 7 8 9 10 11 12
80
50
33
kD

←

FIGURE 1. Western-blot analysis of matrix metalloproteinases present in gingival crevicular fluid (GCF), gingival tissue, and dental plaque extracts from healthy and periodontitis sites. **(A) Lanes 1–4:** GCF samples (5 μg protein) from periodontally healthy sites stained with anti-MMP-8 antibody; **lanes 5–9:** GCF from adult periodontitis (AP) sites stained with anti-MMP-8 antibody; **lanes 10 and 11:** GCF from localized juvenile periodontitis (LJP) sites (clinically healthy) stained with anti-MMP-8 antibody; **lane 12:** periodontally inflamed LJP GCF sample stained with anti-MMP-8-antibody; **lane 13:** neutrophil extract stained with anti-MMP-8-antibody; **lanes 14 and 15:** AP gingival tissue extracts. **(B) Lanes 1–2:** GCF (5 μg protein) from periodontally healthy sites stained with anti-MMP-2; **lane 3:** AP GCF stained with anti-MMP-2; **lanes 4–8:** LJP GCF samples stained with anti-MMP-2. **(C) Lanes 1, 2, and 5:** GCF (5 μg protein) from AP sites stained with anti-MMP-9 antibody; **lanes 3 and 4:** GCF from periodontally healthy sites; **lanes 6 and 7:** GCF from LJP sites; **lanes 8 and 9:** gingival tissue extracts. **(D) Lanes 1–5:** extracts (5 μg protein) and subgingival (*lanes 1–3*) and supragingival (*lanes 4 and 5*) dental plaque stained with anti-MMP-8; **lanes 6–10:** extracts of subgingival (*lanes 6–8*) and supragingival (*lanes 9 and 10*) dental plaque stained with anti-MMP-2 antibody; **lanes 11 and 12:** extracts of sub- and supragingival dental plaque stained with anti-MMP-9. The specificities of the antibodies used are shown in FIGURE 5. Molecular weight markers are indicated.

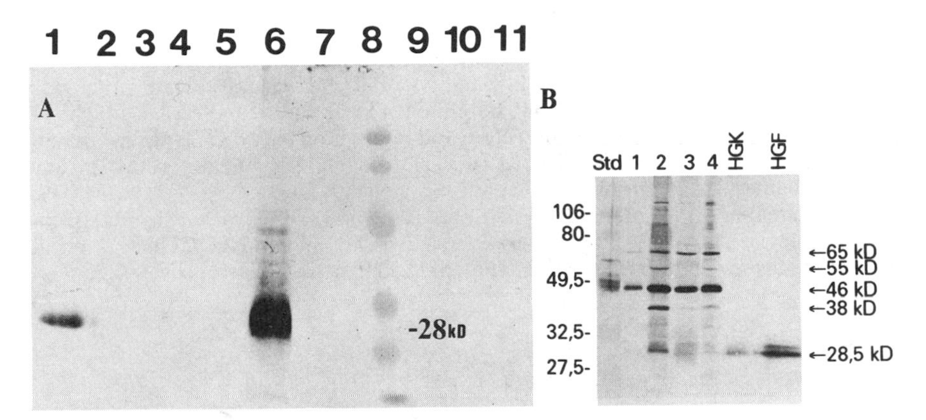

FIGURE 2. (A) Western-blot analysis of gingival crevicular fluid samples (GCF) from adult periodontitis (AP) and localized juvenile periodontitis (LJP) patients, dental plaque extracts, neutrophil (PMN) extract, and control GCF samples with anti-TIMP-1 antibody. **Lane 1** represents conditioned media of cultured gingival fibroblasts (20 μg total protein) showing the clear presence of 28-kD TIMP-1; **lanes 2–5**, concentrated AP GCF samples (20 μg protein); **lane 6**, dental plaque extract (20 μg protein); **lane 7**, human neutrophil (PMN) extract (20 μg protein); **lane 8**, molecular weight standards (not marked because located in the middle of the blot); **lanes 9 and 10**, concentrated LJP GCF samples (20 μg protein); and **lane 11**, concentrated GCF from control person with healthy periodontium (20 μg protein). **(B)** Western blot of saliva of AP and LJP patients with anti-TIMP-1 antibody. **Lane Std** represents molecular weight standards; **lanes 1 and 2**, AP salivary samples (20 μg protein); **lanes 3 and 4**, LJP salivary samples (20 μg protein); lane HGK, conditioned cultured oral keratinocyte media (20 μg protein); **lane HGF**, conditioned human gingival fibroblast culture media—same media also used in FIGURE 2A, lane 1 (20 μg protein). Molecular weight standards are listed on the left side of FIGURE 2B and major polymeric forms of salivary TIMP-1 are listed on the right side of this blot. Mobility of 28-kD TIMP-1 immunoreactivity in FIGURE 2A is indicated.

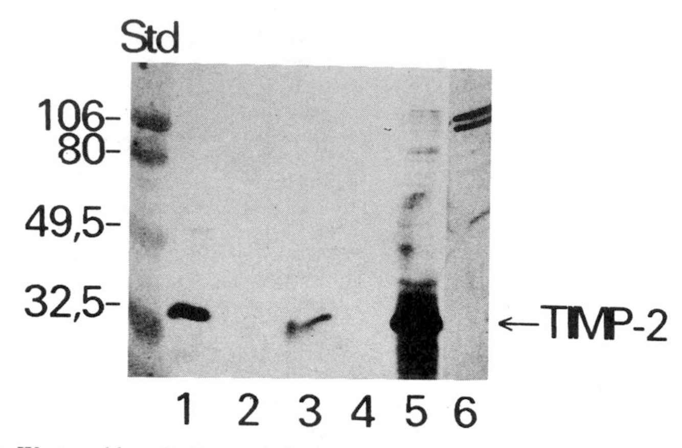

FIGURE 3. Western blot of saliva and gingival crevicular fluid from adult periodontitis (AP) and localized juvenile periodontitis (LJP) patients, dental plaque extracts, and human neutrophil (PMN) extract with anti-TIMP-2 antibody. **Lane 1** represents molecular weight standards; **lane 2**, LJP saliva (20 μg protein); **lane 3**, LJP GCF (20 μg protein); **lane 4**, AP saliva (20 μg protein); **lane 5**, AP GCF (20 μg protein); and **lane 6**, PMN extract (20 μg protein). **Lane Std** indicates molecular weight standards.

and DOXY on purified PMN MMP-8 and MMP-9. NaOCl (10 μM) can activate proMMP-8[16,20] and proMMP-9[34] but no changes in apparent molecular weight of proMMP-8 and proMMP-9 were detected, and interactions of NaOCl, phenylmercuric chloride (PMC), gold thioglucose (GTG), and DOXY did not result in any further changes/fragmentations in apparent molecular weights of these MMPs. Similarly, treatment of proteolytically activated (by *Treponema denticola* chymotrypsin-like proteinase[6]) MMP-8 and MMP-9 with 10 μM DOXY and NaOCl did not result in further fragmentation of these MMPs (FIG. 5D and E, lanes 6–8, respectively).

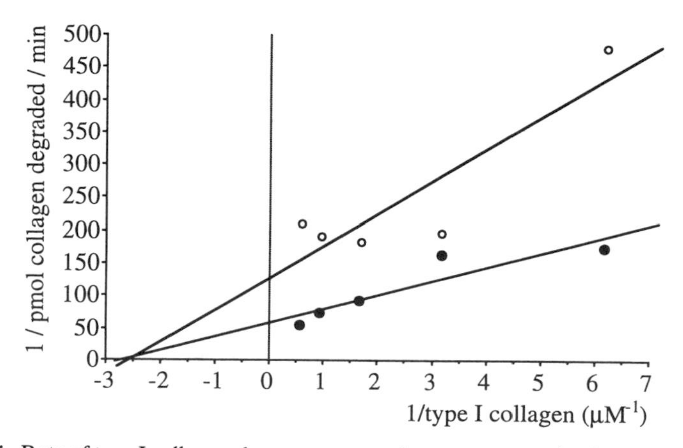

FIGURE 4. Rate of type I collagen cleavage versus collagen concentration by APMA-activated human neutrophil collagenase (MMP-8) without (●) and with (○) pretreatment of 75 μM doxycycline (DOXY): A double-reciprocal analysis.

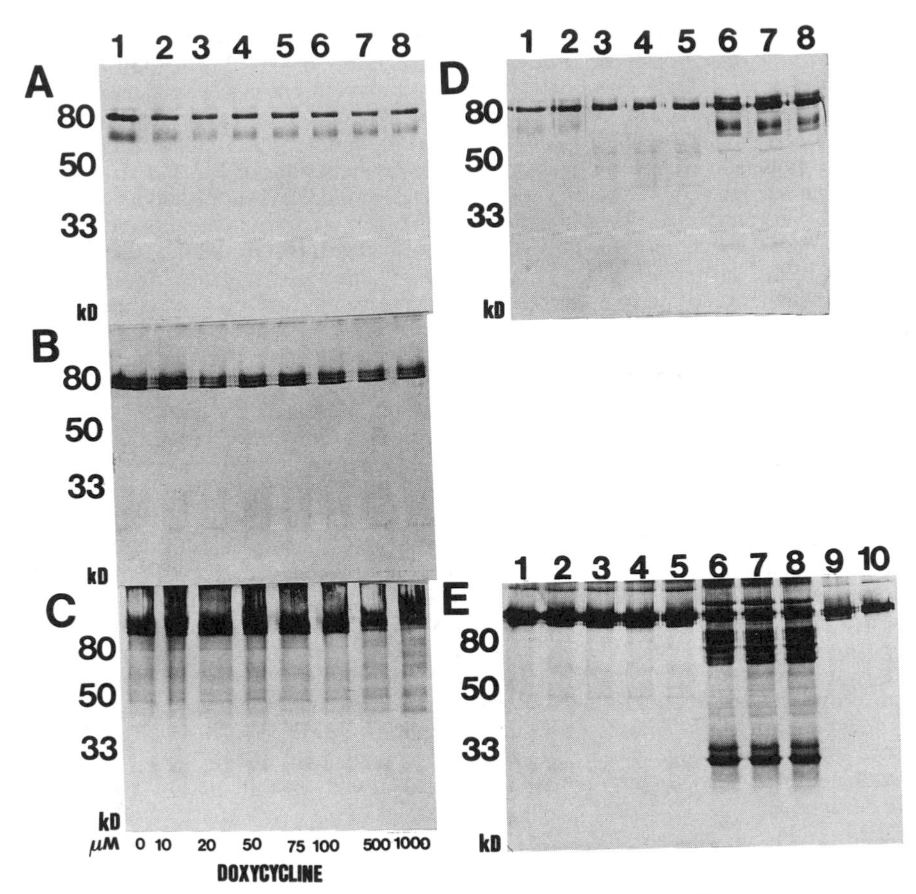

FIGURE 5. Western-blot analysis of the effects of doxycycline (DOXY) on the molecular forms of latent and activated human neutrophil collagenase (MMP-8), MMP-8 present in gingival crevicular fluid (GCF) from adult periodontitis patients, and human neutrophil gelatinase/type IV collagenase (MMP-9). **(A) Lanes 1–8** represent the effects of increasing amounts (0–1000 μM) of DOXY on pro and active MMP-8; indicated amounts of DOXY were incubated with 2 μg MMP-8 for 2 h at 37 °C before Western-blot analysis. **(B)** As in **A** but AP GCF MMP-8 stained with anti-MMP-8. **(C) Lanes 1–8** represent the effects increasing amounts (0–1000 μM) of DOXY on pro and active MMP-9; indicated amounts of DOXY were incubated with 2 μg MMP-9 for 2 h at 37 °C before Western-blot analysis. **(D) Lane 1** represents purified human MMP-8 stained with anti-MMP-8; **lane 2**, as in lane 1 but preincubation with 10 μM DOXY for 2 h at 37 °C; **lane 3**, MMP-8 treated with 10 μM NaOCl; **lane 4**, MMP-8 treated with 10 μM NaOCl and DOXY; **lane 5**, MMP-8 treated with 50 μM gold thioglucose (GTG); **lane 6**, MMP-8 treated with microbial protease (*Treponema denticola* chymotrypsin-like protease); **lane 7**, MMP-8 treated with with microbial protease and 10 μM DOXY; **lane 8**, MMP-8 treated with microbial protease, 10 μM DOXY, and 10 μM NaOCl. **(E) Lane 1** represents purified human neutrophil MMP-9 (2 μg) stained with anti-MMP-9 antibody; **lane 2**, MMP-9 treated with 10 μM DOXY; **lane 3**, MMP-9 treated with 10 μM NaOCl; **lane 4**, MMP-9 treated with 10 μM DOXY and 10 μM NaOCl; **lane 5**, MMP-9; **lane 6**, MMP-9 treated with microbial protease; **lane 7**, MMP-9 treated with microbial protease and 10 μM DOXY; **lane 8**, MMP-9 treated with microbial protease (*Treponema denticola* chymotrypsin-like protease), 10 μM DOXY, and 10 μM NaOCl; **lane 9**, MMP-9 treated with 50 μM GTG; and **lane 10**, MMP-9 treated with 10 μM PMC. Molecular weight markers are indicated.

Medication of reactive arthritis (ReA) patients with a long-term (2-month) regimen of DOXY decreased the activity and levels of MMP-8 in their saliva;[33] however, other salivary proteinase (neutrophil-derived elastase and cathepsin G) and trypsin-like activities were not reduced. Western-blot assessment revealed decreased immunoreactivity for MMP-8 after two months of treatment with DOXY, but no molecular weight changes or further fragmentation of MMP-8 to inactive lower molecular weight species was detected. Also medication of patients with AP with a 2-week regimen of low-dose doxycyline (LDD) is known to result in decreased collagenase activity present in gingival tissue and adjacent GCF.[18,35]

GCF collected from deep (>6 mm) periodontitis sites from 10 untreated patients with AP contained increased PLC activity as compared to periodontally healthy controls (FIG. 6A). *In vitro* PLC (1 mU/mL) was found to be an efficient inducer of isolated PB PMNs to degranulate their collagenase; PLC-induced release of collagenase by PB PMNs was comparable to that seen with PMA (50 ng/mL), a known inducer of PMN degranulation[20,21] (FIG. 6B).

FIGURE 7 shows the immunohistochemical analysis of MMP-8 and MMP-9 in human adult periodontitis gingiva. Actively inflamed areas of human periodontitis gingiva of untreated AP patients were analyzed (FIG. 7A). MMP-8 and MMP-9 are expressed predominantly in neutrophils in the area of acutely inflamed adult human periodontitis gingiva (FIG. 7B and C); the presence of macrophages in the actively/acutely inflamed human adult periodontitis gingiva were identified by anti-CD 68 staining (FIG. 7D). In general, immunohistochemical analysis of the presence of MMP-8 and MMP-9 revealed that both MMP-8 and MMP-9 are expressed in polymorphonuclear leukocytes in the area of active (acute) periodontal inflammation and are scanty in the area of chronic inflammation. MMP-9 was also identified in macrophages and, therefore, especially in chronic areas of periodontal inflammation, the number of MMP-9–positive cells exceeded that of MMP-8–positive cells. Overall, this immunohistochemical data confirmed the involvement of MMP-8 and MMP-9 and neutrophils as their potential sources in the process of periodontal tissue destruction.

DISCUSSION

Periodontal disease is initially characterized by destruction of the extracellular matrix subjacent to the junctional epithelium, including basement membrane, followed by the breakdown of the deeper connective tissue including periodontal ligament and alveolar bone.[1] Birkedal-Hansen[2,3] and Birkedal-Hansen *et al.*[4] have recently reviewed the complex cascade of ECM degradation which includes, according to these reviews,[2–4] an intracellular phagocytic pathway (mediated by acid cathepsins) and three extracellular ("pericellular") pathways, namely, (1) the plasminogen-dependent pathway (serine proteinases), (2) the MMP pathway (note: the plasminogen-dependent pathway may activate some, but evidently not all, precursor forms of MMPs), and (3) the osteoclastic pathway. The MMP family of proteinases not only degrades most, if not all, extracellular matrix macromolecules such as different types of collagens, fibronectin, and proteoglycans but can also inactivate the host's endogenous inhibitors of other classes of proteinases.[8,31] In this regard a number of studies have shown that MMPs can proteolytically inactivate α1-proteinase inhibitor (α-PI), a serum protein and the major inhibitor of PMN elastase, thus further enhancing the connective tissue breakdown.[26,36–38]

Regarding the roles of bacterial toxins, proteinases, and other factors in periodontal ECM degradation, proteolytic enzymes from several periodontopathogens such

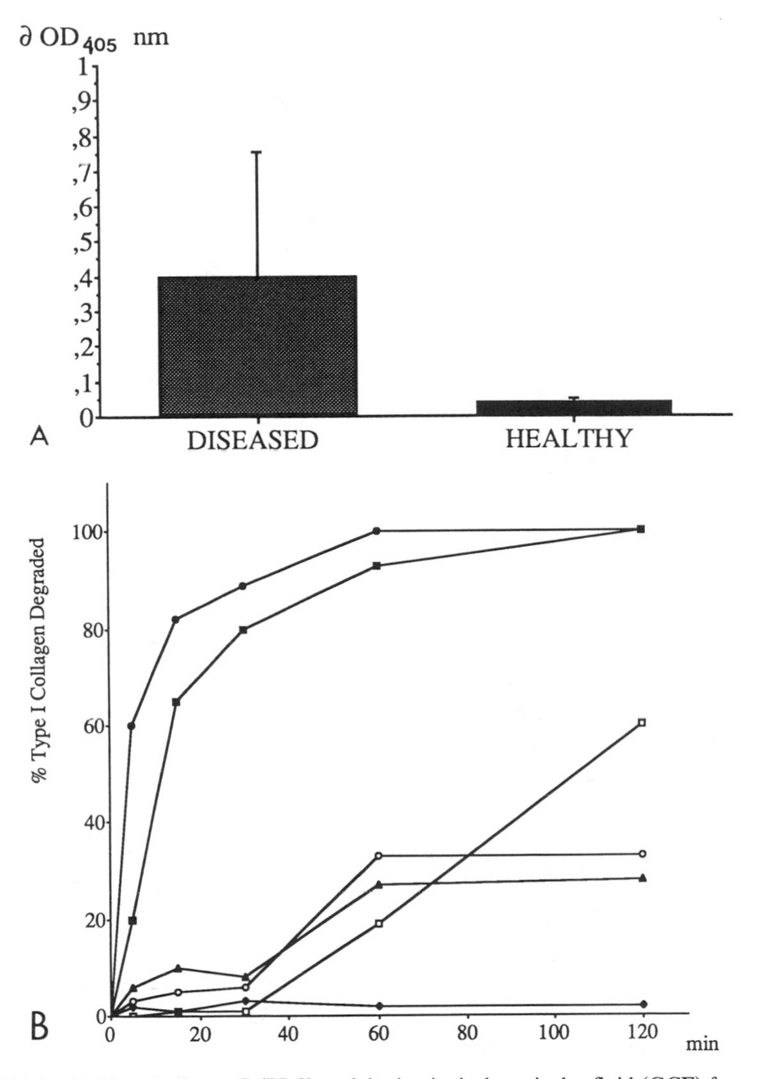

FIGURE 6. (**A**) Phospholipase C (PLC) activity in gingival crevicular fluid (GCF) from adult periodontitis patients and from periodontally healthy controls. (**B**) Release of collagenase by triggered peripheral blood neutrophils (PMNs). Isolated peripheral blood PMNs (1×10^7 cells/mL) were triggered for indicated time periods by phospholipase C (1 mU/mL PLC, indicated by ● and ○), phorbol myristate acetate (50 ng/mL PMA, indicated by ■ and □), and buffer (indicated by ▲ and ◆) and resulting supernatants were assayed for total (1 mM phenylmercuric chloride treatment, PMC+) and endogenously active (PMC−) type I collagenase activities as described in MATERIALS AND METHODS. Collagenase activity is expressed as percent of soluble type I collagen degraded.

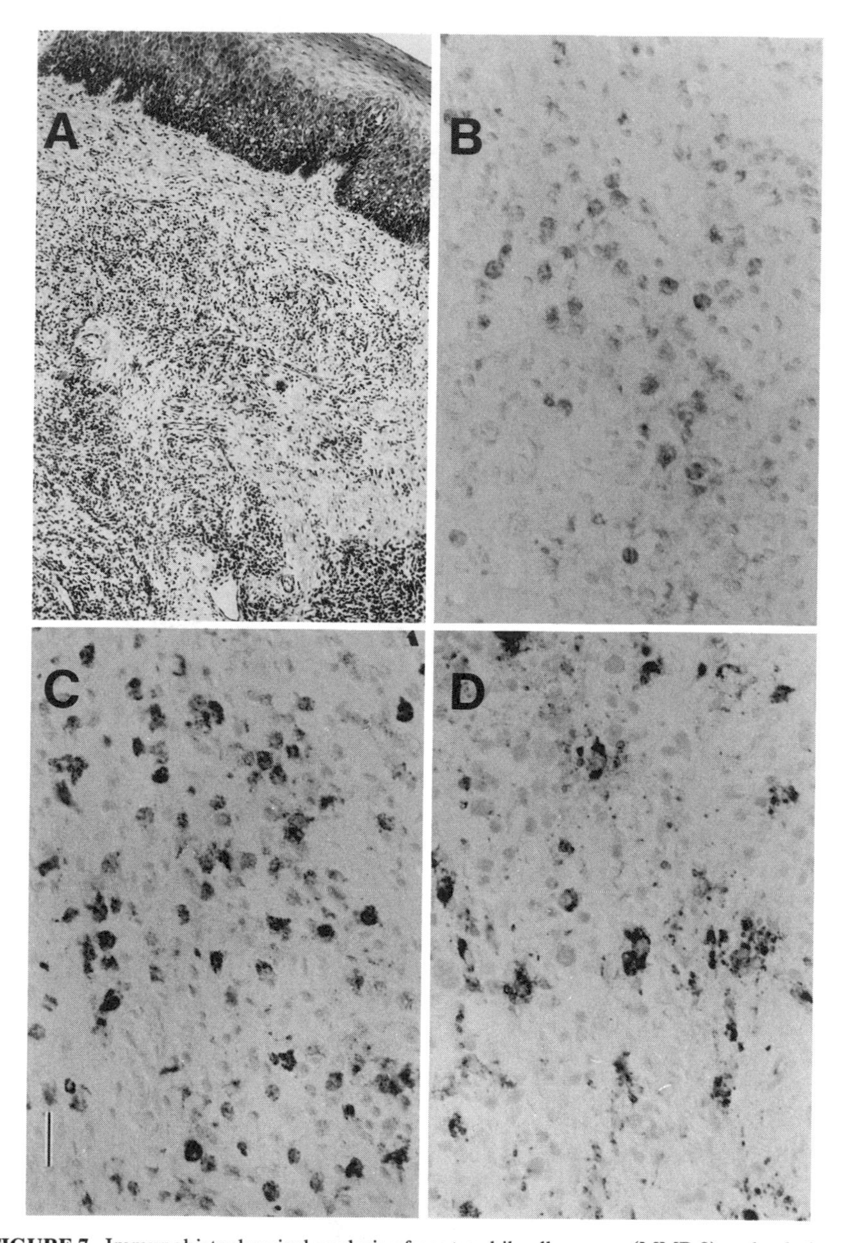

FIGURE 7. Immunohistochemical analysis of neutrophil collagenase (MMP-8) and gelatinase (MMP-9) in inflamed human adult periodontitis gingiva. (**A**) Hematoxylin-eosin staining of adult periodontitis gingiva. A case of acute/active periodontal inflammation is presented, scale = 10 μM. MMP-8 (**B**) and MMP-9 (**C**) are expressed predominantly in neutrophils in the area of acute/active periodontal inflammation; the number of MMP-8 positive cells exceed MMP-9 positive cells, scale = 20 μM. Macrophages (**D**), identified by anti-CD 68 staining, are also abundant in the area inflammation of periodontitis gingiva, scale = 20 μM.

as *Porphyromonas gingivalis, Actinobacillus actinomycetemcomitans,* and *Treponema denticola* are capable of degrading both interstitial and basement membrane collagens, although there is no direct evidence for this pathway in periodontal destruction *in vivo.*[7] However, bacterial proteinases may produce periodontal connective tissue breakdown indirectly by altering the host response.[4–6] Examples include the degradation of immunoglobulins and serpins (the latter include α-PI, also called α-1-antitrypsin) and induction of secretion and direct activation of latent precursor forms of host-derived MMPs.[1–7] Less attention has been focused on the possible role of oral spirochetes in the etiopathogenesis of periodontal tissue destruction. Spirochetes are also known to produce and secrete phosphatidylcholine-hydrolyzing PLC (phosphatidylcholine cholinephophohydrolase).[39] We found here increased activity of PLC in GCF of untreated AP patients when compared to periodontally healthy control GCF samples. Moreover, PLC was found to be an efficient inducer of isolated peripheral blood PMNs to release their collagenase. PLC in GCF can be derived from spirochetes and/or other bacterial species present in diseased periodontal pockets as well as from PMNs.[39] Thus, the initiation of proteolytic cascades of PMN-dependent periodontal tissue destruction may, at least in part, be initiated by hydrolysis of PMN membrane phospholipids by PLC. Alternatively, PLC-dependent hydrolysis of the cell membrane phospholipids can lead to destruction of crevicular epithelial cells, thus releasing arachidonic acid, leukotrienes, and matrix metalloproteinases.[2–4,40] All these host-derived inflammatory mediators can participate in the molecular cascade leading to loss of tooth attachment and bone loss.[1–4]

A number of lines of evidence implicate MMPs as key players in periodontal tissue destruction including the detection of pathologically elevated (compared to normal gingival tissue and oral exudates such as GCF and saliva) activities of MMPs in extracts of inflamed gingival tissues, GCF, and salivary/mouthrinse samples; the activities of these proteinases have been found not only to be positively correlated to the severity of periodontal inflammation and pocket depth at the lesion sites donating these tissue and GCF samples, but also the relative amount of these enzymes recovered in active rather than latent forms appears to increase with greater severity of periodontal disease. Furthermore, the activities of these MMPs and the ratio of active/latent MMPs decreases in periodontitis sites after periodontal instrumentation therapy (scaling and root planing).[1–15] However, the cellular sources of MMPs present in inflamed human gingiva, GCF (inflammatory exudate of inflamed human gingiva), dental plaque, and saliva have been unclear. This information is important not only to understand the pathogenic mechanisms of periodontal tissue destruction but may direct, at least in part, the therapeutic strategy to reduce the pathologically excessive MMP levels associated with different periodontal diseases.[8,14] In this regard, collagenase activity in gingival tissue and adjacent GCF of non-diabetic and diabetic AP patients has been shown to decrease after adjunctive use of either regular or low-dose tetracycline/doxycycline medication, whereas no effect was seen in localized LJP patients.[8,14,15,18] Previous studies have shown that neutrophil collagenase (MMP-8) is sensitive to doxycycline inhibition (IC_{50} = 10–20 μM) whereas fibroblast collagenase (MMP-1) is relatively resistant to doxycycline inhibition (IC_{50} = 280–300 μM).[8,15,23,24] Accordingly, based on these [8,15,23,24] and other studies,[9,10] it was concluded that the source of GCF collagenase in non-diabetic[9] and diabetic AP[15] was PMNs, and in LJP GCF it was probably fibroblasts, epithelial cells and/or monocyte/macrophages.[10,24] The current results revealing the Western-blot assessment of MMPs involved in periodontal ECM degradation showed that major type of collagenase present in GCF of untreated AP patients was MMP-8. In LJP GCF the presence of MMP-8 was dependent upon the degree of inflammation of periodontitis sites studied. In clearly inflamed LJP GCF, MMP-8 could be detected but not in noninflamed periodontitis LJP GCF (noninflamed LJP periodontitis sites are, however, active in disease progression) and some of these noninflamed LJP GCF samples contained also slight amounts of immunoreactivity for MMP-1 and MMP-2; according to Birkedal-Hansen *et al.*[2–4] these MMPs have not been indentified previously in GCF. They may, however, reflect the activation of gingival stromal cells[2–4] during periodontal inflammation, but their amounts present in inflamed human periodontium are evidently not comparable to the amounts of PMN-derived MMPs. GCF samples of periodontally healthy controls did not show the presence of MMP-8, MMP-1, or MMP-2; only MMP-9 was detected by the current Western-blot assessment in these samples. However, anti-MMP-9 antibody detected 92-kD MMP-9 derived both from PMNs and oral keratinocytes.[40] The current Western-blot assessment revealed clear immunoreactivities for both MMP-8 and MMP-9 in extracts of inflamed human gingiva from AP patients, whereas rather slight immunoreactivities were detected for MMP-1 and MMP-2. Based on the current immunological results as well as previous ones concerning the functional characterization of MMPs[8–13] involved in periodontal tissue destruction, PMNs evidently are the major source of MMPs in periodontitis gingiva (inflamed soft periodontal connective tissues), and PMN MMPs are responsible for degradation of soft periodontal tissues, inasmuch as not much MMP-1 and MMP-2 is present in periodontitis gingiva or in oral inflammatory exudates (GCF and saliva). Tonetti *et al.*[41] have recently found that transcripts of MMP-8 are found in cells in inflamed human gingiva, although previous studies indicated that transcription of PMN MMP genes appears to be essentially completed before PMNs emigrate from the blood vessels.[2–4,34] This is in agreement with our recent immunohistochemical and Western-blot findings showing the presence of MMP-8 in inflamed but not in noninflamed human gingival tissue.[41A] It remains to be seen which type of PMN triggering factors can induce solely PMN MMP degranulation and/or separate/concomitant transcription as well as expression of MMPs by PMNs. Furthermore, the specific roles of these two MMP-release/production pathways by PMNs in different forms of periodontitis and tissue destructive diseases remain to be studied. Future studies will also show whether MMP-8 transcripts can be found in other cells than PMNs; nonetheless, our recent PUR/RNA analyses using MMP-8-specific primers have shown that 522 bp–MMP-8-transcripts can also be detected in cells other than PMNs (T. Sorsa, T. Salo, M. Kylmäniemi *et al.*, in preparation). MMP-9 can also be derived from epithelial cells,[40] and periodontitis-associated alveolar bone destruction may involve the action of collagenase/matrix metalloproteinases produced by bone cells.[17] Interestingly, in dental plaque, which is composed of masses of oral bacteria, different glycoproteins, and polysaccharides and is, without doubt, the main cause of periodontal disease, MMP-8 existed in 58-kDa form and MMP-2 and MMP-9 were fragmented low molecular weight species (20–50 kDa) evidently reflecting proteolytic activation and fragmentation of these collagenases and gelatinases, respectively.[22] Of note, MMP-8 and MMP-9 exist in periodontitis gingiva and GCF predominantly in 70–75-kDa and 90–200-kDa forms representing proforms and/or polymeric forms, respectively, rather than proteolytically activated or fragmented forms. As shown in this study, oxidant NaOCl in physiologically relevant concentrations[16,20,34] can activate proMMP-8 and proMMP-9 without changes in apparent molecular weights of the PMN proMMPs. These findings indicate that PMN proMMPs may well be activated by PMN-generated reactive oxygen species[20,34] during periodontal inflammation in inflamed gingiva and GCF, but evidently proteolytically by the action of other host-derived proteases (tryptases, cathepsin G, etc.)[1–5,7,19] and/or proteases from potent periodontopathogenic bacteria[6] in dental plaque and saliva. Reactive oxygen species can also activate fibroblast proMMP-1 and bone-cell–derived procollagenase.[17] Importantly, tetracycline/doxycycline and CMTs can prevent oxidative activation of proMMP-8 and bone-cell–derived procollagenase (probably proMMP-1).[16,17]

The activity of host-cell–derived MMPs is also regulated by their major endogenous inhibitors, tissue inhibitors of matrix metalloproteinases (TIMP-1 and TIMP-2).[1–4] Previously the presence and molecular forms of TIMP-1 and -2 present in dental plaque, GCF, and saliva of periodontitis patients have not been addressed by Western blotting. Drouin *et al.*[42] have partially purified and characterized TIMP-1 from human saliva and they found, in accordance with our present results, higher molecular weight forms of TIMP-1 in saliva in addition to 28-kDa TIMP-1. To confirm and further extend their observations[42] we found that TIMP-1 exists as multiple molecular weight form complexes probably representing polymeric and/or MMP-bound forms of TIMP-1 in saliva of periodontitis patients. No TIMP-1 immunoreactivity was detected in peripheral blood PMN extracts or, surprisingly, in concentrated GCF samples from healthy controls, AP and LJP patients as analyzed by Western-blot technique. Among the multiple molecular weight forms/polymeric (28–120 kDa) of salivary TIMP-1 species, the 48-kDa TIMP-1 was predominant and may represent TIMP-1 bound to MMP-1 and/or MMP-3. In saliva TIMP-1 may also to some extent be bound to PMN-derived 70–80-kDa MMP-8 and/or 92-kDa MMP-9; these complexes of TIMP-1 are not very stable and may dissociate under SDS-PAGE conditions. However, in dental plaque extracts a strong 28-kDa TIMP-1 immunoreactivity was detected. Thus PMNs, being the major source of MMPs in periodontitis gingiva and GCF, apparently do not contain TIMP-1 (in case PMNs do contain TIMP-1 the amounts of TIMP-1 in PMNs is not comparable to their MMPs).[2–4,34] This explains also the relatively low TIMP levels in GCF[2–4] when compared to GCF PMN-type MMP levels, especially during active phases of periodontal disease[8–12] and, accordingly, saliva, not GCF, may be the major source of TIMP-1 present in dental plaque. The presence of intact 28-kDa TIMP-1 in dental plaque extracts supports our recent findings that the proteases from potent periodontopathogenic bacteria evidently do not degrade and/or alter significantly the MMP-inhibitory action of TIMP-1.[6] TIMP-2 (21 kDa) was also found to be present in saliva and dental plaque extracts but not detectable in GCF by current Western blot analysis. Similarly to TIMP-1, TIMP-2 was not fragmented by dental plaque. In addition to PMN extracts the high molecular weight 92-kDa doublet representing TIMP-2 immunoreactivity was also detected in extracts of inflamed human gingiva (T. Sorsa and T. Salo, unpublished results); the nature of this immunoreactivity remains unclear. In fact, dental plaque may act as a reservoir and site of activation of MMPs and TIMPs during the progression of periodontal disease.[22]

Progress in elucidating the roles and regulation of MMPs in periodontal ECM degradation has led to a concept involving chemotherapeutic inhibition of MMPs. Golub and co-workers have reported that specially formulated LDD inhibits the gingival tissue and adjacent GCF collagenase activities.[18] Other studies have also shown, using these LDD formulations as adjuncts to periodontal instrumentation therapy, that progression of periodontal disease (gingival attachment loss) can also be inhibited (for recent reviews see Ingman *et al.*[8] and Golub *et al.*[14]). However, it was not clear whether this *in vivo* effect was direct or indirect. We recently found that DOXY inhibits collagenase extracted from inflamed human gingival tissue (IC_{50} = 10–20 μM) as efficiently as GCF collagenase (IC_{50} = 10–20 μM) and PMN MMP-8 (IC_{50} = 10–20 μM), significantly differing from MMP-1 (IC_{50} = 280–300 μM) independent of the presence of APMA.[32] Also gingival tissue gelatinase was inhibited by DOXY at concentrations comparable to those efficient for MMP-9 from PMNs and GCF gelatinase; the IC_{50} for these gelatinases ranged between 20–50 μM. It appears that DOXY can inhibit *in vivo* activated gingival tissue collagenase and gelatinase directly at therapeutically attainable levels.[32] The molecular interactions involved in inhibition of native, proteolytically and oxidatively activated PMN MMPs, as well as PMN MMPs present in gingiva, GCF, and saliva, by DOXY were studied by Western blotting. It was found that treatment of latent and activated MMP-8 and MMP-9 (either purified or *in vivo* forms present in inflamed gingiva and oral exudates) did not result in the fragmentation of latent and activated forms of these MMPs which could result from autoproteolysis, denaturation or inactivation. DOXY was also found to inhibit MMP-8 in a noncompetitive manner. It has been shown that excess Ca^{2+} can overcome the inhibition of collagenase by doxycycline (or other tetracyclines); however, this inhibition occurs very efficiently *in vitro* and *in vivo* in the presence of physiologically relevant Ca^{2+} concentrations, and recent studies have also shown that Zn^{2+} is more potent than Ca^{2+} in preventing the inhibition of collagenase and gelatinase.[14] Furthermore, the present Michaelis-Menten kinetics data in the present study, demonstrating the noncompetitive nature of the inhibition of collagenase by doxycycline, suggest that this drug interacts with Zn^{2+} at a secondary site in the enzyme, rather than the active-Zn^{2+} site, to mediate this inhibitory effect; such observations would thus be consistent with the recent crystallographic data of Lovejoy *et al.* (Science **263:** 375–377, 1994) describing this secondary Zn^{2+} site in collagenase. When patients with ReA and AP were medicated either with a long-term (2-month) regular doxycycline regimen[33] or 2 weeks of LDD medication,[18,35] respectively, and samples (gingival tissue, GCF, and saliva) were analyzed for collagenase activity by functional assay and for the molecular forms of MMP-8 by Western blotting, the results revealed that latent and endogenously active collagenase activities in gingiva, GCF, and saliva, respectively, statistically significantly reduced and decreased immunoreactivity—but not fragmentation—of MMP-8. Another possible mechanisms, in addition to the direct anticollagenolytic effect and prevention of the oxidative activation latent collagenase/MMP, may be the modification of PMN functions,[43] because doxycycline/tetracycline can be taken up by PMNs and concentrated intracellularly at levels 4–10 times greater than concentrations of these drugs determined extracellularly. Therefore one explanation for the reduction of gingival tissue, GCF, and salivary collagenases in long-term regular and/or low-dose doxycycline medication may be the interaction of DOXY with MMP-8 intracellularly in PMNs in addition to direct extracellular inhibitory actions. We thus suggest that DOXY orally administered, producing a blood peak level about 10 μM (however, even 40-μM tetracycline/doxycycline levels can be reached in GCF as a result of regular medication[8,14,22]), becomes—especially during long-term medication—hyperconcentrated within PMNs and reacts/binds with proMMPs. During degranulation the proMMPs are released from PMNs into the extracellular matrix, where DOXY blocks oxidative activation of proMMPs,[16] and inhibits already-active PMN MMPs.[15,23,24] Other important factors to regulate proteolysis are proteinase inhibitors.[1–4,34] Because both α-PI, an endogenous inhibitor of serine proteinases (such as PMN elastase) and substrate for MMPs[34] and TIMP-1[1–4] can be oxidatively inactivated by reactive oxygen species,[34,44] the therapeutic levels of doxycycline/tetracycline/CMTs—especially during long-term medication—may prevent general proteolytic events (in addition to rather specific collagenolysis) by protecting the proteinase inhibitor shield.[37,45] In summary, we conclude that (1) PMN MMPs are important in periodontal tissue destruction, (2) DOXY can directly inhibit MMP activity (the same IC_{50} obtained ± APMA), (3) DOXY can prevent the activation of proMMPs to MMP by NaOCl, (4) we found no evidence in our *in vitro* and *in vivo* studies for the DOXY-induced denaturation and/or fragmentation of proMMPs to inactive lower molecular weight fragments (yet this can be another mechanism for reducing MMP activity by DOXY), (5) the apparent low levels of TIMPs in periodontitis GCF and PMNs imply that the important regulation step for PMN MMPs in the extracellular milieu is proenzyme activation, and (6) DOXY has the potential to prevent serpins (i.e., α-PI) from proteolytic and oxidative inactivation.[32] Therefore, direct inhibition and prevention of the oxidative activation of PMN MMPs by DOXY and/or CMTs address the therapeutic importance of these drugs (and in future their derivatives) as "host modulators" in controlling tissue destruction events associated with periodontal and other diseases.

SUMMARY

The characterization and regulation of matrix metalloproteinases (MMPs) have been studied to determine their role(s) in periodontal tissue destruction. Progress in elucidating the roles of MMPs in periodontal tissue destruction has led to a new concept involving the chemotherapeutic inhibition on MMPs, a therapeutic strategy which less than a decade ago was considered "a difficult and perhaps impossible task." Tetracyclines/doxycycline (DOXY) and their chemically modified nonantimicrobial derivatives (CMTs) are known to inhibit the matrix metalloproteinases, especially preferring human neutrophil collagenase (MMP-8), and prevent the oxidative activation of procollagenases. We characterized by Western blotting the molecular forms and cellular sources of gingival tissue, dental plaque, gingival crevicular fluid (GCF), and salivary MMPs associated with periodontitis. Also the molecular forms of tissue inhibitors of matrix metalloproteinases (TIMP-1 and TIMP-2) in periodontitis were studied by Western blot. Neutrophil (PMN)-derived MMPs were found to predominate in periodontitis, and phospolipase C present in increased amounts in periodontitis sites was found to be a potential inducer of PMN degranulation. We further studied the effects of DOXY on molecular forms of different latent and active MMPs purified from different cellular sources (PMNs, fibroblasts, keratinocytes) and present *in vivo* in oral exudates (gingival extracts, GCF, and saliva). DOXY inhibition of activated (oxidatively or proteolytically) MMPs were not associated with MMP fragmentation. Michaelis-Menten plots of initial rates of degradation of soluble type I collagen revealed an apparent K_m value of 0.3–0.6 μM for MMP-8, and 75 μM DOXY inhibited MMP-8 in a manner which did not result in changes in apparent K_m value but did prevent the initial degradation reaching V_{max} providing evidence for noncompetitive inhibition. Treatment of patients with long-term DOXY medication results in decreased MMP-8 activities/levels in gingival tissue, crevicular fluid, and saliva, but not fragmentation of MMP-8 *in vivo.* These data further support and extend the key role of PMN-MMPs in periodontitis, and the activities of these PMN MMPs can be inhibited directly by therapeutic levels of DOXY.

REFERENCES

1. SODEK, J. & C. M. OVERALL. 1992. Matrix **12 (Suppl. 1):** 352–362.
2. BIRKEDAL-HANSEN, H. 1993. J. Periodontal Res. **28:** 500–510.
3. BIRKEDAL-HANSEN, H. 1993. J. Periodontol. **64:** 474–484.
4. BIRKEDAL-HANSEN, H., W. G. I. MOORE, H. K. BODDEN, L. M. WINDSOR, B. BIRKEDAL-HANSEN, A. DECARLO & J. A. ENGLER. 1993. Crit. Revs. Oral Biol. Med. **4:** 197–250.
5. UITTO, V-J., H. LARJAVA, J. HEINO & T. SORSA. 1988. Infect. Immun. **57:** 213–218.

6. SORSA, T., T. INGMAN, K. SUOMALAINEN, M. HAAPASALO, Y. T. KONTTINEN, O. LINDY, H. SAARI & V-J. UITTO. 1992. Infect. Immun. **60:** 4491–4495.
7. UITTO, V-J., K. T. TRYGGVASON & T. SORSA. 1987. Proc. Finn. Dent. Soc. **83:** 119–130.
8. INGMAN, T., T. SORSA, K. SUOMALAINEN, S. HALINEN, O. LINDY, A. LAUHIO, H. SAARI, Y. T. KONTTINEN & L. M. GOLUB. 1993. J. Periodontol. **64:** 82–88.
9. SORSA, T., V-J. UITTO, K. SUOMALAINEN, M. VAUHKONEN & S. LINDY. 1988. J. Periodontal Res. **23:** 386–393.
10. SUOMALAINEN, K., T. SORSA, L. SAXEN, M. VAUHKONEN & V-J. UITTO. 1991. Oral Microbiol. Immunol. **6:** 6–14.
11. INGMAN, T., M. KÖNÖNEN, Y. T. KONTTINEN, H. S. SIIRILÄ, K. SUOMALAINEN & T. SORSA. 1994. J. Clin. Periodontol. **21:** 301–307.
12. UITTO, V-J., K. SUOMALAINEN & T. SORSA. 1990. J. Periodontal Res. **25:** 135–142.
13. INGMAN, T., T. SORSA, O. LINDY, H. KOSKI & Y. T. KONTTINEN. 1994. J. Clin. Periodontol. **21:** 26–31.
14. GOLUB, L. M., K. SUOMALAINEN & T. SORSA. 1992. Curr. Opin. Dent. **2:** 80–89.
15. SORSA, T., T. INGMAN, K. SUOMALAINEN, S. HALINEN, H. SAARI, Y. T. KONTTINEN, V-J. UITTO & L. M. GOLUB. 1992. J. Clin. Periodontol. **19:** 146–149.
16. LAUHIO, A., T. SORSA, O. LINDY, K. SUOMALAINEN, H. SAARI, L. M. GOLUB & Y. T. KONTTINEN. 1992. Arthritis Rheum. **36:** 195–198.
17. RAMAMURTHY, N. S., A. T. VERNILLO, R. A. GREENWALD, H-M. LEE, T. SORSA, L. M. GOLUB & B. R. RIFKIN. 1993. J. Bone Miner. Res. **8:** 1247–1253.
18. GOLUB, L. M., S. CIANCIO, N. S. RAMAMURTHY, M. LEUNG & T. F. MCNAMARA. 1990. J. Periodontal Res. **25:** 321–330.
19. SORSA, T. 1987. Scand. J. Rheumatol. **16:** 167–175.
20. WEISS, S. J., G. PEPPIN, X. ORTIZ, C. RAGSDALE & S. T. TEST. 1985. Science **227:** 747–749.
21. KONTTINEN, Y. T., O. LINDY, P. KEMPPINEN, H. SAARI, K. SUOMALAINEN, M. VAUHKONEN, S. LINDY & T. SORSA. 1991. Matrix **11:** 296–301.
22. SORSA, T., V. KNÄUPER, U. WESTERLUND, K. SUOMALAINEN, T. INGMAN, H. SAARI, O. LINDY, H. TSCHESCHE & Y. T. KONTTINEN. 1993. J. Dent. Res. **72** (Special Issue): IADR Abstr. 1517: 293.
23. SUOMALAINEN, K., T. SORSA, L. M. GOLUB, H-M. LEE, N. S. RAMAMURTHY, V-J. UITTO, H. SAARI & Y. T. KONTTINEN. 1992. Antimicrob. Agents Chemother. **36:** 227–229.
24. SUOMALAINEN, K., T. SORSA, T. INGMAN, O. LINDY & L. M. GOLUB. 1992. Oral Microbiol. Immunol. **8:** 121–123.
25. TURTO, H., S. LINDY, V-J. UITTO, O. WEGELIUS & J. UITTO. 1977. Anal. Biochem. **83:** 557–569.
26. MICHAELIS, J., M. C. M. VISSERS & C. C. WINTERBOURN. 1990. Biochem. J. **270:** 809–814.
27. BURNETTE, W. N. 1981. Anal. Biochem. **112:** 195–203.
28. KURIOKA, S. & M. MATSUDA. 1976. Anal. Biochem. **75:** 281–289.
29. SHI, S. R., M. E. KEY & K. L. KALRA. 1991. J. Histochem. Cytochem. **39:** 741–748.
30. SHIN, R-W., T. IWAKI, T. KITAMOTO & J. TATEISHI. 1991. Lab. Invest. **64:** 693–702.
31. OHTANI, H., S. NAKAMURA, Y. WATANABE, T. MIZOI, T. SAKU & H. NAGURA. 1993. Lab. Invest. **68:** 520–527.
32. GOLUB, L. M., T. SORSA, H.-M. LEE, S. CIANCIO, D. SORBI, B. GRUBER, N. S. RAMAMURTHY, T. SALO & Y. T. KONTTINEN. 1994. J. Clin. Periodontol. In press.
33. LAUHIO, A., Y. T. KONTTINEN, H. TSCHESCHE, D. NORDSTRÖM, T. SALO, J. LÄHDEVIRTA, L. M. GOLUB & T. SORSA. 1994. Antimicrob. Agents Chemother. **38:** 400–402.
34. WEISS, S. J. 1990. N. Engl. J. Med. **320:** 365–379.
35. LEE, H-M., T. SORSA, Y. DING, L. M. GOLUB, S. CIANCIO, N. S. RAMAMURTHY, Y. T. KONTTINEN & T. SALO. 1994. J. Dent. Res. **73** (Spec. Issue): IADR Abstr. 1224: 255.
36. KNÄUPER, V., H. REINKE & H. TSCHESCHE. 1990. FEBS Lett. **263:** 355–357.
37. SORSA, T., O. LINDY, Y. T. KONTTINEN, K. SUOMALAINEN, T. INGMAN, H. SAARI, S. HALINEN, H-M. LEE, L. M. GOLUB, J. HALL & S. SIMON. 1993. Antimicrob. Agents Chemother. **37:** 592–594.
38. DESCROCHES, P. E. J., J. J. JEFFREY & S. J. WEISS. 1991. J. Clin. Invest. **87:** 2258–2265.
39. SIBOO, R., W. AL-JOBURI, M. GORNISTKY & E. C. CHAN. 1989. J. Clin. Microbiol. **27:** 568–570.

40. SALO, T., J. G. LYONS, F. RAHEMTULLA, H. BIRKEDAL-HANSEN & H. LARJAVA. 1991. J. Biol. Chem. **266:** 11436–11441.
41. TONETTI, M. S., K. FREIBURGHAUS, N. P. LANG & M. BICKEL. J. Periodontal Res. **28:** 511–513.
41A. INGMAN, T., *et al.* 1994. Scand. J. Dental Res. In press.
42. DROUIN, L., C. M. OVERALL & J. SODEK. J. Periodontal Res. **23:** 370–377.
43. GABLER, W. L., J. SMITH & N. TSUKUDA. 1992. Res. Commun. Chem. Pathol. Pharmacol. **78:** 151–160.
44. STRICKLIN, G. P. & J. HOIDART. 1992. Matrix **12 (Suppl. 1):** 325–326.
45. LAUHIO, A., T. SORSA, O. LINDY, K. SUOMALAINEN, H. SAARI, L. M. GOLUB & Y. T. KONTTINEN. 1993. Arthritis Rheum. **36:** 1335–1336.

Clinical Experiences with Tetracyclines in the Treatment of Periodontal Diseases

SEBASTIAN G. CIANCIO

Departments of Periodontology and Pharmacology, and
Center for Clinical Dental Studies
State University of New York at Buffalo
Buffalo, New York 14214-3008

Periodontal disease (periodontitis) is the major cause of tooth loss over age 35. Periodontitis destroys tissues in the region where they attach to the root surface (periodontal attachment). As a result of this destruction the normal crevice around a tooth deepens and becomes a periodontal pocket. The objectives of periodontal therapy are to halt the progression of the active disease process, to reduce or eliminate the periodontal pocket, to stabilize the region of the periodontal tissue attachment, and, when possible, to increase the level of the periodontal tissue attachment. Therapy is aimed at reducing the level of etiological factors at the site of disease below the threshold capable of producing continuing breakdown, allowing repair of the affected region. In recent years it has been found that regeneration of lost periodontal structures can be enhanced. Although many variables responsible for predictable regeneration are unknown, much current research is focused on this area.

Periodontal disease can be treated on a site-specific basis by a variety of methods including surgical and nonsurgical modalities. Pharmacological therapy may play an adjunctive role in the management of at least several types of periodontitis. Numerous investigators have evaluated the use of antibiotics to halt the progression of periodontitis, with some benefit demonstrated when these medications are incorporated into the treatment protocol.[1–4] Also, adjunctive use of antibiotics may be indicated in patients who do not respond favorably to mechanical debridement procedures or for specific types of periodontitis.[5–7] Continued refinement of local delivery systems for antimicrobials may eventually allow controlled and predictable use at refractory sites.[8–9]

Available comparative longitudinal studies indicate that therapeutic approaches aimed at root surface debridement are successful in halting the progression of periodontitis in most patients.[10–11] Antibiotic therapy may be indicated for patients, or sites within patients, that are nonresponsive to root surface debridement. The establishment of other treatment objectives or presence of specific host conditions will also influence the practitioner in the selection of a nonsurgical and/or a surgical treatment plan.

Increased knowledge about the infectious nature of periodontal diseases has prompted the use of adjunctive antibiotics in the treatment of patients with periodontitis. Since a specific plaque hypothesis was proposed, there have been several studies on plaque infection. It has been shown that the anaerobic microorganisms in the complex microflora associated with advanced periodontitis comprises 90% of cultivable bacteria. In active periodontal lesions, investigators isolated higher proportions of *Campylobacter rectus, Prevotella intermedia, Bacteroides forsythus, Porphyromonas gingivalis,* and *P. intermedia,* which have been found to be more prevalent in localized active disease and *Peptostreptococcus micros* and *C. rectus* in generalized active disease. On the other hand, facultative anaerobic or capnophilic microorganisms such as *Actinobacillus actinomycetemcomitans* (*Aa*) or *Eikenella corrodens* may play an important role in the pathogenesis of localized juvenile periodontitis and aggressive adult forms of periodontitis.[12–14]

Tetracyclines have an excellent profile of activity against these bacteria. Additionally, because collagen destruction occurs in periodontal disease, reduction of host-derived collagenase activity may increase their therapeutic value beyond that of antibacterial activity.

In our early work with tetracyclines in humans, we decided to evaluate the effect of minocycline therapy on this disease. Minocycline HCl is a semisynthetic tetracycline that was first introduced in 1967. It is effective against a broad spectrum of bacteria affected by most tetracyclines and identified as periodontal pathogens. Minocycline HCl also has a longer serum half-life and a lower urinary excretion rate than most other tetracyclines which permits the use of smaller and less frequent doses. Also, it is the most lipid soluble of all commercially available tetracyclines and therefore may diffuse into more body fluids than other tetracyclines.

The objective of our first study was to determine the passage into and concentration of minocycline in gingival crevicular fluid (GCF) and the relationship between its concentration in saliva, GCF, serum, and changes in periodontal health.[15]

Twenty adults, males and females, between the ages of 22 and 55, with gingivitis and/or periodontitis were included in this study.

TABLE 1. Concentration of Minocycline in Body Fluids—Group 1[a]

	Day 1	Day 2	Day 3	Day 5	Day 8
Saliva	0.068 ± 0.017	0.104 ± 0.02	0.269 ± 0.099[b]	0.184 ± 0.047[b]	0.175 ± 0.028[b]
Serum	1.02 ± 0.10	1.82 ± 0.14	2.20 ± 0.18[b]	2.76 ± 0.20[b]	3.26 ± 0.30[b]
GCF	5.26 ± 0.95	5.95 ± 0.59	9.90 ± 1.10[b]	10.38 ± 1.93[b]	15.5 ± 2.63[b]

[a]Mean ± SEM (μg/mL). $N = 9$ for each value except GCF on day 5.
[b]Statistically significant difference from day 1, Student's *t* test.

During the 8 days of this study, 10 subjects (Group 1) received 100 mg of minocycline (Minocin) in the morning and again in the evening. The remaining 10 subjects received 50 mg in the morning and 100 mg in the evening (Group 2). They were told to take their morning medication 1 h before their appointment.

Nine subjects completed the study in Group 1 and 10 in Group 2. The dropout in Group 1 was due to a death in the family.

GCF and serum samples were obtained from each subject according to the following schedule:

Day 1: At 1, 3, 5, and 7 h after taking the medication.

Day 2: Same as day 1.

Days 3, 5, and 8: At 1, 5, and 7 h after taking the medication.

At the termination of the study, all parameters were measured. No subject was discharged unless oral health was equivalent to or better than upon entry into the study.

The results of the analysis of minocycline in serum, saliva, and gingival crevicular fluid are presented in TABLES 1 and 2. The tables show that the concentration of Minocin in saliva is far below that in serum, and the concentration in GCF is far above that in serum. Also, the antibiotic levels in serum and GCF reached bacteriostatic concentrations on day 1 and remained bacteriostatic throughout the study.

TABLE 2. Concentration of Minocycline in Body Fluids—Group 2[a]

	Day 1	Day 2	Day 3	Day 5	Day 8
Saliva	0.079 ± 0.010	0.115 ± 0.014[b]	0.139 ± 0.020[b]	0.141 ± 0.014[b]	0.166 ± 0.027[b]
Serum	1.5 ± 0.13	2.53 ± 0.26[b]	3.05 ± 0.24[b]	1.80 ± 0.17	2.02 ± 0.11[b]
GCF	3.98 ± 0.62	11.05 ± 4.32[b]	6.63 ± 0.83[b]	15.89 ± 3.12[b]	10.61 ± 2.80[b]

[a]Mean ± SEM (μg/mL). $N = 10$ with the following exceptions: Day 1, saliva (9), GCF (4); Day 2, GCF (9); Day 5, saliva (8), GCF (7); Day 8, saliva (8), GCF (8).
[b]Statistically significant difference from day 1, Student's *t* test.

The Gingival Index (GI) was evaluated on a daily basis, and the results are presented in TABLE 3. These data show that the GI scores were markedly reduced during this study.

The Plaque Index (PI) was evaluated on a daily basis, and the results were presented in TABLE 4. From these data one notes a marked reduction in the scores.

Although Group 1 and Group 2 received different dosages of minocycline, the differences for all indices between the groups were not statistically significant. All data were statistically analyzed by a *t* test of significance. No significant changes in pocket depth were noted, which was an expected finding from a short-term study.

The results of this study showed that minocycline (1) is effective against oral microorganisms based on plaque reduction scores, (2) is present in serum at therapeutically effective levels when given in doses of either 200 or 150 mg per day, (3) is concentrated in gingival crevicular fluid at levels five times as high as serum, and (4) produces an improvement in gingival health.

In a later study we compared the clinical and microbiological effects of (1) periodontal scaling and root planing; (2) periodontal scaling, root planing, and adjunctive systemic minocycline therapy; and (3) systemic minocycline therapy alone in adults with moderate to severe periodontitis. The results of treatment were evaluated at frequent intervals over a period of 70 days.[16]

Twenty-six volunteers, 35 to 65 years of age, and generally with moderate periodontal disease, participated in this study. The clinical trial was a double-blind, split-mouth design. The subjects were randomly divided into two groups of 13 members each. One group received daily doses of 200 mg of minocycline for 7 days; during this time the other group received placebo on a daily basis (both the minocycline and the placebo were provided by Lederle Laboratories). The investigators were not aware of the content of the tablets during the experimental phase of the study. Serum and gingival crevicular fluid concentrations of minocycline were determined at day 7 as described in our previous study using a modification of a method by Bennett *et al.*[17]

In each patient, two quadrants of the mouth, each randomly selected from the upper and lower jaw, received periodontal scaling and root planing (root surface debridement). The other half of the mouth was left unscaled. The periodontal

TABLE 3. Daily Gingival Index Scores[a]

	Day 1	Day 8
Group 1	1.88 ± 0.08	1.19 ± 0.13[b]
Group 2	1.76 ± 0.10	1.16 ± 0.11[b]

[a]Mean ± SEM.
[b]Statistically significant difference from day 1, Student's *t* test.

TABLE 4. Daily Plaque Index Scores[a]

	Day 1	Day 8
Group 1	1.91 ± 0.10	1.03 ± 0.12[b]
Group 2	1.70 ± 0.15	1.27 ± 0.09[b]

[a]Mean ± SEM.
[b]Statistically significant difference from day 1, Student's *t* test.

scalings and root planings were carried out by two dental hygienists. The periodontal scaling of each patient averaged 45 to 60 min. No attempt was made to instruct the patients in oral hygiene procedures, although they were provided with a soft toothbrush and dentifrice.

The following clinical measurements were carried out: (1) Löe and Silness' Gingival Index,[18] (2) Silness and Löe's Plaque Index,[19] and (3) periodontal pocket depth, assessed with a calibrated periodontal probe with markings at each millimeter except for the fourth and sixth millimeters.

Gingival crevicular fluid flow was measured at the same time each day on the mesial surfaces of the maxillary left central and mandibular right lateral incisors. Clinical measurements were obtained at days 0 (immediately before treatment), 7, 14, 35, 49, and 70. The clinical indices were obtained by the same periodontist who, at the time of examination, did not know which quadrants had been scaled.

The mean serum and gingival crevicular fluid concentrations of minocycline in those patients receiving the medication were 2.58 ± 0.32 and 8.03 ± 1.64 μg/mL, respectively, on day 7. These findings showed that the patients were compliant and that effective antimicrobial levels were obtained in gingival crevicular fluid. (Prior studies in our laboratory showed that concentrations of minocycline of 5 μg/mL inhibited growth of 98% of all periodontal isolates.)

The effect of minocycline on the total subgingival microflora is presented in TABLE 5. In all four study groups, the total subgingival cell counts were lower throughout the 70-day experimental period than at the initiation of the study. The largest decrease in cell counts was seen in the minocycline-scaled group, in which the total cell number immediately after therapy decreased approximately 25-fold and remained below half of the pretreatment counts during the 70-day study period. In the minocycline-unscaled group, the total cell count decreased 10-fold after therapy. A more rapid return in subgingival cell counts occurred in this latter group, with half of the pretreatment level being reached as early as day 49. In the two placebo groups, the total cell counts were reduced only 4-fold (scaled group) and 2-fold (unscaled group) at day 7. However, long-term suppression of the subgingival microflora was also seen with the scaled placebo group, in which, at day 70, the total cell count was still less than half of that found prior to treatment.

TABLE 5. Effect of Minocycline/Placebo Therapy on the Subgingival Total Bacterial Counts

	Days after Therapy					
Patient Group	0	7	14	35	49	70
Minocycline-scaled	49[a]	2	3	9	17	19
Minocycline-unscaled	65	7	5	15	33	43
Placebo-scaled	85	19	31	17	39	31
Placebo-unscaled	68	34	23	30	38	41

[a]Geometric means of the total cell counts per milliliter multiplied by 10^6.

The Gingival Index measurements are presented in TABLE 6. Minocycline therapy resulted in improved gingival health in both the scaled and unscaled groups; however, the greatest reduction in the gingivitis score was found when minocycline therapy was combined with scaling. The scaled and unscaled placebo groups showed little or no change in gingival inflammation during the study.

The gingival crevicular fluid data are presented in TABLE 7. In agreement with the findings for the Gingival Index, a reduction in gingival crevicular fluid flow was seen in the minocycline groups for up to 35 days after treatment, whereas no major changes were found for the placebo groups.

The findings from this study showed that the minocycline-scaled group responded most favorably, with improved gingival health for at least 49 days and with marked reductions in total bacterial counts and proportions of spirochetes for at least 70 days (termination of the study). Minocycline administration with no periodontal scaling and root planing also resulted in major, long-lasting shifts in the subgingival microflora. Scaling alone was least effective in changing the microflora. This study established the fact that mechanical debridement of periodontal pockets (scaling and root planing), in conjunction with minocycline therapy offered benefits from a microbiological viewpoint which were better than mechanical debridement alone.

Systemic versus topical administration of antibiotics in the treatment of periodontal disease is still a matter for further study. The systemic route was used in the present investigation in the thought that it might allow the antibiotic, which enters the periodontal pocket with the gingival crevicular fluid, to affect microorganisms that would be difficult to reach with topical administration. Also, an oral dose that is effective in inhibiting the growth of most periodontal bacterial species is readily achieved by systemic administration in all periodontal pockets for the entire treatment period. On the other hand, topical administration through an appropriate delivery system has the potential advantage of producing subgingival concentrations that significantly exceed those achievable by systemic administration.

Several investigators have evaluated the use of locally placed tetracycline HCl impregnated fibers of ethylene vinyl acetate copolymer (Actisite®) in the treatment of adult periodontitis. The studies have shown that the fibers can produce mean tetracycline concentrations in gingival crevicular fluid in excess of 1300 μg/mL with minimal side effects.[20–21] Furthermore, significant reductions in levels of microbes associated with active periodontal disease and improvements in pocket depth and attachment levels over that achieved by scaling and root planing alone have been noted.[22–24]

Recently we evaluated the concentration of tetracycline in gingival tissue adjacent to periodontal pockets packed with these fibers.[25] The study population consisted of 10 patients with at least two pockets in both maxillary quadrants of ≥ 5 mm in depth and exhibiting bleeding on probing. After an initial scaling and root planing, placebo or tetracycline fibers were randomly assigned by quadrant to two nonadjacent pockets. Fibers were removed at the time of surgery; that is, day 8, and periodontal surgery was performed utilizing a flap incision that allowed biopsy of one interdental papilla from each of the two test sites in each quadrant. One biopsy was analyzed for tetracycline concentrations by high-performance liquid chromatography (HPLC). The second biopsy was examined by both light and ultraviolet fluorescence microscopy to determine the location of residual tetracycline and the intensity of inflammatory cell infiltrates. Results showed that the tissue concentration of the antibiotic in tetracycline-treated sites was 64.4 ± 7.01 ng/mg (ng of tetracycline/mg tissue weight), which corresponds to 43 μg/mL of tetracycline, and was below levels of accurate measurement in placebo-treated sites. Tetracycline tissue concentrations corresponded to the ultraviolet fluorescence microscopy with a Pearson correlation

TABLE 6. Gingival Index Measurements in Periodontitis Patients Treated with Minocycline or Placebo

Patient Group	Days after Therapy					
	0	7	14	35	49	70
Minocycline-scaled	1.60 ± 0.25[a]	1.35 ± 0.21[b]	1.35 ± 0.25[b]	1.36 ± 0.22[b]	1.26 ± 0.18[b]	1.43 ± 0.27
Minocycline-unscaled	1.59 ± 0.20	1.42 ± 0.25[b]	1.43 ± 0.28[b]	1.47 ± 0.25	1.31 ± 0.23	1.34 ± 0.24[b]
Placebo-scaled	1.68 ± 0.17	1.62 ± 0.25	1.64 ± 0.28	1.53 ± 0.26	1.46 ± 0.25	1.52 ± 0.24
Placebo-unscaled	1.59 ± 0.19	1.64 ± 0.20	1.68 ± 0.24	1.56 ± 0.24	1.44 ± 0.19	1.51 ± 0.20

[a]Arithmetic mean of gingival inflammatory index ± S.D.
[b]Statistically significant $p < 0.05$, compared to both placebo groups as determined by Student's t test.

TABLE 7. Gingival Crevicular Fluid Flow after Therapy with Minocycline or Placebo in Patients with Periodontal Disease

Patient Group	Days after Therapy					
	0	7	14	35	49	70
Minocycline-scaled	0.038 ± 0.006[a]	0.016 ± 0.004[b,c]	0.019 ± 0.008[b,c]	0.028 ± 0.007[b]	0.038 ± 0.008	0.034 ± 0.007
Minocycline-unscaled	0.042 ± 0.006	0.021 ± 0.004[b,c]	0.034 ± 0.008	0.027 ± 0.007	0.032 ± 0.008	0.040 ± 0.007
Placebo-scaled	0.049 ± 0.006	0.028 ± 0.004[c]	0.038 ± 0.008	0.046 ± 0.007	0.045 ± 0.008	0.040 ± 0.007
Placebo-unscaled	0.046 ± 0.006	0.024 ± 0.003[c]	0.037 ± 0.008	0.031 ± 0.007[c]	0.051 ± 0.008	0.050 ± 0.007

[a]Arithmetic mean of gingival crevicular fluid flow (μL/min) ± S.E. of the mean.
[b]Statistically significant ($p < 0.05$) compared to the placebo groups as determined by Student's t test.
[c]Statistically significant ($p < 0.05$) compared to day 0 as determined by Student's t test.

coefficient of r = 0.92. Tetracycline fluorescence was noted in the soft tissue wall ranging from 1 to 20 μm. This study showed that these fibers could deliver levels of antibiotic into tissue adjacent to a periodontal pocket at concentrations that could kill all suspected periodontal pathogens. It is noteworthy that these tissue concentrations could also reduce host-derived collagenase activity.[26]

Although research relative to the role of tetracyclines as adjuncts to traditional treatment methods has been evaluated from a microbiological point of view, interest has now begun to focus on their roles as inhibitors of collagenase activity. Indeed, it may be that the combination of antimicrobial activity with "anticollagenase" activity is the reason why the tetracyclines are the antibiotics that have given the most consistent therapeutic response, compared to other antibiotics, for the treatment of some forms of periodontal diseases.[27–30]

REFERENCES

1. KORNMAN, K. S. & E. H. KARL. 1982. The effect of long-term low-dose tetracycline therapy on subgingival microflora in refractory adult periodontitis. J. Periodontol. **53:** 604–610.
2. HAFFAJEE, A. D., J. L. DZINK & S. S. SOCRANSKY. 1988. Effect of modified Widman flap surgery and systemic tetracycline on the subgingival microbiota of periodontal lesions. J. Clin. Periodontol. **15:** 255–262.
3. SODER, P. O., L. FRITHIOF, S. WIKNER *et al.* 1990. The effect of systemic metronidazole after non-surgical treatment in moderate and advanced periodontitis in young adults. J. Periodontol. **61:** 281–288.
4. OKUDA, K., L. WOLFF, R. OLIVER *et al.* 1992. Minocycline slow-release formulation effect on subgingival bacteria. J. Periodontol. **63:** 73–79.
5. VAN WINKELHOFF, A. J., C. J. TIJHOF & J. DEGRAFF. 1992. Microbiological and clinical results of metronidazole plus amoxicillin therapy in *Actinobacillus actinomycetemcomitans*-associated periodontitis. J. Periodontol. **63:** 52–57.
6. MAGNUSON, I., W. B. CLARK, S. B. LOW, J. MARUNIAK, R. G. MARKS & C. B. WALKER. 1989. Effect of non-surgical periodontal therapy combined with adjunctive antibiotics in subjects with "refractory" periodontal disease. J. Clin. Periodontol. **16:** 647–653.
7. KORNMAN, K. S. & P. B. ROBERTSON. 1985. Clinical and microbiological evaluation of therapy for juvenile periodontitis. J. Periodontol. **56:** 443–446.
8. ECKLES, T. A., R. A. REINHARDT, J. K. DYER, C. J. TUSSING, W. M. SZYDLOWSKI & L. M. DUBOIS. 1990. Intracrevicular application of tetracycline in white petrolatum for treatment of periodontal disease. J. Clin. Periodontol. **17:** 454–462.
9. GOODSON, J. M., S. OFFENBACHER, D. H. FARR & P. E. HOGAN. 1985. Periodontal disease treatment by local drug delivery. J. Periodontol. **56:** 265–272.
10. KALDAHL, W., K. KALKWARF & K. PATIL. 1991. Evaluation of four periodontal therapies following five years of maintenance. J. Dent. Res. **70 (Spec. issue):** 556 (Abstr. 2323).
11. KALDAHL, W. B., K. L. KALKWARF & K. D. PATIL. 1992. Evaluation of periodontally treated sites during six years of maintenance. J. Dent. Res. **71 (Spec. issue):** 739 (Abstr. 1789).
12. SLOTS, J. 1977. The predominant cultivable microflora of advanced periodontitis. Scand. J. Dent. Res. **85:** 114–121.
13. DZINK, J. L., A. C. R. TANNER, A. D. HAFFAJEE & S. S. SOCRANSKY. 1985. Gram-negative species associated with active destructive periodontal lesions. J. Clin. Periodontol. **12:** 648–653.
14. HAFFAJEE, A. D., S. S. SOCRANSKY, C. SMITH & S. DIBART. 1991. Relation of baseline microbial parameters to future periodontal attachment loss. J. Clin. Periodontol. **18:** 744–752.
15. CIANCIO, S. G., M. L. MATHER & J. A. MCMULLEN. 1980. An evaluation of minocycline in patients with periodontal disease. J. Periodontol. **51:** 530–534.
16. CIANCIO, S. G., J. SLOTS, H. S. REYNOLDS, J. J. ZAMBON & J. D. MCKENNA. 1982. The

effect of short-term administration of minocycline HCl on gingival inflammation and subgingival microflora. J. Periodontol. **53:** 557–561.

17. BENNETT, J. B., J. C. BRODIE, E. J. BENNER & W. M. M. KIRBY. 1966. Simplified accurate method for antibiotic assay of clinical specimens. Appl. Microbiol. **14:** 170–175.
18. LÖE, H. & J. SILNESS. 1959. Periodontol disease in pregnancy. I. Prevalence and severity. Acta Odontol. Scand. **30:** 533–551.
19. SILNESS, J. & H. LÖE. 1964. Periodontal disease in pregnancy. II. Correlation between oral hygiene and periodontal condition. Acta Odontol. Scand. **22:** 121–135.
20. TONETTI, M., M. A. CUGINI & J. M. GOODSON. 1990. Zero-order delivery with periodontal placement of tetracycline-loaded ethylene vinyl acetate fibers. J. Periodontal Res. **25:** 243–249.
21. HEIJL, L., G. DAHLEN, Y. SUNDIN, A. WENANDER & J. M. GOODSON. 1991. A 4-quadrant comparative study of periodontal treatment using tetracycline-containing drug delivery fibers and scaling. J. Clin. Periodontol. **18:** 111–116.
22. GOODSON, J. M., P. E. HOGAN & S. L. DUNHAM. 1985. Clinical responses following periodontal treatment by local drug delivery. J. Periodontol. **56(Suppl.):** 81–87.
23. GOODSON, J. M., M. A. DUGINI, R. L. KENT, G. C. ARMITAGE, *et al.* 1991. Multi-center evaluation of tetracycline fiber therapy: II. Clinical response. J. Periodontal Res. **26:** 371–379.
24. MORRISON, S. L., C. M. COBB, G. M. KAZAKOS & W. J. KILLOY. 1992. Root surface characteristics associated with subgingival placement of monolithic tetracycline-impregnated fibers. J. Periodontol. **63:** 137–143.
25. CIANCIO, S. G., C. M. COBB & M. LEUNG. 1992. Tissue concentration and localization of tetracycline following site-specific tetracycline fiber therapy. J. Periodontol. **63:** 849–852.
26. GOLUB, L. M., N. S. RAMAMURTHY, T. MCNAMARA, R. A. GREENWALD & B. R. RIFKIN. 1991. Tetracyclines inhibit connective tissue breakdown: New therapeutic implications for an old family of drugs. Crit. Rev. Oral Biol. Med. **2:** 297–322.
27. LINDHE, J. 1982. Treatment of localized juvenile periodontitis. *In* Host-Parasite Interactions in Periodontal Diseases. R. Genco, S. Mergenhagen, Eds.: 382. American Society for Microbiology, Washington, DC.
28. LINDHE, J. & B. LILJENBERG. 1984. Treatment of localized juvenile periodontitis. Results after 5 years. J. Clin. Periodontol. **11:** 399–410.
29. KORNMAN, K. S. & P. B. ROBERTSON. 1985. Clinical and microbiological evaluation of therapy for juvenile periodontitis. J. Periodontol. **56:** 443–446.
30. MANDELL, R. L., L. S. TRIPODI, E. SAVITT, J. M. GOODSON & S. S. SOCRANSKY. 1986. The effect of treatment on *Actinobacillus actinomycetemcomitans* in localized juvenile periodontitis. J. Periodontol. **57:** 94–99.

Doxycycline and Chemically Modified Tetracyclines Inhibit Gelatinase A (MMP-2) Gene Expression in Human Skin Keratinocytes

VELI-JUKKA UITTO,[a,b] JAMES D. FIRTH,[a] LESLIE NIP,[a] AND LORNE M. GOLUB[c]

[a]*Department of Oral Biology*
University of British Columbia
2199 Wesbrook Mall
Vancouver, B.C., V6T 1Z3, Canada

[c]*Department of Oral Biology and Pathology*
State University of New York at Stony Brook
Stony Brook, New York 11794

Tetracyclines have been used both systemically and locally in the treatment of various infections caused by gram-negative bacteria. During recent years it has been established that tetracyclines exert biological functions entirely independent of their antimicrobial property.[1,2] Therefore, tetracyclines have been found beneficial in the treatment of conditions that do not have a microbial etiology such as epidermolysis bullosa, rosacea, α-1-antitrypsin-deficiency panniculitis, pyoderma gangrenosum, and other ulcerative or bullous skin diseases.[3–7] It has been proposed that the anti-inflammatory mechanisms are due in part to the ability of tetracyclines to inhibit leukocyte proliferation and activity, and to scavenge hypochlorous acid and superoxide radicals produced by phagocytes.[8–10] Furthermore, several investigations involving both *in vitro* and *in vivo* animal studies have shown that tetracycline antibiotics and their chemically modified analogues with no antimicrobial activity can inhibit mammalian collagenase activity and collagen breakdown.[1,2,11–14] Tetracyclines, therefore, may have direct suppressive effects on tissue destruction during inflammatory diseases such as dermatitis, rheumatoid and osteoarthritis, and periodontitis. Little is known about the effects of tetracyclines on epithelial cells. In a recent study, we examined the effects of tetracyclines and non-antimicrobial analogues of tetracycline on mammalian epithelial cells. Cultures of porcine cell rests of Malassez, the nonkeratinized epithelial cells of periodontal ligament, served as the model system for our investigations of tetracyclines on cell viability, cell morphology, and matrix metalloproteinase activity. We found that doxycycline and chemically modified tetracyclines CMT-1 and CMT-6 had direct inhibitory effects on both 92-kDa (MMP-9) and 72-kDa (MMP-2) gelatinases.[15] In this study we examine the gelatinase gene expression of the epithelial cells treated with tetracyclines.

[b]Corresponding author.

MATERIALS AND METHODS

Tetracyclines

Doxycycline, 4-de-dimethylaminotetracycline (CMT-1), and 6-deoxy 5-hydroxy 4-de-dimethylaminotetracycline (CMT-8) were obtained from Drs. Lorne Golub and Thomas McNamara, State University of New York at Stony Brook, New York.

Cell Culture

Oral epithelial cells derived from porcine rests of Malassez were isolated from periodontal ligament and cultured by the method described by Brunette *et al.*[16] These cells exhibit a homogeneous nonkeratinizing epithelial cell population with a cytokeratin pattern (K 4, 5, 6, 14, 16, 19) typical of basal cell-like, undifferentiated, hyperproliferative characteristics. The cells of 5–8 passages were cultured in alpha minimum essential medium supplemented with antibiotics and 15% fetal calf serum (Flow Laboratories, McLean, VA). Human foreskin epithelial cells (NHEK), obtained from Clonetics, San Diego, CA, were cultured in serum free keratinocyte medium (KGM) containing 25 μg/mL of pituitary extracts and 0.03 mM $CaCl_2$. All cultures were grown on plastic dishes to 70–80% confluency until used in the assays with tetracyclines. Each plate contained about 2×10^6 cells, as counted by a Coulter counter.

Cytotoxicity Assay

The periodontal ligament epithelial cells were cultured on a 96-well plate to near confluency. Conversion of tetrazolium salt into blue formazan by viable cells was determined using the "Cell titer 96" test kit of Promega Corporation (Madison, WI), which is based on a modification of the method of Mosmann.[17]

Gelatinase Assay

Enzymography using Laemmli's discontinous buffer system[18] was performed using a 7.5% polyacrylaminde gel that had been cast with 1.0 mg/mL 2-methoxy-2,4-dipheny-3(2H)-furanone-labeled gelatin according to O'Grady *et al.*[19] After each run, gels were washed at room temperature for 30 min sequentially with two buffers: (1) 50 mM Tris, 2.5% Triton X-100, 0.02% NaN_3, pH 7.5 and (2) 50 mM Tris, 2.5% Triton X-100, 0.02% NaN_3, 5 mM $CaCl_2$, 1 μM $ZnCl_2$, pH 7.5. The gels were incubated in the latter buffer without Triton X-100 for up to 24 h at 37 °C. The gelatinolysis was monitored by long-wave ultraviolet (UV) light. Zones of enzymatic activity were subsequently visualized by negative Coomassie brilliant blue staining. Photographs of the gels were scanned by an image-digitized optical scanner (Silverscan, LaCie, Beaverton, OR) and analyzed by a computer software (Image 1.4, NIH).

Total RNA Extraction

Cellular RNA was prepared by a guanidinium thiocyanate (GT) extraction method,[20] modified as described below. Cultures in triplicate at different time points were washed three times in cold PBS before the cells were lysed by 3 mL of guanidinium thiocyanate buffer (GT buffer, containing 4 M guanidinium thiocyanate, 0.025 M $Na_3Citrate(H_2O)_2$, pH 7.0, 0.5% (w/v) sarcosyl, 70 mM β-mercaptoethanol, and 5 mM vanadyl-ribonucleoside complex (VRC, BRL Inc.) as a RNAse inhibitor. Cell lysates were immediately transferred into a polypropylene centrifuge tube. After brief vigorous vortexing, 0.3 mL 2 M NaAc, pH 4.1, 3 mL of RNAse-free water-saturated phenol and 0.6 mL of chloroform-isoamylalcohol were added separately, with vortexing between each addition. After incubation on ice for 15 min and centrifugation at 12,000 *g* for 20 min at 4 °C, RNA was collected from the upper aqueous phase, and the aqueous phase was precipitated overnight at −20 °C in 1 volume of ice-cold isopropanol. The precipitated RNA pellet was rinsed with 80% cold ethanol, vacuum dried, and dissolved in 100 μL RNase-free water. The total RNA yields were determined for each sample by spectroscopic analysis of one-tenth of the final sample volume and calculated per 10^6 cells.

Northern Hybridization

Aliquots of extracted cell RNA (5 μg) were prepared in a loading solution containing 2.2 M formaldehyde and ethidium bromide (40 μg/mL), incubated at 65 °C for 15 min, chilled on ice, then fractionated on 1.2% (w/v) agarose gels containing final concentrations of 2.2 M formaldehyde and 20 mM 3-N-morpholinolpropanesulfonic acid (MOPS, pH 7.0), and transferred onto Hybond-N nylon membrane (0.45 μm pore size, Amersham Corp.) using a Posiblot Pressure Blotter (Stratagene, CA).[21] After 3 min of UV cross-linkage, the blots were prehybridized at 68 °C for 2 h in 5% (w/v) SDS, 50 mM PIPES, 0.1 M sodium chloride, 50 mM Na_2HPO_4, 50 mM NaH_2PO_4, 1 mM ethylenediaminetetraacetic acid tetrasodium salt (EDTA), pH 7.0. Hybridization was performed for 18 h at 68 °C with [^{32}P]dCTP-labeled (MMP-2) cDNA probe in 5 mL of hybridization solution.[22] The full-length human 72-kD gelatinase cDNA probe was kindly provided by Drs. Pirkko Huhtala and Karl Tryggvason, Department of Biochemistry, University of Oulu, Finland, and labeled by random priming with [^{32}P]dCTP ($>$3,000 Ci/mmol, Amersham Corp.) to a specific activity of $\sim 1.8 \times 10^9$ cpm/μg cDNA. After hybridization, the blots were washed in 1 × SSC (standard saline citrate), 5% (w/v) SDS at room temperature for 10 min, followed by a change of the same solution at 68 °C for 25 min. Blots were autoradiographed at −70 °C using double emulsion Cronex 4 X-ray film with two thulium Quanta Detail intensifier screens (Dupont). The films were then scanned and analyzed as described above for zymographs. The data were normalized for the amount of RNA loaded as determined from photographs of the ethidium bromide-stained bands.

Scanning Electron Microscopy

After culture, cells were fixed with 2.5% glutaraldehyde, dehydrated, and processed by the tannic-acid technique of Katsumoto *et al.*[23] Cells were fixed with 2.5% glutaraldehyde in 0.1 M phosphate buffered saline, pH 7.4 for 30 min at room temperature and postfixed with 1% osmium tetroxide in the same buffer for 20 min at room temperature. The specimens were then treated with 2% tannic acid and 1% OsO_4, respectively, for 20 min. The specimens were dehydrated through a graded series of ethanol solutions, then dried using the critical-point drying (CPD) method with dry ice; the samples were mounted with silver dag, DC-sputtered with platinum and examined with a Cambridge Stereoscan microscope (Cambridge, U.K.)

RESULTS

Studies with periodontal ligament epithelial cells showed that doxycycline and the chemically modified tetracycline (CMT-1) at concentrations from 5–250 μg/mL were not toxic up to 24 h of culture (FIG. 1). At 48 and 72 h, however, viability of the cells treated with 100 μg/mL of the drugs was reduced in cultures with both 0.2 and 0.05 mM calcium (FIG. 2).

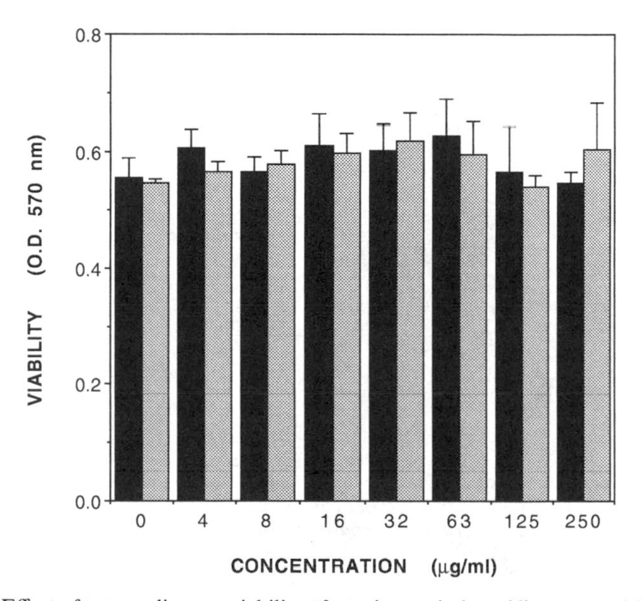

FIGURE 1. Effect of tetracyclines on viability of porcine periodontal ligament epithelial cells *in vitro.* Cells were cultured to subconfluency in α-MEM plus 15% serum. The medium was changed to keratinocyte growth medium (KGM, 0.2 mM Ca^{2+}) and cultures were continued for 24 h with doxycycline (*solid bars*) or 4-de-dimethylaminotetracycline (CMT-1; *stippled bars*). Cell viability was measured with the tetrazolium/formazan assay. Values are mean ± S.D. of triplicate samples.

Activity of gelatinolytic proteases of the periodontal ligament epithelial cells' culture medium was measured by the enzymography technique. These cells secreted both a 72-kDa (gelatinase A, MMP-2) and 92-kDa (gelatinase B, MMP-9) gelatinase into the medium. When the gelatin containing gels were incubated in the presence of different doxycycline concentrations, about 50 μg/mL of the drug was required to inhibit both gelatinase activities by 50%. However, when the cells were cultured for 18 h in the presence of doxycycline, concentrations about ten times smaller resulted in the same inhibition of the gelatinase activity.

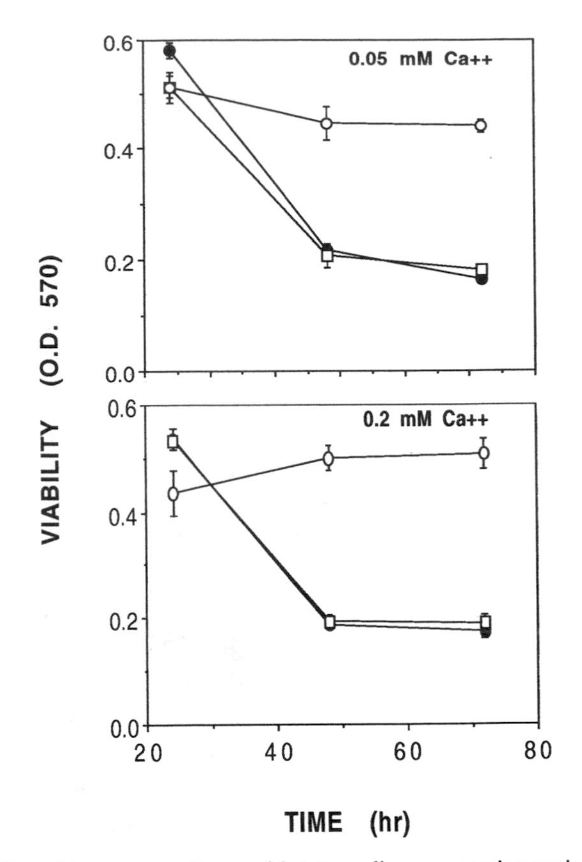

FIGURE 2. Effect of long-term cultures with tetracyclines on porcine periodontal ligament epithelial cell viability. Either 100 μg/mL of doxycycline (*closed circles*) or 4-de-dimethylamino-tetracycline (*squares*) were added to subconfluent epithelial cell cultures in KGM medium containing either 0.05 or 0.2 mM Ca^{2+}. Cells without drugs (*open circles*) were cultured as controls. After 24, 48 or 72 h the viability was measured with the tetrazolium/formazan assay. Values are mean ± S.D. of triplicate samples.

In order to study if the transcription of the gelatinases is affected by tetracyclines, human skin keratinocytes were cultured in serum-free keratinocyte growth medium (KGM) to semiconfluency. Thereafter, the cultures were continued in the presence of doxycycline, or modified tetracyclines (CMT-1 or CMT-8). The medium was collected for gelatin enzymography, and the cells were processed for RNA analysis as described in MATERIALS AND METHODS. Under these conditions MMP-2 was the major gelatinase detected in the culture medium. Culture with 50 μg/mL of doxycycline, CMT-1, and CMT-8 for 24 h inhibited the gelatinase activity by about 74, 63 and 71%, respectively. Culture for 6 h at 50 μg/mL or at 5 μg/mL for 24 h inhibited the activity by 20 to 52%. Northern hybridization using a human MMP-2 cDNA probe of the cells exposed to 5 μg/mL of doxycycline for 6 h showed that the gelatinase mRNA levels were about 25% of the control (FIG. 3 and TABLE 1). Interestingly, after 24-h exposure at 5 μg/mL of doxycycline, the mRNA levels were higher, that is, about 37% of control. In cells cultured for 24 h with 50 μg/mL of doxycycline the mRNA levels were 11% of control. Similar inhibition of the MMP-2 expression was observed by CMT-1, the effect being strongest in 24-h cultures. CMT-8 exerted somewhat less dramatic inhibition than doxycycline. Cells cultured for 24 h in the presence of 50 μg/mL of doxycycline, CMT-1, and CMT-8 had their total RNA content reduced by 39, 33, and 33%, respectively. The lower drug concentration did not have any marked effects on the total RNA yield (TABLE 1).

Under scanning electron microscope the control keratinocytes were flat and well attached to each other. The cells incubated for 24 h with 5 μg of doxycycline, CMT-1 or CMT-8 appeared to have lost their cell-cell contacts. The cell-substrate adhesion seemed to be less affected (FIG. 4B, E, and G). Similar results were observed in 6 h cultures at a 50 μg/mL concentration. The effect was accentuated in cultures for 24 h with 50 μg/mL of the drugs. Although doxycycline caused rounding-up of most keratinocytes, many of the cells cultured with chemically modified tetracyclines appeared spindle-shaped and disorganized (FIG. 4C, D, F, and H).

DISCUSSION

Several inflammatory diseases of skin and other tissues involve activation and increased proliferation of epithelial cells. It is well documented that activated epithelial cells may behave aggressively and secrete several proteolytic enzymes. High-collagenase activities have been detected in epithelium during wound healing,[24,25] in gingival inflammation,[26,27] and in proliferating mucosal keratinocytes in

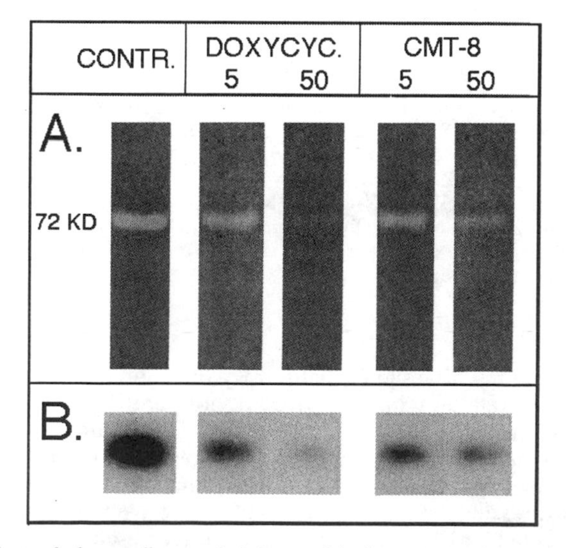

FIGURE 3. Effect of doxycycline and 6-deoxy 5-hydroxy 4-de-dimethylaminotetracycline (CMT-8) on gelatinase A (MMP-2) gene expression in human skin keratinocytes. The drugs were added at 5 or 50 μg/mL to subconfluent keratinocyte cultures in KGM medium for 24 h. Medium was assayed for gelatinase activity with gelatin zymography technique (**A**) and total RNA extracted from the cells was analyzed by Northern hybridization using [^{32}P]dCTP-labeled human MMP-2 cDNA probe (**B**).

culture.[28] In addition, a recent study indicates that migrating periodontal epithelial cells produce collagenase and gelatinases capable of connective tissue destruction.[29] Epithelial invasion into connective tissue has been shown to coincide with expression of metalloproteinases during tissue development[30] and cancer growth.[31,32] In addition to collagenase, epithelial cells may produce, under some conditions, gelatinase A (MMP-2) and gelatinase B (MMP-9).[33] These proteinases have been found to degrade denatured interstitial collagens and native type IV, V, and VII collagens. In addition, stromelysin-2 (MMP-10) has been found to be produced by epithelial cells.[34] These cells, therefore, have the potential to degrade basement membranes as well as connective tissue matrix. However, most of the information on matrix metalloproteinases has been obtained from cell culture studies. The expression of the different matrix degrading enzymes by epithelial cells *in vivo* and their specific roles in different conditions can only be speculated on at the present time.

TABLE 1. Content of Total RNA and Gelatinase (MMP-2) mRNA in Human Skin Keratinocytes Cells[a]

	Total RNA/10⁶ Cells µg (% of control)		MMP-2 mRNA (O.D., % of control)		MMP-2 Activity in Medium (O.D., % of control)	
	6 h	24 h	6 h	24 h	6 h	24 h
Control	60 ± 3 (100)	62 ± 6 (100)	100 ± 12	100 ± 13	100	100
Doxycycline, 5 µg/mL	59 ± 9 (98)	59 ± 4 (96)	25 ± 11[b]	37 ± 8[b]	83	80
Doxycycline, 50 µg/mL	53 ± 2 (88)	37 ± 3[b] (61)	20 ± 7[b]	11 ± 1[b]	48	26
CMT-1, 5 µg/mL	62 ± 1 (102)	62 ± 2 (102)	61 ± 10[b]	27 ± 2[b]	116	77
CMT-1, 50 µg/mL	51 ± 4 (85)	41 ± 3[b] (67)	51 ± 8[b]	8 ± 2[b]	61	37
CMT-8, 5 µg/mL	54 ± 3 (89)	52 ± 11 (84)	42 ± 5[b]	56 ± 8	109	69
CMT-8, 50 µg/mL	58 ± 5 (97)	41 ± 10 (67)	38 ± 6[b]	32 ± 7[b]	84	29

[a]Cells cultured for 6 or 24 h in the presence of 5 or 50 µg/mL of doxycycline or chemically modified tetracyclines (CMT-1 and CMT-8). After culture, the total cellular RNA was extracted and 5-µg aliquots were analyzed by Northern hybridization with [^{32}P]dCTP-labeled MMP-2 cDNA probe. MMP-2 signal levels were quantified by computer analysis of the digitized optical scans of the exposed X-ray film. Medium was analyzed for gelatinase (MMP-2) activity by computer analysis of the digitized optical scans of the 72-kDa band of the gelatin substrate gel enzymographs. Values are means ± S.D. of three samples, except for medium gelatinase activity, which is mean of two samples.

[b]Significant difference from control (Student's *t* test, $p < 0.05$).

Many earlier studies have shown that tetracyclines and their chemically modified forms can inhibit metalloproteinases derived from several cell types. In those studies the inhibition was primarily attributed to binding of Zn^{2+} and Ca^{2+} that is required for the activity of the proteases. In this study we compared the effects of doxycycline, its chemical modification, CMT-8, and a similarily modified tetracycline, CMT-1, on the gelatinase expression in human skin keratinocytes. We found that the amount of gelatinase A secreted into the culture medium was decreased in 24 h cultures with all three tetracyclines. Further, doxycycline drastically decreased the gelatinase A gene expression within 6 h even at a low drug concentration. The chemically modified tetracyclines also significantly inhibited the gelatinase gene expression, but the effect appeared later in culture. This is possibly due to the relatively slow transport of the

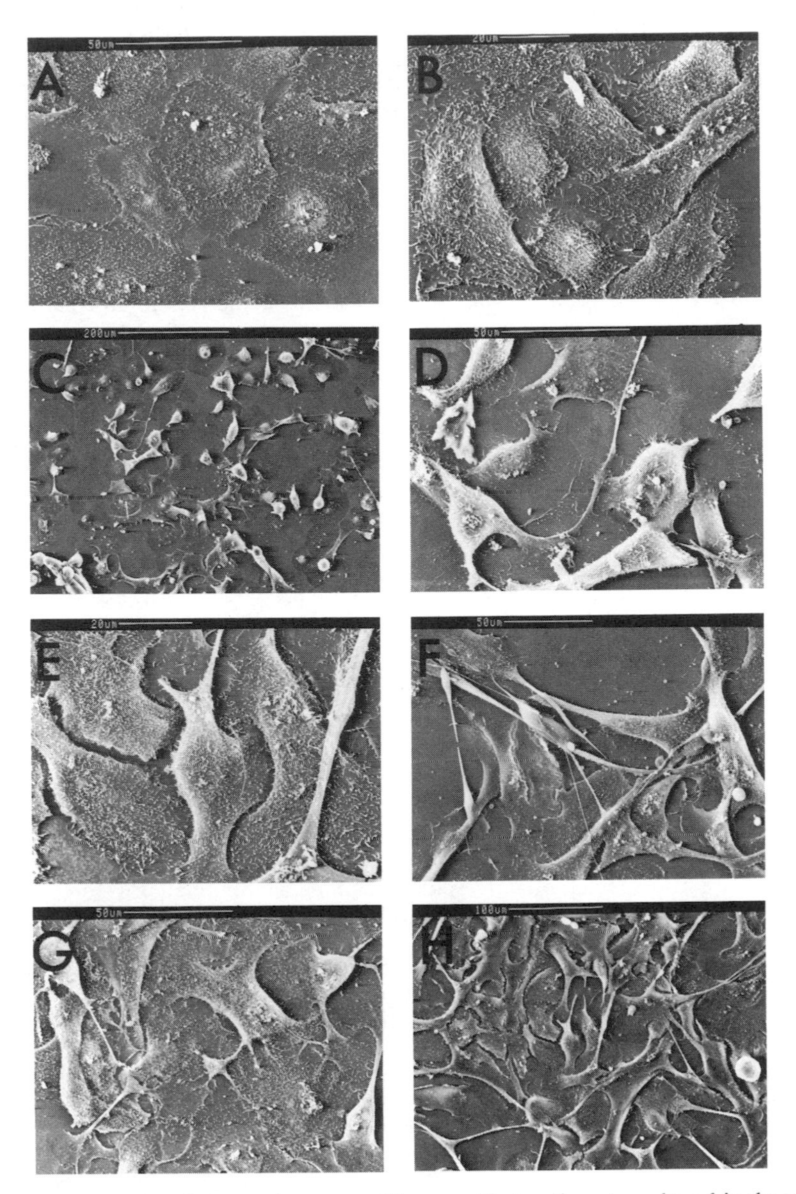

FIGURE 4. Scanning electron microscopy of human skin keratinocytes cultured in the presence of doxycycline or chemically modified tetracyclines (CMT). Subconfluent epithelial cell cultures in KGM medium were cultured for 24 h in the presence of the drugs. (**A**) control; (**B**) doxycycline, 5 μg/mL; (**C** and **D**) doxycycline, 50 μg/mL; (**E**) CMT-1, 5 μg/mL; (**F**) CMT-1, 50 μg/mL; (**G**) CMT-8, 5 μg/mL; (**H**) CMT-8, 50 μg/mL.

chemically modified tetracyclines into the cells. Cultured neutrophils and red blood cells have been shown to concentrate doxycycline intracellularly more effectively than CMT-1.[35] The antibacterial action of tetracyclines results from its binding to the 30S subunit of prokaryote ribosomes and inhibition of binding of aminoacyl tRNAs. These drugs have not been previously reported to act on the transcription or translation of eukaryotic cells. The mechanism of the inhibition of the gelatinase gene expression in epithelial cells remains to be explored. Even though 4-de-dimethylamino doxycycline (CMT-8) appeared to be a somewhat less effective inhibitor of the gelatinase gene expression than doxycycline, both the non-antimicrobial CMT-1 and CMT-8 clearly reduced the gelatinase mRNA levels. This indicates that the mechanism of epithelial cell inhibition is different from that which inhibits the bacterial protein synthesis. It is possible that binding of cations by tetracyclines may result in derangement of some transcriptional and posttranscriptional steps. For instance RNA polymerases require Mg^{2+} or Mn^{2+} for their activity. It may also be that translational inhibition of some factors necessary for matrix metalloproteinase transcription may result in gene down-regulation.

Other possible mechanisms may involve binding of pericellular divalent cations and lead indirectly to altered gene expression. Electron microscopy of the cultured epithelial cells showed that the cell-cell contacts were dissociated even at 5 μg/mL of the drugs. This may be due to binding of Ca^{2+} or Mg^{2+} that is needed for proper function of cell adhesion molecules.[36] Cell contact with matrix and other cells has been found to be an important factor in the regulation of the protease genes. In general, loss of cell contacts and cell rounding have been found to be associated with increased matrix metalloproteinase gene expression.[37,38] In 24 h cultures with tetracyclines, the cells were markedly more spherical than after 6 h. This may explain the fact that the gelatinase expression in cells treated with 5 μg/mL of doxycycline or CMT-8 for 24 h was not lower than at 6 h. The enzyme induction associated with the cell-shape change possibly competed with the inhibitory effect of the tetracyclines. Nevertheless, the longer tetracycline treatment did result in clearly more diminished gelatinase secretion than the short treatment. Even though the amount of the secreted gelatinase levels was generally related to the levels of the gene expression, the relationship was not always clear, particularly in the case of CMT-1.

It is possible that, besides gene expression, other aspects of protein synthesis and secretion are also affected by tetracyclines. One possibility related to the protease inhibitory action of tetracyclines was provided by a recent study revealing that singlet oxygen induces the expression of collagenase in human fibroblasts.[39] Sodium azide, like tetracyclines, is a potent quencher of $^1O_2$ and was found to reverse the collagenase gene induction. At a high concentration of tetracyclines (50 μg/mL) the total RNA was also decreased in the epithelial cells. Even at high concentrations the tetracyclines did not show significant toxic effects on the epithelial cells within 24 h of culture, when measured with the formazan/tetrazolium assay. However, in the 2- and 3-day cultures at 100 μg/mL of the drugs, the cell viability clearly decreased. This suggests that long-term use of locally delivered tetracyclines resulting in high tissue concentrations[40] may be toxic to human cells. At this point it is not known whether the inhibition of gene expression is specific for the gelatinase. In fact, some studies have shown that tetracyclines at pharmacological doses increase skin collagen synthesis and RNA levels in diabetic animals.[41] Moreover, our preliminary studies indicate that total protein synthesis in the cultured epithelial cells is not decreased by tetracyclines. Studies on expression of genes encoding various structural and secreted proteins will provide answers to this question. Gelatinase A is found to be constitutively produced in culture by many cell types, for example, epithelial cells, but it is only moderately inducible by growth factors. Gelatinase B has been found to be much more responsive to induction and is believed to play an important role in disease conditions involving epithelial cell activation. It is important, therefore, to determine whether gelatinase B induction can also be inhibited by tetracyclines.

In conclusion, doxycycline and chemically modified tetracyclines inhibit epithelial cell matrix metalloproteinases. This may result partly from inhibition of the protease gene expression and partially from binding of the cations essential for the enzyme activity. These results provide support for the view that tetracyclines, delivered either systemically or locally, can be valuable in treatment of conditions involving epithelial-mediated tissue destruction, for example, cancer, epidermolysis bullosa, corneal ulcers, and some forms of dermatitis, and periodontitis.

SUMMARY

The mechanism of tetracycline-induced inhibition of matrix metalloproteinases (MMP) was studied by measuring the MMP secretion and MMP-2 mRNA levels in unkeratinizing periodontal ligament epithelial cells and skin keratinocytes cultured in the presence of doxycycline or chemically modified tetracyclines (CMT) lacking antimicrobial activity. Doxycycline, CMT-1, and CMT-8 exerted a direct dose-dependent inhibition of porcine periodontal ligament epithelial cell medium MMP activity as assayed by gelatin enzymography. Both the 92-kDa (MMP-9) and 72-kDa (MMP-2) gelatinases were inhibited by the tetracyclines added to the conditioned medium. Culturing the cells in the presence of the tetracyclines required considerably smaller concentrations to reduce the secreted MMP activity. The drugs were not toxic to the epithelial cells at concentrations from 4 to 250 μg/mL up to 24 h of culture. Tetracycline effects on the MMP-2 mRNA levels were studied in human skin keratinocytes using Northern hybridization analysis with a specific cDNA probe. A marked inhibition in the MMP-2 gene expression was observed by 6 h with 5 μg/mL of doxycycline, CMT-1 or CMT-8. Doxycycline inhibition was somewhat stronger than the two other tetracyclines. After 24 h of culture with 50 μg/mL of the drugs, the total RNA levels also decreased by 33 to 40%. The 72-kDa gelatinase activity in culture medium of the keratinocytes followed roughly the pattern of inhibition of the gene expression. We conclude that doxycycline and the chemically modified tetracyclines, in addition to inhibiting the MMP activity may also reduce the enzyme expression at the transcriptional level.

REFERENCES

1. GOLUB, L. M., N. S. RAMAMURTHY & T. F. MCNAMARA. 1991. Tetracyclines inhibit connective tissue breakdown: New therapeutic implications for an old family of drugs. Crit. Rev. Oral Biol. Med. **2:** 297–322.
2. GOLUB, L. M., K. SUOMALAINEN & T. SORSA. 1992. Host modulation with tetracyclines and their chemically modified analogues. Curr. Opin. Dent. **2:** 80–90.
3. LYNCH, W. S. & W. F. BERGFELD. 1978. Pyoderma gangrenosum responsive to minocycline hydrochloride. CUTIS **21:** 535–538.
4. PLEWIG, G. & F. SCHOPF. 1975. Anti-inflammatory effects of antimicrobial agents: An in vivo study. J. Invest. Dermatol. **65:** 532–537.
5. WHITE, J. E. 1989. Minocycline for dystrophic epidermolysis bullosa. Lancet **1:** 966.
6. HUMBERT, P., R. A. RENAUD, R. LAURENT & P. AGACHE. 1989. Tetracyclines for dystrophic epidermolysis bullosa. Lancet **1:** 277.
7. HUMBERT, P., B. FAIVRE, R. GIBEY & P. AGACHE. 1991. Use of anti-collagenase properties

of doxycycline in treatment of alpha1-antitrypsin deficiency panniculitis. Acta Dermato-Venerol. **71:** 189–194.
8. PREUS, H., T. TOLLEFSEN & B. MÖRLAND. 1987. Effect of tetracycline on human monocyte phagocytosis and lymphocyte proliferation. Acta Odontol. Scand. **45:** 297–302.
9. WASIL, M., B. HALLIWELL & C. P. MOOREHOUSE. 1989. Scavenging of hypochlorous acid by tetracycline, rifampicin and some other antibiotics: A possible antioxidant action of rifampicin and tetracycline? Biochem. Pharmacol. **37:** 775–778.
10. VAN BAAR, H. M. J., P. C. M. VAN DE KERKHOF, P. D. MIER & R. HOPPLE. 1987. Tetracyclines are potent scavengers of superoxide radicals. Br. J. Dermatol. **117:** 131–134.
11. GOLUB, L. M., H. M. LEE, G. LEHRER, A. NEMIROFF, T. F. MCNAMARA, R. KAPLAN & N. S. RAMAMURTHY. 1983. Minocycline reduces gingival collagenolytic activity during diabetes: Preliminary observations and a proposed new mechanism of action. J. Periodontal Res. **18:** 516–526.
12. GOLUB, L. M., T. F. MCNAMARA, G. D'ANGELO, R. A. GREENWALD & N. S. RAMAMURTHY. 1987. A non-antibacterial chemically-modified tetracycline inhibits mammalian collagenase activity. J. Dent. Res. **66:** 1310–1314.
13. GOLUB, L. M., N. S. RAMAMURTHY, T. F. MCNAMARA, R. A. GREENWALD & B. R. RIFKIN. 1991. Tetracyclines inhibit connective tissue breakdown: New therapeutic implications for an old family of drugs. Crit. Rev. Oral Biol. Med. **2:** 297–322.
14. SUOMALAINEN, K., T. SORSA, L. M. GOLUB, N. RAMAMURTHY, H.-M. LEE, V.-J. UITTO, H. SAARI & Y. T. KONTTINEN. 1992. Specificity of the anticollagenase action of tetracyclines: Relevance to their anti-inflammatory potential. Antimicrob. Agents Chemother. **36:** 227–229.
15. NIP, L. H., V.-J. UITTO & L. M. GOLUB. 1993. Inhibition of epithelial cell matrix metalloproteinases by tetracyclines. J. Periodontal Res. **28:** 379–385.
16. BRUNETTE, D. M., A. H. MELCHER & H. K. MOE. 1976. Culture and origin of epithelial-like and fibroblast-like cells from porcine periodontal ligament explant and cell suspensions. Arch. Oral Biol. **2:** 393–400.
17. MOSMANN, T. 1983. Rapid colorimetric assay for cellular growth and survival: Application to proliferation and cytotoxicity assays. J. Immunol. Methods **65:** 55–63.
18. LAEMMLI, U. K. 1970. Cleavage of structural proteins during the assembly of the head of bacteriophage T4. Nature (Lond.) **227:** 680–685.
19. O'GRADY, R. L., A. NETHERY & N. HUNTER. 1984. A fluorescent screening assay for collagenase using collagen labelled with 2-methoxy-2, 4-dipheny-3(2H)-furanone. Anal. Biochem. **140:** 490–494.
20. CHOMCZYNSKI, P. & N. SACCHI. 1987. Single-step method of RNA isolation by acid guanidinium thiocyanatephenol-chloroform extraction. Anal. Biochem. **162:** 156–159.
21. KHANDJIAN, E. W. 1987. Optimized hybridization of DNA blotted and fixed to nitrocellulose and nylon membranes. Bio Tech. **5:** 165–167.
22. THOMAS, P. S. 1980. Hybridization of denatured RNA and small DNA fragments transferred to nitrocellulose. Proc. Natl. Acad. Sci. USA **69:** 1408–1412.
23. KATSUMOTO, T., T. NAGURO, A. IINO & A. TAKAGI. 1984. The effect of tannic acid on the preservation of tissue culture cells for scanning electron microscopy. J. Electron Microsc. **30:** 177–182.
24. GRILLO, H. & J. GROSS. 1967. Collagenolytic activity during mammalian wound repair. Dev. Biol. **15:** 300–317.
25. DONOFF, R. B., J. E. MCLENNAN & H. C. GRILLO. 1969. Preparation and properties of collagenases from epithelium and mesenchyme of healing mammalian wounds. Biochim. Biophys. Acta **227:** 639–653.
26. UITTO, V-J., R. APPELGREN & P. J. ROBINSON. 1981. Collagenase activity in extracts of inflamed human gingiva. J. Periodontal Res. **16:** 417–424.
27. OVERALL, C. M., O. W. WIEBKIN & J. C. THONARD. 1987. Demonstration of tissue collagenase activity in vivo and its relationship to inflammation severity in human gingiva. J. Periodontal Res. **22:** 81–88.
28. LIN, H-Y, B. R. WELLS, R. E. TAYLOR & H. BIRKEDAL-HANSEN. 1987. Degradation of type I collagen by rat mucosal keratinocytes. J. Biol. Chem. **14:** 6823–6831.

29. SALONEN, J., V.-J. UITTO, Y-M. PAN & D. ODA. 1991. Proliferating oral epithelial cells in culture are capable of both extracellular and intracellular degradation of interstitial collagen. Matrix **11:** 43–55.
30. WERB, Z., C. M. ALEXANDER & R. R. ADLER. 1992. Expression of matrix metalloproteinases in development. *In* Matrix Metalloproteinases and Inhibitors. H. Birkedal-Hansen, Z. Werb, H. G. Welgus & H. E. Van Wart, Eds.: 337–343. Matrix (Spec. Suppl. No. 1.) Gustav Fisher. Stuttgart.
31. DRESDEN, M. H., S. A. HEILMAN & J. D. SMIDT. 1972. Collagenolytic enzymes in human neoplasms. Cancer Res. **32:** 993–996.
32. TRYGGVASON, K., M. HÖYHTYÄ & T. SALO. 1987. Proteolytic degradation of extracellular matrix in tumor invasion. Biochim. Bioph. Acta **907:** 191–217.
33. BIRKEDAL-HANSEN, H., W. G. I. MOORE, M. K. BODDEN, L. J. WINDSOR, B. BIRKEDAL-HANSEN, A. DECARLO & J. A. ENGLER. 1993. Matrix metalloproteinases: A review. Crit. Rev. Oral Biol. Med. **4:** 197–250.
34. L. J. WINDSOR, H. GRENETT, M. K. BODDEN, B. BIRKEDAL-HANSEN, J. A. ENGLER & H. BIRKEDAL-HANSEN. 1993. Cell type-specific regulation of SL-1 and SL-2 genes. J. Biol. Chem. **268:** 17341–17347.
35. GABLER, W. L., J. SMITH & N. TSUKUDA. 1992. Comparison of doxycycline and a chemically modified tetracycline inhibition of leukocyte functions. Res. Commun. Chem. Pathol. Pharmacol. **78:** 151–159.
36. ALBELDA, S. M. & C. A. BUCK. 1990. Integrins and other cell adhesion molecules. FASEB J. **4:** 2868–2880.
37. AGGELER, J., S. M. FRISH & Z. WERB. 1984. Changes in cell shape correlate with collagenase gene expression. J. Cell Biol. **98:** 1662–1671.
38. HONG, H. L. & D. M. BRUNETTE. 1987. Effect of cell shape on proteinase secretion by epithelial cells. J. Cell Sci. **87:** 259–267.
39. SCHARFETTER, M., M. WLASCHEK, K. BRIVIBA & H. SIES. 1993. Singlet oxygen induces collagenase expression in human skin fibroblasts. FEBS Lett. **33:** 304–306.
40. CIANCIO, S. G., C. M. COBB & M. LEUNG. 1992. Tissue concentration and localization of tetracycline following site-specific tetracycline fibre therapy. J. Periodontol. **63:** 849–853.
41. SCHNEIR, M., N. RAMAMURTHY & L. GOLUB. 1990. Minocycline-treatment of diabetic rats normalizes skin collagen production and mass: Possible causative mechanisms. Matrix **10:** 112–123.

Collagenolytic Enzymes in Gingival Crevicular Fluid as Diagnostic Indicators of Periodontitis

C. A. G. McCULLOCH

Faculty of Dentistry
University of Toronto
124 Edward Street, Room 430
Toronto, Ontario M5G 1G6, Canada

IS THERE A NEED FOR DIAGNOSTIC TESTS OF PERIODONTAL DISEASES?

Periodontal diseases are heterogeneous and include a variety of infections and inflammatory lesions. Notably, periodontitis is a prevalent disease of man that features degradation of connective tissues and that may result in tooth loss. Differential diagnosis of the currently accepted subtypes of periodontitis poses no major difficulty assuming the use of an accurate and complete history and examination (TABLE 1). The presenting features of prepubertal, juvenile, rapidly progressive, adult, and refractory periodontitis, as well as the forms of periodontitis associated with a variety of systemic diseases are sufficiently striking to permit an accurate diagnosis based on traditional clinical criteria alone.[1] Indeed, diagnostic data obtained by a thoughtful history and by the performance of a disciplined clinical examination are usually much more powerful than anything that can be currently gleaned from the diagnostic laboratory.

Yet there is a need for diagnostic information that goes beyond the traditional examination. The alternating chronic and episodic nature of periodontitis, the wide range of disease severities affecting different teeth within the same subject, and the requirement for repeated treatments over a long period of time indicate that diagnostic information extending beyond the traditional classification of disease would be very useful for improved clinical management (TABLE 2). Currently, there is considerable interest in the development, testing, and commercial application of novel diagnostic tests. Experimental studies of biologic onset in gingivitis,[2] longitudinal studies of the natural history of periodontitis,[3] as well as the retrospective analysis of treated cases,[4] initiated a series of investigations that, collectively, led to significant changes in our understanding of the periodontal diseases. First, gingival inflammation without bone loss (i.e., gingivitis) does not necessarily proceed to destructive periodontitis; second, assuming rigorous longitudinal monitoring of the level of the connective tissue around individual teeth, early diagnosis may closely follow biologic onset in some cases; third, only a small fraction of the population exhibits progressive periodontitis; fourth, lack of clinical intervention may not necessarily lead to progressive disease; fifth, because of measurement errors, it is very difficult to assess with any degree of certainty the natural history or clinical course of the periodontal diseases. As indicated in TABLE 2, appropriately applied diagnostic tests could significantly improve the clinical management of patients with periodontal diseases by putting diagnosis on a more prospective footing and by

TABLE 1. Types of Periodontitis[a]

• Prepubertal
• Localized or generalized juvenile
• Rapidly progressive
• Adult
• Associated with systemic disease
• Necrotizing ulcerative
• Refractory

[a]The differential diagnosis of these subtypes of peridontitis (as defined by clinical appearance) can be made with relative ease on the basis of a good history and a careful clinical and radiographic examination. However, physical examination provides little insight into either the natural history or the clinical course of these infections.

reducing the reliance on retrospective, physical measurements of periodontal connective tissue attachment that are currently used (FIG. 1).

RATIONALE FOR GINGIVAL CREVICULAR FLUID SAMPLING

If efficacious diagnostic tests for periodontitis are to be developed, it is important to be able to sample molecules that may reflect either pathogenic mechanisms or those that are elements of the host response. In this context, the sampling of gingival crevicular fluid exhibits several useful features, and, indeed, the diagnostic potential of gingival crevicular fluid (GCF) has been widely recognized.[1,5,6] For practical

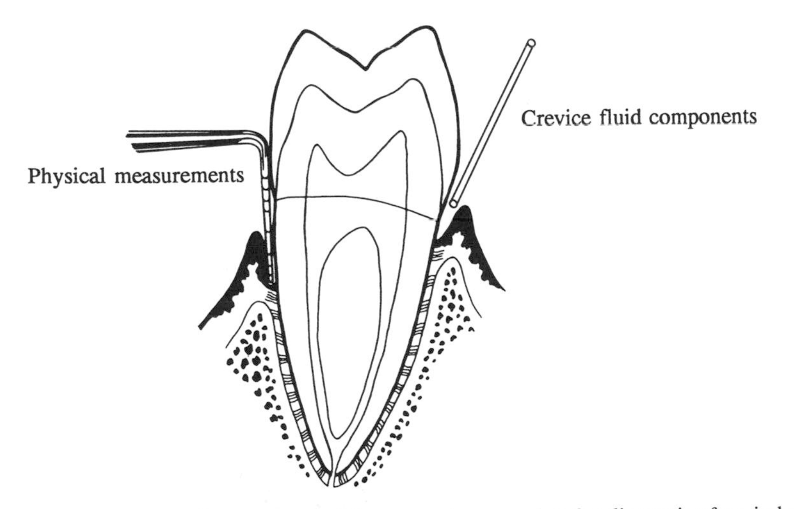

FIGURE 1. Diagram to illustrate two of the current approaches for diagnosis of periodontal diseases. Physical methods such as the periodontal probe rely on measurement of lost gingival connective tissue and bone to assess the cumulative destruction of tissue. Although direct, these methods are retrospective and can only detect disease after it has occurred. In contrast, measurement of molecules in gingival crevicular fluid (e.g., host enzymes such as collagenase) may provide insight into risk of disease and indicate response to therapy (e.g., inhibitors of matrix metalloproteinases).

TABLE 2. Uses of Diagnostic Tests

- Detect presence of disease and identify patients at risk before biologic onset
- Evaluate severity of disease
- Predict subsequent clinical course and prognosis
- Reduce time between biologic onset and usual clinical diagnosis
- Estimate responsiveness to treatment before additional therapy
- Assess actual response to treatment after completion

purposes, gingival crevicular fluid is an inflammatory exudate from the gingival microcirculation that crosses inflamed periodontal tissues and *en route* collects molecules of potential interest from the local inflammatory reaction.[7] In terms of application to diagnostic tests, sampling of crevicular fluid provides advantages that are analogous to the drawing of blood by the physician (TABLE 3). However, the accurate and reproducible sampling of fluid is not trivial; a number of studies have demonstrated that the choice of collection device,[8] the length of collection time,[9] the number of sample repetitions,[10,11] and the calculation of data as absolute measures or as flow rates[12] contribute in a large measure to overall variability. Due to these variables and to expected fluctuations based on the operator-sensitive use of collection devices (e.g., depth of insertion of filter-paper strip or capillary tube), potential depletion of sample by prolonged collection,[13] potential contamination by serum components,[14] and loss of sample from the collection device,[15] no consensus exists on which sampling protocols exhibit the lowest bias, the highest reproducibility, and the strongest validity. Perhaps of even greater consequence to the question of diagnostic validity are the unknown dynamics of the binding, metabolism, and redistribution of fluid components that occur during their traverse of the periodontal tissues and their fate upon entry into the crevice. For example the kinetics of neutrophil collagenase activation by microbial enzymes in the crevice may play a critical role in the accurate assessment of this host enzyme activity.[16] Currently there is no definitively superior method to overcome these basic and intrinsic problems with collection and analysis of crevicular fluid.

HOST ENZYMES IN GINGIVAL CREVICULAR FLUID

Inflammation is associated with the release of a wide variety of hydrolytic enzymes from stromal cells, epithelial cells, and cells of the hemopoietic lineage in periodontal tissues. Many of these enzymes can be detected in GCF (TABLE 4). Enzymes synthesized and secreted in the tissues are carried by GCF to the crevice where they are augmented by release of enzymes from bacteria and host cells present in the pocket. For any particular enzyme, it is not possible presently to discriminate

TABLE 3. Crevicular Fluid Sampling

Advantages	*Uncertainties*
Ease of access	Type of collection device
Atraumatic	Length of collection time
Rapid equilibration with whole pool	Number of repetitions
Can vary time and site	Absolute or relative values
Repeated sampling possible	Contamination by serum components
	Loss of sample over time

between intracrevicular and intratissue release of enzymes. This central problem intrinsically limits our ability to determine whether GCF enzyme data directly assess inflammatory mechanisms *within* the tissue. However, of equal consequence is the importance of distinguishing host from bacterial origins. Many enzymes (e.g., hyaluronidase, gelatinase, collagenase) are produced by both host and bacterial cells, and indeed, as described above, some bacterial proteases may activate host enzymes (see ref. 16). Consequently laboratory-based or chairside assays for the measurement of a specific enzyme should be able to measure specifically the host enzyme with high selectivity. Further, if an important synergistic relationship exists between bacterial proteases and host enzymes, then some consideration should be given to the simultaneous measurement of both host and bacterial enzymes.

TABLE 4. Partial List of Host Enzymes That Are Detectable in Gingival Crevicular Fluid

Alkaline phosphatase
Arylsulfatase
Aspartate aminotransferase
Beta glucuronidase
Cathepsin-like proteases
Elastase
Hyaluronidase
Lactate dehydrogenase
Lysozyme
Matrix metalloproteinases (collagenases, gelatinases)
Myeloperoxidase

MATRIX METALLOPROTEINASES IN PERIODONTAL DISEASES

These enzymes are a family comprising several different gene products that are synthesized by a number of cell types in the periodontal tissues.[17] Unlike bacterial enzymes that make multiple cuts along the collagen molecule, mammalian collagenases initiate degradation by making a single cut. Definitive demonstration of host versus bacterial origins can be provided by assays that discriminate the single versus multiple cuts,[18,19] and this feature is likely to be important for development and validation of collagenolytic enzymes as diagnostic tests. Subsequent degradation of the denatured collagen molecule can be mediated (with somewhat less specificity) by the fibroblast-type gelatinase (MMP-2; 72 kDa) or the neutrophil gelatinase (MMP-9; 92 kDa), and multiple forms of gelatinases have been detected in GCF;[20] however, the diagnostic potential of these different forms is uncertain.

In periodontitis, the collagenase detected in GCF is derived largely from emigrating polymorphonuclear leukocytes.[18,19] The human neutrophil collagenase (MMP-8) is not as well characterized as the fibroblast enzyme (MMP-1) but appears to be a separate gene product.[21] Although multiple forms of the neutrophil collagenase have been identified,[21,22] MMP-8 has different substrate kinetics,[23] different sequence specificity,[24] and is also antigenically distinct[25] from MMP-1. Neutrophil collagenase is released in a latent form from secretory granules,[26] and both the latent and active forms are detectable in the GCF.[27] Activation of the proenzyme may occur by a variety of mechanisms that include catalytic cleavage and conformational change,[28,29] but the regulation of these processes is not well understood *in situ*[30] and it is not possible to discriminate between intratissue and intracrevicular activation. Some preliminary data have suggested that tetracyclines can be used to distinguish between MMP-1 and MMP-8[31] by differential inhibition *in vitro* but this supposition has not been rigorously tested by independent laboratories.

COLLAGENOLYTIC ENZYMES AS DIAGNOSTIC TESTS

The rationale for the measurement of collagenolytic activity is that collagens are the primary structural proteins of the periodontium and that their degradation is a central feature of progressive lesions. Early studies[32,33] found collagenase activity in GCF and Golub and co-workers[33] suggested that ". . . collagenase activity in gingival fluids could reflect active tissue destruction and be of diagnostic value." Currently, there is no definitive proof that collagenases are central to the destruction of the periodontal connective tissues.

Previous studies have demonstrated that using SDS-PAGE fluorography and radiolabeled substrate, GCF collagenolytic activity can be measured with considerable sensitivity[34] and with reasonable reproducibility.[35] The concentration of total enzyme has been positively associated with the volume of exudate[33,36] and with the amount of cumulative destruction.[11,37] However, the study by Villela and colleagues[11] failed to demonstrate convincing associations between collagenase activity and attachment loss, in part because of the large error terms. Because total (i.e., latent plus active) enzyme activity was measured and efforts were not made to discriminate between active and latent activities, it is conceivable that diagnostic information was lost. Other studies have strongly implicated the role of the active enzyme in progressive lesions in contrast with the elevation of latent enzyme in tissues with inflammation alone.[27,38,39] Further, after pharmacologic reduction of infection (e.g., antibiotics) or after mechanical debridement, the rate of connective tissue destruction is reduced and the collagenolytic activity decreases.[27,40] Collectively, these reports indicate that *active* collagenase may be more related to tissue destruction than simple measurement of the total enzyme activity. However, as suggested by Birkedal-Hansen,[17] measurement of the active enzyme may be confounded by activation artifacts *in vitro,* particularly when assay times are prolonged. The detection of active enzyme may also reflect inclusion of neutrophils in the GCF sample that release enzyme during the assay. Further, it is unclear how active enzyme can be detected in GCF when inhibitors such as α-2-macroglobulin are present at such high concentrations. Recently, Gustafsson and co-workers[41] showed that in contrast to healthy sites, GCF from sites of destruction contains reduced inhibitor concentrations, suggesting that during active lesions, the inhibitor is degraded and the protease-inhibitor complex is quickly eliminated. These results provide some insight into why active collagenase can be detected at sites of active destruction even when inhibitors may be present.

GCF gelatinase activity has been measured by conventional micropipette sampling[19] and by a whole mouth rinse method.[40,42] A human longitudinal study has demonstrated good reproducibility on replicate measures, and during episodes of attachment loss of 2 mm or greater, whole mouth gelatinase activity was doubled.[40] Individuals at risk of developing attachment loss exhibited gelatinase levels ~40% higher than individuals with stable periodontal lesions. When whole mouth gelatinase was used as a screening test for recurrent periodontitis and the threshold was set at ~71,000 units, sensitivity was 0.79 and specificity was 0.88. These data indicate that the use of a screening test prior to the application of site-specific assays could overcome the prevalence problem (and hence low positive predictive values) that are seen with single tests.[13]

To assess directly the relationship between connective tissue degradation and the concentration of active and latent MMP-8 expression in the inflammatory exudates of human periodontal lesions, a study was conducted[43] which utilized three diagnostically well-defined subject groups, each of which exhibited either (1) subjects with inflammation and progressive loss of bone and connective tissue detachment; (2) subjects with inflammation and bone loss but without progressive connective tissue detachment; or (3) subjects with inflammation but without connective tissue detachment from the tooth and without bone loss. Examination of these groups permitted discrimination between the enzymes found in lesions with only inflammation and in lesions with both inflammation and net connective tissue destruction. Pooled data of active collagenase activity were respectively five- and sixfold higher in the subjects with progressive loss of connective tissue compared to subjects with either inflamed tissues alone or subjects with inflammation and previous bone loss. In contrast, latent collagenase activity was increased up to twofold in subjects with inflammation but no bone loss compared to subjects with progressive lesions. The ratio of active to total collagenase activity was 50% higher in subjects with progressive disease. In all subjects, replicate measurements of active collagenase obtained one month apart demonstrated wide variation ($r < 0.50$). However, only in subjects with progressive lesions was there continuing increase of active collagenase with time ($\sim$900 units per day). Enzyme levels peaked sharply before loss of connective tissue. At the time of detection of connective tissue attachment loss, a 40% increase of active collagenase activity was found in subjects with progressive loss of connective tissue compared to prebreakdown sampling times. These data provide strong *in vivo* evidence for a direct role of active collagenase in pathological connective tissue destruction and also indicate the potential diagnostic efficacy of this test. Further, this study emphasizes the contention that individuals with recurrent periodontitis would be good candidates to enter into studies of efficacy of diagnostic tests of periodontitis inasmuch as they in particular have a good deal to gain by early diagnosis. A commercially available test kit based on neutral protease activity in GCF against a collagen substrate has been developed,[44] but independent testing of the kit in a ligature-induced monkey model failed to provide any positive test results.[45] The kit did, however, generate a positive test result after incubation with bacterial collagenase.

SCREENING ASSAYS

Collagenolytic activity has been measured by SDS-PAGE fluorography of radiolabeled substrate degradation,[27] by enzymography,[42] or by incubation with radiolabeled fibrils and subsequent measurement of soluble collagen radioactivity by scintillation counting.[11] However, these assays are not particularly applicable to the analysis of large numbers of samples and are costly, labor intensive, and time consuming. Development of fluorogenic assays based on peptides that mimic the cleavage site of neutrophil collagenase and that incorporate a fluorescent leaving group would facilitate measurements of gelatinase in site-specific or whole mouth rinse samples. Appropriate use of such an assay as a screening test for the presence of disease in the whole mouth would facilitate periodontal diagnosis.[40] For example, only subjects with above-threshold values would be examined by further, site-specific tests (e.g., elastase, β-glucuronidase, collagenase). The strategy of a battery of tests could help improve the positive predictive value of single tests by increasing the pretest prevalence.

We developed a fluorogenic assay based on a methoxycoumarin (MC) containing septapeptide analog of the collagen cleavage site developed by Knight and co-workers.[46] This compound, Mca-Pro-Leu-Gly-Leu-Dpa-Ala-Arg-NH_2, may potentially provide a more efficient alternative to electrophoretic analysis of collagen degradation fragments. Enzymatic cleavage between the Mca containing Pro residue and the quenching amino acid analog Dpa causes a measurable increase in fluorescence intensity (FI; FIG. 2). Human crevicular fluid obtained by mouth rinse, conditioned media from human gingival and periodontal ligament fibroblast cultures, and neutrophil supernatant were assayed by fluorogenic substrate cleavage and by gelatin enzymograms. Samples separated on 12% polyacrylamide gels containing 400 μg/mL gelatin were incubated in buffer, stained, and quantified by optical scanning and densitometry software. Cleavage of the fluorogenic substrate by the samples was followed at timed intervals for up to 48 h and compared to a standard curve of MC. The rate of liberation of cleaved fluorescent substrate was constant for up to 72 h whereas FI was 2 to 10 times background, corresponding to 1–5% of the substrate.[47] The rate of fluorescence increase versus the amount of gelatin digested in the enzymogram bands was linearly related ($r = 0.95$) by the equation: Gelatinase activity (ng h^{-1}) = 1192 × FI (nM h^{-1}). The substrate reported activity due to both 72-kDa and 92-kDa gelatinase, interstitial collagenase, and pancreatic elastase, and detected total collagenolytic activity in crevicular fluid samples obtained by mouth rinse (45,000-fold dilution). Elastase activity was blocked by addition of 1 mM PMSF, whereas the other collagenolytic enzymes were insensitive to it. These data indicate that the fluorogenic assay reliably parallels data obtained by electrophoretic methods, indicating its potential usefulness in a screening program for periodontitis.

To test the diagnostic accuracy of the fluorogenic peptide in a clinical study, patients were grouped into recurrent periodontitis (R), stable periodontitis (S), or untreated disease (U) groups.[48] Complete clinical measurements were obtained to ensure that patients fit their respective disease groups and 5 mL mouth rinse crevicular fluid samples were taken. PMSF (1 mM) and NaN_3 (3 mM) were added to the samples to prevent degradation of the enzymes of interest by proteolytic enzymes or bacterial action, and to eliminate any cleavage due to elastase. Samples were incubated in assay buffer containing the fluorogenic substrate at 1 μM and FI was measured at timed intervals. The fluorescence was quantified from a MC standard curve using a Hitachi F-2000 spectrofluorimeter calibrated against an ovalene crystal, and the rate of the reaction was expressed in nM of MC equivalents per hour. The samples examined so far indicate that the cleavage rates were highest in the U group (0.77 ± 0.09 nM h^{-1}, $n = 22$) and lowest in the S group (0.13 ± 0.02 nM h^{-1}, $n = 16$). The R (0.35 ± 0.28 nM h^{-1} $n = 30$) group exhibited a wide range overall, but there was good individual consistency between visits ($r = 0.78$). These preliminary results indicate that regular monitoring with the fluorogenic substrate assay may be a useful early indicator of changes in periodontal disease status.

Mca-Pro-Leu-Gly↕Leu-Dpa-Aka-Arg-NH_2

FIGURE 2. Diagram to illustrate fluorogenic septapeptide that mimics the collagen cleavage site as devised by Knight *et al.*[46] Cleavage of the Gly-Leu bond (*arrow*) results in increased fluorescence of the Mca leaving group when excited at 328 nm and emission collected at 393 nm. For gelatinase, k_{cat}/k_m ($M^{-1} \cdot s^{-1}$) = 629,000. Samples of gingival crevicular fluid containing collagenase can be measured with this system as described in the text.

TABLE 5. Constraints of Collagenolytic Enzymes as Diagnostic Indicators

• Binding, metabolism, and redistribution
• Activation mechanisms, potential for artifact
• Inhibitors
• Other enzymes in sample fluid
• Cellular source
• Drainage area size—influence of pocket depth

ASSESSING EFFICACY OF COLLAGENOLYTIC ENZYMES AS DIAGNOSTIC TESTS

Available data from collagenolytic assays of GCF indicate promise for improved diagnosis of periodontitis. The evidence indicates that collagenolytic enzymes can be readily measured in GCF and will likely provide a good reflection of the global host response in the periodontium. Yet several hurdles need to be overcome before these enzymes are likely to be very effective at the chairside (TABLE 5). First, what is the most appropriate choice for sampling of GCF? Site specific sampling by micropipettes or filter-paper strips are both useful but there is no commonly accepted method for depth of insertion, time of sampling or the most appropriate expression and computation of enzyme activity (i.e., total amount versus concentration). More work is needed to resolve these basic issues of sampling techniques and data management. Second, what does collagenase activity in GCF actually measure—net collagen destruction, collagen turnover, or simply level of inflammation (that is, PMN count)? Third, how can the problems of enzyme activation and assay artifact be overcome? What is the source of the rather large variations on replicate samplings (see refs. 11 and 35)? Fourth, if studies are to be conducted on matrix metalloproteinase inhibitors as treatments for periodontitis, then the large sample-to-sample variation and patient-to-patient variation must be addressed. For example, as shown in FIGURE 3 which presents data from an earlier, randomized placebo-controlled trial of doxycycline,[35] healthy sites exhibit very low levels of collagenase activity compared to sites with advanced destruction. After pharmacological treatment with doxycycline, there was no statistically significant difference between the drug and placebo groups at the deep sites. This lack of ability to detect a difference may reflect in part the significant error terms that are present in the assay procedures.

Several criteria should be examined carefully before any single test is embraced for clinical use (TABLE 6; ref. 49). First, is there a commonly accepted "gold standard" with which to compare the test results? Although longitudinal measurement of attachment loss has been used for the last decade as the best available standard against which to compare other tests, is this in fact the standard we should be using? Second, if a test is to be of clinical use, it should be able to perform accurately over a wide range of clinical conditions and of disease severity. If, for example a test can only differentiate gingivitis from periodontitis, its usefulness would seem to be limited. Third, can the test be used in an office environment and perform as reproducibly and as accurately as in a clinical research center? Fourth, is the test able to discriminate between patients who are and who are not responding well to treatment? This would likely be the most important feature for clinicians.

To improve the potential diagnostic accuracy of any test, particularly when prevalence is low, the most obvious strategy would be to use multiple tests, preferably in a battery. For example, the first step in the diagnostic procedure might be to screen patients for disease, possibly with an assay such as use of the fluorogenic

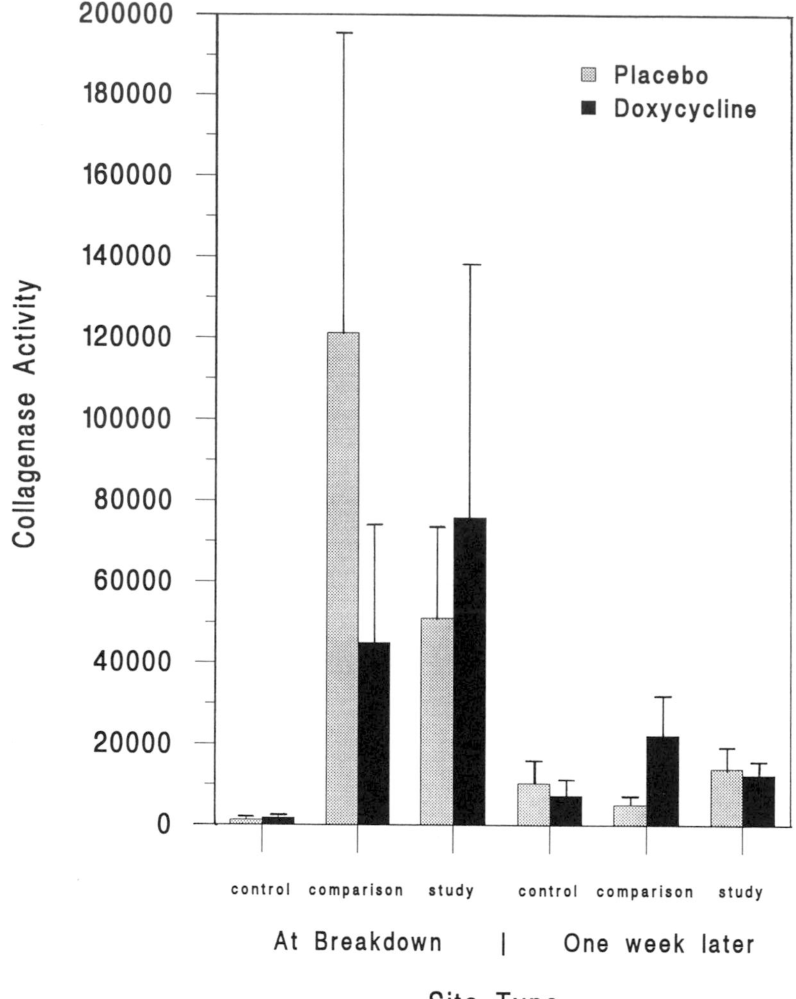

FIGURE 3. Measurement of mean collagenase activity in gingival crevicular fluid in patients with recurrent periodontitis entered in a randomized placebo-controlled trial of doxycycline (Lee *et al.*[35]). Samples were obtained by micropipette sampling, and collagenase activity was measured by a soluble, radioactive collagen assay and SDS-PAGE fluorography. Active collagenase was measured at the time of detection of lost periodontal attachment and one week later. Patients with progressive periodontitis were randomly assigned to either placebo or doxycycline (100 mg per day for 3 weeks), and both groups received deep scaling of the teeth. Samples from teeth with pockets < 3 mm and no inflammation (control) or from teeth with pockets > 5 mm and with inflammation and periodontal destruction were measured at one week after the administration of either doxycycline or placebo. Although there were large reductions of the mean collagenase activities after doxycycline treatment, equally large reductions were found after treatment with the placebo. At the time of breakdown, collagenase activities were markedly lower in control sites compared to the sites with deep pockets. Data are displayed as mean and standard error of the mean.

peptide assay described above. It would be particularly useful to measure all sites simultaneously with a whole-mouth sampling technique described earlier.[42] Subsequently, if an above-threshold screening test result were obtained, then other, independent tests of specific sites could be performed. By computing the likelihood ratios for each test (LI = sensitivity/1 − specificity), it is possible to calculate the combined likelihood of disease after use of two or three tests serially. If a positive test result is obtained for the first and second tests, then the joint likelihood ratio is the product of the separate ratios, assuming that the tests are reasonably independent. Employment of this strategy would help address the problem of low prevalence (~4–8%) that is commonly seen in the general population.

The ultimate criterion for the use of a diagnostic test is whether the patient is any better off as a result of the application of the test.[49] Early diagnosis is theoretically an advantage, but only if therapies are available that are less costly, more effective, and less painful if applied early. Further, the clinician must be able to appropriately interpret the findings of the test and provide reasonable advice on treatment. If early diagnosis can be provided well in advance of the usual clinical diagnosis, then the outcome for the patient may well be improved. In the final analysis, the usefulness of collagenolytic enzymes as diagnostic indicators will need to be examined in randomized controlled trials in which the question is asked: Are patients better off as a result of testing?

TABLE 6. Criteria for Evaluating the Clinical Usefulness of a Diagnostic Test[a]

1. Was comparison made with "gold standard" of diagnosis?
2. Were evaluations performed with a wide spectrum of disease?
3. Was the clinical setting of test appropriate?
4. Was prevalence artificially high in study (i.e., filter effect)?
5. Was reproducibility evaluated?
6. Was observer variation measured?
7. Was test appropriately responsive to changes effected by therapy?
8. Was evaluation of test in a battery of tests described?
9. Can test be reasonably conducted in an office setting?
10. Has usefulness of test been examined?
11. Was prognostic value evaluated?

[a]Adapted from Sackett *et al.*[49]

SUMMARY

Periodontal diseases are associated with the production of several families of enzymes that are detectable in gingival crevicular fluid and that are released by stromal, epithelial or inflammatory cells. Measurement of collagen degrading enzymes in crevicular fluid could contribute to insights into pathogenesis of periodontal diseases and also provide a rational basis for the development of novel diagnostic tests. However, similar to the development of other diagnostic tests, the appropriate validation of collagenolytic enzymes as diagnostic indicators is dependent on clear-cut demonstrations of the identity of the enzyme in the assay, and the reproducibility, diagnostic accuracy, and clinical utility of the test. If collagenolytic enzymes are to be of clinical usefulness, they should be easily measured over a broad range of disease severities and in varied clinical settings. Ideally, the diagnostic test should assay for an essential component of proposed pathogenic mechanisms. Neutrophil collagenase and gelatinase are promising enzymes for diagnostic tests because of (1) their apparently central role in periodontal attachment loss and disease progression; (2) demonstrations of positive associations between enzyme levels and attachment loss and inflammation; and (3) availability of sensitive and specific assays to quantify these enzymes. However, much less data exist on reproducibility, diagnostic accuracy, and clinical use in longitudinal studies. In the future, more emphasis must be placed on the importance of appropriate study design for establishing the efficacy of collagenolytic enzymes as diagnostic tests.

REFERENCES

1. PAGE, R. C. 1992. Host response tests for diagnosing periodontal diseases. J. Periodontol. **63:** 356–366.
2. LOE, H., E. THEILADE & S. B. JENSEN. 1965. Experimental gingivitis in man. J. Periodontol. **35:** 177–178.
3. GOODSON, J., A. C. R. TANNER, A. D. HAFFAJEE, G. C. SORNBERGER & S. S. 1982. Patterns of progression and regression of advanced destructive periodontal disease. J. Clin. Periodontol. **9:** 472–481.
4. HIRSCHFELD, L. & B. WASSERMAN. 1978. A long-term survey of tooth loss in 600 treated periodontal patients. J. Periodontol. **49:** 225–237.
5. JOHNSON, N. W. 1991. Crevicular fluid-based diagnostic tests. Curr. Opin. Dent. **1:** 52–65.
6. LAMSTER, I. B. 1992. The host response in gingival crevicular fluid: Potential applications in periodontitis clinical trials. J. Periodontol. **63:** 1117–1123.
7. CIMASONI, G. 1983. *In* Crevicular Fluid Updated. Monographs in Oral Science. G. Cimasoni, Ed. Vol. 12: 24–28. Karger. Basel.
8. EGELBERG, J. & R. ATTSTROM. 1973. Comparison between orifice and intracrevicular methods of sampling gingival crevicular fluid. J. Periodontal Res. **8:** 384–388.
9. PERSSON, G. R., T. A. DEROUEN & R. C. PAGE. 1990. Relationship between gingival crevicular fluid levels of aspartate aminotransferase and active tissue destruction in treated chronic periodontitis patients. J. Periodontal Res. **25:** 81–87.
10. BINDER, T. A., J. M. GOODSON & S. S. SOCRANSKY. 1987. Gingival fluid levels of acid and alkaline phosphatase. J. Periodontal Res. **22:** 14–19.
11. VILLELA, B., R. B. COGEN, A. A. BARTOLUCCI & H. BIRKEDAL-HANSEN. 1987. Crevicular fluid collagenase activity in healthy, gingivitis, chronic adult periodontitis and localized juvenile periodontitis patients. J. Periodontal Res. **22:** 209–211.
12. LAMSTER, I. B., R. I. VOGEL, L. J. HARTLEY, C. A. DEGEORGE & J. M. GORDON. 1985. Lactate dehydrogenase. β-glucuronidase and arylsulfatase in gingival crevicular fluid associated with experimental gingivitis in man. J. Periodontol. **56:** 139–146.
13. PERSSON, G. R. & R. C. PAGE. 1990. Effect of sampling time and repetition on gingival crevicular fluid and aspartate aminotransferase activity. J. Periodontal Res. **25:** 235–242.
14. LAMSTER, I. B., R. D. MANDELL & J. M. GORDON. 1985. Lactate hydrogenase activity in gingival crevicular fluid collected with filter paper strips: Analysis in subjects with non-inflamed and mildly inflamed gingiva. J. Clin. Periodontol. **12:** 153–161.
15. ROSSOMANDO, E. F., J. E. KENNEDY & J. HADJIMICHAEL. 1990. Tumour necrosis factor alpha in gingival crevicular fluid as a possible indicator of periodontal disease in humans. Arch. Oral Biol. **35:** 431–434.
16. SORSA, T., T. INGMAN, K. SUOMALAINEN, M. HAAPASALO, Y. T. KONTTINEN, O. LINDY, H. SAARI, V.-J. UITTO. 1992. Identification of proteases from periodontopathic bacteria as activators of latent human neutrophil and fibroblast-type interstitial collagenases. Infect. Immun. **60:** 4491–4495.
17. BIRKEDAL-HANSEN, H. 1993. Role of matrix metalloproteinases in human periodontal diseases. J. Periodontol. **64:** 474–484.
18. SORSA, T., V.-J. UITTO, K. SUOMALAINEN, M. VAUHKONEN & S. LINDY. 1988. Comparison of interstitial collagenases from human gingiva, sulcular fluid and polymorphonuclear leukocytes. J. Periodontal Res. **23:** 386–393.
19. OVERALL, C. M., J. SODEK, C. A. G. MCCULLOCH & P. BIREK. 1992. Evidence for

polymorphonuclear leukocyte collagenase and 92-kilodalton gelatinase in gingival crevicular fluid. Infect. Immun. **59:** 4687–4692.

20. INGMAN, T., T. SORSA, O. LINDY, H. KOSKI & Y. T. KONTTINEN. 1994. Multiple forms of gelatinases/type IV collagenases in saliva and gingival crevicular fluid of periodontitis patients. J. Clin. Periodontol. **21:** 26–31.
21. VAN WART, H. 1992. Human neutrophil collagenase. Matrix (Suppl. No. 1): 31–36.
22. MURPHY, G., J. J. REYNOLDS, U. BRETZ & M. BAGGIOLINI. 1982. Partial purification of collagenase and gelatinase from human polymorphonuclear leukocytes. Analyses of their actions on soluble and insoluble collagens. Biochem. J. **203:** 209–221.
23. HASTY, K. A., J. J. JEFFREY, M. S. HIBBS & H. G. WELGUS. 1987. The collagen substrate specificity of human neutrophil collagenase. J. Biol. Chem. **262:** 10048–10052.
24. NETZEL-ARNETT, S., G. B. FIELDS, H. BIRKEDAL HANSEN & H. E. VAN WART. 1991. Sequence specificities of human fibroblast and neutrophil collagenases. J. Biol. Chem. **266:** 6747–6755.
25. HASTY, K. A., M. S. HIBBS, A. H. KANG & L. MAINARDI. 1984. Heterogeneity among human collagenases demonstrated by a monoclonal antibody that selectively recognizes and inhibits human neutrophil collagenase. J. Exp. Med. **159:** 1455–1463.
26. HASTY, K. A., M. S. HIBBS, A. H. KANG & L. MAINARDI. 1986. Secreted forms of human neutrophil collagenase. J. Biol. Chem. **261:** 5645–5650.
27. LARIVEE, J., J. SODEK & J. M. FERRIER. 1986. Collagenase and collagenase inhibitor activities in crevicular fluid of patients receiving treatment for localized juvenile periodontitis. J. Periodontal Res. **21:** 702–715.
28. GRANT, G. A., A. Z. EISEN, B. L. MARMER, W. T. ROSWIT & G. I. GOLDBERG. 1987. The activation of human skin fibroblast procollagenase. Sequence identification of the major conversion products. J. Biol. Chem. **262:** 5886–5889.
29. WHITMAN, S. E., G. MURPHY, P. ANGEL, H.-J. RAHMSDORF, B. J. SMITH, A. LYONS, T. J. R. HARRIS, J. J. REYNOLDS, P. HERRLICH & A. J. P. DOCHERTY. 1986. Comparison of human stromelysin and collagenase by cloning and sequence analysis. Biochem. J. **2402:** 913–916.
30. WEISS, S. J. 1989. Tissue destruction by neutrophils. N. Engl. J. Med. **320:** 365–376.
31. SUOMALAINEN, K., T. SORSA, T. INGMAN, O. LINDY & L. M. GOLUB. 1992. Tetracycline inhibition identifies the cellular origin of interstitial collagenases in human periodontal diseases *in vivo.* Oral Microbiol. Immun. **7:** 121–123.
32. OHLSSON, K., I. OLSSON & G. TYNELIUS-BRATTHALL. 1973. Neutrophil leukocyte collagenase, elastase and serum protease inhibitors in human gingival crevices. Acta Odontol. Scand. **31:** 51–59.
33. GOLUB, L. M., K. SIEGEL, N. S. RAMAMURTHY & I. D. MANDEL. 1976. Some characteristics of collagenase activity in gingival crevicular fluid and its relationship to gingival diseases in humans. J. Dent. Res. **55:** 1049–1057.
34. OVERALL, C. M. & J. SODEK. 1987. Initial characterization of a neutral metalloproteinase, active on native 3/4 collagen fragments, synthesized by ROS 17/2.8 osteoblastic cells, periodontal fibroblasts, and identified in gingival crevicular fluid. J. Dent. Res. **66:** 1271–1282.
35. LEE, W., S. AITKEN, G. V. KULKARNI, P. BIREK, C. M. OVERALL, J. SODEK & C. A. G. MCCULLOCH. 1991. Collagenase activity in recurrent periodontitis: Relationship to disease progression and doxycycline therapy. J. Periodontal Res. **26:** 479–485.
36. KOWASHI, Y., F. JACCARD & G. CIMASONI. 1979. Increase of free collagenase and neutral protease activities in the gingival crevice during experimental gingivitis in man. Arch. Oral Biol. **24:** 645–650.
37. HAKKARAINEN, K., V.-J. UITTO & J. AINAMO. 1988. Collagenase activity and protein content of sulcular fluid after scaling and occlusal adjustment of teeth with deep periodontal pockets. J. Periodontal Res. **23:** 204–210.
38. KRYSHTALSKYJ, E., J. SODEK & J. M. FERRIER. 1986. Correlation of collagenolytic enzymes and inhibitor in gingival crevicular fluid with clinical and microscopic changes in experimental periodontitis in the beagle dog. Arch. Oral Biol. **31:** 21–31.
39. KRYSHTALSKYJ, E. & J. SODEK. 1987. Nature of collagenolytic enzyme and inhibitor

activities in gingival crevicular fluid from chronic adult periodontitis patients and experimental gingivitis subjects. J. Periodontal Res. **25:** 69–73.
40. TENG, Y. T., J. SODEK & C. A. G. MCCULLOCH. 1992. Gingival crevicular fluid gelatinase and its relationship to periodontal disease in human subjects. J. Periodontal Res. **27:** 544–552.
41. GUSTAFSSON, A., B. ASMAN & K. BERGSTROM. 1994. Altered relation between granulocyte elastase and α-2-macroglobulin in gingival crevicular fluid from sites with periodontal destruction. J. Clin. Periodontol. **21:** 17–21.
42. GANGBAR, S., C. M. OVERALL, C. A. G. MCCULLOCH & J. SODEK. 1990. Identification of polymorphonuclear leukocyte collagenase and gelatinase activities in mouthrinse samples: Correlation with periodontal disease activity in adult and juvenile periodontitis. J. Periodontal Res. **25:** 257–267.
43. LEE, W., S. AITKEN, J. SODEK & C. A. G. MCCULLOCH. 1994. Evidence of a direct relationship between neutrophil collagenase activity and periodontal destruction *in vivo*: Role of active enzyme in human periodontal disease. J. Periodontal Res. In press.
44. BOWERS, J. E. & R. T. ZAHRADNIK. 1989. Evaluation of a chairside protease test for use in periodontal diagnosis. J. Clin. Dent. **1:** 106–109.
45. BIREK, P., C. A. G. MCCULLOCH & C. M. OVERALL. 1989. Measurements of probing velocity with an automated periodontal probe and the relationship with experimental periodontitis in the Cynomolgus monkey (*Macaca fascicularis*). Arch. Oral Biol. **34:** 793–801.
46. KNIGHT, C. G., R. WILLENBROCK & G. MURPHY. 1992. A novel coumarin-labelled peptide for sensitive continuous assays of the matrix metalloproteinases. FEBS Lett. **296:** 263–266.
47. NGUYEN, L. P., L. C. SMITH & C. A. G. MCCULLOCH. 1994. Validation of a rapid fluorogenic assay for oral fluid collagenolytic activity. J. Dent. Res. **73:** 1630 (Abstr.).
48. SMITH, L. C., L. NADEAU, P. BIREK & C. A. G. MCCULLOCH. 1994. Classification of periodontal disease status using a fluorescent collagenolytic assay. J. Dent. Res. **73:** 1631 (Abstr.).
49. SACKETT, D. L., R. B. HAYNES & P. TUGWELL. 1985. Clinical Epidemiology. A Basic Science for Clinical Medicine: 47–57. Little Brown. Toronto.

Modulation of Bone Resorption by Tetracyclines[a]

BARRY R. RIFKIN,[b,c] ANTHONY T. VERNILLO,[b]
LORNE M. GOLUB,[d] AND
NUNGAVARUM S. RAMAMURTHY[d]

[b]*New York University College of Dentistry*
Division of Basic Sciences
345 East 24th Street
New York, New York 10010

[d]*State University of New York at Stony Brook*
Department of Oral Biology and Pathology
School of Dental Medicine
Stony Brook, New York 11794

Tetracyclines (TCs) were first introduced in 1948 as broad spectrum antibiotics and have gained wide therapeutic use since that time.[1] TCs have been used in dentistry, particularly as adjuncts to periodontal therapy, an approach based on three perceived advantages: (1) their effectiveness in suppressing Gram-negative anaerobic periodontopathogenic organisms in the subgingival plaque;[2,3] (2) their ability to concentrate in the gingival crevicular fluid (GCF) at levels substantially greater than in serum;[4,5] and (3) their ability to bind to the tooth surface and then be slowly released in an active form, a property which prolongs therapeutic effectiveness.[6]

In the early to mid-1980s, Golub and co-workers first made the key observation that TCs can also inhibit the activity of mammalian collagenase and that this inhibition was *unrelated* to the antimicrobial efficacy of these drugs. This latter finding was supported by the fact that the TC effect was seen in germ-free as well as conventional animals,[7] in uninfected tissues and in sterile *in vitro* systems.[7-10] Some analogs of TC (i.e., chemically modified TC, CMT) were chemically modified to *lose* their antimicrobial activity, but still retained their ability to inhibit collagenase.[8] In their earlier clinical studies, Golub *et al.* found that regular doses of minocycline, doxycycline, or tetracycline administered to both nondiabetics and diabetics with adult periodontitis reduced the excessive collagenase activity in the GCF.[7,9,10] During the same time period, Gomes *et al.*[11] showed that TCs also inhibited parathyroid hormone-induced bone resorption in organ culture; several years later, Rifkin *et al.*[12] demonstrated inhibition of bone resorption with nonantimicrobial chemically modified tetracycline. Therefore, TCs might be therapeutically useful in the management of diseases characterized by pathologic collagen breakdown.[13,14] That excessive collagenase activity might also be associated with bone resorption in periodontal disease led to investigations to determine whether TCs had an effect on bone cell metabolism.

[a]This work was supported by the National Institute of Dental Research grants R01DE-09576 and R37DE-03987 and Collagenex Inc., Wayne, Pennsylvania 19107.

[c]Corresponding author.

ROLE OF COLLAGENASE IN BONE RESORPTION

Collagen is the major component of the bone matrix; it represents at least 90% of the organic matrix of bone.[15] Destruction of the connective tissue matrix is an essential step in the pathogenesis of the lytic bone diseases, including periodontitis. Interstitial collagenase (matrix metalloproteinase-1, MMP-1) activity, as well as the activity of other MMPs (e.g., gelatinase, stromelysin), may be important in both the normal and pathologic remodeling or resorption of the bone collagen extracellular matrix.[16] The activity of the bone cell MMPs is regulated *in vivo* by the tissue inhibitors of MMPs (TIMPs).[17] Bone resorbing hormones (e.g., parathyroid hormone, PTH) act through receptors found on osteoblasts (primary bone-forming cells) that synthesize bone matrix proteins and initiate mineralization.[18] Multinucleated osteoclasts are responsible for the resorption of the mineralized matrix.[19] However, the functions of these two principal cells in resorption have been interrelated.

Walker *et al.*[20] provided the first experimental data suggesting that PTH stimulated the release of collagenolytic activity from bone tissue. The discovery of a latent, trypsin-activated form of collagenase (i.e., procollagenase) in tadpole tissue cultures by Harper *et al.*[21] gave impetus to subsequent studies on the regulation of collagenase. Sakamoto *et al.*[22] first reported in 1975 a correlation between bone resorption in organ culture and the release of collagenase by the addition of parathyroid hormone extract. In 1976, Puzas and Brand first showed collagenolytic activity from enzymatically isolated bone cells (an osteoblast-enriched preparation).[23] These investigators subsequently demonstrated PTH stimulation of collagenase secretion by an osteoblast-enriched preparation isolated from fetal rat calvaria.[24]

That osteoblasts appeared to be the target cells for bone resorbing hormones (e.g., PTH) led to the hypothesis, as proposed by Sakamoto and Sakamoto in 1982, that osteoblasts played a pivotal role in bone resorption;[25] osteoblasts were apparently the collagenase-synthesizing cells as demonstrated by immunocytochemistry (i.e., anti-mouse bone collagenase antibody), whereas osteoclasts, the principal resorbing cells of bone, were not reactive for MMP-1.[26] Therefore, under the influence of bone resorption-stimulating hormones, it was the osteoblast and not the osteoclast interstitial collagenase (MMP-1) that would degrade the surface type I collagen (osteoid) and initiate resorption.[27] A role for the osteoblast-derived collagenase in bone resorption was further supported by the observations of Chambers and Fuller[28] whereby isolated osteoclasts cultured on calvarial explants did not resorb the mineralized matrix unless the osteoid layer was first removed, either by pretreatment with osteoblasts or collagenase.

In addition to their role as target cells for calciotropic hormones (i.e., PTH), osteoblasts presumably interacted with osteoclasts to mediate the resorption of bone. For example, McSheehy and Chambers[29] showed that PTH was without effect on isolated and cultured osteoclasts, but if osteoblasts and osteoclasts were cultured together, then PTH enhanced osteoclastic bone resorption. Indeed, the studies of Chambers on the model of disaggregated osteoclasts were particularly important in establishing such an osteoblast-mediated hormonal stimulation of osteoclastic activity.[30,31] Similarly, additional bone resorption inducers such as 1,25-dihydroxyvitamin D_3,[32] interleukin-1, IL-1,[33] or TNF (tumor necrosis factor)-α and TNF-β,[34] which had no effect on disaggregated osteoclasts alone, caused a significant increase in osteoclastic resorption when cocultured with osteoblasts. Finally, Rodan and Martin[35] also suggested an initiator role (from a morphological point of view) for the osteoblast; in response to PTH, contracted osteoblasts might then yield a greater surface for recruiting osteoclasts to continue the resorptive process.

Information on bone metabolism rapidly accrued with the availability of cloned rodent osteoblastic cell lines (e.g., UMR 106, ROS 17/2). Agents that had been known to stimulate bone resorption (e.g., PTH; 1,25-dihydroxyvitamin D_3; prostaglandin) in osteoblast-osteoclast cocultures also up-regulated the production of collagenase from isolated clones of osteoblastic cells.[36] In 1983, Partridge *et al.*[37] initially characterized the ultrastructural and biochemical properties of four clonal osteogenic osteosarcoma lines (transformed cells). These osteoblastic clones (UMR cells) exhibited a stable phenotype through many passages in culture (e.g., high alkaline phosphatase activity and PTH activation of adenylate cyclase) and are still used widely to study osteoblast structure and function. Otsuka *et al.*[38] assayed collagenase and TIMP from ROS 17/2 cells in culture.

The mechanism of PTH-stimulated collagenase synthesis by osteoblastic clones was further examined in UMR 106-01 cells. Partridge *et al.*[36] assayed MMP-1 activity from culture media after limited trypsinization (to activate the procollagenase); in response to PTH, the cells produced not only significant amounts of enzyme (12–48 h) but also TIMP (72–96 h). The levels of collagenase were markedly curtailed after the appearance of TIMP. These data,[36] therefore, suggested a complex pattern in the regulation of collagenase and its inhibitor. PTH also caused a substantial increase in MMP-1 mRNA via transcription of the collagenase gene in UMR 106-01 cells;[39] such PTH-mediated induction of transcription required the synthesis of protein factors and occurred through cyclic AMP. Recently, human osteoblasts were also shown to synthesize the MMPs, collagenase, gelatinase B (92 kDa), and stromelysin when treated with PTH.[40] This particular finding gives the putative role of collagenase additional relevance in the human bone diseases.

Work from the laboratory of Partridge[36,37,39,41] had expanded the Chambers and Fuller model[28] in which the osteoblast played a key role in the *initiation* of bone resorption. Thus, the osteoblast is initially activated by a calciotropic hormone, such as PTH, which up-regulates the synthesis and secretion of procollagenase. Procollagenase may, in turn, be activated through the plasminogen activator/plasmin pathway.[41] Consequently, the activation of the latent collagenase initiates the degradation of osteoid, followed by migration of osteoclasts into the osteoid-free areas to resorb the mineralized matrix. These models and data strongly supported the concept of the osteoclast-mediated degradation of the calcified bone matrix; the osteoblast was viewed as the initiator of resorption.

Despite such vigorous investigations to delve into the mechanisms of collagenase synthesis and secretion, the possible correlations between this MMP and bone resorption still remain scarce and contradictory. Initial studies[22] to examine such a correlation using bone organ culture systems were inconclusive. A similar correlation was not observed by Eilon and Raisz,[42] Delaisse *et al.*,[43] or Lenaers-Claeys and Vaes,[44] even when collagenase was assayed immunochemically. Sellers *et al.*[45] observed that vitamin A and 1,25-dihydroxyvitamin D_3 increased the production of collagenase and reduced TIMP in the conditioned medium from cultured mouse calvaria; however, these observations were difficult to interpret. In contrast to and nearly a decade after these earlier studies, Delaisse *et al.*[46] demonstrated that *bone matrix* not only contained procollagenase, but resorption of fetal mouse calvariae *in vitro* also proceeded along with the accumulation of procollagenase, primarily within the nonmineralized matrix; these data suggested strongly that the activity of osteoblast collagenase was required (perhaps, more definitively) for the initial resorption of intact, embryonic mammalian bone. Very recently, further support for the role of collagenase has been provided.[47] Hill *et al.* inhibited bone resorption induced by either PTH or 1,25-dihydroxyvitamin D_3 in cultured neonatal mouse calvariae with recombinant human TIMP-1 or TIMP-2.[47] Their results suggested that endogenous TIMPs played a central role in regulating both physiological and pathological bone resorption.

Nonetheless, these experiments[42–47] collectively neither excluded nor provided conclusive evidence between bone resorption and the release of collagenase. Furthermore, these investigations did not definitively conclude that collagenase was involved in the resorption of bone matrix, or that TIMP controlled collagenase activity in resorption.

Unlike the study of osteoblast metabolism, investigative work in the field of osteoclast biology lagged because it was difficult to obtain large numbers of purified osteoclasts. However, techniques[48,49] were developed to isolate and purify large numbers of avian osteoclasts and, consequently, salient progress has been made in the last several years substantiating the central role of osteoclasts in the resorption of bone.

The resorption of bone requires the removal of both the mineral and the organic matrix components. This degradative process occurs in an extracellular compartment (the subosteoclastic resorption zone) acidified by proton transport at the osteoclast ruffled border membrane.[50,51] The protons are generated by an electrogenic vacuolar proton pump identical to that found in the intercalated cells of renal tubules.[51] However, it has been most recently established that this pump is probably unique to the osteoclast and, therefore, may not be identical to that found in the kidney.[52] Nonetheless, the low pH (4–5) generated by such a pump likely allows dissolution of the mineral phase, exposes the organic matrix, and favors the degradative action of lysosomal enzymes.[53,54] Blair *et al.*[55] clearly showed that isolated avian osteoclasts resorbed both the organic and inorganic components of bone. In a very recent study, Blair *et al.*[56] also isolated and characterized an avian osteoclast acid collagenase that resembled the mammalian lysosomal enzyme, cathepsin B. Mannose-6-phosphate receptors for the trafficking of lysosomal enzymes were also recently reported.[57] In this acid milieu, the lysosomal cysteine proteinases, directionally secreted by the osteoclast into its bone-resorbing compartment, are not only involved in the degradation of collagen but could also be sufficient for its complete degradation without requiring the participation of MMP-1. Such an implication that cysteine proteinases (i.e., cathepsins) and, not collagenase, might be sufficient to degrade bone matrix collagen received additional support from the work of Delaisse *et al.*;[43] these investigators showed that the addition of TIMP (the inhibitor of MMP-1) did not inhibit bone resorption *in vitro.* More specifically, Rifkin *et al.*[58] showed that a selective inhibitor of cathepsin L significantly inhibited bone resorption when added to disaggregated rat osteoclasts seeded onto cortical bovine bone. Furthermore, cathepsin L activity in isolated avian osteoclasts was significantly greater than cathepsin B, implying that the former may play a key role in resorption.[58] Of particular interest was the observation that cathepsin B activity was significantly higher than cathepsin L activity from freshly isolated marrow cells.[59] In contrast, authentic osteoclasts and marrow generated giant cells (GCs) stimulated with osteoblast-conditioned media had significantly higher cathepsin L than B activity and, thereby, demonstrated a reversal of cathepsin B:L activity noted in isolated marrow cells.[59] The implication of such findings is intriguing: cathepsin L activity may be a developmental marker for differentiated bone-resorbing cells such as osteoclasts.

However, collagenase[60,61] and gelatinase[62] were also ultimately identified in the osteoclast; these latter findings were of enormous significance because the osteoclast, apart from any possible participation from osteoblast collagenase, might be enzymatically self-sufficient as a bone-resorbing cell. Nonetheless, the exact roles for the osteoclast collagenase and its cysteine proteinases (i.e., cathepsins B and L) are not yet clear.

Cooperative modes of interaction have been proposed between these above enzymes.[19,63] Conceivably, the collagenolytic action of the cysteine proteinases, optimal at pH 4–5, could be exerted preferentially in the most acid portion of the bone-resorption lacuna and in the immediate vicinity of the ruffled border. In contrast, collagenase (active between pH 6–7.5) could be predominantly deeper in the lacuna at the interface between demineralized and mineralized matrix where the pH is likely more neutral due to the buffering capacity of the dissolved salts. It is even likely that the concerted, sequential action of both enzymes may render collagen degradation much more efficient compared to the isolated action of each separately.[19] Collagenase could also degrade the fringes of yet undegraded, but already demineralized collagen, likely remaining at the base of the resorption lacuna when the osteoclast detaches[19] and, thereby, allow a sudden neutralization of the pH at that site. This already demineralized collagen could also be denatured at an acidic pH and at physiologic temperature. Moreover, denatured collagen at this fringe could be degraded by osteoclast gelatinase. If the pH were suddenly neutralized, then the acid-requiring cysteine proteinases would be inactive, creating a permissive milieu for the action of neutral MMP-1 (and possibly neutral gelatinase) and the completion of the resorbing process. Despite its reputation as a neutral proteinase, bone collagenase is still active in an acidic environment; at pH 6, it retains nearly 100% of its activity on reconstituted collagen gels and approximately 50% at pH 5.5.[64] Therefore, the sites for the action of collagenase could be even more widely distributed. Collagenase could act not only in the deeper part of the lacunae but also in the immediate vicinity of the osteoclast ruffled border at the bone interface. However, Fuller *et al.* have very recently shown that novel, potent inhibitors against collagenase did not affect parameters of bone resorption including the area, depth, or demineralized fringe thickness of excavations formed by isolated osteoclasts.[65] Such a finding reinforces the ongoing controversy regarding the exact role of collagenase in bone resorption.[42–47] Nonetheless, Fuller *et al.*[65] have very recently suggested that collagenase could still play a significant role because the secretion of it from the osteoclast could facilitate its movement through the osteoid layer and, thereby, its subsequent attachment to the underlying mineralized bone.

The precise mechanisms for the activation of latent bone collagenase are also not yet completely understood. For example, procollagenase is activated by the acid cysteine proteinase, cathepsin B;[66] therefore, it is conceivable that bone cell (e.g., osteoclast) procollagenase could also be similarly activated, providing yet another possible mode of interaction between these two enzyme systems in the remodeling of bone. Apart from such enzyme-to-enzyme interactions, additional physiologic mechanisms include the activation of bone procollagenase through reactive oxygen species, ROS. Fallon *et al.*[67] and Oursler *et al.*[68] showed that the osteoclast produced superoxide anion. An ROS mechanism was demonstrated in neutrophils (PMNs) that produced HOCl from hydrogen peroxide through the myeloperoxidase system.[69] Activation of PMN progelatinase and procollagenase was demonstrated with HOCl.[70,71] Our group first showed HOCl activation of rat osteoblastic procollagenase *in vitro.*[72] Further investigation is needed not only to characterize more extensively osteoclast ROS, but also to define more precisely the role of these reactive molecules in bone resorption.

Nonetheless, the mechanisms for the removal of osteoid, prior to osteoclastic resorption, are still unclear; consequently, the precise role of collagenase and its significance in bone resorption have not yet been completely established. However, the osteoclast has a potentially singular role in the degradation of bone. Therefore, it can be a primary therapeutic target, particularly in the medical management of the human lytic bone diseases.

OSTEOCLAST FUNCTION

Aside from the matrix metalloproteinases and cysteine proteinases and their potential roles in resorption, the osteoclast has other properties for which characterization has also been relatively recent. The effect of tetracyclines on the above osteoclast enzymes has been established preliminarily; other parameters of osteoclast structure and function (as discussed below) in response to TC exposure are now under intensive study.

The multinucleated osteoclast is the principal cell involved in the resorption of bone; however, the precise cellular and molecular mechanisms by which these cells resorb bone are incompletely understood. The attachment of the osteoclast through specialized, filamentous, adhesion structures containing F-actin (i.e., podosomes)[73] consequently creates a sealed environment (i.e., subosteoclastic resorption zone) which may be viewed as a secondary lysosome. The formation of podosomes may be regulated through changes in the intracellular calcium concentration.[74] In fact, podosomes are dynamic structures that appear during resorption and disappear when resorption is inhibited, leading to the retraction of the cell from the bone surface.[75] The osteoclast ruffled border is the active site of bone resorption,[19] and it is surrounded by an area substantially devoid of organelles, the "clear zone."[76] Lysosomal enzymes (e.g., cathepsin L[58]) are directionally secreted from the ruffled border[57] into the subjacent subosteoclastic resorption zone; the generation (through an electrogenic pump in the ruffled border) of an acidic pH likely optimizes the activity of these enzymes.[53,54] In addition to the morphologic changes in podosomes, the osteoclast ruffled border at the bone interface can undergo dynamic change (i.e., reduction in its surface area) in response to physiologic agents that inhibit bone resorption.[30] Finally, acidification can also be affected by physiologic agents that inhibit resorption[77,78] and, therefore, is another parameter of osteoclast function that can be exploited as a therapeutic target. Thus, the effects of TCs on osteoclast parameters such as attachment, ruffled border area, lysosomal enzyme secretion, and the production of acid have become an exciting and clinically relevant area for investigations (FIG. 1); these effects of TC are presented in a later section.

EFFECT OF TETRACYCLINES ON BONE MATRIX METALLOPROTEINASE ACTIVITY

Tetracyclines reduced the excessive collagenase activity in the GCF of diabetics and nondiabetics with adult periodontitis;[7,9,10] those clinical findings, nearly a decade ago, provided a catalyst for recent studies to examine the effect of TCs and its chemically modified analogs (i.e., CMTs) on extracellular MMP activity from clonal osteoblastic cells (UMR 106-01 and ROS 17/2.8) *in vitro*.[79,80] Responsiveness to bone resorbing hormones (i.e., secretion of collagenase) had been documented for the osteoblast;[22–26] however, collagenase had not yet been identified in osteoclasts. Therefore, the osteoblast was chosen for the initial experiments to examine the effects of TCs on MMP activity in bone cells. When added directly to the cultures, minocycline and CMT-1 (one of several chemically modified TC analogs) inhibited the activity of already secreted collagenase 64% and 90%, respectively, from PTH-treated UMR 106-01 cells;[79] these data first showed that TCs inhibited extracellular MMP activity from osteoblastic cells. Nonetheless, additional experiments must determine whether TCs affect intracellular pathways (e.g., transcription, synthesis) for MMPs in bone cells. In that regard, our preliminary data indicated that doxycycline and CMT-1 inhibited PTH-stimulated collagenase synthesis in UMR 106-01 cells 63% and 78%, respectively, as demonstrated by ELISA against rat collagenase (unpublished observations).

The inhibitory effect of TCs on extracellular MMP activity ostensibly differed between cell types. For example, the drug concentration (doxycycline or CMT-1) required to inhibit 50% of the enzyme activity (i.e., IC_{50}) was 15 to 30 μM for collagenases (i.e., MMP-8) from human PMNs and GCF. In contrast, human fibroblast collagenase (MMP-1) was relatively resistant to TC inhibition; the IC_{50} for doxycycline and CMT-1 were 280 μM and 510 μM, respectively.[81] ROS 17/2.8 gelatinase was not inhibited with TC exposure, whereas doxycycline or CMT-1 had an approximate IC_{50} of 40–60 μM against UMR 106-01 gelatinase activity.[80] This latter finding is interesting because osteoblastic, like fibroblast, collagenase is an interstitial enzyme (i.e., MMP-1) but its inhibitory profile to TC more closely approximated the PMN enzyme (MMP-8). Clearly, there is a need to determine the inhibitory profile of TCs against the MMPs from the various cell types because TC inhibition is not solely dependent upon the type of MMP. In fact, the differential sensitivity of PMN and fibroblast collagenases to TC treatment may have substantial therapeutic benefits. In this regard, pharmacologic concentrations might inhibit the activity of PMN, but not the fibroblast enzyme. Such a selective inhibition might reduce collagenolytic activity and tissue degradation during inflammation, but not the normal collagen turnover required for the maintenance of tissue integrity.

The TC molecule was chemically modified to determine whether its anticollagenase and antimicrobial properties could be dissociated from each other. Removal of the dimethylamino group from carbon-4 of the TC molecule (i.e., CMT-1) did not eliminate its anticollagenase activity, even though it no longer had effective antimicrobial activity.[14] Several CMTs (1 to 5) were then obtained and modified at different sites to ascertain their effects not only on osteoblast collagenase (MMP-1) and gelatinase (TABLE 1) activity but also bone resorption.[82] CMT-1 (4-dedimethylamino tetracycline) and CMT-3 (6-demethyl 6-deoxy 4-dedimethylamino tetracycline) inhibited both extracellular osteoblast collagenase (clonal UMR 106-01) activity (culture media) and bone resorption (disaggregated rat osteoclast pit assay). In contrast, CMT-2 (tetracyclinonitrile), CMT-4 (7-chloro 4-dedimethylamino tetracycline), and CMT-5 (tetracyclinpyrazole) affected neither osteoblast collagenase activity nor bone resorption.[82] It is of interest that CMT-2 and CMT-4 did inhibit PMN collagenase selectively, but not the osteoblast enzyme; CMT-5 inhibited neither PMN nor osteoblast collagenase. These findings, particularly with CMT-1 and CMT-3, suggested that both collagenase activity and bone resorption might be coupled processes; further discussion of this implication is presented in the next section.

Nonetheless, the exact mechanisms by which MMPs are activated and their precise roles *in vivo* are still unknown. *In vitro* studies had shown that HOCl activated PMN procollagenase and progelatinase.[70,71] Our laboratory extended these studies to bone cells and first showed that HOCl also activated UMR 106-01 extracellular procollagenase.[72] However, it was known that TCs scavenge ROS.[83] Therefore, our laboratory also examined the ability of TCs to inhibit the activation of osteoblastic procollagenase in the presence of HOCl. Both doxycycline and CMT-1 significantly inhibited the HOCl activation of osteoblastic (UMR 106-01) procollagenase, presum-

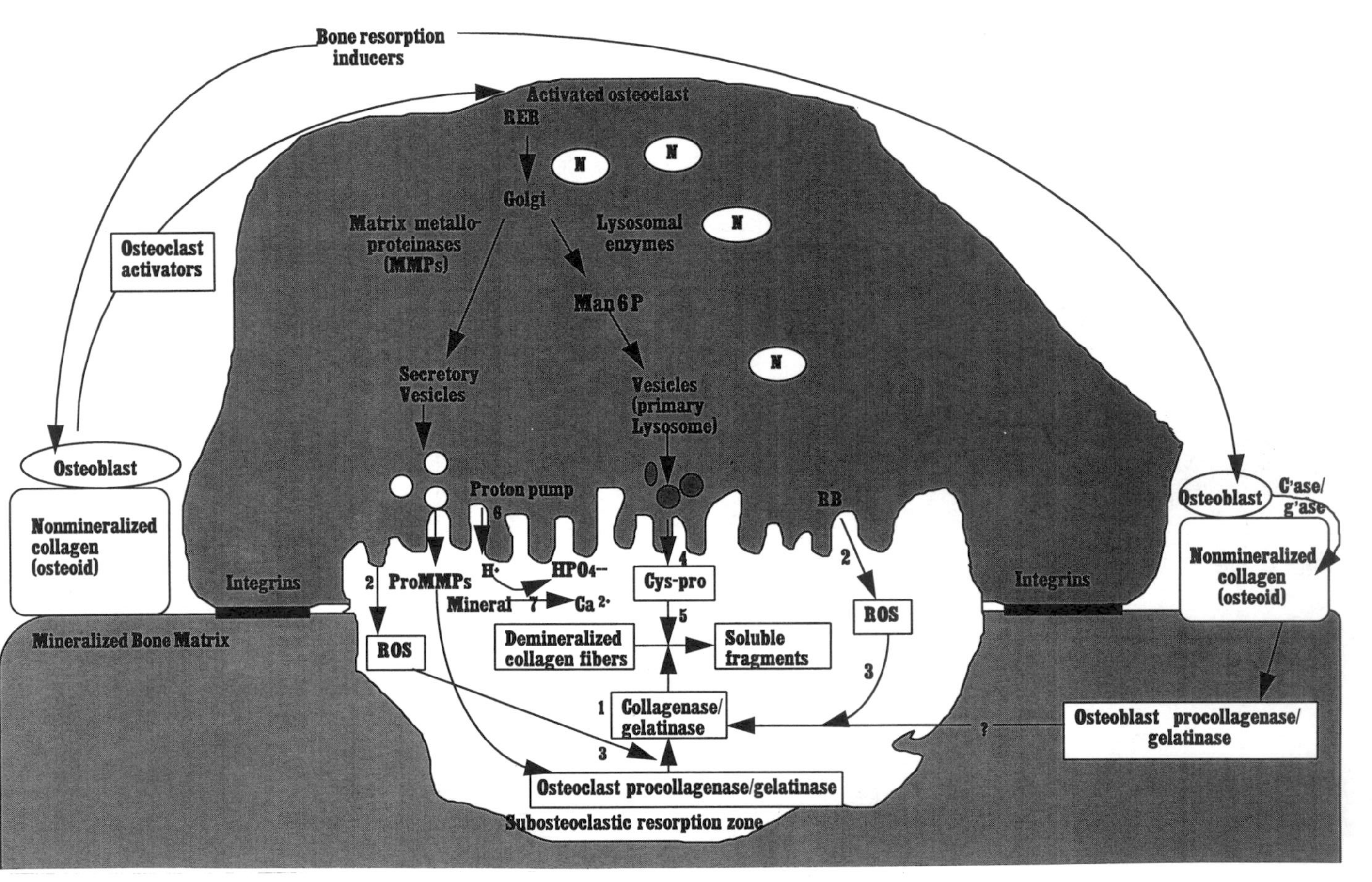
Bone resorption inducers
Osteoclast activators
Activated osteoclast
RER
Golgi
Matrix metallo-proteinases (MMPs)
Lysosomal enzymes
Man 6 P
Secretory Vesicles
Vesicles (primary Lysosome)
Proton pump
RB
Osteoblast
Nonmineralized collagen (osteoid)
Integrins
Mineralized Bone Matrix
ProMMPs
H+
HPO4--
Mineral
Ca 2+
Cys-pro
ROS
Demineralized collagen fibers
Soluble fragments
Collagenase/ gelatinase
Osteoclast procollagenase/gelatinase
Subosteoclastic resorption zone
Osteoblast procollagenase/ gelatinase
C'ase/ g'ase

TABLE 1. Effect of Doxycycline and CMTs on UMR 106-01 Gelatinase Activity from Conditioned Media[a]

Inhibitor	Concentration	Gelatinase Activity (μg gelatin degraded/hour)	% Inhibition
None	0	225	0
EDTA	25 mM	0	100
Doxycycline	50 μM	0	100
CMT-1	50 μM	0	100
CMT-2	50 μM	210	7
CMT-3	50 μM	72	68
CMT-4	50 μM	210	7
CMT-5	50 μM	232	0

[a]Cultures were treated with 10^{-7} M b-PTH-(1-34), and media were analyzed for gelatinase activity as previously described.[36,80] Values represent duplicate determinations. CMT, chemically modified tetracyclines.

ably by reducing the available concentration of HOCl.[72] As suggested by the earlier work of Chambers and Fuller,[28] the activation of secreted osteoblast procollagenase in tissue might facilitate the subsequent osteoclastic resorption of bone; thus, TCs could play a key role in modulating osteoclast function by blocking such an ROS activation of osteoblast procollagenase.

The very recent demonstration of collagenase and gelatinase in osteoclasts[60–62] has led to studies from our laboratory on the anti-MMP effects of TCs using avian multinucleated preosteoclasts.[84] These cells are presumably generated from osteoclast precursors in the bone marrow and have osteoclast-like characteristics such as cytochemical staining for tartrate-resistant acid phosphatase, significant cathepsin L activity, and calcitonin receptors. These cells have been generated in significant numbers and with high purity.[48,49] Our laboratory recently showed that CMT-1 (50–400 μM) inhibited the extracellular activity (culture media) of MMP-1 from such avian preosteoclasts (FIG. 2, lanes 5–8); even 5–25 μM CMT-1 (FIG. 2, lanes 9–11) showed a partial inhibition of collagenase activity.[84] Furthermore, the IC_{50} for CMT-1 against the avian preosteoclast collagenase was 50–70 μM (unpublished observations), comparable to the inhibitory concentration against clonal osteoblastic gelatinase.[80] Such an IC_{50} (especially at the lower range) may be relevant *in vivo* because TCs bind to bone mineral and may be highly concentrated at the resorption site. Therefore, it is likely that TCs may also inhibit the activity of MMPs from bone-resorbing osteoclasts *in vivo*. Significant extracellular gelatinase activity has

←

FIGURE 1. Potential sites for tetracycline (TC) or chemically modified tetracycline (CMT) molecule action on bone cell metabolism. TCs and CMTs have both anti-matrix metalloproteinase (anti-MMP) and antiosteoclast properties. Tetracyclines and their chemically modified analogs may act (1) to inhibit directly active extracellular collagenase/gelatinase (C′ase/g′ase); (2) to reduce the available concentration of osteoclast-generated superoxide radicals (reactive oxygen species [ROS]), and thereby (3) inhibit the superoxide radical conversion of extracellular procollagenase/progelatinase (proMMPs) to active enzyme; (4) to reduce the secretion of the lysosomal osteoclast cysteine proteinases (Cys-pro) such as cathepsin L, and thereby (5) reduce resorption of bone collagen; (6) to reduce the secretion of acid (protons, H^+) from the osteoclast ruffled border (RB), and thereby alter the pH optima (pH 4–5) for lysosomal enzymes; and (7) to inhibit acid solubilization of bone mineral. N, nucleus; RER, rough endoplasmic reticulum; Man 6 P, mannose-6-phosphate.

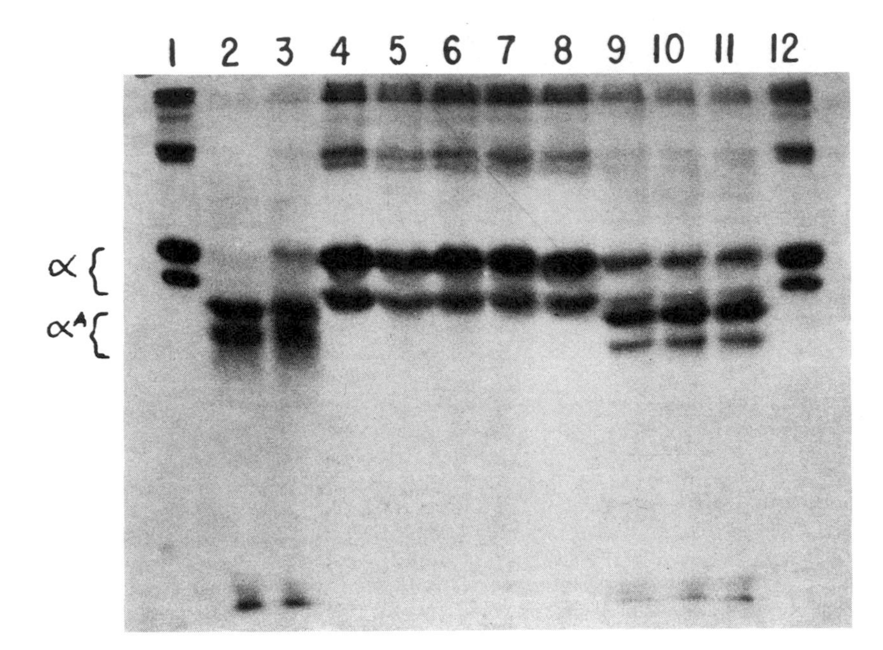

FIGURE 2. Effects of chemically modified tetracycline-1 (CMT-1) on avian preosteoclast (preOC) collagenase activity. Pooled media from untreated cultures were lyophilized, reconstituted with collagenase buffer, and subjected to 5% sodium dodecylsulfate polyacrylamide gel electrophoresis (SDS-PAGE) and fluorography.[9,82] The media alone were treated by the addition of CMT-1 prior to SDS-PAGE. **Lane 1:** a standard internal control representing undegraded, intact collagen alone; **lanes 2 and 3:** collagen plus preOC media; **lane 4:** preOC media plus 50 mM ethylenediaminetetraacetic acid (EDTA); **lane 5:** preOC media plus 400 μM CMT-1; **lane 6:** preOC media plus 200 μM CMT-1; **lane 7:** preOC media plus 100 μM CMT-1; **lane 8:** preOC media plus 50 μM CMT-1; **lane 9:** preOC media plus 25 μM CMT-1; **lane 10:** preOC media plus 10 μM CMT-1; **lane 11:** preOC media plus 5 μM CMT-1; **lane 12:** same as lane 1. Alpha (α) represents the intact α_1 and α_2 chains of collagen; α^A represents the breakdown products of the α chains to $\alpha_1{}^A$ and $\alpha_2{}^A$ products. (From Vernillo *et al.*[84] Reproduced, with permission, from *Current Science.*)

also been assayed from avian preosteoclasts (unpublished observations); however, the effect of TCs on this preosteoclast gelatinase awaits further investigation. Finally, doxycycline and CMT-1 significantly inhibited extracellular gelatinase activity from authentic avian osteoclasts; furthermore, HOCl activated avian preosteoclast procollagenase secreted into the culture media (unpublished observations). Clearly, additional investigations will include HOCl activation of osteoclast procollagenase and progelatinase and the effects of TCs.

EFFECT OF TETRACYCLINES ON BONE RESORPTION AND OSTEOCLAST FUNCTION

The effects of TCs on bone resorption were investigated shortly after the discovery of their anticollagenase properties because collagenase activity is a critical step in bone resorption. Therapeutic concentrations of TCs inhibited PTH stimulation of calcium and hydroxyproline release from fetal rat long bone cultures.[12] Our laboratory showed that this inhibition was reversible; removal of TCs after 48 h resulted in a resumption of bone resorption.[85] Furthermore, no morphological or biochemical cytotoxicity was noted at 20 μg/mL (minocycline or CMT-1). Inhibition of bone resorption occurred without significant change in osteoclast numbers, but the ruffled border area of osteoclasts was reduced. Such an effect on the osteoclast ruffled border suggested that TCs might also directly inhibit osteoclast function.[13,85–87]

CMTs have been tested in both fetal long bone and disaggregated rat osteoclast (pit assay) systems; both bone resorption systems have shown comparatively good correlation.[82] In the presence of various TCs, isolated rat osteoclasts resorbed significantly less bone than controls. CMT-1 and CMT-3 comparably inhibited resorption in fetal long bone cultures and in the pit assay;[82] whereas CMT-2 and CMT-4 did not inhibit PTH stimulated bone resorption in fetal long bones or in the pit assay. Furthermore, the antimicrobial TC, doxycycline, and CMT-1 and CMT-3 inhibited extracellular osteoblast collagenase activity, whereas CMTs-2, 4, and 5 were without effect on this enzyme.[82] Such findings reinforce the concept that resorption, as assayed independently by two systems, may be coupled with the activity of collagenase. More recently, our laboratory examined the effects of the newly derived CMTs on bone resorption. Two of these CMTs, CMT-7 (12 alpha deoxy-4-dedimethylamino tetracycline) and CMT-8 (6 alpha deoxy-5 hydroxy-4-dedimethylamino tetracycline), are extremely potent and have an IC_{50} less than 1 μg/mL. Of significant interest is the fact that these recently synthesized CMTs (CMT-7 and CMT-8) may be more potent inhibitors of bone resorption than CMTs developed earlier (TABLE 2).

The osteoclast ruffled border is the active site from where lysosomal enzymes are secreted into the resorption lacunae.[57] As stated previously, isolated osteoclasts in the presence of TCs resorbed significantly less bone than controls. Tetracyclines (doxycycline and CMT-1) also caused a significant decrease in the extracellular activities of two key osteoclast enzymes: tartrate-resistant acid phosphatase (unpublished data), a thoroughly investigated lysosomal marker enzyme, and cathepsin L,[88] an important cysteine proteinase (TABLE 3). However, doxycycline and CMT-1 had no direct inhibitory effect on already secreted lysosomal enzyme activities; the treated osteoclast lysates showed no difference in comparison to controls. Thus, TCs might inhibit bone resorption by affecting intracellular pathways involved in the secretion of key lysosomal enzymes, especially the cysteine proteinases. In contrast,

TABLE 2. Potency of Tetracyclines as Inhibitors of Bone Resorption *in Vitro*

Tetracycline	IC_{50} (μg/mL)
Minocycline[a,b]	15–20
Doxycycline[a,b]	6
CMT-1[a,b]	8
CMT-2[a,b]	No inhibition
CMT-3[a,b]	<2
CMT-4[a,b]	No inhibition
CMT-5[a,b]	No inhibition
CMT-6[a,b]	2–4
CMT-7[b]	0.5–1.0
CMT-8[b]	0.1–0.5

NOTE: Based on studies using 19-day fetal rat long bones[a] and disaggregated neonatal rat osteoclasts[b] cultured on devitalized cortical bovine bone.

and perhaps not unexpectedly, osteoclasts cultured on bone slices with CMT-4 showed no change in cathepsin L activity; furthermore, CMT-4 did not inhibit resorption or osteoblast collagenase activity.[82]

The apical or ruffled border membrane is also the site of proton extrusion for acidification of the bone resorbing compartment. Acidification occurs via a vacuolar-like, adenosine triphosphate (ATP)-driven, proton pump; protons are generated intracellularly by carbonic anhydrase II.[89] The acidic pH not only likely optimizes the action of lysosomal enzymes but also dissolves the hydroxyapatite mineral. Acridine orange, a fluorescent dye, has been used in studies of osteoclast function to determine whether agents increase or decrease its acid production.[89] With such techniques, tetracyclines (doxycycline, minocycline and CMT-1) caused a dose-dependent reduction in acid formation by avian osteoclasts;[90] a possible effect on the proton pump was investigated. Plasma membrane vesicles were then isolated from avian osteoclasts and examined for their ability to pump protons in the presence or absence of CMT-1: CMT-1 had no effect on the proton pump. Since carbonic

TABLE 3. Effect of Doxycycline on Secreted Osteoclast Cathepsin L Activity[a]

Treatment	Time (h)	Activity (M AFC released/mL/min)
Control	3	$15.15 \times 10^{-8} \pm 1.11$
Doxycycline		
2 μg/mL	3	$8.50 \times 10^{-8} \pm 0.67$[b]
6 μg/mL	3	$5.29 \times 10^{-8} \pm 1.16$[c]
10 μg/mL	3	$1.99 \times 10^{-8} \pm 0.50$[c]
Control	6	$23.75 \times 10^{-8} \pm 1.57$
Doxycycline		
2 μg/mL	6	$10.25 \times 10^{-8} \pm 0.89$[b]
6 μg/mL	6	$7.39 \times 10^{-8} \pm 1.16$[c]
10 μg/mL	6	$0.66 \times 10^{-8} \pm 0.57$[c]

[a]Disaggregated rat osteoclasts were cultured and treated for 3 or 6 h. Release of the fluoroprobe AFC was used as a measure of enzyme activity.[58]
[b]Significantly different from control, $p < 0.01$.
[c]Significantly different from control, $p < 0.001$.
AFC, 7-amino-(4-trifluoromethyl) coumarin.

anhydrase II provides protons for acidification and also requires zinc for activity, we hypothesized that TC, a metal chelator, might also interfere with its activity; however, TCs had no effect on it.

The osteoclast clear zone contains specialized, discrete adhesion sites called podosomes;[73] these are foot-like, conically shaped, vertical protrusions from this zone. Each podosome consists of a core of F-actin containing microfilaments associated with α-actinin, fimbrin, and the Ca^{2+}-binding protein, gelsolin.[76] The core of the podosome along with α-actinin, fimbrin, and gelsolin is in turn linked through vinculin and talin to an integrin receptor; this receptor recognizes and binds to extracellular matrix proteins. Podosomes also stain with rhodamine-conjugated phalloidin and appear as a broad ring or band of punctate dots. We have used rhodamine phalloidin to study the expression of F-actin containing podosomes in cultured avian osteoclasts treated with two different TC analogs, doxycycline and CMT-1. After 3–4 days in culture, approximately 60% of control osteoclasts expressed such podosomes; in contrast, treated osteoclasts showed a significant reduction of them.[85] Podosome assembly in osteoclasts is controlled by cytosolic calcium.[76] Increases in [Ca^{2+}] result in decreased podosome expression and bone resorption. Therefore, podosomes are viewed as calcium-regulated adhesion structures, and the significant effect on them with TC may result from an alteration in [Ca^{2+}].

Of particular interest are recent data showing that TCs altered [Ca^{2+}] in isolated rat osteoclasts.[91] Osteoclasts possess, on the plasma membrane, a Ca^{2+} "sensor" or "receptor."[92] In the osteoclast, a change in the extracellular calcium concentration is monitored and transduced into a cytosolic calcium signal resulting in marked functional inhibition. Thus, a number of correlates of osteoclast function are consequently reduced and include cell matrix adhesion, cell spreading, podosome expression, enzyme secretion, and bone resorption.[92,93] Our laboratory further investigated the pharmacologic basis for the action of TCs on the osteoclast: (1) Minocycline and doxycycline strongly attenuated, in a dose- and time-dependent manner, the rise of cytosolic calcium resulting from calcium receptor activation by calcium or nickel.[94] Nickel is used as a surrogate agonist of the calcium receptor.[93] Similar results were obtained when CMT-1 inhibited nickel-induced calcium receptor activation. However, TC did not alter the responsiveness of osteoclasts to calcitonin, a hormone that interacts with its own cell surface receptor to elicit a cytosolic calcium signal. Therefore, the inhibitory effects of TC may be localized to the transduction pathway of triggered calcium activation. (2) Minocycline elevated cytosolic calcium in isolated osteoclasts and well within the therapeutic range (1–4 mg/L).[94] Indeed, physiologic and pharmacologic agents that elevated cytosolic calcium also inhibited bone resorption. (3) Minocycline and CMT-1 produced marked retraction in isolated osteoclasts. Microspectrofluorimetric and morphometric data showed that TC acted on the transduction pathways triggered by divalent cation exposure.[94,95] The exciting possibility exists that TCs might interact with the osteoclast calcium receptor—a singular mechanism to shutdown osteoclast function.

CONCLUDING REMARKS

The anticollagenase property of tetracycline and CMTs has enormous medical and dental potential. Osseous destructive diseases associated with excessive mammalian collagenase activity include the lytic bone diseases such as periodontitis. However, apart from its anticollagenase effect, TCs are also potent inhibitors of osteoclast function. TCs can affect parameters of osteoclast function and, consequently, inhibit bone resorption by (1) elevating intracellular calcium and interacting with the putative calcium receptor; (2) decreasing ruffled border area; (3) diminishing acid production; (4) diminishing the secretion of cysteine proteinases (e.g., cathepsin L); (5) inducing cell retraction by affecting podosomes; (6) inhibiting osteoclast MMPs, collagenase and gelatinase; and (7) scavenging reactive oxygen species which activate extracellular osteoclast (and osteoblast) proMMPs. Future investigations should further elucidate the mechanisms of tetracycline action in bone.

REFERENCES

1. GILMAN, A. G., L. S. GOODMAN, T. W. RALL & F. MURAD, EDS. 1985. Goodman and Gilman's The Pharmacologic Basis of Therapeutics. 7th edit. Macmillan Publishing Company. New York.
2. SLOTS, J. & B. G. ROSLING. 1983. J. Clin. Periodontol. **10:** 465–468.

3. SLOTS, J., P. MASHIMO, M. J. LEVINE & R. J. GENCO. 1979. J. Periodontol. **50:** 495–509.
4. CIANCIO, S. G., M. L. MATHER & J. A. MCMULLEN. 1980. J. Periodontol. **51:** 530–534.
5. GOODSON, J. M. & S. S. SOCRANSKY. 1981. J. Periodontol. **52:** 609–618.
6. BAKER, P., R. EVANS, R. COBURN & R. GENCO. 1983. J. Periodontol. **54:** 580–586.
7. GOLUB, L. M., H. M. LEE, G. LEHRER, A. NEMIROFF, T. F. MCNAMARA, R. KAPLAN & N. S. RAMAMURTHY. 1983. J. Periodontal Res. **18:** 516–526.
8. GOLUB, L. M., T. F. MCNAMARA, G. D'ANGELO, R. A. GREENWALD & N. S. RAMAMURTHY. 1987. J. Dent. Res. **66:** 1310–1314.
9. GOLUB, L. M., M. WOLFF, H. M. LEE, T. F. MCNAMARA, N. S. RAMAMURTHY, J. ZAMBON & S. G. CIANCIO. 1985. J. Periodontol. Res. **20:** 12–23.
10. GOLUB, L. M., N. RAMAMURTHY, T. F. MCNAMARA, B. GOMES, M. WOLFF, A. CASINO, A. KAPOOR, J. ZAMBON, S. CIANCIO, M. SCHNEIR & H. PERRY. 1984. J. Periodontal Res. **19:** 651–655.
11. GOMES, B. C., L. M. GOLUB & N. S. RAMAMURTHY. 1984. Experentia **40:** 1273–1275.
12. RIFKIN, B., B. GOMES, N. S. RAMAMURTHY, T. F. MCNAMARA & L. M. GOLUB. 1987. J. Cell Biol. **105:** 216A.
13. GOLUB, L. M., S. CIANCIO, N. S. RAMAMURTHY, M. LEUNG & T. F. MCNAMARA. 1990. J. Periodontal Res. **25:** 321–330.
14. GOLUB, L. M., N. S. RAMAMURTHY, T. F. MCNAMARA, R. A. GREENWALD & B. R. RIFKIN. 1991. Crit. Rev. Oral Biol. Med. **2:** 297–322.
15. PIEZ, K. A. & A. H. REDDI, EDS. 1984. Extracellular Matrix Biochemistry. Elsevier. Amsterdam.
16. HARRIS, E. D., H. G. WELGUS & S. M. KRANE. 1984. Collagen Relat. Res. **4:** 493–499.
17. BIRKEDAL-HANSEN, H., W. G. I. MOORE, M. K. BODDEN, L. J. WINDSOR, B. BIRKEDAL-HANSEN, A. DECARLO & J. A. ENGLER. 1993. Crit. Rev. Oral Biol. Med. **4:** 197–250.
18. KAHN, A. J. & N. C. PARTRIDGE. 1987. Am. J. Otolaryngol. **8:** 258–264.
19. DELAISSE, J.-M. & G. VAES. 1992. The Biology and Physiology of the Osteoclast. B. R. Rifkin & C. V. Gay, Eds. CRC Press. Boca Raton, FL.
20. WALKER, D. G., C. M. LAPIERE & J. GROSS. 1964. Biochem. Biophys. Res. Commun. **15:** 397–402.
21. HARPER, E., K. J. BLOCH & J. GROSS. 1971. Biochemistry **10:** 3035–3041.
22. SAKAMOTO, S., M. SAKAMOTO, P. GOLDHABER & M. GLIMCHER. 1975. Biochem. Biophys. Res. Commun. **63:** 172–178.
23. PUZAS, J. E. & J. S. BRAND. 1976. Biochim. Biophys. Acta **429:** 964–974.
24. PUZAS, J. E. & J. S. BRAND. 1979. Endocrinology **104:** 559–562.
25. SAKAMOTO, S. & M. SAKAMOTO. 1982. J. Periodontal Rcs. **17:** 523–526.
26. SAKAMOTO, S. & M. SAKAMOTO. 1984. Biomed. Res. **5:** 29–38.
27. PECK, W. A., ED. 1986. Bone and Mineral Research. Vol. 5. Elsevier. Amsterdam.
28. CHAMBERS, T. J. & K. FULLER. 1985. J. Cell Sci. **76:** 155–166.
29. MCSHEEHY, P. M. J. & T. J. CHAMBERS. 1986. Endocrinology **118:** 824–828.
30. CHAMBERS, T. J. 1982. J. Cell Sci. **57:** 247–252.
31. CHAMBERS, T. J., N. A. ATHANASOU & K. FULLER. 1984. Endocrinology **102:** 281–286.
32. MCSHEEHY, P. M. J. & T. J. CHAMBERS. 1987. J. Clin. Invest. **80:** 425–431.
33. THOMSON, B. M., J. SAKLATVALA & T. J. CHAMBERS. 1986. J. Exp. Med. **164:** 104–110.
34. THOMPSON, B. M., G. R. MUNDY & T. J. CHAMBERS. 1987. J. Immunol. **138:** 775–881.
35. RODAN, G. A. & T. J. MARTIN. 1981. Calcif. Tissue Int. **33:** 349–351.
36. PARTRIDGE, N. C., J. J. JEFFREY, L. S. EHLICH, S. L. TEITELBAUM, C. FLISZAR, H. G. WELGUS & A. J. KAHN. 1987. Endocrinology **120:** 1956–1962.
37. PARTRIDGE, N. C., D. ALCORN, V. P. MICHELANGELI, G. RYAN & T. J. MARTIN. 1983. Cancer Res. **43:** 4308–4314.
38. OTSUKA, K., J. SODEK & H. F. LIMEBACK. 1984. Calcif. Tissue Int. **36:** 722–724.
39. SCOTT, D. K., K. D. BRAKENHOFF, J. C. CLOHISY, C. O. QUINN & N. C. PARTRIDGE. 1992. Mol. Endocrinol. **6:** 2153–2159.
40. MEIKLE, M. C., S. BORD, R. M. HEMBRY, J. COMPSTON, P. I. CROUCHER & J. J. REYNOLDS. 1992. J. Cell Sci. **103:** 1093–1099.
41. HAMILTON, J. A., S. LINGELBACH, N. C. PARTRIDGE & T. J. MARTIN. 1985. Endocrinology **116:** 2186–2192.

42. EILON, G. & L. G. RAISZ. 1978. Endocrinology **103:** 1969–1975.
43. DELAISSE, J. M., A. BOYDE, E. MACONNACHIE, N. N. ALI, C. SEAR, Y. EECKHOUT, G. VAES & S. J. JONES. 1987. Bone **8:** 305–313.
44. LENAERS-CLAEYS, G. & G. VAES. 1979. Biochim. Biophys. Acta **584:** 375–388.
45. SELLERS, A., M. C. MEIKLE & J. J. REYNOLDS. 1980. Calcif. Tissue Int. **31:** 35–43.
46. DELAISSE, J. M., Y. EECKHOUT & G. VAES. 1988. Endocrinology **123:** 264–276.
47. HILL, P. A., J. J. REYNOLDS & M. C. MEIKLE. 1993. Biochim. Biophys. Acta **1177:** 71–74.
48. OURSLER, M. J., P. COLLIN-OSDOBY, F. ANDERSON, L. LI, D. WEBBER & P. OSDOBY. 1991. J. Bone Miner. Res. **4:** 375–385.
49. COLLIN OSDOBY, P., M. J. OURSLER, D. WEBBER & P. OSDOBY. 1991. J. Bone Miner. Res. **6:** 1353–1356.
50. BARON, R., L. NEFF, D. LOUVARD & P. COURTOY. 1985. J. Cell Biol. **101:** 2210–2222.
51. BLAIR, H. C., S. L. TEITELBAUM, R. GHISELLI & S. GLUCK. 1989. Science **245:** 855–857.
52. CHATTERJEE, D., L. NEFF, M. CHAKRABORTY, M. LEIT, S. JAMSA-KELLOKUMPU & R. FUCHS. 1992. Proc. Natl. Acad. Sci. USA **89:** 6257–6262.
53. BARON, R. 1989. Anat. Rec. **224:** 317–324.
54. VAES, G. 1988. Clin. Orthop. Relat. Res. **231:** 239–271.
55. BLAIR, H. C., A. J. KAHN, E. C. CROUCH, J. J. JEFFREY & S. TEITELBAUM. 1986. J. Cell Biol. **102:** 1164–1172.
56. BLAIR, H. C., S. L. TEITELBAUM, L. E. GROSSO, D. L. LACEY, H.-L. TAN, D. W. MCCOURT & J. J. JEFFREY. 1993. Biochem. J. **294:** 873–884.
57. BARON, R., L. NEFF, W. BROWN, P. J. COURTOY, D. LOUVARD & M. G. FARQUHAR. 1988. J. Cell Biol. **106:** 1863–1872.
58. RIFKIN, B. R., A. T. VERNILLO, A. P. KLECKNER, J. M. AUSZMANN, L. R. ROSENBERG & M. ZIMMERMAN. 1991. Biochem. Biophys. Res. Commun. **179:** 63–69.
59. AUSZMANN, J., P. COLLIN-OSDOBY, R. GALVIN, P. OSDOBY & B. RIFKIN. 1992. J. Bone Miner. Res. **7:** 884A.
60. OKAMURA, T., H. SHIMOKAWA, Y. TAKAGI, H. ONO & S. SASAKI. 1993. Calcif. Tissue Int. **52:** 325–330.
61. BARON, R., Y. EECKHOUT, L. NEFF, C. FRANCOIS-GILLET, P. HENRIET, J.-M. DELAISSE & G. VAES. 1990. J. Bone Miner. Res. **5:** 203.
62. WUCHERPFINNIG, A. L., Y. P. LI, W. G. STETLER-STEVENSON, A. E. ROSENBERG & P. STASHENKO. 1994. J. Bone Miner. Res. **9:** 549–556.
63. EVERTS, V., J.-M. DELAISSE, W. KORPER, A. NICHOF, G. VAES & W. BEERTSEN. 1992. J. Cell. Physiol. **150:** 221–231.
64. VAES, G. 1972. Biochem. J. **126:** 275–289.
65. FULLER, K., S. KOMIYA, B. C. BRETON, J. T. L. THONG & T. J. CHAMBERS. 1993. J. Bone Miner. Res. **8:** 1081A.
66. EECKHOUT, Y. & G. VAES. 1977. Biochem. J. **166:** 21–31.
67. FALLON, M., S. SILVERTON, P. SMITH, C. MOSKAL, R. CONSTANTINESCU, E. FELDMAN, E. GOLUB & I. SHAPIRO. 1986. J. Bone Miner. Res. **1:** 1A.
68. OURSLER, M. J., P. COLLIN-OSDOBY, L. LI, E. SCHMITT & P. OSDOBY. 1991. J. Cell. Biochem. **46:** 331–334.
69. BABIOR, B. M. 1984. Blood **64:** 956–966.
70. PEPPIN, G. J. & S. J. WEISS. 1986. Proc. Natl. Acad. Sci. USA **83:** 4322–4326.
71. WEISS, S. J., G. PEPPIN, X. ORTIZ, C. RAGSDALE & S. T. TEST. 1985. Science **277:** 747–749.
72. RAMAMURTHY, N. S., A. T. VERNILLO, R. A. GREENWALD, H.-M. LEE, T. SORSA, L. M. GOLUB & B. R. RIFKIN. 1993. J. Bone Miner. Res. **8:** 1247–1253.
73. ZAMBONIN-ZALLONE, A., A. TETI, A. CARANO & P. C. MARCHISIO. 1988. J. Bone Miner. Res. **3:** 517–523.
74. ZAIDI, M., H. K. DATTA, A. PATCHELL, B. MOONGA & I. MACINTYRE. 1989. Biochem. Biophys. Res. Commun. **163:** 1461–1465.
75. AUBIN, J. E., J. N. M. HEERSCHE, J. KANEHISA, A. OKUDA, K. TURKSEN & M. OPAS. 1992. *In* The Biological Mechanisms of Tooth Movement and Craniofacial Adaptation. Z. Davidovitch, Ed. EBSCO Media. Birmingham, AL.
76. TETI, A. & A. ZAMBONIN-ZALLONE. 1992. *In* Biology and Physiology of the Osteoclast. B. R. Rifkin & C. V. Gay, Eds. CRC Press. Boca Raton, FL.

77. ANDERSON, R. E., D. M. WOODBURY & W. S. S. JEE. 1986. Calcif. Tissue Int. **39:** 252–258.
78. HUNTER, S. J., H. SCHRARER & C. V. GAY. 1988. J. Bone Miner. Res. **3:** 297–303.
79. RAMAMURTHY, N. S., A. T. VERNILLO, H.-M. LEE, L. M. GOLUB & B. R. RIFKIN. 1990. Res. Commun. Chem. Pathol. Pharmacol. **70:** 323–335.
80. VERNILLO, A. T., N. S. RAMAMURTHY, H.-M. LEE, S. MALLYA, J. AUSZMANN, L. M. GOLUB & B. R. RIFKIN. 1993. J. Dent. Res. **73:** 367A.
81. SUOMALAINEN, K., T. SORSA, L. M. GOLUB, N. RAMAMURTHY, H.-M. LEE, V.-J. UITTO, H. SAARI & Y. T. KONTTINEN. 1992. Antimicrob. Agents Chemother. **36:** 227–229.
82. RIFKIN, B. R., L. M. GOLUB, F. SANAVI, A. T. VERNILLO, A. P. KLECKNER, T. F. MCNAMARA, J. M. AUSZMANN & N. S. RAMAMURTHY. 1992. *In* The Biological Mechanisms of Tooth Movement and Craniofacial Adaptation. Z. Davidovitch, Ed. EBSCO Media. Birmingham, AL.
83. WASIL, M., B. HALLIWELL & C. P. MOORHOUSE. 1988. Biochem. Pharmacol. **37:** 775–778.
84. VERNILLO, A. T., N. S. RAMAMURTHY, L. M. GOLUB & B. R. RIFKIN. 1994. Curr. Opin. Periodontol. **2:** 111–118.
85. RIFKIN, B. R., A. T. VERNILLO & L. M. GOLUB. 1993. J. Periodontol. **64:** 819–827.
86. RIFKIN, B. R., B. C. GOMES, A. KLECKNER, N. S. RAMAMURTHY & L. M. GOLUB. 1986. J. Dent. Res. **65:** 517A.
87. RIFKIN, B., F. SANAVI, A. KLECKNER, B. GOMES, N. S. RAMAMURTHY & L. GOLUB. 1988. J. Dent. Res. **67:** 2068A.
88. AUSZMANN, J., A. KLECKNER, A. VERNILLO & B. RIFKIN. 1992. J. Dent. Res. **72:** 1928A.
89. GAY, C. V. 1992. *In* Biology and Physiology of the Osteoclast. B. R. Rifkin & C. V. Gay, Eds. CRC Press. Boca Raton, FL.
90. RIFKIN, B. R., N. S. RAMAMURTHY & C. V. GAY. 1992. J. Dent. Res. **71:** 1038A.
91. DONAHUE, H. J., K. IIJIMA, M. S. GOLIGORSKY, C. T. RUBIN & B. R. RIFKIN. 1992. J. Bone Miner. Res. **7:** 1313–1318.
92. ZAIDI, M., A. S. M. T. ALAM, C. L. H. HUANG, M. PAZIANAS, C. M. R. BAX, B. E. BAX, B. S. MOONGA, P. J. R. BEVIS & V. S. SHANKAR. 1993. Cell Calcium **214:** 271–277.
93. ZAIDI, M., A. S. M. T. ALAM, V. S. SHANKAR, B. E. BAX, C. M. R. BAX, B. S. MOONGA, P. J. R. BEVIS, C. STEVENS, D. R. BLAKE, M. PAZIANAS & C. L. H. HUANG. 1993. Biol. Rev. **68:** 197–264.
94. BAX, C. M. R., V. S. SHANKAR, A. S. M. T. ALAM, B. E. BAX, B. S. MOONGA, C. L. H. HUANG, M. ZAIDI & B. R. RIFKIN. 1993. Biosci. Rep. **13:** 169–174.
95. ZAIDI, M., B. S. MOONGA, C. L. H. HUANG, A. S. M. T. ALAM, V. S. SHANKAR, M. PAZIANAS, J. B. EASTWOOD, H. K. DATTA & B. R. RIFKIN. 1993. Biosci. Rep. **13:** 175–182.

Treatment of Destructive Arthritic Disorders with MMP Inhibitors

Potential Role of Tetracyclines

ROBERT A. GREENWALD

Division of Rheumatology
Long Island Jewish Medical Center
Room 337, Faculty Practice
New Hyde Park, New York 11042

Although there are over 100 different rheumatic afflictions, three relatively common disorders—rheumatoid arthritis, osteoarthritis, and psoriatic arthritis—account for virtually all of the naturally occurring (i.e., nontraumatic) joint destruction that afflicts the human species. Although the last three decades have seen great strides in our understanding of the mechanisms and control of joint inflammation, rheumatologists must sadly admit that the greatest advance in the treatment of destructive arthritis has been the orthopedic joint implant rather than a treatment strategy to control the inexorable progression of arthritis—strong testimony to our failure to identify and impede the mechanisms leading to collagen and proteoglycan degradation.

Hundreds of physiologic species have been implicated in the pathogenesis of both osteoarthritis and rheumatoid arthritis—immunologic elements, cell types of many lineages, cytokines, prostaglandins, reactive oxygen species, etc.; every national meeting brings forth yet more players to the game. But in the simplest analysis, it seems quite clear that both collagen and proteoglycan destruction must be primarily enzymatic in nature. Although we may not have fully identified all of the enzymes capable of this process, surely the known matrix metalloproteinases (MMPs) represent a strong starting point for a pharmacologic approach. Identification of agents that might inhibit MMPs *in vivo* has long been a therapeutic goal for modulation of connective tissue degradation in rheumatoid arthritis and related disorders.[1–3] In this paper, I review some of the data on pathologically excessive MMP activity in human and animal model arthritis, some strategies for monitoring connective tissue destruction, and the results to date with MMP inhibition therapy in animal models and in patients. I also suggest some directions for future research along these lines.

MMP ACTIVITY IN ARTHRITIS

Space does not permit a fully documented review of all aspects of MMP involvement in animal models of arthritis nor in human disease. A number of reviews are available.[4–7] MMPs are clearly capable of degrading the macromolecules of connective tissue matrix, and it is widely believed that these enzymatic actions play an important role in destruction of cartilage, tendon, ligaments, and bone.

Rheumatoid Arthritis

In inflammatory arthritides such as rheumatoid arthritis (RA), the enzymes that might participate in these reactions include (but are not limited to) interstitial collagenase (MMP-1), 72-kDa gelatinase A (MMP-2), stromelysin (MMP-3), neutrophil collagenase (MMP-8), and 92-kDa gelatinase B (MMP-9). These enzymes have been identified in homogenates and cultures of rheumatoid synovium,[8–11] detected in inflammatory synovial fluids (SFs),[12–14] and localized immunologically and by *in situ* hybridization in proliferative pannus and synovium.[15–18] In addition, the breakdown products of collagenase action have also been detected in rheumatoid synovium.[19] (For a review of MMP in synoviocyte cell cultures, see Brinckerhoff.[5])

Walakovits *et al.*[14] compared stromelysin (MMP-3) and collagenase (MMP-1) levels in SFs from patients with either RA or with traumatic arthritis using a sensitive and specific double-antibody sandwich ELISA. There were modest elevations of both enzymes in the posttraumatic samples, and impressive elevations of both enzymes in RA. In both cases, however, the MMP-3 levels were much higher than the MMP-1 levels, and most of the MMP-3 was proenzyme. Of great interest was the calculation that the level of tissue inhibitor of metalloproteinase (TIMP) known to occur in the SF would at best be only 20% of that needed to inhibit this enzyme level if all of the pro-MMPs were activated. Such data suggest that a "cocktail" of MMP inhibitors might be required to truly protect all connective tissues from degradation.

The therapeutic potential for use of MMP inhibitors in RA is clearly established by the pathologic excess of collagenase and stromelysin, because the primary connective tissue substrates for these enzymes—collagen and proteoglycan—are known to be degraded during the course of the disease. The recent findings of aggrecan degradation by collagenases (MMP-1 and MMP-8) reinforce the potential benefits of an inhibitor(s) of excess MMP action.[20]

Osteoarthritis

Osteoarthritis (OA) presents a pathophysiologic conceptual problem which differs from that of rheumatoid arthritis. In the latter, one can imagine a self-sustaining immunologic process being initiated by some as yet unidentified stimulus, perhaps an infectious agent; the various inflammatory cells react and interact to produce enzymes that destroy matrix over a time period that might be as short as one to two years. In OA, however, the inciting agent is totally unknown, and the disease process is believed to transpire over a period of years, if not a decade(s), with only minimal inflammation. Only by constructing a scenario involving a subtle excess of MMP to inhibitor can one imagine matrix degradation taking place so slowly.

Nevertheless, numerous studies, going back to 1977, have identified all three major MMPs as probable contributors to collagen and proteoglycan degradation in OA. Recent reviews by Dean,[21] Carney,[22] and others[23] summarize many of these findings. MMPs have been found in excess in synovium, in SF, and directly within cartilage. In a series of recent studies, using mRNA analysis, OA tissues showed consistent elevation of MMP-1, MMP-3, and TIMP as compared to controls; as with RA, MMP-3 levels often exceed MMP-1 rather substantially.[24] Another recent study has shown the presence of MMP-9 in OA cartilage.[25] It remains to be determined, for both RA and for OA, whether a selective inhibitor of MMP-1, MMP-8, MMP-9, or MMP-3 would be a better treatment for prevention of matrix degradation.

ANIMAL MODELS

Scientists who have not had hands-on experience in the care of patients with RA or OA often underestimate the virulent biology of these diseases. Despite polypharmacy with as many as six or eight different agents administered simultaneously to patients with RA, the disease usually proceeds inexorably to connective tissue destruction. Even the use of agents optimistically called disease modifying antirheumatic agents (DMARDs) generally fails to stop these destructive processes.

Part of the problem lies, perhaps, in that we have no good *in vitro* systems for studying joint erosion and cartilage injury. Despite systems such as interleukin-stimulated cartilage degradation, it is not at all clear that agents which block this process in an 18-h culture will have a salutary effect in a human afflicted with a 20-year-long illness. Unfortunately, there also are no ideal animal models for the major human rheumatic diseases. All of the RA models have both virtues and deficits. For OA the situation is worse because most rapid models require drastic surgical intervention as an inducing agent, and the alternative—spontaneous disease in a susceptible species—often requires experiments of more than 12 months' duration. (For reviews see Greenwald[7] and Burton-Wurster *et al.*[23])

MMP Activity in Animal Models

Despite the widespread use of the adjuvant arthritis (AA) model, surprisingly little attention has been paid to direct study of connective tissue degradation; most work has focused on anti-inflammatory effects and on the immunology of the system. Pannus formation with neutrophil accumulation and loss of metachromasia has been shown in the AA model.[26] Several papers have identified collagenase as a factor that might mediate cartilage breakdown. Kuberasampath and Bose[27] reported increased levels of collagenase (as well as several cathepsins) in skin, liver, bone, kidney, and spleen. Suppression of AA with prednisolone reversed the enzyme changes. It was also reported that serum collagenase levels in AA fell with duration of disease, in association with induction of natural inhibitors of neutral proteinases (α1-antiproteinase and α2-macroglobulin).[28] Transin, the rat equivalent of stromelysin, has been immunolocalized in tissues from rats with streptococcal cell wall arthritis.[29]

Coffey and Salvador[30] measured collagenase activity directly in inflamed rat paws. They noted a 30-fold increase at day 26, with substantial amounts of latent enzyme that could be trypsin activated. Induction of collagenase paralleled the inflammatory index. Brinckerhoff *et al.*[31] reported that they cultured adherent cells from the arthritic synovium and measured elaboration of collagenase and PGE_2; both moieties were readily expressed in culture, and oral treatment of the animals with 13-*cis*-retinoic acid suppressed the changes.

Confirmation of induction of both collagenase and gelatinase in AA paw tissue has been obtained in my laboratory using both conventional fibril assays as well as SDS-PAGE fluorography. Activity of both enzymes is readily inhibitable either *in vivo* or *in vitro* by tetracyclines.[32–33] Interestingly, the combination of tetracycline treatment with a nonsteroidal anti-inflammatory drug (NSAID)—either flurbiprofen, ibuprofen or tenidap, the agents tested so far—appears to have a marked salutary effect on radiographic bone destruction as well. Ramamurthy *et al.* have extended these findings of joint protection in the AA model from the paw to the temporomandibular joint, again showing that the combination of an NSAID and a tetracycline-derived MMP inhibitor prevented radiographic destruction in concert with suppression of enzyme activity (see paper in this volume).

An alternative to the adjuvant rat is the induction of arthritis by injection of purified type II collagen, known as CIA (collagen-induced arthritis). Surprisingly little data exists on collagenase activity in the collagen arthritis model; stromelysin (MMP-3) has attracted the most attention. Hasty *et al.*[34] confirmed the loss of safranin O staining in CIA and immunolocalized MMP-3 to the cartilage-pannus junction and to the proliferating synovium. Additional staining was noted in chondrocytes and near erosions, but the enzyme was absent from the cartilage matrix. Collagenase was found only at the cartilage-pannus junction. Neither enzyme was found in nonarthritic controls.

Two studies have attempted to find the MMPs responsible for cartilage degradation in rabbit immune-mediated synovitis (the Dumonde-Glynn model). Collagenase was identified in cartilage fragments (directly extracted, without culture) from chronic synovitis knees, exclusively in latent form, readily activatible with trypsin or APMA; normal tissue contained no enzyme.[35] In a contrasting study, another group found substantial MMP production (collagenase, stromelysin) from the synovium of such animals, but very little activity from explants of cartilage in culture.[36] Similar data were obtained in two additional rabbit models of arthritis.[37]

THERAPEUTIC USE OF MMP INHIBITORS IN ANIMAL MODELS

Most studies of disease progression in the AA model have focused on evidence of radiologic bone/joint destruction. The published direct evidence of an anticollagenase effect consists of a few studies in which HyPro excretion was monitored, our previously reported findings on MMP activity in inflamed rat paw tissues,[33] and our recent work on urinary collagen-derived cross-links (see below). In the AA model, many conventional NSAIDs, for example, piroxicam, appear to be "remittive" agents in that radiologic evidence of bone destruction (demineralization or bone loss but not excretion of collagen cross-links) is inhibited;[38] under some circumstances, histologic evidence of cartilage destruction can also be reversed. Bone and cartilage destruction in AA appear to be much more prostaglandin-dependent in rats than in human RA. This makes interpretation of radiologic changes most difficult and mandates the careful use of controls.

Beyond various X-ray techniques and histology, only three obvious biochemical markers seem practical for testing the effect of MMP inhibitors in AA and similar animal models: urinary HyPro, tissue MMP levels, and urinary excretion of other collagen-breakdown products. Reddy and Dhar showed that urinary HyPro excretion was increased 28% after 29 days of AA, but the value went up to 112% of control at day 49.[39] By comparing the total and specific activities of labeled HyPro in various pools, they concluded that AA was characterized by enhanced collagen degradation. Direct tissue measurement of enzymes is one of the techniques that we have employed, and urinary excretion of the pyridinium-derived collagen cross-links is the latest marker to become available (see below and poster papers in this volume).

A review of the literature for evidence of anti-MMP effect in animal models reveals only a handful of studies. In the Brinckerhoff study,[31] oral administration of high-dose (40–160 mg/kg) 13-*cis*-retinoic acid reduced collagenase secretion by cultured synovial cells from treated animals; a parallel effect on paw swelling was noted, but no measures of tissue damage were assessed. A conventional NSAID, naproxen, was reported to decrease cartilage MMP activity in the canine anterior cruciate ligament model.[40] Tenidap, a new NSAID not yet on the market, has been reported to inhibit the release of activated neutrophil collagenase as well as being anti-inflammatory in AA. Carmichael *et al.*[41] administered TIMP intraperitoneally to

DBA/1 mice in which collagen arthritis had been induced; details were not provided, but they reported an overall reduction by 50% in clinical, radiologic, and histologic severity.

On the specific question of bone loss, most studies have employed known inhibitors of osteoclastic bone resorption. For example, calcitonin given subcutaneously to AA rats markedly ameliorated the usual loss of bone density,[42] an effect that was enhanced when the hormone treatment was combined with an NSAID. A diphosphonate showed similar effects.[43] By use of a sensitive but complex radiologic system for assessing bone density, cyclosporin A has also been shown to ameliorate bone damage in AA, with parallel findings on undecalcified histology.[44]

Exciting MMP-inhibition data for an animal model of OA come from the laboratories of Brandt and his colleagues at Indiana University (described in detail in this volume). The Brandt group uses an aggressive osteoarthritis model in dogs (surgical intervention in the knee combined with sectioning of the dorsal root ganglion, i.e., the afferent nerves), and they have noted that prophylactic administration of doxycycline (DOXY), a semisynthetic tetracycline (TET), produces substantial amelioration of the gross and histologic features of the model, correlating with decreased collagenase and gelatinase activity as well. This group is continuing their studies with treatment (as opposed to prevention) protocols and with further *in vitro* studies on the mechanism of action of the effect.

Inhibition of MMPs in Adjuvant Arthritis by Tetracyclines—Histologic Findings

Our concept of the use of tetracycline-derived MMP inhibitors focuses on MMP-mediated events, such as connective tissue degradation, rather than an anti-inflammatory effect, which is easily attained with a host of other compounds of much greater potency. In our 1992 paper,[33] we clearly saw a separation between the ability of drugs to reduce paw swelling and joint score as opposed to reduction in MMP level and radiologic deterioration. The effect of the TETs on tissue MMP level *in vivo* was sometimes limited, and this turned out to be due to failure of the drug to penetrate the hard tissues within the inflamed limbs. When an NSAID was given simultaneously, the combination usually resulted in complete elimination of MMP activity as well as substantial, if not total, joint preservation on X-ray.

In further related studies not previously reported except as an abstract,[45] we have examined the rat paws using histomorphometry applied to undecalcified sections. These studies were performed by Dr. Webster S. S. Jee of the University of Utah and his colleagues, Q. Zeng and H. Ke. Ankle joints from AA rats were cleaned, stained with Villaneuva stain for one week, destained, embedded in methyl methacrylate, hardened for 5–7 days, cut serially into 230-μm sections, and then ground to 100 μm. Microradiographs were taken and the sections reground to 30 μm for histology (toluidine blue) and histomorphometry by semiautomated Digitizing Image Analyzer. Light and/or fluorescent microscopes were used to project the images onto a digitizing pad which were then quantified on Macintosh computers using stereology software.

The sensitive parameters included cortical bone area, trabecular bone area, resorption area, and measures of internal and external woven bone (periosteal new bone). Adjuvant disease produces dramatic changes in all of these parameters especially in the distal tibia, talus, calcaneus, cuneiform, and cuboid. Flurbiprofen (FBP) by itself partially corrected many of these abnormalities, but the combination of FBP plus chemically modified tetracycline (CMT)-1 normalized several of them rather dramatically; due to the small number of subjects for each group, however,

TABLE 1. Histomorphometric Measurements on Adjuvant Rat Paws after Treatment with CMT-1 Alone, Flurbiprofen Alone, or a Combination

Parameter	Normal	No Rx	FBP	CMT	Combo
DTM, tba	64.6	30.0	53.1	37.7	56.6
DTM, cba	58.3	41.9	53.6	45.1	57.2
DTM, ra	0	5.8	2.7	3.0	1.5
DTE, iwba	0	0.30	0.16	0.40	0.10
TAL, iwba	0	0.20	0.08	0.14	0.03
TAL cart resorption	0%	32%	2.4%	27%	0%
CAL iwba	0	2.9	0.9	1.12	0.44
CAL cart	0%	48%	20%	33%	7.2%
CUB total bone area	82%	51%	48%	56%	48%

NOTE: All measurements in mm^2 except where % shown. Most of the differences between AA/No Rx and normal, and between FBP and AA/No Rx, are statistically significant, but the trends between combination treatment and FBP alone are not.

Abbreviations: FBP, flurbiprofen; CMT, chemically modified tertracycline; DTM, distal tibial metaphysis; tba, total bone area; cba, cortical bone area; ra, resorption area; DTE, distal tibial ephiphysis; iwba, internal woven bone area; TAL, talus; cart, cartilage; CAL, calcaneus; CUB, cuboid.

statistical significance versus FBP alone was not achieved. Internal new woven bone formation was the parameter most beneficially affected by FBP and/or CMT (TABLE 1); a synergistic benefit of the combination was apparent at multiple anatomic sites. The preservation of bone integrity by the combination treatment is readily appreciated histologically (FIG. 1) and radiologically (FIG. 2).

In a parallel, electron microscopic study performed in collaboration with Dr. Takahisa Sasaki of Showa University Dental School, Tokyo, also previously published as an abstract,[46] treatment regimens utilizing FBP, CMT or both were used to assess effects on bone ultrastructure. FBP reduced paw swelling and joint score but CMT had no such effects; CMT markedly reduced tissue MMP activity, but FBP had no such effect. The combination abolished MMP activity without loss of anti-inflammatory efficacy. Radiologically, FBP produced some amelioration of bone loss, CMT had little effect, and the combination, acting synergistically, virtually normalized the radiographs. Light microscopy and backscattered electron microscopy of AA tibias showed prominent osteoclastic bone resorption in both compact and trabecular bones; FBP produced some lessening of the resorption. The combination of FBP + CMT ameliorated bone resorption more significantly and also stimulated new trabecular bone formation. Osteoblasts from normal rats contained abundant cytoplasmic organelles including secretory granules of procollagen and were adjacent to osteoid; osteoclasts (OCs) exhibited well-developed ruffled borders contiguous with resorptive bone surfaces. In AA, osteoblasts (OBs) still had abundant cytoplasmic organelles but osteoid was missing from the OB-bone interface, and OCs still showed ruffled borders. Some OBs in AA rats were abundant in lysosomes and showed phagocytosis of intact collagen fibers. Quantitative X-ray

→

FIGURE 1. Photomicrographs of histologic sections of rat ankles (see text for details). **Top panel,** normal rat; **center panel,** untreated adjuvant arthritis; **lower panel,** after treatment with combination of flurbiprofen and chemically modified tetracycline-1.

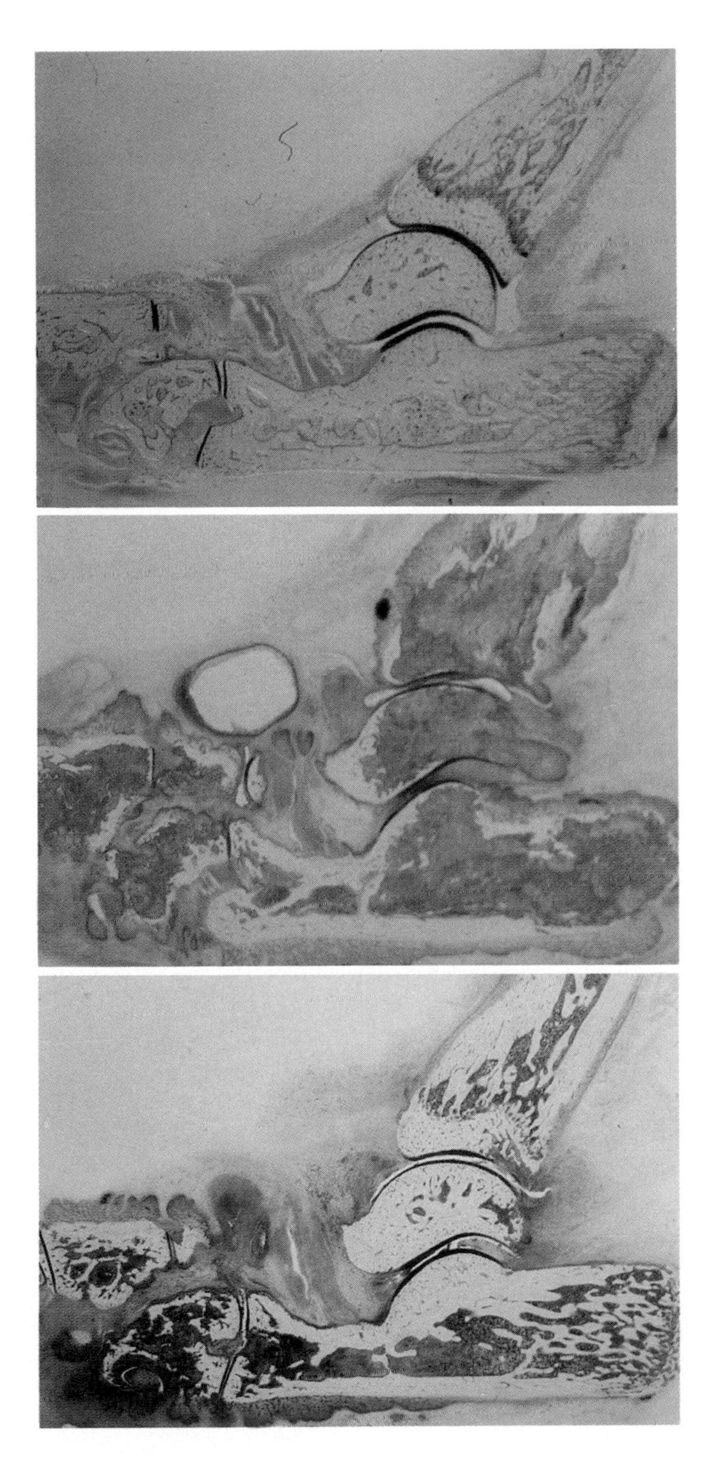

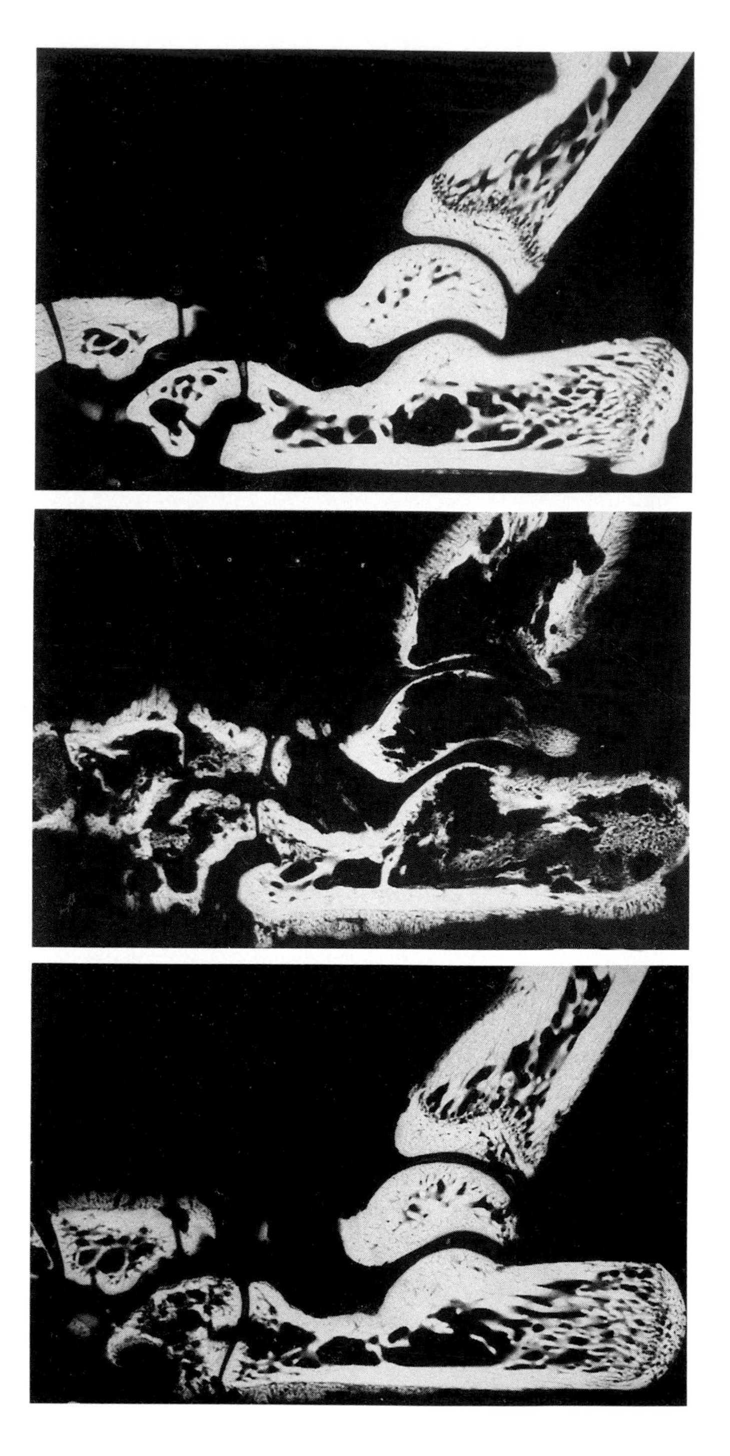

TABLE 2. Comparison of PYD and DPYD Measurements in Urine of Adjuvant Arthritic Rats Collected at Day 21 Using Two Different Assay Systems—ELISA (Metra Biosystems) or HPLC (Nichols Institute)

	ELISA[a]	HPLC PYD	HPLC DPYD	Arthritis Score
Normals				
1	322	82	72	n/a
2	184	66	60	n/a
3	318	74	64	n/a
4	275	72	64	n/a
Mean	275	72	64	
Arthritic				
5	508	180	132	9
6	414	239	218	8
7	520	307	278	13
8	545	161	108	9
9	470	163	112	4
10	278	152	107	6
11	744	202	159	12
Mean	497	202	159	
% Normal	180%	280%	248%	

[a]ELISA assay measures only PYD.
Abbreviations: HPLC, high-performance liquid chromatography; PYD, pyridinoline; DPYD, deoxypyridinoline.

microanalysis of weight% Ca, P, and Ca/P ratio in normal bone revealed 28.24% Ca, 13.35% P, and Ca/P = 1.63; in combination therapy in AA bone, the values were 22.46, 10.84, and 1.60. These data could be interpreted to indicate new bone formation rather than inhibition of resorption. Alternatively, increased production of MMP-1 by OBs, leading to osteoid loss in AA, is blocked by the combination drug treatment. The radiologic improvement seen with these drug regimens correlated with restoration of normal OB and OC histology and functional relationships.

Inhibition of MMPs by Tetracyclines—Biochemical

In data presented at the 1993 San Antonio meeting of the American College of Rheumatology,[47] we showed that AA in rats resulted in enhanced excretion of pyridinoline (PYD), detectable either by HPLC or by an ELISA (Metra Biosystems) designed for human use but which cross-reacts well with the rat antigen. In an exploratory experiment, urine was collected from small groups of both normal and untreated arthritic rats, and PYD/DPYD excretion measured by both methods. Pathologically excessive excretion of collagen cross-links was readily demonstrable by both assays, although the absolute amounts were not equivalent (TABLE 2).

←

FIGURE 2. Microradiographs of rat ankles. **Top panel,** normal rat; **center panel,** untreated adjuvant arthritis; **lower panel,** after treatment with combination of flurbiprofen and chemically modified tetracycline-1.

We then initiated treatment with various CMTs in groups of arthritic rats, with parallel groups receiving either FBP, ibuprofen, or both. Two CMTs, known as CMT-1 (4-dedimethylaminotetracycline) and CMT-4 (7-chloro-4-dedimethylamino-tetracycline), virtually normalized the pathologically excessive excretion of PYD in these animals (FIG. 3). The FBP treatment alone did not affect the PYD excretion, demonstrating a selective effect of MMP inhibition not shared by NSAID. Ganu *et al.* have obtained similar data using DOXY (see paper in this volume).

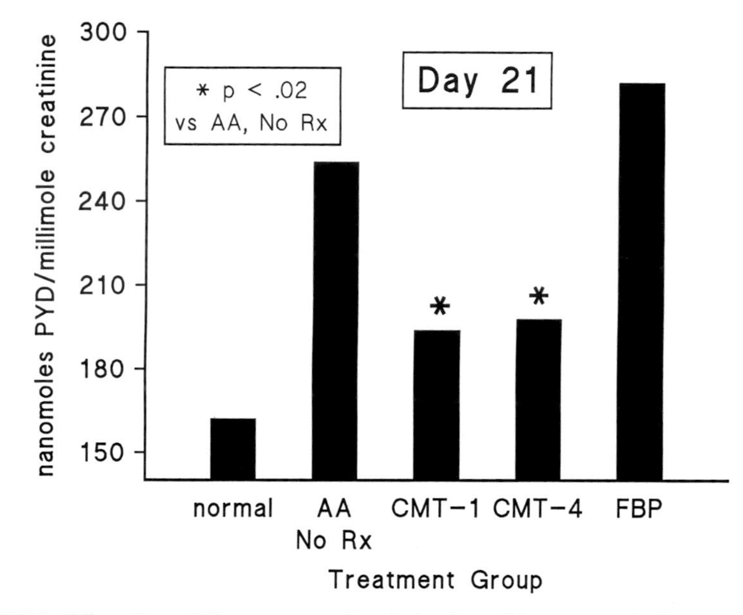

FIGURE 3. Effect of two different tetracycline derivatives with potent anticollagenase activity on urinary excretion of pyridinoline (PYD) in rats with adjuvant arthritis at day 21. At day 14, lesser elevation of PYD, but the same pattern of inhibition was observed. Like flurbiprofen, ibuprofen also had no effect on the PYD excretion.

The Guinea Pig Model

Virtually all animal models of OA depend on an initiating surgical lesion; the arthropathy which then develops has features of normal human OA but is generally more aggressive and rapid. Spontaneous knee OA develops in guinea pigs (GPs) fed *ad libitum* after 9–12 months and is characterized by typical histologic changes[48–49] and by the appearance of the proteoglycan epitope 3-B-3(−) that can be detected in extracts of articular cartilage and/or meniscus in a reparative phase.[50] Suppression of this marker can be used to monitor the efficacy of chondroprotective drugs in slowing the disease process. We therefore administered DOXY chronically to a cohort of GPs to determine if such treatment could suppress the advent of spontaneous OA.

Eighteen male albino Hartley GPs (starting weight 689 g) were fed *ad libitum* for 12 (n = 10) or 18 (n = 8) months. Two-thirds of the animals received DOXY (20 mg/kg, recalculated monthly) orally in the cheek pouch every weekday; 6 served as controls (no therapy). At sacrifice, the joints were examined grossly for features of

OA, radiographs were performed, and tissues were harvested for histologic and immunologic studies (performed in the laboratory of Dr. Bruce Caterson, University of North Carolina, Chapel Hill). Articular cartilage and/or meniscus was dissected from one knee (and some hips) of each GP, extracted in standard manner in 4M GuCl containing 10 mM DTT, alkylated with iodoacetate, dialyzed, lyophilized, reconstituted in 8M urea, fractionated by agarose/acrylamide composite gel electrophoresis, transferred to nitrocellulose, and subjected to immunolocalization with monoclonal antibodies recognizing the 3-B-3(−) epitope, and quantitated by scanning densitometry. Results were expressed as a ratio of 3B3−/3B3+ with or without chondroitinase pretreatment; a ratio less than 0.30 meant suppression of epitope expression.

At 12 months, the mean weight of both groups of GPs was 1090 g; at 18 months, there was a slight but equivalent weight decline in both groups. At sacrifice at 12 months, gross evidence of OA (cartilage erosion, osteophytes) was rated as severe in 2 of 4 controls, mild in 1, and no OA in 1; in the DOXY animals, one animal had moderately severe OA and the remaining 5 had no or trivial lesions. Histologically (toluidine blue), the DOXY specimens generally showed preservation of cartilage thickness with no loss of metachromasia, fewer (or no) osteophytes, and preservation of cellularity (FIG. 4). Radiographs of the knees of 8 animals at 18 months showed osteophyte formation/joint space narrowing in only 1 of 6 DOXY animals. Histology at 18 months showed further evidence of reduced osteophyte formation and of preservation of toluidine blue metachromasia.

In articular cartilage extracts from 3 of 5 DOXY-treated animals (12 months), the appearance of the 3-B-3(−) epitope was suppressed below 0.30; this result was noted in only 1 of 3 controls. In the meniscus extracts, 3 of 6 DOXY-treated samples showed suppressed 3-B-3(−), but only 1 of 4 controls was low. Immunohistologic staining of sections from two animals confirmed the presence of the 3-B-3(−) epitope in the superficial layer of the meniscus from an untreated animal; in sections from a DOXY-treated GP, no epitope was detected. At 18 months, the appearance of the 3B3 epitope was clearly suppressed in most of the tissues that could be examined. When viewed in the context of the results obtained by Brandt and his group in Indiana using the canine ACL model, the data strongly suggest that prophylactic administration of DOXY or a similar TET-based MMP inhibitor in early human osteoarthritis might be beneficial.

MONITORING DISEASE PROGRESSION

Animal Studies

Documentation of MMP inhibition in animal models has generally depended upon direct enzyme measurement in extracted tissues correlated with gross and/or histologic changes. The two papers in this volume from my laboratory and by Ganu *et al.* are the first demonstrations using PYD excretion in the AA model. This will probably serve as the "gold standard" for many future studies.

PYD excretion abnormalities have been reported in four other studies involving animals. Ovariectomy in rats resulted in increased excretion of both PYD and DPYD at seven weeks and correlated with reduced residual collagen content of the bones (immunohistologic staining with antibodies to type I collagen).[51] In contrast, feeding alcohol to rats for six weeks reduced the DPYD excretion with no effect on PYD.[52] Infusion of parathyroid hormone into normal rats substantially increased excretion of both markers,[53] suggesting, to this author, that an MMP inhibitor, for example,

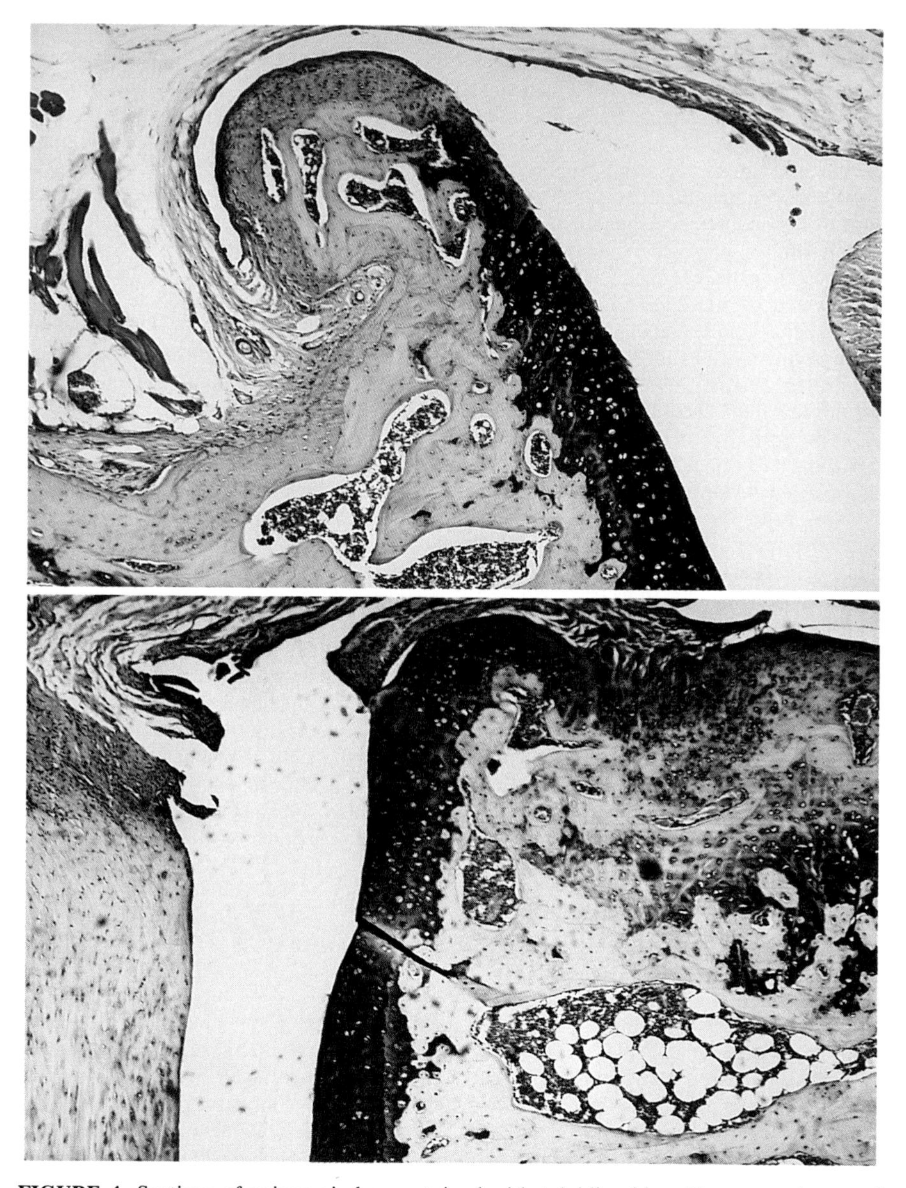

FIGURE 4. Sections of guinea-pig knees stained with toluidine blue. **Upper panel,** control guinea pig showing loss of metachromatic staining (blue), osteophytes, and cartilage narrowing; **lower panel,** guinea pig after 12 months of doxycycline treatment, showing preservation of metachromasia (purple), no osteophytes, and increased cartilage thickness.

DOXY, might be of benefit in chronic renal failure patients who invariably develop renal osteodystrophy and secondary hyperparathyroidism. Finally, the observation has been extended to rabbits, where injection of chymopapain into knees also resulted in enhanced PYD excretion.[54]

Human Disease

Animal models have the advantage of uniformity, that is, most of the subjects will get the same disease at the same time, progressing at the same rate. Human disease variability is orders of magnitude greater. Tight inclusion criteria are needed when designing a study of human RA or OA in order to attain homogeneity of the

TABLE 3. Parameters of MMP Inhibition for Human Disease

Body Fluids That Can Be Sampled	
Serum	Plasma
Joint fluid	Urine
GCF	Neutrophils
Serous cavity fluid	
Markers That Might Be Measured in Body Fluids	
Enzymes	TIMP
α_1-Antitrypsin	α_2-Macroglobulin
HyPro	PYD/DPYD
PG fragments (aggrecan, KS, etc.)	
Tissue Markers	
Residual matrix molecule content (e.g., collagen, GAG)	
Functional assays of MMP activity in tissue extracts (spontaneous activity, after activation)	
Immunohistologic localization of MMP enzymes	
mRNA for MMPs—*in situ* hybridization	
Degradation products, e.g., degraded collagen detected immunologically	
Markers of regeneration, e.g., 3-B-3	
Markers of the Whole Organism	
Radiologic—conventional erosion counts or bone loss, bone density, MRI	
Clinical evidence of tissue integrity: teeth, bone, gums, tendons, joint stability	
Survival (e.g., metastatic cancer)	
Arthroscopic assessment	

treatment and control (or placebo) groups, and large numbers of subjects must be enrolled in order to attain a sufficient number of subjects at the conclusion of the trial, because there will always be logistic, toxic, or personal dropouts. When one factors in the time needed to see changes in parameters of connective tissue degradation (e.g., X-rays), the problems of doing an MMP inhibitor trial in human arthritis become rather formidable.

The search for surrogates has therefore been given a great deal of attention by those interested in this problem. TABLE 3 shows a list of possible parameters by which an MMP inhibitor might be evaluated in humans. The list encompasses both collagenase- and stromelysin-mediated events; there seems to be no obvious gelatinase breakdown product which one could assess, although the enzyme itself can be readily measured.

TREATMENT OF HUMAN ARTHRITIS WITH TETRACYCLINES

Interest in the use of antibiotics in general, and TETs in particular, for treatment of human arthritides, has been sustained for several decades. Three rationales have been cited for such an approach: (1) antibiotic elimination of an as yet unidentified, putative, causative organism; (2) anti-inflammatory benefit; and (3) inhibition of excessive MMP activity.

The antibiotic approach has no apparent scientific validity. No organism has yet been identified in RA tissue, let alone a TET-sensitive one, despite years of search. No controlled antimicrobial trial has been published in a peer-reviewed journal. Anecdotal data from persons utilizing widely varying antibiotic regimens in no consistent, or even controlled, pattern can hardly be construed as scientific data.

Dermatologists have long known that certain TETs have mild anti-inflammatory properties; these were summarized in a earlier paper from my laboratory.[33] Two decades of intense pharmaceutical research have brought forth over 30 currently marketed NSAIDs, all of which have anti-inflammatory powers far in excess than those of any TET. Because most patients with RA are generally receiving one or more NSAIDs plus low-dose prednisone, and because their disease proceeds inexorably despite what would appear from *in vitro* work to be doses of anti-inflammatory medication sufficient to quench even the most zealous neutrophil, it seems irrational to expect a substantive additional *anti-inflammatory* effect from addition of a tetracycline. The same argument holds true in OA, where years of NSAID therapy does not prevent disease progression.

Despite the above caveats, there have been a number of trials of tetracycline treatment for RA. The first was the Skinner trial of 1971[55] in which 30 patients were given either placebo or tetracycline HCl, 250 mg/day, for one year. Using standard evaluation parameters of inflammatory disease, no statistically significant difference was found between treatment and control. The dose chosen was low, the drug used was not a good MMP inhibitor compared to other TETs, and compliance rates were not assessed. However, it is of interest to note that in the treatment group, only 1 patient out of 5 showed progression of hand X-ray erosions, whereas 3 out of 7 in the placebo group worsened.

After publication of our studies of minocycline (MIN) inhibition of rheumatoid collagenase, as well as other studies, the Dutch and Israeli groups performed additional studies,[56–58] both of which were updated with presentations at the San Antonio meeting of the American College of Rheumatology (November, 1993). The second Dutch MIN trial showed statistically, but not clinically, significant changes in a few minor indices but no overall or clinically useful improvement, as well as minor changes in laboratory findings that were more impressive statistically but also were of no clinical import. No radiologic improvement or arrest was noted (6 months). The Israeli experience is primarily open-label, and they enthusiastically report that despite a high incidence of skin discoloration from MIN, a quarter of their patients have continued to take the drug; however, another quarter dropped out for lack of efficacy, and another quarter for toxicity (including the skin pigmentation and oral candidiasis.)

Based partly on public pressure and partly on the scientific background regarding MIN inhibition of collagenase, the National Institutes of Health sponsored a double-blind, placebo-controlled trial known as MIRA (minocycline in RA) which was presented in San Antonio in preliminary and abbreviated fashion (the authors were granted only 5 min). Although the results have not been published as of this writing, partial data is known. The patients had very mild RA, as evidenced, for example, by a mean hematocrit of 40 and a mean ESR of 33. (In even moderate RA, mild anemia is common and an ESR above 40 is the general rule.) The mild overall disease pattern was confirmed by several other parameters as well.

The investigators relied heavily on a parameter called "meaningful improvement in joint swelling," which is defined as 50% or more reduction in number of swollen or tender joints. This result was reported to have been attained in more than 50% of the MIN patients as opposed to only about 40% of the placebo cases, $p < 0.05$. However, this was apparently the only clinical index of inflammatory disease activity that improved significantly. As with the Dutch studies, overall improvement on subjective scales—global assessment of disease activity by either patient or examiner—could not be demonstrated. As in all previous trials, some toxicity, notably dizziness, was noted.

It is the thesis of the current author that although there may be great potential benefit from use of tetracyclines in RA, as well as in OA, studies that focus only on parameters of inflammation rather than on tissue destruction do not address the major problem at hand and do not exploit the ultimate potential of the drug. The best analogy for potential use of a tetracycline in RA is to that of parenteral gold injections. In successful gold treatment, patients do eventually feel better, with subsidence of swelling, morning stiffness, etc., but they also have the potential for a disease-modifying effect. If addition of a TET to an RA regimen improves symptom levels, the result is desirable, but the primary motivation should be to forestall tissue destruction.

I believe that the following principles should control future trial design:

1. The major rationale that should be pursued is that of MMP inhibition, in which regard a tetracycline would be an adjunct to treatment with multiple other agents. Inasmuch as no known agent stops the progression of either RA or OA, it makes no sense to design a trial in which a TET is substituted for another drug(s).

2. Doxycycline, the most potent MMP inhibitor of the approved TETs, works at low doses on MMP activity, and the regimen of choice is therefore on the order of 40–80 mg/day given for several months, if not years. When and if a CMT of greater potency is approved for human use, then there may be a better agent which would avoid even the low risk of microbial side effects (e.g., the oral candidiasis seen in the Israeli experience).

3. The primary parameters to be assessed should be those which MMP inhibition might rationally affect, for example, those in TABLE 3. I present our data on a preliminary trial with 13 RA patients measuring urinary PYD excretion (this volume). If patients in such a trial feel better, all well and good, but in this preliminary study, they were carefully advised that no effect on symptoms was anticipated, and that the end point would be demonstration of a biochemical effect pertaining to collagen breakdown. They readily accepted this rationale.

If we are to make headway in the battle against connective tissue degradation, novel approaches are required; clearly, the conventional drugs fall far short. The tetracyclines have the advantage of proven safety, and I believe that the time is at hand for a long-term trial of adjunctive doxycycline therapy in both rheumatoid arthritis and osteoarthritis to determine if indeed this goal can be attained.

ACKNOWLEDGMENTS

Susan Moak, my research assistant for 14 years, performed all of the original studies reported herein, and has been invaluable in all aspects of these scientific endeavors. Dr. Majeedul H. Chowdhury assisted with the guinea pig studies. Drs.

L. M. Golub and N. S. Ramamurthy provided much insight and support for this work as well. Assisting with the work on the guinea pigs at the University of North Carolina were John Morrison, Beth Crossley, and Hao Wang.

REFERENCES

1. Harris, E. D., Jr, H. G. Welgus & S. M. Krane. 1984. Regulation of the mammalian collagenases. Collagen Relat. Res. **4:** 493–512.
2. Brinckerhoff, C. E. 1991. Joint destruction in arthritis: Metalloproteinases in the spotlight. Arthritis Rheum. **34:** 1073–1075.
3. Cawston, T. 1993. Blocking cartilage destruction with metalloproteinase inhibitors: A valid therapeutic target? Ann. Rheum. Dis. **52:** 769–770.
4. Mainardi, C. L. 1985. Collagenase in rheumatoid arthritis. Ann. N. Y. Acad. Sci. **460:** 345–354.
5. Brinckerhoff, C. E. 1990. Retinoids and rheumatoid arthritis: Modulation of extracellular matrix by controlling expression of collagenase. Methods Enzymol. **190:** 175–188.
6. Nagase, H. & J. F. Woessner. 1993. Role of endogenous proteinases in the degradation of cartilage matrix. *In* Joint Cartilage Degradation: Basic and Clinical Aspects. J. F. Woessner & D. S. Howell, Eds.: 159–186. Marcel Dekker. New York.
7. Greenwald, R. A. 1993. Cartilage degradation in animal models of inflammatory joint disease. *In* Joint Cartilage Degradation: Basic and Clinical Aspects. J. F. Woessner & D. S. Howell, Eds.: 385–408. Marcel Dekker. New York.
8. Sorsa, T., Y. T. Kontinnen, O. Lindy, C. Ritchlin, H. Saari, K. Suomalainen, K. Eklund & S. Santavirta. 1992. Collagenase in synovitis of rheumatoid arthritis. Semin. Arthritis Rheum. **22:** 44–53.
9. Wooley, D. E., C. E. Brinckerhoff & C. L. Mainardi. 1977. Collagenase production by rheumatoid synovial cells. Ann. Rheum. Dis. **38:** 262–270.
10. Bauer, E. A., A. Z. Eisen & J. J. Jeffrey. 1991. Studies on purified rheumatoid synovial collagenase in vitro and in vivo. J. Clin. Invest. **50:** 2056–2064.
11. Dayer, J. M., S. M. Krane & R. O. G. Russell. 1976. Production of collagenase and prostaglandins by isolated adherent rheumatoid synovial cells. Proc. Natl. Acad. Sci. USA **73:** 945–949.
12. Al-Haik, N., D. A. Lewis & G. Struthers. 1984. Neutral protease, collagenase, and elastase activities in synovial fluids from arthritic patients. Agents Actions **15:** 436–442.
13. Gysen, P., M. Malaise & S. Gaspar. 1985. Measurement of proteoglycans, elastase, collagenase, and protein in synovial fluid in inflammatory and degenerative arthropathies. Clin. Rheum. **4:** 39–50.
14. Walakovits, L. A., V. L. Moore, N. Bhardwaj, G. S. Gallick & M. W. Lark. 1992. Detection of stromelysin and collagenase in synovial fluid from patients with rheumatoid arthritis and posttraumatic knee injury. Arthritis Rheum. **35:** 35–42.
15. Wooley, D. E., M. J. Crossley & J. M. Evanson. 1977. Collagenase at sites of cartilage erosion in the rheumatoid joint. Arthritis Rheum. **20:** 1231–1239.
16. Krane, S. M. 1982. Collagenase and collagen degradation. J. Invest. Dermatol. **79:** 83s–86s.
17. Gravallese, E. M., J. M. Darling, A. L. Ladd, J. N. Katz & L. M. Glimcher. 1991. In situ hybridization studies of stromelysin and collagenase messenger RNA expression in rheumatoid synovium. Arthritis Rheum. **34:** 1076–1084.
18. Macachren, S. S. 1991. Expression of metalloproteinases and metalloproteinase inhibitor in human arthritic synovium. Arthritis Rheum. **34:** 1085–1093.
19. Okada, Y., Y. Gonoji, I. Nakanishi, H. Nagase & T. Hayakawa. 1990. Immunohistochemical demonstration of collagenase and tissue inhibitor of metalloproteinases (TIMP) in synovial lining cells of rheumatoid synovium. Virchows Arch. B Cell Pathol. **59:** 305–312.
20. Fosang, A. J., K. Last, V. Knauper, P. J. Neame, G. Murphy, T. E. Hardingham, H. Tschesche & J. A. Hamilton. 1993. Fibroblast and neutrophil collagenases cleave at two sites in the cartilage aggrecan interglobular domain. Biochem. J. **295:** 273–276.

21. Dean, D. D. 1991. Proteinase-mediated cartilage degradation in osteoarthritis. Semin. Arthritis Rheum. **20:** 2–11.
22. Carney, S. L. 1991. Cartilage research, biochemical, histologic, and immunohistochemical markers in cartilage, and animal models of osteoarthritis. Curr. Opin. Rheum. **3:** 669–675.
23. Burton-Wurster, N., R. J. Todhunter & G. Lust. 1993. Animal models of osteoarthritis. *In* Joint Cartilage Degradation: Basic and Clinical Aspects. J. F. Woessner & D. S. Howell, Eds.: 347–384. Marcel Dekker. New York.
24. Zafarullah, M., J. P. Pelletier, J. M. Cloutier & J. Marcel-Pelletier. 1993. Elevated metalloproteinases and tissue inhibitor of metalloproteinase mRNA in human osteoarthritic synovia. J. Rheumatol. **20:** 693–697.
25. Mohtai, M., R. L. Smith, D. J. Schurman, Y. Tauji, F. M. Torti, N. I. Hutchinson, W. G. Stetler-Stevenson & G. I. Goldberg. 1993. Expression of 92-kD Type IV collagenase/gelatinase (gelatinase B) in osteoarthritic cartilage and its induction in normal human articular cartilage by interleukin 1. J. Clin. Invest. **92:** 179–185.
26. Mohr, W., A. Wild & H. P. Wolf. 1981. Role of polymorphs in inflammatory cartilage destruction in adjuvant arthritis of rats. Ann. Rheum. Dis. **40:** 171–176.
27. Kuberasampath, T. & S. M. Bose. 1980. Effect of adjuvant arthritis on collagenase and certain lysosomal enzymes in relation to the catabolism of collagen. Agents Actions **10:** 78–84.
28. Al-Haik, N., D. P. Parrott, D. A. Lewis. 1984. Collagenase and elastase in the serum of adjuvant arthritic rats. Agents Actions **14:** 688–693.
29. Case, J. P., H. Sano, R. Lafyatis, E. F. Remmers, G. K. Kumkumian & R. L. Wilder. 1989. Transin/stromelysin expression in the synovium of rats with experimental erosive arthritis. J. Clin. Invest. **84:** 1731–1740.
30. Coffey, J. W. & R. A. Salvador. 1981. Levels of collagenolytic activity, β-glucuronidase and collagen prolyl hydroxylase in paws from rats with developing adjuvant arthritis. Biochim. Biophys. Acta **677:** 243–252.
31. Brinckerhoff, C. E., J. W. Coffey & A. C. Sullivan. 1983. Inflammation and collagenase production in rats with adjuvant arthritis reduced with 13-*cis*-retinoic acid. Science **221:** 756–758.
32. Greenwald, R. A., L. M. Golub, N. S. Ramamurthy & T. McNamara. 1990. Direct detection of collagenase and gelatinase in periarticular tissue from adjuvant arthritis rats: Inhibition by tetracyclines and potential amelioration of bone destruction. Trans. Orthop. Res. Soc. **15:** 270.
33. Greenwald, R. A., S. A. Moak, N. S. Ramamurthy & L. M. Golub. 1992. Tetracyclines suppress matrix metalloproteinase activity in adjuvant arthritis and in combination with flurbiprofen ameliorate bone damage. J. Rheumatol. **19:** 927–938.
34. Hasty, K. A., R. A. Reife, A. H. Kang & J. M. Stuart. 1990. The role of stromelysin in the cartilage destruction that accompanies inflammatory arthritis. Arthritis Rheum. **33:** 388–397.
35. Ciosek, C. P. & T. W. Harrity. 1980. Cartilage associated collagenolytic activity in rabbits with antigen induced chronic synovitis. J. Lab. Clin. Med. **96:** 460–469.
36. Henderson, B., E. R. Pettipher & G. Murphy. 1990. Metalloproteinases and cartilage proteoglycan depletion in chronic arthritis. Arthritis Rheum. **33:** 241–246.
37. Hembry, R. M., M. R. Bagga, G. Murphy, B. Henderson & J. J. Reynolds. 1993. Rabbit models of arthritis: Immunolocalization of matrix metalloproteinases and tissue inhibitor of metalloproteinase in synovium and cartilage. Am. J. Pathol. **143:** 628–642.
38. Otterness, I. G., D. L. Larson & J. G. Lombardino. 1982. An analysis of piroxicam in rodent models of arthritis. Agents Actions **12:** 308–312.
39. Reddy, G. K. & S. C. Dhar. 1992. Metabolic studies on connective tissue collagens in bone and tendon of adjuvant arthritic rat. Calcif. Tissue Int. **50:** 320–326.
40. Ratcliffe, A., M. P. Rosenwasser, F. Mahmud, P. A. Glazer, N. F. Saed, N. Lane & V. C. Mow. 1993. The in vivo effects of naproxen on canine experimental osteoarthritic articular cartilage: Composition, metalloproteinase activity, and metabolism. Agents Actions (Suppl.) **39:** 207–211.
41. Carmichael, D. F., G. P. Stricklin & J. M. Stuart. 1989. Systemic administration of

TIMP in the treatment of collagen-induced arthritis in mice. Agents Actions **27:** 378–379.
42. Bobalik, G. R., J. P. Aldred, R. P. Kleszynski, R. K. Stubbs, R. A. Zeedyk & J. W. Bastian. 1974. Effect of salmon calcitonin and combination drug therapy on rat adjuvant arthritis. Agents Actions **4:** 364–369.
43. Francis, M. D., K. Hovancik & R. W. Boyce. 1989. NE-58095: A diphosphonate which prevents bone erosion and preserves joint architecture in experimental arthritis. Int. J. Tissue React. **11:** 239–252.
44. del Pozo, E., M. Graeber, P. Elford & T. Payne. 1990. Regression of bone and cartilage loss in adjuvant arthritic rats after treatment with cyclosporin A. Arthritis Rheum. **33:** 247–252.
45. Greenwald, R. A., S. Moak, L. M. Golub, N. S. Ramamurthy, Q. Zeng, D. Ke & W. S. S. Jee. 1991. CMT, a metalloproteinase inhibitor, prevents bone resorption in adjuvant arthritis. Arthritis Rheum. **34 (Suppl. 9):** S66 (Abstr. A6).
46. Sasaki, T., R. A. Greenwald, S. A. Moak, N. S. Ramamurthy & L. M. Golub. 1992. Osteoblast (OB) and osteoclast (OC) ultrastructure in adjuvant arthritis (AA): Effects of flurbiprofen (FBP) and a metalloproteinase (MMP) inhibitor. Arthritis Rheum. **35 (Suppl. 9):** S140.
47. Greenwald, R. A., S. A. Moak, M. H. Chowdhury & L. M. Golub. 1993. Metalloproteinase (MMP) inhibitors suppress pathologically excessive collagen crosslink excretion in adjuvant arthritis. Arthritis Rheum. **36 (Suppl. 9):** S45 (Abstr. 40).
48. Bendele, A. M. & J. F. Hulman. 1988. Spontaneous cartilage degeneration in guinea pigs. Arthritis Rheum. **31:** 561–565.
49. Bendele, A. M. & J. F. Hulman. 1991. Effects of body weight restriction on the development and progression of spontaneous osteoarthritis in guinea pigs. Arthritis Rheum. **31:** 1180–1184.
50. Hughes, C. E., B. Caterson, R. J. White, P. J. Roughley & J. S. Mort. 1992. Monoclonal antibodies recognizing protease-generated neoepitopes from cartilage proteoglycan degradation. J. Biol. Chem. **267:** 16011–16014.
51. Black, D., C. Farquharson & S. P. Robins. 1989. Excretion of pyridinium cross-links of collagen in ovariectomized rats as urinary markers for increased bone resorption. Calcif. Tissue Int. **44:** 343–347.
52. Preedy, V. R., R. A. Sherwood, C. I. O. Akpoguma & D. Black. 1991. The urinary excretion of the collagen degradation markers pyridinoline and deoxypyridinoline in an experimental rat model of alcoholic bone disease. Alcohol & Alcohol. **26:** 191–198.
53. Jerome, C. P., A. Colwell, R. Eastell, R. G. G. Russell & U. Trechsel. 1992. The effect of rat parathyroid hormone (1-34) infusion on urinary 3-hydroxypyridinium crosslink excretion in the rat. Bone Miner. **19:** 117–125.
54. Uebelhart, D., E. J. Thonar, D. W. Pietryla & J. M. Williams. 1993. Elevation in urinary levels of pyridinium cross-links of collagen following chymopapain induced degradation of articular cartilage in the rabbit knee provides evidence of metabolic changes in bone. Osteoarthritis Cart. **1:** 185–192.
55. Skinner, M., E. S. Cathcart, J. A. Mills & R. S. Pinals. 1971. Tetracycline in the treatment of rheumatoid arthritis. Arthritis Rheum. **14:** 727–732.
56. Breedveld, F. C., B. A. C. Dijkmans & H. Mattie. 1990. Minocycline treatment for rheumatoid arthritis: An open dose finding study. J. Rheumatol. **17:** 43–46.
57. Kloppenburg, M., F. C. Breedveld, A. M. Miltenburg & B. A. C. Dijkmans. 1993. Antibiotics as disease modifiers in arthritis. Clin. Exp. Rheumatol. **11 (Suppl.):** S113–S115.
58. Langevitz, P., I. Bank, D. Zemer, M. Book & M. Pras. 1992. Treatment of resistant rheumatoid arthritis with minocycline: An open study. J. Rheumatol. **19:** 1502–1504.

Insights into the Natural History of Osteoarthritis Provided by the Cruciate-Deficient Dog

An Animal Model of Osteoarthritis

KENNETH D. BRANDT

Rheumatology Division and
Multipurpose Arthritis and Musculoskeletal Diseases Center
Indiana University School of Medicine
541 Clinical Drive, Room 492
Indianapolis, Indiana 46202-5103

No effective, noninvasive practical techniques are available today that permit recognition of the earliest stages of osteoarthritis (OA). Even if they were available, ethical considerations would preclude sampling of weight-bearing articular cartilage for direct study at that stage. For this reason, animal models have been used extensively to gain insight into the underlying pathogenetic mechanisms in OA. The biochemistry, metabolism, and biomechanical alterations in articular cartilage from the unstable knee of the cruciate-deficient dog[1,2] have been characterized more extensively than those in other animal models.[3–5]

THE NATURAL HISTORY OF OSTEOARTHRITIS IN THE CANINE CRUCIATE-DEFICIENCY MODEL

It is important to recognize that until recently, because of an apparent absence of progressive changes and lack of full-thickness cartilage ulceration (as seen in human OA), the cruciate-deficient dog was widely viewed with skepticism as a valid model of OA. Studies by Marshall several years ago had suggested that the articular cartilage changes in dogs followed for as long as two years after ligament transection were no more severe than those seen only a few months after surgery.[6] It was considered that mechanical factors, for example, capsular fibrosis and buttressing osteophytes, stabilized the cruciate-deficient knee and prevented progressive cartilage breakdown. The lack of progressive cartilage changes led a number of investigators to contend that the model represented cartilage injury and repair, but not OA.

Recent studies that we carried out in collaboration with Dr. Mark Adams, however, clearly validate the cruciate-deficient dog as a model of OA.[7] They emphasize, furthermore, the tremendous capacity of the chondrocyte for repair in the earlier stages of OA.

In the earliest metabolic studies of articular cartilage from the cruciate-deficient knee, McDevitt, Muir and colleagues showed that an increase in proteoglycan (PG) synthesis occurs within days after ligament transection.[3–5] Adams and I found that in both foxhounds and mongrels, cartilage from the OA knee, although showing typical biochemical, metabolic, and histological changes of OA, remained thicker than normal for as long as 64 weeks after knee instability was created.[7] Notably, this thickening showed a persistent increase in PG synthesis and increases in both the content and concentration of PGs in the OA cartilage.

By magnetic resonance imaging (MRI),[8] we showed that this hypertrophic cartilage was maintained for as long as three years after cruciate transection, but that progressive loss of articular cartilage then occurred in the unstable knee, so that by 45 months after cruciate ligament transection no articular cartilage remained over extensive areas of the joint surface.[9] Studies of dogs sacrificed 54 months after cruciate ligament transection provided pathological confirmation of our MRI observations. Thus, if one merely waits long enough, the changes of OA in this model are progressive. This is indeed a model of OA.

The phenomenon of hypertrophic repair of articular cartilage in OA is illustrated particularly well by longitudinal study of the canine cruciate deficiency model, but is by no means unique to that model. It was recognized initially in human OA cartilage by Bywaters[10] and subsequently by Johnson,[11] and has been detected also in OA cartilage from rabbits that have undergone partial meniscectomy[12] and from Rhesus macaques developing OA spontaneously.[13,14]

Contemporary descriptions of the pathology of OA emphasize the progressive loss of articular cartilage, but fail to take into account evidence of increases in cartilage thickness and PG synthesis in the earlier stages of OA, which are consistent with a phase of compensated, stabilized OA. This phase, which may persist for a lengthy period, may keep pace with cartilage breakdown and maintain the joint in a reasonably functional state for years.[8] The repair tissue, however, often does not hold up as well to mechanical stresses as normal hyaline cartilage. Eventually, at least in some cases, the rate of PG synthesis falls off, abrogating the ability of the cells to maintain the matrix. End-stage OA then develops, with full-thickness loss of articular cartilage.[8,9]

EFFECTS OF ANTIRHEUMATIC DRUGS ON ARTICULAR CARTILAGE

Several years ago we showed that when dogs with cruciate-deficient knees were fed aspirin daily in clinical relevant anti-inflammatory doses, the slowly progressive and "compensated" OA characteristic of this model, described above, progressed rapidly, with destruction of articular cartilage within only a few months.[15] *In vitro* studies showed that salicylates reversibly inhibited PG synthesis in normal canine and human articular cartilage in a concentration-dependent fashion[16] and that this effect was much more marked in OA cartilage than in normal cartilage.[17] Additional *in vitro* studies showed that suppression of cartilage PG synthesis was caused not only by salicylates but also by several other nonsteroidal anti-inflammatory drugs (NSAIDs), although not by all NSAIDs.[18] The effect of salicylate could be attributed to inhibition of enzymes involved in PG biosynthesis, for example, glucuronyl transferase,[19] and was independent of inhibition of prostaglandin biosynthesis.[20] Studies in other laboratories, using other animal models of OA, have confirmed the acceleration of OA changes induced by salicylate administration, although there is no evidence that salicylates in reasonable doses adversely affect normal articular cartilage.

On the basis of such observations, the question has arisen whether chronic therapy of OA with NSAIDs—albeit effective in relieving pain—might accelerate the progression of OA. This has led to initiation of clinical trials which are currently in progress, aimed at comparing the long-term effects of an NSAID with those of a pure analgesic (e.g., acetaminophen) on the natural history of OA in man. Indeed, the observations in animal models of OA awakened interest also in the question of whether NSAIDs, although clearly more effective than placebo, are superior in the treatment of OA to pure analgesics having no anti-inflammatory effect.[21]

RELATION OF SYNOVITIS TO ARTICULAR CARTILAGE CHANGES IN OSTEOARTHRITIS

Little is known today about the contribution of synovitis to the progression of cartilage breakdown in this disorder. However, the canine cruciate-deficiency model has provided some interesting insights into this issue.

To some extent, the biochemical and metabolic changes in OA cartilage reported by various investigators using the model have been variable. The basis for this variability is unclear, but could be due to such factors as the age or sex of the dog, level of postoperative load-bearing on the unstable knee, etc. We recently considered the possibility that synovial inflammation might be an important variable. Indeed, much interest exists today in the possibility that cytokine mediators, such as interleukin-1 (IL-1), released from inflamed synovium, drive the progression of cartilage damage in OA, and considerable effort is being directed toward discovery of inhibitors of IL-1 or of matrix-degrading enzymes produced by chondrocytes.

Some recent data, however, suggest that IL-1 may not be a major factor in OA. First, IL-1 *inhibits* proteoglycan synthesis by articular cartilage both *in vitro* and *in vivo*[22,23] whereas, as indicated above, PG synthesis in OA cartilage is increased. Second, adult articular cartilage, especially in the dog, is highly resistant to both the anti-anabolic and the catabolic effects of IL-1.[24] Therefore, it is possible that IL-1 does not play a significant role in driving cartilage breakdown in this canine model of OA.

Synovial inflammation, characterized by extensive synovial infiltration with mononuclear cells, lining cell hyperplasia and lymphoid aggregates, routinely develops in the unstable canine knee after cruciate ligament transection.[25] By paying meticulous attention to hemostasis at the time of ligament transection, however, we produced a cruciate-deficiency model with virtually no synovitis.[25] When intra-articular bleeding was controlled by electrocautery, and the joint was copiously lavaged prior to closure after ligament transection, inflammatory changes were minimal. This difference is explained by the fact that arteries lying on the surface of the cruciate ligament inevitably are torn when the ligament is transected, whether via arthrotomy or through a blind stab incision (the Pond-Nuki model). With a blind stab wound, furthermore, blood vessels in the plica synovialis also are severed, increasing intra-articular bleeding. In such cases, marked iron deposition in the synovium is associated with inflammatory changes.

We recently compared the biochemical, metabolic, and morphological changes in articular cartilage from the unstable knees of two groups of dogs, both of which underwent cruciate ligament transection. In one group hemostasis was controlled; in the other, no effort was made to control bleeding at the time of surgery. Our results showed that in the first group, marked synovial inflammatory changes were present, whereas in the second group, synovitis was minimal. Notably, except for cloning of chondrocytes, which was more marked when no attempt had been made to control hemostasis (probably due to the presence of growth factors in the intra-articular blood), changes in articular cartilage from the two groups of animals were indistinguishable.[25]

Thus, in this model, at least in the initial months, the mechanical changes induced by cruciate ligament transection appear to be of much greater importance than synovial inflammation in driving the cartilage changes of OA. This suggestion is supported by the observation that daily oral administration of prednisone, 0.1 mg/kg, provided no protection against the cartilage changes of OA after cruciate ligament transection,[26] although that dose was sufficient to have inhibited IL-1 production.

NEUROGENIC ACCELERATION OF OSTEOARTHRITIS

Charcot arthropathy is marked by severe, chaotic joint destruction, with osteochondral fractures, loose bodies, effusions, ligament instability, and formation of new bone and new cartilage within the joint. The general concept of the pathogenesis of Charcot arthropathy is that deafferentation of the extremity by a neurologic disease (e.g., tabes dorsalis or syringomyelia) deprives the central nervous system of afferent input from nociceptive or proprioceptive nerve fibers, leading to recurrent episodes of joint trauma and, ultimately, joint breakdown. It is impossible, however, to produce Charcot arthropathy experimentally by neurosurgical procedures that deafferentate an extremity. Thus, we were unable to produce joint pathology in normal dogs subject to unilateral L4-S1 dorsal root ganglionectomy (which extensively deafferentated the ipsilateral limb).[27]

On the other hand, when the ipsilateral anterior cruciate ligament was transected in dogs that had previously undergone dorsal root ganglionectomy,[27] articular cartilage in the knee of the deafferented extremity broke down within only three weeks! This stands in sharp contrast to the very slowly progressive changes seen in the neurologically intact dog with a cruciate-deficient knee, in which the articular cartilage remains hypertrophic for as long as three years.

MODIFICATION OF OSTEOARTHRITIS BY ORAL DOXYCYCLINE THERAPY

Acceleration of the pathologic changes of OA by deafferentation of the ipsilateral limb prior to transection of the cruciate ligament markedly increases the utility of the model for assessing the effectiveness of putative disease-modifying drugs. We have recently shown, for example, that oral administration of doxycycline inhibits the breakdown of articular cartilage in the accelerated model.[28,29] The effect was most striking when the drug was administered prophylactically,[28] but reduction in the severity of articular cartilage degeneration was noted even when treatment was delayed until cartilage degeneration had already been established.[29] The protective effect of doxycycline was accompanied by striking reductions in the levels of total and active collagenase and gelatinase in extracts of the OA cartilage. Studies of the mechanism of action of doxycycline, presented in detail by Dr. Gerald Smith elsewhere in this volume, strongly suggest that the drug altered the conformation of recombinant human neutrophil procollagenase, rendering it more susceptible to proteolysis and resulting in irreversible loss of enzyme protein.[30] The fact that doxycycline administration in our canine model of OA reduced the level of total gelatinase as well as of total collagenase in extracts of the articular cartilage suggests that it may inhibit activation of several matrix metalloproteinases.

With respect to the demonstrated chondroprotective effect of doxycycline in the canine OA model, our findings are relevant to the recent observation[31] that neutrophil collagenase uniquely exhibits "aggrecanase" activity, that is, that it specifically cleaves the interglobular domain of the proteoglycan core protein between glu-373 and ala-374.[32,33] If chondrocytes secrete an enzyme similar to neutrophil collagenase, doxycycline treatment could inhibit *in vivo* catalysis of both proteoglycan and collagen in articular cartilage.

PERIARTICULAR MUSCLE

Several years ago we showed that when the hind limb of a normal dog was immobilized in an orthopedic cast, the articular cartilage developed striking atrophic changes within only a few weeks, and exhibited thinning, a decrease in proteoglycan concentration, reduction in net proteoglycan synthesis, and a defect in proteoglycan aggregation.[34] Notably, if the unstable knee is immobilized immediately after cruciate ligament transection, OA does not develop but changes of cartilage atrophy are seen,[35] which are identical to those noted after immobilization of dogs with intact cruciate ligaments, emphasizing the importance of altered joint mechanics in the genesis of the cartilage changes in this model of OA.

Studies aimed at elucidating the pathogenesis of the cartilage changes seen with immobilization indicated that these were not due principally to a decrease in oscillatory motion of the joint, but to reduction in use of the periarticular muscles that span the joint (e.g., hamstrings, quadriceps) and normally contract to stabilize the limb during stance.[36]

SUBCHONDRAL BONE

Some investigators have suggested that stiffening of the subchondral bone might be of primary importance in the etiopathogenesis of OA in man.[37] Recent data shed some light on bone changes in the canine cruciate-deficiency model.

Radiographic studies of dogs that have undergone cruciate ligament transection reveal typical bony changes of OA, including subchondral sclerosis, by 24 months after surgery.[38] However, direct examination of the subchondral plate and the mass of subchondral trabeculae by computerized tomographic microdensitometry in samples from dogs maintained for as long as 72 weeks after cruciate transection showed no increase in thickness of the subchondral plate, but osteopenia, with an increase in the intertrabecular distance.[39] Later, however, a trend was noted for thickening of the subchondral plate. This loss of bone in the subchondral trabeculae is presumably due to the decrease in load-bearing after cruciate ligament transection; vertical forces generated by the unstable limb are only about 60% of control values.[40]

Thus, our observations show that typical articular cartilage changes of OA may occur in the presence of osteopenia. Clearly, stiffening of subchondral bone is not a requisite for *initiation* of the early cartilage changes in this OA model. Indeed, the loss of subchondral bone could theoretically increase mechanical strain in the overlying articular cartilage, leading to degeneration.[41] Thickening of the subchondral plate, a relatively late phenomenon in the canine model, could, however, contribute to the failure of intrinsic repair mechanisms and *progression* of cartilage breakdown.

REFERENCES

1. Pond, M. J. & G. Nuki. 1973. Experimentally-induced osteoarthritis in the dog. Ann. Rheum. Dis. **32:** 387–388.

2. ADAMS, M. E. & J.-P. PELLETIER. 1988. The canine anterior cruciate ligament transection model of osteoarthritis. *In* Handbook of Models of Arthritis Research. R. A. Greenwald & H. S. Diamond, Eds.: 265–297. CRC Press. Boca Raton, FL.
3. MCDEVITT, C. A., E. GILBERTSON & H. MUIR. 1977. An experimental model of osteoarthritis; early morphological and biochemical changes. J. Bone Jt. Surg. Br. **57:** 24–35.
4. GILBERTSON, E. M. M. 1975. Development of periarticular osteophytes in experimentally induced osteoarthritis in the dog. Ann. Rheum. Dis. **34:** 15–25.
5. MCDEVITT, C. A., H. MUIR & M. J. POND. 1983. Canine articular cartilage in naturally and experimentally-induced osteoarthritis. Biochem. Soc. Trans. **1:** 287–289.
6. MARSHALL, J. L. & S.-E. OLSSON. 1971. Instability of the knee. A long-term experimental study in dogs. J. Bone Jt. Surg. Am. **53:** 1561–1570.
7. ADAMS, M. E. & K. D. BRANDT. 1991. Hypertrophic repair of canine articular cartilage in osteoarthritis after anterior cruciate ligament transection. J. Rheumatol. **18:** 428–435.
8. BRAUNSTEIN, E. M., K. D. BRANDT & M. ALBRECHT. 1990. MRI demonstration of hypertrophic articular cartilage repair in osteoarthritis. Skeletal Radiol. **19:** 335–339.
9. BRANDT, K., E. BRAUNSTEIN, D. VISCO *et al.* 1991. Anterior cruciate ligament transection (ACLT): A bona fide model of canine osteoarthritis (OA), not merely of cartilage injury and repair. J. Rheumatol. **18:** 436–446.
10. BYWATERS, E. G. L. 1937. Metabolism of joint tissue. J. Pathol. Bacteriol. **44:** 247–268.
11. JOHNSON, L. C. 1959. Kinetics of osteoarthritis. Lab. Invest. **8:** 1223–1238.
12. VIGNON, E., M. ARLOT, D. HARTMAN, B. MOYER & G. VILLE. 1983. Hypertrophic repair of articular cartilage in experimental osteoarthrosis. J. Rheumatol. **142:** 82–88.
13. CHÂTEAUVERT, J., K. P. H. PRITZKER, M. J. KESSLER & M. D. GRYNPAS. 1989. Spontaneous arthritis in Rhesus macaques. I. Chemical and biochemical studies. J. Rheumatol. **16:** 1098–1104.
14. CHÂTEAUVERT, J., K. P. H. PRITZKER, M. J. KESSLER & M. D. GRYNPAS. 1990. Spontaneous arthritis in Rhesus macaques. II. Characterization of disease and morphometric studies. J. Rheumatol. **17:** 73–83.
15. PALMOSKI, M. & K. BRANDT. 1983. *In vivo* effect of aspirin on canine osteoarthritic cartilage. Arthritis Rheum. **26:** 994–1001.
16. PALMOSKI, M. & K. BRANDT. 1979. Effect of salicylate on proteoglycan metabolism in normal canine articular cartilage *in vitro.* Arthritis Rheum. **22:** 746–754.
17. PALMOSKI, M. J., R. A. COLYER & K. D. BRANDT. 1980. Marked suppression by salicylate of the augmented proteoglycan synthesis (? matrix repair) in osteoarthritic cartilage. Arthritis Rheum. **23:** 83–91.
18. BRANDT, K. D. & M. J. PALMOSKI. 1984. The effects of salicylates and other nonsteroidal anti-inflammatory drugs on articular cartilage. Am. J. Med. **77:** 65–69.
19. HUGENBERG, S. T., K. D. BRANDT & C. A. COLE. 1993. Effect of sodium salicylate, aspirin and ibuprofen on enzymes required by the chondrocyte for synthesis of chondroitin sulfate. J. Rheumatol. **20:** 2128–2133.
20. PALMOSKI, M. & K. BRANDT. 1984. Effects of salicylate and indomethacin on glycosaminoglycan and prostaglandin E_2 synthesis in intact canine knee cartilage *ex vivo.* Arthritis Rheum. **27:** 398–403.
21. BRADLEY, J. D., K. D. BRANDT, B. P. KATZ, L. A. KALASINSKI & S. I. RYAN. 1991. Comparison of an anti-inflammatory dose of ibuprofen, an analgesic dose of ibuprofen, and acetaminophen in the treatment of patients with osteoarthritis of the knee. N. Engl. J. Med. **325:** 87–91.
22. SAKLATVALA, J., L. M. C. PILSWORTH, S. J. SARSFIELD, J. FAVRILOVIC & J. K. HEATH. 1984. Pig catabolin is a form of interleukin-1. Cartilage and bone resorb, fibroblasts make prostaglandin and collagenase, and thymocyte proliferation is augmented in response to one protein. Biochem. J. **224:** 461–466.
23. PETTIPHER, E. R., G. A. HIGGS & B. HENDERSON. 1986. Interleukin-1 induces leukocyte infiltration and cartilage proteoglycan degradation in the synovial joint. Proc. Natl. Acad. Sci. USA **83:** 8749–8753.
24. BAYLISS, M. T., V. VILIM, T. E. HARDINGHAM & H. MUIR. 1989. Age-related changes in the biosynthetic response of human articular cartilage to interleukin-1. Trans. Orthop. Res. Soc. **14:** 329.

25. MYERS, S., K. D. BRANDT, B. O'CONNOR, D. VISCO & M. ALBRECHT. 1990. Synovitis and osteoarthritis changes in canine articular cartilage after cruciate ligament transection. Effect of surgical hemostasis. Arthritis Rheum. **33:** 1406–1415.
26. MYERS, S. L., K. D. BRANDT & B. L. O'CONNOR. 1991. "Low-dose" prednisone treatment does not reduce the severity of osteoarthritis in dogs after anterior cruciate ligament transection. J. Rheumatol. **18:** 1856–1862.
27. O'CONNOR, B., M. PALMOSKI & K. BRANDT. 1985. Neurogenic acceleration of degenerative joint lesions. J. Bone Jt. Surg. Am. **67:** 562–572.
28. YU, L. C., JR., G. N. SMITH, JR., K. D. BRANDT, S. L. MYERS, B. O'CONNOR & D. A. BRANDT. 1992. Reduction of the severity of canine osteoarthritis by prophylactic treatment with oral doxycycline. Arthritis Rheum. **35:** 1150–1159.
29. YU, L. C., JR., G. N. SMITH, JR., K. D. BRANDT, B. L. O'CONNOR & S. L. MYERS. 1993. Therapeutic administration of doxycycline (DOXY) slows the progression of cartilage destruction in canine osteoarthritis (OA). Trans. Orthop. Res. Soc. **18:** 724. (Abstr.).
30. SMITH, G. N., JR., K. D. BRANDT & K. A. HASTY. 1994. Procollagenase is reduced to inactive fragments upon activation in the presence of doxycycline. Ann. N.Y. Acad Sci. This volume.
31. FOSANG, A. J., K. LAST, P. J. NEAME, C. E. HUGHES, B. CATERSON, T. E. HARDINGHAM, V. KNAUPER, G. MURPHY & H. TSCHESCHE. 1994. Neutrophil collagenase cleaves at the aggreganase site in the interglobular region of aggrecan. Trans. Orthop. Res. Soc. **19:** 48.
32. SANDY, J. D., C. R. FLANNERY, P. J. NEAME & L. S. LOHMANDER. 1992. The structure of aggrecan fragments in human synovial fluid. Evidence for the involvement of osteoarthritis of a novel proteinase which cleaves the glu 373-ala 374 bond of the interglobular domain. J. Clin. Invest. **89:** 1512–1516.
33. SANDY, J. D., P. J. NEAME, R. E. BOYNTON & C. R. FLANNERY. 1991. Catabolism of aggrecan in cartilage explants. Identification of a major cleavage site within the interglobular domain. J. Biol. Chem. **266:** 894–902.
34. PALMOSKI, M., E. PERRICONE & K. D. BRANDT. 1979. Development and reversal of a proteoglycan aggregation defect in normal canine knee cartilage after immobilization. Arthritis Rheum. **22:** 508–517.
35. PALMOSKI, M. & K. BRANDT. 1982. Immobilization of the knee prevents osteoarthritis after anterior cruciate ligament transection. Arthritis Rheum. **25:** 1201–1208.
36. PALMOSKI, M., R. COLYER & K. BRANDT. 1980. Joint motion in the absence of normal loading does not maintain normal articular cartilage. Arthritis Rheum. **23:** 325–334.
37. RADIN, E. L. 1975. Mechanical aspects of osteoarthritis. Bull. Rheum. Dis. **26:** 862.
38. WIDMER, W., D. VISCO, B. O'CONNOR, W. BLEVINS & K. BRANDT. 1990. Radiographic features of experimental osteoarthritis (OA) in dogs with unstable stifle (knee) joints. Trans. Orthop. Res. Soc. **15:** 569.
39. DEDRICK, K. D., S. A. GOLDSTEIN, K. D. BRANDT, B. L. O'CONNOR, R. W. GOULET & M. ALBRECHT. 1993. A longitudinal study of subchondral plate and trabecular bone in cruciate-deficient dogs with osteoarthritis followed for up to 54 months. Arthritis Rheum. **36:** 1460–1467.
40. O'CONNOR, B. L., D. M. VISCO, D. HECK & K. D. BRANDT. 1989. Gait alterations in dogs following transection of the anterior cruciate ligament. Arthritis Rheum. **32:** 1142–1147.
41. BROWN, T. D., E. L. RADIN, R. B. MARTIN & D. B. BURR. 1984. Finite element studies of some juxtarticular stress changes due to localized subchondral stiffening. J. Biomech. **17:** 11–24.

Application of Peptide-Based Matrix Metalloproteinase Inhibitors in Corneal Ulceration[a]

ROBERT D. GRAY[b] AND CHRISTOPHER A. PATERSON[c]

Departments of Biochemistry[b] and of Ophthalmology and Visual Sciences[c]
University of Louisville School of Medicine
Louisville, Kentucky 40292

A variety of pathological conditions, including chemical and thermal burns, eye infections involving certain bacteria, viruses and fungi, various "dry eye" syndromes, and severe vitamin A deficiency, can lead to ulceration of the cornea. In addition, corneal ulceration is also associated with diseases such as Mooren's ulcer, lupus erythematosus, Wegener's granulomatosis, and rheumatoid arthritis.[1] Although the detailed pathophysiological mechanisms associated with these ulcerative states remain to be fully elucidated, a common theme is proteolytic degradation of the structural macromolecules of the cornea. One approach to minimizing or preventing the loss of vision associated with corneal ulceration is the application of inhibitors of the proteolytic processes that are a prerequisite for ulceration. Toward this end, we prepared a number of metalloproteinase inhibitors. The purpose of this paper is to review briefly our studies to evaluate the impact of the inhibitors in animal models of two metalloproteinase-dependent ulcerative conditions: the alkali-burned cornea and *Pseudomonas* keratitis in the rabbit.

To aid in understanding the mechanism of ulceration, it may be useful to review some aspects of corneal anatomy. As illustrated in FIGURE 1, the human cornea consists of five distinct layers: an epithelium five to six cells in thickness, a basal lamina called Bowman's layer, a stromal layer, a basement membrane (Descemet's membrane), and last, a single layer of endothelial cells whose posterior aspect contacts the aqueous humor. The major structural components of the cornea have been summarized by Berman.[2] Collagen and proteoglycan comprise 12–15% and 1–3%, respectively, of the wet weight of the tissue. Bowman's membrane contains collagen types IV and VII, laminin, dermatan sulfate, and chondroitin sulfate. The stroma, which comprises about 90% of the total thickness of the cornea, consists largely of collagen fibrils embedded in a matrix of proteoglycan. The fibers are composed predominantly of type I collagen, with smaller amounts of types III and V collagen. The fibers are arranged in lamellae in which parallel fibers of one lamella lie at an oblique angle to those of adjacent lamellae. Human and bovine cornea, but not rabbit cornea, also contain significant amounts of type VI collagen. The proteoglycan ground substance consists mainly of keratan sulfate, dermatan sulfate, and chondroitin sulfate; fibroblast-like cells, the keratocytes, are widely dispersed within the stromal layer.

[a]This work was supported by U.S. Public Health Service grants AR 39573 (R.D.G.), EY 06918 (C.A.P.), 1F32-EY-06048, the Kentucky Lions Eye Foundation, and an unrestricted grant from Research to Prevent Blindness, Inc. C.A.P. is a Research to Prevent Blindness Senior Scientific Investigator.

Injuries resulting from exposure to corrosive chemicals are among the most difficult of corneal traumas to treat successfully.[3] Alkalis such as sodium hydroxide, calcium hydroxide or ammonia rapidly penetrate the cornea, resulting in destruction of the epithelial cells and stromal keratocytes as well as the endothelial cells; in severe burns, the stromal layer becomes cloudy and other structures of the eye are damaged. The course of healing after an alkali wound has been described.[4] At least two clinical courses are possible for the ulcerating cornea; in less severe cases, the ulcer may heal with the formation of scar tissue and resultant decrease in visual acuity, whereas in severe cases, the ulcer may perforate with loss of vision and possibly loss of the entire eye. The complete healing process involves regrowth of the epithelium over the wound, which can take many weeks. Polymorphonuclear leukocytes (PMNs) usually infiltrate the stroma, and peripheral vascularization often occurs. Re-epithelialization of the wound is important because ulceration is associated with a defective epithelial layer—the longer the delay in epithelial healing, the greater the likelihood of ulceration.

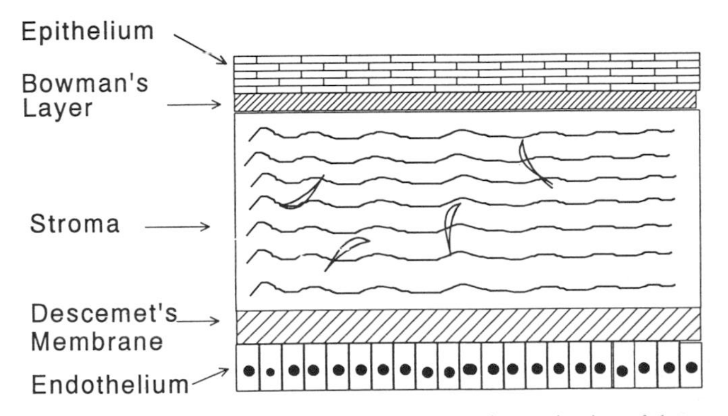

FIGURE 1. Schematic diagram showing the structural organization of the cornea.

The identity and source of the proteolytic enzymes responsible for ulceration remain the subject of debate. Early studies of Brown *et al.*[5] demonstrated the presence of collagenase in ulcerating corneas. Later immunocytochemical studies implicated both resident stromal cells[6] and infiltrating PMNs[7] as sources of the collagenolytic activity. Collagenases from either source probably can mediate the ulcerative process; which enzyme is present may depend on details such as the degree of inflammation and the viability of the resident corneal cells at the site of injury.

Matsubara *et al.*[8] recently provided evidence for the possible importance of the gelatinolytic MMPs in corneal ulceration and wound healing. In thermally wounded rat corneas, 92-kDa gelatinase (MMP-9), which is capable of degrading basement (Bowman's) membrane collagens (types IV, V, and VII), was detected by zymography and by specific antibodies. In addition, 72-kDa gelatinase (MMP-2) was present in healing corneas. The authors suggested that MMP-9 might destroy Bowman's membrane, thereby allowing invasion of the stroma by epithelial cells. This scenario is of interest because Johnson-Muller and Gross showed that normal corneal epithelium stimulates the production of collagenase by stromal fibroblasts in culture.[9]

Bacterial infections of the eye can also result in corneal ulceration. Superficial damage to the corneal epithelium sets up ideal conditions for colonization by the common environmental pathogen, *Pseudomonas aeruginosa.*[10] Such infections are frequently associated with the use of contaminated cosmetics[11] or contact lenses.[12] The organism is capable of secreting two metalloproteinases, elastase and alkaline proteinase. Both bacterial and host metalloproteinases may contribute to corneal damage in *Pseudomonas* keratitis;[13] bacterial elastase inhibitors have shown some efficacy in treating experimental *Pseudomonas* keratitis.[14–16]

Since our initial report describing the design and synthesis of synthetic MMP inhibitors,[17] we and others have prepared MMP inhibitors of high potency.[18–24] Structurally, these inhibitors are generally modeled on captopril, an inhibitor of the metallopeptidase angiotensin-converting enzyme.[25] In this type of inhibitor a metal-binding amino acid analogue capable of coordinating to the active site zinc is coupled to a substrate analogue peptide that can also interact with the extended substrate binding pocket of the proteinase (FIG. 2). One of our more potent MMP inhibitors, $HSCH_2CH[CH_2CH(CH_3)_2]CO$-Phe- Ala-$NH_2$, can be considered a derivative of the tripeptide Leu-Phe-Ala-NH_2 in which the α-amino group is replaced by a β-mercaptomethyl ($HSCH_2$-) group. Designated SIMP, this compound was found to inhibit porcine fibroblast collagenase and 72-kDa gelatinase *in vitro* with IC_{50} values of 20 nM and <1 nM, respectively.[26] SIMP also inhibited rabbit corneal MMPs;[27] it has been extensively tested *in vivo* in models of corneal melting[28,29] due to an alkali burn and in bacterial keratitis.[30] Schultz *et al.*[31] also utilized the alkali-burn model to demonstrate the efficacy of a hydroxamic acid-based metalloproteinase inhibitor.

These two corneal systems provide a convenient means of evaluating the potential therapeutic utility of metalloproteinase inhibitors. Both models develop along predictable and reproducible courses that can readily be evaluated. The severity of the pathological state can be altered by changing the initial conditions. Furthermore, because the cornea is an external structure, topical application of the inhibitor is appropriate, thereby minimizing concerns relating to metabolic stability and drug availability. Finally, corneal ulceration is a significant public health problem, so the results obtained are directly applicable to clinical medicine.

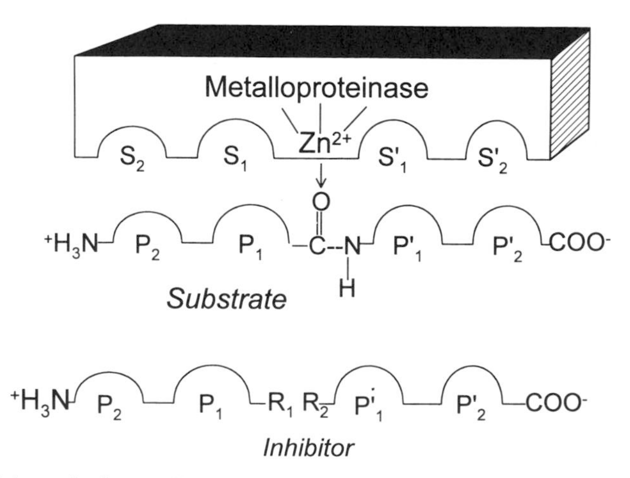

FIGURE 2. Schematic diagram illustrating the interaction between a metalloproteinase and a substrate or a substrate-based zinc-binding inhibitor. For SIMP, $R_2 = HSCH_2-$.

TABLE 1. Clinical Criteria for Evaluation of Corneal Ulceration after an Experimental Alkali Burn

Score	Clinical Observation
0	No corneal pathology
1	Superficial ulcer (depth to anterior one-third)
2	Moderate ulcer (depth to middle one-third)
3	Deep ulcer (depth to posterior one-third)
4	Descemetocele
5	Perforation

ALKALI-INDUCED CORNEAL ULCERATION

Materials and Methods

Detailed procedures for generating an alkali burn in the rabbit cornea and treatment protocols have been published.[28] Briefly, one cornea of an anesthetized New Zealand Dutch strain albino rabbit was exposed to 2 N NaOH for 60 s by pipeting the solution into a plastic well held firmly against the cornea. After irrigation of the eye with saline and application of erythromycin ophthalmic ointment, animals were randomly assigned to treatment (SIMP) or control (vehicle) groups. Unless otherwise noted, topical treatment with one drop of inhibitor or vehicle was initiated after the burn at 8 AM and continued at two-hour intervals until 6 PM. At 8 PM, a subconjunctival injection of 0.5 mL of inhibitor solution or vehicle was administered. At 9 PM, erythromycin ointment was applied to the burned eye. For the first ten days of the study, each of the burned eyes was examined by slit lamp for the presence of corneal defects using the clinical scoring system summarized in TABLE 1. For the remaining 11 days of the study, detailed examinations were conducted on alternate days. At the conclusion of the studies, corneas were removed, fixed in formalin, embedded in paraffin and stained with hematoxylin-eosin prior to examination by light microscopy.

SIMP was dissolved at a concentration of 25–30 mM in 95% ethanol containing 10 mM acetic acid and stored at −20 °C to minimize oxidation to the disulfide form. A working formulation of SIMP for therapeutic use was prepared at one- to two-day intervals by dilution of the stock solution to a concentration of 1 mM in Adsorbotear supplied without EDTA or thimerosol by Alcon, Inc., Forth Worth, Texas; this solution was stored on ice between uses.

Results

The experimental conditions used in these studies resulted in severe corneal burns characterized by opacity extending to the limbus of each alkali-treated eye. The most significant result of the study is that SIMP strongly inhibited corneal ulceration and prevented perforation. The data summarized in TABLE 2 show that each of the ten vehicle-treated eyes ulcerated; eight of these defects were observed to be "deep" ulcers. By the termination of the experiment at the end of three weeks, seven of the vehicle-treated corneas had perforated. In contrast, none of nine SIMP-treated corneas remaining in the study perforated, and only four showed any sign of ulceration. Of these, only one ulcer was "deep." Neither the SIMP nor the vehicle-treated eyes showed signs of infection, necrosis, or corneal vascularization.

TABLE 2. Effect of SIMP on the Degree of Corneal Ulceration and Perforation Subsequent to an Experimental Alkali Burn[a]

Treatment Group	Total Number of Ulcers	Total Number of Deep Ulcers	Total Number of Perforations
Vehicle-treated (n = 10)	10	8	7
SIMP-treated (n = 9)	4	1	0

[a]Modified from Burns *et al.*[28] Used with permission of *Investigative Ophthalmology and Visual Sciences.*

The time course of ulceration is shown in FIGURE 3. Even though fewer of the SIMP-treated animals developed ulcers during the three weeks of the experiment, the average time for ulceration for both groups was not significantly different (vehicle-treated: 14 ± 4 days versus 16 ± 1 days for the SIMP group). However, once established, ulcers in the vehicle-treated group developed earlier and progressed more rapidly to descemetocele formation and perforation.

Histological examination of the burned corneas at the conclusion of the treatment period revealed that re-epithelialization had not occurred in either group. A variable degree of peripheral vascularization was observed that did not differ appreciably between the two treatment groups. Both showed a loss of stromal keratocytes, but the architecture of the stroma was preserved to a greater extent in the SIMP-treated corneas, as would be expected if there were less degradation of the collagen fibers and/or ground substance. Inflammatory cells such as PMNs were

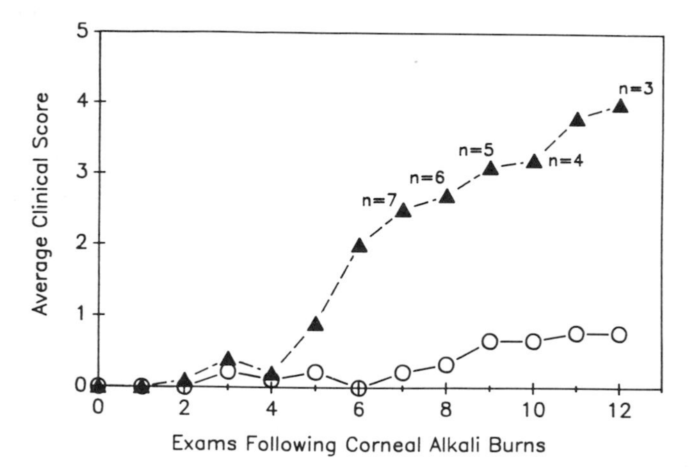

FIGURE 3. Progression of clinical changes in alkali-burned corneas treated with vehicle (▲) or SIMP (○). Corneas were evaluated using the criteria of TABLE 1 on alternate days from examinations 1–5; from examinations 6–12, the corneas were evaluated on a daily basis. The day of the alkali burn was designated exam 0; exam 12 corresponds to day 20 post-burn. Initially, 10 animals were assigned to the vehicle-treated group and 9 to the SIMP-treated group. The n values in the control group decreased over the course of the experiment because animals were removed from the study upon corneal perforation. The two groups differed when comparing the severity of ulceration using the Mann-Whitney U test ($p < 0.01$). (Chart modified from ref. 28 and used with permission from *Investigative Ophthalmology and Visual Sciences.*)

found surrounding the ulcers or perforations in the corneas in the vehicle-treated group. In the SIMP-treated group, the four corneas with ulcers had PMN infiltration; however, the non-ulcerating corneas were devoid of PMNs.

To determine if SIMP could arrest the progress of an established ulcer or accelerate the rate of corneal wound healing after an alkali burn, we carried out a series of experiments in which treatment with inhibitor was delayed until ulceration had progressed either to clinical stage 2 (mid-stromal level) or to clinical stage 1 (superficial level). The experimental procedure was essentially identical to that described above.[29] When treatment was delayed until a mid-stromal ulcer had formed, SIMP exhibited no significant effect on either the time course or end state of ulceration (data not shown). However, when topical treatment with SIMP was initiated at clinical stage 1 (superficial ulceration), approximately 90% of the corneas in the inhibitor-treated group showed reversal or cessation of the ulcerative process (FIG. 4). In contrast, the vehicle-treated corneas progressed over the two-week

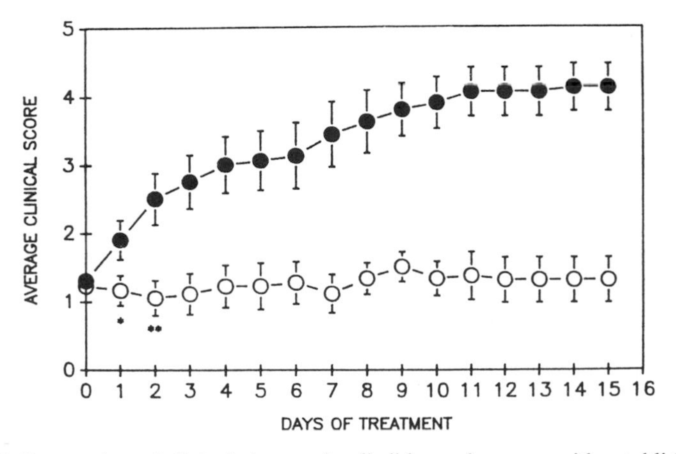

FIGURE 4. Progression of clinical changes in alkali-burned corneas with established ulcers treated with vehicle (●) or SIMP (○). Corneas were evaluated using the criteria of TABLE 1. When a cornea attained a clinical score of 1 (superficial ulcer), treatment with SIMP or vehicle was initiated. Statistical analysis was performed using the two-tailed Student's t test, with a significant difference in the severity and progression of ulceration between the two groups noted after one day of treatment ($^*p < 0.01$; $^{**}p < 0.005$). (Chart modified from ref. 29 and used with permission from *Investigative Ophthalmology and Visual Sciences.*)

period of evaluation to an average clinical score of 4 (descemetocele formation). Gross examination of the corneas at the conclusion of the experiment revealed no significant differences. About half of the vehicle-treated corneas were vascularized, whereas none of the SIMP-treated corneas showed vascularization. Histological examination at the conclusion of the second experiment revealed a lack of epithelium in the corneas from both treatment regimens. The stromal architecture of the SIMP-treated corneas again appeared to be largely preserved, whereas that of the vehicle-treated corneas was highly disrupted. In addition, the SIMP-treated corneas had few infiltrating PMNs, whereas in the ulcerating, vehicle-treated corneas, PMNs had infiltrated to the posterior regions of the stroma.

Discussion

The results of these experiments show that a metalloproteinase inhibitor, SIMP, inhibits ulceration and prevents perforation of an alkali-burned animal cornea if treatment is initiated before the ulcerative process has progressed beyond a superficial stage. Inasmuch as SIMP inhibits all of the MMPs, these studies shed no light on the relative importance of the various members of this family in corneal ulceration. Because interstitial collagen is the major structural component of the stroma and because light microscopy of treated corneas revealed that the fibrous nature of the stroma is largely preserved, inhibition of collagenase was probably an important aspect of the action of SIMP. However, this does not necessarily rule out the importance of other members of the MMP family such as basement membrane-degrading 72-kDa or 92-kDa gelatinases.

An additional interesting mechanistic aspect of corneal melting is the relative role of interstitial collagenases from PMNs and fibroblasts. PMNs were absent from SIMP-treated, non-ulcerating corneas. Since SIMP did not inhibit the chemotactic response of PMNs *in vitro,*[32] presumably the lack of PMN infiltration results from the decrease in collagen degradation in the treated corneas. Collagen fragments have been reported to be chemotactic for PMNs;[33] therefore, the absence of collagenolysis in the SIMP-treated corneas could account for the absence of these cells. Because some of the ulcerating corneas also lacked PMNs, we suggest that the presence of either type of collagenase is sufficient to initiate degradation of collagen fibrils and consequent corneal melting.

From a therapeutic standpoint, the results on delayed treatment are important because they indicate that some leeway in the initiation of treatment is possible. By limiting or preventing the ulcerative process, corneal scarring is minimized, thereby decreasing the loss of visual acuity that occurs in chemical wounding of the eye. These results clearly indicate that SIMP should be valuable in treating ulcers in humans that result from chemical injury to the eye.

Assuming that the same spectrum of metalloproteinases contributes to corneal melting in other diseases such as Mooren's ulcer and rheumatoid arthritis, SIMP may offer a reasonable treatment regimen for controlling corneal degeneration in these syndromes as well.

PSEUDOMONAS KERATITIS

Materials and Methods

Detailed procedures for induction of *Pseudomonas* keratitis in rabbits have been described.[31] Briefly, a 4-mm section of corneal epithelium was removed from one cornea of anesthetized New Zealand Dutch albino rabbits by gently scraping the surface with a scalpel blade. A known number of *Pseudomonas aeruginosa* (strain PA-28, previously demonstrated to produce both elastase and alkaline proteinase and to be susceptible to gentamicin) was injected intrastromally into the center of the cornea. In initial experiments, the bacteria were suspended in growth medium containing both proteinases, whereas in subsequent experiments they were washed free of extracellular proteinases and suspended in sterile saline prior to injection. The treatment regimen consisted of topical administration of 1 mM SIMP beginning 15 min after infection and continuing at 15-min intervals for 4 h, followed by 1 drop

TABLE 3. Clinical Criteria for Evaluation of Corneal Ulceration in Experimental *Pseudomonas* Keratitis

Score	Clinical Observation
0	No corneal pathology
1	Corneal infiltrate without melting
2	Superficial ulcer (depth to anterior one-third)
3	Moderate ulcer (depth to middle one-third)
4	Deep ulcer (depth to posterior one-third)
5	Descemetocele formation/severe bulging
6	Perforation

administered topically every 30 min for 20 h post-inoculation. Corneas were evaluated as indicated in TABLE 3.

Results

The development of corneal pathology subsequent to intrastromal injection of *P. aeruginosa* in growth medium containing elastase and alkaline proteinase is shown in FIGURE 5. In this extremely aggressive model of bacterial keratitis, the vehicle-treated corneas on average reached stage 4 (descemetocele formation) within 4 h of infection and perforation ensued within 10 h. In contrast, the SIMP-treated corneas remained at stage 1 (superficial ulceration) for 10–12 h after infection. However, by 24 h after infection, most of the corneas developed descemetoceles, although by this time, on average, fewer perforated corneas were observed in the inhibitor-treated

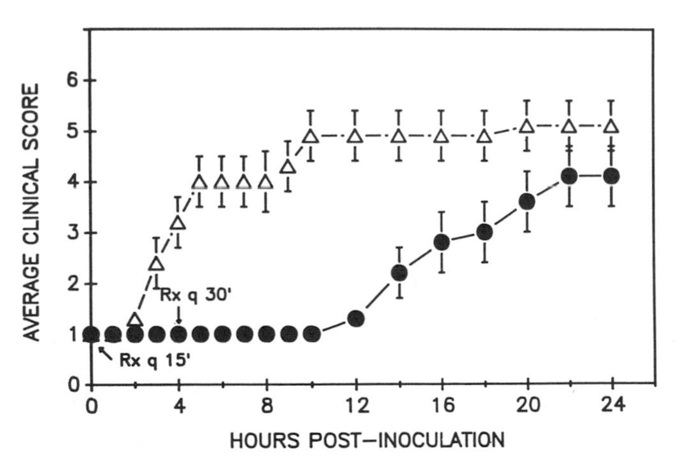

FIGURE 5. Clinical progression of *Pseudomonas* keratitis in a SIMP-treated (●, $n = 9$) and vehicle-treated (△, $n = 10$) corneas. The clinical scoring system used is given in TABLE 3. The clinical scores were significantly lower in the corneas treated with SIMP from 3 h ($p < 0.02$) through 18 h ($p < 0.05$) post-inoculation. Error bars represent the standard error of the mean. Rx q 15′, inhibitor given every 15 min; Rx q 30′, inhibitor given every 30 min. (Chart modified from ref. 30 and used with permission of the American Society for Microbiology.)

group. In a similar experiment, gentamicin-treated and vehicle-treated corneas exhibited identical clinical courses that did not differ significantly from that depicted in FIGURE 5 for the vehicle-treated animals. This experiment suggests that in the model of keratitis tested, the corneal melting resulted predominantly, if not exclusively, from bacterial proteinases present in the injection medium, rather than from proteinases secreted by the organism after infection, or from proteinases released intracorneally by the host.

In a second experiment, SIMP had no significant effect on corneal melting when *P. aeruginosa* were washed free of extracellular proteinases prior to intrastromal injection. A possible explanation for the lack of effectiveness of SIMP in this model of keratitis is that stroma-degrading proteinases were present that were not susceptible to the inhibitor. An obvious candidate is *Pseudomonas* alkaline proteinase; when SIMP was tested *in vitro* against this metalloproteinase, it was found to be only a weak inhibitor (data not shown).

Discussion

The experiments discussed above modeling *Pseudomonas* keratitis illustrate that SIMP delayed the onset of corneal melting under conditions where secreted bacterial proteinases are injected. The susceptibility of *Pseudomonas* elastase, but not *Pseudomonas* alkaline proteinase to SIMP inhibition *in vitro,* suggests that the efficacy of SIMP may be related to its ability to halt the action of the elastase; however, ulceration eventually occurred as a result of multiplication of the infecting organisms and, possibly, the action of alkaline proteinase.

ACKNOWLEDGMENTS

The authors thank the following colleagues for their essential contributions to these studies: F. R. Burns, J. S. Wentworth, A. F. Spatola, and K. Darlak.

REFERENCES

1. BERMAN, M. B. 1980. Collagenase and corneal ulceration. *In* Collagenase in Normal and Pathological Connective Tissues. D. E. Woolley & J. M. Evanson, Eds.: 141–174. John Wiley and Sons. Chichester, England.
2. BERMAN, E. R. 1991. Biochemistry of the Eye.: 101–129. Plenum Press. New York.
3. BURNS, F. R., J. S. WENTWORTH & C. A. PATERSON. 1990. Chemical eye injuries: Pathogenesis and emergency treatment. Chem. Safety **1:** 22–29.
4. ARFFA, R. C. 1991. Chemical injuries. *In* Grayson's Diseases of the Cornea. 3rd edit.: 649–665. Mosby Yearbook. St. Louis, MO.
5. BROWN, S. I., C. A. WELLER & S. AKIYA. 1970. Pathogenesis of ulcers of the alkali-burned cornea. Arch. Ophthalmol. **83:** 205–208.
6. GORDON, J. M., E. A. BAUER & A. Z. EISEN. 1980. Collagenase in the human cornea. Arch. Ophthalmol. **98:** 341–345.
7. KAO, W. W.-Y., J. EBERT, C. W.-C. KAO, H. COVINGTON & C. CINTRON. 1986. Development of monoclonal antibodies recognizing collagenase from rabbit PMN; the presence of this enzyme in ulcerating corneas. Curr. Eye Res. **5:** 801–815.
8. MATSUBARA, M., K. D. ZOESLE & M. E. FINI. 1991. Mechanism of basement membrane

dissolution preceding corneal ulceration. Invest. Ophthalmol. Visual Sci. **32:** 3221–3237.
9. JOHNSON-MULLER, B. & J. GROSS. 1978. Regulation of corneal collagenase production: Epithelial-stromal cell interactions. Proc. Natl. Acad. Sci. USA **75:** 4417–4421.
10. MCDONNELL, P. J. & W. R. GREEN. 1990. Keratitis. *In* Principles and Practice of Infectious Diseases. 3rd edit. G. G. Mandell, R. G. Douglas & J. E. Bennett, Eds.: 981–987. Churchill-Livingstone, Inc. New York.
11. ALFONSO, E., S. MANDELBAUM, M. J. FOX & R. K. FORSTER. 1986. Ulcerative keratitis associated with contaminated eye mascaras. Am. J. Ophthalmol. **84:** 112–119.
12. WILSON, L. A. & D. G. AHEARN. 1977. *Pseudomonas*-induced corneal ulcers associated with contact lens wear. Am. J. Ophthalmol. **101:** 429–433.
13. KESSLER, E., H. E. KENNAH & S. I. BROWN. 1977. The corneal response to *Pseudomonas aeruginosa:* Histopathological and enzymatic characterization. Invest. Ophthalmol. Visual Sci. **16:** 488–497.
14. KAWAHARAJO, K., J. Y. HOMMA, T. AOYAGI & H. UMEZAWA. 1982. Effect of phosphoramidon on protection against corneal ulcer caused by elastase and protease from *Pseudomonas aeruginosa.* Jpn. J. Exp. Med. **52:** 271–272.
15. KESSLER, E., A. SPIERER & S. BLUMBERG. 1983. Specific inhibition of *Pseudomonas aeruginosa* elastase injected intracorneally in rabbit eyes. Invest. Ophthalmol. Visual Sci. **24:** 1093–1097.
16. KESSLER, E. & A. SPIERER. 1984. Inhibition by phosphoramidon of *Pseudomonas aeruginosa* elastase injected intracorneally in rabbit eyes. Curr. Eye Res. **3:** 1075–1078.
17. GRAY, R. D., H. S. SANEII & A. F. SPATOLA. 1983. Metal binding peptide inhibitors of vertebrate collagenase. Biochem. Biophys. Res. Commun. **101:** 1251–1258.
18. GRAY, R. D., R. B. MILLER & A. F. SPATOLA. 1986. Inhibition of mammalian collagenases by thiol-containing peptides. J. Cell. Biochem. **32:** 71–77.
19. MOOKHTIAR, K. A., C. K. MARLOWE, P. A. BARTLETT & H. VAN WART. 1987. Phosphonamidate inhibitors of human neutrophil collagenase. Biochemistry **26:** 1962–1965.
20. DARLAK, K., R. B. MILLER, M. S. STACK, A. F. SPATOLA & R. D. GRAY. 1990. Thiol-based inhibitors of mammalian collagenase: Substituted amide and peptide derivatives of the leucine analogue, 2-[(*R,S*)-mercaptomethyl]-4-methylpentanoic acid. J. Biol. Chem. **265:** 5199–5205.
21. JOHNSON, W. H., N. A. ROBERTS & N. BORKAKOTI. 1987. Collagenase inhibitors: Their design and potential therapeutic use. J. Enzyme Inhibition **2:** 1–22.
22. SCHWARTZ, M. A., S. VENKATARAMAN, M. A. GHAFFARI, A. LIBBY, K. A. MOOKHTIAR, S. K. MALLYA, H. BIRKEDAL-HANSEN & H. VAN WART. 1991. Inhibition of human collagenases by sulfur-based substrate analogs. Biochem. Biophys. Res. Commun. **176:** 173–179.
23. GROBELNY, D., L. PONCZ & R. E. GALARDY. 1992. Inhibition of human skin collagenase, thermolysin, and *Pseudomonas aeruginosa* elastase by peptide hydroxamic acids. Biochemistry **31:** 7152–7154.
24. BESZANT, B., J. BIRD, L. M. GASTER, G. P. HARPER, I. HUGHES, E. H. KARRAN, R. E. MARKWELL, A. J. MILES-WILLIAMS & S. A. SMITH. 1993. Synthesis of novel modified dipeptide inhibitors of human collagenase: β-mercapto carboxylic acid derivatives. J. Med. Chem. **36:** 4030–4039.
25. CUSHMAN, D. W., H. S. CHEUNG, E. F. SABO & M. A. ONDETTI. 1977. Design of potent competitive inhibitors of angiotensin-converting enzyme. Carboxyalkanoyl and mercaptoalkanoyl amino acids. Biochemistry **16:** 5484–5491.
26. STACK, M. S. & R. D. GRAY. 1989. Comparison of vertebrate collagenase and gelatinase using a new fluorogenic synthetic substrate. J. Biol. Chem. **264:** 4277–4281.
27. BURNS, F. R., M. S. STACK, R. D. GRAY & C. A. PATERSON. 1989. Inhibition of purified collagenase from alkali-burned rabbit corneas. Invest. Ophthalmol. Visual Sci. **30:** 1569–1575.
28. BURNS, F. R., R. D. GRAY & C. A. PATERSON. 1990. Inhibition of alkali-induced corneal ulceration and perforation by a thiol peptide. Invest. Ophthalmol. Visual Sci. **31:** 107–114.

29. WENTWORTH, J. S., C. A. PATERSON & R. D. GRAY. 1992. Effect of a metalloproteinase inhibitor on established corneal ulcers after an alkali burn. Invest. Ophthalmol. Visual Sci. **33:** 2174–2179.
30. BURNS, F. R., C. A. PATERSON, R. D. GRAY & J. T. WELLS. 1990. Inhibition of *Pseudomonas aeruginosa* elastase and *Pseudomonas* keratitis using a thiol-based peptide. Antimicrob. Agents Chemother. **34:** 2065–2069.
31. SCHULTZ, G. S., S. STRELOW, G. A. STERN, N. CHEGINI, M. B. GRANT, R. E. GALARDY, D. GROBELNY, J. J. ROWSEY, K. STONECIPHER, V. PARMLEY & P. T. KHAW. 1992. Treatment of alkali-injured rabbit corneas with a synthetic inhibitor of matrix metalloproteinases. Invest. Ophthalmol. Visual Sci. **33:** 3325–3331.
32. WENTWORTH, J. S. Unpublished data.
33. PFISTER, R. R., J. L. HADDOX, R. W. DOSDON & L. E. HARKINS. 1987. Alkali-burned collagen produces a locomotory and metabolic stimulant to neutrophils. Invest. Ophthalmol. Visual Sci. **28:** 295–304.

Clinical Trials of a Low Molecular Weight Matrix Metalloproteinase Inhibitor in Cancer

PETER D. BROWN

British Bio-technology Ltd.
Clinical Research & Development
Watlington Road
Cowley, Oxford OX4 5LY, United Kingdom

Matrix metalloproteinases (MMPs) first received broad recognition as the proteinases thought to be responsible for joint degeneration in rheumatoid arthritis,[1,2] and it was with this end point in mind that many pharmaceutical companies established research programs in the early 1980s with the goal of obtaining clinically useful low molecular weight MMP inhibitors. It is therefore perhaps a little surprising that one of the first indications to be pursued in the clinic would be the treatment of malignant ascites. This paper reviews some of the developments that have led to the clinical trial of a low molecular weight MMP inhibitor in cancer.

MATRIX METALLOPROTEINASES AND TUMOR PROGRESSION

A key development in the field of MMP research was the establishment of the genetic identity of MMPs and the MMP gene family. In a period of 2–3 years the genes for collagenase,[3] stromelysin,[4] 72-kDa type IV collagenase,[5] 92-kDa type IV collagenase,[6] and pump-1[7] were sequenced and cloned. At about this time several correlative studies pointed to a role for MMPs in tumor progression. It had long been established that the breakdown of basement membranes and other tissue structures was a cardinal feature of malignant disease. These correlative studies now suggested that MMPs might be responsible for this tissue breakdown. High levels of fibrillar collagenase activity were detected at the invading edge of gastric carcinomas,[8] and immunohistochemical analysis showed an increased staining for collagenase in the connective tissue stroma of colorectal carcinomas as compared with the staining in adenomas and normal mucosa.[9] A high level of mRNA encoding for stromelysin was detected in carcinoma of the lung.[10] Immunohistochemical analysis also showed the expression of 72-kDa type IV collagenase to be correlated with the progression of colorectal, gastric, and breast carcinomas,[11] and mRNA *in situ* hybridization studies for pump-1 revealed a high level of expression in the majority of gastric and colorectal carcinomas.[12]

This correlative evidence in support of a role for MMP activity in tumor progression was strengthened by studies with two native inhibitors of the matrix metalloproteinase family, TIMP-1[13] and TIMP-2.[14] Intraperitoneal administration to mice of recombinant TIMP-1 was shown to block effectively the colonization of lungs by both B16-F10 mouse melanoma cells[15] and ras-transfected rat embryo 4R cells.[16] In subsequent experiments with 4R cells that had been stably transfected with TIMP-2 cDNA it was shown that production of TIMP-2 markedly reduced tumor growth following subcutaneous implantation.[17] TIMP-1 was also shown to inhibit tumor-induced vascularization of the rabbit cornea,[18] presumably by blocking the remodeling of the extracellular matrix that precedes new capillary growth. As a result of these and other studies, a model has been proposed in which MMPs facilitate tumor progression by (1) disrupting local tissue architecture to allow tumor growth, (2) breaking down basement membrane barriers to allow metastatic spread, and (3) remodeling extracellular matrices to permit tumor neovascularization (FIG. 1). Theoretically, MMP inhibitors could act to block tumor progression at any or all of these steps.

MMP INHIBITORS AS ANTI-CANCER DRUGS

Research on MMP inhibitors and cancer soon extended to studies of low molecular weight (M_r < 600) synthetic inhibitors. One of those studied was batimastat (BB-94), a broad spectrum MMP inhibitor with inhibitory activity against collagenase, stromelysin, type IV collagenases, and pump-1 in the low nanomolar range. Although classed as a broad spectrum inhibitor, batimastat is selective for the MMP family of metalloenzymes and does not show appreciable inhibitory activity against other metalloproteinases such as enkephalinase. The basis of the inhibitory activity is the hydroxamate group in the compound which chelates the zinc atom at the active site of the metalloproteinase. Selectivity for MMP over other metalloproteinases is conferred by the structures surrounding this group.

Batimastat was tested in several different animal cancer models and was shown to inhibit both tumor growth and metastatic spread.[19] In one of the models studied, batimastat inhibited the development of malignant ascites.[20] In that model a human ovarian carcinoma xenograft, HU, is introduced directly into the peritoneums of nude mice where it develops rapidly as a malignant ascites. The survival of animals bearing this xenograft is typically in the range of 16–25 days. In experiments with

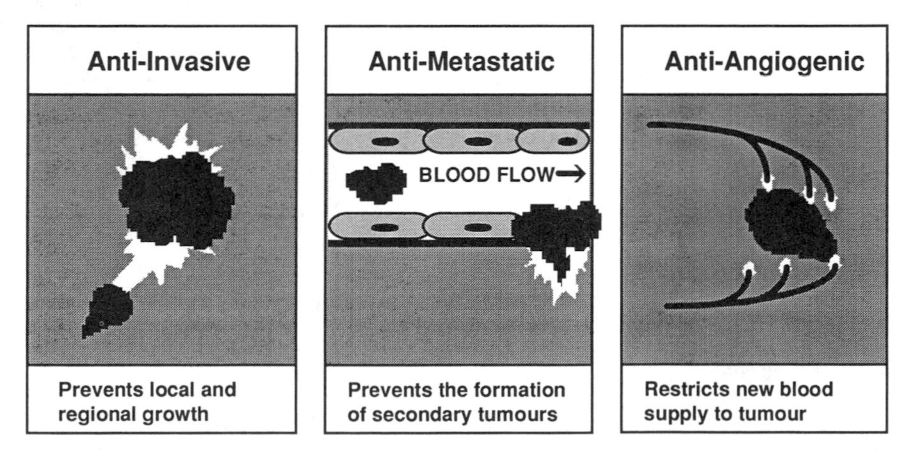

FIGURE 1. The potential sites of action of MMPs in tumor progression and the points at which MMP inhibitors might be expected to exert anti-cancer activity. Several studies have now shown that the proportion of active MMPs overwhelms the local inhibitory activity surrounding an invasive tumor. This net MMP activity facilitates the metastatic spread of malignant cells, the invasive growth of solid tumors, and the ingrowth of new blood vessels. Preclinical studies indicate that inhibition of MMP activity at any of these sites can inhibit tumor progression.

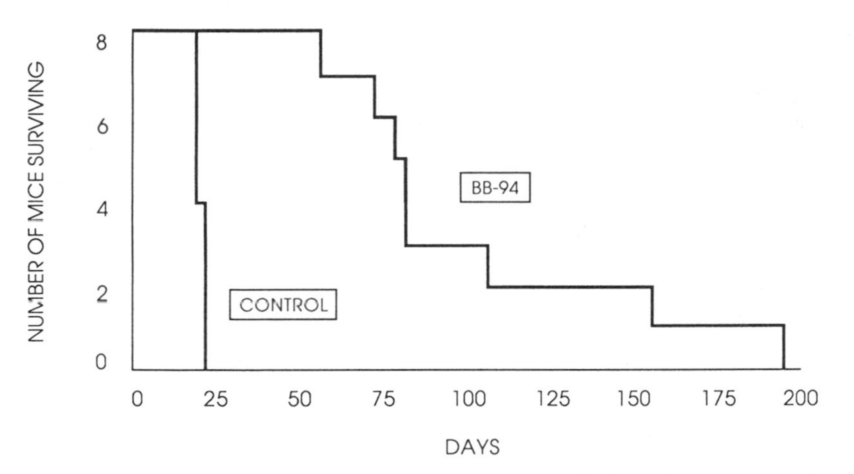

FIGURE 2. Resolution of malignant ascites by a synthetic MMP inhibitor. The synthetic hydroxamate MMP inhibitor, batimastat, was administered, daily for 14 days, to nude mice bearing HU human ovarian carcinoma malignant ascites. The ascites was resolved into small avascular nodules containing islands of tumor cells surrounded by dense stromal tissue. This resolution of ascitic disease led to a 4.4-fold increase in survival.

batimastat nude mice were injected with ascitic HU tumor (approximately 1×10^6 cells) on day 0. Batimastat was initially administered by the intraperitoneal route at 40 mg/kg, once daily from day 3 to day 21, as a 2.5 mg/mL suspension in phosphate-buffered saline containing 0.01% Tween-20 pH 7.4. This treatment resulted in a marked increase in survival from a median of 18.5 days in the control group to 120 days in the treated group, a 6.5-fold increase. An improvement in survival was also observed when batimastat was administered from days 7–20, with a median survival in the BB-94 treated group of 81 days compared to 18.5 days in the group treated with diluent alone (FIG. 2). This was significant because by day 7 the mice had approximately 1–2 mL of ascites, and this seemed to resolve during treatment with batimastat.

The effects of more restricted schedules of batimastat were also investigated. Batimastat is poorly soluble in aqueous media and forms a depot when administered intraperitoneally. Therefore, it seemed likely that daily administration was not necessary. This was confirmed using a schedule in which only a single dose of batimastat (40 mg/kg) was administered on day 3. The median survival of the treated group was 49 days compared to 18 days for the group treated with diluent alone. *In vitro* studies with these and other tumor cell lines have shown that batimastat is not directly cytotoxic even at saturating concentrations.[19,20]

Visual and histological analysis of the peritoneal cavity of mice on day 14 and day 21 revealed that in the presence of batimastat the malignant ascites resolved into small avascular nodules comprising a dense stromal capsule surrounding small islands of tumor cells. Histopathological evidence of necrosis in these tumor islands was present by day 21. Treatment of tumor-bearing mice with BB-1268, the poorly active diastereomer of batimastat, did not reduce the rate and extent of ascites formation. Inasmuch as this compound has similar physical properties to batimastat, it seems unlikely that the latter is exerting its antitumor activity by any nonspecific action. Rather, the histopathological finding of stromal encapsulation indicates that batimastat is blocking the turnover or remodeling of matrix by the ovarian cancer cells, allowing a dense wall of tissue to develop around them.

Earlier experiments with the HU xenograft model had shown that the daily administration of recombinant human tumor necrosis factor (rhTNF), from day 7 onwards, would resolve the ascites and double the survival of the animals.[21] Autopsy of the mice revealed that the rhTNF had induced the formation of solid invasive tumors that were well vascularized. These tumors were quite different in appearance from the "benign" nodules that formed after treatment with batimastat, and appeared to be responsible for the death of the rhTNF-treated mice. This is of particular interest because a recent Phase I/II clinical trial, in patients with refractory malignant ascites, showed that intraperitoneal administration of rhTNF caused a complete resolution of ascites in approximately 55% of patients and a partial resolution in a further 21%, although there did not appear to be an effect on survival.[22]

BATIMASTAT: CLINICAL TRIALS IN CANCER

On the basis of these and other studies in models of human colorectal carcinoma, batimastat has progressed to clinical trial in patients with malignant ascites. At this early stage of the clinical research program patients with ascites from any primary malignancy are eligible for inclusion in the trial, an open study of rising single intraperitoneal doses of batimastat. Clinical trials of batimastat have also been started in patients with malignant pleural effusion, where the drug is administered via an intercostal drain, and in patients with metastatic bone disease, where the drug is administered orally.

Batimastat will almost certainly be joined by other low molecular weight MMP inhibitors in a range of cancer trials and possibly by therapies involving one of the native MMP inhibitors, TIMP-1 or TIMP-2. MMP inhibitors will also be tested against a range of diseases including the one that gave rise to so much early research interest, rheumatoid arthritis. Only then will it be possible to judge the extent to which this novel class of compound will be of therapeutic value.

REFERENCES

1. DINGLE, J. T. 1978. Articular damage in arthritis and its control. Ann. Int. Med. **88:** 821.
2. KRANE, S. M. 1979. Mechanisms of tissue destruction in rheumatoid arthritis. *In* Arthritis and Allied Conditions. 9th edit. D. J. McCarty, Ed.: 449. Lea & Febiger. Philadelphia, PA.
3. GOLDBERG, G. I., S. M. WILHELM, A. KRONBERGER, E. A. BAUER, G. A. GRANT & A. Z. EISEN. 1986. Human fibroblast collagenase: Complete primary structure and homology to an oncogene transformation-induced rat protein. J. Biol. Chem. **261:** 6600–6605.
4. MATRISIAN, L. M., P. LEROY, C. RUHLMANN, M.-C. GESNEL & R. BREATHNACH. 1986. Isolation of the oncogene and epidermal growth factor-induced transin gene: Complex control in rat fibroblasts. Mol. Cell. Biol. **6:** 1679–1686.
5. COLLIER, I. E., S. M. WILHELM, A. Z. EISEN, B. L. MARMER, G. A. GRANT, J. L. SELTZER, A. KRONBERGER, C. HE, E. A. BAUER & G. I. GOLDBERG. 1988. H-ras oncogene-transformed human bronchial epithelial cells (TBE-1) secrete a single metalloprotease capable of degrading basement membrane collagen. J. Biol. Chem. **263:** 6579–6587.
6. WILHELM, S. M., I. E. COLLIER, B. L. MARMER, A. Z. EISEN, G. A. GRANT & G. I. GOLDBERG. 1989. SV40-transformed human lung fibroblasts secrete a 92-kDa type IV collagenase which is identical to that secreted by normal human macrophages. J. Biol. Chem. **264:** 17213–17221.

7. QUANTIN, B., G. MURPHY & R. BREATHNACH. 1989. Pump-1 cDNA codes for a protein with characteristics similar to those of classical collagenase family members. Biochemistry **28:** 5325–5334.
8. KUBOCHI, K. 1990. New direct assay method of type IV collagenase in tissue homogenate and biochemical role of collagenase against type I and IV collagens to the invasion of the stomach and lung cancer. Nippon Geka. Gakkai Zasshi **91:** 174–183.
9. HEWITT, R. E., I. H. LEACH, D. G. POWE, I. M. CLARK, T. E. CAWSTON & D. R. TURNER. 1991. Distribution of collagenase and tissue inhibitor of metalloproteinases (TIMP) in colorectal tumours. Int. J. Cancer **49:** 666–672.
10. MULLER, D., R. BREATHNACH, A. ENGELMANN, R. MILLON, G. BRONNER, H. FLESCH, P. DUMONT, M. EBER & J. ABECASSIS. 1991. Expression of collagenase-related metalloproteinase genes in human lung or head and neck tumours. Int. J. Cancer **48:** 550–556.
11. D'ERRICO, A., S. GARBISA, L. A. LIOTTA, V. CASTRONOVO, W. G. STETLER-STEVENSON & W. F. GRIGIONI. 1991. Augmentation of type IV collagenase, laminin receptor, and Ki67 proliferation antigen associated with human colon, gastric and breast carcinoma progression. Mod. Pathol. **4:** 239–246.
12. MCDONNELL, S., M. NAVRE, R. J. COFFEY & L. M. MATRISIAN. 1991. Expression and localization of the matrix metalloproteinase pump-1 (MMP-7) in human gastric and colon carcinomas. Mol. Carcinogenesis **4:** 527–533.
13. DOCHERTY, A. J. P., A. LYONS, B. J. SMITH, E. M. WRIGHT, P. E. STEPHENS & T. J. R. HARRIS. 1985. Sequence of human tissue inhibitor of metalloproteinases and its identity to erythroid-potentiating activity. Nature **318:** 66–69.
14. STETLER-STEVENSON, W. G., H. C. KRUTZSCH & L. A. LIOTTA. 1989. Tissue inhibitor of metalloproteinase (TIMP-2): A new member of the metalloproteinase inhibitor family. J. Biol. Chem. **264:** 17374–17378.
15. SCHULTZ, R. M., S. SILBERMAN, B. PERSKY, A. S. BAJKOWSKI & D. F. CARMICHAEL. 1988. Inhibition by human recombinant tissue inhibitor of metalloproteinases of human amnion invasion and lung colonization by murine B16-F10 melanoma cells. Cancer Res. **48:** 5539–5545.
16. ALVAREZ, O. A., D. F. CARMICHAEL & Y. A. DECLERCK. 1990. Inhibition of collagenolytic activity and metastasis of tumour cells by a recombinant human tissue inhibitor of metalloproteinases. J. Natl. Cancer Inst. **82:** 589–595.
17. DECLERCK, Y. A., N. PEREZ, H. SHIMADA, T. C. BOONE, K. E. LANGLEY & S. M. TAYLOR. 1992. Inhibition of invasion and metastasis in cells transfected with an inhibitor of metalloproteinases. Cancer Res. **52:** 701–708.
18. MOSES, M. A., J. SUDHALTER & R. LANGER. 1990. Identification of an inhibitor of neovascularization from cartilage. Science **248:** 1408–1410.
19. CHIRIVI, R. G. S., A. GAROFALO, M. J. CRIMMIN, P. D. BROWN & R. GIAVAZZI. 1994. Inhibition of the metastatic spread and growth of B16-BL6 murine melanoma by a synthetic matrix metalloproteinase inhibitor. Int. J. Cancer. In press.
20. DAVIES, B., P. D. BROWN, M. CRIMMIN, N. EAST & F. R. BALKWILL. 1993. A synthetic metalloproteinase inhibitor decreases tumour burden and prolongs survival of mice bearing human ovarian carcinoma xenografts. Cancer Res. **53:** 2087–2091.
21. MALIK, S. T. A., D. B. GRIFFIN, W. FIERS & F. R. BALKWILL. 1989. Paradoxical effects of tumour necrosis factor in experimental ovarian cancer. Int. J. Cancer **44:** 918–925.
22. RATH, U., M. KAUFMANN, H. SCHMID & J. HOFMANN. 1991. Effect of intraperitoneal rhTNFα on malignant ascites. Eur. J. Cancer **27:** 121–125.

Matrix Metalloproteinases and Their Inhibitors in Tumor Progression[a]

YVES A. DE CLERCK,[b,c] HIROYUKI SHIMADA,[d]
SHIRLEY M. TAYLOR,[e] AND KEITH E. LANGLEY[f]

[b]Division of Hematology-Oncology
Departments of Pediatrics and of [d]Pathology
University of Southern California
Los Angeles, California 90027

[e]Department of Microbiology and Immunology
Medical College of Virginia
Richmond, Virginia 23298

[f]Amgen, Inc.
Thousand Oaks, California 91320

A characteristic feature of malignant cells is their ability to invade the surrounding tissues and to form new metastatic tumors in distant organs. Metastasis is the major life-threatening aspect of cancer and the major cause of failure to maintain long-term disease-free remission in cancer patients. When a cancer has disseminated, it is almost always resistant to conventional treatment. Over the last two decades significant efforts have been made to understand the molecular and cellular mechanisms involved in tumor dissemination, with the anticipation that this information may lead to the identification of innovative approaches to control this process. Studies have shown that the metastatic process is under genetic control separate from tumorigenesis and that alteration of several genes is required. Emphasis has been placed on the interaction between tumor cells and the extracellular matrix (ECM), and on the many changes that perturb the harmony between cells and their extracellular environment during tumor invasion (reviewed in ref. 1). Degradation of matrix proteins is an important step that allows tumor cells to penetrate the ECM, to intravasate into blood vessels, and to extravasate at distant sites. In addition, because the ECM is also a reservoir of many factors, including growth factors, it is anticipated that proteolytic degradation of the matrix will significantly affect the equilibrium between cells, matrix proteins to which they are attached, and matrix bound growth factors for which they have specific membrane receptors. Since 1972, when collagenase activity in cancer was first described,[2,3] a growing number of reports have pointed to the key role of matrix metalloproteinases (MMPs) in this regard.[4,5] These proteases, which now comprise a large family, are secreted as inactive proenzymes which can subsequently become activated (reviewed in ref. 6). Several lines of evidence have implicated the MMPs in cancer cell tumorigenicity, invasion, and metastasis, as outlined in TABLE 1. This manuscript reviews these lines of evidence,

[a]This work was supported in part by grant BE84 from the American Cancer Society, and by grant CA 42919 from the National Institutes of Health, Department of Health and Human Services to Y.A.D.

[c]Address correspondence to Yves A. De Clerck, M.D., Division of Hematology-Oncology Childrens Hospital Los Angeles, 4650 Sunset Boulevard, MS #54, Los Angeles, California 90027.

TABLE 1. MMPs in Cancer—Some Milestones

Year	Observation
1972	Collagenase expression in cancer cells
1979	Description of a type IV collagen-degrading enzyme in invasive tumor cells
1980–1986	Positive correlations between type IV collagenase production and metastatic potential in tumor cell lines
1985	Transin, a gene isolated from polyoma virus transformed rat cells, is homologous to human stromelysin
1987–1989	Cloning of MMP cDNAs from cDNA libraries derived from transformed cells
1988–1992	TIMPs inhibit tumor invasion and metastasis
1990–1993	MMPs are expressed in many human tumor tissues

emphasizing the importance of the imbalance between MMPs and their inhibitors as a key determinant in malignant invasion.

EVIDENCE FOR A CAUSAL ROLE OF MMPS IN TUMOR PROGRESSION

In recent years our understanding of the role of MMPs in tumor invasion and metastasis has evolved from suggestive evidence to proof of a causal role. The available data can be grouped in five major categories (TABLE 2). First, a large number of studies have shown a positive correlation between the production of MMPs by established mammalian cell lines and invasive and metastatic behavior *in vitro* and *in vivo*.[4,7–9] Second, studies of the mechanisms regulating MMP production have shown that growth factors such as epidermal growth factor, fibroblast growth factor, and platelet-derived growth factor that stimulate cell proliferation also up-regulate the production of several MMPs, suggesting regulatory pathways common to MMP production and cell proliferation.[10,11] Third, the binding of ECM proteins, including laminin and vitronectin, to their receptors on tumor cells has been shown to stimulate the production of several MMPs and also to enhance the invasive behavior of the cells.[12,13] Fourth, an essential approach showing causality for MMPs has involved genetic manipulations that have induced or enhanced MMP expression in cells, and also induced or enhanced invasion and metastasis. Transfection of nontumorigenic cells with c-Ha-ras induced a metastatic phenotype and the secretion of type IV collagenase;[8,14] however, other changes associated with the expression of the proto-oncogene could not be eliminated in these experiments. More recently cDNAs for specific MMPs have been transfected into mammalian

TABLE 2. Role of MMPs in Tumor Cell Lines. From Correlation to Causality

1. Positive correlation between MMP production and invasive/metastatic behavior of tumor cell lines
2. Factors that promote tumor growth also induce production of several MMPs
3. Stimulation of extracellular matrix receptors that promote invasion also induce production of MMPs
4. Genetic manipulations that induce/enhance MMP production also induce/enhance invasive and metastatic behavior
5. Inhibitors of MMPs inhibit tumor invasion and metastasis

cells such that overexpression of the MMPs in positive transfectants has been shown to enhance the metastatic phenotype. For example, Powell *et al.* have reported that matrilysin overexpression in DU-145 human prostate cancer cells dramatically enhanced their ability to invade the diaphragm when cells were injected intraperitoneally in mice.[15] Further experiments performed with other members of the MMP family will allow us to better define the role of specific MMPs in tumor progression. Fifth, down-regulation of MMP activity in tumor cells by exogenously added synthetic as well as natural MMP inhibitors can inhibit invasion of tumor cells *in vitro* and metastasis *in vivo,*[16–21] and overexpression of the natural inhibitors, tissue inhibitors of metalloproteinases (TIMPs), in B16 melanoma cells or c-Ha-ras transfected cells has had similar effects.[22–24]

EXPRESSION OF MMPS IN HUMAN CANCER

Adding to the evidence showing a causal association between MMPs and invasive behavior of tumor cell lines, several investigators have demonstrated the expression of MMPs in human tumor tissues. As monoclonal antibodies against all subclasses of

TABLE 3. Expression of MMPs in Human Tumor Tissues

MMP	Cancer	Preferential Expression	Reference
Collagenase	Colon, lung, head and neck, squamous cell carcinoma	Stroma and malignant epithelium	25–28, 37
Gelatinases	Skin, breast, thyroid, prostate, colon, lung	Malignant epithelium	29–35, 46
Stromelysins	Head and neck, breast, basal cell carcinoma, lung	Adjacent stroma	27, 32, 36–40
Matrilysin	Stomach, colon, lung, prostate	Malignant epithelium	26, 41, 42

MMPs and human cDNA probes for these proteases became available, studies have taken advantage of several methods of analysis, in particular immunohistochemistry, Northern blot analysis of tumor RNA, and *in situ* hybridization. These studies have demonstrated the presence in human tumor tissue of most members of the MMP family including interstitial collagenase, gelatinases A and B, stromelysins 1, 2 and 3, and matrilysin (TABLE 3). Although some MMPs are preferentially expressed in particular cancers, such as matrilysin in gastric cancers and stromelysin 3 in breast cancer, there is no unique and specific association between an MMP and a particular type of neoplasm.[25–42] In the majority of these studies a positive correlation between the level of MMP detected in tumor tissues and the degree of local recurrence, lymph node metastasis, distant metastasis, and adverse clinical outcome was documented.

With the use of *in situ* hybridization and immunohistochemistry, it has also become apparent that in many cases, MMPs were not produced by the tumor cells but rather by adjacent stromal cells.[27,28,37,39] This is particularly the case for interstitial collagenase and stromelysins which were detected in stromal fibroblasts directly adjacent to malignant tumor masses. On the other hand, matrilysin and gelatinase A seem to be preferentially produced by malignant epithelial cells.[32,41,42] These studies indicate significant differences in mechanisms involved in the production of MMPs in human cancers. They also suggest that some tumor cells can produce factors that stimulate the production of MMPs by the adjacent stromal cells. One such factor, designated tumor cell-derived collagenase stimulatory factor,[43] has been recently characterized and partially sequenced.[44] This 58-kDa protein released in tumor cell conditioned media and associated with tumor cell membranes stimulates the production by fibroblasts of several MMPs, including interstitial collagenase, stromelysin, and gelatinase A.[45] In some cases, an apparent discrepancy exists between information derived from *in situ* hybridization and that from immunohistochemistry.[31,32,46] For example, in a study of colon carcinoma, the transcripts for gelatinase A were predominantly localized in stromal fibroblasts, whereas by immunohistochemistry this protease was localized in the malignant epithelium. This observation suggests that some tumor cells may to able to concentrate gelatinase A produced by normal cells at their surface via a specific receptor. In support of this concept, a cell surface binding protein for gelatinase A has in fact been described recently in two human breast cancer cell lines, MCF7 and MDA-MB-231.[47] Thus much evidence shows that MMPs play an essential and causal role in the progression of human neoplasms. Our ability to control the activity of these proteases may therefore allow us to regain control over tumor progression.

REGULATION OF MMP IN THE EXTRACELLULAR MATRIX

In the extracellular matrix, the activity of MMPs is tightly regulated by a family of natural inhibitors known as tissue inhibitors of metalloproteinases (TIMPs). Three members of this family have been identified so far. The prototype TIMP is a 28.5-kDa ubiquitous glycoprotein initially isolated from rabbit bone[48] and human fibroblasts.[49] It can inhibit the activity of all MMPs by forming a tight (Ki $< 10^{-9}$) 1:1 stoichiometric noncovalent complex with the activated enzymes.[50] The second member, designated metalloproteinase inhibitor (MI) or TIMP-2, was described by us[51] and others.[52,53] It is a 21.5-kDa nonglycosylated protein, often present as a complex with progelatinase A.[52–54] A third member, designated TIMP-3, has been recently reported in human breast tumors.[55] In contrast to TIMP-1 and TIMP-2 which are present in a soluble form, this latter inhibitor is insoluble and binds to components of the ECM.[56] From structure-function analyses of TIMPs, it has become clear that they exert a complex multilevel control on the extracellular activity of MMPs. A first level of control is exerted on the proenzyme. We have demonstrated that TIMP-2 can form a SDS stable complex with proMMP-1 and can block the autoproteolytic activation of the proenzyme. Similarly, TIMP-1 blocks the activation of progelatinase B[57] by stromelysin, and TIMP-2 blocks the activation of progelatinase A by a membrane-bound activator.[58] As mentioned above, a second level of control is directed toward the activated MMPs with which TIMP-1 and TIMP-2 form tight 1:1 stoichiometric complexes. Finally, TIMPs have been shown to block the autoproteolytic degradation of several MMPs. For example, human PMN collagenase undergoes autoproteolytic degradation into a 40-kDa N-terminal active fragment and a 27-kDa C-terminal inactive fragment that includes the hemopexin-like domain and binds to native collagen.[59] The 40-kDa fragment is active but has lost its substrate specificity. Autoproteolytic degradation into these two fragments is inhibited in the presence of a stoichiometric amount of TIMPs, and TIMPs may therefore also play an important role in maintaining the integrity and substrate specificity of MMPs. Thus TIMPs are the major regulators of MMP activity in the ECM, and a balance between MMPs and TIMPs is a critical determinant in maintaining the homeostasis and the integrity of the ECM.

MMP-TIMP BALANCE IN TISSUE REMODELING AND IN HUMAN CANCERS

Subtle changes in the MMP-TIMP balance happen during many physiological processes associated with tissue penetration by normal migratory cells and ECM remodeling. For example, proteolytic degradation of matrix proteins occurs during trophoblastic implantation,[60,61] neovascularization,[62] endometrial proliferation,[63,64] embryogenesis,[65–67] and mammary gland or lung development.[68,69] These physiological processes involve a controlled invasion of the ECM by normal cells including maintenance of the MMP-TIMP balance to limit the degradation of matrix proteins. Disruption of this balance can have dramatic effects on these processes. For example, excessive production of MMPs inhibits branching morphogenesis during lung and mammary gland development, and addition of TIMP restores normal development.[68,69] Transgenic mice expressing the human collagenase transgene in the lungs show histological evidence of pulmonary emphysema.[70] Targeted disruption of TIMP-1 by homologous recombination in mouse embryonic stem cells promotes their invasiveness.[71] Thus, disruption of the MMP-TIMP balance during development results in profound changes in the structure of the ECM and in cell behavior that is suggestive of that observed during neoplastic invasion.

During tumor progression, the breakdown of the ECM is excessive and control of the MMP-TIMP balance appears to be lost. As mentioned, this concept has been supported by experiments in which TIMPs were used to down-regulate MMP activities. Such experiments have utilized recombinant TIMPs to test anti-invasive and antimetastatic activities *in vitro* and *in vivo* or genetic manipulation of TIMP expression in cells. rTIMP-1 has been shown to inhibit the *in vitro* invasion of a human amniotic membrane by tumor cells, as well as lung colonization of mice injected intravenously with metastatic B16 melanoma cells.[21] We have similarly shown that rTIMP-1 and rTIMP-2 inhibit the *in vitro* ECM degradation by invasive and metastatic c-Ha-ras transformed mouse embryo cells and experimental metastasis *in vivo* in nude mice.[19,72]

Genetic manipulation experiments have recently brought several additional observations confirming the important role of MMP-TIMP balance in tumor progression. Khokha *et al.* have down-regulated TIMP-1 expression in mouse 3T3 cells by transfection of antisense TIMP-1 RNA. This resulted in increased tumorigenicity and enhanced ability to invade a human amnion membrane and to form metastatic tumors in nude mice.[73] We have obtained transfectant clones of the invasive and metastatic c-Ha-ras transfected mouse embryo cells overexpressing TIMP-2.[22] These clones showed a marked decrease in proteolytic activity for ECM *in vitro,* and in metastatic potential *in vivo.* Furthermore, when injected subcutaneously in nude mice, the clones formed only small tumors that were encapsulated in a thick layer of connective tissue and failed to invade the adjacent muscle layer (FIG. 1). In parallel experiments, subcutaneous injection of rTIMP-2 in these tumors also induced a local fibrotic reaction that blocked invasion of the adjacent muscle. Confirming these observations, Khokha *et al.* showed that up-regulation of TIMP-1 in metastatic B16 F10 mouse melanoma cells inhibits invasion *in vitro* and experimental metastasis *in vivo.*[23,24] These data clearly indicate the importance of the MMP-TIMP balance, and suggest that TIMPs can play a tumor suppressor role.

A few studies have been made on the expression of TIMPs in human cancers. TIMP-1 expression in colon carcinoma,[74,75] breast tumors,[32] and head and neck carcinoma,[38] and TIMP-2 expression in colon carcinoma[46] and head and neck carcinoma[38] have been reported. In contrast to TIMP-1, which was found in both

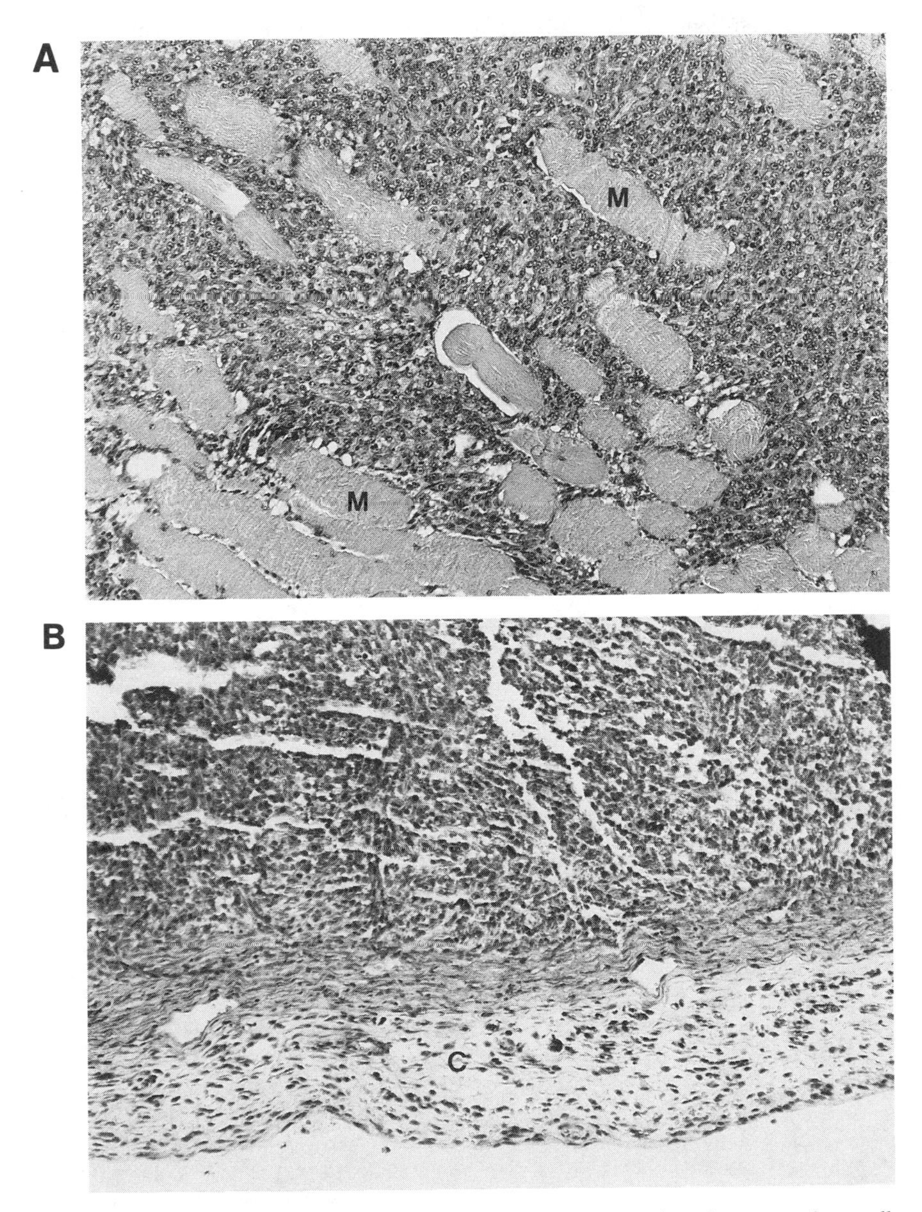

FIGURE 1. Cross section of tumors obtained from c-Ha-Ras transfected mouse embryo cells injected subcutaneously in nude mice. c-Ha-ras transfected rat embryo cells (**A**) and a clone overexpressing TIMP-2 (**B**) were injected (10^5 cells) subcutaneously in the flank of nude mice. After 14 days the tumors and the adjacent muscle were resected and processed for thin sectioning and stained with hematoxylin-eosin. Note the invasion of the muscle (M) in parent cells, and the absence of invasion and the presence of a connective tissue capsule (C) in TIMP-2 expressing transfectant (original ×400; reduced to 67%).

tumor and non-tumor cells, TIMP-2 was detected mainly in endothelial cells and sparse stromal cells. Although a reverse correlation between TIMP expression and clinical aggressiveness of tumor cells was reported,[76] there are other reports of a direct correlation between TIMP expression and clinical aggressiveness of tumors.[74,77] It is important to note that these studies (as well as most of the MMP tumor localization studies referred to above) lack any insights into the relative levels of and interactions between MMPs and TIMPs.

In the context of the direct correlations between TIMP expression and tumor aggressivity, it is noteworthy that there continue to be provocative reports of *in vitro* growth promotion by TIMPs, toward erythroid precursor cells (erythroid potentiating activity)[78,79] and toward a variety of normal and tumor cells.[80] Such effects may involve a membrane receptor,[81] although no such receptor has yet been fully characterized. TIMP-1 is also homologous to phorbin, a phorbol-inducible protein whose expression in human colon carcinoma has been shown to correlate with increased invasiveness.[75] TIMP-3 potentiates the malignant transformation of cells.[56] These data raise the possibility that TIMPs are multifunctional proteins having, in addition to an antimetalloproteinase activity, a direct growth-promoting effect on cells.

CONCLUSION

The data reviewed in this article support the concept that matrix metalloproteinases play an important and active part in the invasive and metastatic behavior of human cancers. They also show that TIMPs can control the activity of MMPs and preserve the integrity of the ECM, allowing the ECM to maintain its control over neoplastic progression. The challenge ahead will be to develop innovative approaches to apply these observations to human cancer. The fact that TIMPs are natural inhibitors offers the advantage that genetic manipulation is applicable. As new methods to alter gene expression and to deliver genes to specific tissues become available, it is our hope that they can be used to modify the MMP-TIMP balance in human neoplasms and arrest their progression in patients.

REFERENCES

1. Liotta, L. A., P. S. Steeg & W. G. Stetler-Stevenson. 1991. Cancer metastasis and angiogenesis: An imbalance of positive and negative regulation. Cell **64:** 327–336.
2. Harris, E. D., C. S. Faulkner & S. Wood. 1972. Collagenase in carcinoma cells. Biochem. Biophys. Res. Commun. **48:** 1247–1253.
3. Yamanishi, Y., M. K. Dabbous & K. Hashimoto. 1972. Effect of collagenolytic activity in basal cell epithelioma of the skin on reconstituted collagen and physical properties and kinetics of the crude enzyme. Cancer Res. **32:** 2551–2560.
4. Liotta, L. A., K. Tryggvason, S. Garbisa, I. Hart, C. M. Foltz & S. Shafie. 1980. Metastatic potential correlates with enzymatic degradation of basement membrane collagen. Nature **284:** 67–68.
5. Stetler-Stevenson, W. G., S. Aznavoorian & L. Liotta. 1993. Tumor cell interactions with the extracellular matrix during invasion and metastasis. Annu. Rev. Cell Biol. **4:** 541–573.
6. Matrisian, L. M. 1990. Metalloproteinases and their inhibitors in matrix remodeling. Trends Genet. **6:** 121–125.
7. Nakajima, M., D. R. Welch, P. N. Belloni & G. L. Nicolson. 1987. Degradation of basement membrane type IV collagen and lung subendothelial matrix by rat mammary

adenocarcinoma cell clones of differing metastatic potentials. Cancer Res. **47:** 4869–4876.
8. POZZATTI, R., R. MUSCHEL, J. WILLIAMS, R. PADMANABHAN, B. HOWARD, L. LIOTTA & G. KHOURY. 1986. Primary rat embryo cells transformed by one or two oncogenes show different metastatic potentials. Science **232:** 223–227.
9. STETLER-STEVENSON, W. G. 1990. Type IV collagenases in tumor invasion and metastasis. Cancer Metastasis Rev. **9:** 289–303.
10. MATRISIAN, L. M. & B. L. HOGAN. 1990. Growth factor-regulated proteases and extracellular matrix remodeling during mammalian development. Curr. Top. Dev. Biol. **24:** 219–259.
11. MATRISIAN, L. M. 1992. The matrix degrading metalloproteinases. Bioessays **14:** 455–463.
12. SEFTOR, R. B., E. A. SEFTOR, K. R. GEHLSEN, W. G. STETLER-STEVENSON, P. D. BROWN, E. RUOSLAHTI & M. J. C. HENDRIX. 1992. Role of the $\alpha_v\beta_3$ integrin in human melanoma invasion. Proc. Natl. Acad. Sci. USA **89:** 1557–1561.
13. TURPEENNIEMI-HUJANEN, T., U. P. THORGEIRSSON, C. N. RAO & L. A. LIOTTA. 1986. Laminin increases the release of type IV collagenase from malignant cells. J. Biol. Chem. **261:** 1883–1889.
14. GARBISA, S., A. NEGRO, T. KALEBIC, R. POZZATTI, R. MUSCHEL, U. SAFFIOTTI & L. A. LIOTTA. 1988. Type IV collagenolytic activity linkage with the metastatic phenotype induced by ras transfection. Adv. Exp. Med. Biol. **233:** 179–186.
15. POWELL, W. C., J. D. KNOX, M. NAVRE, T. M. GROGAN, J. KITTELSON, R. B. NAGLE & G. T. BOWDEN. 1993. Expression of the metalloproteinase matrilysin in DU-145 cells increases their invasive potential in severe combined immunodeficient mice. Cancer Res. **53:** 417–422.
16. PERSKY, B., L. E. OSTROWSKI, P. PAGAST, A. AHSAN & R. M. SCHULTZ. 1986. Inhibition of proteolytic enzymes in the in vitro amnion model for basement membrane invasion. Cancer Res. **46:** 4129–4134.
17. REICH, R., B. STRATFORD, K. KLEIN, G. R. MARTIN, R. A. MUELLER & G. C. FULLER. 1988. Inhibitors of collagenase IV and cell adhesion reduce the invasive activity of malignant tumour cells. Ciba Found. Symp. **141:** 193–210.
18. REICH, R., E. W. THOMPSON, Y. IWAMOTO, G. R. MARTIN, J. R. DEASON, G. C. FULLER & R. MISKIN. 1988. Effects of inhibitors of plasminogen activator, serine proteinases, and collagenase IV on the invasion of basement membranes by metastatic cells. Cancer Res. **48:** 3307–3312.
19. ALVAREZ, O. A., D. F. CARMICHAEL & Y. A. DECLERCK. 1990. Inhibition of collagenolytic activity and metastasis of tumor cells by a recombinant human tissue inhibitor of metalloproteinases. J. Natl. Cancer Inst. **82:** 589–595.
20. ALBINI, A., A. MELCHIORI, L. SANTI, L. A. LIOTTA, P. D. BROWN & W. G. STETLER-STEVENSON. 1991. Tumor cell invasion inhibited by TIMP-2. J. Natl. Cancer Inst. **83:** 775–779.
21. SCHULTZ, R. M., S. SILBERMAN, B. PERSKY, A. S. BAJKOWSKI & D. F. CARMICHAEL. 1988. Inhibition by human recombinant tissue inhibitor of metalloproteinases of human amnion invasion and lung colonization by murine B16-F10 melanoma cells. Cancer Res. **48:** 5539–5545.
22. DECLERCK, Y. A., N. PEREZ, H. SHIMADA, T. C. BOONE, K. E. LANGLEY & S. M. TAYLOR. 1992. Inhibition of invasion and metastasis in cells transfected with an inhibitor of metalloproteinases. Cancer Res. **52:** 701–708.
23. KHOKHA, R., M. J. ZIMMER, S. M. WILSON & A. F. CHAMBERS. 1992. Up-regulation of TIMP-1 expression in B16-F10 melanoma cells suppresses their metastatic ability in chick embryo. Clin. Exp. Metastasis **10:** 365–370.
24. KHOKHA, R., M. J. ZIMMER, C. H. GRAHAM, P. K. LALA & P. WATERHOUSE. 1992. Suppression of invasion by inducible expression of tissue inhibitor of metalloproteinase-1 (TIMP-1) in B16-F10 melanoma cells. J. Natl. Cancer Inst. **84:** 1017–1022.
25. GRAY, S. T., K. YUN, T. MOTOORI & Y. M. KUYS. 1993. Interstitial collagenase gene expression in colonic neoplasia. Am. J. Pathol. **143:** 663–671.
26. URBANSKI, S. J., D. R. EDWARDS, A. MAITLAND, K. J. LECO, A. WATSON & A. E.

KOSSAKOWSKA. 1992. Expression of metalloproteinases and their inhibitors in primary pulmonary carcinoma. Br. J. Cancer **66:** 1188–1194.

27. MULLER, D., R. BREATHNACH, A. ENGELMANN, R. MILLON, G. BRONNER, H. FLESCH, P. DUMONT, M. EBER & J. ABECASSIS. 1991. Expression of collagenase-related metalloproteinase genes in human lung or head and neck tumours. Int. J. Cancer **48:** 550–556.
28. GRAY, S. T., R. J. WILKINS & K. YUN. 1992. Interstitial collagenase gene expression in oral squamous cell carcinoma. Am. J. Pathol. **141:** 301–306.
29. PYKE, C., E. RALFKIAER, P. HUHTALA, T. HURSKAINEN, K. DANO & K. TRYGGVASON. 1992. Localization of messenger RNA for Mr 72,000 and 92,000 type IV collagenases in human skin cancers by *in situ* hybridization. Cancer Res. **52:** 1336–1341.
30. BROWN, P. D., R. E. BLOXIDGE, E. ANDERSON & A. HOWELL. 1993. Expression of activated gelatinase in human invasive breast carcinoma. Clin. Exp. Metastasis **11:** 183–189.
31. POULSOM, R., A. M. HANBY, M. PIGNATELLI, R. E. JEFFERY, J. M. LONGCROFT, L. ROGERS & G. W. STAMP. 1993. Expression of gelatinase A and TIMP-2 mRNAs in desmoplastic fibroblasts in both mammary carcinomas and basal cell carcinomas of the skin. J. Clin. Pathol. **46:** 429–436.
32. CLAVEL, C., M. POLETTE, M. DOCO, I. BINNINGER & P. BIREMBAUT. 1992. Immunolocalization of matrix metallo-proteinases and their tissue inhibitor in human mammary pathology. Bull. Cancer (Paris) **79:** 261–270.
33. CAMPO, E., M. J. MERINO, L. LIOTTA, R. NEUMANN & W. STETLER-STEVENSON. 1992. Distribution of the 72-kd type IV collagenase in nonneoplastic and neoplastic thyroid tissue. Hum. Pathol. **23:** 1395–1401.
34. BOAG, A. H. & I. D. YOUNG. 1993. Immunohistochemical analysis of type IV collagenase expression in prostatic hyperplasia and adenocarcinoma. Mod. Pathol. **6:** 65–68.
35. LEVY, A. T., V. CIOCE, M. E. SOBEL, S. GARBISA, W. F. GRIGIONI, L. A. LIOTTA & W. G. STETLER-STEVENSON. 1991. Increased expression of the Mr 72,000 type IV collagenase in human colonic adenocarcinoma. Cancer Res. **51:** 439–444.
36. MULLER, D., C. WOLF, J. ABECASSIS, R. MILLON, A. ENGELMANN, G. BRONNER, N. ROUYER, M. C. RIO, M. EBER, G. METHLIN, P. CHAMBON & P. BASSET. 1993. Increased stromelysin 3 gene expression is associated with increased local invasiveness in head and neck squamous cell carcinomas. Cancer Res. **53:** 165–169.
37. POLETTE, M., C. CLAVEL, D. MULLER, J. ABECASSIS, I. BINNINGER & P. BIREMBAUT. 1991. Detection of mRNAs encoding collagenase I and stromelysin 2 in carcinomas of the head and neck by in situ hybridization. Invasion Metastasis **11:** 76–83.
38. POLETTE, M., C. CLAVEL, P. BIREMBAUT & Y. A. DE CLERCK. 1993. Localization by *in situ* hybridization of mRNAs encoding stromelysin 3 and tissue inhibitors of metalloproteinases TIMP-1 and TIMP-2 in human head and neck carcinomas. Pathol. Res. Pract. **189:** 1052–1057.
39. WOLF, C., N. ROUYER, Y. LUTZ, C. ADIDA, M. LORIOT, J. P. BELLOCQ, P. CHAMBON & P. BASSET. 1993. Stromelysin 3 belongs to a subgroup of proteinases expressed in breast carcinoma fibroblastic cells and possibly implicated in tumor progression. Proc. Natl. Acad. Sci. USA **90:** 1843–1847.
40. WOLF, C., M. P. CHENARD, P. DURAND DE GROSSOUVRE, J. P. BELLOCQ, P. CHAMBON & P. BASSET. 1992. Breast-cancer-associated stromelysin-3 gene is expressed in basal cell carcinoma and during cutaneous wound healing. J. Invest. Dermatol. **99:** 870–872.
41. MCDONNELL, S., M. NAVRE, R. COFFEY, JR. & L. M. MATRISIAN. 1991. Expression and localization of the matrix metalloproteinase pump-1 (MMP-7) in human gastric and colon carcinomas. Mol. Carcinog. **4:** 527–533.
42. PAJOUH, M. S., R. B. NAGLE, R. BREATHNACH, J. S. FINCH, M. K. BRAWER & G. T. BOWDEN. 1991. Expression of metalloproteinase genes in human prostate cancer. J. Cancer Res. Clin. Oncol. **117:** 144–150.
43. MURAOKA, K., K. NABESHIMA, T. MURAYAMA, C. BISWAS & M. KOONO. 1993. Enhanced expression of a tumor-cell-derived collagenase-stimulatory factor in urothelial carcinoma: Its usefulness as a tumor marker for bladder cancers. Int. J. Cancer **55:** 19–26.
44. NABESHIMA, K., W. S. LANE & C. BISWAS. 1994. Partial sequencing and characterization of the tumor cell-derived collagenase stimulatory factor. Arch. Biochem. Biophys. **285:** 90–96.
45. KATAOKA, H., R. DECASTRO, S. ZUCKER, AND C. BISWAS. 1993. Tumor cell-derived

collagenase-stimulatory factor increases expression of interstitial collagenase, stromelysin, and 72-kDa gelatinase. Cancer Res. **53:** 3154–3158.

46. POULSOM, R., M. PIGNATELLI, W. G. STETLER-STEVENSON, L. A. LIOTTA, P. A. WRIGHT, R. E. JEFFERY, J. M. LONGCROFT, L. ROGERS & G. W. STAMP. 1992. Stromal expression of 72 kda type IV collagenase (MMP-2) and TIMP-2 mRNAs in colorectal neoplasia. Am. J. Pathol. **141:** 389–396.
47. EMONARD, H. P., A. G. REMACLE, A. C. NOEEL, J. A. GRIMAUD, W. G. STETLER-STEVENSON & J. M. FOIDART. 1992. Tumor cell surface-associated binding site for the M(r) 72,000 type IV collagenase. Cancer Res. **52:** 5845–5848.
48. CAWSTON, T. E., A. W. GALLOWAY, E. MERCER, G. MURPHY & J. J. REYNOLDS. 1981. Purification of rabbit bone inhibitor of collagenase. Biochem. J. **195:** 159–165.
49. STRICKLIN, G. P. & H. G. WELGUS. 1983. Human skin fibroblast collagenase inhibitor. Purification and biochemical characterization. J. Biol. Chem. **258:** 12252–12258.
50. WELGUS, H. G., J. J. JEFFREY, A. Z. EISEN, W. T. ROSWIT & G. P. STRICKLIN. 1985. Human skin fibroblast collagenase: Interaction with substrate and inhibitor. Collagen Relat. Res. **5:** 167–179.
51. DE CLERCK, Y. A., T. D. YEAN, B. J. RATZKIN, H. S. LU & K. E. LANGLEY. 1989. Purification and characterization of two related but distinct metalloproteinase inhibitors secreted by bovine aortic endothelial cells. J. Biol. Chem. **264:** 17445–17453.
52. STETLER-STEVENSON, W. G., H. C. KRUTZSCH & L. A. LIOTTA. 1989. Tissue inhibitor of metalloproteinase (TIMP-2). A new member of the metalloproteinase inhibitor family. J. Biol. Chem. **264:** 17374–17378.
53. GOLDBERG, G. I., B. L. MARMER, G. A. GRANT, A. Z. EISEN, S. WILHELM & C. S. HE. 1989. Human 72-kilodalton type IV collagenase forms a complex with a tissue inhibitor of metalloproteases designated TIMP-2. Proc. Natl. Acad. Sci. USA **86:** 8207–8211.
54. HOWARD, E. W. & M. J. BANDA. 1991. Binding of tissue inhibitor of metalloproteinases 2 to two distinct sites on human 72-kDa gelatinase. Identification of a stabilization site. J. Biol. Chem. **266:** 17972–17977.
55. URIA, J. A., A. A. FERRANDO, G. VELASCO, J. M. P. FREIJE & C. LOPEZ-OTIN. 1994. Structure and expression in breast tumors of human TIMP-3, a new member of the metalloproteinase inhibitor family. Cancer Res. **54:** 2091–2094.
56. YANG, T. T. & S. P. HAWKES. 1992. Role of the 21-kDa protein TIMP-3 in oncogenic transformation of cultured chicken embryo fibroblasts. Proc. Natl. Acad. Sci. USA **89:** 10676–10680.
57. GOLDBERG, G. I., A. STRONGIN, I. E. COLLIER, L. T. GENRICH & B. L. MARMER. 1992. Interaction of 92-kDa type IV collagenase with the tissue inhibitor of metalloproteinases prevents dimerization, complex formation with interstitial collagenase, and activation of the proenzyme with stromelysin. J. Biol. Chem. **267:** 4583–4591.
58. BROWN, P. D., D. E. KLEINER, E. J. UNSWORTH & W. G. STETLER-STEVENSON. 1993. Cellular activation of the 72 kDa type IV procollagenase/TIMP-2 complex. Kidney Int. **43:** 163–170.
59. KNAUPER, V., A. OSTHUES, Y. A. DECLERCK, K. E. LANGLEY, J. BLASER & H. TSCHESCHE. 1993. Fragmentation of human polymorphonuclear-leucocyte collagenase. Biochem. J. **291:** 847–854.
60. LALA, P. K. & C. H. GRAHAM. 1990. Mechanisms of trophoblast invasiveness and their control: The role of proteases and protease inhibitors. Cancer Metastasis Rev. **9:** 369–379.
61. YAGEL, S., R. S. PARHAR, J. J. JEFFREY & P. K. LALA. 1988. Normal nonmetastatic human trophoblast cells share in vitro invasive properties of malignant cells. J. Cell. Physiol. **136:** 455–462.
62. MIGNATTI, P., R. TSUBOI, E. ROBBINS & D. B. RIFKIN. 1989. In vitro angiogenesis on the human amniotic membrane: Requirement for basic fibroblast growth factor-induced proteinases. J. Cell Biol. **108:** 671–682.
63. MARBAIX, E., J. DONNEZ, P. J. COURTOY & Y. EECKHOUT. 1992. Progesterone regulates the activity of collagenase and related gelatinases A and B in human endometrial explants. Proc. Natl. Acad. Sci. USA **89:** 11789–11793.
64. RODGERS, W. H., K. G. OSTEEN, L. M. MATRISIAN, M. NAVRE, L. C. GIUDICE & F. GORSTEIN. 1993. Expression and localization of matrilysin, a matrix metalloproteinase,

in human endometrium during the reproductive cycle. Am. J. Obstet. Gynecol. **168:** 253–260.

65. BRENNER, C. A., R. R. ADLER, D. A. RAPPOLEE, R. A. PEDERSEN & Z. WERB. 1989. Genes for extracellular-matrix-degrading metalloproteinases and their inhibitor, TIMP, are expressed during early mammalian development. Genes Dev. **3:** 848–859.
66. EDWARDS, D. R., J. K. HEATH, B. L. HOGAN, S. NOMURA & A. J. WILLS. 1992. Expression of TIMP in fetal and adult mouse tissues studied by in situ hybridization. Matrix (Suppl. 1): 286–293.
67. FLENNIKEN, A. M., C. E. CAMPBELL & B. R. WILLIAMS. 1992. Regulation of TIMP gene expression in cell culture and during mouse embryogenesis. Matrix (Suppl. 1): 275–280.
68. TALHOUK, R. S., J. R. CHIN, E. N. UNEMORI, Z. WERB & M. J. BISSELL. 1991. Proteinases of the mammary gland: Developmental regulation in vivo and vectorial secretion in culture. Development **112:** 439–449.
69. GANSER, G. L., G. P. STRICKLIN & L. M. MATRISIAN. 1991. EGF and TGF alpha influence in vitro lung development by the induction of matrix-degrading metalloproteinases. Int. J. Dev. Biol. **35:** 453–461.
70. D'ARMIENTO, J., S. S. DALAL, Y. OKADA, R. A. BERG & K. CHADA. 1992. Collagenase expression in the lungs of transgenic mice causes pulmonary emphysema. Cell **71:** 955–961.
71. ALEXANDER, C. M. & Z. WERB. 1992. Targeted disruption of the tissue inhibitor of metalloproteinases gene increases the invasive behavior of primitive mesenchymal cells derived from embryonic stem cells in vitro. J. Cell Biol. **118:** 727–739.
72. DECLERCK, Y. A., T. D. YEAN, D. CHAN, H. SHIMADA & K. E. LANGLEY. 1991. Inhibition of tumor invasion of smooth muscle cell layers by recombinant human metalloproteinase inhibitor. Cancer Res. **51:** 2151–2157.
73. KHOKHA, R. & D. T. DENHARDT. 1989. Matrix metalloproteinases and tissue inhibitor of metalloproteinases: A review of their role in tumorigenesis and tissue invasion. Invasion Metastasis **9:** 391–405.
74. LU, X. Q., M. LEVY, I. B. WEINSTEIN & R. M. SANTELLA. 1991. Immunological quantitation of levels of tissue inhibitor of metalloproteinase-1 in human colon cancer. Cancer Res. **51:** 6231–6235.
75. GUILLEM, J. G., M. F. LEVY, L. L. HSIEH, M. D. JOHNSON, P. LOGERFO, K. A. FORDE & I. B. WEINSTEIN. 1990. Increased levels of phorbin, c-myc, and ornithine decarboxylase RNAs in human colon cancer. Mol. Carcinog. **3:** 68–74.
76. TESTA, J. E. 1992. Loss of the metastatic phenotype by a human epidermoid carcinoma cell line, HEp-3, is accompanied by increased expression of tissue inhibitor of metalloproteinase 2. Cancer Res. **52:** 5597–5603.
77. KOSSAKOWSKA, A. E., S. J. URBANSKI, S. A. HUCHCROFT & D. R. EDWARDS. 1992. Relationship between the clinical aggressiveness of large cell immunoblastic lymphomas and expression of 92 kDa gelatinase (type IV collagenase) and tissue inhibitor of metalloproteinases-1 (TIMP-1) RNAs. Oncol. Res. **4:** 233–240.
78. GASSON, J. C., D. W. GOLDE, S. E. KAUFMAN, C. A. WESTBROOK, R. M. HEWICK, R. J. KAUFMAN, G. G. WONG, P. A. TEMPLE, A. C. LEARY, E. L. BROWN, E. C. ORR & S. C. CLARK. 1985. Molecular characterization and expression of the gene encoding human erythroid-potentiating activity. Nature **316:** 768–771.
79. STETLER-STEVENSON, W. G., N. BERSCH & D. W. GOLDE. 1992. Tissue inhibitor of metalloproteinase-2 (TIMP-2) has erythroid-potentiating activity. FEBS Lett. **296:** 231–234.
80. HAYAKAWA, T., K. YAMASHITA, J. KISHI & K. HARIGAYA. 1990. Tissue inhibitor of metalloproteinases from human bone marrow stromal cell line KM 102 has erythroid-potentiating activity, suggesting its possibly bifunctional role in the hematopoietic microenvironment. FEBS Lett. **268:** 125–128.
81. AVALOS, B. R., S. E. KAUFMAN, M. TOMONAGA, R. E. WILLIAMS, D. W. GOLDE & J. C. GASSON. 1988. K562 cells produce and respond to human erythroid-potentiating activity. Blood **71:** 1720–1725.

TIMP-2 Expression Modulates Human Melanoma Cell Adhesion and Motility

JILL M. RAY[a] AND WILLIAM G. STETLER-STEVENSON

Extracellular Matrix Pathology Section
Laboratory of Pathology
National Cancer Institute
National Institutes of Health
Building 10, Room 2A33
Bethesda, Maryland 20892

In order to traverse connective tissue barriers, invasive cells must create localized defects in the extracellular matrix (ECM). This process is thought to be critically regulated in the pericellular milieu by the balance of proteases and protease inhibitors. During angiogenesis this balance has been shown to modulate the morphological outcome of endothelial cell invasion. Excess protease activity results in decreased invasiveness and the formation of saccular structures instead of the normal tubular network.[1] Proteases from all four classes of endopeptidases have been correlated with invasive cell behavior. The matrix metalloproteinases (MMPs) consist of a family of at least nine zinc-dependent enzymes, including the collagenases, gelatinases, and stromelysins, that collectively proteolyze virtually every component of the ECM. Gelatinase A is synthesized and secreted at low levels by a variety of normal cells, such as myoepithelial cells and endothelial cells. In contrast, this protease is produced at high levels in many invasive and malignant cancers.[2] Increased levels of the 72-kDa type IV collagenase (gelatinase A) have been closely correlated with increased tumor cell invasion, and down-regulation of this enzyme activity has been shown to cause loss of the invasive phenotype.[3–6]

Protease inhibitors have been useful in defining the enzymes utilized in invasive cell processes and have allowed progression from correlation to functional analysis. Tissue inhibitors of metalloproteinase (TIMPs) will block all active members of the MMP family with varying affinity. Three forms of TIMPs have been isolated to date. Although they have overlapping inhibitory activities, they are located on different chromosomes and their expression is differentially regulated. TIMP-1 is a 28.5-kDa glycoprotein that preferentially forms a 1:1 complex with activated interstitial collagenase, stromelysin-1, and the 92-kDa type IV collagenase (gelatinase B).[7,8] TIMP-2, a nonglycosylated 21-kDa protein, has the highest affinity for progelatinase A, and will form a 1:1 complex with either the latent or activated forms of the enzyme. TIMP-3, a 24-kDa unglycosylated protein is distinct from TIMP-1 and TIMP-2 in its ability to bind preferentially to the extracellular matrix.[9] TIMP-2 inhibits both the gelatinase A type IV collagenolytic and gelatinolytic activities,[10] and will block the hydrolytic activity of all MMPs.[10,11] The balance between the levels of activated MMPs and free TIMPs determines the net MMP activity. Altering this equilibrium affects the metastatic process. Overexpression of TIMP-2 in ras-transformed rat embryo fibroblasts results in reduction of *in vivo* growth rate and locally invasive character when the transfected cells producing TIMP-2 are injected subcutaneously, as well as loss of lung colony formation when these cells were

[a] Corresponding author.

injected intravenously in nude mice.[12] TIMP-1 has also been shown to inhibit *in vitro* invasion of human amniotic membranes,[13,14] and *in vivo* metastasis in animal models.[14,15] Disruption of TIMP-1 by homologous recombination results in increased invasive behavior in embryonic stem cells. This effect was reversed by the addition of exogenous TIMP-1.[16] In addition, down-regulation of TIMP-1 using an antisense construct transfected into NIH3T3 cells causes enhanced invasion of human amniotic membranes and formation of tumors in athymic mice.[17] Therefore, both TIMP-1 and TIMP-2 may function as natural suppressors of cellular invasion.

In the present study we examine the effects of altered TIMP-2 production on some of the events of the invasion cascade in the A2058 human melanoma cell line. We discuss the finding that altering the balance of gelatinase A and TIMP-2 through genetic manipulation of TIMP-2 expression modulates not only extracellular matrix proteolysis but also cell attachment to extracellular matrix components and motility of cells on matrix components.

MATERIALS AND METHODS

Fibronectin (Mr 440,000) was purchased from Collaborative Research, Inc. (Bedford, MA). Nucleopore membranes (polyvinyl-pyrrolidone-free) and the 48-well chemotaxis chambers were purchased from Neuro Probe, Inc. (Cabin John, MD). Fatty acid-free BSA was purchased from Sigma Chemical Co. (St. Louis, MO).

Cell Culture Methods and Growth Assays

PA317 cells were obtained from American Type Tissue Culture Collection (Rockville, MD). A2058 human melanoma cells[18] were cultured in Dulbecco's modified Eagle's medium containing 10% (v/v) fetal bovine serum, penicillin (100 units/mL), streptomycin (100 units/mL), and when appropriate Geneticin (400 μg/mL) (Gibco/BRL, Gaithersburg, MD). For growth assays, cells were plated in triplicate in 60-mm dishes at 7.5×10^4 cells/dish. The cells were trypsinized and counted at various time points after plating by trypan blue staining using a hemocytometer.

Construction of the Retroviral Expression Vectors Containing TIMP-2 and Infection of A2058 Cells

The retroviral expression vector pLXSN was a generous gift from D. Miller.[19] pLXSN contains the murine leukemia retroviral 5′ LTR that serves as a constitutive promoter for sequences inserted into a downstream multiple cloning site. This plasmid also contains a neo gene which confers G418 resistance for selection of positive infectants. The TIMP-2 containing plasmids pLXSNT2 and pLXSNT2R were constructed using conventional recombinant DNA technology. A 1062 base pair Eco*RI* fragment of *timp*-2 cDNA was isolated from pT2-MO1[20] and ligated into the Eco*RI* site of pLXSN. The 1062 base pair fragment encodes the pro-TIMP-2 protein of 220 amino acids, including a 26-residue signal peptide sequence and mature TIMP-2 encoded by 194 amino acids. The fragment also contains 270 base pairs of sequence upstream of the start site for translation of the signal peptide and a 3′-untranslated sequence of 130 nucleotides which includes a polyadenylation signal.

Recombinant plasmids were screened for orientation of the *timp*-2 insert relative to the retroviral vector promoter. A plasmid was selected that contains *timp*-2 in the proper orientation relative to the LTR promoter pLXSN for transcription and translation of the protein and was named pLXSNT2. A plasmid containing *timp*-2 in the antisense orientation relative to the LTR promoter was named pLXSNT2R.

Production of infective virus particles and infection of the virus particles into A2058 cells were carried out as previously described.[19,21] Briefly, pLXSN, pLXSNT2 or pLXSNT2R was transiently transfected into PA317, an amphoteric packaging cell line using standard calcium phosphate methodology (Bethesda Research Laboratories, Gaithersburg, MD). PA317 cells were plated at 5×10^5 cells per 60-mm dish for 24 h, were transfected and exposed to the DNA precipitate for 18 h, then the medium was aspirated and fresh medium was added. The virus-containing medium was removed 24 h later, centrifuged at $3000 \times g$ for 5 min to remove cells and debris, and used to infect A2058 cells. A2058 cells were plated at 5×10^5 cells per 60-mm dish for 24 h, then infected with 1, 10 or 100 μL of virus-containing medium and incubated for 24 h. Fresh medium was added for another 24 h, and the cells were then trypsinized and split 1:20–1:100 onto 100-mm dishes containing complete media and 400 μg/mL G418. The medium was changed every 3–4 days until colonies formed, and when individual colonies were large enough they were isolated using cloning rings. Clones that were expanded were given an isolate number added onto the end of A-13 when containing pLXSN vector DNA, T2-1 when overproducing TIMP-2, or T2R-7 when expressing antisense *timp*-2 mRNA.

Determination of Levels of TIMP-2 Expression in G418-Resistant A2058 Clones by Northern and ELISA Analyses

Clones were screened for expression of TIMP-2 at the RNA level by Northern analysis[22–24] using the 1062 base pair EcoR1 fragment from pT2-MO1 as a probe. Total RNA was prepared from subconfluent cultures by the method of Gough.[25] RNA was normalized to levels of GAPDH. Antisense TIMP-2 production was detected by Northern analysis using riboprobes (Stratagene, La Jolla, CA) of sense and antisense TIMP-2 mRNA as probes against polyadenylated RNA.[23] To quantitate TIMP-2 protein production conditioned media was collected from cells grown to 80% confluence in complete media. The cells were washed with serum-free Dulbecco's modified Eagle's medium, incubated with serum-free medium for 1 h, then incubated for an additional 24 h. The conditioned medium was removed from the cell monolayer, centrifuged at $800 \times g$ at room temperature for 5 min, decanted from the cell debris, then Brij-35 was added to 0.02% (w/v), and the conditioned medium was stored at −20 °C. The cell monolayer was trypsinized and counted for normalization of conditioned medium to cell number. TIMP-2 concentrations were quantitated in a sandwich ELISA assay using a monoclonal TIMP-2 antibody capture phase and polyclonal antisera detection phase (A. N. Murphy and W. G. Stetler-Stevenson, in preparation).

Adhesion Assays

Maxisorb 96-well plates (Nunc, Naperville, IL) were coated either with 4 μg/mL fibronectin or 1 μg/mL type IV collagen, gelatin or vitronectin in PBS (total volume 100 μL) at room temperature for 2 h. Excess matrix component was removed by aspiration and the wells were then blocked with 1 mg/mL fatty acid-free BSA (Sigma

Chemical Co., St. Louis, MO) in PBS for 30 min. Subconfluent monolayers (80% confluent) of cells were trypsinized, resuspended in DMEM containing 10% fetal bovine serum, counted, allowed to recover for 1 h, then centrifuged at 800 × *g* for 5 min and resuspended in DMEM containing 1 mg/mL fatty acid-free BSA to a concentration of 3 × 10^5 cells/mL. The blocking agent was removed from the wells and 100 μL of cells were plated per well. The plate was incubated at 37 °C for 1–4 h depending on the experiment. After incubation the medium was removed and the wells were washed gently with 100 μL of PBS, then fixed and stained with Diff-Quik Fixative and Diff-Quik Solution II (Baxter Scientific, McGaw Park, IL), respectively, for 3 min each. Excess stain was removed with water; then the cells were either photographed and counted or the dye was extracted in 10% methanol, 5% acetic acid and read at 650 nm on an ELISA plate reader (Molecular Devices, Mountain View, CA). At 3 h, optimal adhesion was achieved, and a linear and reproducible correlation between cell number and absorbency at 650 nm was observed for all of the infectants. As a result, standard adhesion assays were carried out for 3 h and quantitated at OD_{650}.

Cell Motility and Chemotaxis Assays

Chemotaxis assays were carried out as previously described.[26] In short, chemotaxis was assayed in triplicate using 48-well microchemotaxis chambers with 8-μm or 12-μm pore size Nucleopore filters. Filters were precoated by soaking overnight in a solution of fibronectin (10 mg/mL in CAPS buffer). Conditioned medium (1X A2058) from cells grown in serum-free DMEM and containing 0.1% BSA was used as the chemoattractant in the lower chamber. Trypsinized, recovered cells (1.1 × 10^5) were placed in the upper compartment, and chambers were incubated for 4 h at 37 °C. The filters were then removed, fixed, stained with DiffQuik, mounted on glass slides and analyzed by scanning the stained filters into a computer using an Arcus Scanner (Agfa, Wilmington, MA) and quantitated using Image 1.41 (public domain software, Twilight Clone, BBS, Silver Spring, MD). Chemokinesis was assayed using serum-free DMEM, 0.1% BSA in the lower chamber. When appropriate, 500 μg/mL purified TIMP-2 was added either to the cells or to the conditioned media.

RESULTS

Construction, Selection, and Characterization of Expression/Antisense Constructs and Stable Infectant Cell Lines

To examine the effects of altering the ratio of TIMP-2 to gelatinase A we infected highly invasive human melanoma A2058 cells with recombinant human TIMP-2. The TIMP-2 cDNA was ligated in both orientations into a retroviral expression vector, pLXSN, that placed the gene under the control of a high-level constitutive retroviral LTR promoter. The construct containing TIMP-2 ligated in the sense direction relative to the promoter was named pLXSNT2. The antisense construct was called pLXSNT2R. Stable infectant lines were selected that exhibited enhanced TIMP-2 mRNA expression on Northern blot analysis. In addition, serum-free conditioned media from positive clones were screened for overexpression of TIMP-2 protein using anti-TIMP-2 antibodies in ELISA assays. TABLE 1 shows the level of TIMP-2 secreted protein from the selected clone, as measured by ELISA assays, compared to parental A2058 cells and a control cell line (A-13) that was infected with vector only. Secreted TIMP-2 protein levels in T2-1 cells were five times the level secreted by vector control cells. An antisense clone (T2R-7) was selected that exhibits increased expression of antisense TIMP-2 mRNA as shown by Northern analysis using an RNA probe to the sense strand (data not shown) and by decreased expression of TIMP-2 as measured by ELISA analysis. TIMP-2 protein expression in this cell line was decreased to 50% of control infected or parental A2058 cell expression. Gelatinase A expression and secretion was not altered in either T2-1 or T2R-7 cells when compared with A2058 infected controls (data not shown).

Amount of TIMP-2 Produced Modulates Cell Morphology

Alterations in TIMP-2 production correlated with changes in the morphology of the infectant cell lines (data not shown). The T2-1 clone, expressing elevated levels of TIMP-2, was significantly more spread and had more numerous sites of peripheral cell attachment than A-13 controls or A2058 parental cells. The T2-1 clone was also larger than the parental cell line both when spread on a plate and when rounded after trypsinization. In contrast, the T2R-7 clone expressing antisense TIMP-2 mRNA and decreased TIMP-2 protein was smaller, elongated, and spindled in appearance and had fewer attachment sites. These data suggest that altered TIMP-2 expression in the cell has an effect on either a cell attachment or a cell-spreading phenomenon.

TABLE 1. TIMP-2 Production in A2058 Parental Cells and Clonal Lines Containing LXSN, LXSNT2, and LXSNT2R[a]

Cell Line	ng TIMP-2 per 10,000 Cells per 24 h
A2058	1.4
A-13	1.5
T2-1	6.8
T2R-7	0.7

[a]TIMP-2 concentrations were determined by ELISA analysis as described in MATERIALS AND METHODS and were normalized to cell number. Vector control: A-13 = pLXSN infected into A2058 cells. Cell line overproducing TIMP-2: T2-1 = pLXSNT2 infected into A2058 cells. Cell line underexpressing TIMP-2: T2R-7 = pLXSNT2R infected into A2058 cells.

Changes in TIMP-2 Secretion Do Not Affect in Vitro *Cell Growth*

Overproduction of TIMP-2 had no significant effect on log phase growth of the cell lines tested (FIG. 1). At confluence, the parental cells continued to grow logarithmically while cells producing greater or lesser amounts of TIMP-2 than the parental cells stopped or slowed their doubling rate. In addition, the parental cell line could be seen to form large numbers of deep foci. T2R-7 cells underexpressing TIMP-2 grew more slowly during the late log phase than the T2-1 infected or control cells, and did not appear to form foci as the parental cells or reach stationary phase as seen with the T2-1 cells. Both the T2-1 cell lines and cells expressing antisense TIMP-2 mRNA, T2R-7, appeared to grow only in a monolayer and showed no evidence of foci formation. These results suggest that modulation of TIMP-2

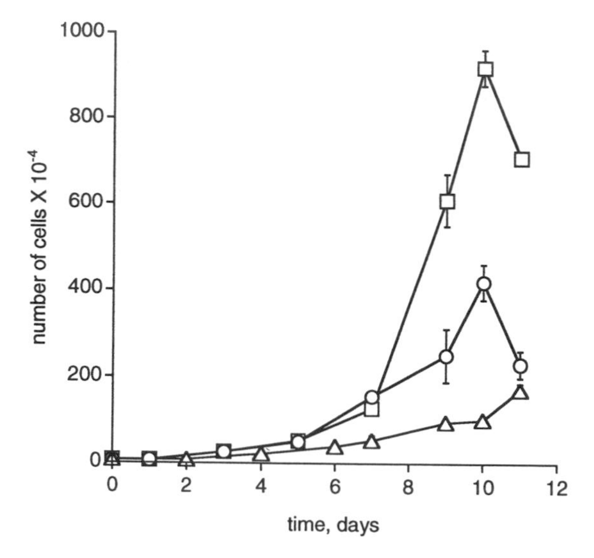

FIGURE 1. Time course of growth of infected A2058 cells *in vitro.* Equal numbers of log-phase cells (7.5×10^4) were plated in triplicate on 60-mm dishes in complete medium for each cell line. At time zero, the plates were approximately 40% confluent. Cells were trypsinized and counted at timed intervals as described in MATERIALS AND METHODS. A-13, *squares*; T2-1, *circles*; T2R-7, *triangles.*

production has an effect on foci formation. However, no significant effects on the growth rate were seen in early log phase growth for any of the cell lines examined.

Adhesive Properties of Infected A2058 Cells Producing Altered Levels of TIMP-2

While isolating and screening sense and antisense infectants we also found a correlation between the concentration of TIMP-2 secreted by the cells and the difficulty with which the cells could be released from tissue culture plastic by trypsinization. This phenomenon was examined by measuring the adhesion of the cell lines to fibronectin. FIGURES 2 and 3 show a time course of adhesion of A-13 (vector control), T2-1 (overexpressing TIMP-2), and T2R-7 (antisense TIMP-2 isolate) cells. Adhesion was analyzed directly by cell counting (FIG. 2A) and also by spectrophotometric determination of cell density (FIG. 2B). At 0.5 h, essentially the same number of T2-1 and A-13 cells had attached to the culture well (FIG. 2A), though the T2-1 cells appeared better spread (FIG. 3) and demonstrated greater absorbency at OD_{650} than the control and antisense cell lines (FIG. 2B). After 1 h the difference in the number of adherent cells determined by cell counting was more pronounced between the three cell lines, and the relationship between the number of cells counted and the OD_{650} became linear. These results indicate that TIMP-2 does not have an effect on the initial attachment of the cells to the matrix components, but may influence either the spreading properties or the long-term adhesive capability of the cells. Those cells that do not spread as well are more likely to be released from the matrix attachment sites during cell-washing procedures. Adhesion to gelatin, type IV collagen, and vitronectin was also measured to determine whether the effect of alterations in TIMP-2 levels were specific for fibronectin (FIG. 4). In all cases, the infectant overexpressing TIMP-2 was 2–3 times more stably adherent than the vector control A-13 cells, and the cell line depleted of TIMP-2 was 25–50% less adherent than A-13 cells.

Altered Production of TIMP-2 Results in Decreased Motility of A2058 Cells

Alterations in TIMP-2 levels in infected A2058 cells were found to negatively affect melanoma cell migration toward a chemoattractant. Cell lines that overexpress or underexpress TIMP-2 were found to migrate at a much slower rate than A2058 or vector control infected cells on fibronectin in response to concentrated A2058 conditioned media (FIG. 5). T2-1 cells exhibited a 50% decrease in chemotaxis, whereas T2R 7 cells displayed a 85 90% decrease. These results suggest that there is a critical balance of MMP activity that is required for optimal cell attachment and migration.

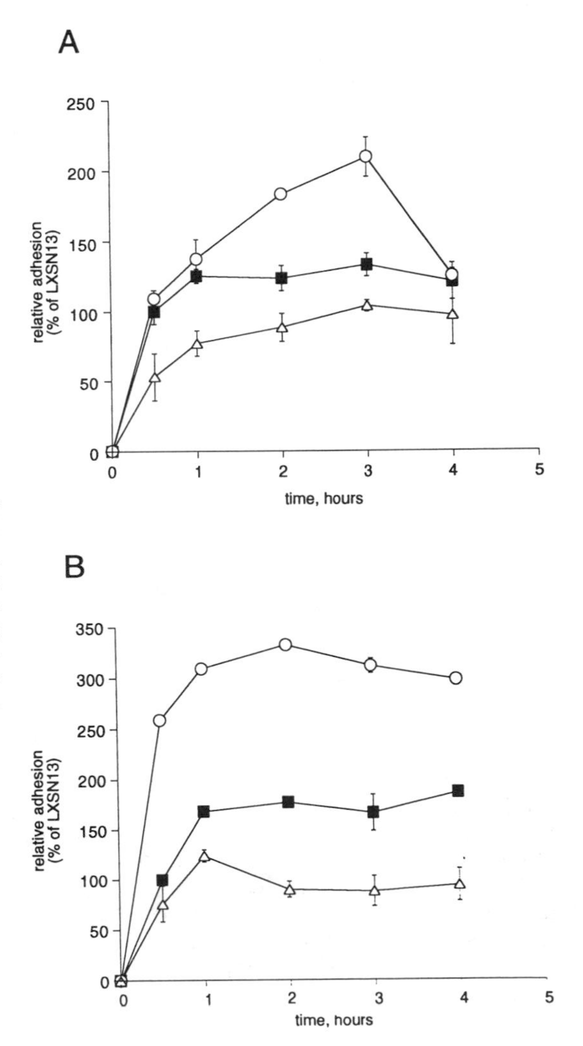

FIGURE 2. Time course of adhesion to fibronectin to A2058 cells with altered levels of TIMP-2 expression. Adhesion was assayed in triplicate and in at least two separate experiments as described in MATERIALS AND METHODS. Data are presented using adhesion of A-13 control cells at 0.5 h as 100% adhesion. The time-course data were analyzed both by counting three fields of cells for each time point (**A**), and then by solubilizing the fixed, stained cells and reading the OD_{650} spectrophotometrically (**B**). A-13, *squares*; T2-1, *circles*; T2R-7, *triangles.*

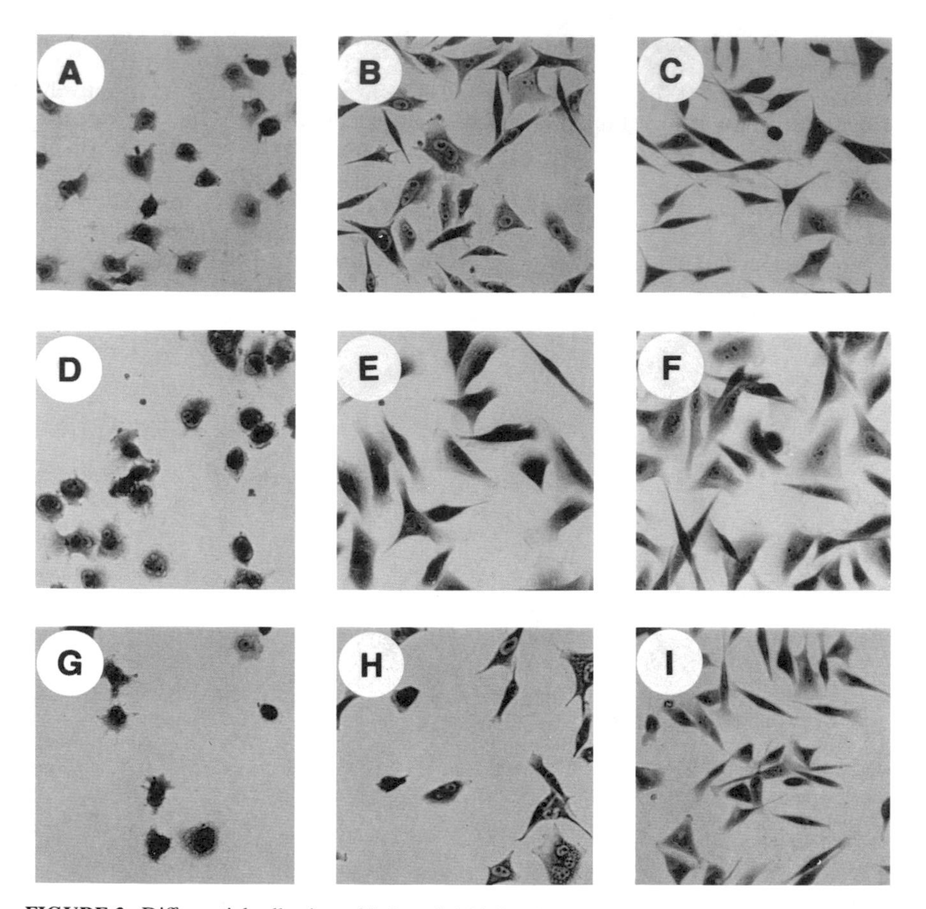

FIGURE 3. Differential adhesion of infected A2058 cells on fibronectin. Cells (3×10^4) were plated in wells of Maxisorb microtiter plates coated with fibronectin. At each time point nonadherent cells were removed by washing the wells with PBS. The adherent cells were fixed and stained and then photographed. (**A–C**) A-13; (**D–F**) T2-1; (G–I) T2R-7. (**A, D,** and **G**) 0.5 h; (**B, E,** and **H**) 1 h; (**C, F,** and **I**) 3 h.

DISCUSSION

This study was designed to characterize the effects of altered production of TIMP-2 on the processes involved in A2058 human melanoma cell invasion. These cells produce the MMPs progelatinase A and interstitial procollagenases well as the MMP inhibitors TIMP-2 and TIMP-1.[20] The MMPs produced by A2058 cells can be inhibited by TIMP-2 following activation. Progelatinase A, which has the unique ability to bind TIMP-2 to the C-terminus of the proenzyme, is widely distributed in a variety of normal and transformed cell types[27] and has been shown to be important during tumor cell invasion and normal cell development.[28] Inhibitors of metalloproteinases in general and specific inhibition of gelatinase A blocks tumor cell invasion *in vitro* and *in vivo,* as well as invasion of cytotrophoblasts *in vitro.*[3,12,14,15,29–33]

Maintenance of the balance between the MMPs and their inhibitors is required for homeostasis of the extracellular matrix. Thus far, work in this field has focused on how alterations of MMP-inhibitor balance affects ECM proteolysis.[16,20,24,34–36] Excessive MMP activity has been implicated to be involved in the matrix degradative step of the cellular invasion cascade. Overexpression of TIMPs has been shown to block local tissue invasion in ras-transfected rat embryo fibroblasts, presumably by directly blocking matrix degradation.[12]

In the present study, we have used a retroviral infection vector to over- or underproduce TIMP-2 in human melanoma A2058 cells. Our results indicate that altering the production of TIMP-2 modulates not only proteolysis of the ECM but also the adhesive and migratory properties of A2058 cells, and also results in changes in cell morphology and contact-inhibited growth.

Overproduction of TIMP-2 by A2058 cells increased the adhesive capacity of the cells and underproduction of TIMP-2 correlated with decreased adhesion. The

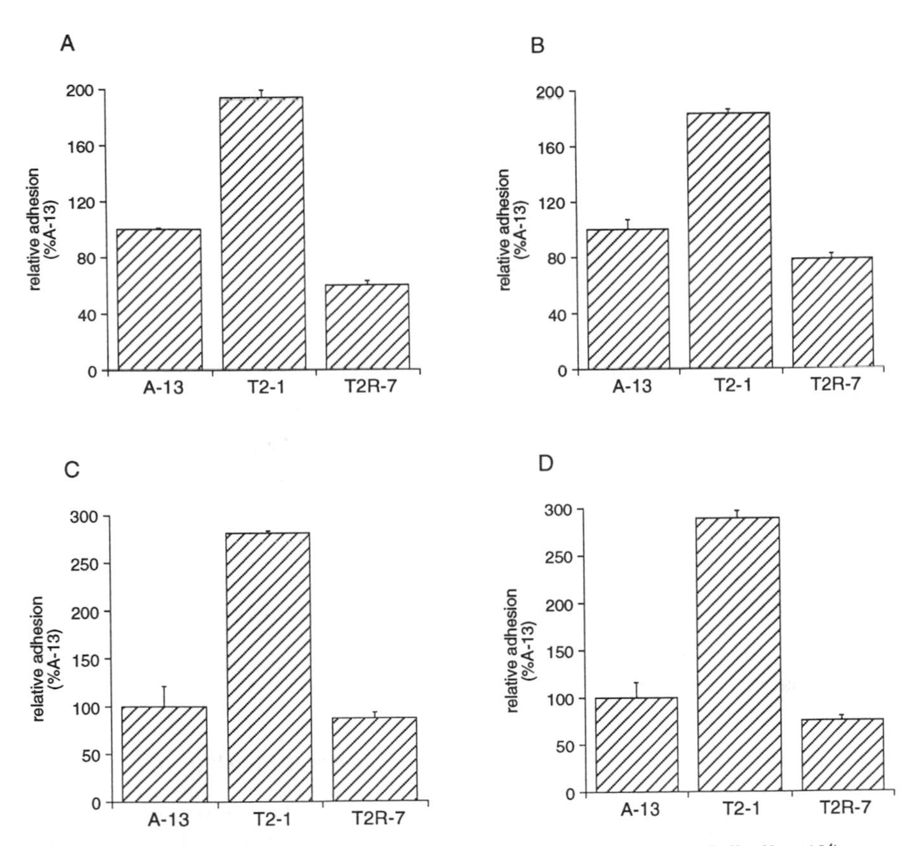

FIGURE 4. Adhesion of infected cells to various matrix components. Cells (3×10^4) were plated in wells of Maxisorb microtiter plates coated with either (**A**) gelatin, (**B**) type IV collagen, (**C**) fibronectin or (**D**) vitronectin. After 3 h, the cells were rinsed with PBS to remove nonadherent cells, then were fixed, stained, solubilized, and read at 650 nm on an ELISA plate reader. Each bar represents the mean of three experiments done in triplicate. Data are expressed as a percentage of attachment of A-13 control cells.

process of cellular adhesion requires initial cell attachment to the matrix followed by cell spreading in combination with stable cell attachment. Our data indicate that cells producing higher TIMP-2 concentrations show similar rates of initial attachment, as demonstrated by counting the number of cells attached after half an hour. At later time points, cells overproducing TIMP-2 showed enhanced attachment compared to controls. These cells also spread more quickly than the controls or underproducing clones. This suggests that TIMP-2 is modulating cell spreading or stable adhesion rather than initial attachment. Our results also demonstrate that

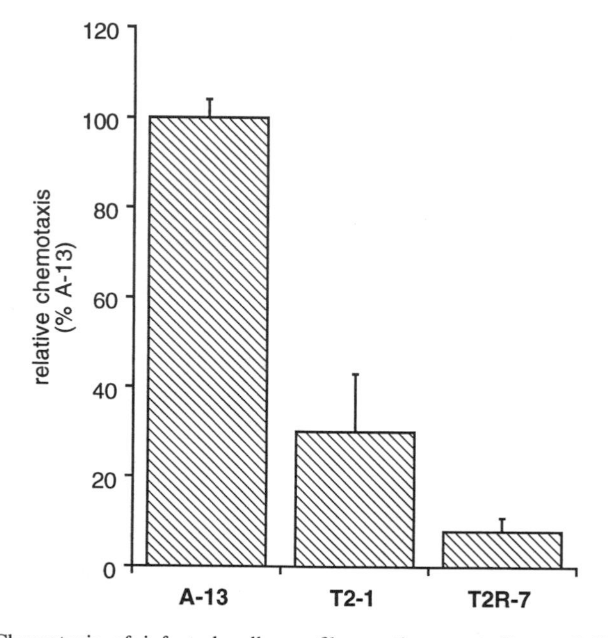

FIGURE 5. Chemotaxis of infected cells on fibronectin-coated filters. Cell motility was assessed in 48-well microchemotaxis chambers. Polycarbonate filters were precoated on both sides with fibronectin. After the addition of 1.1×10^5 cells in serum-free DMEM, 0.1% BSA to the upper compartments, chambers were incubated at 37 °C for 4 h. Control = chemotaxis to 1X A2058 conditioned media. Chemokinesis (random migration in response to DMEM, 0.1% BSA) was subtracted from raw chemotaxis data (migration toward A2058 conditioned media). Results are expressed as a percentage of A-13 chemotaxis. Data represent the mean of triplicate determination for a single experiment of migration. All experiments were repeated three to four times.

alterations in the level of TIMP-2 expression modulates cell attachment and spreading on a number of extracellular matrix macromolecules. This suggests that the mechanism of action of TIMP-2 in this process may be downstream of cell surface receptor-mediated cell attachment, that is, cell spreading. A2058 adhesion may be analogous to the fibroblast system, which appears to use an RGD-dependent mechanism for initial cell attachment, yet spreads by utilizing the heparin sulfate binding site on fibronectin.[37]

Based on the influence of TIMP-2 production on adhesion of A2058 infectants, it could be predicted that there would also be an effect on cell motility. Increased adhesion could result in increased traction on the matrix, which could either facilitate or block migration. Alternatively, decreased adhesion could aid in cell detachment, which also could either enhance or block migration. Interestingly, we found that alteration in TIMP-2 levels in either direction decreased both cell motility on fibronectin-coated filters. Thus, the balance of proteases and protease inhibitors can influence cell attachment and in turn cell migration, both of which are critical for cell invasion.

A2058 cells producing higher or lower concentrations of TIMP-2 appear to grow at the same rate until the cells reach confluence. At this point, the cells under- or overproducing TIMP-2 slow their growth rate and remain in a monolayer, while A2058 parental cells continue doubling at the same rate and form foci composed of several cell layers. This apparent contact inhibition of growth was also observed by Khokha and co-workers in NIH3T3 cells overexpressing antisense mRNA to TIMP-1.[17] These results may indicate that a relationship exists between the constraints on migration of cells with altered TIMP-2 production and the ability of cells to continue growing in a confined space by crawling over one another to form three-dimensional foci. Alternatively, altered TIMP-2 expression may result in partial reversion of the transformed phenotype in which cells regain contact-inhibition of growth.

Modification in the ratio of metalloproteinase to inhibitor alters not only the invasive phenotype but also affects both cell shape and cell-to-matrix contact. Clones that overproduce TIMP-2 are larger than control cells, have more focal adhesion contacts, and are better spread. Cells underproducing TIMP-2 are smaller, appear to contact the matrix at fewer sites, are more refractile and quite elongated. Similar morphological changes were observed with overproduction of antisense mRNA to TIMP in NIH3T3 cells.[17] The changes in cell shape do not result from changes in the cytoskeleton because β-tubulin and F-actin staining showed similar patterns in all the cell types (data not shown). These alterations in the shape of cells and number of contact sites with the matrix may be influenced by the amount of active metalloproteinase available. Spacial distribution or compartmentalization of the enzyme and inhibitor may also affect the net proteolytic activity.[38,39]

In addition to transcriptional regulation of metalloproteinase and inhibitor production, control of activation of MMPs may be critical for monitoring proteolysis spatially and temporally. It has been demonstrated that activation of gelatinase A requires interaction of the C-terminus with the cell surface.[40] Gelatinase A activator has been characterized at the cell surface in HT1080 fibrosarcoma cells and WI-38 fibroblasts.[41–44] Activation of gelatinase A at the cell surface would imply that proteolysis would be limited to the immediate vicinity and regulation of matrix degradation would be under stringent cellular control.

DeClerck and co-workers[45] saw no effect of exogenous TIMP-2 on adhesion or motility of HT1080 fibrosarcoma cells or ras-transfected rat embryo 4R cells.[29] In contrast, addition of exogenous TIMP-2 has been shown to inhibit A2058 invasion *in vitro*.[3,29] The invasion assay could be more responsive to the effects of exogenous TIMP-2, because inhibition of any step of the invasive cascade (adhesion, matrix degradation or motility) would disrupt the process.

Recently, the TIMPs have been implicated to have direct inhibitory effects on angiogenesis not mediated by inhibition of proteolysis.[46,47] The effects of TIMP-2 on cell growth, morphology, adhesion, and motility on A2058 cells that we describe here appear to be mediated at least to some extent by inhibition of proteolysis. It is also possible that some of the effects are either receptor mediated or are due to interaction of TIMP-2 with other receptors or cell-signaling components, such as fibronectin, laminin, or vitronectin receptors.[46,47] Current experimental work is directed at exploring these possibilities.

The data presented demonstrate that the invasive cascade of human melanoma cells is composed of a series of interrelated events. Furthermore, these results implicate the involvement of MMPs and their inhibitors in all aspects of the cellular invasion cascade. These data also support the hypothesis that highly invasive cell lines establish a balance of MMPs and inhibitors that is optimal for invasion, and alteration of this balance in either direction results in perturbation of the invasive phenotype. Overproduction of TIMP-2 may enhance cell attachment and spreading, thereby blocking release of cells from the matrix as well as inhibiting proteolysis of the matrix components, both of which would compromise the process of cellular invasion. Protease activity in excess of the critical level for facilitation of invasion may result in uncontrolled matrix degradation and disruption of cell-matrix interactions required for migration and invasion. These results demonstrate a direct role for MMPs and their inhibitors in cell attachment, spreading and migration, and provide insight into the interrelated nature of these events with regard to tumor cell invasion.

SUMMARY

Invasion of cells across extracellular matrix barriers requires attachment of cells to the matrix, creation of a proteolytic defect in the matrix, and migration of the cells through the defect. To date, alterations in the balance between matrix metalloproteinases (MMPs) and their inhibitors have been shown to alter cellular invasion only through effects on matrix degradation. We used a retroviral infection system to over- and underproduce tissue inhibitor of metalloproteinase-2 (TIMP-2) in human A2058 melanoma cells. Our results indicate that altering the balance of MMPs and TIMP-2 through genetic manipulation of TIMP-2 production modulates not only proteolysis of the extracellular matrix but also cell attachment to the extracellular matrix and motility of cells through matrix components. Altering the production of TIMP-2 also results in the ability of cells to form foci. These results implicate the MMPs and their inhibitors in all aspects of the cellular invasion cascade. This supports the hypothesis that highly invasive cell lines establish a balance of MMPs and inhibitors that is optimal for invasion, and alteration of this balance in either direction results in perturbation of the invasive phenotype.

ACKNOWLEDGMENTS

The authors wish to thank A. N. Murphy for thoughtful scientific insight and critical reading of the manuscript, A. D. Miller for the generous gift of the plasmid pLXSN, and R. Bird of Oncologics for rTIMP-2.

REFERENCES

1. Montesano, R., M. S. Pepper, U. Mohle-Steinlein, W. Risau, W. F. Wagner & L. Orci. 1990. Increased proteolytic activity is responsible for the aberrant morphogenetic behavior of endothelial cells expressing the middle T oncogene. Cell **62:** 435–445.
2. Stetler-Stevenson, W. G., S. Aznavoorian & L. A. Liotta. 1993. Tumor cell interactions with the extracellular matrix during invasion and metastasis. Annu. Rev. Cell Biol. **9:** 541–573.
3. Albini, A., A. Melchiori, L. Santi, L. A. Liotta, P. D. Brown & W. G. Stetler-

STEVENSON. 1991. Tumor cell invasion inhibited by TIMP-2. J. Natl. Cancer Inst. **83:** 775–779.

4. LIOTTA, L. A., K. TRYGGVASON, S. GARBISA, I. HART, C. M. FOLTZ & S. SHAFIE. 1980. Metastatic potential correlates with enzymatic degradation of basement membrane collagen. Nature (Lond.) **284:** 67–68.
5. NAKAJIMA, M., D. WELCH, P. N. BELLONI & G. L. NICOLSON. 1987. Degradation of basement membrane type IV collagen and lung subendothelial matrix by rat mammary adenocarcinoma cell clones of differing metastatic potentials. Cancer Res. **47:** 4869–4876.
6. NAKAJIMA, M., D. LOTAN, M. M. BAIG, R. M. CARRALERO, W. R. WOOD, M. J. C. HENDRIX & R. LOTAN. 1989. Inhibition by retinoic acid of type IV collagenolysis and invasion through reconstituted basement membrane by metastatic rat mammary adenocarcinoma cells. Cancer Res. **49:** 1698–1706.
7. WILHELM, S. M., I. E. COLLIER, B. L. MARMER, A. Z. EISEN, G. A. GRANT & G. I. GOLDBERG. 1989. SV40-transformed human lung fibroblasts secrete a 92-kDa type IV collagenase which is identical to that secreted by normal human macrophages. J. Biol. Chem. **264:** 17213–17221.
8. DOCHERTY, A. J. P., A. LYONS, B. J. SMITH, E. M. WRIGHT, P. E. STEPHENS, T. J. R. HARRIS, G. MURPHY & J. J. REYNOLDS. 1985. Sequence of human tissue inhibitor of metalloproteinases and its identity to erythroid-potentiating activity. Nature **318:** 66–69.
9. STASKUS, P. W., F. R. MASIARZ, L. J. PALLANCK & S. P. HAWKES. 1991. The 21-kDa protein is a transformation-sensitive metalloproteinase inhibitor of chicken fibroblasts. J. Biol. Chem. **266:** 449–454.
10. STETLER-STEVENSON, W. J., H. C. KRUTZSCH & L. A. LIOTTA. 1989. Tissue inhibitor of metalloproteinase (TIMP-2), a new member of the metalloproteinase family. J. Biol. Chem. **264:** 17374–17378.
11. GOLDBERG, G. I., B. L. MARMER, G. A. GRANT, A. Z. EISEN, A. WILHELM & C. HE. 1989. Human 72-kilodalton type IV collagenase forms a complex with a tissue inhibitor of metalloproteinases designated TIMP-2. Proc. Natl. Acad. Sci. USA **86:** 8207–8211.
12. DECLERCK, Y. A., N. PEREZ, H. SHIMADA, T. C. BOONE, K. E. LANGLEY & S. M. TAYLOR. 1992. Inhibition of invasion and metastasis in cells transfected with an inhibitor of metalloproteinases. Cancer Res. **52:** 701–708.
13. MIGNATTI, P., E. R. ROBBINS & D. B. RIFKIN. 1986. Tumor invasion through the human amniotic membrane: Requirement for a proteinase cascade. Cell **47:** 487–498.
14. SCHULTZ, R. M., S. SILBERMAN, B. PERSKY, A. S. BAJKOWSKI & D. F. CARMICHAEL. 1988. Inhibition by human recombinant tissue inhibitor of metalloproteinases of human amnion invasion and lung colonization by murine B-16F10 melanoma cells. Cancer Res. **48:** 5539–5545.
15. ALVAREZ, O. A., D. F. CARMICHAEL & Y. A. DECLERCK. 1990. Inhibition of collagenolytic activity and metastasis of tumor cells by a recombinant human tissue inhibitor of metalloproteinases. J. Natl. Cancer Inst. **82:** 589–595.
16. ALEXANDER, C. M. & Z. WERB. 1992. Targeted disruption of the tissue inhibitor of metalloproteinases gene increases the invasive behavior of primitive mesenchymal cells derived from embryonic stem cells in vitro. J. Cell Biol. **118:** 727–739.
17. KHOKHA, P., P. WATERHOUSE, S. YAGEL, P. K. LALA, C. M. OVERALL, G. NORTON & D. T. DENHARDT. 1989. Antisense RNA-induced reduction in metalloproteinase inhibitor causes mouse 3T3 cells to become tumorigenic. Science **243:** 947–950.
18. TODARO, G. J., C. FRYLING & J. E. DELARCO. 1980. Transforming growth factors produced by certain human tumor cells: Polypeptides that interact with epidermal growth factor receptors. Proc. Natl. Acad. Sci. USA **77:** 5258–5262.
19. MILLER, A. D. & G. J. ROSMAN. 1989. Improved retroviral vectors for gene transfer and expression. Biotechniques **7:** 980–982.
20. STETLER-STEVENSON, W. G., P. D. BROWN, M. ONISTO, A. T. LEVY & L. A. LIOTTA. 1990. Tissue inhibitor of metalloproteinases-2 (TIMP-2) mRNA expression in tumor cell lines and human tumor tissues. J. Biol. Chem. **265:** 13933–13938.

21. KRIEGLER, M. 1990. Gene transfer and expression: A laboratory manual. Stockton Press. New York.
22. LEVY, A. T., V. CIOCE, M. E. SOBEL, S. GARBISA, W. F. GRIGIONI, L. A. LIOTTA & W. G. STETLER-STEVENSON. 1991. Increased expression of the Mr 72,000 type IV collagenase in human colonic adenocarcinoma. Cancer Res. **51:** 439–444.
23. SAMBROOK, J., E. F. FRITSCH & T. MANIATIS. 1989. Molecular cloning: A laboratory manual. Cold Spring Harbor Laboratory Press. Cold Spring Harbor, NY.
24. ALEXANDER, C. M. & Z. WERB. 1991. Extracellular matrix degradation. *In* Cell Biology of the Extracellular Matrix. 2nd edit. E. D. Hay, Ed.: 255–302. Plenum Publishing Corp. New York.
25. GOUGH, N. M. 1988. Rapid and quantitative preparation of cytoplasmic RNA from small numbers of cells. Anal. Biochem. **173:** 93–95.
26. AZNAVOORIAN, S., M. L. STRACKE, H. KRUTZSCH, E. SCHIFFMANN & L. A. LIOTTA. 1990. Signal transduction for chemotaxis and haptotaxis by matrix molecules in tumor cells. J. Cell Biol. **110:** 1427–1438.
27. BIRKEDAL-HANSEN, H., W. G. I. MOORE, M. K. BODDEN, L. J. WINDSOR, B. BIRKEDAL-HANSEN, A. DECARLO & J. A. ENGLER. 1993. Matrix metalloproteinases: A review. Crit. Rev. Oral Biol. Med. **4:** 197–250.
28. LIOTTA, L. A., P. S. STEEG & W. G. STETLER-STEVENSON. 1991. Cancer metastasis and angiogenesis: An imbalance of positive and negative regulation. Cell **64:** 327–336.
29. DECLERCK, Y. A., T. D. YEAN, D. CHAN, H. SHIMADA & K. E. LANGLEY. 1991. Inhibition of tumor invasion of smooth muscle cell layers by recombinant human metalloproteinase inhibitor. Cancer Res. **51:** 2151–2157.
30. GRAHAM, C. H. & P. K. LALA. 1992. Mechanisms of placental invasion of the uterus and their control. Biochem. Cell Biol. **70:** 867–874.
31. HOYHTYA, M., E. HUJANEN, T. TURPEENNIEMI-HUJANEN, U. THORGEIRSSON, L. A. LIOTTA & K. TRYGGVASON. 1990. Modulation of type-IV collagenase activity and invasive behavior of metastatic human melanoma (A2058) cells in vitro by monoclonal antibodies to type-IV collagenase. Int. J. Cancer **46:** 282–286.
32. MELCHIORI, A., A. ALBINI, J. M. RAY & W. G. STETLER-STEVENSON. 1992. Inhibition of tumor cell invasion by a highly conserved peptide sequence from the matrix metalloproteinase enzyme prosegment. Cancer Res. **52:** 2353–2356.
33. LIBRACH, C. L., Z. WERB, M. L. FITZGERALD, K. CHIU, N. M. CORWIN, R. A. ESTEVES, D. GROBELNY, R. GALARDY, C. H. DAMSKY & S. J. FISHER. 1991. 92-kD type IV collagenase mediates invasion of human cytotrophoblasts. J. Cell Biol. **113:** 437–449.
34. MATRISIAN, L. M. 1990. Metalloproteinases and their inhibitors in matrix remodeling. Trends Genet. **6:** 121–125.
35. WERB, Z., P. M. TREMBLE, O. BEHRENDSTEN, E. CROWLEY & C. H. DAMSKY. 1989. Signal transduction through the fibronectin receptor induces collagenase and stromelysin gene expression. J. Cell Biol. **109:** 877–889.
36. WERB, Z. 1989. Proteinases and matrix degradation. *In* Textbook of Rheumatology. 3rd edit. W. N. Kelley, E. D. Harris, Jr., S. Ruddy & C. B. Sledge, Eds.: 300–321. W. B. Saunders. Philadelphia, PA.
37. AKIYAMA, S. K., H. LARJAVA & K. M. YAMADA. 1990. Differences in the biosynthesis and localization of the fibronectin receptor in normal and transformed cultured human cells. Cancer Res. **50:** 1601–1607.
38. UNEMORI, E. N., K. S. BOUHANA & Z. WERB. 1990. Vectorial secretion of extracellular matrix proteins, matrix-degrading proteinases, and tissue inhibitor of metalloproteinases by endothelial cells. J. Biol. Chem. **265:** 445–451.
39. MOSCATELLI, D. & D. B. RIFKIN. 1988. Membrane and matrix localization of proteinases: A common theme in tumor cell invasion and angiogenesis. Biochim. Biophys. Acta **948:** 67–85.
40. MURPHY, G., R. WARD, J. GAVRILOVIC & S. ATKINSON. 1992. Physiological mechanisms for metalloproteinase activation. Matrix **1:** 224–230.
41. BROWN, P. D., A. T. LEVY, I. M. MARGULIES, L. A. LIOTTA & W. G. STETLER-STEVENSON. 1990. Independent expression and cellular processing of Mr 72,000 type IV collagenase and interstitial collagenase in human tumorigenic cell lines. Cancer Res. **50:** 6184–6191.

42. BROWN, P. D., R. E. BLOXIDGE, N. S. A. STUART, K. C. GATTER & J. CARMICHAEL. 1993. Association between expression of activated 72-kilodalton gelatinase and tumor spread in non-small-cell lung carcinoma. J. Natl. Cancer Inst. **85:** 574–578.
43. OVERALL, C. M. & J. SODEK. 1990. Concanavalin A produces a matrix-degradative phenotype in human fibroblasts. Induction and endogenous activation of collagenase, 72-kDa gelatinase, and Pump-1 is accompanied by the suppression of the tissue inhibitor of matrix metalloproteinases. J. Biol. Chem. **265:** 21141–21151.
44. STRONGIN, A. Y., B. L. MARMER, G. A. GRANT & G. I. GOLDBERG. 1993. Plasma membrane-dependent activation of the 72-kDa type IV collagenase is prevented by complex formation with TIMP-2. J. Biol. Chem. **268:** 14033–14039.
45. DECLERCK, Y. A., T. YEAN, B. J. RATZKIN, H. S. LU & K. E. LANGLEY. 1989. Purification and characterization of two related but distinct metalloproteinase inhibitors secreted by bovine aortic endothelial cells. J. Biol. Chem. **264:** 17445–17453.
46. MOSES, M. A., J. SUDHALTER & R. LANGER. 1990. Identification of an inhibitor of neovascularization from cartilage. Science **248:** 1408–1410.
47. MURPHY, A. N., E. J. UNSWORTH & W. G. STETLER-STEVENSON. 1993. Tissue inhibitor of metalloproteinases-2 inhibits bFGF-induced human microvascular endothelial cell proliferation. J. Cell. Physiol. **157:** 351–358.

Plasma Assay of Matrix Metalloproteinases (MMPs) and MMP-Inhibitor Complexes in Cancer

Potential Use in Predicting Metastasis and Monitoring Treatment[a]

STANLEY ZUCKER,[b–d] RITA M. LYSIK,[b]
HOSEIN M. ZARRABI,[b,c] UTE MOLL,[c] SIMON P. TICKLE,[e]
WILLIAM STETLER-STEVENSON,[f] TERRY S. BAKER,[e]
AND ANDREW J. P. DOCHERTY[e]

[b]*Departments of Research and Medicine*
Veterans Affairs Medical Center
Northport, New York 11768

[c]*Departments of Medicine and Pathology*
State University of New York at Stony Brook
Stony Brook, New York 11794

[e]*Celltech Limited*
Slough SLI4EN, Berkshire, United Kingdom

[f]*Department of Pathology*
National Cancer Institute
Bethesda, Maryland 20892

INTRODUCTION

Tumor Markers

Tumor markers are defined as cancer-related substances that can be identified in the bloodstream of patients with cancer. The concentration of tumor markers in blood generally reflects the body burden of cancer rather than the biological behavior of the cancer. In clinical medicine, tumor markers have been employed to (1) screen a population for the presence of cancer, (2) distinguish between malignant tumor types or between benign and malignant tumors, (3) assess prognosis, and (4) follow the clinical course of a cancer patient over time in response to treatment. Classically, tumor markers were considered to be synthesized by the tumor, but they may also be produced by host tissues in response to invasion by cancer cells.[1]

In this report, we propose to expand the concept of tumor markers to include substances released into the bloodstream during cancer dissemination which can be used as markers of cancer invasion and metastasis. Based on the observation that matrix metalloproteinases (MMPs) are produced in excess by cancer cells and can be

[a]This research was supported by a Merit Review grant from the Department of Veterans Affairs, the Ann Shermerhorn Cancer Foundation, and Dianon Systems, Inc.

[d]Address correspondence to Stanley Zucker, M.D., VA Medical Center, Mail Code 151, Northport, New York 11768.

detected in plasma and serum, we propose that specific MMPs may be increased in the blood of patients with metastatic or potentially metastatic cancer.

Matrix Metalloproteinase Production in Cancer

MMPs are considered to play a crucial role in cancer dissemination. Numerous studies have found a correlation between secretion of gelatinase A (72-kDa type IV collagenase, MMP-2) and gelatinase B (92-kDa type IV collagenase, MMP-9) by cancer cell lines and the invasive and metastatic potential of these cells in model systems.[2–5] The concentration of gelatinases in selected human cancer biopsy specimens has also been shown to correlate with tumor grade, invasiveness, and metastasis.[6,7] It has been proposed that the ratio of active gelatinase to latent proenzyme may provide a better correlation with the invasive behavior of human cancer cells than the overall expression of enzyme.[7,8] Immunohistochemical studies of surgical specimens using specific antibodies to gelatinase A have confirmed that human breast, prostate, and gastrointestinal cancer cells contain high levels of cytoplasmic gelatinase A; the percentage of positively stained cells was reported to correlate with evidence of metastasis.[9,10]

The dogma that cancer cells alone are responsible for the increased MMP production in the tumor environment has been challenged.[11] Recent observations suggest that peritumoral fibroblasts surrounding breast cancers and colon cancers are actually responsible for producing of stromelysin-3 and gelatinase A rather than the cancer cells themselves.[7,12–14] Likewise, peritumoral macrophages appear to be the cells responsible for producing gelatinase B. Other reports have described a lower percentage of positive immunostaining for gelatinase A in breast cancer and esophageal cancer with staining in both tumor cells and in adjacent fibroblasts.[6,15] Irrespective of the origin of MMPs in cancer, the bulk of evidence with most types of cancers suggests that MMPs are increased in cancer tissue and play an important role in the metastatic process. In human mesenchymal tumors, however, there may be no correlation between synthesis of gelatinases and biological behavior of the tumor, suggesting that other factors must be involved in the spread of tumor cells into neighboring tissues.[16]

Identification of Matrix Metalloproteinase Activity in Plasma and Serum

Gelatinase/type IV collagenase activity was initially identified in plasma inadvertently as a proteinase contaminant during the isolation of fibronectin.[17] Gelatinase A and gelatinase B were subsequently identified as individual components of normal plasma using gelatin zymography which depicted these MMPs as negatively staining bands of gelatinolytic activity in gelatin-impregnated SDS polyacrylamide gels following electrophoresis, substitution of the detergent, and incubation in a calcium-containing buffer.[18]

MMP activity in human plasma has been measured in various physiological and disease states using various collagen substrate assays. By use of a specific substrate assay, serum collagenase levels have been reported to be increased in term pregnancy; the concentration of collagenase correlated with ripening of the cervix prior to delivery.[19] By use of a substrate degradation assay which presumably measures the combination of gelatinase A/B activity, type IV collagenase (gelatinase) was reported to be increased in the serum of a small subgroup of patients with metastatic hepatocellular carcinoma and thrombosis.[20] The spurious results obtained using serum specimens for measurement of gelatinase B will be discussed in a later section.

Formation of MMP:TIMP Complexes

The demonstration that many types of cancer cells and normal cells produce both tissue inhibitors of metalloproteinases (TIMPs) as well as MMPs and frequently secrete noncovalent complexes of these proteins has complicated interpretation of the interaction of these proteins in extracellular matrix turnover (FIG. 1). Not only do

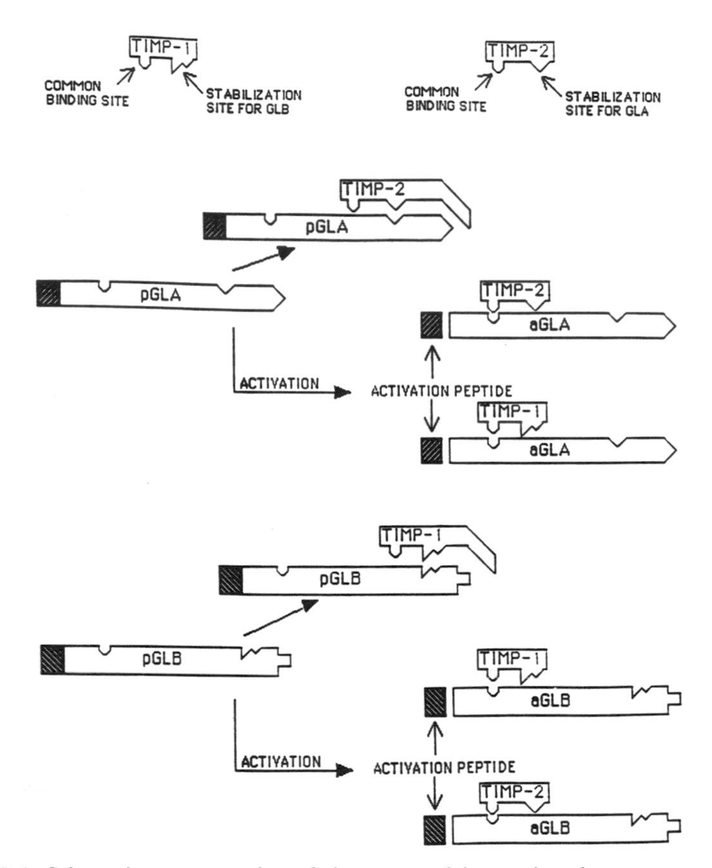

FIGURE 1. Schematic representation of the proposed interactions between progelatinase (pGLA, pGLB), activated gelatinases (aGLA, aGLB), and tissue inhibitor of metalloproteinases (TIMP-1, TIMP-2). Activation peptides are released from the proenzyme after activation. The common binding site on both TIMP-1 or TIMP-2 can bind to the activated forms of both gelatinase A or gelatinase B, whereas the stabilization site on TIMP-1 is specific for progelatinase B and the stabilization site on TIMP-2 is specific for progelatinase A.

TIMPs form complexes with all activated MMPs in the amino terminal portion of the enzymes, but TIMP-1 forms a specific complex with latent gelatinase B, and TIMP-2 forms a specific complex with latent gelatinase A in the carboxy terminal component of the enzyme. These unique complexes lead to stabilization of the enzyme activation mechanism, thereby preventing autoactivation of gelatinases.[21–24] Inasmuch as TIMP:

gelatinase complexes are secreted by cells *in vitro,* it was not surprising to find that these complexes also leach into the bloodstream.

RESULTS AND DISCUSSION

Characterization of Metalloproteinases and TIMPs in Plasma

Moutsiakis *et al.* have identified and characterized the spectrum of metalloproteinases and MMP:TIMP complexes in human plasma.[25] Fractionation of plasma by gelatin Sepharose chromatography (which binds gelatinases and gelatinase:TIMP complexes) provided a simple, one-step procedure for the isolation of gelatinases and gelatinase:TIMP complexes (FIG. 2). Based on its high plasma concentration (approximately 400 ng/mL), it was possible to purify gelatinase A to homogeneity from plasma by molecular sieve chromatography as a single protein band of 72 kDa on SDS-PAGE. Antibodies to the amino terminal peptides of each proenzyme were used to demonstrate that the vast majority of plasma gelatinases circulate as latent enzymes. TIMP-1 and TIMP-2 were also isolated from gelatin Sepharose-bound plasma, which is consistent with their presence in gelatinase:TIMP complexes because free TIMPs do not bind to gelatin Sepharose. In nonreduced, nonheated SDS gels, gelatin Sepharose-bound TIMP-1 was identified by immunostaining with rabbit anti-TIMP-1 in complexes with gelatinase B at 92 kDa as well as in the free form at 28 kDa (FIG. 3). This result suggests that a portion of the complex dissociated in SDS; heating and reduction of the complex with β-mercaptoethanol resulted in complete dissociation of gelatinase and TIMP. In molecular sieve chromatography, the slower migrating component of gelatinase B (eluting at 11.5 mL) had a lesser component of associated TIMP-1, which is consistent with the delayed elution of free gelatinase B. In contrast to TIMP-1, TIMP-2 isolated from gelatin Sepharose was represented on nonreduced SDS-PAGE by at least 10 immunostained bands migrating between 20–120 kDa. This suggests that TIMP-2 may be bound to lower molecular weight degradation products of gelatinase A or other proteinases in plasma. Comparison of the concentration of plasma TIMPs bound to gelatin Sepharose to the unbound fraction shows that the proportion of bound to unbound TIMPs in plasma varies considerably between different individuals; approximately 70% of TIMP-2 in plasma is complexed to gelatinase A, and 20% of TIMP-1 is complexed with gelatinase B (data not shown).

Stromelysin-1 was readily isolated and identified in human plasma as a 57-kDa band in the chromatography fraction which did not bind to gelatin Sepharose, but which subsequently bound to Matrex green A.[25] A previously unidentified MMP has recently been isolated from normal human plasma in the Matrex green A bound, red Sepharose bound fraction that was further purified by molecular sieve chromatography (FIG. 4). This 34–36-kDa proteinase degraded [^{3}H]carboxymethylated transferrin and α casein, but not gelatin, type I collagen or type IV collagen, and was inhibited by EDTA, but not TIMP-1. Whether this enzyme represents a breakdown product of an established enzyme or a new species of metalloproteinase must await amino acid sequence data on the purified enzyme. The fact that the 34–36-kDa metalloproteinase did not react on immunoblotting with antibodies to either gelatinases, collagenase, stromelysin-1, or matrilysin (PUMP-1) suggests that this may be a new enzyme species (Zucker *et al.*, unpublished data).

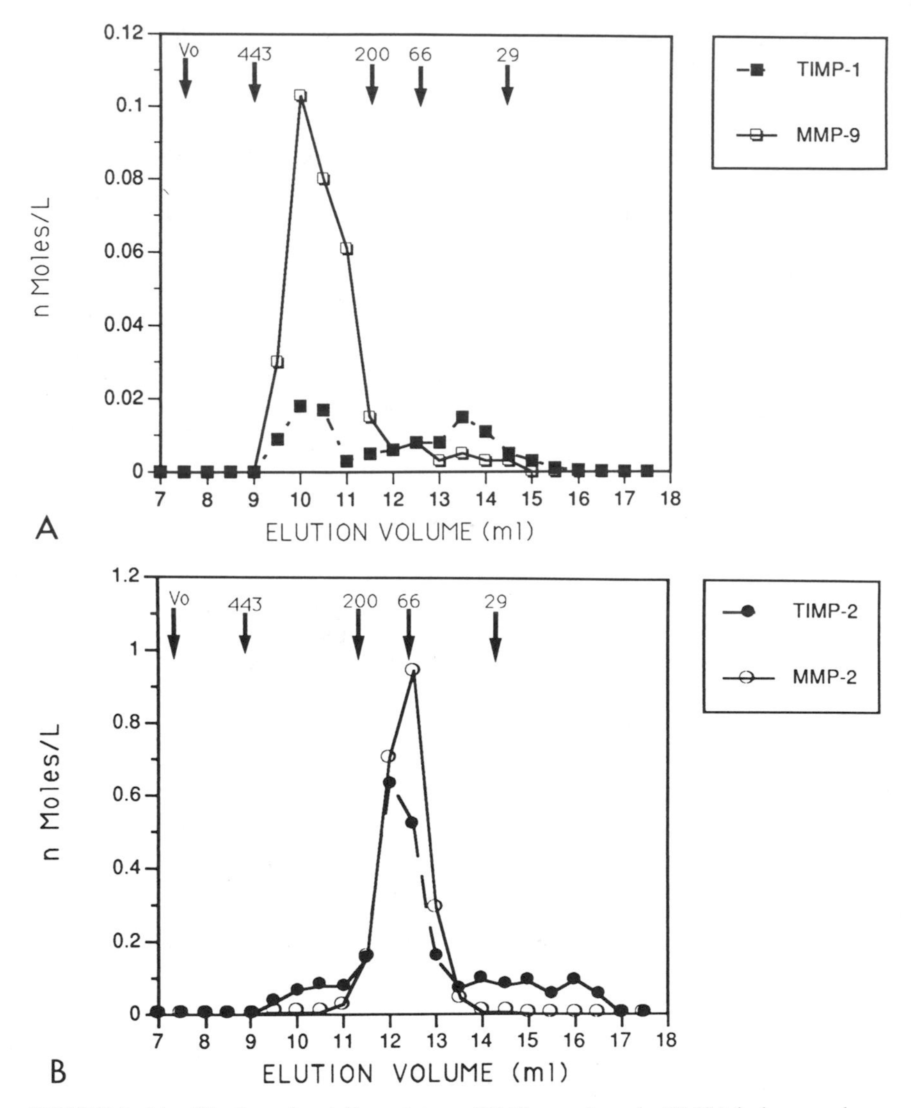

FIGURE 2. Identification of metalloproteinase:TIMP complexes by ELISA in human plasma fractions isolated by gelatin Sepharose chromatography followed by molecular sieve chromatography. Gelatinase and gelatinase:TIMP complexes were isolated from human plasma by gelatin-Sepharose affinity chromatography (eluted with 4% DMSO) followed by molecular sieve chromatography. Fractions eluted from a Superose 12 column (Pharmacia) were analyzed by ELISA for gelatinase B (MMP-9) (*open squares*) and TIMP-1 (*filled squares*) (**A**) and gelatinase A (MMP-2) (*open circles*) and TIMP-2 (*filled circles*) (**B**). The concentration of MMPs and TIMPs was calculated as nmoles/L using linear regression analysis of standard curves of purified antigen. Molecular weights, as determined with marker proteins, are indicated at the top of the figures with arrows. As noted, gelatinase B and TIMP-1 levels peaked in fractions between 10–11.5 mL (>200 kDa); gelatinase A and TIMP-2 levels peaked in fractions between 12–13 mL (70–130 kDa), suggesting the presence of complexes. The molar concentration of gelatinase A is approximately 10-fold higher than that of gelatinase B.

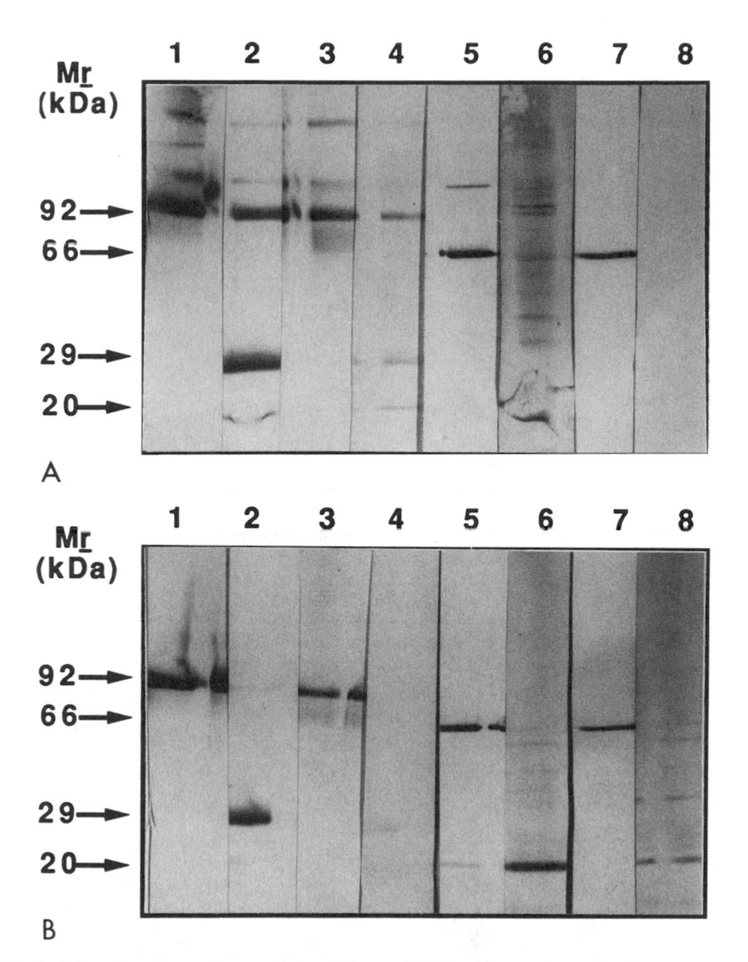

FIGURE 3. Identification of metalloproteinase:inhibitor complexes in human plasma by immunoblotting. Gelatinases and TIMPs were isolated by molecular sieve chromatography from human plasma as described in FIGURE 2. **Lanes 1 and 2** were eluted at 10 mL; **lanes 3 and 4,** at 11.5 mL; **lanes 5 and 6,** at 12 mL, and **lanes 7 and 8,** at 13 mL. Proteins electrophoresed under nonreducing conditions (not heated) are visualized in **A,** whereas proteins electrophoresed under reducing conditions are visualized in **B. Lanes 1 and 3,** probed with a rabbit polyclonal antihuman gelatinase B antibody, show bands of reactivity at high molecular weight as well as at 92 kDa. After reduction a single band is present at 92 kDa (**B**). **Lanes 2 and 4,** probed with rabbit polyclonal anti-TIMP-1, demonstrate bands of reactivity at 92 kDa, indicating the presence of TIMP-1 in a metalloproteinase:inhibitor complex (**A**). In addition, a TIMP-1 band is also seen at 28 kDa. Weak staining for TIMP-1 is also noted at higher molecular weights. After reduction, a single TIMP-1 band is present at 28 kDa (**B**). **Lanes 5 and 7** demonstrate strong reactivity with rabbit polyclonal antigelatinase A at 66 kDa and a weaker band at 130 kDa. **Lane 6,** probed with a rabbit polyclonal anti-TIMP-2 antibody, demonstrates multiple bands with molecular weights between 20–200 kDa, only one of which coincides with the gelatinase A band at 66 kDa (**A**). After reduction, only a singe TIMP-2 band at 22 kDa was visualized with the anti-TIMP-2 antibody (**B**). *Interpretation of data*: Normal human plasma contains both gelatinase A:TIMP-2 and gelatinase B:TIMP-1 complexes. It is possible that the multiple high molecular weight bands seen when proteins are probed with anti-TIMP-2 represent, in part, the presence of multimers of TIMP-2, and/or alternatively, that TIMP-2 is binding to gelatinase A or gelatinase B fragments that are not detected with the antigelatinase A or antigelatinase B antibodies used as probes.

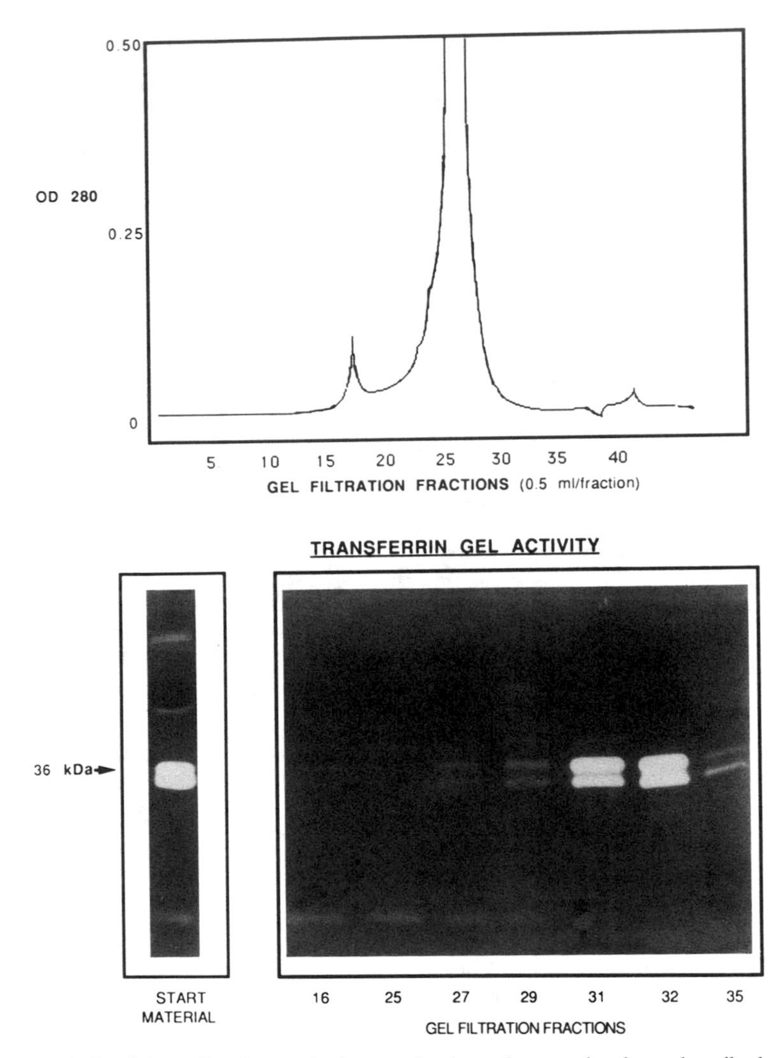

FIGURE 4. Partial purification and characterization of a previously undescribed matrix metalloproteinase from human plasma. The unbound fraction of human plasma (stored at −20 °C) after gelatin Sepharose affinity chromatography was bound to Matrex green A, eluted with 0.3M NaCl, bound to red Sepharose, eluted with 0.5M KCl, concentrated and treated by molecular sieve chromatography (*top panel*) on a Superose 12 column (Pharmacia). The fractions (0.5 mL) were individually assayed by 2-day incubation in a carboxymethylated transferrin[42] gel (0.3%) prepared similarly to a gelatin SDS-polyacrylamide gel with carboxymethylated transferrin replacing gelatin in the recipe (*bottom panel*). The negatively staining bands of transferrin degradation at 34 and 36 kDa eluted at fractions 31–32, which is consistent with their lower molecular weight. These active fractions had no activity on gelatin zymography. The procedure for the preparation of the proteinase-sensitive carboxymethylated transferrin substrate was previously reported by Okada *et al.*[42]

Immunoassay of Plasma and Serum Metalloproteinases and Inhibitors

Immunoassays have been developed to quantitate gelatinases, stromelysin-1, and collagenase, as well as TIMP-1 and TIMP-2, in plasma, serum, and body fluids.[26–32] Although numerous antibodies have been used for the identification of MMPs in tissue, it has become apparent that most of these antibodies lack the characteristics required for assay of these antigens in blood (unpublished results).

After screening a large number of antibodies to gelatinases, Zucker *et al.*[30] developed a sensitive and specific sandwich-type ELISA for measuring gelatinase A in human plasma which employed the combination of rabbit polyclonal antibodies and mouse monoclonal antibodies to human gelatinase A. Affinity-purified polyclonal antibodies to gelatinase A were quite effective in binding gelatinase A in the capture step of an ELISA and did not interfere with the subsequent effectiveness of the monoclonal detecting antibody. Color was developed using a biotinylated goat anti-mouse Ig, followed by an alkaline phosphatase-streptavidin conjugate, and finally the addition of a *p*-nitrophenyl phosphate substrate. The concentration of gelatinase A, purified and used as a reference standard or present in plasma, showed a linear correlation with optical density (405 nm) on a log:log scale.

To determine whether physiologic conditions associated with increased connective tissue turnover are accompanied by increased gelatinase A levels in plasma, Zucker *et al.*[30] demonstrated that plasma gelatinase A (mean ± SD) was significantly increased during the second half of pregnancy (650 ± 312 ng/mL) as compared to early pregnancy (356 ± 139 ng/mL) or the nonpregnant state (387 ± 193 ng/ml). Because of the wide distribution of plasma gelatinase concentrations in the normal population, however, considerable overlap in results was noted between plasma specimens from pregnant and nonpregnant women. Plasma gelatinase A levels in 43 healthy individuals and 231 patients with gastrointestinal, breast, genitourinary, and lung cancers, and lymphomas-leukemias were then measured with the expectation that some groups of cancer patients would have increased gelatinase A levels. The results, however, demonstrated that plasma gelatinase A levels were not increased in cancer patients, even in disseminated malignancy.[33] In an effort to explain this data, gelatinase A secretion by human umbilical vein endothelial cells and lung cancer cells passaged as cell lines were compared. Contrary to expectations, endothelial cells secreted higher levels of gelatinase A than lung cancer cell lines. Based on these findings it was hypothesized that blood vessel lining cells and other normal cells make a sizable contribution to plasma levels of gelatinase A in the intact animal, and may thereby obfuscate detection of increased levels of enzyme originating from solid tumors. In contrast to these initial findings in 18 patients with lung cancer (primarily stage I-II), Garbisa *et al.*[34] reported significantly increased levels of serum gelatinase A in a larger group of patients with metastatic lung cancer; patients not responding to chemotherapy had significantly higher levels of gelatinase A than those who responded to treatment. Gabrisa proposed that monitoring serum gelatinase levels has the potential for forecasting tumor aggressiveness and response to therapy. To readdress the issue of stage of cancer and plasma gelatinase A levels, Zucker and Soriano have subsequently measured gelatinase A levels in plasma of an additional 40 patients with advanced lung cancer and found no significant increase in gelatinase A levels in this group (unpublished data).

Zucker *et al.*[29] also developed a specific sandwich ELISA for measurement of human gelatinase B in plasma, which employed a pair of monoclonal antibodies to gelatinase B developed in Dr. Quigley's laboratory at SUNY, Stony Brook, NY. In this assay, the first monoclonal antibody was highly effective in capturing gelatinase B in solution. The second monoclonal antibody recognized a different epitope on gelatinase B and was employed as a biotinylated detecting antibody. In this study, no significant group differences were noted in plasma concentrations of gelatinase B between 60 healthy subjects (9 ± 11 ng/mL), 136 hospitalized patients without cancer, and 179 patients with cancer of the lung, genitourinary tract or lymphomas-leukemias. In contrast, plasma gelatinase B was significantly increased in 122 patients with gastrointestinal tract cancer and breast cancer (18 ± 23 ng/mL and 21 ± 22 ng/mL, respectively). However, there was no correlation between elevations of plasma gelatinase B and stage of disease or in the level of carcinoembryonic antigen (CEA) in patients with colorectal cancer. In fact, an inverse relationship was identified between plasma gelatinase B and CEA suggesting that production of gelatinase B (possibly by tumor macrophages) occurs primarily in colorectal cancers with low CEA production. It remains to be determined whether gelatinase B is a marker for metastasis or other aspects of prognosis in this disease. In these studies, plasma gelatinase B levels did not appear to correlate with the magnitude of the local inflammatory response noted on pathological examination of cancer tissues. Combining both CEA and gelatinase B plasma measurements improved the sensitivity of detection of colon cancer with little overlap between these tests.[29] In a similar type of study in experimental mammary adenocarcinoma, Nakajima *et al.* reported sevenfold higher levels of plasma gelatinase B (measured by substrate degradation) in rats with metastasis to the lungs from fat pad tumors as compared to nonmetastatic tumors.[35]

Although glossed over in some reports,[35] serum concentrations of gelatinase B have been reported to be more than 200-fold higher than simultaneously collected plasma concentrations, which suggests that during *in vitro* blood clotting neutrophils degranulate and release stored gelatinase B.[29] This observation may explain why Hashimoto *et al.* detected increased levels of serum type IV collagenase (combined gelatinase A and B) only in the small subset of cancer patients with thrombosis of the portal vein.[20] Furthermore, a correlation between plasma gelatinase B and the neutrophil count in blood has been reported;[32] this observation awaits confirmation.

Plasma levels of stromelysin-1 have also been measured in patients with breast cancer (41 ± 43 ng/mL), lung cancer (85 ± 12 ng/mL), and genitourinary tract cancer (prostate, 53 ± 45 ng/mL); levels of this MMP were not significantly increased in any of these cancer groups as compared to healthy control subjects (61 ± 41 ng/mL) (TABLE 1). In contrast to this negative data in cancer, plasma stromelysin-1 levels were increased approximately threefold in patients with rheumatoid arthritis (187 ± 102 ng/mL), which is consistent with the extremely elevated levels of stromelysin noted in the synovial fluid of patients with rheumatoid arthritis.[36] Although these stromelysin assay results in arthritis lend support to the general concept that plasma levels of MMPs are reflective of the tissue concentration of these enzymes in diseased states, they bring into question the actual levels of latent gelatinase A and gelatinase B in cancer tissue as compared to the much larger contribution to the gelatinase plasma pool by normal tissues.

It is noteworthy that the absolute plasma concentration of MMPs in normal individuals is quite different with gelatinase A > stromelysin-1 > gelatinase B. This is consistent with the observation that gelatinase A is produced constitutively in many tissues in the body;[37] stromelysin-1 production appears limited to selected mesenchymal tissues; and gelatinase B production is restricted primarily to macrophages and keratinocytes. Likewise, physiologic and pathologic regulation of MMP production varies considerably with gelatinase B and stromelysin-1 production increased in response to cytokines such as IL-1 and TNF, whereas gelatinase A production appears to be independent of control by most cytokines.[38] Another confounding factor affecting clinical MMP measurements is that MMPs are capable of binding to connective tissue matrix;[39] hence, increased local secretion of MMPs in disease may not necessarily be translated into increased plasma levels.

Immunoassay of Plasma Latent MMP:TIMP Complexes

Based on the observation that gelatinases circulate in plasma in complexes with TIMPs as well as in the free form, we have measured the concentration of gelatinase B:TIMP-1 in human plasma of patients with various forms of cancer. These immunoassays were developed using the same principles and reagents utilized for the independent assay of gelatinases and TIMPs. Monoclonal antibodies (MAC-015) were utilized to capture the TIMP-1 component[28] of the complex from plasma, followed by a biotinylated antigelatinase B antibody used as the detecting antibody.[29] Gelatinase B:TIMP-1 complexes (TABLE 1) were found to be significantly increased ($p < 0.05$) in the plasma of patients with colorectal cancer and female genitourinary tract cancer. With the range of normal values based on ±2 standard deviations around the mean of healthy individuals, 23% of patients with colorectal cancer and

TABLE 1. Plasma Gelatinase B, GLB:TIMP-1 Complexes, and Stromelysin-1 Levels in Control Subjects and Patients in Various Cancer Groups[a]

Group	*N*	GLB:TIMP-1 (units[b])	GLB (ng/mL)	Stromelysin-1 (ng/mL)
Control	60	3.4 ± 1.7	9.0 ± 10.7	61.3 ± 40.6
GI Cancer	94	5.5 ± 3.9[c]	18.4 ± 23.0[c]	79.8 ± 45.4
GU Cancer	23	7.6 ± 5.3[c]	20.2 ± 18.6[c]	52.9 ± 45.8
Breast Cancer	30	4.2 ± 2.5	21.4 ± 22.4[c]	41.3 ± 43.6
Lung Cancer	8	4.3 ± 1.7	4.5 ± 3.3	84.7 ± 11.6

[a]Data is listed as X ± S.D.
[b]GLB:TIMP-1 levels are expressed as arbitrary units.
[c]Significant difference when compared to control value ($p < 0.05$).
Abbreviations: GLB, plasma gelatinase B; TIMP, tissue inhibitor of metalloproteinase; GI, gastrointestinal; GU, female genitourinary cancer or males with prostate cancer (stromelysin-1 only).

52% of patients with female genitourinary tract cancer had increased plasma levels of gelatinase B:TIMP-1 complexes. When used in combination, abnormal levels of either gelatinase B:TIMP-1 complexes, gelatinase B, or both were found in 37% of patients with colorectal cancer and 65% of patients with genitourinary tract cancer. No correlation was found between the plasma concentration of gelatinase B:TIMP-1 complexes and CEA in patients with colorectal cancer. In contrast to these results, measurement of latent gelatinase A:TIMP-2 complexes in plasma revealed that cancer patients did not have increased levels of this complex (data not shown).

Immunoassay of Plasma Activated Gelatinases and Activated Gelatinase:TIMP Complexes

From a theoretical point of view, the detection of activated gelatinases[8] or complexes of activated gelatinase with the noncorresponding TIMP (gelatinase A:TIMP-1 complex or gelatinase B:TIMP-2 complex; see FIG. 1) should provide a better plasma test for monitoring metastasis in patients with cancer than the assay of latent enzymes and latent MMP:TIMP complexes because activation of MMPs is ultimately required for matrix degradation. Improvement in the sensitivity of these ELISAs will be required before their full potential can be recognized, because the concentration of activated gelatinase complexes in plasma appears to make up less than 1% of the level of the latent enzymes. Thus, the concentration of gelatinase B:TIMP-2 complexes in plasma is probably less than 0.1 ng/mL, which is below the level of detection using currently available antibodies. The production of antibodies that exclusively recognize activated gelatinase A or gelatinase B in plasma, but not the latent form of these enzymes, would provide an extremely useful reagent for monitoring tissue MMP activation and presumably the associated degradation of matrix proteins. Production of these types of exclusive antiactivated gelatinase antibodies for clinical use has proven to be a difficult proposition, probably because of the likelihood that an activation epitope within the gelatinase molecule would immediately be blocked by the binding effect of circulating TIMPs in plasma. Other avenues for exploration would be to (1) identify a unique epitope on TIMP that appears after the formation of a complex with activated gelatinase (a similar type of epitope has been identified on α2-macroglobulin after trapping of a proteinase[40]), (2) develop a plasma assay to detect the activation peptide of MMPs (see FIG. 1), (3) assay plasma levels of unique matrix protein degradation products.

Recently we employed the combination of a rabbit polyclonal antibody to gelatinase A[30] as the capture antibody and a mouse monoclonal antibody to TIMP-1 as the detecting antibody[28] in an ELISA for detection of activated gelatinase A:TIMP-1 complexes. Recombinant human TIMP-1 and aminophenyl mercuric acetate activated gelatinase A were utilized to produce the MMP:TIMP complexes employed in the reference curve. The reference data produced with recombinant complexes fit the log:log relationship noted with our other ELISAs. Plasma specimens from healthy individuals and cancer patients provided ELISA data that fit within the lower end of the reference curve (1–2 ng/mL). The results of this study indicated that plasma gelatinase A:TIMP-1 complexes are within the normal range in patients with colorectal cancer, genitourinary tract cancer, and breast cancer (FIG. 5). These data confirm our previous suspicion that production of gelatinase A and formation of gelatinases A:TIMP-1 complexes is ubiquitous in normal tissues, hence limiting the usefulness of these assays in diagnosis of disease. In contrast, plasma measurement of MMPs and MMP:TIMP complexes of more restricted tissue origin, which fluctuate more widely in response to tissue injury, may prove to be more useful as surrogate markers of disease; stromelysin-3 and matrilysin are likely prospects in this regard.

MMP-TIMP Assays for Monitoring Cancer Treatment with Anti-Metastatic Drugs

Another area of potential use for plasma assays of activated gelatinase:TIMP complexes would be in monitoring the response to therapy of cancer patients treated with pharmacologic agents designed to inhibit the MMP activation process. Pharmacologic inhibition of gelatinase activation in the tissues would presumably be associated with decreased plasma levels of gelatinase B:TIMP-2 and gelatinase A:TIMP-1 complexes, thus permitting the measurement of these complexes in plasma to serve as surrogates of drug effect at the tissue level. Since the cancer metastatic process can take years before clinical recognition even by sophisticated radiologic procedures, a blood test that could determine the potential effectiveness of an anti-metastatic drug would be highly valuable. Although the demonstration that prevention of MMP activation in the tissues would not guarantee that a drug would be useful in preventing metastasis, the reverse statement is likely to be correct.

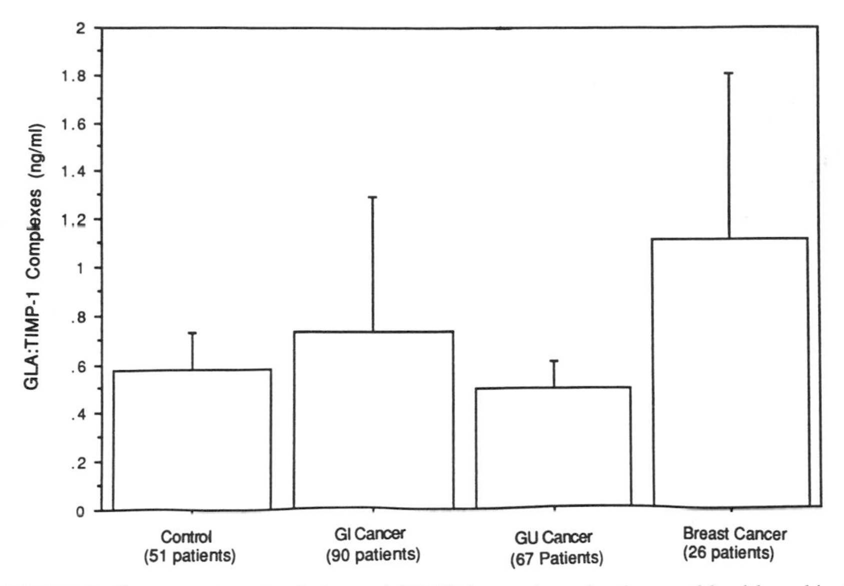

FIGURE 5. Concentration of gelatinase A:TIMP-1 complexes in plasma of healthy subjects and patients with cancer. Because progelatinase A does not form a complex with TIMP-1, this assay is a measure of the activated gelatinase A:TIMP-1 complex. As noted, none of the cancer groups demonstrated increased plasma levels of the MMP activation complex.

In other words, a pharmacologic agent designed to inhibit MMP activation in tissues that does not lower the plasma concentration of activated MMPs or activated MMP:TIMP complexes in a drug-treated patient is likely to be of no clinical usefulness.

Comparison of Assay Techniques for Measurement of Gelatinases in Plasma and Tissues

To address the issue of whether the conclusions reached in various studies of cancer may be due to differences in techniques employed to measure gelatinases, Zucker *et al.*[41] did a comparative study of methods employed for the assay of gelatinases. Immunoassays and substrate degradation assays yielded parallel-type data. In addition to being highly specific for individual metalloproteinases, the ELISAs for gelatinase A and gelatinase B were shown to be more than 100-fold more sensitive than the [^{3}H]gelatin degradation assay and threefold more sensitive than gelatin zymography. Gelatin zymography, however, has the distinct advantage of permitting the distinction between activated and latent gelatinases. In this regard, gelatin zymography has been useful in identifying activated forms of gelatinases in breast cancer specimens from patients with more invasive cancer.[8]

CONCLUSION

We conclude that the assay of metalloproteinases and their complexes with TIMPs in the plasma of patients with cancer may be clinically useful in identifying early metastatic events in certain types of cancer. Standardization of the laboratory tests and long-term clinical follow-up of cancer patients is required before the clinical usefulness of these tests can be determined. Development of high-affinity antibodies for the measurement of activated MMP:TIMP complexes and antibodies to matrilysin and stromelysin-3 will be useful to fully explore the clinical applications of plasma MMP measurements in cancer.

REFERENCES

1. BATES, S. E. & D. L. LONGO. 1987. Use of serum tumor markers in cancer diagnosis and management. Semin. Oncol. **14:** 102–138.
2. LIOTTA, L. A. 1992. Cancer cell invasion and metastasis. Sci. Am. 34–41.
3. GARBISA, S., R. POZZATTI, R. J. MUSCHEL, U. SAFFIOTTI, M. BALLIN, R. H. GOLDFARB, G. KHOURY & L. A. LIOTTA. 1987. Secretion of type IV collagenolytic protease and metastatic phenotype: Induction by transfection with c-Ha-ras but not c-Ha-ras plus Ad2-E1a. Cancer Res. **47:** 1523–1528.
4. BERNHARD, E. J., R. J. MUSCHEL & E. N. HUGHES. 1990. Mr 92,000 gelatinase release correlates with the metastatic phenotype in transformed rat embryo cells. Cancer Res. **50:** 3872–3877.
5. STEARNS, M. E. & M. WANG. 1993. Type IV collagenase (Mr 72,000) expression in human prostate: Benign and malignant tissue. Cancer Res. **53:** 878–883.
6. SHIMA, I., Y. SASAGURI, J. KUSUKAWA, H. YAMANA, H. FUJITA, T. KAKEGAWA & M. MORIMATSU. 1992. Production of matrix metalloproteinase-2 and metalloproteinase-3 related to malignant behavior of esophageal carcinoma. Cancer **70:** 2747–2753.
7. DAVIES, B., D. W. MILES, L. C. HAPERFIELD, M. S. NAYLOR, L. G. BOBROW, R. D. RUBENS & F. R. BALKWILL. 1993. Activity of type IV collagenases in benign and malignant breast disease. Br. J. Cancer **67:** 1126–1131.
8. BROWN, P. D., R. E. BLOXIDGE, E. ANDERSON & A. HOWELL. 1993. Expression of activated gelatinase in human invasive breast cancer. Clin. Exp. Metastasis **11:** 183–189.
9. MONTEAGUDO, C., M. J. MERINO, J. SAN-JUAN, L. A. LIOTTA & W. G. STETLER-STEVENSON. 1990. Immunohistochemical distribution of type IV collagenase in normal, benign, and malignant breast tissue. Am. J. Pathol. **136:** 585–592.
10. LEVY, A. T., V. CIOCE, M. E. SOBEL, S. GABRISA, W. F. GRIGIONI, L. A. LIOTTA & W. G. STETLER-STEVENSON. 1991. Increased expression of the Mr 72,000 type IV collagenase in human adenocarcinoma. Cancer Res. **51:** 439–444.
11. BASSET, P., J. P. BELLOCQ, C. WOLF, I. STOLL, P. HUTIN, J. M. LIMACHER, O. L. PODHAJCER, M. P. CHENARD, M. C. RIO & P. CHAMBON. 1990. A novel metalloproteinase gene specifically expressed in stromal cells of breast carcinomas. Nature **348:** 699–704.
12. POULSOM, R., M. PIGNATELLI, W. G. STETLER-STEVENSON, L. A. LIOTTA, P. A. WRIGHT, T. E. JEFFREY, J. M. LONGCROFT, L. ROGERS & G. W. H. STAMP. 1992. Stromal expression of 72 kDa type IV collagenase (MMP-2) and TIMP-2 mRNAs in colorectal neoplasia. Am. J. Pathol. **141:** 389–394.
13. PYKE, C., E. RALFKIAER, K. TRYGGVASON & K. DANO. 1993. Messenger RNA for two types of type IV collagenases is located in stromal cells in human colon cancer. Am. J. Pathol. **142:** 359–365.
14. AUTIO-HARMAINEN, H., T. KARTTUNEN, T. HURSKAINEN, M. HOYHTYA, A. KAUPPILA & K. TRYGGVASON. 1993. Expression of 72 kDa type IV collagenase (gelatinase A) in benign and malignant ovarian tumors. Lab. Invest. **69:** 312–321.
15. CLAVEL, C., M. POLETTE, M. DOCO, I. BINNINGER & P. BIREMBAUT. 1992. Immunolocalization of matrix metalloproteinases and their inhibitor in mammary pathology. Bull. Cancer **79:** 261–270.
16. SOINI, Y., T. SALO, A. OIKARINEN & H. AUTO-HARMAINEN. 1993. Expression of 72 kilodalton and 92 kilodalton type IV collagenase in malignant fibrous histiocytomas and dermatofibromas. Lab. Invest. **69:** 305–311.

17. JOHANSSON, S. & B. SMEDSROD. 1986. Identification of a plasma gelatinase in preparations of fibronectin. J. Biol. Chem. **261:** 4363–4366.
18. VARTIO, T. & M. BAUMANN. 1989. Human gelatinase/type IV collagenase is a regular plasma component. FEBS Lett. **155:** 285–289.
19. GRANSTROM, L. M., G. E. EKMAN, A. MALSTROM, U. ULMSTEM & F. WOESSNER. 1992. Serum collagenase levels in relation to the state of the human cervix during pregnancy and labor. Am. J. Obstet. Gynecol. **167:** 1284–1288.
20. HASHIMOTO, N., M. IKOBAYASHI & T. TSUJI. 1988. Serum type IV collagen-degrading enzyme in hepatocellular carcinoma with metastasis. Acta Med. Okayama **42:** 1–6.
21. WARD, R. V., S. J. ATKINSON, P. M. SLOCOMBE, A. J. P. DOCHERTY, J. J. REYNOLDS & G. MURPHY. 1991. Tissue inhibitor of metalloproteinases-2 inhibits the activation of 72 kDa progelatinase by fibroblast membranes. Biochim. Biophys. Acta **1079:** 242–246.
22. HOWARD, E. W., E. C. BULLEN & M. J. BANDA. 1991. Regulation of the autoactivation of human 72-kDa progelatinase by tissue inhibitor of metalloproteinases-2. J. Biol. Chem. **266:** 13064–13069.
23. GOLDBERG, G. I., A. STRONGIN, I. E. COLLIER, T. GENRICH & B. L. MARMER. 1992. Interaction of 92 kDa type IV collagenase with tissue inhibitor of metalloproteinases prevents dimerization, complex formation with interstitial collagenase, and activation of the proenzyme with stromelysin. J. Biol. Chem. **267:** 4583–4591.
24. FRIDMAN, R., T. FUERST, R. E. BIRD, M. HOYHTYA, M. OELKUCT, S. KRAUS, D. KORAREK, L. A. LIOTTA, M. L. BERMAN & W. G. STETLER-STEVENSON. 1992. Domain structure of human 72-kDa gelatinase/type IV collagenase. J. Biol. Chem. **267:** 15398–15405.
25. MOUTSIAKIS, D., P. MANCUSO, H. KRUTZSCH, W. STETLER-STEVENSON & S. ZUCKER. 1992. Characterization of metalloproteinases and tissue inhibitors of metalloproteinases in human plasma. Connect. Tissue Res. **28:** 213–230.
26. WACHER, M. P., H. C. KRUTZSCH, L. A. LIOTTA & W. G. STEVENSON. 1990. Development of a novel substrate capture assay for the detection of a neutral metalloproteinase capable of degrading membrane (type IV) collagen. J. Immunol. Methods **126:** 239–246.
27. BERGMANN, U., J. MICHAELIS, R. OBERHOFF, V. KNAUPER, R. BECKMANN & H. TSCHESCHE. 1989. Enzyme linked immunosorbent assay (ELISA) for the quantitative determination of human leukocyte collagenase and gelatinase. J. Clin. Chem. Clin. Biochem. **27:** 351–359.
28. COOKSLEY, S., J. P. HIPKISS, S. P. TICKLE, E. HOLMES-LEVERS, A. J. P. DOCHERTY, G. MURPHY & A. D. G. LAWSON. 1990. Immunoassays for the detection of human collagenase, stromelysin, tissue inhibitor of metalloproteinases (TIMP) and enzyme-inhibitor complexes. Matrix **10:** 285–291.
29. ZUCKER, S., R. M. LYSIK, M. H. ZARRABI & U. MOLL. 1993. Mr 92,000 type IV collagenase is increased in plasma of patients with colon cancer and breast cancer. Cancer Res. **53:** 140–146.
30. ZUCKER, S., R. M. LYSIK, M. GURFINKEL, M. H. ZARRABI, W. G. STETLER-STEVENSON, L. A. LIOTTA, H. BIRKEDAL-HANSEN & W. MANN. 1992. Immunoassay of type IV collagenase/gelatinase (MMP-2) in human plasma. J. Immunol. Methods **148:** 189–198.
31. CLARK, I. M., J. K. WRIGHT, T. E. CAWSTON & B. L. HAZLEMAN. 1992. Polyclonal antibodies against human fibroblast collagenase and the design of an enzyme-linked immunosorbent assay to measure TIMP-collagenase complex. Matrix **12:** 108–115.
32. KJELDSEN, L., O. W. BJERRUN, D. HOVGAARD, A. H. JOHNSEN, M. SEHESTED & N. BORREGAARD. 1992. A marker for circulating neutrophils. Purification and quantification by enzyme linked immunosorbent assay. Eur. J. Haematol. **49:** 180–191.
33. ZUCKER, S., R. M. LYSIK, M. H. ZARRABI, W. STETLER-STEVENSON, L. A. LIOTTA, H. BIRKEDAL-HANSEN, W. MANN & M. FURIE. 1992. Type IV collagenase/gelatinase (MMP2) is not increased in plasma of patients with cancer. Cancer Epidemiol. Biomarkers Prevention **1:** 475–479.
34. GARBISA, S., G. SCAGLIOTTI, L. MASIERO, C. DIFRANCESCO, C. CAENAZZO, M. ONISTO, M. MICELA, W. G. STETLER-STEVENSON & L. A. LIOTTA. 1992. Correlation of serum metalloproteinase levels with lung cancer metastasis and response to therapy. Cancer Res. **52:** 4548–4549.

35. NAKAJIMA, M., D. R. WELCH, D. M. WYNN, T. TSURUO & G. L. NICHOLSON. 1993. Serum and plasma Mr 92,000 progelatinase levels correlate with spontaneous metastasis of rat 13762 NF mammary adenocarcinoma. Cancer Res. 5802–5807.
36. ZUCKER, S., R. M. LYSIK, M. H. ZARRABI, R. A. GREENWALD, B. GRUBER, S. P. TICKLE, T. S. BAKER & A. J. P. DOCHERTY. 1994. Elevated plasma stromelysin levels in arthritis. J. Rheumatol. In press.
37. REPONEN, P., C. SAHLBERG, P. HUHTALA, T. HURSKAINEN, I. THESLEFF & K. TRYGGVASON. 1992. Molecular cloning of murine 72-kDa type IV collagenase and its expression during mouse development. J. Biol. Chem. **267:** 7856–7862.
38. MACKAY, A. R., M. BALLIN, M. D. PELINA, A. R. FARINA, A. M. NASON, J. L. HARTZLER & U. P. THORGEIRSSON. 1992. Effect of phorbol ester and cytokines on matrix metalloproteinase and tissue inhibitor of metalloproteinase expression in tumor and normal cell lines. Invasion Metastasis **12:** 168–184.
39. MOSCATELLI, D. & D. B. RIFKIN. 1988. Membrane and matrix localization of proteinases: A common theme in tumor invasion and angiogenesis. Biochim. Biophys. Acta **948:** 67–84.
40. ZUCKER, S., R. M. LYSIK, M. H. ZARRABI, J. J. FIORE & D. K. STRICKLAND. 1991. Proteinase-alpha2 macroglobulin complexes are not increased in plasma of patients with cancer. Int. J. Cancer **48:** 399–403.
41. ZUCKER, S., P. MANCUSO, B. DIMASSIMO, R. M. LYSIK, C. CONNER & C.-L. WU. 1994. Comparison of techniques for measurement of gelatinase/type IV collagenases: Enzyme-linked immunoassays versus substrate degradation assays. Clin. Exp. Metastasis. In press.
42. OKADA, Y., H. NAGASE & E. D. HARRIS, JR. 1986. A metalloproteinase from human rheumatoid synovial fibroblasts that digests connective tissue matrix components. Purification and characterization. J. Biol. Chem. **261:** 14245–14255.

Inhibition of Tumor Angiogenesis[a]

ERIC P. SIPOS,[b] RAFAEL J. TAMARGO,[b]
JON D. WEINGART,[b] AND HENRY BREM[b-d]

*Departments of Neurological Surgery,[b] Oncology,[c]
and Ophthalmology[d]
Brain Tumor Research Center, Hunterian 817
The Johns Hopkins University School of Medicine
725 North Wolfe Street
Baltimore, Maryland 21205*

The dependence of solid tumor growth upon neovascularization provides a therapeutic opportunity to modulate the growth of neoplasms by inhibiting tumor-induced angiogenesis.[1-3] Although the experimental and clinical evidence that solid tumor growth is angiogenesis-dependent has been summarized in detail elsewhere,[4-7] a brief review of this topic will introduce potential applications for the treatment of cancer.

Angiogenesis and Solid Tumor Growth

Solid tumor growth occurs in two phases: an avascular phase and a vascular phase. During the initial avascular phase of growth, tumors exist as small aggregates of neoplastic cells supported by simple diffusion of oxygen and nutrients. Tumor volume increases at a slow linear rate to a maximum of less than 3 cubic millimeters. This volume is determined by the limits of diffusion;[8] no further three-dimensional growth occurs until the tumor becomes vascularized. When deprived of a vascular supply, as in the anterior chamber of the eye, tumors remain viable for prolonged periods, but they are dormant from a growth perspective.[9] A balance exists between cell proliferation at the periphery and cell death at the center of the tumor.

Tumors produce diffusable angiogenic factors that induce host capillary endothelial cells to proliferate, migrate, and form new vessels which supply the neoplastic cells. Once vascularized, tumors are maintained by perfusion and their growth becomes rapid. During this phase, tumor volume increases exponentially.[9-11]

INHIBITION OF TUMOR ANGIOGENESIS

Within the last two decades, a variety of seemingly unrelated classes of drugs have been found to inhibit angiogenesis *in vivo* and *in vitro*. These have been summarized elsewhere[6,7] and are presented in TABLE 1.[12-59] A few examples are reviewed in greater detail below.

[a]This work was supported by the Preuss Foundation, National Institutes of Health Cooperative Agreement UO1 CA52857, National Institutes of Health grant P20 NS31081, and a gift from the Poole and Kent Foundation. Drs. Jon Weingart and Eric Sipos are National Research Service Awardees (CA-09574).

Cartilage

Cartilage, a normally avascular tissue resistant to invasion by most tumors,[60] develops from a vascularized embryonic form that subsequently loses its blood vessels in the early neonatal period.[61,62] These observations suggested to Brem and Folkman the possibility that neonatal cartilage may contain a factor that inhibits vessels, a hypothesis that was confirmed in 1975. With use of the rabbit cornea angiogenesis assay, it was shown that cartilage produces a potent diffusible inhibitor of tumor-induced angiogenesis.[12] Once isolated and purified,[13] the cartilage-derived factor delivered by selective intra-arterial infusions was found to modulate angiogenesis and tumor growth in rabbits and mice.[63] Further purification and characteriza-

TABLE 1. Reported Inhibitors of Angiogenesis

Inhibitor	Refs.	Inhibitor	Refs.
Extracts from avascular tissues		**Agents that bind heparin**	
• Cartilage-derived factor	12–15	• Protamine	45
• Vitreous extract	16, 17	• Platelet factor-4	46
Steroid hormones		**Other modulators of collagen biosynthesis**	
• Cortisone, hydrocortisone, tetrahydrocortisol & others	18–21	• Proline analogues	47
• Medroxyprogesterone	22, 23	• α,α-Dipyridyl	47
• Heparin-steroid conjugates	24	• β-Aminoproprionitrile	47
Antibiotics		• Tricyclodecan-9-yl-xanthate (D609)	48
• Minocycline	25, 26	• GPA1734	49
• AGM-1470 and other fumagillin derivatives	27–30	• Tissue inhibitors of metalloproteinases (TIMP and TIMP-2)	50
• Herbimycin A	31	• Interferons	51–53
• 15-Deoxyspergualin	32, 33	**Miscellaneous inhibitors of angiogenesis**	
• Eponemycin	34	• Thrombospondin	54–56
Antirheumatic agents		• Site-specific mutants of angiogenin	57
• D-Penicillamine	35, 36	• α-Difluoromethylornithine (DMFO)	58
• Inhibitors of prostaglandin synthesis	37–40	• Anti-bFGF monoclonal antibody	59
• Gold compounds	41, 42		
Vitamins and derivatives			
• Vitamin D_3 analogues	43		
• Retinoids	44		

tion of the factor revealed that it is a protein with an approximate molecular weight of 24,000, potent anticollagenase properties, and a high sequence homology to a collagenase inhibitor isolated from cultured human skin fibroblasts.[14,15]

Angiostatic Steroids

The observation that heparin administered in combination with cortisone has potent antiangiogenic effects, causes tumor regression, and inhibits formation of metastases[18] led to the discovery of a new class of steroids with antiangiogenic properties. These corticosteroids inhibit angiogenesis independent of their relative mineralocorticoid or glucocorticoid activity and have been termed "angiostatic steroids."[19]

The antitumor potential of angiostatic steroids and heparin was confirmed in a series of *in vivo* studies. Heparin and hydrocortisone administered orally inhibited angiogenesis and tumor growth in nude mice with human neurofibrosarcoma implants.[64] Heparin and cortisone, when incorporated into a biodegradable polyanhydride controlled-release polymer, had an antiangiogenic effect against tumor-induced neovascularization in the rabbit cornea model and inhibited tumor growth when implanted into 9L-gliosarcomas growing in the flanks of adult rats.[65] This demonstrated the feasibility of using implantable controlled-release polymers for the interstitial delivery of angiostatic agents[66] to inhibit tumor growth. Given the limitations imposed by the blood-brain barrier upon drug penetration into the central nervous system,[67] this drug-delivery system has become useful for studying the effects of angiogenesis inhibitors upon the growth of intracranial neoplasms.

Although it is possible that the systemic toxicity of cortisone could limit its clinical use as an inhibitor of angiogenesis, potentially less toxic alternatives have been found to have antiangiogenic properties. Tetrahydrocortisol, a major metabolite of cortisol devoid of glucocorticoid or mineralocorticoid activity, is the most potent of the known angiostatic steroids. This is only one of several examples of angiostatic steroids with no other apparent biological activity.[21]

Similar concerns that the anticoagulant effect of heparin might limit the clinical use of this potential cancer therapy prompted the discovery of heparin fragments[18] and synthetic heparin substitutes such as β-cyclodextrin tetradecasulfate[20,68] that potentiate the antiangiogenic properties of angiostatic steroids without anticoagulation. More recently, Thorpe *et al.* developed new angiogenesis inhibitors by covalently linking angiostatic steroids to a non-anticoagulant heparin derivative, thus achieving antitumor effects in mice with a compound less likely to have the side effects of heparin and cortisone.[24]

Fumagillin Derivatives

Another class of newly discovered angiogenesis inhibitors has been derived from fumagillin, an antibiotic purified from cultures of *Aspergillus fumigatus* fresenius, which inhibits endothelial cell proliferation *in vitro.*[27] Although the parent compound is systemically toxic, many less toxic and more potent antiangiogenic derivatives have been synthesized.[27,29,30] The most potent of the known fumagillin analogues, O-(chloroacetylcarbamoyl)fumagillol or AGM-1470, effectively reduces angiogenesis in several standard assays[27–30] and inhibits tumor growth and metastasis in a wide variety of mouse and rat tumors with little apparent toxicity.[27,69] When delivered systemically in nude mice, this agent inhibits tumor angiogenesis and growth of human schwannoma, neurofibroma, and neurofibrosarcoma xenografts.[70] It causes endothelial cell rounding and thus prevents endothelial proliferation. AGM-1470 is cytotoxic *in vitro* only in high concentrations; therefore its antineoplastic effects are thought to be due to inhibition of angiogenesis. This drug is currently being evaluated clinically for treatment of the highly vascular Kaposi's sarcoma in AIDS patients.

Inhibitors of Collagen Metabolism

As more is learned about the varied biological effects of different classes of antiangiogenic compounds, a common feature in their mechanisms of action is the modulation of collagen, extracellular matrix, and basement membrane metabolism. The inhibitor of angiogenesis derived from cartilage, for example, is a protein that inhibits mammalian collagenase.[14,15] Similarly, the angiostatic steroids modulate collagen metabolism,[22,71,72] induce basement membrane dissolution,[73] and inhibit synthesis of plasminogen activator, an enzyme involved in remodeling the extracellular matrix and the capillary basement membrane.[23,74]

Capillary retraction, endothelial cell rounding, inhibition of endothelial proliferation, and vascular involution are all associated with dissolution of the basement membrane.[47,73] Regression of growing capillaries in the chick chorioallantoic membrane and potentiation of the antiangiogenic effects of heparin and steroid combinations have been achieved with a variety of agents, including proline analogues and an inhibitor of prolyl hydroxylase, that induce structural alterations in the extracellular matrix by interfering with collagen metabolism.[47] Agents that affect collagen metabolism, either anabolically or catabolically, may inhibit neovascularization and therefore have potential roles in the treatment of cancer and other angiogenic diseases. The feasibility of this therapeutic approach has been demonstrated clinically with the successful use of recombinant interferon α_{2a} to treat pulmonary hemangiomatosis, a rare angiogenic disease that is usually relentlessly progressive and fatal.[53] We have investigated other clinically available agents that modulate collagen metabolism in a search for alternative inhibitors of angiogenesis, as indicated below.

Tetracycline Derivatives

Tetracyclines, which specifically block mitochondrial protein synthesis, have long been used clinically for their broad-spectrum antibiotic properties. In 1984, Kroon *et al.* reported that tumor growth can be inhibited by tetracycline derivatives and hypothesized that this cytostatic effect was due to inhibition of mitochondrial translation in neoplastic cells.[75]

Another recently recognized characteristic of these drugs is the inhibition of extracellular matrix metalloproteinases, including type IV collagenase.[76–78] Collagenase inhibition appears to be unrelated to the antibiotic action of these compounds because chemically modified derivatives of tetracycline that lack antimicrobial activity may retain their anticollagenase properties.[79,80]

Among the commercially available tetracyclines already in clinical use as antibiotics, the semisynthetic derivative minocycline is the most lipid soluble, has the best tissue penetration,[81,82] and is a potent collagenase inhibitor.[83] We therefore began laboratory investigations to test our hypothesis that minocycline is an inhibitor of tumor angiogenesis.[25]

The rabbit cornea, normally clear and avascular, provides an important model for study of putative angiogenesis inhibitors incorporated into biocompatible polymers capable of sustained drug release because it permits direct observation and quantification of vessel growth.[12,66] This model was used to study minocycline impregnated into ethylene-vinyl acetate copolymer (EVAc), an inert nonbiodegradable controlled-release polymer well-suited for local drug delivery.[84–86] Similar polymers were fabricated with a combination of heparin and cortisone or with cortisone alone. The polymers were then implanted with rabbit VX2 carcinoma into the rabbit cornea. Minocycline significantly inhibited tumor angiogenesis with results comparable to those achieved with combinations of heparin and cortisone.

To elucidate the mechanisms of antiangiogenic action by tetracycline derivatives, we explored the *in vitro* effects of minocycline upon some of the major cell types involved in the structure and regulation of the cerebral microvasculature.[26] These experiments demonstrated that minocycline selectively inhibits endothelial cell growth at concentrations that have minimal effect on astrocytes and pericytes. DNA and protein synthesis, as measured by incorporation of tritiated thymidine and tritiated leucine, were also inhibited selectively in endothelial cells. When other tetracycline derivatives were tested, it was found that the degree of endothelial cell growth inhibition correlated with the potency of anticollagenase activity and not antimicrobial action for each of the compounds studied.

Teicher *et al.* and Sotomayor *et al.* have reported the successful use of minocycline as an adjunct to standard cancer therapies against subcutaneously implanted Lewis lung carcinoma in mice.[87,88] In this model, the inhibition of tumor growth achieved with cyclophosphamide, melphalan, or radiation treatment was enhanced by the concurrent administration of systemic minocycline. A significant decrease in the number and size of vascularized pulmonary metastases was also demonstrated in these studies when minocycline was administered as an adjunct to cisplatin, melphalan, or cyclophosphamide. Animals treated with the combination of minocycline and cyclophosphamide showed the greatest response.

Weingart *et al.* demonstrated that interstitial administration of minocycline by controlled-release polymers for treatment of brain tumors in the rat 9L-gliosarcoma model significantly inhibited tumor growth and prolonged survival.[89] Furthermore, local delivery of minocycline, administered as an adjunct to systemic carmustine (BCNU) therapy, resulted in synergistic prolongation of survival in this brain tumor model.[90,91]

In vitro studies demonstrating that minocycline at very high concentrations is minimally cytotoxic against cultured EMT-6 murine mammary tumor cells[88] are consistent with the hypothesis that the minocycline effect is due to inhibition of angiogenesis, presumably related to its anticollagenase properties.

Further studies aimed at elucidating the mechanism of antiangiogenic action of minocycline and other chemically modified tetracycline derivatives are in progress. Minocycline shows considerable promise as a biological response modifier for use as an adjunct to cytotoxic agents in the treatment of solid tumors. If this therapeutic potential is confirmed, minocycline is an ideal candidate for clinical trials as an inhibitor of tumor neovascularization because it is readily available commercially and has for several years been safely used as a systemic antibiotic.

CONCLUSION

Antiangiogenic drugs have emerged as important potential therapeutic agents to be evaluated either independently or in combination with other approaches for the treatment of cancer and other angiogenic diseases. The variety of compounds reported to inhibit angiogenesis suggests that a broad range of antiangiogenic mechanisms exists. This diversity of therapeutic options could be exploited by using multiple different inhibitors together to more effectively control neovascularization.

SUMMARY

The exponential growth of solid tumors depends upon induction of new vessel growth, a process mediated by diffusable angiogenic factors produced by tumor cells. By inhibiting angiogenesis, it is now possible to modulate tumor growth and metastasis in laboratory animals. The first described inhibitor of angiogenesis was a protein derived from cartilage. Other important classes of antiangiogenic agents include angiostatic steroids combined with heparin or heparin derivatives, and the synthetic derivatives of fumigallin. As the mechanisms of action of these and other angiostatic agents are being elucidated, it is becoming apparent that many modulators of collagen metabolism inhibit angiogenesis and may offer clinically useful anticancer treatments. Minocycline and other tetracycline derivatives with anticollagenase properties have been shown to be potent inhibitors of angiogenesis. These agents, when administered with other standard cancer therapies, help prolong survival in laboratory animals with solid tumors. Further studies of these biologic response modifiers of tumor progression are under way in the hope that they will offer effective new treatments for cancer in humans.

ACKNOWLEDGMENTS

The authors thank Drs. Pamela Talalay and Reid Thompson for reviewing the manuscript.

REFERENCES

1. Folkman, J. 1971. Tumor angiogenesis: Therapeutic implications. N. Engl. J. Med. **285:** 1182–1186.
2. Folkman, J. 1972. Anti-angiogenesis: New concept for therapy of solid tumors. Ann. Surg. **175:** 409–416.
3. Folkman, J. 1975. Tumor angiogenesis: A possible control point in tumor growth. Ann. Int. Med. **82:** 96–100.
4. Folkman, J. & M. Klagsbrun. 1987. Angiogenic factors. Science **235:** 442–447.
5. Folkman, J. 1990. What is the evidence that tumors are angiogenesis dependent? J. Natl. Cancer Inst. **82:** 4–6.
6. Folkman, M. J. 1991. Antiangiogenesis. *In* Biologic Therapy of Cancer. V. DeVita, S. Hellman & S. Rosenberg, Eds. J. B. Lippincott Co. Philadelphia, PA.
7. Folkman, J. 1993. Tumor angiogenesis. *In* Cancer Medicine. J. Holland, E. Frei, R. Bast, D. Kufe, D. Morton & R. Weichselbaum, Eds. Lea and Febiger. Philadelphia, PA.
8. Folkman, J. & M. Hochberg. 1973. Self-regulation of growth in three dimensions. J. Exp. Med. **138:** 745–753.
9. Gimbrone, M., S. Leapman, R. Cotran & J. Folkman. 1972. Tumor dormancy in vivo by prevention of neovascularization. J. Exp. Med. **136:** 261–276.
10. Gimbrone, M., R. Cotran, S. Leapman & J. Folkman. 1974. Tumor growth and neovascularization: An experimental model using the rabbit cornea. J. Natl. Cancer Inst. **52:** 413–419.
11. Knighton, D., D. Ausprunk, D. Tapper & J. Folkman. 1977. Avascular and vascular phases of tumour growth in the chick embryo. Br. J. Cancer **35:** 347–356.
12. Brem, H. & J. Folkman. 1975. Inhibition of tumor angiogenesis mediated by cartilage. J. Exp. Med. **141:** 427–439.
13. Langer, R., H. Brem, K. Falterman, M. Klein & J. Folkman. 1976. Isolation of a cartilage factor that inhibits tumor neovascularization. Science **193:** 70–72.
14. Moses, M. A., J. Sudhalter & R. Langer. 1990. Identification of an inhibitor of neovascularization from cartilage. Science **248:** 1408–1410.
15. Murray, J. B., K. Allison, J. Sudhalter & R. Langer. 1986. Purification and partial amino acid sequence of a bovine cartilage-derived collagenase inhibitor. J. Biol. Chem. **261:** 4154–4159.
16. Preis, I., R. Langer, H. Brem, J. Folkman & A. Patz. 1977. Inhibition of neovascularization by an extract derived from vitreous. Am. J. Ophthalmol. **84:** 323–328.
17. Lutty, G. A., D. C. Thompson, J. Y. Gallup, R. J. Mello, A. Patz & A. Fenselau.

1983. Vitreous: An inhibitor of retinal extract-induced neovascularization. Invest. Ophthalmol. Visual Sci. **2:** 52–56.
18. FOLKMAN, J., R. LANGER, R. LINHARDT, C. HAUDENSCHILD & S. TAYLOR. 1983. Angiogenesis inhibition and tumor regression caused by heparin or a heparin fragment in the presence of cortisone. Science **221:** 719–725.
19. CRUM, R., S. SZABO & J. FOLKMAN. 1985. A new class of steroids inhibits angiogenesis in the presence of heparin or a heparin fragment. Science **230:** 1375–1378.
20. FOLKMAN, J., P. B. WEISZ, M. M. JOULLIE, W. W. LI & W. R. EWING. 1989. Control of angiogenesis with synthetic heparin substitutes. Science **243:** 1490–1493.
21. FOLKMAN, J. & D. E. INGBER. 1987. Angiostatic steroids. Method of discovery and mechanism of action. Ann. Surg. **206:** 374–383.
22. GROSS, J., R. AZIZKHAN, C. BISWAS, R. BRUNS, D. HSIEH & J. FOLKMAN. 1981. Inhibition of tumor growth, vascularization, and collagenolysis in the rabbit cornea by medroxyprogesterone. Proc. Natl. Acad. Sci. USA **78:** 1176–1180.
23. ASHINO-FUSE, H., Y. TAKANO, T. OIKAWA, M. SHIMAMURA & T. IWAGUCHI. 1989. Medroxyprogesterone acetate, an anti-cancer and anti-angiogenic steroid, inhibits the plasminogen activator in bovine endothelial cells. Int. J. Cancer **44:** 859–864.
24. THORPE, P. E., E. J. DERBYSHIRE, S. P. ANDRADE, N. PRESS, P. P. KNOWLES, S. KING, G. J. WATSON, Y. C. YANG & B. M. RAO. 1993. Heparin-steroid conjugates: New angiogenesis inhibitors with antitumor activity in mice. Cancer Res. **53:** 3000–3007.
25. TAMARGO, R. J., R. A. BOK & H. BREM. 1991. Angiogenesis inhibition by minocycline. Cancer Res. **51:** 672–675.
26. GUERIN, C., J. LATERRA, T. MASNYK, L. M. GOLUB & H. BREM. 1992. Selective endothelial growth inhibition by tetracyclines that inhibit collagenase. Biochem. Biophys. Res. Commun. **188:** 740–745.
27. INGBER, D., T. FUJITA, S. KISHIMOTO, K. SUDO, T. KANAMARU, H. BREM & J. FOLKMAN. 1990. Synthetic analogues of fumagillin that inhibit angiogenesis and suppress tumour growth. Nature **348:** 555–557.
28. KUSAKA, M., K. SUDO, T. FUJITA, S. MARUI, F. ITOH, D. INGBER & J. FOLKMAN. 1991. Potent anti-angiogenic action of AGM-1470: Comparison to the fumagillin parent. Biochem. Biophys. Res. Commun. **174:** 1070–1076.
29. MARUI, S., F. ITOH, Y. KOZAI, K. SUDO & S. KISHIMOTO. 1992. Chemical modification of fumagillin. I. 6-O-acyl, 6-O-sulfonyl, 6-O-alkyl, and 6-O-(N-substituted-carbamoyl)fumagillols. Chem. Pharm. Bull. (Tokyo) **40:** 96–101.
30. MARUI, S. & S. KISHIMOTO. 1992. Chemical modification of fumagillin. II. 6-Amino-6-deoxyfumagillol and its derivatives. Chem. Pharm. Bull. (Tokyo) **40:** 575–579.
31. OIKAWA, T., K. HIROTANI, M. SHIMAMURA, H. ASHINO-FUSE & T. IWAGUCHI. 1989. Powerful antiangiogenic activity of herbimycin A (named angiostatic antibiotic). J. Antibiot. **42:** 1202–1204.
32. OIKAWA, T., M. SHIMAMURA, H. ASHINO-FUSE, T. IWAGUCHI, M. ISHIZUKA & T. TAKEUCHI. 1991. Inhibition of angiogenesis by 15-deoxyspergualin. J. Antibiot. **44:** 1033–1035.
33. OIKAWA, T., M. HASEGAWA, I. MORITA, S. MUROTA, H. ASHINO, M. SHIMAMURA, A. KIUE, R. HAMANAKA, M. KUWANO, M. ISHIZUKA *et al.* 1992. Effect of 15-deoxyspergualin, a microbial angiogenesis inhibitor, on the biological activities of bovine vascular endothelial cells. Anticancer Drugs **3:** 293–299.
34. OIKAWA, T., M. HASEGAWA, M. SHIMAMURA, H. ASHINO, S. MUROTA & I. MORITA. 1991. Eponemycin, a novel antibiotic, is a highly powerful angiogenesis inhibitor. Biochem. Biophys. Res. Commun. **181:** 1070–1076.
35. MATSUBARA, T., R. SAURA, K. HIROHATA & M. ZIFF. 1989. Inhibition of human endothelial cell proliferation in vitro and neovascularization in vivo by D-penicillamine. J. Clin. Invest. **83:** 158–167.
36. BREM, S. S., D. ZAGZAG, A. M. C. TSANACLIS, S. GATELY, M. P. ELKOUBY & S. E. BRIEN. 1990. Inhibition of angiogenesis and tumor growth in the brain. Suppression of endothelial cell turnover by penicillamine and the depletion of copper, an angiogenic cofactor. Am. J. Pathol. **137:** 1121–1142.
37. COOPER, C. A., M. V. W. BERGAMINI & I. H. LEOPOLD. 1980. Use of flurbiprofen to inhibit corneal neovascularization. Arch. Ophthalmol. **98:** 1102–1105.

38. PETERSON, H. 1983. Effects of prostaglandin synthesis inhibitors on tumor growth and vascularization: Experimental studies in the rat. Invasion Metastasis **3:** 151–159.
39. PETERSON, H. 1986. Tumor angiogenesis inhibition by prostaglandin synthetase inhibitors. Anticancer Res. **6:** 251–254.
40. MILAS, L., N. HUNTER, Y. FURUTA, I. NISHIGUCHI & S. RUNKEL. 1991. Antitumor effects of indomethacin alone and in combination with radiotherapy: Role of inhibition of tumour angiogenesis. Int. J. Radiat. Biol. **60:** 65–70.
41. MATSUBARA, T. & M. ZIFF. 1987. Inhibition of human endothelial cell proliferation by gold compounds. J. Clin. Invest. **79:** 1440–1446.
42. KOCH, A. E., M. CHO, J. BURROWS, S. J. LEIBOVICH & P. J. POLVERINI. 1988. Inhibition of production of macrophage-derived angiogenesis activity by the anti-rheumatic agents gold sodium thiomalate and auranofin. Biochem. Biophys. Res. Commun. **154:** 205–212.
43. OIKAWA, T., K. HIROTANI, H. OGASAWARA, T. KATAYAMA, O. NAKAMURA, T. IWAGUCHI & A. HIRAGUN. 1990. Inhibition of angiogenesis by vitamin D_3 analogues. Eur. J. Pharmacol. **178:** 247–250.
44. OIKAWA, T., K. HIROTANI, O. NAKAMURA, K. SHUDO, A. HIRAGUN & T. IWAGUCHI. 1989. A highly potent antiangiogenic activity of retinoids. Cancer Lett. **48:** 157–162.
45. TAYLOR, S. & J. FOLKMAN. 1982. Protamine is an inhibitor of angiogenesis. Nature **297:** 307–312.
46. MAIONE, T. E., G. S. GRAY, J. PETRO, A. J. HUNT, A. L. DONNER, S. I. BAUER, H. F. CARSON & R. J. SHARPE. 1990. Inhibition of angiogenesis by recombinant human platelet factor-4 and related peptides. Science **247:** 77–79.
47. INGBER, D. & J. FOLKMAN. 1988. Inhibition of angiogenesis through modulation of collagen metabolism. Lab. Invest. **59:** 44–51.
48. MARAGOUDAKIS, M. E., E. MISSIRLIS, M. SARMONIKA, M. PANOUTSACOPOULOU & G. KARAKIULAKIS. 1990. Basement membrane biosynthesis as a target to tumor therapy. J. Pharmacol. Exp. Ther. **252:** 753–757.
49. MARAGOUDAKIS, M. E., M. SARMONIKA & M. PANOUTSACOPOULOU. 1988. Inhibition of basement membrane biosynthesis prevents angiogenesis. J. Pharmacol. Exp. Ther. **244:** 729–733.
50. TAKIGAWA, M., Y. NISHIDA, F. SUZUKI, J. KISHI, K. YAMASHITA & T. HAYAKAWA. 1990. Induction of angiogenesis in chick yolk-sac membrane by polyamines and its inhibition by tissue inhibitors of metalloproteinases (TIMP and TIMP-2). Biochem. Biophys. Res. Commun. **171:** 1264–1271.
51. SIDKY, Y. A. & E. C. BORDEN. 1987. Inhibition of angiogenesis by interferons: Effects on tumor- and lymphocyte-induced vascular responses. Cancer Res. **47:** 5155–5161.
52. TSURUOKA, N., M. SUGIYAMA, Y. TAWARAGI, M. TSUJIMOTO, T. NISHIHARA, T. GOTO & N. SATO. 1988. Inhibition of in vitro angiogenesis by lymphotoxin and interferon-γ. Biochem. Biophys. Res. Commun. **155:** 429–435.
53. WHITE, C. W., H. M. SONDHEIMER, E. C. CROUCH, H. WILSON & L. L. FAN. 1989. Treatment of pulmonary hemangiomatosis with recombinant interferon alfa-2a. N. Engl. J. Med. **320:** 1197–1200.
54. RASTINEJAD, F., P. J. POLVERINI & N. P. BOUCK. 1989. Regulation of the activity of a new inhibitor of angiogenesis by a cancer suppressor gene. Cell **56:** 345–355.
55. BAGAVANDOSS, P. & J. W. WILKS. 1990. Specific inhibition of endothelial cell proliferation by thrombospondin. Biochem. Biophys. Res. Commun. **170:** 867–872.
56. IRUELA-ARISPE, M. L., P. BORNSTEIN & H. SAGE. 1991. Thrombospondin exerts an antiangiogenic effect on cord formation by endothelial cells in vitro. Proc. Natl. Acad. Sci. USA **88:** 5026–5030.
57. SHAPIRO, R. & B. L. VALLE. 1989. Site-directed mutagenesis of histidine-13 and histidine-114 of human angiogenin. Alanine derivatives inhibit angiogenin-induced angiogenesis. Biochemistry **28:** 7401–7408.
58. TAKIGAWA, M., M. ENOMOTO, Y. NISHIDA, H. PAN, A. KINOSHITA & F. SUZUKI. 1990. Tumor angiogenesis and polyamines: α-Difluoromethylornithine, an irreversible inhibitor of ornithine decarboxylase, inhibits B16 melanoma-induced angiogenesis in ovo and the proliferation of vascular endothelial cells in vitro. Cancer Res. **50:** 4131–4138.

59. HORI, A., R. SASADA, E. MATSUTANI, K. NAITO, Y. SAKURA, T. FUJITA & Y. KOZAI. 1991. Suppression of solid tumor growth by immunoneutralizing monoclonal antibody against basic fibroblast growth factor. Cancer Res. **51:** 6180–6184.
60. EISENSTEIN, R., N. SORGENTE, L. W. SOBLE, A. MILLER & K. E. KUETTNER. 1973. The resistance of certain tissues to invasion: Penetrability of explanted tissues by vascularized mesenchyme. Am. J. Pathol. **73:** 765–774.
61. HARALDSSON, S. 1962. The vascular pattern of a growing and fullgrown human epiphysis. Acta Anat. **48:** 156–167.
62. BLACKWOOD, H. J. J. 1965. Vascularization of the condylar cartilage of the human mandible. J. Anat. **99:** 551–563.
63. LANGER, R., H. CONN, J. VACANTI, C. HAUDENSCHILD & J. FOLKMAN. 1980. Control of tumor growth in animals by infusion of an angiogenesis inhibitor. Proc. Natl. Acad. Sci. USA **77:** 4331–4335.
64. LEE, J. K., B. CHOI, R. A. SOBEL, E. A. CHIOCCA & R. L. MARTUZA. 1990. Inhibition of growth and angiogenesis of human neurofibrosarcoma by heparin and hydrocortisone. J. Neurosurg. **73:** 429–435.
65. TAMARGO, R. J., K. W. LEONG & H. BREM. 1990. Growth inhibition of the 9L glioma using polymers to release heparin and cortisone acetate. J. Neuro-Oncol. **9:** 131–138.
66. LANGER, R. & J. MURRAY. 1983. Angiogenesis inhibitors and their delivery systems. Appl. Biochem. Biotechnol. **8:** 9–24.
67. TAMARGO, R. J. & H. BREM. 1992. Drug delivery to the central nervous system: A review. Neurosurg. Q. **2:** 259–279.
68. WEISZ, P. B., H. C. HERMANN, M. M. JOULLIÉ, K. KUMOR, E. M. LEVINE, E. J. MACARAK & D. B. WEINER. 1992. Angiogenesis and heparin mimics. Exs **61:** 107–117.
69. YAMAOKA, M., T. YAMAMOTO, T. MASAKI, S. IKEYAMA, K. SUDO & T. FUJITA. 1993. Inhibition of tumor growth and metastasis of rodent tumors by the angiogenesis inhibitor O-(chloroacetyl-carbamoyl)fumagillol (TNP-470; AGM-1470). Cancer Res. **53:** 4262–4267.
70. TAKAMIYA, Y., R. M. FRIEDLANDER, H. BREM, A. MALIK & R. L. MARTUZA. 1993. Inhibition of angiogenesis and growth of human nerve-sheath tumors by AGM-1470. J. Neurosurg. **78:** 470–476.
71. MARAGOUDAKIS, M., M. SARMONIKA & M. PANOUTSACOPOULOU. 1989. Antiangiogenic action of heparin plus cortisone is associated with decreased collagenous protein synthesis in the chick chorioallantoic membrane system. J. Pharmacol. Exp. Ther. **251:** 679–682.
72. HARADA, I., T. KIKUCHI, Y. SHIMOMURA, M. YAMAMOTO, H. OHNO & N. SATO. 1992. The mode of action of anti-angiogenic steroid and heparin. Exs **61:** 445–448.
73. INGBER, D., J. MADRI & J. FOLKMAN. 1986. A possible mechanism for inhibition of angiogenesis by angiostatic steroids: Induction of capillary basement membrane dissolution. Endocrinology **119:** 1768–1775.
74. BLEI, F., E. L. WILSON, P. MIGNATTI & D. B. RIFKIN. 1993. Mechanism of action of angiostatic steroids: Suppression of plasminogen activator activity via stimulation of plasminogen activator inhibitor synthesis. J. Cell. Physiol. **155:** 568–578.
75. KROON, A. M., B. H. J. DONTJE, M. HOLTROP & C. VAN DEN BOGERT. 1984. The mitochondrial genetic system as a target for chemotherapy: Tetracyclines as cytostatics. Cancer Lett. **25:** 33–40.
76. GOLUB, L. M., N. RAMAMURTHY, T. F. MCNAMARA, B. GOMES, M. WOLFF, A. CASINO, A. KAPOOR, J. ZAMBON, S. CIANCIO, M. SCHNEIR, *et al.* 1984. Tetracyclines inhibit tissue collagenase activity. A new mechanism in the treatment of periodontal disease. J. Periodontal Res. **19:** 651–655.
77. GOLUB, L. M., M. WOLFF, H. M. LEE, T. F. MCNAMARA, N. S. RAMAMURTHY, J. ZAMBON & S. CIANCIO. 1985. Further evidence that tetracyclines inhibit collagenase activity in human crevicular fluid and from other mammalian sources. J. Periodontal Res. **20:** 12–23.
78. GOLUB, L. M., N. S. RAMAMURTHY, T. F. MCNAMARA, R. A. GREENWALD & B. R. RIFKIN. 1991. Tetracyclines inhibit connective tissue breakdown: New therapeutic implications for an old family of drugs. Crit. Rev. Oral Biol. Med. **2:** 297–321.

79. GOLUB, L. M., T. F. MCNAMARA, G. D'ANGELO, R. A. GREENWALD & N. S. RAMAMURTHY. 1987. A non-antibacterial chemically-modified tetracycline inhibits mammalian collagenase activity. J. Dent. Res. **66:** 1310–1314.
80. YU, Z., M. K. LEUNG, N. S. RAMAMURTHY, T. F. MCNAMARA & L. M. GOLUB. 1992. HPLC determination of a chemically modified nonantimicrobial tetracycline: Biological implications. Biochem. Med. Metab. Biol. **47:** 10–20.
81. ZBINOVSKY, V. & G. P. CHREKIAN. 1977. Minocycline. *In* Analytical Profiles of Drug Substances. K. Florey, Ed. Academic Press. New York.
82. SANDE, M. A. & G. L. MANDELL. 1990. Antimicrobial agents: Tetracyclines, chloramphenicol, erythromycin, and miscellaneous antibacterial agents. *In* Goodman and Gilman's The Pharmacological Basis of Therapeutics. A. G. Gilman, T. W. Rall, A. S. Nies & P. Taylor, Eds. Pergamon Press. New York.
83. GOLUB, L. M., H. M. LEE, G. LEHRER, A. NEMIROFF, T. F. MCNAMARA, R. KAPLAN & N. S. RAMAMURTHY. 1983. Minocycline reduces gingival collagenolytic activity during diabetes. Preliminary observations and a proposed new mechanism of action. J. Periodontal Res. **18:** 516–526.
84. LANGER, R. & J. FOLKMAN. 1976. Polymers for the sustained release of proteins and other macromolecules. Nature **263:** 797–800.
85. RHINE, W. D., D. S. T. HSIEH & R. LANGER. 1980. Polymers for sustained macromolecule release: Procedures to fabricate reproducible delivery systems and control release kinetics. J. Pharm. Sci. **69:** 265–270.
86. LANGER, R., H. BREM & D. TAPPER. 1981. Biocompatibility of polymeric delivery systems for macromolecules. J. Biomed. Mater. Res. **15:** 267–277.
87. TEICHER, B. A., E. A. SOTOMAYOR & Z. D. HUANG. 1992. Antiangiogenic agents potentiate cytotoxic cancer therapies against primary and metastatic disease. Cancer Res. **52:** 6702–6704.
88. SOTOMAYOR, E. A., B. A. TEICHER, G. N. SCHWARTZ, S. A. HOLDEN, K. MENON, T. S. HERMAN & E. FREI. 1992. Minocycline in combination with chemotherapy or radiation therapy in vitro and in vivo. Cancer Chemother. Pharmacol. **30:** 377–384.
89. WEINGART, J., R. TAMARGO & H. BREM. 1992. Minocycline, an angiogenesis inhibitor, inhibits brain tumor growth (Abstract). Proc. Am. Assoc. Cancer Res. **33:** 75.
90. WEINGART, J. & H. BREM. 1992. Minocycline, an angiogenesis inhibitor, acts synergistically with carmustine (BCNU) to control 9L gliosarcoma growth (Abstract). Mol. Biol. Cell **3(s):** A21.
91. WEINGART, J. D., E. P. SIPOS & H. BREM. 1994. The role of minocycline in treatment of intracranial 9L glioma. J. Neurosurg. In press.

Guidelines for Clinical Trial Design for Evaluation of MMP Inhibitors

ROBERT A. GREENWALD

Division of Rheumatology
Long Island Jewish Medical Center
New Hyde Park, New York 11042

As pointed out elsewhere in this volume, matrix metalloproteinase (MMP) inhibitors have been "on the shelf" at several pharmaceutical firms for at least two decades, if not longer. Simultaneously, clinicians in several fields have recognized that MMPs play a pathologic role in a variety of disorders and that inhibitor therapy might have an obvious beneficial effect. Why then, one might ask, is there no commercially available MMP inhibitor that was brought to market especially for that purpose?

The superficial answer is that drug design scientists want to have the best available agent before committing the tens of millions of dollars needed to bring an agent to market. However, an alternative explanation might be that the burgeoning interest in MMP inhibitor design and *in vitro* testing has sped far ahead of concepts for clinical evaluation of such agents. Hence the pharmaceutical industry has been coaxed into endlessly perfecting each agent for lack of an obvious clinical evaluation paradigm. Clearly, an MMP inhibitor, to be marketed specifically for that purpose, would represent a unique category of drug for which clinical testing guidelines are not readily available.

With many of the world's experts, both pharmacologic and clinical, gathered together in Tampa, I convened a round-table discussion group for the express purpose of gathering ideas on how one might present a clinical evaluation protocol to an agency such as the U.S. Food and Drug Administration. Three main areas were selected for concentration—periodontal disease, arthritis, and cancer—but discussion in every field was encouraged. From notes taken during the discussion, a typed transcript of the proceedings, and material supplied by several speakers, I have prepared the following summary of the discussion which ensued. The comments have been attributed as much as possible to the individual speakers when they were identifiable; speakers from the audience were usually unidentifiable. I have written this summary both in the first and third person, but the exact wording of each speaker's comments has been edited, and if I have distorted the meaning of anyone's comments, I take responsibility for any inadvertent changes.

STEPHEN KRANE (*Boston, MA*): Krane commented on the difference between the power of modern biology to understand structure at a sophisticated level versus the pragmatic need to help people on an empirical basis. He also pointed out that several drugs for treatment of rheumatoid arthritis (as an example) were introduced for the wrong reasons, for example, gold as an antituberculous agent, penicillamine to destroy rheumatoid factor, etc. If a drug works but we don't understand exactly how, we should still take advantage of its power.

ROBERT GREENWALD (*New Hyde Park, NY*): As mentioned in an earlier session, I believe that if an MMP inhibitor is to be tested in arthritis patients, the primary parameters for evaluation should be those that an MMP inhibitor might influence, which is probably far removed from the features of a standard anti-inflammatory agent. The case in point, of course, is the MIRA trial. If indeed joint counts, grip strength, and ESR are the wrong parameters, then it behooves me as a critic to suggest what I think is a better approach.

I have compiled a list of some things we might measure in an MMP inhibitor trial (see TABLE 3, page 193 in this volume). The two major categories are the enzymes themselves and the breakdown products of their action. One must decide in which body fluid and/or tissue(s) these will be assessed, and in an ideal study, one would like to have both pretreatment and posttreatment samples for comparison, even if it means repeated biopsies. Bone density, tissue integrity (corneas, ruptured tendons), and survival are perhaps easiest to measure and hardest to influence.

KENNETH BRANDT (*Indianapolis, IN*): There already have been animal studies showing prevention of osteoarthritis, and Brandt urged that anyone doing such studies, especially in industry, publish their data in peer-reviewed journals so as to enhance the scientific underpinnings of proposed clinical trials. To perform a trial at reasonable cost, he suggested that a high-risk population be selected based on epidemiologic data, and his choice would be overweight women over the age of 50 with unilateral osteoarthritis (OA) of the knee. This group is at high risk for development of contralateral OA; epidemiologic studies suggest that 50% will develop radiologic disease within two years. Patients with far advanced disease would obviously be a poor choice for proper testing of a preventative agent. A population of 150–200 patients studied for two years might be a good starting point.

However, he also pointed out that all of the markers are surrogates—even joint space narrowing on a radiograph. Even with MRI, there are major problems of reproducibility, and many of the other markers have yet to be validated. A normal bone scan, however, might be used to exclude certain joints which have a low probability (less than 20%) of developing OA and therefore avoid low-risk cases. Arthroscopy might help, but there is too much interobserver variability and this needs to be standardized. Finally, Brandt urged that the FDA start developing these concepts now, because a drug trial may be just around the corner. If the FDA insists on an outcome measure such as need for arthroplasty or disability, then such studies will be very difficult. The FDA worked with industry to develop standards for an osteoporosis drug(s), and it should do the same for MMP inhibitors in arthritis.

KRANE: Methotrexate, one of our most powerful drugs, has not clearly been shown to prevent progressive bone erosions, which is a crucial target for any remittive agent. Bone resorption and osteoclast function in rheumatoid arthritis (RA) are valid targets; biphosphonates may be useful adjuncts. Tissues harvested for assessment of enzyme activity probably need to be cultured rather than merely extracted; the "grind 'em up" school may well measure neutrophil enzyme, but it will not reflect activity generated by mesenchymal cells. Finally, synovial fluid may not be a good marker of tissue events, and use of synovial fluid (SF) could be quite misleading.

AUDIENCE: A speaker from the floor made three points. First, a measure must be quantifiable. Scoring of radiographs and MRIs is generally nonparametric at this time; DEXA and QCT are better. Secondly, tissue inhibitor of metalloproteinases (TIMP) and other natural inhibitors are probably irrelevant if testing a collagenase inhibitor. Thirdly, acute phase reactants should be avoided if we want to be pure about not mixing inflammation with anti-MMP effect.

AUDIENCE: Is it practical to consider direct injection of an MMP inhibitor into a joint?

KRANE: We have a 40-year experience with steroid injections, and any new agent will have to compete with steroids—which are cheap, effective, and low in toxicity. One must also consider that in RA, the patient has involvement of multiple joints.

GREENWALD: Remember the experience with superoxide dismutase (SOD), a cure that has been in search of a disease for over three decades. In one OA study, SOD injections had no effect until 14 weeks after treatment had terminated, long after any biologic effect could have conceivably occurred. Bizarre things can happen when studying injectables.

BRANDT: With regard to a "gold standard," we need to make the distinction between symptomatic relief and disease modification. In a slowly evolving disease like OA, it takes time to demonstrate the latter. In a two-year OA trial, a certain number of injections might be acceptable to patients if the treatment had disease-modifying expectations.

MICHAEL LARK (*Merck, Rahway, NJ*): Dr. Brandt, how would you define end-stage disease in order to exclude such patients from a trial? Joint space narrowing or outcome measure?

BRANDT: By outcome measure rather than joint space narrowing. There is no reliable scale for the latter. X-rays are fraught with error.

AUDIENCE: Since we may not know what enzyme we are inhibiting, perhaps we should concentrate on the chondroprotective aspect and the matrix components, and ignore the proteinases themselves.

GREENWALD: Good point. My patients are actually worried about their collagen, not their collagenase. I'm treating a 32-year-old woman with a two-year history of RA who has already ruptured three finger tendons; she is on low-dose doxycycline, and if she has no more ruptures, she'll consider the trial a success even if my biochemical measures don't change. Remember that nonsteroidal anti-inflammatory drugs (NSAIDs) are discovered by screening drugs for inhibitory activity against prostaglandin synthetase, but in clinical trials of NSAIDs, we don't measure prostaglandins, we measure inflammatory indices. I don't care if you screen your compound with an *in vitro* assay and then bring it to the clinic and don't want to measure MMP levels. I just want to know what to measure instead?

AUDIENCE: Measuring actual MMP level can be a guide to dosing.

KRANE: Let's move on to cancer. First you get arthritis, then you get cancer. Actually, that can really happen, depending on the drugs used.

STANLEY ZUCKER (*Stony Brook, NY*): Measurement in cancer is much easier than in arthritis—survival! Time of survival is easily quantified.

ANDREW DOCHERTY (*Celltech, Slough, UK*): Our goal has been to make antimetastatic drugs, and although there is an important need for such agents, demonstrating an effect may be quite difficult as most patients already have metastases when they present. We have used animal models to try to test measurable parameters applicable in clinical conditions. We have envisioned four scenarios in which such agents might be employed (TABLE 1); Dr. A. M. Sopwith of Celltech shares the credit for enunciating these principles.

Containment therapy is designed as primary therapy, that is, to reduce established tumors, inhibit growth and angiogenesis, prevent invasion, prevent additional metastases, and perhaps prevent bone resorption at the sites of bone metastases. These are ambitious goals. Bladder cancer might be a good example; perhaps an MMP inhibitor could keep it at the noninvasive stage.

Adjuvant therapy involves use of an MMP inhibitor after primary tumor debulking. The goals are to prevent local recurrence, further metastases, and bone resorption. Lung, colorectal, breast, or prostate cancers might be so treatable. This is perhaps the most likely usage.

A third scenario would be to prevent iatrogenic spread during surgery. MMP inhibitors might prevent neovascularization and/or extravasation and prevent secondary spread. Based on what we know about mechanism, this might be a reasonable

TABLE 1. Clinical Scenarios for Evaluation of an MMP Inhibitor in Cancer Patients[a]

Containment therapy—Reduce the malignancy of established tumors
Inhibit growth through the antiangiogenic properties of MMP inhibitors
Prevent tumor invasion into surrounding stroma
Prevent or delay further development of occult metastases by the above mechanisms
Prevent tumor-induced bone resorption at sites of bone metastases
Example: Prevent progression of bladder cancer
Adjuvant therapy—After primary debulking therapy
Prevent or delay local recurrence
Prevent or delay further development of occult metastases
Prevent bone resorption at sites of bone metastasis
Examples: Lung, colorectal, breast, prostate, liver metastases
Reduce iatrogenic spread
Inhibit dissemination through the ability of MMP inhibitors to reduce extravasation
Inhibit neovascularization at secondary sites of growth
Examples: Colorectal, breast (with inhibitor administered at time of surgery)
Reduce morbidity
Resolve pleural effusions through MMP inhibitor-induced fibrosis
Resolve ascites through inhibitor-induced fibrosis
Examples: Lung, breast, ovarian

[a]Compiled by Andrew Docherty and A. M. Sopwith of Celltech, Slough, United Kingdom.

goal if the inhibitor were administered at the time of surgery. Finally, such agents might reduce morbidity, for example, by reducing effusions.

Finally, because these drugs lower the enzyme activity but do not alter their synthesis, measuring enzyme levels may be quite irrelevant. We need surrogate markers—events that are eliminated by inhibiting the enzymes. The candidates might include type IV collagen fragments from tumor invasion or type I fragments from bone resorption. We also need to demonstrate actual MMP inhibition within the tumor biochemically in order to convince clinicians to start long-term trials.

PETER BROWN (*British Bio-technology, Oxford, UK*): Many companies have avoided studying their MMP inhibitors because of the complexity of the required clinical trials. They are also worried about long-term toxicity, for example, liver fibrosis. In cancer trials, quality of life and survival will be the most important markers. Regulatory agencies will want that kind of data.

WILLIAM STETLER-STEVENSON (*National Institutes of Health, Bethesda, MD*): Tumors that lose basement membrane continuity have a poor prognosis—they have converted to invasiveness. We may not have sufficient levels of matrix degradation fragment to allow measurements, as can be done in arthritis. Enzyme and drug effect within the tissue will be critical.

HENRY BREM (*Johns Hopkins, Baltimore, MD*): Forty percent of admissions for cancer are for pain control, and most of that pain is from bone metastases, so pain relief is another measurable goal of cancer therapy. Our approach is to identify drugs that might have anti-MMP or antiangiogenic properties and then work out the dosage for local and/or systemic administration; we study the kinetics of drug release from polymers. We also evaluate synergy with other agents. We also would rely on human tumors in nude mice as a test system before going to human trial.

GREENWALD: I would like to comment about synergy. RA is a terribly difficult disease to control. My average RA patient is on five or six different drugs, usually two different NSAIDs (a short-acting drug during the day and a long-acting drug at bedtime), two different second line agents (e.g., hydroxychloroquine and gold), low-dose prednisone, and a supplemental analgesic, and despite all of this, most of them are still doing poorly. The FDA has a very narrow focus concerning drug combinations, except perhaps in oncology. For example, we have reported very exciting animal data when certain tetracyclines were combined with flurbiprofen, but Upjohn was not interested in any effect which required a drug combination. They were concerned that FDA approval would be impossible to attain. Several papers at this meeting have dealt with effective drug combinations. Since an MMP inhibitor will not show anti-inflammatory effect, the FDA will have to consider such drugs in combination or all hope for MMP inhibition is doomed.

ZUCKER: Studying advanced disease is a mistake; the chance of detecting useful effects is too low. I think that the specific aims of a cancer trial (TABLE 2) should be based on a cancer with a high short-term probability of metastasis, for example, small-cell cancer of the lung, malignant melanoma greater than 4 mm thickness, or breast cancer with more than 10 nodes. I would also select cancers where distant metastases occur despite good control of the primary. I would exclude cases where lack of control of the primary produces too much morbidity, for example, head and neck patients.

TABLE 2. Concepts for Development of Human Trials for Testing MMP Inhibitors in Cancer[a]

Goal of the trial: To determine if an MMP inhibitor is able to decrease the incidence of metastasis and/or prolong survival in patients with cancer.

Specific aims in planning a human trial:

1. Select cancer types which have a high probability of metastasis and in which there is experimental evidence that specific MMPs are involved in the mechanism of cancer dissemination.
2. The concentration of specific MMP (protein, mRNA) of interest should have been shown to be increased in human cancers of this type. Ideally, the primary tumor from each patient should be analyzed for MMP activity/content.
3. Select a cancer type and disease stage in which the likelihood of distant metastasis over a short-term follow-up is high and relatively predictable. *Examples:* Localized small cell lung cancer, malignant melanoma >4 mm thickness, breast cancer with >10 positive nodes.
4. Select a cancer type in which distant metastasis occurs in spite of good local control of the primary tumor mass. *Examples:* Small-cell lung cancer after chemotherapy, malignant melanoma after resection of the primary. Avoid cancers in which the lack of local control of the primary will produce considerable morbidity/mortality even if metastatic disease is prevented, e.g., unresectable head and neck cancers.
5. Permit the patients to be treated with the best standard treatment modality (e.g., surgery for localized non-small-cell lung cancer or melanoma, chemotherapy for small-cell lung cancer) prior to or simultaneously with anti-MMP treatment. Total treatment regimen must be standardized, not just the MMP inhibitor component.
6. Begin treatment soon after diagnosis and continue treatment until the likelihood of metastasis has diminished (e.g., localized small-cell lung cancer tends to recur within a year; breast cancer recurrence is less predictable.)
7. Begin treatment before evidence of distant metastasis is demonstrated. In cancer of the breast, axillary node metastasis should not exclude the patient from the trial; likewise for hilar nodes in lung cancer.
8. Drug dosage and frequency of administration should be tailored to achieve tissue levels of drug that have been demonstrated to have the desired inhibitory effect on MMPs.
9. Consider trying to achieve higher local MMP inhibitor levels in specific tumors by local administration, e.g., intraperitoneal injection for ovarian cancer.

[a]As formulated by Stan Zucker.

AUDIENCE: An ideal agent would inhibit MMP production, inhibit adherence, impede cellular proliferation, and even be toxic to the cell. There may be common pathways and it might be possible to target drugs to control multiple transcription processes.

AUDIENCE: Screening with an angiogenesis model might be as effective as using MMP inhibitor assays.

AUDIENCE: As an ENT surgeon, I disagree with Dr. Zucker. I think that head and neck cancer patients are ideal candidates. The super radical surgery of the last 30 years has not altered mortality rates. Current thinking is to refer the patients earlier to medical oncologists and to consider surgery only for a selected population. Head and neck tumors are easy to monitor radiologically with CT or MRI and quality of life can also be assessed easily. Tumor invasion can be quantified, and because the expected life span is not long, the end point is practical.

KRANE: The final topic for discussion is dental disease.

LORNE GOLUB (*Stony Brook, NY*): The classic clinical parameter used by the periodontist is the depth of the periodontal pocket, from gingival margin to base, which is thought to reflect disease progression. However, the disease is episodic in nature, and pocket depth may not indicate current disease activity. Inactive lesions cannot be distinguished from active ones by traditional clinical indices. We believe that biochemical markers need to be used to identify sites which might be expected to respond to anti-MMP therapy. One might measure (in gingival crevicular fluid [GCF]) prostaglandin E, collagenase, gelatinase, β-glucuronidase, or whatever is relevant to the study.

In one of our studies, we screened 200 patients to find 75 for inclusion, and each of the 75 had 10 or more pockets—some active, some stagnant. If you include the latter in the trial, any agent has little chance of success. We therefore measured collagenase in the GCF several times before baseline and only accepted into the study those sites with high enzyme levels on three separate occasions two weeks apart—about two sites per patient. Fortunately for us, the placebo-treated group did deteriorate from baseline over 9 months, assessed by detecting progressive increase in pocket depth (actually loss of gingival attachment), so the treatment could be shown to be effective; the treatment tested was our low-dose doxycycline formulation previously found to inhibit gingival and GCF collagenase activity. I should also mention that the patients were not randomized into the treatment groups, they were stratified by enzyme level so that each group had the same mean initial collagenase levels at baseline (time zero).

CHRISTOPHER MCCULLOCH (*Toronto, Ontario, Canada*): I believe that one should strive to reduce heterogeneity by selecting a particular patient type, preferably one that will progress rapidly so that the trial can be expedited and made more efficient. I personally would use patients with refractory periodontitis who have failed with conventional treatment. Inasmuch as the new treatment—for example, with CMT—is to be layered onto conventional therapy as an adjunct, once again it is the patient who has not responded to the usual treatment who stands to benefit the most. I also believe that subjects should meet rigid inclusion criteria. Finally, as to outcome criteria, the traditional dental measures, such as radiographic bone height and gingival attachment, should be supplemented by quality of life items which have recently received some attention in periodontal research.

AUDIENCE: I didn't realize that standard periodontal therapy was so effective. Having paid handsomely for my wife's treatment, I'm pleased to hear that.

GREENWALD: The periodontist cleans out your pockets and fills his!

AUDIENCE: Can you design a periodontal trial separating antibacterial effect from anti-MMP?

McCULLOCH: In our doxycycline study, we did not find a great deal of decrease in collagenase levels, and the relative risk reduction for attachment loss was only about 50%. However, when the subjects were then treated with metranidazole, a conventional antibiotic, the relative risk reduction went to 95%. Thus combined treatment, perhaps due to MMP inhibition plus antibacterial, was quite useful.

GOLUB: Remember that with low-dose doxycycline, the blood levels attained have no antimicrobial effect and do not change the flora. The peak blood level from 20 mg orally b.i.d. is only about 0.3 μg/mL. It should also be remembered that the subgroup referred to by Chris McCullough, that is, refractory periodontitis, is perhaps only 15%, or at most 25%, of the patients. Our treatment approach concentrates on garden-variety periodontal disease.

TIMO SORSA (*Helsinki, Finland*): We know that doxycycline treatment of patients with adult periodontitis reduces GCF collagenase levels as well as prevents inactivation of α-1 antiproteinase. The MMPs in this dental condition arise primarily from neutrophils, and PMN collagenase is very sensitive to inhibition by doxycycline. Tenidap, an NSAID, has been reported to prevent the release and oxidative activation of PMN procollagenase. NSAIDs were shown by Greenwald *et al.* to work synergistically with tetracyclines in reducing bone destruction in arthritic rats. Golub and his colleagues at Stony Brook are testing this approach for periodontal disease in collaboration with our Helsinki group, and "combination" trials are under way. I believe that such combination approaches may prove enlightening and should be encouraged. Periodontal researchers should agree on standardized sample collection and assay procedures. I would analyze gingival biopsies as well as GCF for both fibroblast and neutrophil collagenase, as well as for TIMPs, cytokines, and serpins.

GREENWALD: The arthritic joint, the metastatic lesion, and the periodontal pocket all have something in common. After patiently explaining all of the treatment options to a patient, it is not unusual to hear, "Well doctor, should I start the treatment now?" My usual response is that I have yet to identify a disease which is better treated late than early. Since most of our animal work is done prophylactically, I don't think we should wait until the bitter end to start treatment, and perhaps that argument applies to refractory periodontal disease.

KRANE: Thank you all.

Concluding thoughts. When the idea for this conference first occurred to me in the winter of 1991, there were, to the best of my knowledge, no major human clinical trials under way for any MMP inhibitor. Since then, the British Bio-technology compound has entered Phase II/III testing for cancer in the United Kingdom, several periodontal disease trials have been completed using tetracyclines, and a pilot study in human rheumatoid arthritis has been performed. Obviously, considerable progress has been made.

There was no consensus from the panel concerning parameters to measure that might apply to every trial. In periodontal disease, where GCF is readily sampled and where MMP activity can be used to select sites which might benefit from treatment, sequential assays of MMPs during longitudinal trials clearly seems indicated. In cancer and arthritis, where tissue enzyme level cannot easily be accessed, surrogates such as collagen or proteoglycan degradation products must be utilized, along with MMP-relevant clinical indices, for example, the appearance of a new metastasis. Regardless of what is chosen, the panelists and audience seemed to agree that MMP inhibitors should be evaluated by a rational method centered on their unique capabilities and that study design should be based upon one or more MMP-mediated events that can be quantified serially.

Inhibition of Angiogenesis by Anthracyclines and Titanocene Dichloride[a]

MICHAEL E. MARAGOUDAKIS, PLATON PERISTERIS, ELEFTHERIA MISSIRLIS, ALEXIS ALETRAS, PARASKEVI ANDRIOPOULOU, AND GEORGE HARALABOPOULOS

Department of Pharmacology
University of Patras Medical School
26110 Patras, Greece

Type IV collagenases belong to the metalloproteinase family of enzymes, whose activity is considered to be a key rate-limiting step in extracellular matrix degradation.[1] In recent years it has been recognized that metalloproteinases have significant roles in a number of biological processes such as embryonic development,[2,3] renal disease,[4] cancer metastasis,[5,6] and angiogenesis.[7,8] In the multistep cascade of angiogenesis, the first event requires the disruption of basement membrane surrounding the preexisting vessels, allowing endothelial cells to traverse the vessel wall. This local dissolution of basement membrane is correlated with the production of proteolytic enzymes such as collagenases and plasminogen activator by the endothelial cells after an angiogenic stimulus.[9] It has also been shown that inhibition of these proteolytic systems prevents both neovascularization of cartilage[10] and invasion of human amnionic membrane by bovine endothelial cells.[11] Furthermore, Moses and Langer[12] have demonstrated that a preparation from cartilage inhibits collagenase, angiogenesis, and tumor growth; they suggested inhibition of metalloproteinases as a mechanism for inhibiting angiogenesis.

In the present series of experiments we investigate the ability of the antitumor agents daunorubicin, doxorubicin, epirubicin, and titanocene dichloride to suppress angiogenesis in relation to their ability to inhibit a type IV collagenase from rat Walker 256 carcinosarcoma reported previously.[13,14]

MATERIALS AND METHODS

The hydrochloride salts of doxorubicin, daunorubicin, and epirubicin were obtained from Farmitalia, Carlo Erba. Titanocene dichloride (cyclopentadienyl titanium dichloride), $(C_5H_5)_2TiCl_2$, was synthesized by the method of Wilkinson and Birmingham[15] and recrystalized in boiling toluene. Fresh fertilized eggs were obtained locally and kept at 10 °C before incubation at 37 °C. The plastic discs were 13-mm round tissue culture cover slips from Nunc Inc. (Naperville, IL). Collagenase type VII from clostridium histolyticum, cortisone acetate, collagen type IV from human placenta, hematoxylin, EDTA, and gelatin were obtained from SIGMA

[a]This work was supported in part by grants from the Greek Ministry of Industry and Technology and NATO collaborative grant No. SA5-2-05 (CRG 920005) 336/92/Jarc-501.

Chemical Co. (Poole, UK). [^{3}H]Acetic anhydride, specific activity 9.8 Ci/mmol, was obtained from Amersham International. L-[U-^{14}C]Proline (specific activity 273 mCi/mmol) was obtained from New England Nuclear (Boston, MA). Fetal bovine serum, ECGS, M199, DMEM, penicillin, streptomycin, L-glutamine, gentamycin, heparin, and trypsin-EDTA were obtained from ICN Flow Laboratories (UK). Matrigel was a kind gift from Dr. Hynda Kleinman of the National Institutes of Health (Bethesda, MD).

Preparation of [^{3}H]Labeled Collagen

[^{3}H]labeled type IV collagen was prepared by acetylation of type IV collagen from human placenta with [^{3}H]acetic anhydride as previously described.[6]

Assay for Collagen Type IV Degrading Activity

The method for measuring type IV collagen degrading activity was according to Karakiulakis *et al.*[6,13] Assays were performed with 200 μL (15 nM) of [^{3}H]acetylated type IV collagen and solution of test substance made up to a volume of 280 μL with 50 mM Tris-HCl, 200 mM NaCl, 3 mM $CaCl_2$ (pH 7.4). NaN_3 (0.01% final) was included in all incubations to prevent contamination with microorganisms that may elaborate proteinase activity. The reaction was started with the addition of 20 μL of collagen type IV degrading enzyme solution (1 mg/mL), prepared from Walker 256 carcinoma as described by Karakiulakis *et al.*[6] Reaction was stopped with the addition of 150 μL of 6% trichloroacetic acid per 0.3% tannic acid. Tubes were transferred to 4 °C for 20 min, and undigested substrate was centrifuged out at 6500 $\times$ *g* for 20 min. An aliquot of the supernatants (200 μL) was transferred to vials containing 4.5 mL of Lumagel scintillation mixture. Radioactivity was measured in a Beckman LS 1801 liquid scintillation spectrometer. Radioactivity released without tumor enzyme was considered as "blank" and subtracted from each value. Radioactivity released in the presence only of tumor enzyme without inhibitors was considered as "control." Collagen type IV degrading activity was determined as cpm released in the supernatant. Results were expressed as % of mean of cpm ± S.E. released by the tumor enzyme as compared to total cpm added in the reaction mixture and analyzed by unpaired *t* test.

Chick Chorioallantoic Membrane System

The *in vivo* chick chorioallantoic membrane (CAM) angiogenesis model, initially described by Folkman[16] and modified as previously reported,[17] was used. Briefly, fresh fertilized eggs were incubated for 4 days at 37 °C when a window was opened on the egg shell exposing the CAM. The window was covered with sterile cellophane tape, and the eggs were returned to the incubator until day 9 when the test materials were applied. The test materials or vehicle and 0.5 μCi [U-^{14}C]labeled proline were placed on sterile plastic discs and were allowed to dry under sterile conditions. The control discs (containing vehicle and radiolabeled proline) were placed on the CAM one centimeter away from the disc containing the test material. A sterile solution of cortisone acetate (100 μg/disc) was routinely incorporated in all discs in order to prevent an inflammatory response. The loaded and dried discs were inverted and placed on the CAM, the windows were covered, and the eggs incubated until day 11 when assessment of angiogenesis took place.

Biochemical Evaluation of Angiogenesis

Collagenous proteins represent about 10% of the total proteins synthesized by the CAM under the conditions described above. About 80% of these collagenous proteins synthesized by the CAM represent type IV collagen as shown previously.[18] The extent of collagenous protein biosynthesis reaches a maximum between days 8 and 11 and coincides with the stage of maximal angiogenesis in the CAM as shown by morphological evaluation of vascular density. Furthermore, at day 10, collagenous protein biosynthesis is 11-fold higher than that of day 15 when angiogenesis has reached a plateau.[17]

Biochemical evaluation of newly formed vessels was performed by determining the extent of collagenous protein biosynthesis in the CAM lying directly under the discs. Briefly, the area under the disc was cut off and placed in an appropriate buffer, and protein biosynthesis was stopped by addition of cycloheximide and dipyridyl.[17] Non-protein-bound radioactivity was removed by washing with trichloroacetic acid. Pellets containing protein-bound radioactivity were resuspended and subjected to clostridial collagenase (type VII) digestion. The resulting radiolabeled tripeptides corresponding to basement membrane collagen and other collagenous material synthesized by the CAM from [U-^{14}C]proline were counted and expressed as cpm/mg protein.

In order to assess the rate of release of the test material from the discs, 50 μL solution containing 10^6 cpm of [U-^{14}C]proline was placed on discs, dried, and placed on the CAM. The radioactivity remaining on the discs was measured at various time intervals after placement on the CAM. It was shown that within 60 min more than 90% of the radioactivity had disappeared from the discs. Collagenous protein biosynthesis under the discs containing the test materials or vehicle were calculated as cpm/mg protein. For each egg, collagenous protein biosynthesis under the disc containing the test materials was expressed as % of the value obtained for the control. The results were analyzed by paired *t* test and were expressed as mean of % change of control ± SE.

Morphological Evaluation of Angiogenesis

Eggs were treated as above in the absence of radiolabeled proline. At day 11, the eggs were flooded with 10% buffered formalin, the plastic discs were removed, and the eggs were kept at 37 °C until dissection. A large area around the disc was removed and placed on a glass slide, and the vascular density index was measured by the method of Harris-Hooker *et al.*[19] Vascular density was determined as the number of vessels intersecting three concentric circles of 4-, 5- and 6-mm diameter, respectively. For each egg the vascular density under the disc containing the test materials was expressed as % of the value obtained for the control. The results were analyzed by paired *t* test and were expressed as mean of % change of control ± SE.

Human Umbilical Vein Endothelial Cells Tube Formation Assay and Assessment of the Area

Human umbilical vein endothelial cells (HUVEC) were isolated from freshly obtained umbilical cords by 1% collagenase digestion according to the method previously described by Jaffe *et al.*[20] Cells were maintained at 37 °C in medium (supplemented with ECGS, FBS, and antibiotics) and 5% CO_2.

The Matrigel tube assay was performed as has been previously described.[21] Briefly, 300 μL Matrigel per well was used to coat 24-well Limbro cluster plates at 4 °C. After the Matrigel had formed a gel at 37 °C, approximately 40,000 cells were seeded into each well in medium containing the test substances and 10% fetal bovine serum. After 18 h incubation at 37 °C, the cells were fixed and stained, and the tube area was quantified using the MCID Image Analysis system from Brock University, Ontario, Canada, previously described by Grant *et al.*[22] In all experiments with HUVECs on Matrigel, the results were analyzed by unpaired *t* test and were expressed as % inhibition ± SE of tube area as compared to control wells.

Endothelial Cells on Plastic

HUVECs were used at passage 6. Approximately 40,000 cells per well were put in 24-well culture plates under similar conditions as the tube assay but without Matrigel. After incubation for 18 h, the medium was removed and cells were tested for viability and proliferation using trypan blue. Subsequently, cells were fixed and stained in the plates. The attached cells in three random areas in each well were counted using the 10× objective.

Cell Cultures

Chick skin fibroblasts were isolated and cultured as described previously.[23] The Walker 256 cell line was maintained *in vivo* as ascites tumor within adult Sprague-Dawley male rats. The isolation and culture of tumor cells were performed according to Shaugnessy *et al.*[24]

Conditioned medium from both cultures was used as source of collagenases in zymography.

Gelatin Zymography

Gelatinase zymography in sodium dodecyl sulfate (SDS) gels was performed using an 11% gel, which had been cast in the presence of gelatin (1 mg/mL).[25] After electrophoresis the gels were washed, incubated in the presence or absence of test compounds, and stained as previously described.[26]

The number of samples, eggs or wells used for each treatment is indicated in each figure legend.

RESULTS

Effect of Anthracyclines on Angiogenesis in the Chorioallantoic Membrane in Vivo

The anthracyclines, daunorubicin, doxorubicin and epirubicin, caused a dose-dependent inhibition of angiogenesis in the CAM as evidenced by a decrease in collagenous protein biosynthesis (FIG. 1). The degree of inhibition, compared to control, ranged from 11.6 ± 4.2 to 76.6 ± 4.7% for daunorubicin (6 to 50 μg/disc), 26 ± 6.3 to 60 ± 3.6% for doxorubicin (3 to 25 μg/disc), and 27.6 ± 7.1 to 67.6 ± 3.6% for epirubicin (2 to 20 μg/disc). Collagenous protein biosynthesis has been

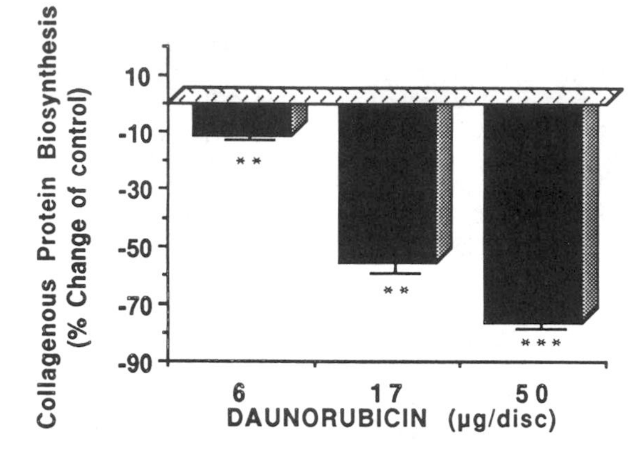

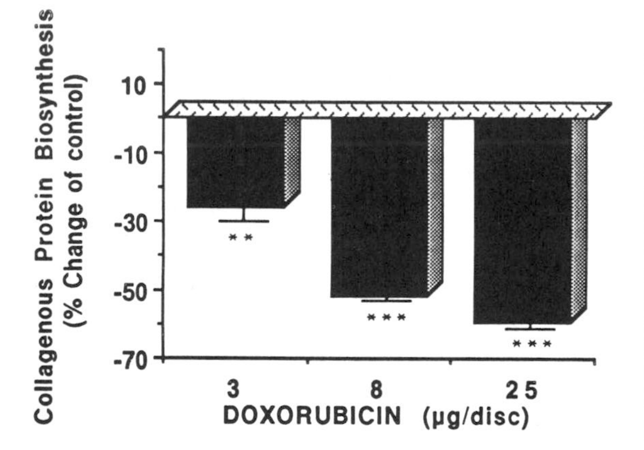

FIGURE 1. Dose-related effect of anthracyclines, daunorubicin, doxorubicin and epirubicin, on angiogenesis in the chick chorioallantoic membrane (CAM) *in vivo,* expressed as collagenous protein biosynthesis. Results are expressed as mean ± S.E. of the mean % of control and are compared by paired t test. Asterisks denote a statistical difference from the control which is taken as 0%. $^{**}p < 0.01$, $^{***}p < 0.001$. Eight to 12 eggs were used per treatment group.

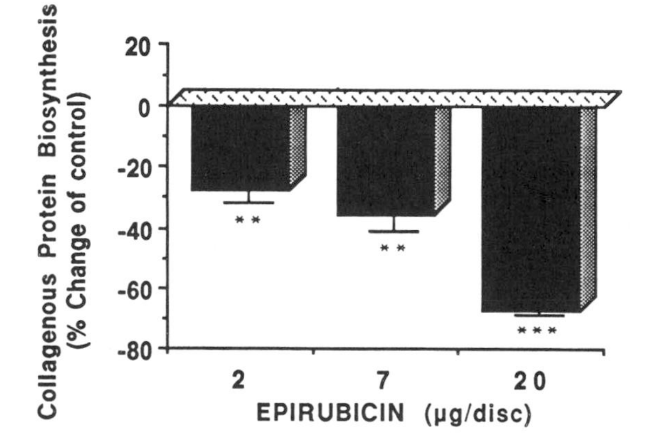

shown to be a sensitive and reliable biochemical index of angiogenesis.[7] In experiments using CAM morphological evaluation performed by the method of Harris-Hooker *et al.*,[19] the vascular density under the discs containing 17 μg of daunorubicin, 8 μg of doxorubicin or 7 μg of epirubicin was reduced by 38 ± 4.9, 21 ± 1.2, and 30 ± 2.9%, respectively (FIG. 2).

Effect of Anthracyclines on the Matrigel Tube Formation Assay System

When HUVECs were plated on Matrigel, in the presence of anthracyclines, tube formation expressed as relative tube area was inhibited in a dose-dependent manner (FIG. 3). None of the three anthracyclines caused any cytotoxic effects at the first two doses used (> 95% viability). Only epirubicin and daunorubicin at the highest dose used presented a ~35% cytotoxicity (data not shown). Furthermore, with the low doses of anthracyclines used in the tube formation assay, no inhibition of endothelial cell proliferation for the time period of the assay was observed (data not shown). It should be noted that tube formation does not require cell proliferation of HUVECs, but a change in cell phenotype. Anthracyclines also had no effect on cell viability or proliferation of Walker 256 carcinosarcoma cells after 24 h incubation at doses compared to those used for inhibition of tube formation (data not shown).

Effect of Titanocene Dichloride on Type IV Collagen Degrading Activity

Titanocene dichloride is an experimental antitumor agent with unknown mechanism of action. We have previously shown that titanocene dichloride retards Walker

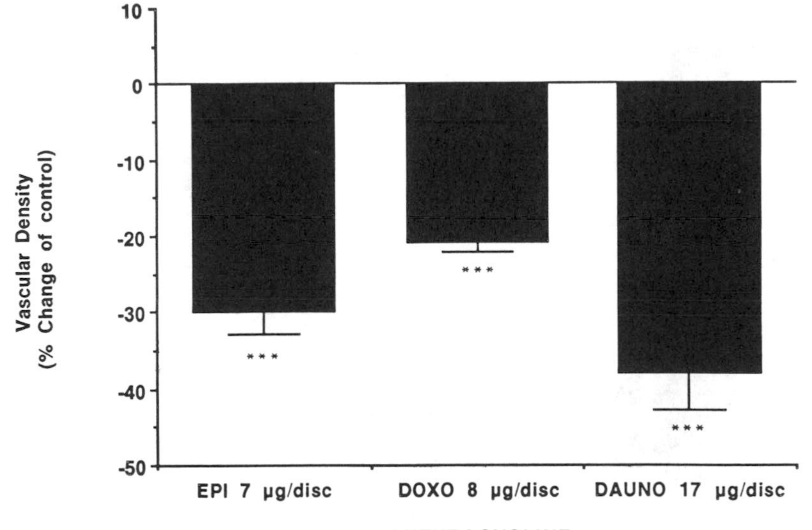

FIGURE 2. Effect of anthracyclines, daunorubicin (DAUNO), doxorubicin (DOXO) and epirubicin (EPI), on vascular density in the chick chorioallantoic membrane (CAM) *in vivo*. Results are expressed as mean ± S.E. of the mean % of control and are compared by paired *t* test. Asterisks denote a statistical difference from the control which is taken as 0%. $^{***}p < 0.001$. Six to eight eggs were used per treatment group.

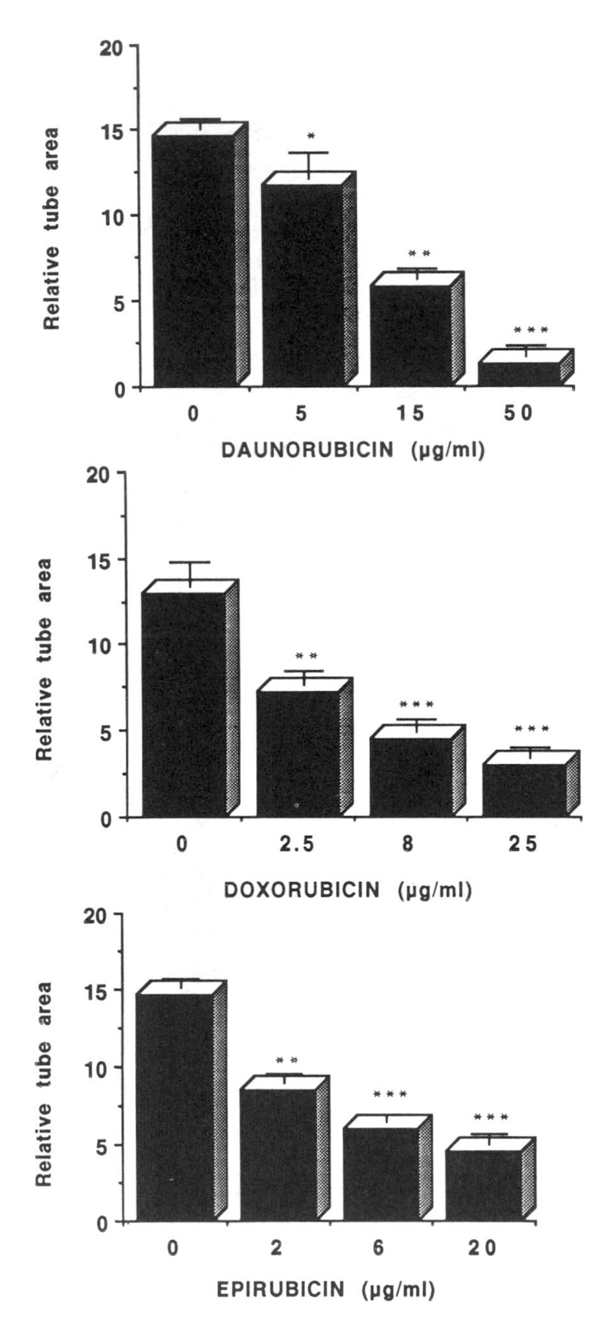

FIGURE 3. Dose-related effect of anthracyclines, daunorubicin, doxorubicin and epirubicin, on tube formation by endothelial cells plated on Matrigel. Results are expressed as mean ± S.E. of the mean relative tube area. Asterisks denote a statistical difference from the control (in the absence of test substance). $^{*}p < 0.05$, $^{**}p < 0.01$, and $^{***}p < 0.001$. Each treatment group represents 8 wells.

256 carcinosarcoma growth and metastasis *in vivo.*[14] This agent does not affect the growth and viability of HUVECs, Walker 256 carcinosarcoma cells or cells of human A549 lung adenosarcoma at doses up to 30 μg/mL.[14] As shown in FIGURE 4 it inhibited in a dose-dependent manner collagen type IV degrading activity from

Walker 256 carcinosarcoma. Inhibitions ranged from 19.5 ± 1.5 to 80 ± 1% with respective concentrations from 0.01 to 0.6 mM.

Effect of Titanocene Dichloride on Angiogenesis in the Chorioallantoic Membrane in Vivo

Titanocene dichloride inhibited in a dose-dependent manner angiogenesis in the CAM as evidenced by a decrease in collagenous protein biosynthesis (FIG. 5). Inhibitions, compared to controls, ranged from 18 ± 5 to 38 ± 4% with 10–100 μg/disc of titanocene dichloride. In this same system morphological evaluation performed by the method of Harris-Hooker *et al.*[19] showed that the vascular density under the discs containing 10–100 μg/disc was reduced from 9.5 ± 0.9 to 49 ± 4% compared to controls (FIG. 6).

Effect of Titanocene Dichloride on Relative Tube Area on the Matrigel Tube Formation Assay

Titanocene dichloride inhibited, in a dose-dependent manner, tube formation by HUVECs on Matrigel, starting at 1 μg/mL and causing 95% inhibition at 50 μg/mL.

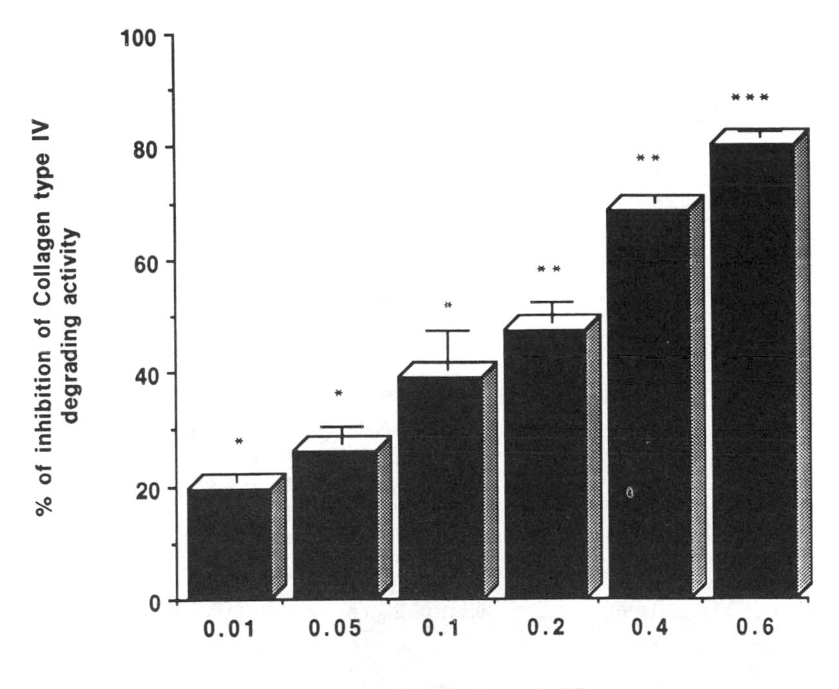

FIGURE 4. Dose-related inhibition of titanocene dichloride on type IV collagen degrading activity of rat Walker 256 carcinoma. Results are expressed as mean ± S.E. of the mean % of control which is taken as 0% and are compared by unpaired t test. Asterisks denote a statistical difference from the control which is taken as 0%. $^{*}p < 0.05$, $^{**}p < 0.01$, and $^{***}p < 0.001$. All experiments were conducted in quadruplicate.

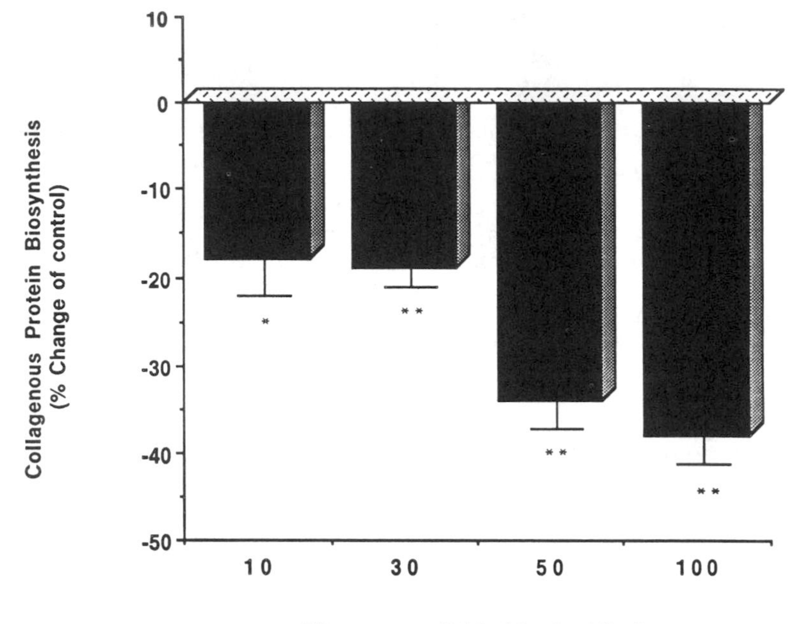

FIGURE 5. Dose-related effect of titanocene dichloride on angiogenesis in the chick chorioallantoic membrane (CAM) *in vivo,* expressed as collagenous protein biosynthesis. Results are expressed as mean ± S.E. of the mean % of control and are compared by paired t test. Asterisks denote a statistical difference from the control which is taken as 0%. $^{*}p < 0.05$, $^{**}p < 0.01$. For each dose of titanocene dichloride, 13–32 eggs were used.

At these doses titanocene did not present any toxic effect on HUVECs (data not shown).

Effect of Anthracyclines and Titanocene Dichloride on Zymograms of 72-kDa and 92-kDa Gelatinases

It has been shown that cultured chick skin fibroblasts release into the medium a 72-kDa gelatinase;[23] tumor Walker 256 cells release a 92-kDa gelatinase.[24] Conditioned medium from these two types of culture was analyzed in zymograms. Anthracyclines at a concentration of 200 mM failed to inhibit either gelatinase whereas titanocene dichloride at 1 mM inhibited completely the activity of these two enzymes (FIG. 7). EDTA (10 mM), a metal chelator, also inhibited the two enzymes.

DISCUSSION

Considerable progress has been made in recent years in demonstrating the involvement of collagenases in angiogenesis. Schnaper *et al.*[27] have shown in the Matrigel system of *in vitro* angiogenesis, used in this study, that antibodies against type IV 72- and 92-kDa gelatinases/collagenases decrease the area of tube formation. Similarly, both tissue inhibitors of metalloproteinases, TIMP-1 and TIMP-2, decrease tube formation when added to cultures of HUVECs in Matrigel. Moses and Langer,[12] using a cartilage-derived collagenase inhibitor, have demonstrated in the CAM system of angiogenesis that metalloproteinases have a key role in the process of neovascularization.

In the present series of experiments, we investigated the inhibition of angiogenesis by the antitumor agent titanocene dichloride and the anthracycline antibiotics in relation to their ability to inhibit collagenases. Titanocene dichloride is an experimental antitumor substance with significantly less hepatotoxic effects than *cis*-platinum.[28] The mechanism of the antitumor effect has not been elucidated. We have shown in this report that titanocene dichloride inhibits collagenase type IV from rat Walker 256 carcinosarcoma with IC_{50} ~0.2 mM. It has also been demonstrated that this agent is a potent inhibitor of angiogenesis both in the Matrigel system and in the CAM system.[14] The antiangiogenic effect of titanocene in the CAM is evident both by morphometric evaluation of blood vessels using the method of Harris-Hooker, which is the most rigorous test of angiogenesis in the CAM, and also by measuring collagenous protein biosynthesis in the presence and absence of this agent. Collagenous protein biosynthesis in the CAM has been shown to be a reliable and sensitive index of angiogenesis.[7]

Titanocene dichloride inhibits angiogenesis in a dose-dependent fashion which is

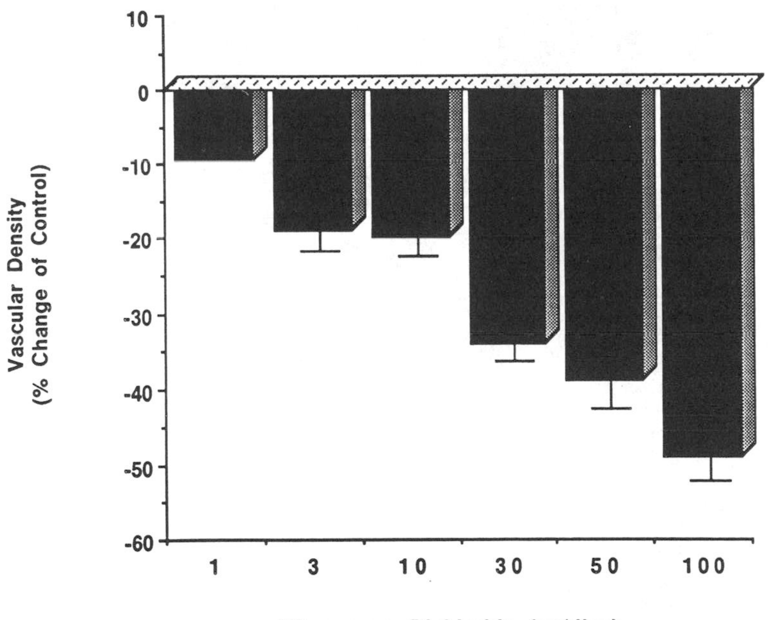

FIGURE 6. Effect of titanocene dichloride on vascular density in the chick chorioallantoic membrane (CAM) *in vivo.* Results are expressed as mean ± S.E. of the mean % of control and are compared by paired t test. At all doses test groups presented a $p < 0.001$ statistical difference from the control, which is taken as 0%. For each dose of titanocene dichloride, 6–12 eggs were used.

evident at very low concentrations. The antiangiogenic effect is specific because other related drugs such as cis-platinum are without effect.[14] The findings are not a result of cytotoxicity since the agent has no effect on endothelial cell proliferation and attachment, nor does it affect the viability of tumor cells at comparable concentrations. The antiangiogenic effect of titanocene dichloride is unlikely to be explained by inhibition of metalloproteinases, because the IC_{50} for this effect is quite high although both the 72-kDa and 92-kDa proteolytic enzymes with specificity for type IV collagen are inhibited in zymograms.

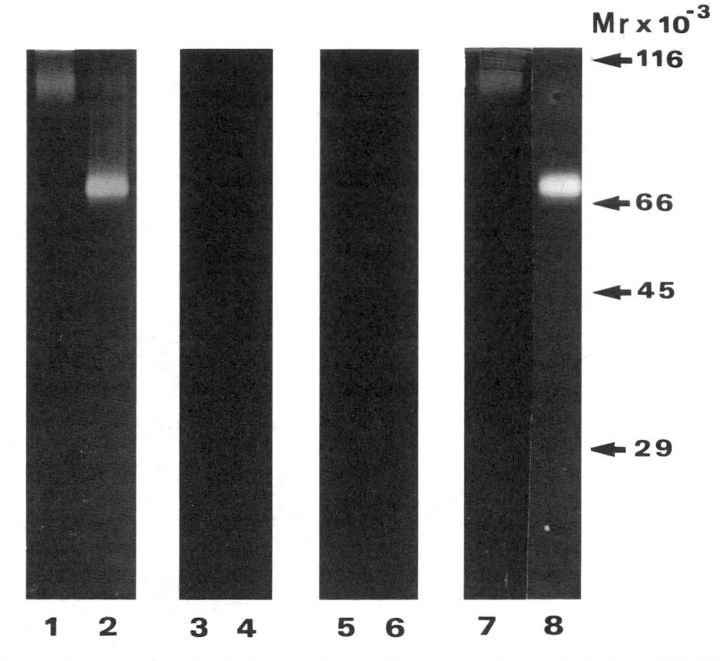

FIGURE 7. Zymography of gelatinases from cell cultures. **Lanes 1, 3, 5, and 7,** Walker 256 tumor cells and **lanes 2, 4, 6, and 8** chick skin fibroblast conditioned culture medium, respectively. After electrophoresis in nonreducing conditions, the gels were incubated at 37 °C in the absence (*lanes 1 and 2*) or in the presence of 10 mM EDTA (*lanes 3 and 4*), 1 mM titanocene dichloride (*lanes 5 and 6*), and 0.2 mM daunorubicin (*lanes 7 and 8*). M_r's of standard proteins run under reducing conditions are indicated. Similar results were obtained with doxorubicin and epirubicin (data not shown), as with daunorubicin.

It has been previously reported the anthracyclines, daunorubicin, doxorubicin, and epirubicin, inhibit in a noncompetitive and reversible manner collagen type IV degrading activity from homogenates of murine Walker 256 carcinoma. The K_i were 92, 49, and 40 μM for daunorubicin, doxorubicin, and epirubicin, respectively.[13] However, this inhibitory effect is probably unrelated to the suppression of angiogenesis reported in this paper. None of the anthracyclines exhibited inhibitory effect on the 72-kDa and 92-kDa collagenase at concentrations up to 200 μM in zymograms. Furthermore, the inhibitory effect obtained by anthracyclines both in the Matrigel system of angiogenesis and in the CAM is attainable at much lower concentrations than that reported for collagenase type IV inhibition (~40–90 μM). In the CAM the effect on angiogenesis is dose-dependent as shown by the inhibition of collagenous protein biosynthesis (FIG. 1). Similarly, in the Matrigel system the disruption of tube formation by 50% is obtained at about 2.5 μg/mL for epirubicin and doxorubicin and about 15 μg/mL for daunorubicin. These effects cannot be attributed to cytotoxic effects of anthracyclines on the endothelial cells. Parallel experiments have shown that at the first two low doses used, cells were viable to an extent of more than 95%. In the Matrigel system there is no proliferation of endothelial cells, but a change into an angiogenic phenotype. In experiments where anthracyclines were added at comparable concentrations to proliferating endothelial cells no inhibition of growth was shown for the same time period (18 h). Also, in primary cultures of Walker 256 carcinosarcoma, no cytotoxicity was evident in the presence of comparable concentrations of anthracyclines.

We conclude from these experiments that anthracyclines and titanocene dichloride have a specific effect in modulating the angiogenic phenotype of endothelial cells by a mechanism that is not related to inhibition of metalloproteinases. The suppression of angiogenesis by anthracyclines is obtained at therapeutically attainable concentrations and may explain, at least in part, their effectiveness in the treatment of different malignancies.[29]

SUMMARY

The anthracycline antibiotics, daunorubicin, doxorubicin, and epirubicin, which are widely used for treatment of malignancies, have been evaluated for their effect on angiogenesis in relation to the inhibition of collagenase type IV reported previously.[6] In the chick chorioallantoic membrane (CAM) system of angiogenesis, anthracyclines inhibited vascular density at doses of 5–20 μg/disc as well as collagenous protein biosynthesis, which is a reliable index of angiogenesis. Similarly, all three anthracyclines inhibited tube formation in the *in vitro* system of angiogenesis using human umbilical vein endothelial cells (HUVECs) plated on Matrigel. The inhibition was dose-dependent and caused 50% inhibition at concentrations of 2.5–15 μg/mL. At concentrations of anthracyclines which prevented tube formation and angiogenesis, there were no cytotoxic effects, as evidenced by methylene blue uptake, and the growth of these endothelial cells was not inhibited. The experimental antitumor agent titanocene dichloride inhibited collagenase type IV from Walker 256 carcinosarcoma with IC_{50} approximately 0.2 mM. Titanocene also prevented angiogenesis in the CAM and tube formation by HUVECs on Matrigel at concentrations that were without effect on growth or cytotoxicity of endothelial cells or Walker 256 cells in culture.

The antiangiogenic effect of the aforementioned antitumor agents at therapeutically attainable concentrations may explain, at least in part, their antitumor properties because angiogenesis is an essential process for tumor growth and metastasis. The antiangiogenic effect is, however, unrelated to metalloproteinase inhibition because higher concentrations are required for that effect than for inhibition of angiogenesis.

ACKNOWLEDGMENTS

We thank Dr. N. Klouras for the synthesis of titanocene dichloride.

REFERENCES

1. MURPHY, G. J. P., G. MURPHY & J. J. REYNOLDS. 1991. The origin of matrix metalloproteinases and their familial relationship. FEBS Lett. **284:** 4–7.
2. BRENNER, C. A., R. R. ADLER, D. A. RAPPOLEE, R. A. PEDERSEN & Z. WERB. 1989. Genes for extracellular matrix-degrading metalloproteinases and their inhibitor, TIMP, are expressed during early mammalian development. Genes Dev. **3:** 848–854.
3. NOMURA, S., B. L. M. HOGAN, A. J. WILLS, J. K. HEATH & D. R. EDWARDS. 1989. Developmental expression of tissue inhibitor of metalloproteinases (TIMP) mRNA. Development **105:** 575–583.
4. BARICOS, W. H. & S. V. SHAH. 1991. Proteolytic enzymes as mediators of glomerular injury. Kidney Int. **40:** 161–173.
5. SALO, T., L. A. LIOTTA, J. KESKI-OJA & T. TURPEENNIEMI-HUJANEN. 1982. Secretion of basement membrane collagen degrading enzyme and plasminogen activator by transformed cells—Role in metastasis. Int. J. Cancer **30:** 669–673.
6. KARAKIULAKIS, G., E. MISSIRLIS, A. ALETRAS & M. E. MARAGOUDAKIS. 1988. Degradation of intact basement membranes by human and murine tumor enzymes. Biochim. Biophys. Acta **967:** 163–175.
7. MISSIRLIS, E., G. KARAKIULAKIS & M. E. MARAGOUDAKIS. 1990. Angiogenesis is associated with collagenous protein synthesis and degradation in the chick chorioallantoic membrane. Tissue & Cell **22:** 419–426.
8. WEIDNER, N., J. P. SEMPLE, W. R. WELCH & J. FOLKMAN. 1991. Tumor angiogenesis and metastasis—Correlation in invasive breast carcinoma. N. Engl. J. Med. **324:** 1–8.
9. GROSS, T. L., D. MOSCATELLI & D. B. RIFKIN. 1983. Plasminogen activator and collagenase production by cultured capillary endothelial cells. Proc. Natl. Acad. Sci. USA **80:** 2623–2637.
10. MOSES, M. A., J. SUDHALTER & R. LANGER. 1990. Identification of an inhibitor of neovascularization from cartilage. Science **248:** 1408–1410.
11. MIGNOTTI, P., R. TSUBOI, E. ROBBINS & D. B. RIFKIN. 1989. In vitro angiogenesis on the human amniotic membrane: Requirement for basic fibroblast growth factor-induced proteinases. J. Cell Biol. **108:** 671–682.
12. MOSES, M. A. & R. LANGER. 1993. Metalloproteinases inhibition as a mechanism for inhibition of angiogenesis. *In* Angiogenesis. Key Principles, Science, Technology and Medicine. R. Steiner, P. B. Weisz & R. Langer, Eds. Vol. **1:** 146–151. Birkhauser Verlag. Basel.
13. KARAKIULAKIS, G., E. MISSIRLIS, A. ALETRAS & M. E. MARAGOUDAKIS. 1990. Basement membrane collagen degrading activity from a malignant tumour is inhibited by anthracycline antibiotics. Biochim. Biophys. Acta **1035:** 218–222.
14. BASTAKI, M., E. MISSIRLIS, N. KLOURAS, G. KARAKIULAKIS & M. E. MARAGOUDAKIS. 1994. Suppression of angiogenesis by the antitumor agent titanocene dichloride. Eur. J. Pharmacol. **251**:263–269.
15. WILKINSON, J. & J. M. BIRMINGHAM. 1954. *Bis*-cyclopentadienyl compounds of Ti, Zr, V, Nb and Ta. J. Am. Chem. Soc. **76:** 42–81.
16. FOLKMAN, J. 1985. Tumour angiogenesis. Adv. Cancer Res. **43:** 172–203.
17. MARAGOUDAKIS, M. E., M. PANOUTSAKOPOULOU & M. SARMONICA. 1988. Rate of basement membrane biosynthesis as an index to angiogenesis. Tissue & Cell **20:** 531–539.
18. MARAGOUDAKIS, M. E., E. MISSIRLIS, G. KARAKIULAKIS, M. SARMONICA, M. BASTAKI & N. TSOPANOGLOU. 1993. Basement membrane biosynthesis as a target for developing inhibitors of angiogenesis with anti-tumor properties. Kidney Int. **143:** 147–150.
19. HARRIS-HOOKER, S. A., C. M. GAJDUSEK, T. N. WIGHT & S. M. SCHWARTZ. 1983. Neovascularization responses induced by cultured endothelial cells. J. Cell. Physiol. **114:** 302–310.
20. JAFFE, E. A., R. L. NACHMAN, C. G. BECKER & C. R. MINICK. 1973. Culture of human endothelial cells derived from umbilical veins—Identification and immunological criteria. J. Clin. Invest. **52:** 2745–2756.
21. KUBOTA, Y., H. K. KLEINMAN, G. R. MARTIN & T. J. LAWLEY. 1988. Role of laminin and

basement membrane in the morphological differentiation of human endothelial cells into capillary-like structures. J. Cell Biol. **107:** 1589–1598.

22. GRANT, D. S., K. I. TASHIRO, B. SEGUI-REAL, Y. YAMADA, G. R. MARTIN & H. K. KLEINMAN. 1989. Two different laminin domains mediate the differentiation of human endothelial cells into capillary-like structures in vitro. Cell **58:** 933–943.
23. GRAIG, F. M., C. W. ARCHER & G. MURPHY. 1991. Isolation and characterization of a chicken gelatinase (type IV collagenase). Biochim. Biophys. Acta **1074:** 243–250.
24. SHAUGNESSY, S. G., M. WHALEY, R. M. LATRENIE & F. W. ORR. 1993. Walker 256 tumour cell degradation of extracellular matrices and latent gelatinase activated by reactive oxygen species. Arch. Biochem. Biophys. **304:** 314–321.
25. HEUSSEN, G. & E. B. DOWDLE. 1980. Electrophoretic analysis of plasminogen activators in polyacrylamide gels containing sodium dodecyl sulfate and copolymerized substrates. Anal. Biochem. **102:** 192–202.
26. ZUCKER, S., J. WIEMAN, R. M. LYSIK, B. IMHOF, H. NAYASE, N. RAMAMURTHY, L. A. LIOTTA & L. M. GOLUB. 1989. Gelatin-degrading type IV collagenase isolated from human small cell lung cancer. Invasion Metastasis **9:** 167–181.
27. SCHNAPER, H. W., D. S. GRANT, W. G. STETLER-STEVENSON, R. FRIDMAN, G. D'ORAZI, A. N. MURPHY, R. E. BIRD, M. HOYTHYA, T. R. FUERST, D. L. FRENCH, J. P. QUIGLEY & H. K. KLEINMAN. 1993. Type IV collagenase(s) and TIMPs modulate endothelial cells morphogenesis in vitro. J. Cell. Physiol. **156:** 235–246.
28. KÖPF-MAIER, P., W. WAGNER, B. HESSE & H. KÖPF. 1981. Tumour inhibition by metallocenes: Activity against leukemias and detection of the systemic effect. Eur. J. Cancer **57:** 665–669.
29. DIMARCO, A., F. ARCAMONE & F. ZUNINO. 1975. *In* Antibiotics—Mechanism of Action of Antimicrobial and Antitumour Agents. J. W. Cockran & F. H. Hahn, Eds. Vol. 3. Springer. Heidelberg.

Interaction of α_2-Macroglobulin with Matrix Metalloproteinases and Its Use for Identification of Their Active Forms[a]

HIDEAKI NAGASE,[b] YOSHIFUMI ITOH, AND SHARON BINNER

Department of Biochemistry and Molecular Biology
University of Kansas Medical Center
3901 Rainbow Boulevard
Kansas City, Kansas 66160-7421

The activities of matrix metalloproteinases (MMPs) are regulated by two major groups of endogenous inhibitors: (1) tissue inhibitors of metalloproteinases (TIMPs) secreted from a number of cell types and (2) a general plasma proteinase inhibitor, α_2-macroglobulin (α_2M) and its related proteins. TIMPs probably control MMP activities immediately around the cell in the tissues, whereas α_2M functions as a regulator of MMPs in body fluids, especially in the inflammatory sites. More than 95% of anticollagenase activity in plasma is due to α_2M.[1]

Human α_2M is a tetrameric glycoprotein of 725 kDa. It consists of four identical subunits of 185 kDa that are linked in pairs by disulfide bonds and assembled noncovalently.[2,3] The inhibition mechanism of α_2M is unique. It inhibits almost all endopeptidases regardless of their specificities.[2,4] The binding of α_2M and a proteinase is initiated by proteolytic attack of the enzyme on the particular locus, so-called bait region, located near the middle of the subunit. The cleavage of the bait region triggers a conformational change in the α_2M molecule that in turn entraps the enzyme without blocking the active site of the enzyme.[4] The enzyme within the complex remains free to hydrolyze low molecular mass substrates but is restricted from reactions with large protein substrates. Inactive proteinases do not bind to α_2M.[4] Another molecular feature of α_2M is that the inhibitor contains an unusual intrachain β-cysteinyl-γ-glutamyl thiolester.[2,3] Upon proteolysis of the bait region of α_2M, the thioester bond becomes immediately available for nucleophilic attack by the proteinase, and a large portion of the trapped enzymes are covalently linked to the α_2M subunit.[2,3]

The rate of complex formation with α_2M differs among proteinases. In general, proteinases with limited specificity react slowly with α_2M. We therefore investigated the rate of interaction of α_2M with MMP-1 (interstitial collagenase), which readily digests the triple helical region of interstitial collagens I, II, and III, but has limited activity on other proteins. Kinetic studies indicated that α_2M is about 150 times better substrate for MMP-1 than collagen I, indicating it is a major inhibitor of collagenase activity.[5] We also show that the unique trapping reaction of α_2M that occurs only with active enzymes is useful for identification of catlytically active forms of MMPs. This feature was used to investigate the activation processes of the proMMP-2 (progelatinase A)–TIMP-2 and the proMMP-9 (progelatinase B)–TIMP-1 complexes upon treatment with 4-aminophenyl mercuric acetate (APMA) or trypsin.

[a]This work was supported by National Institutes of Health grants AR 39189 and AR 40994.
[b]Corresponding author.

RAPID BINDING OF MMP-1 TO α_2-MACROGLOBULIN

MMP-1 specifically cleaves collagen types I, II, and III at a Gly-Ile (or Leu) bond located in the triple helices approximately three-quarters away from their N-terminals.[6-8] When these substrates are heat denatured to gelatin, they become relatively resistant to proteolysis by MMP-1, indicating that the helical structure of collagen molecules are important for enzyme recognition. MMP-1 also digests aggrecan core protein,[9] cartilage link protein,[10] reduced-carboxymethylated transferrin,[11] but its activity on these substrates are much weaker compared to that on interstitial collagens. These observations suggest that the helical structure collagen molecules are important for favorable interaction with the enzyme. The binding kinetic studies of MMP-1 to α_2 M, however, indicated that α_2M is the best substrate for MMP-1 that has been characterized to date.[5]

The rate of human MMP-1 binding to α_2M was determined by incubating [^{125}I]labeled MMP-1 with a 2–20 molar excess human α_2M at 25 °C in 20 μL of 50 mM Tris-HCl (pH 7.5), 0.15 M NaCl, 5 mM $CaCl_2$. The reaction was terminated by adding 40 μL of 50 mM 1,10-phenanthroline and 25 mM EDTA in 10% (v/v) ethanol. The samples were then run on 5% polyacrylamide gels, and the amount of [^{125}I]labeled MMP-1 bound to α_2M was measured by incorporation of radioactivity.

The anticipated reaction scheme for formation of irreversible enzyme-inhibitor complex (EI*) is described as follows:

$$E + I \underset{k_{-1}}{\overset{k_1}{\rightleftarrows}} EI \overset{k_2}{\rightarrow} EI^* \quad (1)$$

$$K_i = \frac{k_{-1}}{k_1} \quad (2)$$

where K_i is the dissociation constant of the reversible complex (EI) and k_2 the first order rate constant for the formation of irreversible complex (EI*). This reaction scheme has been simplified: two steps of the initial cleavage of α_2M by a proteinase and the subsequent entrapment of the enzyme are combined because the experimental procedure described here does not distinguish between the two reactions. The progressive binding of [^{125}I]labeled MMP-1 to α_2M with time showed pseudo-first order kinetics in accordance with equation 3.

$$\ln \frac{EI^*_\infty - EI^*_t}{EI^*_\infty} = -k_{app} \cdot t \quad (3)$$

where EI^*_∞ represents the initial concentration of the enzyme and EI^*_t is the amount of EI* formed at time t. EI^*_∞ was measured from the total EI* formed at infinite time. According to the method of Kitz and Wilson,[12] the apparent pseudo-first order rate constant, k_{app} was described as a function of the concentration of inhibitor, I (equation 4).

$$\frac{1}{k_{app}} = \frac{1}{k_2} + \frac{K_i}{k_2} \cdot \frac{1}{I} \quad (4)$$

The K_i value and the k_2 value for the complex formation of α_2M and MMP-1 were 0.171×10^{-6} M and 0.48 s^{-1}, respectively. The k_2/K_i values of 2.8×10^6 M^{-1} s^{-1}

TABLE 1. Kinetic Parameters for Hydrolysis of Macroglobulins and Collagens by Human MMP-1 and MMP-3

Enzyme	Substrate	K_i or K_m (μM)	k_2 or k_{cat} ($s^{-1} \times 10^3$)	k_2/K_i or k_{cat}/K_m ($M^{-1}s^{-1} \times 10^{-3}$)	Ref.
MMP-1	α_2M (human)	0.17	483	2,800	5
	Ovostatin (chicken)	0.32	0.58	1.8	5
	Type I collagen				
	Human	0.8	15	18	13
	Guinea pig	0.9	6.0	7.0	13
	Type II collagen				
	Human	2.1	0.28	0.13	13
	Rat	1.1	1.3	1.1	13
	Type III collagen				
	Human	1.4	160	110	13
	Guinea pig	0.7	5.0	7.2	13
MMP-3	α_2M (human)	0.10	5.8	56	5

indicates that α_2M is approximately 150-fold better as a substrate than collagen I[13] (TABLE 1).

Ovostatin, an α_2M-like inhibitor in egg white,[14] also binds to MMP-1, but the binding kinetic of human MMP-1 with chicken ovostatin is considerably slower than α_2M (TABLE 1).

The kinetic parameters for the complex formation of α_2M and MMP-3 (stromelysin-1) indicates that MMP-3 binds to α_2M much more slowly than MMP-1 in spite of its stronger general proteolytic activity. The K_i and k_2 values were 0.104×10^{-6} M and $5.8 \times 10^{-3}\ s^{-1}$ ($k_2/K_i = 5.6 \times 10^4\ M^{-1}\ s^{-1}$), respectively. The reaction of MMP-3 with chicken ovostatin was even slower: the reaction of a stoichiometric amount of ovostatin and MMP-3 was not completed even after incubation at 37 °C for 2 h. This indicates that chicken ovostatin is not an effective inhibitor of human MMP-3.

MMP-1 cleaved the Gly^{679}-Leu^{680} bond of the α_2M subunit,[5,15] and MMP-3 at the Gly^{679}-Leu^{680} and at the Phe^{684}-Tyr^{685} bonds in the bait region.[5] The high susceptibility of the Gly-Leu bond in the Gly-Pro-Glu-Gly-Leu-Arg-Val-Gly sequence to MMP-1 is likely to be controlled by the local conformation of the α_2M bait region, but secondary structural predictions of this region did not show any dominant secondary structure of a conventional type.

The rapid binding of MMP-1 to α_2M suggests that α_2M is a major regulator of collagenolysis especially in inflammatory lesions. This is also supported by the work of Cawston and Mercer,[16] which demonstrated that MMP-1 preferentially bound to α_2M when it was reacted with an equinolar amount of TIMP-1 and α_2M. The concentration of α_2M in rheumatoid synovial fluids is 0.7–1.0 mg/mL, about one-third of the normal plasma level.[17,18] Finding of the collagenase-α_2M complexes in synovial fluid[17] and the disposition of α_2M in the rheumatoid synovial lining cells[19] further suggest the importance of α_2M in regulating tissue destruction by proteinases.

USE OF α_2M FOR IDENTIFICATION OF ACTIVE MMPs

MMPs are secreted from the cells as inactive zymogens (proMMPs) and their activation occurs, in most cases, in a stepwise manner.[11] During activation processes endogenous inhibitors TIMPs may interact with intermediate forms before proMMPs are converted to their fully active final products. Such a possibility may occur more readily for the proMMP-2–TIMP-2 and the proMMP-9–TIMP-1 complexes. Both proMMPs are considered to be activated even complexed with respective TIMPs by treatment with APMA, although their proteolytic activities are considerably reduced.[20–24] We investigated whether the decreased activity (usually less than 10% the activity of the activated TIMP-free enzyme) is due to the decrease in specific activity of the activated complex or whether it is derived from a small amount of TIMP-free zymogen present in the complex preparation. α_2M is useful to distinguish between the active and inactive species of proteinases inasmuch as it complexes only with enzymically active enzymes.

ProMMP-2 and proMMP-9 were purified as described by Okada *et al.*[25] and Morodomi *et al.*,[26] respectively. The proMMP-2–TIMP-2 complex was purified from the culture medium of human uterine cervical fibroblasts by gelatin-Sepharose affinity chromotography and gel filtration on Sephacryl S-200. The amount of TIMP-2 in the complex was estimated by titration of the known amount of MMP-1. Some preparations of the complex contained a small portion of free proMMP-2 ($< 10\%$). The proMMP-9–TIMP-1 complex was purified from the culture medium of HT-1080 cells treated with 25 ng/mL phorbol myristate acetate by a similar procedure described for the proMMP-2–TIMP-2 complex. The proMMP-9–TIMP-1 preparation was virtually free from free proMMP-9 and free TIMP-1.

When proMMP-9 and the proMMP-9–TIMP-1 complex were treated with 10 μg/mL trypsin or 1 mM APMA at 37 °C for various periods of time, only proMMP-9 exhibited gelatinolytic activity (TABLE 2). Zymographic analysis of the trypsin-activated products after electrophoresis under nonreducing conditions indicated that proMMP-9 in both samples was converted to 80 kDa, 74 kDa and a minor species of 66 kDa (FIG. 1). The treatment of proMMP-9 with 1 mM APMA resulted in formation of the 80-kDa and 68-kDa species, whereas the proMMP-9 complexed with TIMP-1 was converted only to 80 kDa (FIG. 1B). However, it is difficult to assess which molecular species are proteolytically active by this method because even the 88-kDa proMMP-9 exhibits gelatinolysis due to autoactivation induced by molecular perturbation in SDS during electrophoresis. Furthermore, the TIMP-MMP complex may partially dissociate in SDS and exhibit the enzymic activity. To identify the active species, the activated samples were treated with a four-molar excess of α_2M for 1 h at

TABLE 2. Gelatinolytic Activity of proMMP-9 and the proMMP-9–TIMP-1 Complex upon Treatment with APMA and Trypsin[a]

Treatment	MMP-9 Activity	
	ProMMP-9 (%)	ProMMP-9–TIMP-1 (%)
Nontreatment	0	0
APMA (1 mM)		
24 h	45	1
48 h	100	6
Trypsin (10 μg/mL)		
15 min	55	0
30 min	95	0
60 min	100	0

[a]Activation reactions were carried out in 50 mM Tris-HCl (pH 7.5), 0.15 M NaCl, 10 mM $CaCl_2$, 0.02% NaN_3, 0.05% Brij 35 at 37 °C.

37 °C, and then the samples were subjected to zymography. Only the proteolytically active forms of the enzyme bind to α_2M, and they move to the top of the gel together with α_2M under nonreducing conditions.

As shown in FIGURE 1, proMMP-9 without activation did not bind to α_2M, but the gelatinolytic zones exhibited by the trypsin-activated MMP-9 shifted to the top of the gel. In contrast, although proMMP-9 complexed with TIMP-1 was also processed to the seemingly active forms by trypsin or APMA, none of the gelatinolytic zones was removed with α_2M, indicating that all of these species do not possess proteolytic activity (FIG. 1B). These observations, together with the results in TABLE 2, suggest that upon activation of the zymogen, TIMP-1 binds to the catalytic domain of the enzyme and blocks the active site. This is also supported by the observation that the

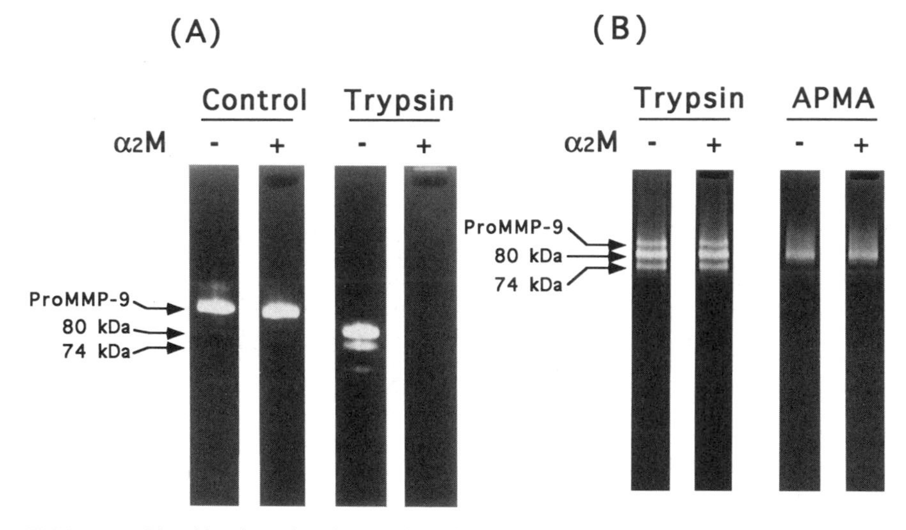

FIGURE 1. Identification of active species of MMP-9 by complex formation with α_2M. (**A**) proMMP-9 and the trypsin-activated MMP-9 were subjected to zymographic analysis on SDS/polyacrylamide gel containing gelatin (0.8 mg/mL) before and after complexing with a 4-molar excess of α_2M. Active MMP-9 was complexed with α_2M and the zones of gelatinolysis were observed at the top of the gel together with α_2M dimer. (**B**) The proMMP-9–TIMP-1 complex was treated with trypsin (10 μg/mL) for 15 min or with 1 mM APMA for 24 h at 37 °C. Trypsin activity was terminated with 2 mM diisopropyl fluorophosphate. The samples were then subjected to zymography before and after complexing with a 4-molar excess of α_2M.

APMA-treated proMMP-9–TIMP-1 complex could no longer inhibit other MMPs, whereas the untreated complex inhibited MMP-1 stoichiometrically.

When proMMP-2 and the proMMP-2-TIMP-2 complex were treated with APMA, gelatinolytic activity was observed only with proMMP-2 (TABLE 3). SDS/PAGE analysis showed that the free proMMP-2 was converted to 69 kDa and 67 kDa by APMA and both species could react with α_2M (FIG. 2A). A small amount of the band remained at 67 kDa, but it is not clear whether it is a portion of MMP-2 dissociated from the α_2M–MMP-2 complex or whether it is the MMP-2–TIMP-2 complex that is resistant from binding to α_2M. In contrast, when the proMMP-2–TIMP-2 complex was treated with APMA at 37 °C, no significant amount of proMMP-2 was shifted to the lower molecular species even after 24 h (FIG. 2B). The APMA-treated complex

TABLE 3. Gelatinolytic Activity of proMMP-2 and the proMMP-2–TIMP-2 Complex upon Treatment with APMA[a]

Treatment	MMP-2 Activity	
	ProMMP-2 (%)	ProMMP-2–TIMP-2 (%)
Nontreatment	0	0
APMA (1 mM)		
30 min	100	4
60 min	70	2
240 min	30	0

[a]Activation reactions were carried out as in TABLE 2.

failed to react with α_2M, indicating that they were not proteolytically active. These results suggest that when the proMMP-2–TIMP-2 complex is treated with APMA, the catalytic center of the zymogen becomes unmasked, possibly by dissociating the interaction of the zinc atom at the catalytic site and the unique cysteine in the propeptide; however, the activated zymogen is readily inhibited by TIMP-2 prior to propeptide being processed. Upon APMA treatment, TIMP-2 in the complex lost its

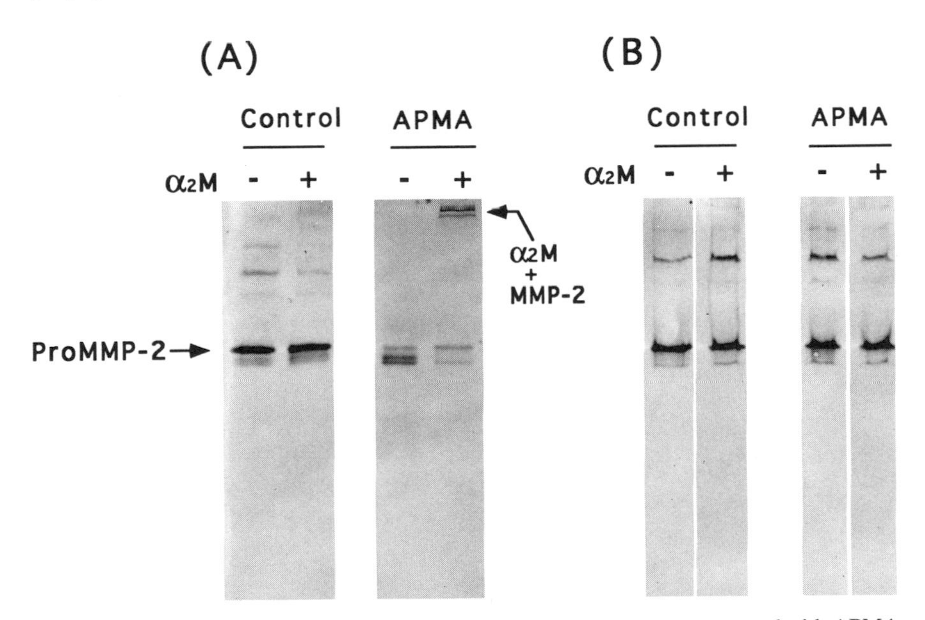

FIGURE 2. Inability to bind to α_2M of the proMMP-2–TIMP-2 complex treated with APMA. (**A**) proMMP-2 (control) and proMMP-2 treated with 1 mM APMA for 60 min at 37 °C were subjected to Western blotting analysis before and after complexing with a 4-molar excess of α_2M. The proteins were visualized using rabbit anti-(human MMP-2) antibody. The majority of the APMA-activated MMP-2 bound to α_2M. (**B**) The proMMP-2–TIMP-2 complex was treated with 1 mM APMA for 4 h at 37 °C, and then the sample was subjected to Western blotting analysis for MMP-2 before and after reacting with a 4-molar excess of α_2M. ProMMP-2 of the complex was neither converted to a low molecular mass species nor bound to α_2M even after APMA treatment. Control, sample without APMA treatment.

inhibitory capacity against other MMPs, as in the case of the proMMP-9–TIMP-1 complex.

On the other hand, when the proMMP-2–TIMP-2 complex contained a small amount of free proMMP-2, proMMP-2 was slowly converted to the 67-kDa species. However, the majority of the 67-kDa species was not able to complex with α_2M (FIG. 3), suggesting that the 67-kDa form is inhibited by TIMP-2. The slow conversion of the 72-kDa proMMP-2 to the 67-kDa MMP-2 is probably mediated by free proMMP-2 that was rapidly activated to the 67-kDa MMP-2 by APMA. The latter form can then readily bind to TIMP-2 of the proMMP-2–TIMP-2 complex and may release an activated proMMP-2 which is then converted to 67 kDa. The repeat of this cycle may eventually result in the conversion of proMMP-2 to a 67-kDa form, but it is inhibited by TIMP-2. Another possibility is that MMP-2 directly removes the propeptide from the activated proMMP-2 complexed with TIMP-2. To examine the possibility that active MMP-2 facilitates the conversion of proMMP-2 in the complex, the proMMP-2–TIMP-2 complex containing approximately 10% free proMMP-2 was treated with APMA in the absence or in the presence of α_2M (FIG. 4). Without α_2M, proMMP-2 was slowly converted to the 67-kDa species. In the presence of α_2M, however, a majority of proMMP-2 remained as pro-form even after a 24-h treatment with 1 mM APMA at 37 °C, and neither the 72-kDa proMMP-2 or the 67-kDa species bound to α_2M. This indicates that the active MMP-2 was rapidly complexed with α_2M and could not participate in the processing of the proMMP-2 in the complex.

Our studies using α_2M as a reagent to identify active MMPs indicated that both the proMMP-9–TIMP-1 and the proMMP-2–TIMP-2 complexes do not exhibit an appreciable proteolytic activity upon activation by trypsin or APMA. Both zymogens are activated by these treatments, but activated enzyme is readily inhibited by the TIMP that is bound to the proenzyme. However, the modes of interaction of the

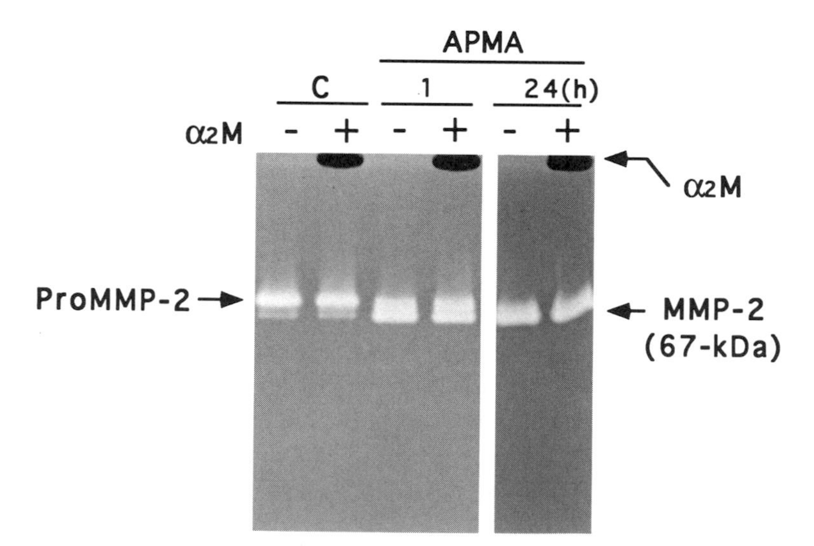

FIGURE 3. Processing of the proMMP-2–TIMP-2 complex by APMA in the presence of a small amount of free proMMP-2. The proMMP-2–TIMP-1 complex containing about 10% proMMP-2 was treated with 1 mM APMA for 1 h and 24 h. The samples were subjected to zymography before and after reacting with a 4-molar excess of α_2M, as in FIGURE 1. C, sample without APMA treatment.

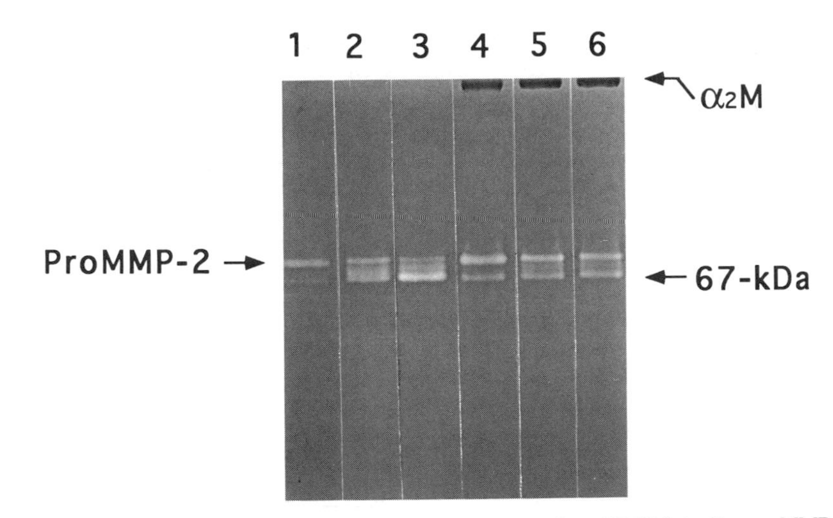

FIGURE 4. The lack of the APMA-induced conversion of proMMP-2 in the proMMP-2–TIMP-2 complex in the presence of α_2M. The proMMP-2–TIMP-2 complex preparation containing about 10% proMMP-2 was treated with 1 mM APMA for various periods of time at 37 °C in the absence (*lanes 1–3*) and presence (*lanes 4–6*) of a 4-molar excess of α_2M. The samples were subjected to zymographic analysis as in FIGURE 1. **Lanes 1 and 4,** without APMA treatment; **lanes 2 and 5,** treated with APMA for 1 h; **lanes 3 and 6,** treated with APMA for 24 h.

activated gelatinases with respective TIMPs are different. In the case of proMMP-9, the zymogen is processed to low molecular mass species, whereas the processing of proMMP-2 is inhibited by TIMP-2 upon activation. These differences may be due to how the progelatinases are complexed with respective TIMPs and difference in the rate of interaction of the activated zymogen/enzyme and the inhibitor.

ACKNOWLEDGMENTS

We thank Lori Blunt for typing the manuscript.

REFERENCES

1. WOOLLEY, D. E., D. R. ROBERTS & J. M. EVANSON. 1976. Nature **261:** 325–327.
2. BARRETT, A. J. 1981. Methods Enzymol. **80:** 737–754.
3. SOTTRUP-JENSEN, L. 1989. J. Biol. Chem. **264:** 11539–11542.
4. BARRETT, A. J. & P. M. STARKEY. 1973. Biochem. J. **133:** 709–724.
5. ENGHILD, J. J., G. SALVESEN, K. BREW & H. NAGASE. 1989. J. Biol. Chem. **264:** 8779–8785.
6. MILLER, E. J., E. D. HARRIS, JR., E. CHUNG, J. E. FINCH, JR., P. A. MCCROSKERY & W. T. BUTLER. 1976. Biochemistry **16:** 781–792.
7. HOFMANN, H., P. P. FIETZEK & K. KÜHN. 1978. J. Mol. Biol. **125:** 137–165.
8. DIXIT, S. N., C. L. MAINARDI, J. M. SEYER & A. H. KANG. 1979. Biochemistry **18:** 5416–5422.
9. FOSANG, A. J., K. LAST, V. KNÄUPER, P. J. NEAME, G. MURPHY, T. E. HARDINGHAM, H. TSCHESCHE & J. A. HAMILTON. 1993. Biochem. J. **295:** 273–276.

10. NGUYEN, Q., G. MURPHY, C. E. HUGHES, J. S. MORT & P. J. ROUGHLEY. 1993. **295:** 595–598.
11. NAGASE, H., Y. OGATA, K. SUZUKI, J. J. ENGHILD & G. SELVESON. 1991. Biochem. Soc. Trans. **19:** 715–718.
12. KITZ, R. & I. B. WILSON. 1962. J. Biol. Chem. **237:** 3245–3249.
13. WELGUS, H. G., J. J. JEFFREY & A. Z. EISEN. 1981. J. Biol. Chem. **256:** 9511–9515.
14. NAGASE, H. & E. D. HARRIS, JR. 1983. J. Biol. Chem. **258:** 7490–7498.
15. SOTTRUP-JENSEN, L. & H. BIRKEDAL-HANSEN. 1989. J. Biol. Chem. **264:** 393–401.
16. CAWSTON, T. E. & E. MERCER. 1986. FEBS Lett. **209:** 9–12.
17. ABE, S., M. SHINMEI & Y. NAGAI. 1973. J. Biol. Chem. **73:** 1007–1011.
18. VIRCA, G. D., R. K. MALLYA, M. B. PAPYS & H. P. SCHNEBLI. 1982. Adv. Exp. Med. Biol. **167:** 345–353.
19. FLORY, E. D., B. J. CLARRIS & K. D. MURIDEN. 1982. Ann. Rheum. Dis. **41:** 520–526.
20. WILHELM, S. M., I. E. COLLIER, B. L. MARMER, A. Z. EISEN, G. A. GRANT & G. I. GOLDBERG. 1989. J. Biol. Chem. **264:** 17213–17221.
21. MOLL, U. M., G. L. YOUNGLEIB, K. B. ROSINSKI & J. P. QUIGLEY. 1990. Cancer Res. **50:** 6162–6170.
22. GOLDBERG, G. I., B. L. MARMER, G. A. GRANT, A. Z. EISEN, S. M. WILHELM & C. HE. 1989. Proc. Natl. Acad. Sci. USA **86:** 8207–8211.
23. KLEINER, D. E., H. C. KRUTZSCH, E. J. UNSWORTH & W. G. STETLER-STEVENSON. 1992. Biochemistry **31:** 1665–1672.
24. KOLKENBROCK, H., D. ORGEL, A. HECKER-KIA, W. NOACK & N. ULBRICH. 1991. Eur. J. Biochem. **198:** 775–781.
25. OKADA, Y., T. MORODOMI, J. J. ENGHILD, K. SUZUKI, A. YASUI, I. NAKANISHI, G. SALVESEN & H. NAGASE. 1990. Eur. J. Biochem. **194:** 721–730.
26. MORODOMI, T., Y. OGATA, Y. SASAGURI, M. MORIMATSU & H. NAGASE. 1992. Biochem. J. **285:** 603–611.

Interaction of Matrix Metalloproteinases with Serine Protease Inhibitors

New Potential Roles for Matrix Metalloproteinase Inhibitors[a]

SATISH K. MALLYA,[b] JOSEPH E. HALL,[c] HSI-MING LEE,[d] ELIZABETH J. ROEMER,[e] SANFORD R. SIMON,[b,e,f] AND LORNE M. GOLUB[d]

Departments of Biochemistry and Cell Biology,[b] Pediatrics,[c] Oral Biology and Pathology,[d] and Pathology[e]
State University of New York at Stony Brook
Stony Brook, New York 11794

Proteolytic activity is responsible for inflammatory damage to connective tissues in a number of acute and chronic processes involving such diverse organs as the lungs, heart, joints, skin, and periodontium. Regardless of the nature of the inflammatory stimulus, a significant contribution to the tissue damage associated with inflammatory injury appears to come from unregulated degradation by the proteases from activated leukocytes. Neutrophil elastase (NE) is an especially potent protease which can degrade multiple components of the interstitial matrix *in vitro* and which has been implicated in inflammatory tissue damage *in vivo*. In the plasma and the interstitial stroma, the major endogenous defense against excessive NE activity comes from α_1-antitrypsin (α_1-protease inhibitor, α_1-PI), a plasma serine protease inhibitor, or serpin, which is present in interstitial fluid at about half its normal plasma concentration of 1–2 mg/mL. We recently demonstrated that α_1-PI can be bound tightly to a complete interstitial extracellular matrix (ECM) *in vitro* with retention of its antielastase activity.[1] Such ECM-bound antiprotease activity could contribute, along with fluid phase inhibitor, to the overall defenses against proteolytic damage to stromal tissues *in vivo*. Several laboratories have shown that fluid phase α_1-PI can be completely inactivated by proteolytic cleavage of a constrained loop in the native molecule. A number of matrix metalloproteinases (MMPs), such as neutrophil collagenase (MMP-8), can proteolyze the loop in fluid phase α_1-PI at specific sites[2–5] in addition to cleaving matrix proteins. In this communication, we have extended the earlier fluid phase studies to ECM-bound α_1-PI to determine whether matrix binding affects the sensitivity of this serpin to inactivation by MMPs. It is our hypothesis that inhibition of the MMP-8 activity of activated neutrophils could be of therapeutic value in helping to sustain the normal protective serpin levels within the interstitium while also blocking collagenolytic activity. One such class of inhibitors of MMP activity that could be useful for this application is the tetracy-

[a]This work was supported by grants from the Smokeless Tobacco Research Council, SUSB Biotechnology Center (NYS Office of Science and Technology), Collaborative Laboratories, Cortech Inc., and National Institutes of Health (NHLBI R01-HL-14262 to S.R.S. and NIDR R37-DE-03987 to L.M.G.).

[f]Address correspondence to Sanford R. Simon, Ph.D., Department of Pathology, State University of New York at Stony Brook, Stony Brook, New York 11794-8691.

clines.[6] Although evidence has already been presented to support the ability of tetracyclines to inhibit the MMP-mediated inactivation of α_1-PI in solution,[7] the extent of such inhibition within the environment of the interstitial ECM has remained speculative up until now.

We employed the ECM secreted by R22 rat heart smooth muscle cells as a model system to study hydrolysis of native interstitial tissues.[1,8] In previous studies from our laboratory, we found that R22 ECM hydrolysis by purified NE[1] or by viable human neutrophils[9] is markedly inhibited by α_1-PI, suggesting that NE contributes significantly to degradation in this *in vitro* system. In the current study, we present evidence that a combination of α_1-PI and doxycycline is even more effective in preventing the degradation of R22 ECM by activated human neutrophils. Similar results have been obtained when R22 ECM degradation by MMP-8 was measured in the presence of chemical inactivators of any contaminating serine proteases. These findings indicate that doxycycline protects the interstitial matrix from neutrophil-mediated degradation by inhibiting matrix metalloproteinases, including MMP-8. This inhibition of MMPs not only blocks collagenolysis but also prevents α_1-PI from becoming proteolytically inactivated, thereby preserving its anti-NE activity and further protecting the matrix. The findings illustrate the complex interplay between MMPs, serine proteases, and the antiproteases specific for these two classes of enzymes in determining the extent of inflammatory injury to stromal tissues. The results also suggest that protease inhibitors to be employed as therapeutic agents may reduce such injury by affecting this interplay, and may therefore have effects beyond simple targeting of a single protease.

MATERIALS AND METHODS

ECM was from R22 rat smooth muscle cells cultured in 24-well polystyrene microplates in the presence of ascorbic acid and [^{3}H]proline to metabolically label the matrix as described by Jones *et al.*,[10] with our modifications.[1,8] We have previously shown that this ECM contains 37% collagen, 12% elastin, and 51% proteoglycans and noncollagenous glycoproteins. Human neutrophils were obtained from leukocyte concentrates by the method of Kalmar *et al.*[11] and suspended in Hanks' balanced salt solution (HBSS). Neutrophil-mediated ECM hydrolysis was monitored by adding a suspension of 5×10^5 neutrophils/cm^2 ECM surface area (1×10^6 neutrophils/well) to the matrix in each well of a 24-well microplate. Cells were activated by addition of 25 nM phorbol myristate acetate (PMA). Neutrophil-mediated ECM degradation was generally allowed to proceed for 4–6 h at 37 °C, during which time minimal loss of cell viability occurred. Neutrophil lysate was prepared by sonicating 5×10^8 cells in 5 mL of 50 mM Tris, 10 mM $CaCl_2$, 1.0 M NaCl, 0.05% Brij-52, pH 7.5, followed by seven consecutive cycles of freezing in dry ice/acetone and thawing at 37 °C. The mixture was diluted with an additional 5 mL of NaCl-free buffer and incubated with protease-free DNAase to reduce viscosity. The DNAase-treated suspension was centrifuged at $12{,}000 \times g$ for 2 h, and the supernatant was collected for use in the experiments described below. A volume of 150 μL of lysate was diluted with an equal volume of buffer in each well of an ECM-coated microplate. Human MMP-8 was purified as described by Mookhtiar and Van Wart;[12] 150 μL aliquots of the purified enzyme preparation diluted with an equal volume of buffer were added to individual wells of the ECM-coated plates as described for the neutrophil lysates. PMSF was included during the purification of MMP-8 to ensure that no contaminating serine protease activity was present in the purified enzyme preparation.[12] In experiments in which ECM-bound α_1-PI was present, the wells of an ECM-coated plate were first incubated with 600 μg/mL α_1-PI for 4 h at 37 °C, after which all unbound protein was removed with multiple washes prior to addition of neutrophils. We have previously shown that the ECM becomes saturated with about 0.01 mg tightly bound active inhibitor per mg ECM protein after incubation with this concentration of α_1-PI.[1] Fluid phase α_1-PI, when present, was added at a concentration of 200 μg/mL. Supernatants were removed from the wells after incubation with neutrophils, lysates, or purified MMP-8 and centrifuged at 10,000 rpm in a microfuge prior to addition to liquid scintillation fluid for counting. In most cases, the maximum radioactivity released from the matrix over the interval studied (4–18 h) was less than 25% of the total incorporated label, as determined by solubilizing the matrix with 2N NaOH. When results are expressed as the percentage of control, the control was defined as radioactivity released in the absence of any inhibitor, and was normalized to 100%.

RESULTS

Doxycycline in Combination with α_1-PI Inhibits Neutrophil-mediated ECM Solubilization

When activated with PMA, human neutrophils degraded the R22 ECM about twice as rapidly as in the absence of exogenous stimuli, as shown in FIGURE 1A. At a final concentration of 20 μM, doxycycline inhibited the PMA-activated neutrophil-mediated degradation of this ECM only modestly. In a separate experiment, about 25% inhibition of PMA-activated neutrophil-mediated ECM degradation was achieved with 50 μM doxycycline (FIG. 1B). α_1-PI, a physiological inhibitor of NE, was a more potent inhibitor of activated neutrophil-mediated ECM degradation and inhibited ECM solubilization by about 50%, regardless of whether 200 μg/mL fluid phase inhibitor was present or only ECM-bound inhibitor was present after preincubating the matrix with 600 μg/mL α_1-PI and removing fluid phase inhibitor. When relatively low concentrations of doxycycline and α_1-PI were both present, the combination protected the matrix from degradation by at least 60%. In the presence of 20 μM doxycycline and 200 μg/mL fluid phase α_1-PI, the protection afforded by the combination of inhibitors was especially evident and was greater than the sum of the effects of the inhibitors added separately (FIG. 1A).

Effects of Doxycycline on ECM Hydrolysis by Neutrophil Extracts

Neutrophil lysate prepared in the absence of any protease inhibitors contains active MMPs, including collagenase (MMP-8) and gelatinase (MMP-9), and active serine proteases, including NE and cathepsin G. FIGURE 2A shows that even 100 μM doxycycline was not effective in inhibiting R22 ECM hydrolysis by lysate which was not treated with PMSF. This result reflects the great susceptibility of the complete R22 interstitial ECM to degradation by NE (which is resistant to inhibition by doxycycline[10]), and the relatively modest susceptibility to MMP-8–mediated degradation. ECM hydrolysis by purified NE was not inhibited by doxycycline, even at 100 μM (data not shown).

By treating the neutrophil lysate with 1 mM PMSF to inhibit NE and cathepsin G activity, the effect of doxycycline on ECM degradation by metalloproteinases in the extract (which are unaffected by this irreversible inhibitor) could be measured in the absence of contaminating serine proteases. FIGURE 2B shows that doxycycline at a concentration as low as 5 μM inhibited ECM hydrolysis by PMSF-treated neutrophil

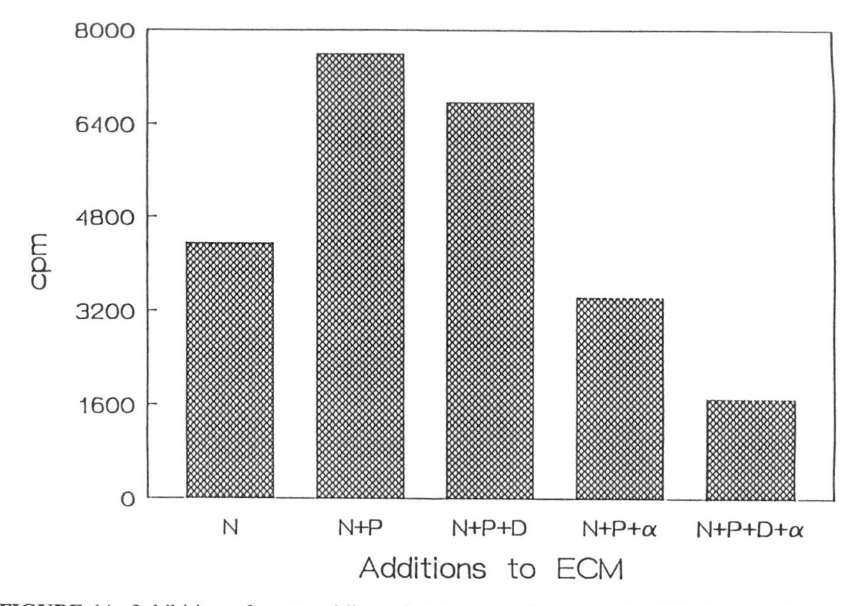

FIGURE 1A. Inhibition of neutrophil-mediated extracellular matrix (ECM) degradation by α_1-PI and doxycycline. Aliquots of 1×10^6 neutrophils in 1 mL HBSS were added to ECM-containing wells for 6 h in the absence of added agents (N), with 125 nM PMA alone (N + P), with PMA + 20 μM doxycycline (N + P + D), with PMA + 200 μg/mL α_1-PI (N + P + α), and with PMA + α_1-PI + doxycycline (N + P + D + α). Supernatants were removed and assayed for radioactivity as described in MATERIALS AND METHODS. Absolute counts are shown. PMA, phorbol myristate acetate; α,-PI, α_1-protease inhibitor.

lysates by 30%, and at higher concentrations (25–50 μM) inhibited degradation by 75–80%. In these experiments, the inhibitor was added to the ECM prior to addition of the neutrophil lysates. The IC_{50} value for doxycycline using this assay lies between 5 and 25 μM, and agrees well with that obtained for the inhibition of hydrolysis of soluble type I collagen by MMP-8. The relatively weak activity of MMP-8 compared to NE in degrading the R22 ECM is evidenced by the fact that approximately the same extent of matrix solubilization was achieved in 24 h with the PMSF-treated neutrophil lysate as was achieved in only 4 h with the same volume of lysate in which serine protease activity was still present.

Inhibition of MMP-8–mediated ECM Degradation by Mixtures of α_1-PI and Doxycycline

At least two reports have suggested that α_1-PI may be as good a substrate for MMP-8 as soluble collagen.[4,13] To see if ECM-bound α_1-PI might compete with the matrix proteins for MMP-8, the R22 ECM was first incubated with 600 μg/mL α_1-PI for 4 h to saturate all the binding sites on the matrix, and the wells were washed five times to remove any unbound α_1-PI. FIGURE 3 shows that ECM-bound α_1-PI could protect the ECM from MMP-8–mediated hydrolysis, reducing the rate of matrix solubilization by 50% even in the presence of 1 mM PMSF. This unexpected result presumably reflects the capacity of α_1-PI to act as a competitive substrate for MMP-8. Addition of 100 μM doxycycline to the α_1-PI–pretreated ECM prior to addition of MMP-8 reduced the rate of matrix solubilization by over 95%, effectively inhibiting all MMP-8–mediated proteolysis. This result demonstrates that doxycycline in combination with α_1-PI is effective in protecting the matrix from attack by purified MMP-8.

DISCUSSION

During the inflammatory response, activated neutrophils and macrophages release proteases that can cause tissue damage in a variety of settings.[14–16] Leukocyte proteolytic activity in the interstitium is controlled under normal conditions partly by protein inhibitors derived from plasma, including α_1-PI (as well as lower amounts of other serine protease inhibitors or "serpins") and very low levels of α_2-macroglobulin (which is too large to pass the endothelial barrier in the absence of inflammation). Tissue inhibitors of metalloproteinases (TIMPs) are released with MMPs by mesenchymal cells within connective tissues and have been visualized in interstitial ECM by immunochemical techniques.[17] It has been argued that a critical balance of proteases and antiproteases in tissues is pivotal for maintaining the architecture of the normal interstitial stroma, and any factor that tips the balance towards unchecked enzyme activity can contribute to tissue injury.[18–20]

Reactive oxygen metabolites, especially HOCl released during the respiratory burst of neutrophils, have been proposed to tip the protease-antiprotease balance, both by oxidizing serpins to less potent species and by activating the precursors of

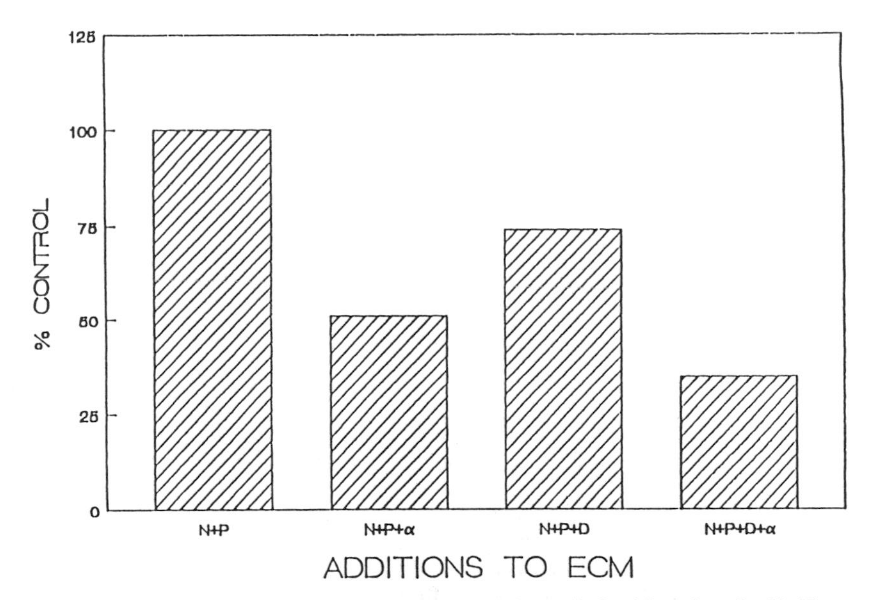

FIGURE 1B. Neutrophils were added to ECM-containing wells for 4 h as described in FIGURE 1A. Except when α_1-PI was present, wells were preincubated with 600 μg/mL α_1-PI for 4 h and then rinsed repeatedly to remove all free inhibitor prior to addition of neutrophils. Neutrophils (1×10^6) in 1 mL HBSS were added with 125 nM PMA alone (N + P), with PMA + ECM-bound α_1-PI (N + P + α), with PMA + 50 μM doxycycline (N + P + D), and with PMA + ECM-bound α_1-PI + doxycycline (N + P + D + α).

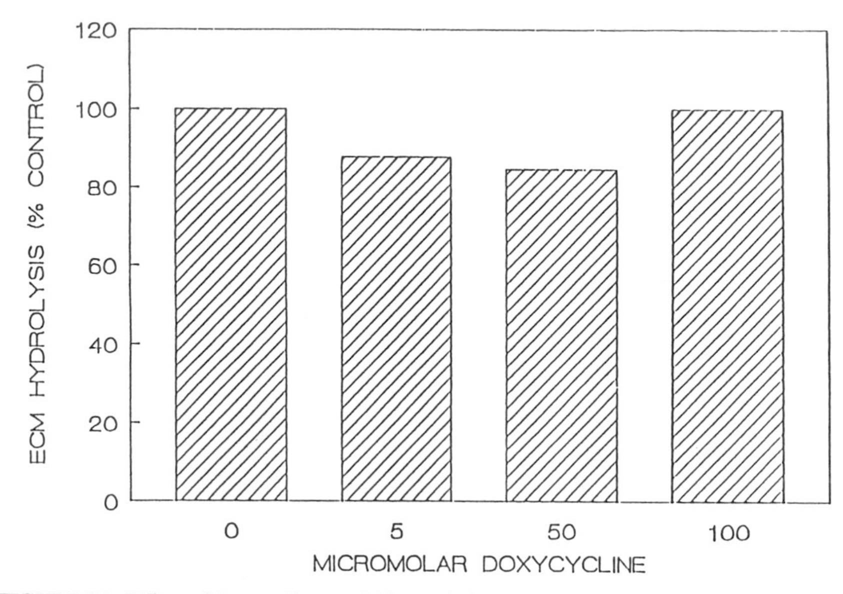

FIGURE 2A. Effect of doxycycline on R22 matrix hydrolysis by neutrophil lysate. Neutrophil lysate (150 μL) was diluted with an equal volume of 50 mM Tris, 10 mM $CaCl_2$, 0.2 M NaCl, 0.05% Brij-52 pH 7.5 and was then incubated in each ECM-containing well for 4 h at 37 °C. The different concentrations of doxycycline were added at the time of addition of the diluted lysates to the wells. ECM, extracellular matrix.

MMPs.[21] More recently, however, some questions have arisen over the significance of oxidative mechanisms in shifting the protease-antiprotease balance, while alternative proteolytic cascade mechanisms have been proposed.[22] Examples of such mechanisms include activation of pro-MMP-8 by cathepsin G and inactivation of TIMP by NE. As an example of shifting the protease-antiprotease balance, the "serpinase" activity of the MMPs, in which the reactive loops of serine protease inhibitors such as α_1-PI, α_1-antichymotrypsin, antithrombin III, and α_2-antiplasmin are cleaved, may be an important part of their normal physiological function, especially within the microenvironment of the ECM, where α_2-macroglobulin, the major plasma MMP inhibitor, is normally excluded.[23] In recent years a number of laboratories have shown that MMPs can inactivate α_1-PI in solution by cleaving it at a site in the reactive loop near Met-358.[3–5,13] Several other examples of serpinase activity by MMPs have also been reported.[23] The MMPs can therefore potentially contribute to tissue damage in two ways, *viz.*, by directly hydrolyzing ECM components as well as by inactivating serpins, thereby facilitating ECM hydrolysis by the serine proteases normally targeted by these inhibitors. Our results indicate that the relative importance of the collagenolytic and serpinolytic activities of MMP-8 may be a function of the nature of the microenvironment in which the enzyme encounters these competing substrates. It has been estimated that these two activities of MMP-8 are approximately equivalent (within a factor of 2) when hydrolysis of type I collagen and α_1-PI in solution are being measured.[4,13] On the other hand, it has been calculated that MMP-8 catalyzes the hydrolysis of insoluble type I collagen fibrils at only one-twentieth its rate of cleavage of soluble collagen into 1/4-length and 3/4-length fragments.[24] Within the interstitial ECM, the collagen in fibrils may be even less accessible to MMP-8 due to the presence of accessory proteins coating the fibril surface.[25] It is therefore possible that even in the presence of a large excess of ECM proteins, the relatively modest fraction of ECM-bound α_1-PI may compete effectively with fibrillar collagen within the matrix as a substrate for MMP-8.

In addition to proteolytic inactivation of ECM-bound α_1-PI by neutrophil-derived MMP activity, we also reported a serpin-degrading activity within the R22 ECM that appears to be MMP-like as well.[1] This activity results in significant reduction in the amounts of intact, active α_1-PI which can be bound to the R22 ECM and the appearance of radiolabel from the α_1-PI in the supernatant medium. The amounts of α_1-PI which remain bound to the ECM are markedly increased if the matrix is pretreated with 1,10-phenanthroline, or with doxycycline. Pretreatment of the matrix with 1,7- or 4,7-phenanthroline (nonchelating isomers of the 1,10-species) does not result in augmented binding of serpins to the R22 ECM, leading us to conclude that the reduced binding capacity and the release of serpin-associated counts into the supernatant medium reflect a zinc-dependent MMP-like activity associated with the matrix. We have not yet characterized this endogenous MMP by zymographic analysis, but it appears to have very little activity against the metabolically labeled components of the ECM itself, because virtually no spontaneous autolysis of the matrix can be detected.

If MMPs can regulate the activity of serine proteases by hydrolyzing the ECM-associated serpins, then inhibition of the MMPs could be of therapeutic value in helping to sustain the normal protective antiprotease levels along with direct inhibition of collagenolytic activity. This appears to be a fundamental underlying mechanism for the therapeutic efficacy of the tetracyclines.[6,7,26] FIGURE 4 illustrates

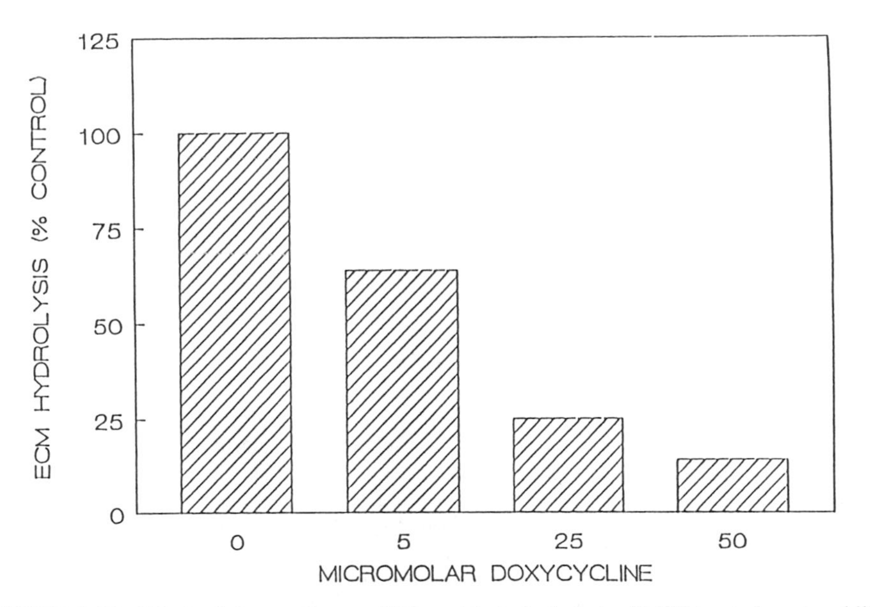

FIGURE 2B. Effect of doxycycline on R22 matrix hydrolysis by PMSF-treated neutrophil lysate. Neutrophil lysate (150 μL) was diluted with an equal volume of 50 mM Tris, 10 mM $CaCl_2$, 0.2 M NaCl, 1 mM PMSF, 0.05% Brij-52 pH 7.5 and was then incubated in each ECM-containing well for 24 h at 37 °C. The different concentrations of doxycycline were added at the time of addition of the diluted PMSF-treated lysates to the wells.

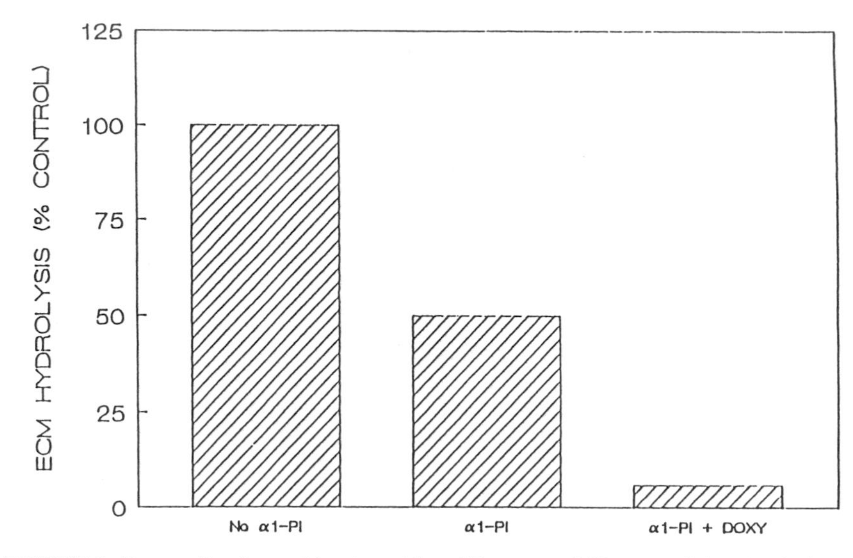

FIGURE 3. Doxycycline in combination with α_1-PI protects R22 extracellular matrix (ECM) from hydrolysis by MMP-8. α_1-PI (600 μg/mL) was preincubated with ECM for 4 h and unbound serpin was rinsed away. MMP-8 was then added in the presence or absence of 100 μM doxycycline, and the extent of ECM hydrolysis after incubation for 18 h was determined. The assays were performed with 150 μL MMP-8 diluted with an equal volume of 50 mM Tris, 10 mM $CaCl_2$, 0.2 M NaCl, 1 mM PMSF, 0.05% Brij-52 in each ECM-containing well at 37 °C.

the potential therapeutic benefit of tetracyclines in restoring the overall protease-antiprotease balance at multiple points under conditions of the neutrophil-mediated inflammatory response. Support for a role of the tetracyclines in restoring the protease-antiprotease balance comes from studies with tetracycline analogues that have no antibiotic activity but are still effective in reducing tissue damage in inflammatory models.[6,26,27] Such tetracycline analogues may also have effects on turnover of ECM components in addition to their capacity to inhibit MMP activity, but evidence such as that reported here suggests that even the antiproteolytic action of tetracyclines may have multifactorial consequences. By inhibiting the MMP-mediated cleavage of α_1-PI, for example, these compounds could reduce formation of the proteolytic fragments of this serpin which have been shown to possess chemotactic activity for neutrophils.[28]

We used the R22 matrix as a model to study ECM degradation during inflammation. This interstitial matrix is an excellent model system because all components are represented in their native form and the ultrastructure closely resembles that of stromal connective tissue.[1,8,10,25] We previously determined that this matrix contains almost 50% collagen (types I and III) and elastin.[1,8] The R22 cells can be cultured in the presence of labeled precursor metabolites to yield ECM in which components may be selectively or preferentially labeled. In these studies, we employed [^{3}H]proline to preferentially label types I and III collagen fibrils. Inclusion of ascorbic acid in the R22 cell culture medium as a cofactor favors production of a relatively collagen-rich matrix which is insoluble even in strong detergents. The hydrolysis of this matrix by neutrophils or purified enzymes can be quantitated by measuring release of radioactivity in the supernatant. Because the connective tissue elements in the R22

ECM retain their native organization, such measurements of degradation reflect more accurately the capacity of cells or enzymes to attack interstitial stroma rather than might be concluded from studies with more artificial systems containing purified individual ECM components stripped of their accessory proteins or with synthetic oligopeptide chromogenic substrates. We stress this distinction because inferences drawn from more classical enzymological studies with single soluble substrates may not always be applicable to the more complex multicomponent organization present within the interstitial ECM.

We have shown that α_1-PI can bind in a saturable quasi-irreversible adsorption process to this matrix and that bound inhibitor retains both the capacity to form a stable complex with NE and to inhibit NE-mediated ECM degradation.[1] In this report we have focused on the effect of ECM-bound α_1-PI in protecting the matrix from degradation by viable neutrophils, rather than by purified NE. Other reports have noted that concentrations of α_1-PI in excess of 250 μg/mL are effective in blocking migration of human neutrophils through an endothelial layer and underlying basement membrane in an *in vitro* "vessel wall" construct.[29] We are currently investigating the effects of ECM-bound α_1-PI and doxycycline, both separately and in combination, on migration of neutrophils through the R22 ECM on porous membrane filters. It is clear from previous work with other *in vitro* models of inflammatory cell invasiveness that predictions based solely on the activity of purified constituent proteases of leukocytes may not be borne out by results with viable cells.[30]

In preliminary studies, we examined the effects of MMPs and doxycycline on binding and degradation of another serpin, plasminogen activator inhibitor-1 (PAI-1), to the R22 ECM. PAI-1 is an effective inhibitor of urokinase and has been implicated in regulating macrophage infiltration and degradative activity, which is plasminogen- (or plasmin-) dependent.[31] Its association with basement membranes is well documented, although binding to interstitial ECM has been less extensively studied.[32] As was the case for α_1-PI, the amounts of PAI-1 that can be bound to the R22 interstitial ECM *in vitro* are markedly augmented if the matrix is pretreated with

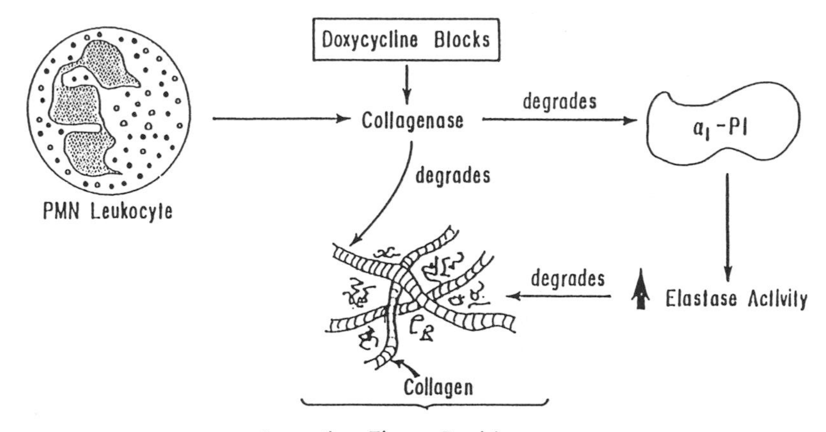

FIGURE 4. A model showing the role of doxycycline in preventing neutrophil-mediated extracellular matrix degradation. Doxycycline inhibits collagenase, thereby preventing it from hydrolyzing the matrix. This inhibition also results in protection of α_1-PI which in turn inhibits hydrolysis of the matrix by neutrophil elastase. α_1-PI, α_1-protease inhibitor.

doxycycline or 1,10-phenanthroline, but nonchelating phenanthroline isomers have no effect on PAI-1 binding. We have interpreted these results, like those on binding of α_1-PI to R22 ECM, as reflecting the action of an endogenous noncollagenolytic MMP-like activity in the ECM which has serpinolytic activity. Unlike α_1-PI, however, PAI-1 appears to be resistant to degradation by MMP-8 or MMP-9, either in fluid phase or when bound to the R22 ECM. Instead, like α_1-antichymotrypsin, PAI-1 is degraded by neutrophil elastase. We hypothesize that preservation of high levels of active interstitial α_1-PI (as could be achieved by inhibition of MMPs with tetracyclines) will also indirectly protect other serpins, such as PAI-1 and α_1-antichymotrypsin, which would otherwise be degraded by unregulated NE-mediated proteolysis.

Extensive evidence for inhibition of MMP activities by tetracyclines in a number of *in vivo* systems has been well documented, but *in vitro* models for these inflammatory processes have been somewhat limited. All our data indicate that doxycycline inhibits R22 interstitial matrix degradation by both activated neutrophils as well as by purified MMP-8. In studies not shown here, we have demonstrated that doxycycline is equally effective in protecting the ECM against degradation by MMP-8 added subsequently ("prophylactic") as it is in inhibiting ongoing MMP-8–mediated ECM degradation ("therapeutic"). This property is not shared by α_1-PI, which is less effective in inhibiting ongoing NE-mediated ECM degradation than it is in protecting the ECM against subsequent addition of NE.[33,34] The detailed molecular mechanism of inhibition by tetracyclines either *in vivo* or *in vitro* has not been elucidated. Previously published accounts have demonstrated that Zn^{2+} can reverse this inhibition,[35,36] and it seems plausible that tetracyclines could bind to the catalytic Zn in the various members of the MMP family. It still remains to be determined, however, whether doxycycline and other tetracyclines may also bind to activated leukocytes or to some elements of extracellular matrices, thereby preventing these proteases from even gaining access to their substrates. We are currently pursuing kinetic studies to resolve some of these questions.

REFERENCES

1. RINEHART, A. R., S. K. MALLYA & S. R. SIMON. 1993. Human α_1-proteinase inhibitor binds to extracellular matrix *in vitro* with retention of inhibitory activity. Am. J. Respir. Cell Mol. Biol. **9:** 666–679.
2. TRAVIS, J., A. DUBIN, J. POTEMA, W. WATOREK & A. KURDOWSKA. 1991. Neutrophil proteinases. Caution signs in designing inhibitors against enzymes with possible multiple functions. Ann. N. Y. Acad. Sci. **624:** 81–86.
3. KNAUPER, V., H. REINKE & H. TSCHESCHE. 1990. Inactivation of human α_1-proteinase inhibitor by human PMN collagenase. FEBS Lett. **263:** 355–357.
4. MICHAELIS, J., M. C. M. VISSERS & C. C. WINTERBOURN. 1990. Human neutrophil collagenase cleaves α_1-antitrypsin. Biochem. J. **270:** 809–814.
5. WINYARD, P. G., Z. CHANG, K. CHIDWICK, D. R. BLAKE, R. W. CARRELL & G. MURPHY. 1991. Proteolytic inactivation of human α_1-antitrypsin by human stromelysin. FEBS Lett. **279:** 91–94.
6. GOLUB, L. M., R. A. GREENWALD, N. S. RAMAMURTHY, T. F. MCNAMARA & B. R. RIFKIN. 1991. Tetracyclines inhibit connective tissue breakdown: New therapeutic implications for an old family of drugs. Crit. Rev. Oral. Biol. Med. **2:** 297–322.
7. SORSA, T., O. LINDY, Y. T. KONTTINEN, K. SUOMALAINEN, T. INGMAN, H. SAARI, S. HALINEN, H.-M. LEE, L. M. GOLUB, J. E. HALL & S. R. SIMON. 1993. Doxycycline in the protection of serum alpha-1-antitrypsin from human neutrophil collagenase and gelatinase. Antimicrob. Agents Chemother. **37:** 592–594.
8. ROEMER, E. J., K. J. STANTON & S. R. SIMON. 1994. *In vitro* assay systems for inflammatory cell-mediated damage to interstitial extracellular matrix. In Vitro Toxicol. **7:** 75–81.

9. RINEHART, A. R. & S. R. SIMON. 1991. Human α_1-proteinase inhibitor binds to extracellular matrix *in vitro* with retention of inhibitory activity. J. Cell. Biochem. Suppl. **15G:** 152.
10. JONES, P. A., T. SCOTT-BURDEN & W. GEVERS. 1979. Glycoprotein, elastin, and collagen secretion by rat smooth muscle cells. Proc. Natl. Acad. Sci. USA **76:** 353–357.
11. KALMAR, J. R., R. R. ARNOLD, M. L. WARBINGTON & M. K. GARDNER. 1988. Superior leukocyte separation with a discontinuous one-step Ficoll-Hypaque gradient for the isolation of human neutrophils. J. Immunol. Methods **110:** 275–281.
12. MOOKHTIAR, K. A. & H. E. VAN WART. 1990. Purification of human neutrophil collagenase to homogeneity. Biochemistry **29:** 10623–10627.
13. DESROCHERS, P. E., K. MOOKHTIAR, H. E. VAN WART, K. A. HASTY & S. J. WEISS. 1992. Proteolytic inactivation of α_1-proteinase inhibitor and α_1-antichymotrypsin by oxidatively inactivated human neutrophil metalloproteinases. J. Biol. Chem. **267:** 5005–5012.
14. CERRA, F. B. 1992. Multiple organ failure syndrome. Disease-a-Month **38:** 847–895.
15. LEE, C. T., A. M. FAIN & M. LIPPMAN. 1981. Elastolytic activity in pulmonary lavage from patients with adult respiratory distress syndrome. N. Engl. J. Med. **304:** 192–196.
16. HORL, W. H. & A. HEIDLAND, ED. 1983. Proteases II: Potential role in health and disease. Adv. Exp. Med. Biol. **240.**
17. HOWARD, E. W. & M. J. BANDA. 1991. Binding of tissue inhibitor of metalloproteinases 2 to two distinct sites on human 72 kDa gelatinase. J. Biol. Chem. **266:** 17972–17977.
18. JANOFF, A. 1985. Elastases and emphysema: Current assessment of the protease-antiprotease hypothesis. Am. Rev. Respir. Dis. **132:** 417–423.
19. CRYSTAL, R. G. 1990. Alpha-1-antitrypsin deficiency, emphysema and liver disease: Genetic bases and strategies for therapy. J. Clin. Invest. **85:** 1434–1452.
20. PRYOR, W. A., M. DOOLEY & D. F. CHURCH. 1984. Inactivation of human plasma α_1-proteinase inhibitor by gas phase cigarette smoke. Biochem. Biophys. Res. Commun. **122:** 676–681.
21. WEISS, S. J. 1989. Tissue destruction by neutrophils. N. Engl. J. Med. **320:** 365–376.
22. BUCURENCI, N., B. R. BLAKE, K. CHIDWICK & P. G. WINYARD. 1992. Inhibition of neutrophil superoxide production by human plasma α_1-antitrypsin. FEBS Lett. **300:** 21–24.
23. MAST, A. E., J. J. ENGHILD, H. NAGASE, K. SUZUKI, S. V. PIZZO & G. SALVESEN. 1991. Kinetics and physiologic relevance of the inactivation of α_1-proteinase inhibitor, α_1-antichymotrypsin and antithrombin III by matrix metalloproteinase-1 (tissue collagenase) and -2 (72 kDa gelatinase/type IV collagenase) and -3 (stromelysin). J. Biol. Chem. **266:** 15810–15816.
24. HASTY, K. A., J. J. JEFFREY, M. S. HIBBS & H. G. WELGUS. 1987. Collagen substrate specificity of human neutrophil collagenase. J. Biol. Chem. **262:** 10048–10052.
25. MECHAM, R. P. & J. HEUSER. 1990. Three-dimensional organization of extracellular matrix in elastic cartilage as viewed by quick freeze deep etch electron microscopy. Connect. Tissue Res. **24:** 83–93.
26. GOLUB, L. M., T. F. MCNAMARA, G. D. D'ANGELO, R. A. GREENWALD & N. S. RAMAMURTHY. 1987. A non-antibacterial chemically modified tetracycline inhibits mammalian collagenase activity. J. Dent. Res. **66:** 1310–1314.
27. SUOMALAINEN, K., T. SORSA, L. M. GOLUB, N. S. RAMAMURTHY, H.-M. LEE, V.-J. UITTO, H. SAARI & Y. T. KONTTINEN. 1992. Specificity of the anticollagenase actions of the tetracyclines: Relevance to their anti-inflammatory potential. Antimicrob. Agents Chemother. **36:** 227–229.
28. BANDA, M. J., A. G. RICE, G. L. GRIFFIN & R. M. SENIOR. 1988. α_1-proteinase inhibitor is a neutrophil chemoattractant after proteolytic inactivation by macrophage elastase. J. Biol. Chem. **263:** 4481–4484.
29. WEISS, S. J. & S. REGIANI. 1984. Neutrophils degrade subendothelial matrices in the presence of α_1-proteinase inhibitor. J. Clin. Invest. **73:** 1297–1303.
30. FURIE, M. B., B. L. NAPRSTEK & S. C. SILVERSTEIN. 1987. Migration of neutrophils across monolayers of cultured microvascular endothelial cells. J. Cell Sci. **88:** 161–175.
31. SALONEN, E. M., A. VAHERI, J. POLLANEN, R. STEPHENS, P. ANDREASEN, M. F. MAYER, K. DANO, J. GALLIT & E. RUOSLAHTI. 1989. Interaction of plasminogen activator inhibitor-1 (PAI-1) with vitronectin. J. Biol. Chem. **264:** 6339–6343.

32. CHAPMAN, H. A., X. YANG, L. Z. SAILOR & D. J. SUGARBAKER. 1990. Developmental expression of plasminogen activator inhibitor type 1 by human alveolar macrophage: Possible role in lung injury. J. Immunol. **145:** 3398–3405.
33. MORRISON, H. M., H. G. WELGUS, R. A. STOCKLEY, D. BURNETT & E. J. CAMPBELL.1990. Inhibition of human leukocyte elastase bound to elastin: Relative ineffectiveness and two mechanisms of inhibitory activity. Am. J. Respir. Cell Mol. Biol. **2:** 263–269.
34. PADRINES, M. & J. G. BIETH. 1991. Elastin decreases the efficiency of neutrophil elastase inhibitors. Am. J. Respir. Cell Mol. Biol. **4:** 187–193.
35. LEE, H. M., J. E. HALL, T. SORSA, S. R. SIMON & L. M. GOLUB. 1992. Doxycycline inhibits PMN-mediated tissue breakdown in culture. J. Dent. Res. **71:** 245.
36. YU, L. P., G. N. SMITH, K. A. HASTY & K. D. BRANDT. 1991. Doxycycline inhibits type XI collagenolytic activity of extracts of human osteoarthritic cartilage and of gelantinase. J. Rheumatol. **18:** 1450–1452.

Low Molecular Weight Inhibitors in Corneal Ulceration[a]

RICHARD E. GALARDY,[b] MARIE E. CASSABONNE, CARLANNE GIESE, JAMES H. GILBERT, FRANCE LAPIERRE, HENRY LOPEZ, MARY E. SCHAEFER, ROBERT STACK, MICHAEL SULLIVAN, BRENT SUMMERS, ROB TRESSLER, DAVE TYRRELL, AND JENNIFER WEE[c]; SCOTT D. ALLEN AND JOHN J. CASTELLOT[d]; JOHN P. BARLETTA AND GREGORY S. SCHULTZ[e]; LEONARDO A. FERNANDEZ[f]; SUSAN FISHER AND TIAN-YI CUI[g]; HARALD G. FOELLMER[h]; DAMIAN GROBELNY[i]; AND WALTER M. HOLLERAN[j]

The matrix metalloproteinases (MMPs) have been implicated in a number of diseases involving inflammation or cellular invasion.[1] GM 6001 (FIG. 1) is an inhibitor of most of these enzymes with K_i's in the low nanomolar range. Though potent *in vitro,* this molecule is short-lived in circulation with a half-life of a few minutes.

We show here that topical GM 6001 prevents the infiltration of inflammatory cells into the alkali-burned rabbit cornea and into phorbol ester–stimulated mouse skin. It thus prevents ulceration in the former and psoriasis-like inflammation and proliferation in the latter. When given systemically it blocks the infiltration of cells into the peritoneal cavity of mice stimulated with thioglycollate. Topical administration of this drug inhibits angiogenesis in the chick chorioallantoic membrane. When given intravenously, it inhibits angiogenesis in rat corneas implanted with a pellet containing tumor extract, a process requiring penetration of vascular basement membrane by endothelial cells. Finally, systemic GM 6001 increases survival of mice in a B16-F10 melanoma metastasis model, presumably by inhibiting cellular invasion or tumor growth.

In addition to the potential for preventing direct destruction of connective tissue

[a] This work was supported in part by National Institutes of Health grant EY05587.

[b] Corresponding author.
[c] Glycomed Incorporated, 860 Atlantic Avenue, Alameda, California 94501.
[d] Department of Anatomy and Cell Biology, Tufts University School of Medicine, Boston, Massachusetts 02111.
[e] University of Florida, Gainesville, Florida 32610.
[f] Department of Diagnostic Radiology, Yale University School of Medicine, New Haven, Connecticut 06510.
[g] Department of Stomatology, University of California, San Francisco, California 94143.
[h] Department of Obstetrics and Gynecology, Yale University School of Medicine, New Haven, Connecticut 06437.
[i] 43 Brassey Avenue, Rosanna 3084, Victoria, Australia.
[j] Department of Dermatology, Veterans Administration Medical Center, 4150 Clement Street, San Francisco, California 94121.

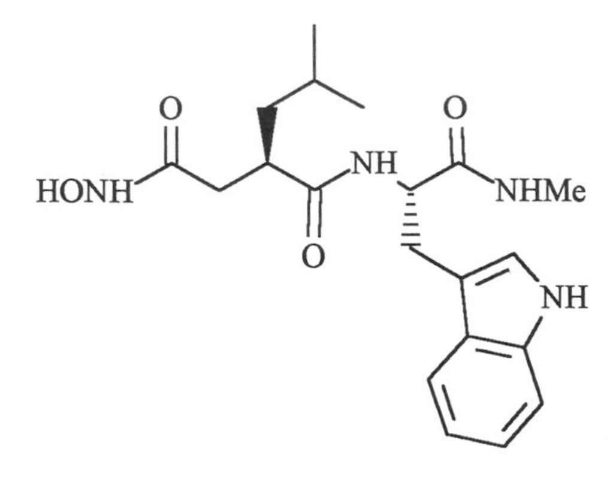

FIGURE 1. Matrix metalloproteinase (MMP) inhibitor GM 6001. The hydroxamic acid group is presumed to act as a bidentate ligand for the active site zinc atom in the MMPs.

by MMPs in inflammatory disease, MMP inhibitors such as GM 6001 appear to prevent infiltration of inflammatory and other cells into the extravascular space. Prevention of extravasation represents a new mechanism of anti-inflammatory activity by this class of drugs. GM 6001 is presently in Phase II/III clinical trials for corneal ulcers resulting from infection by microorganisms.

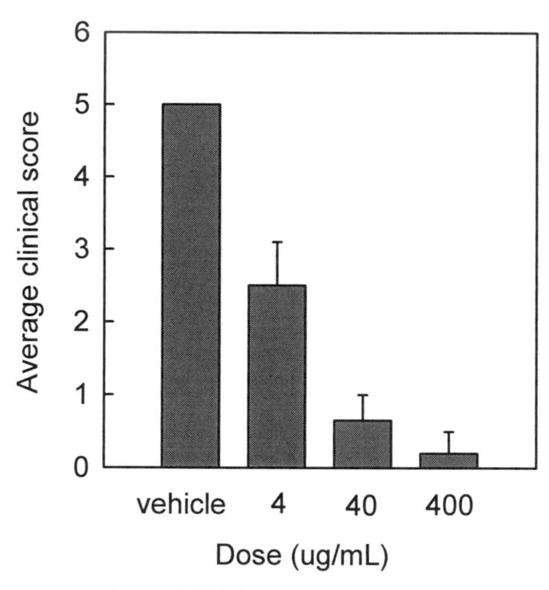

FIGURE 2. Dose response for GM 6001 in the alkali burn model of corneal ulceration.[4] A severe alkali injury (2 N sodium hydroxide for 1 min) was made to corneas of New Zealand white rabbits under anesthesia. Five rabbits in each of four groups were treated with vehicle (50 mM, pH 7.5 HEPES buffer, containing 0.1% in dimethylsulfoxide with 100 U/mL penicillin, 100 μg/mL streptomycin and 0.25 μg/mL amphotericin B), 4, 40, and 400 μg GM 6001 in vehicle every 2 h between 8 AM and 6 PM for 30 days. The corneas were scored 0 (no ulceration) through 5 (perforation). Doses of 40 and 400 μg/mL gave statistically identical results, and all drug treatments were different from vehicle (Mann-Whitney U nonparametric test).

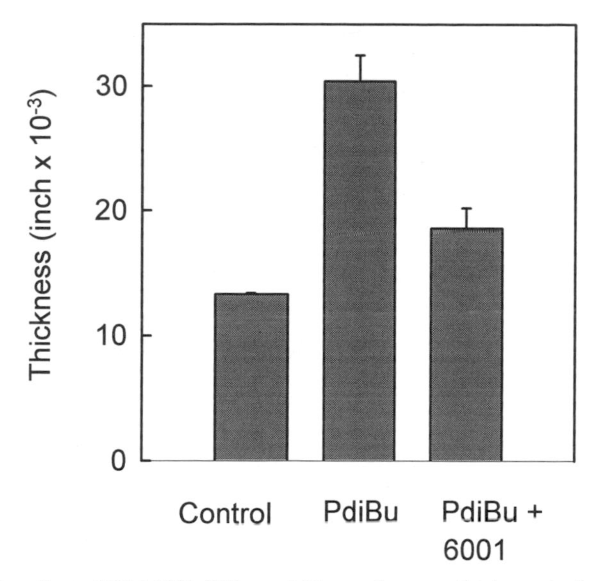

FIGURE 3. The effect of GM 6001 (200 μg, 0.53 μmol) on ear thickness in the phorbol ester model of inflammatory skin disease. Phorbol 12,13-dibutyrate (PdiBu), 20 nM in 20 μL of acetone, was applied to both ears (approximately 1 cm^2 each) of hairless mice (strain h/h). The test compounds in a total volume of 20 μL of ethanol were applied to the right ear within 30 min. The left ear received 20 μL of ethanol vehicle. Test compounds and vehicle were reapplied at 6 and 18 h. At 30 h the animals were anesthetized intraperitoneally with chloral hydrate (0.5 mg/kg), ear thickness was determined with a microcaliper, and a punch biopsy was taken for histology. GM 6001 reduced ear thickness by nearly 60% (n = 8) compared to vehicle (n = 8).

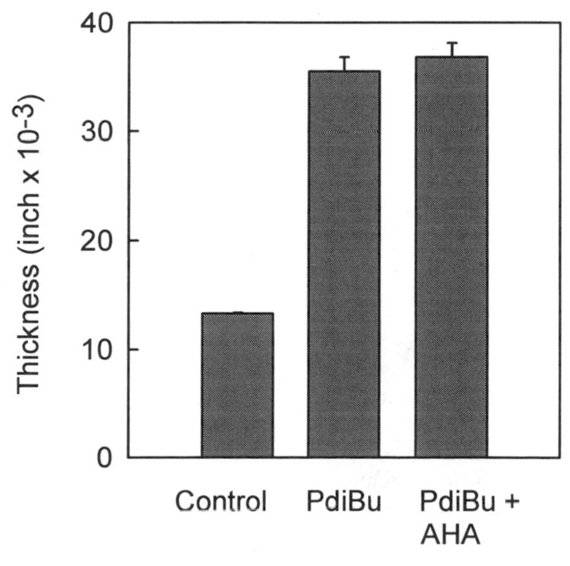

FIGURE 4. The effect of acetohydroxamic acid (AHA, 0.53 μmol) on ear thickness in the phorbol ester model. Experimental conditions are as in FIGURE 3. AHA (n = 6) had no effect on PdiBu-treated ears (n = 6).

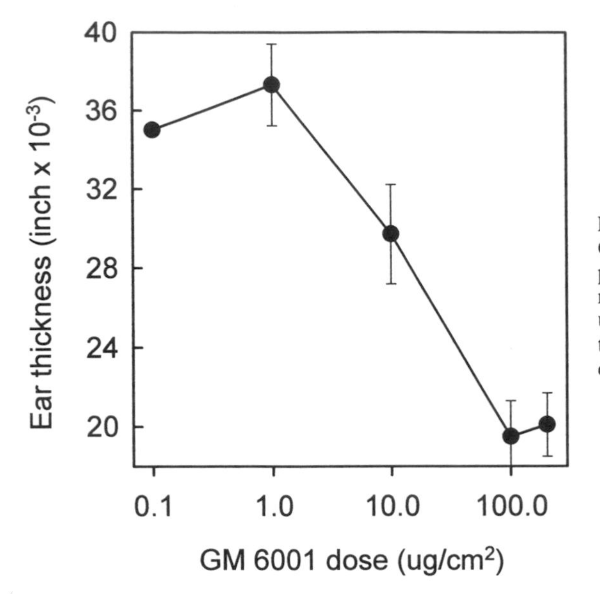

FIGURE 5. Dose response for GM 6001 on ear thickness in the phorbol ester model. Experimental conditions are as in FIGURE 1 except that the concentration of GM 6001 in 20 μL ethanol was varied.

RESULTS AND DISCUSSION

In Vitro *Potency*

GM 6001 has a K_i of 0.4 nM against human skin and human gingival fibroblast collagenase using a synthetic thiolester substrate at pH 6.5.[2] Its K_i is 0.5 nM against 72-kDa gelatinase A, 0.2 nM against 92-kDa gelatinase B, 0.1 nM against human neutrophil collagenase, and 27 nM against human stromelysin using a fluorescent peptide substrate at pH 7.5[3] in an automated 96-well plate assay. GM 6001 inhibits *Pseudomonas aeruginosa* elastase, an extracellular matrix degrading metalloproteinase, with a K_i of 20 nM, using a synthetic substrate.[2]

Pharmacokinetic Properties

GM 6001 has half-life of about 13 min after a single intravenous bolus injection in rabbits and mice. However, intraperitoneal (i.p.) dosing of a suspension of the drug in mice gives micromolar blood levels up to at least 6 h.

Corneal Ulceration

Topical GM 6001 dramatically inhibits corneal ulceration in both the alkali burn model and a model of *Pseudomonas aeruginosa* infection in rabbits. FIGURE 2 shows the dose response in the alkali burn model.[4] Ulceration is almost totally suppressed at doses of 400 and 40 μg/mL of the drug. Histology of the corneas demonstrates a large inflammatory infiltrate in the vehicle-treated eyes, but few inflammatory cells in the high-dose–treated group.

Pseudomonas aeruginosa is a virulent pathogen in human corneal ulceration. It produces a metalloelastase which may be involved in infection and ulceration, and which is inhibited by GM 6001 with a K_i of 20 nM.[2] In this model, 50 μL of organism-free *Pseudomonas aeruginosa* culture broth rich in elastase and exotoxins was injected into 12 rabbit corneas after de-epithelialization with heptanol under anesthesia.[5] The resulting corneal destruction appears to be mediated both by bacterial products and an inflammatory infiltrate. Six of the eyes were treated topically with two drops of vehicle and six with two drops of an aqueous formulation of drug at 800 μg/mL every 15 min for 1 h before and 6 h after injection. A subconjunctival injection of 0.5 mL of inhibitor or vehicle was given just after the last topical dose. After 16 h, the corneal stromas of all six vehicle-treated eyes were liquified, and the corneas themselves grossly distended. The drug-treated corneas were neither liquified nor distended. A similar experiment where the culture broth was concentrated and treated with GM 6001 before injection demonstrated equally dramatic protection of the corneas.[5]

Phorbol Ester Model of Inflammatory Skin Disease

Phorbol esters applied to the skin of an experiental animal cause inflammation and proliferation of the epidermis. This has been used as a model for psoriasis, a

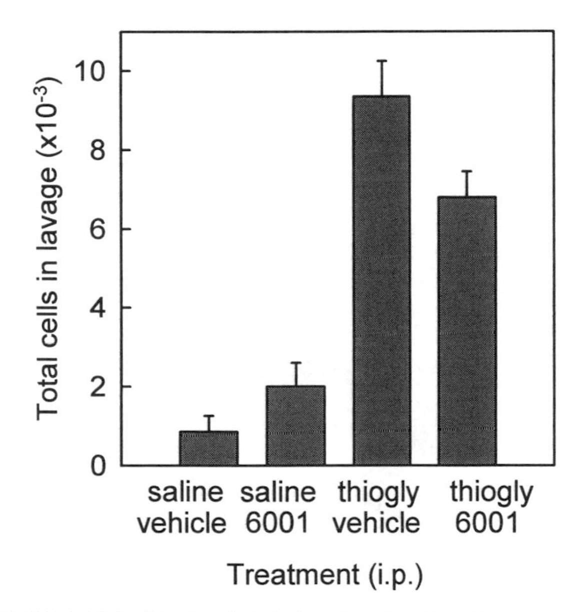

FIGURE 6. GM 6001 inhibits thioglycollate-induced peritonitis in the mouse. At time zero, 1.0 mL of thioglycollate (5.9 mg thioglycollate powder per milliliter of deionized water) was injected intraperitoneally (i.p.) into Balb/c male mice. At 0.5, 1.0, and 2.0 h either 1.0 mL of aqueous GM 6001 at 0.8 mg/mL (n = 10) or 1.0 mL of aqueous vehicle (n = 10) was also injected i.p. At 3.0 h the animals were euthanized by carbon dioxide asphyxiation and 5 mL of lavage fluid (0.1% bovine serum albumin, 5 U heparin per milliliter of saline) was injected i.p., mixed by gentle massage, and removed via syringe (80% recovery). Saline stimulus was used as a negative control for the thioglycollate. GM 6001 treatment after saline caused minimal infiltration of cells. GM 6001 treatment after thioglycollate reduced cellular infiltrate by 27%. Thiogly, thioglycollate.

TABLE 1. Angiostatic Activity of GM 6001 in the Chick Chorioallantoic Membrane[a]

Compound	Dose (μg/pellet)	Score		
		0	+	++
Buffer control	—	45	1	0
GM 6001	40	46	51	8
	20	12	6	3
	12	18	6	0
	4	20	3	0
	2.5	11	0	0
Heparin + hydrocortisone	50	5	11	17

[a]GM 6001 was dissolved in propylene glycol and diluted 1:10 into 0.55% methyl cellulose. Twenty microliters of the methylcellulose solution were dried into 3-mm pellets. The pellets were placed on the chorioallantoic membrane of chicken eggs three days after fertilization. The vascularity around the pellet was scored after 24 h: 0, high vascularity; +, reduced vascularity; ++, completely avascular in an area three-quarters of the way around the pellet.

disease in which neovascularization and then inflammation precede and accompany the epidermal proliferation.[6,7] FIGURE 3 shows the reduction by GM 6001 of ear thickness (edema caused by inflammation) resulting from the application of phorbol 12,13-dibutyrate (PdiBu). FIGURE 4 shows that an equimolar amount of acetohydroxamic acid, containing the same hydroxamate functional group as GM 6001, has no effect. Finally, FIGURE 5 shows that a dose of about 10 $\mu g/cm^2$ gives about 50% of the maximal inhibition. (The maximum observed with GM 6001 is about 70% reduction in thickness, where 100% reduction would give the thickness of a control, untreated ear.) Histology shows proliferation of the epidermis and extravasation of inflammatory cells and red cells into the dermis in vehicle-treated animals. Both extravasation of cells and proliferation are reduced in GM 6001–treated groups.

Thioglycollate-Induced Peritonitis

In this experiment, i.p. injection of thioglycollate causes infiltration of inflammatory cells which are measured by lavage after 3 h. FIGURE 6 shows the 27% inhibition ($p < 0.05$) of cellular infiltration by three i.p. doses of an aqueous solution of GM 6001 (a total of about 100 mg/kg) in a period of 2 h. The drug itself causes minimal infiltration in the absence of thioglycollate. No attempt was made to optimize the GM 6001 concentration or frequency of dosing. GM 6001 presumably inhibits cellular infiltration by preventing extravasation of inflammatory cells.

Angiogenesis

TABLE 1 shows the inhibition of angiogenesis by GM 6001 in the chick chorioallantoic membrane model. The entries in the table are the number of pellets scoring at each level: 0, +, or ++. A clear dose-response relationship is apparent. Inhibition at the maximum dose of 40 μg per pellet is less than that seen with pellets containing heparin and hydrocortisone, known to be a potent angiostatic combination.

Slow-release pellets containing a crude tumor extract cause neovascularization after implantation in the rat cornea. FIGURE 7 shows the reduction in the number of new vessels caused by systemic GM 6001 in rat corneas implanted with an extract of Walker 256 carcinoma. The reduction in the high-dose group (32 mg/kg/day intravenously) is highly significant ($p < 0.001$). The reduction in the low-dose group (1.28 mg/kg/day) is not ($p < 0.13$), possibly because of the small number of animals in this group.

FIGURE 8 shows the same experiment but with evaluation of vessel area using analysis of a video still image. The vessel-area results are similar to the results of the manual analysis of FIGURE 7. Inhibition of neovascularization is significant in the high-dose group ($p < 0.001$), but not in the low-dose group ($p < 0.19$).

Experimental Metastasis

FIGURE 9 shows the effect of GM 6001 on increasing the survival of mice injected intravenously with metastatic B16-F10 melanoma cells. GM 6001 significantly in-

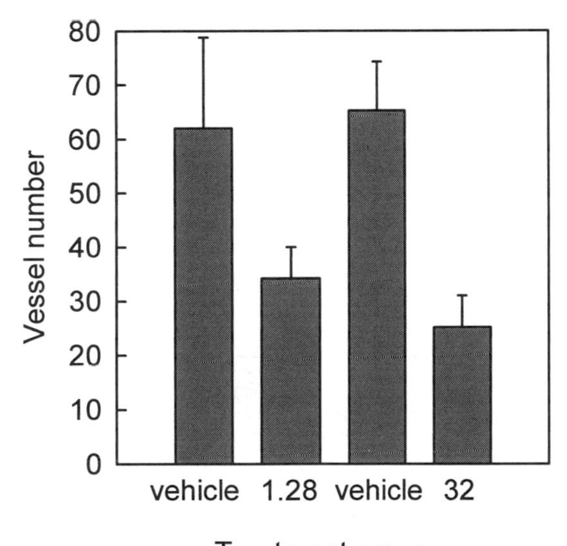

FIGURE 7. GM 6001 reduces the number of new vessels in rat corneas implanted with a neovascularization stimulus. An aqueous extract of Walker 256 rat carcinosarcoma was incorporated into Hydron polymer disks 1.5 mm in diameter. The disks were implanted in the stroma of the corneas of Sprague Dawley rats as described by Gimbrone *et al.*[8] for rabbits. An indwelling catheter was implanted in the inferior vena cava through the right femoral vein and exteriorized from the intercapsular area.[9] Vehicle or drug was continuously infused through the catheter for six days at a rate of 0.034 mL/h using a Harvard syringe pump. The low-dose group ($n = 5$) received GM 6001 (0.4 mg/mL, 1.28 mg/kg/day) in pH 7.5 HEPES buffer containing 0.1% dimethylsulfoxide. The high-dose group ($n = 17$) received GM 6001 (10 mg/mL, 32 mg/kg/day) in 55% dimethylsulfoxide. The individual vehicles were used separately as controls for each group ($n = 6$ and $n = 5$, respectively). The number of new vessels (but not branches) were counted visually.

creased survival compared to vehicle ($p < 0.05$ on days 35–37). The cells were mixed with the drug at 40 μg/mL immediately before injection. Treatment of the animals with the drug began one day before injection of the cells. The growth of cells in tissue culture was not affected by vehicle alone or drug in vehicle at 40 μg/mL. Pretreatment and then treatment with the drug on days one through four only increased survival several weeks later. This suggests interference with an early event in the growth of this tumor such as metastasis. Identical results occurred when the pretreatment of both the cells and the animals was omitted.

Clinical Studies

In a Phase I trial, four concentrations of GM 6001 (50, 200, 400, and 800 μg/mL) were administered in eye drops to normal volunteers. A maximum of 24 drops was administered in 12 h. Dosing continued up to 14 days. No serious events were observed. Ophthalmological examinations revealed no drug-induced abnormalities. Some minor symptoms were noted. However, there was no clear pattern of adverse reactions associated with the drug, and no dose response was observed for any event.

Phase II/III clinical trials are currently in progress for the treatment of patients with frank corneal ulcers caused by infection.

In summary, GM 6001 is active in a number of animal models of inflammation and cellular invasion. A common event in these models is the crossing of vascular basement membrane by inflammatory or other cell types. The drug prevents inflammatory cells from penetrating into the cornea in corneal ulceration models and into the extravascular space in the dermis in a psoriasis model. It also prevents cells from entering the peritoneal cavity in the thioglycollate peritonitis model. Blocking access to the extravascular space is a new mechanism of anti-inflammatory activity for matrix metalloproteinase inhibitors.

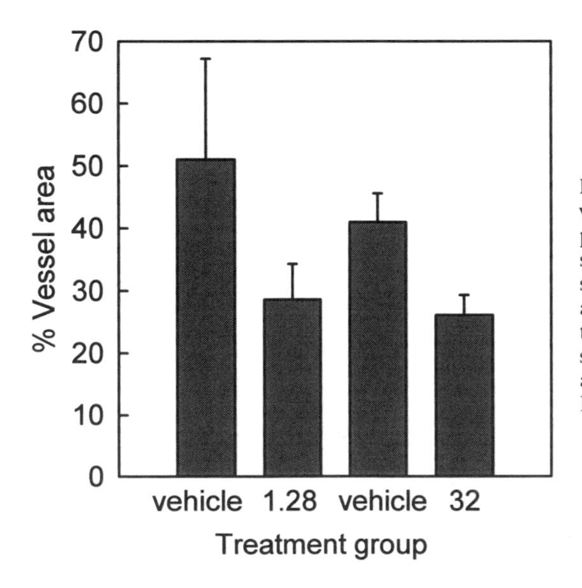

FIGURE 8. GM 6001 reduces new vessel area in rat corneas implanted with a neovascularization stimulus. The experiment is described in FIGURE 7. The % vessel area in a 2-mm^2 rectangle between the pellet and the limbus was measured using a video still camera and Image 1 software (Universal Imaging, Philadelphia, PA).

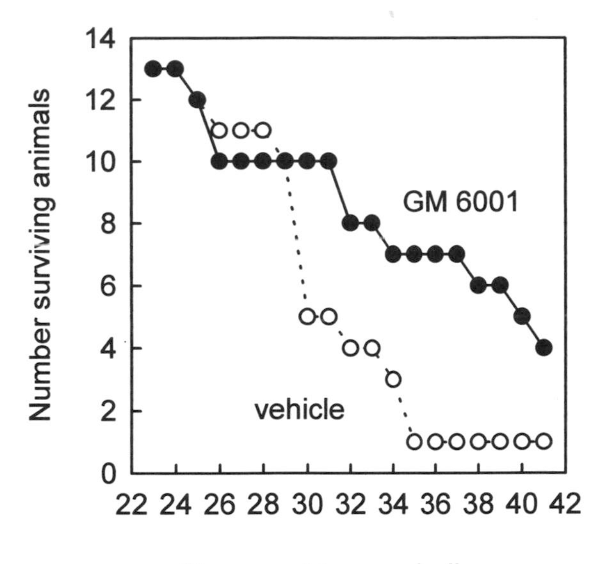

FIGURE 9. GM 6001 increases survival in a B16-F10 experimental metastasis assay in C57BL/6 mice. B16-F10 melanoma (American Type Culture Collection) was cultured, harvested with trypsin-EDTA and suspended at 5×10^5 cells/mL in phosphate-buffered saline for vehicle-treated animals. GM 6001 (40 μg/mL) in the same buffer was added to the cells immediately before injection of drug-treated animals. Cells were injected into the tail vein (5×10^4 cells) in a randomized order. Animals were injected i.p. with GM 6001 (150 mg/kg) suspended in 4% carboxymethylcellulose ($n = 13$) or vehicle alone ($n = 13$) 24 h prior to tumor challenge, 30 min prior to tumor challenge, and on days 2–4. Drug-treated animals demonstrated a significant increase in survival ($p < 0.05$) on days 35–37.

REFERENCES

1. Birkedal-Hansen, H., W. G. I. Moore, M. K. Bodden, L. J. Windsor, B. Birkedal-Hansen, A. DeCarlo & J. A. Engler. 1993. Crit. Rev. Oral Biol. Med. **4:** 197–250.
2. Grobelny, D., L. Poncz & R. E. Galardy. 1992. Biochemistry **31:** 7152–7154.
3. Knight, C. G., F. Willenbrock & G. Murphy. 1992. FEBS Lett. **296:** 263–266.
4. Schultz, G. S., S. Strelow, G. A. Stern, N. Chegini, M. B. Grant, R. E. Galardy, D. Grobelny, J. J. Rowsey, K. Stonecipher, V. Parmley & P. T. Khaw. 1992. Invest. Ophthalmol. Visual Sci. **33:** 3325–3331.
5. Barletta, J. P., K. C. Balch, H. G. Dimova, G. A. Stern, M. T. Moser, G. B. van Setten & G. S. Schultz. 1993. Invest. Ophthal. Visual Sci. **34:** 1058.
6. Braverman, I. M. & J. Silby. 1982. J. Invest. Dermatol. **78:** 12–17.
7. Christophers, E. & G. G. Krueger. 1987. *In* Dermatology in General Medicine, 3rd edit. T. B. Fitzpatrick, A. Z. Eisen, Wolff, I. M. Freedberg & K. F. Austin, Eds.: 461–491. McGraw-Hill. New York.
8. Gimbrone, M. A., R. S. Cotran, S. B. Leapman & J. Folkman. 1974. J. Natl. Cancer Inst. **52:** 413–427.
9. Fernandez, L. A., V. J. Caride, J. Twinckler & R. E. Galardy. 1982. Am. J. Physiol. **243:** H869–H875.

Matrix Metalloproteinase-9 in Tumor Cell Invasion

DEBORAH L. FRENCH, NOEMI RAMOS-DESIMONE, KAREN ROZINSKI, AND GERARD J. NUOVO

Department of Pathology
State University of New York at Stony Brook
Stony Brook, New York 11794-8691

Degradation of the extracellular matrix (ECM) is a critical feature of cancer cell invasion. Accumulating evidence suggests that proteolytic enzymes, including the matrix metalloproteinase (MMP) family, contribute to these multifactorial processes. Important features of these enzymes that exist in normal and neoplastic invasive processes are the induction of proteinase expression and the regulation of proteolytic activity. A common feature of the enzymes cited above is that they are produced as zymogens and require an activation step to become catalytically active. In addition, the expression of these enzymes requires stimuli that induce at the level of transcription whereas others are transcriptionally overexpressed. A correlation between proteolysis and malignant progression has been established. Specific inhibitors of serine and metalloproteinases have been used in *in vitro* and *in vivo* models[1–4] to demonstrate inhibition of tumor cell invasion and metastasis. The exact role of these specific proteinases in the mechanisms involved in the formation of metastatic lesions remains unclear.

The interaction of tumor cells with the ECM and basement membrane occurs at multiple stages in the metastatic cascade. Benign adenoma and *in situ* carcinomas are characterized by a continuous basement membrane that separates the epithelium from the underlying stroma, whereas invasive carcinomas demonstrate zones of basement membrane loss around the invading tumor cells. Attention has been focused on proteolytic enzymes whose substrate specificities include ECM and basement membrane components. Some enzymes that have been implicated in the invasive processes of cancer cells include the MMPs[5] and urokinase-type plasminogen activator (uPA).[6] Natural inhibitors of these enzymes have been identified and include the tissue inhibitors of metalloproteinases-1 and -2 (TIMP-1 and TIMP-2)[7–9] and plasminogen activator inhibitor types-1 and -2 (PAI-1 and PAI-2).[10]

Positive correlations between the expression of MMPs and uPA and the malignant phenotype have been demonstrated *in vitro* using numerous tumor and transformed cell lines.[5–6] These correlative studies have been supported by analyses using specific inhibitors. Inhibitors of MMPs or uPA have been shown to block tumor cell invasion across native or reconstituted basement membranes *in vitro.*[4,11,12] The ability of TIMP to prevent matrix degradation has been demonstrated in cell model systems. Inhibition of tumor cell invasion and metastasis in an *in vivo* animal model has been demonstrated using anti-uPA antibodies[2] or *in vivo* injections of TIMP.[13] A causal role for MMPs in cancer cell invasion was demonstrated using antisense TIMP RNA.[14] These studies demonstrated that a reduction in TIMP levels resulted in a change of the cellular phenotype to one that was invasive in an *in vitro* model and metastatic in an *in vivo* model. These analyses implicate the MMPs in cancer cell invasion.

The MMPs represent a family of structurally related enzymes that are grouped according to their substrate specificities (see reviews, refs. 15–17). A growing body of evidence supports a correlation between elevated levels of MMP-9 or MMP-2 expression and the invasive and metastatic potential of breast cancer cells. The expression of MMP-9 has been detected in malignant breast tumors.[18] By *in situ* hybridization analyses, MMP-9 was detected in some inflammatory and neoplastic cells of malignant tumors, whereas MMP-2 was expressed exclusively by fibroblast cells in both benign and malignant tumors. These data and a transfection study that directly correlated MMP-9 expression with metastasis[19] suggest that MMP-9 may be linked to the metastatic phenotype. A new member of the MMP family, ST-3 or MMP-11, was expressed in stromal cells surrounding malignant epithelial cells of invasive tumor components in all breast carcinomas analyzed.[18,20] This enzyme was not expressed in normal mucosa and has been implicated as a potential mediator of metastatic disease.[18] The MMP-11 protein has recently been expressed, but the substrate specificity of this enzyme is unknown.[21]

The natural or *in vivo* mechanism(s) of activation of MMP-9 is presently unknown. *In vitro* analyses have demonstrated differences in the expression and activation of MMP-2 and MMP-9. MMP-2 is expressed and activated in culture supernatants from a number of different normal and transformed cell lines.[22] Recent studies have demonstrated that MMP-2 activation is localized at the cell surface,[23,24] and a putative receptor has been suggested.[25] Cellular expression of MMP-9 is very different than that of MMP-2[22] in so far as it requires induction by transcriptional activators such as phorbol ester, interleukin-1 (IL-1), or tumor necrosis factor-α (TNF-α).[17] The same conditions that result in the cell surface activation of MMP-2 do not result in the activation of MMP-9.[23] These data suggest a different mechanism of activation. A hypothetical model has been suggested as a proteolytic cascade involving serine and other metalloproteinases.[26] A natural or *in vivo* model needs to be identified to confirm the significance of this mechanism. Additional modes of regulation of these enzymes include the expression of their natural inhibitors or TIMPs. TIMP-1 is secreted as a complex with MMP-9, and TIMP-1 and TIMP-2 can inhibit activated forms of this enzyme.[9] An *in vitro* study has demonstrated that production of TIMP-free MMP-9 may be a mechanism that is critical for the formation of active forms of this enzyme that have a very high specific activity.[26]

We have focused our studies on analyzing the role of MMP-9 in tumor cell invasion. To perform these studies, anti-human MMP-9–specific monoclonal antibodies have been generated.[27] These monoclonal antibodies have been shown to react specifically with MMP-9 by ELISA and immunoblot. They also react by immunoprecipitation with proMMP-9 and the two lower molecular weight forms of this enzyme, as shown in the paper by Ramos-DeSimone and French (this volume). Of significance is the finding that one of the monoclonal antibodies, designated 6-6B, has been shown to inhibit the *in vitro* activation of MMP-9 as demonstrated in a radiolabeled substrate assay using [^{3}H]gelatin.[27] In this study, tumor cell invasion across a reconstituted basement membrane is inhibited by the anti-MMP-9 monoclonal antibodies. In addition, the *in vivo* expression of MMP-9 is analyzed on cytospins of HT1080 fibrosarcoma cells using reverse transcriptase (RT) *in situ* polymerase chain reaction (PCR).

MATERIALS AND METHODS

Methodologies for MMP-9 purification, *in vitro* activation using APMA, silver-stained SDS-PAGE, gelatin zymography, and immunoblot have been described previously.[27]

Reconstituted Basement Membrane Assay Using Matrigel

Polycarbonate filters (8 μm pore size) (Costar, Cambridge, MA) were coated with 7.5 $\mu g/cm^2$ or 10 μg per filter of matrigel (Collaborative Research, Bedford, MA) and solidified with or without antibody (25 μg/mL). HT1080 cells, with or without monoclonal antibodies (25 μg/mL), were added into the upper reservoir of each chamber. The HT1080 cells require a chemoattractant, and formyl-norleucyl-leucyl-phenylalanine (fNLP) was used at a final concentration of 1 mM. The average number of cells, to which no antibody was added, was counted on the underside of the filter and standardized to 100%.

Reverse Transcriptase **in Situ** *Polymerase Chain Reaction*

Tissue sections are fixed with buffered formalin for 4–15 h. This cross-linking fixative is essential to limit the migration of PCR product from the *in situ* site of synthesis. In preparation for cDNA synthesis, tissues are digested with trypsin (2 mg/mL) according to the fixative time:

Buffered Formalin (h)	Trypsin Digestion (min)
4	12
6	20
8	40
>15	120

The trypsin is inactivated by washes in DEPCddH_2O and 100% ethanol. Sections are treated overnight with RNase-free DNase according to manufacturers recommendations (Boehringer Mannheim, Indianapolis, IN). The tissues are incubated directly on the glass slide at 42 °C for 30 min with 10 μL of a solution containing the 3′ primer (1 μM), RT (5–10 U), and RT buffer (RT-PCR kit, Perkin Elmer Corporation, Norwalk, CT). To prevent evaporation, the solution is covered with a plastic coverslip anchored to the slide with a drop of nail polish and overlaid with mineral oil. The glass slide is placed in an aluminum "boat" directly on the block of the thermocycler (Perkin Elmer Corporation, Norwalk, CT). After incubation, the mineral oil is removed with a 5-min wash in xylene, which is removed with a 5-min wash in 100% ethanol. The cDNA is amplified by PCR. The reagents in the amplifying solution include PCR buffer (50 mM KCl, 10 mM Tris-Cl, pH 8.3), 4.5 mM $MgCl_2$, 200 μM dNTPs, 1.0 μM of each primer, and for direct incorporation of label, 10 μM digoxigenin dUTP and 1.2 U of Taq polymerase (AmpliTaq, Perkin Elmer Corporation). The coverslip is anchored to the glass slide using a drop of nail polish and overlaid with mineral oil that has been preheated to 82 °C. The reaction is brought to 94 °C for 3 min followed by 15 cycles of 2-step PCR in which annealing/extension is performed at the appropriate temperature for the primers (50 °–60 °C) for 2 min followed by a denaturing step at 94 °C for 1 min. The slides are treated with xylene and ethanol followed by detection using alkaline phosphatase conjugated antidigoxigenin antibody (Boehringer Mannheim). The chromagen is nitroblue tetrazolium in the presence of 5-bromo-4-chloro-3-indolylphosphate which yields a dark blue color (incubation times are usually 5 min or less).

Controls

Control reactions are performed on the same cells or tissue as the test reaction. Areas of the tissue are blocked off using a hydrophobic pen to prevent mixing of solutions. If the tissue is too small, a serial section must be analyzed under the same coverslip using the same reagents that are used on the experimental tissue. Controls that must be performed include:

1. Either no reverse transcriptase added to the RT *in situ* PCR reaction or primers that are specific for a product that cannot possibly be present in the cell populations that are analyzed (i.e., primers for a viral product) (negative control); and
2. No DNase treatment (positive control).

If a signal is present in the sample in which RT was eliminated or nonspecific primers were added, the probable cause is insufficient DNase treatment because of inadequate protease digestion (see chart above). If no signal or a weak signal is present in the samples in which no DNase treatment was performed, the probable cause is insufficient trypsin treatment of the tissue. The same tissue sample should have a 4+ reaction (at least 75% of cells with an intense signal) without DNase treatment and no signal with nonspecific primers or elimination of RT.

RESULTS AND DISCUSSION

Specific Immunoreactivity of the Monoclonal Antibodies with MMP-9 and Not MMP-2

Monoclonal antibodies—designated 6-6B, 7-11C, and 8-3H—have been generated that react specifically with MMP-9 by ELISA.[27] By immunoblot, the 7-11C monoclonal antibody reacts with the 92-kDa form of the enzyme under reduced and nonreduced conditions.[27] Monoclonal antibodies 6-6B and 8-3H react with the 92-kDa and 83-kDa forms of the enzyme under nonreduced conditions, whereas under reduced conditions the 6-6B loses reactivity to the 83-kDa form of the enzyme and the 8-3H monoclonal antibody loses all reactivity.[27] These analyses suggest that the three IgG_1 monoclonal antibodies recognize different epitopes on purified MMP-9. The monoclonal antibodies were also analyzed by immunoblot on a purified preparation of MMP-2 and a partially purified preparation that contained both MMP-2 and MMP-9 (FIG. 1). A silver-stained SDS-PAGE of the enzyme preparations demonstrate the purity of the MMP-2/TIMP-2 preparation and the mixed enzyme preparation that contains the MMP-2 and MMP-9 proenzymes, their lower molecular forms, and both TIMP-1 and TIMP-2 (FIG. 1, left). By immunoblot analysis, the monoclonal antibodies do not react with MMP-2 which is present in both enzyme preparations, but do react specifically with MMP-9 which is present only in the mixed enzyme preparation (FIG. 1, right). In addition to the IgG_1 6-6B monoclonal antibody, an IgG_{2b} heavy chain switch variant has been isolated. This monoclonal antibody has the same binding immunoreactivity to MMP-9 as the original antibody (FIG. 1, right) and can be used with the IgG_1 monoclonal antibodies in sandwich ELISA detection analyses.

Inhibition of Chemoinvasion of HT1080 Fibrosarcoma Cells by Anti-MMP-9 Monoclonal Antibodies

The monoclonal antibodies were analyzed previously for neutralizing activity, and of significance was the finding that the 6-6B monoclonal antibody inhibited the APMA activation of MMP-9 in a radiolabeled substrate assay.[27] These monoclonal

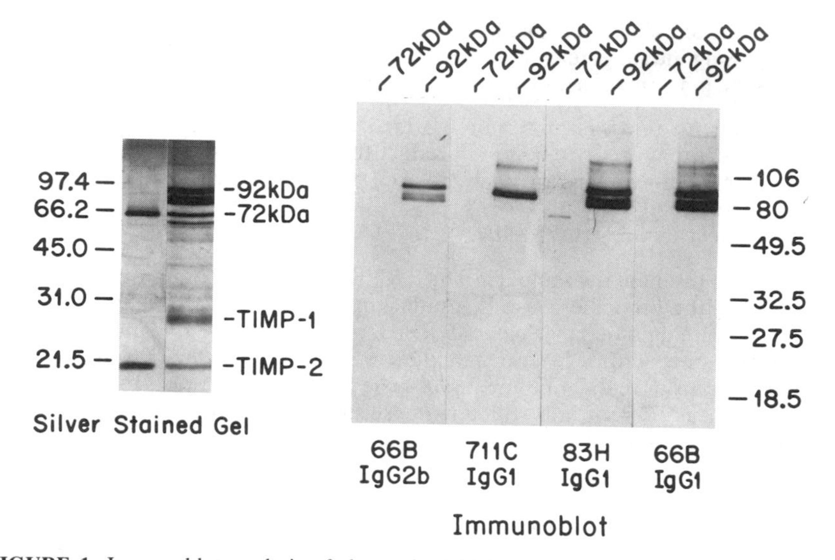

FIGURE 1. Immunoblot analysis of the anti-MMP-9 monoclonal antibodies (mAbs) on a purified preparation of MMP-2 and a partially purified preparation containing both MMP-2 and MMP-9. **Left:** Silver-stained SDS-PAGE of the enzyme preparations showing the purity of the proMMP-2/TIMP-2 preparation and the mixture of pro- and converted MMP-2, pro- and converted MMP-9, TIMP-1, and TIMP-2. **Right:** Immunoblot of the IgG_1 6-6B, 7-11C, and 8-3H mAbs and the IgG_{2b} 6-6B mAb on the MMP-2 and MMP-9 preparations.

antibodies were tested for their ability to inhibit invasion of HT1080 fibrosarcoma cells in an *in vitro* assay in which a reconstituted basement membrane on a filter was used as a tumor cell barrier in a Boyden chamber.[28,29] Matrigel is a solubilized basement membrane preparation extracted from Engelbreth-Holm Swarm (EHS) mouse sarcoma cells.[30] The preparation is not a pure system but a mixture of basement membrane components rather than true basement membrane. This mixture may contain endogenous gelatinases and a potential activator of MMP-9.[31] The role of MMP-9 in HT1080 tumor cell invasion was analyzed in this assay using the 6-6B, 7-11C, and 8-3H monoclonal antibodies (FIG. 2). Approximately 80% inhibition was detected in the presence of the 6-6B and 7-11C monoclonal antibodies. The 8-3H monoclonal antibody inhibited tumor cell invasion by approximately 50%, and the isotype control monoclonal antibody, 23-3, inhibited invasion by approximately 10%. These results suggest that MMP-9 is contributing to tumor cell invasion in this system. Prior studies have been performed using HT1080 cells in the reconstituted basement membrane assay, and complete inhibition of invasion was obtained using TIMP-2 or a polyclonal anti-type IV collagenase antisera.[11] These data suggested that the inhibition of invasion of HT1080 was due to the inhibition of MMP-2. A possible scenerio in this assay system may be that the inhibition of one enzyme is sufficient to inhibit tumor cell invasion. Inasmuch as the HT1080 cells secrete both MMP-2 and MMP-9, a cell line secreting only MMP-9 and not MMP-2 is required for the analysis of the specific role of MMP-9 in tumor cell invasion.

Previous studies have shown that the breast cancer cell line, MDA-MB-231, secretes only MMP-9 and not MMP-2,[32,33] and the same result was implied for another breast cancer cell line, MCF-7.[25] The MDA-MB-231 cell line is a poorly differentiated adenocarcinoma. The MCF-7 cell line is an adenocarcinoma with characteristics of differentiated mammary epithelium that can process estradiol via cytoplasmic estrogen receptors. To evaluate the secretion of MMPs from these cell lines, culture supernatants were tested by gelatin zymography. Briefly, cells were grown to confluency and resuspended in serum-free media with phorbol myristate acetate (PMA) (100 ng/mL). After three days, the supernatants were analyzed by gelatin zymography. MMP-9 was secreted from PMA-treated cells and MMP-2 was not generated (FIG. 3, lanes 2 and 4). Two additional gelatinolytic species of approximate molecular weights of 55,000 and 52,000 were also observed. To determine if these species were MMPs, the culture supernatants were incubated with 1,10-phenanthroline, a specific inhibitor of MMPs. The three gelatinolytic species of 92-kDa, 55-kDa, and 52-kDa were sensitive to this inhibitor (data not shown). To determine if the 55-kDa and 52-kDa species were secreted as zymogens or were potential degradation products of MMP-9, the samples were treated with 2 mM APMA for 3 h and analyzed by gelatin zymography. ProMMP-9 was activated to an 83-kDa form, and the 55-kDa and 52-kDa species generated two additional bands of approximate molecular weights of 45,000 and 42,000, respectively (lane 3). The molecular weights and weak gelatinolytic activity of these additional species are indicative of MMP-3 (stromelysin) and/or MMP-1 (interstitial collagenase). The secretion of MMP-3 and the expression of MMP-1 from the MDA-MB-231 cell line have been demonstrated by casein zymography and Northern blot analysis, respectively.[33] These additional gelatinolytic species are being analyzed by immunoblot, and these cell lines are being used to analyze the specific role of MMP-9 in tumor cell invasion.

MMP-9 Expression in HT1080 Fibrosarcoma Cells Using Reverse Transcriptase in Situ *Polymerase Chain Reaction*

In situ PCR methodology using DNA templates in tissue has proven very effective and highly specific.[34] Extensive work has been done to optimize the *in situ* PCR

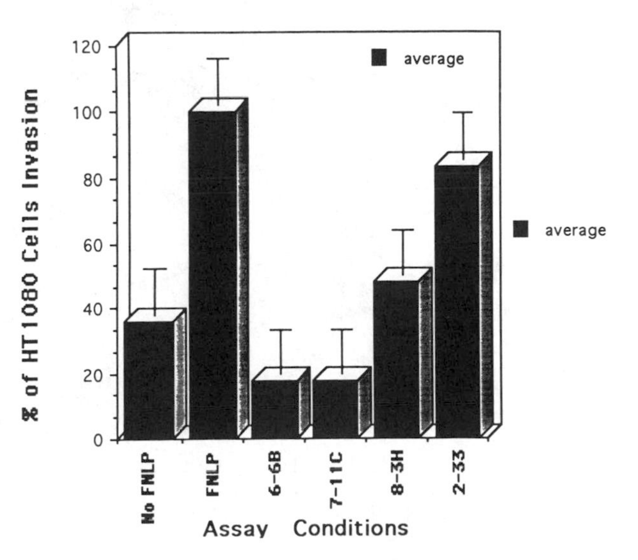

FIGURE 2. Matrigel invasion assay: Chemoinvasion of HT1080 cells in the presence of the anti-MMP-9 monoclonal antibodies (mAbs) in a reconstituted basement membrane assay using matrigel. A total of 15,000 cells per chamber were incubated with or without the mAbs (25 μg/mL) and fNLP (1 mM). The number of invasive cells in the absence of mAbs was standardized to 100%. Each sample was run in triplicate and each bar represents the mean ± standard deviation. fNLP, formyl-norleucyl-leucyl-phenylalanine.

signal, including (1) the elimination of non-cross-linking fixatives, such as ethanol and acetone, because of loss of signal from the cells into the amplifying solution; (2) the use of buffered formalin to eliminate efflux of the PCR product out of the cells, but allow optimal protease digestion to permit PCR reagents into the cells; (3) determining optimal Taq polymerase, $MgCl_2$, and primer concentrations. This methodology has been modified to include reverse transcriptase for the analysis of RNA expression in tissue.[35] The tissue is pretreated with RNase-free DNase which eliminates the native DNA from the amplification reaction. Complete digestion can be determined for each sample by utilizing primers for a ubiquitous human gene, such as bcl-2, and analyzing for the lack of a product signal. Specific PCR fragments are amplified from cDNA and visualized by direct labeling with digoxigenin dUTP. An enzyme-conjugated antibody that is specific for digoxigenin is added to the tissue and the color is developed with a precipitating substrate.

The RT *in situ* PCR technique was used for analyzing the expression of MMP-9 in cytospins of HT1080 cells (FIG. 4). Two cytospins of cells were placed on the same glass slide and fixed for 15 h in 10% buffered formalin. The cells were protease treated with trypsin (2 mg/mL) for only 30 min at 37 °C because cells require shorter protease digestion times than does tissue. One of the samples was pretreated with RNase-free DNase (10 U/cytospin) overnight at 37 °C followed by cDNA synthesis using RT and a 3′ primer for MMP-9. PCR was performed using the same sets of primers on both samples that were under the same coverslip. Digoxigenin-dUTP was incorporated into the PCR amplification reaction. The samples were developed

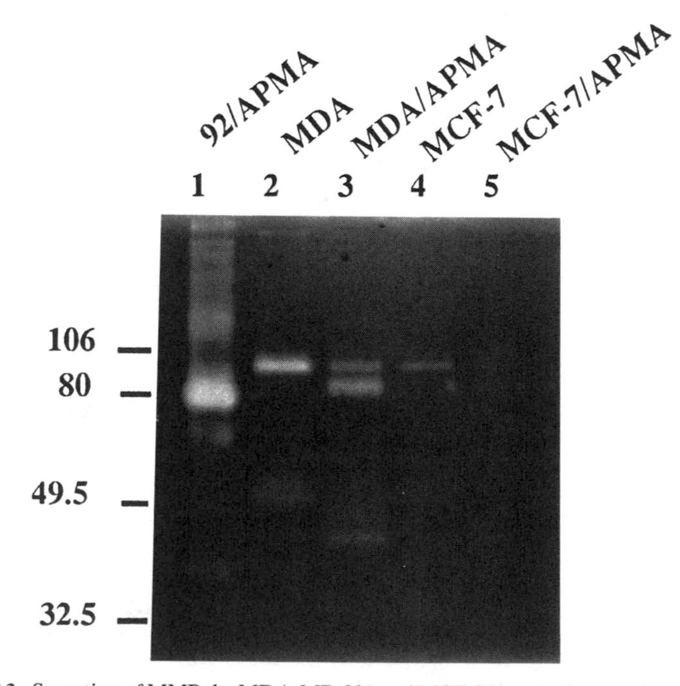

FIGURE 3. Secretion of MMPs by MDA-MB-231 and MCF-7 breast adenocarcinoma cell lines treated with phorbol myristate acetate. Culture supernatants were analyzed by gelatin zymography for the presence of MMPs. **Lane 1:** APMA activated purified MMP-9; **lane 2:** MDA-MB-231 cells untreated; **lane 3:** MDA-MB-231 cells treated with APMA (2 mM) for 3 h; **lane 4:** MCF-7 cells untreated; **lane 5:** MCF-7 cells treated with APMA (2 mM) for 3 h.

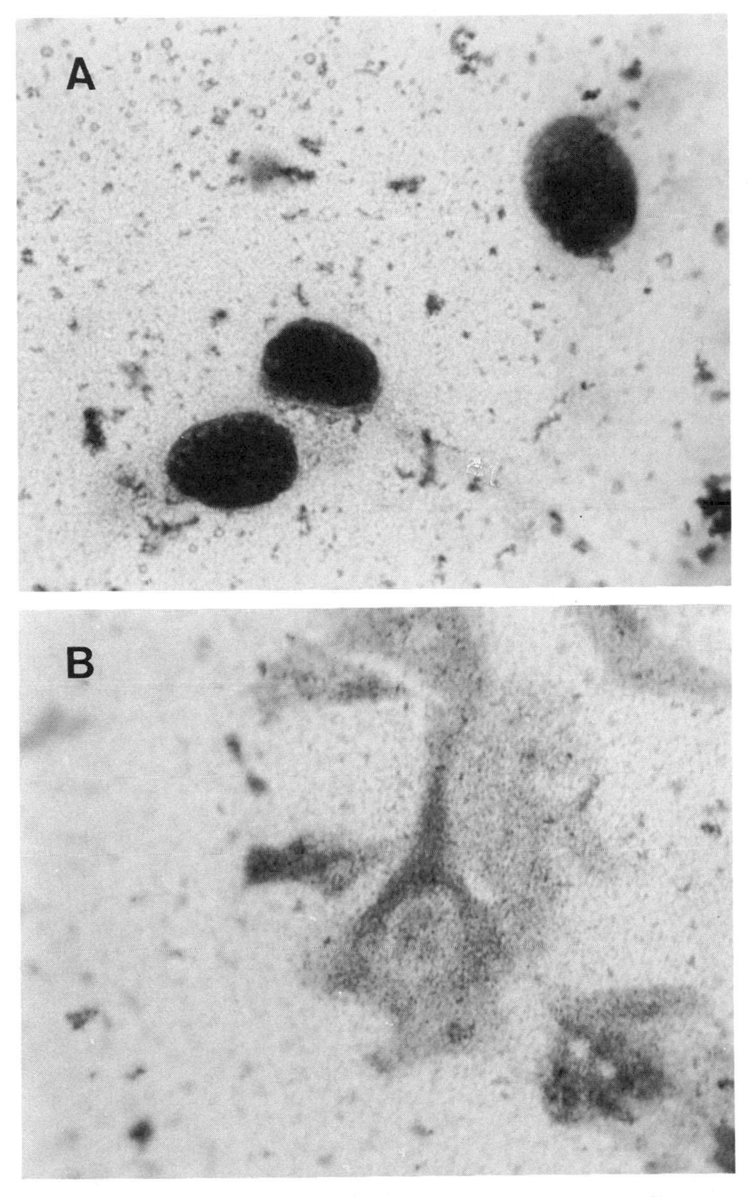

FIGURE 4. Photomicrographs of MMP-9 expression in cytospins of HT1080 fibrosarcoma cells using reverse transcriptase (RT) *in situ* polymerase chain reaction (PCR). (**A**) Cells were not treated with DNase prior to PCR. (**B**) Cells were treated with DNase and followed by RT-PCR.

using enzyme conjugated anti-digoxigenin antibody and precipitating substrate. The results are shown in FIGURE 4A in which no DNase treatment was performed and in FIGURE 4B in which the cells were treated with DNase.

The strong nuclear staining of the non-DNase–treated cells is shown in FIGURE

4A. These cells were not treated with RT and the counterstain (nuclear fast red) does not react with the cytoplasm; thus only the nuclear signal is detectable. A diffuse cytoplasmic staining of the DNase-treated cells is shown in FIGURE 4B in which the stellate appearance of the HT1080 fibrosarcoma cell is visible. A halo of nucleus is visible in these cells due to weak reactivity of the counterstain. The specificity of the reaction in the DNase-treated cells is checked by eliminating RT from the reaction. This control was performed in the experiment and no signal was detected (data not shown because no visible staining is detectable). These results demonstrate the specificity of the RT *in situ* PCR technique and the feasibility of this methodology for analyzing MMP expression in invasive carcinomas.

In summary, we have generated anti-human MMP-9 monoclonal antibodies that are reactive by ELISA, immunoblot, and immunoprecipitation. The 6-6B monoclonal antibody neutralizes the activation of MMP-9. Tumor cell invasion across a reconstituted basement membrane was inhibited by the monoclonal antibodies, and additional studies are in progress using cell lines that secrete MMP-9 and not MMP-2. The cytoplasmic expression of MMP-9 in cytospins of HT1080 cells was shown using the technique of RT *in situ* PCR. We are using the technique to determine the expression of MMP-9 in formalin-fixed breast and cervical cancer tissue from patients with and without metastatic disease.

REFERENCES

1. THORGEIRSSON, U. P., L. A. LIOTTA, T. KALEBIC, I. M. MARGULIES, K. THOMAS, M. RIOS-CALDERONE & R. G. RUSSO. 1982. Effect of natural protease inhibitors and a chemoattractant on tumor cell invasion *in vitro.* J. Natl. Cancer Inst. **69:** 1049–1054.
2. OSSOWSKI, L. & E. REICH. 1983. Antibodies to plasminogen activator inhibit human tumor metastasis. Cell **35:** 611–619.
3. MIGNATTI, P., E. ROBBINS & D. B. RIFKIN. 1986. Tumor invasion through the human amniotic membrane: Requirement for a proteinase cascade. Cell **47:** 487–498.
4. REICH, R., E. THOMPSON, Y. IWAMOTO, G. R. MARTIN, J. R. DEASON, G. C. FULLER & R. MISKIN. 1988. Effects of inhibitors of plasminogen activator, serine proteinases, and collagenase IV on the invasion of basement membrane by metastatic cells. Cancer Res. **48:** 3307–3312.
5. LIOTTA, L. A., P. S. STEEG & W. G. STETLER-STEVENSON. 1991. Cancer metastasis and angiogenesis: An imbalance of positive and negative regulation. Cell **64:** 327–336.
6. DANØ, K., P. A. ANDREASEN, J. GRØNDAHL-HANSEN, P. KRISTENSEN, L. S. NIELSEN & L. SKRIVER. 1985. Plasminogen activators, tissue degradation, and cancer. Adv. Cancer Res. **44:** 139–266.
7. DOCHERTY, A. J. P., A. LYONS, B. J. SMITH, E. M. WRIGHT, P. E. STEPHENS & T. J. R. HARRIS. 1985. Sequence of human tissue inhibitor of metalloproteinases and its identity to erythroid-potentiating activity. Nature **318:** 66–69.
8. STETLER-STEVENSON, W. G., P. D. BROWN, M. ONISTO, A. T. LEVY & L. A. LIOTTA. 1990. Tissue inhibitor of metalloproteinase (TIMP-2). J. Biol. Chem. **265:** 13933–13938.
9. HOWARD, E. W., E. C. BULLEN & M. J. BANDA. 1991. Preferential inhibition of 72- and 92-kDa gelatinases by tissue inhibitor of metalloproteinases-2. J. Biol. Chem. **266:** 13070–13075.
10. ELLIS, V., T. C. WUN, N. BEHRENDT, E. RØNNE & K. DANØ. 1990. Inhibition of receptor-bound urokinase by plasminogen-activator inhibitors. J. Biol. Chem. **265:** 9904–9908.
11. ALBINI, A., A. MELCHIORI, S. PARODI, L. A. LIOTTA, P. D. BROWN & W. G. STETLER-STEVENSON. 1991. TIMP-2 inhibits tumor cell invasion. J. Natl. Cancer Inst. **83:** 775–779.
12. HÖYHTYÄ, M., E. HUJANEN, T. TURPEENNIEMI-HUJANEN, U. THORGEIRSSON, L. A. LIOTTA & K. TRYGGVASON. 1990. Modulation of type-IV collagenase activity and invasive

behavior of metastatic human melanoma (A2058) cells *in vitro* by monoclonal antibodies to type-IV collagenase. Int. J. Cancer **46:** 282–286.

13. McDonnell, S. & L. M. Matrisian. 1990. Stromelysin in tumor progression and metastasis. Cancer Metastasis Rev. **9:** 305–319.
14. Khokha, R., P. Waterhouse, S. Yagel, P. K. Lala, C. M. Overall, G. Norton & D. T. Denhardt. 1989. Antisense RNA-induced reduction in murine TIMP levels confers oncogenicity on Swiss 3T3 cells. Science **243:** 947–950.
15. Matrisian, L. M. 1990. Metalloproteinases and their inhibitors in matrix remodeling. Trends Genet. **6:** 121–125.
16. Woessner, J. F., Jr. 1991. Matrix metalloproteinases and their inhibitors in connective remodeling. FASEB J. **5:** 2145–2154.
17. Birkedal-Hansen, H., W. G. Moore, M. K. Bodden, L. J. Windsor, B. Birkedal-Hansen, A. DeCarlo & J. A. Engler. 1993. Matrix metalloproteinases: A review. Crit. Rev. Oral Biol. Med. **4:** 197–250.
18. Wolf, C., N. Rouyer, Y. Lutz, C. Adida, M. Loriot, J.-P. Bellocq, P. Chambon & P. Basset. 1993. Stromelysin 3 belongs to a subgroup of proteinases expressed in breast carcinoma fibroblastic cells and possibly implicated in tumor progression. Proc. Natl. Acad. Sci. USA **90:** 1843–1847.
19. Bernhard, E. J., S. B. Gruber & R. J. Muschel. Direct evidence linking the expression of MMP-9 to the transformed phenotype in transformed rat cells. Proc. Natl. Acad. Sci. USA. In press.
20. Basset, P., J. P. Bellocq, C. Wolf, I. Stoll, P. Hutin, J. M. Limacher, O. L. Podhajcer, M. P. Chenard, M. C. Rio & P. Chambon. 1990. A novel metalloproteinase gene specifically expressed in stromal cells of breast carcinomas. Nature **348:** 699–704.
21. Murphy, G., J.-P. Segain, M. O'Shea, M. Cockett, C. Ioannou, O. Lefebvre, P. Chambon & P. Basset. 1993. The 28-kDa N-terminal domain of mouse stromelysin-3 has the general properties of a weak metalloproteinase. J. Biol. Chem. **268:** 15435–15441.
22. Moll, U. M., G. L. Youngleib, K. B. Rosinski & J. P. Quigley. 1990. Tumor promoter-stimulated M_r 92,000 gelatinase secreted by normal and malignant human cells: Isolation and characterization of the enzyme from HT1080 tumor cells. Cancer Res. **50:** 6162–6170.
23. Ward, R. V., S. J. Atkinson, P. M. Slocombe, A. J. P. Docherty, J. J. Reynolds & G. Murphy. 1991. Tissue inhibitor of metalloproteinase-2 inhibits the activation of 72kDa progelatinase by fibroblast membranes. Biochim. Biophys. Acta **1079:** 242–246.
24. Brown, P. D., D. E. Kleiner, E. J. Unsworth & W. G. Stetler-Stevenson. 1992. Cellular activation of the 72kDa type IV procollagenase/TIMP-2 complex. Kidney Int. **43:** 163–170.
25. Emonard, J. P., A. G. Remacle, A. C. Noël, J.-A. Grimaud, W. G. Stetler-Stevenson & J.-M. Foidart. 1992. Tumor cell surface-associated binding site for the Mr 72,000 type IV collagenase. Cancer Res. **52:** 5845–5848.
26. Okada, Y., Y. Gonoji, K. Naka, K. Tomita, I. Nakanishi, K. Iwata, K. Yamashita & T. Hayakawa. 1992. Matrix metalloproteinase 9 (92-kDa gelatinase/type IV collagenase) from HT1080 human fibrosarcoma cells. J. Biol. Chem. **267:** 21712–21719.
27. Ramos-DeSimone, N., U. M. Moll, J. P. Quigley & D. L. French. 1993. Inhibition of matrix metalloproteinase 9 activation by a specific monoclonal antibody. Hybridoma **12:** 349–363.
28. Kleinman, H. K., M. L. McGarvey, J. R. Hassell, V. L. Star, F. B. Cannon, G. W. Laurie & G. R. Martin. 1986. Basement membrane complexes with biological activity. Biochemistry **25:** 312–318.
29. Terranova, V. P., E. S. Hujanen, D. M. Loeb, G. R. Martin, L. Thornburg & V. Glushko. 1986. Use of reconstituted basement membrane to measure cell invasiveness and select for highly invasive tumor cells. Proc. Natl. Acad. Sci. USA **83:** 465–469.
30. Kleinman, H. K., M. L. McGarvey, L. A. Liotta, K. Robey, K. Tryggvason & G. R. Martin. 1982. Isolation and characterization of type IV procollagen, laminin, and heparan sulfate proteoglycan from the EHS sarcoma. Biochemistry **21:** 6188–6193.

31. DIAMOND, P. F. 1991. Basement membrane matrigel: Induction of enzymes and protein synthesis. Cell Line **1:** 1.
32. MACKAY, A. R., J. L. HARTZLER, M. D. PELINA & U. P. THORGEIRSSON. 1990. Studies on the ability of 65kDa and 92kDa tumor cell gelatinases to degrade type IV collagen. J. Biol. Chem. **265:** 21929–21934.
33. MACKAY, A. R., M. BALLIN, M. D. PELINA, A. R. FARINA, A. M. NASON, J. L. HARTZLER & U. P. THORGEIRSSON. 1992. Effect of phorbol ester and cytokines on matrix metalloproteinase and tissue inhibitor of metalloproteinase expression in tumor and normal cell lines. Invasion Metastasis **12:** 168–184.
34. NUOVO, G. J. 1992. PCR *in Situ* Hybridizations: Protocols and Applications. Raven Press. New York.
35. NUOVO, G. J., K. LIDONOCCI, P. MACCONNELL & B. LANE. 1993. Intracellular localization of PCR-amplified hepatitis C cDNA. Am. J. Pathol. **17:** 683–690.

Inhibition of Collagenase Gene Expression in Synovial Fibroblasts by All-*trans* and 9-*cis* Retinoic Acid[a]

LUYING PAN[b] AND CONSTANCE E. BRINCKERHOFF[b–d]

Departments of Medicine[b] and Biochemistry[c]
Dartmouth Medical School
Hanover, New Hampshire 03755

A reduction in metalloproteinase levels can be achieved by inhibiting enzyme activity or by controlling enzyme synthesis (reviewed in refs. 1–3). Our laboratory has focused on the latter, with emphasis on mechanisms controlling expression of the gene for interstitial collagenase, matrix metalloproteinase (MMP)-1, in normal fibroblasts. Collagenase is synthesized as preprocollagenase by several cell types,[1–3] and is secreted as soon as it is synthesized.[4] Activation of the latent enzyme is accomplished by proteolytic cleavage of procollagenase.[1–3]

A number of compounds influence collagenase synthesis in these cells.[1–3] Cytokines such as interleukin-1α and -β, tumor necrosis factor-α, polypeptide growth factors such as epidermal growth factor and platelet-derived growth factor, and phorbol esters increase MMP-1 synthesis, whereas compounds such as transforming growth factor-β (TGF-β), glucocorticoid hormones, and vitamin A analogues (retinoids) suppress it.[1–3] When these cells are induced, collagenase is a major gene product, perhaps representing as much as 2% of the mRNA population (ref. 5 and references therein).

Increased synthesis results, at least in part, from increased transcription of the collagenase gene and is mediated by a series of elements within the promoter.[5–10] One element, the 8 bp sequence 5′-TGAGTCAC-3′, is located at approximately −77 in the rabbit gene and −70 in the human gene,[6–10] it is called the activator protein-1 (AP-1) site. It binds the transcription factors Fos and Jun.[10] When gene expression is induced by the addition of phorbol esters, these proteins transactivate transcription. Although the AP-1 site is important in this activation, additional sequences located upstream in the collagenase promoter are also important, and even necessary.[6,7]

Repression of metalloproteinase synthesis by TGF-β,[11] glucocorticoids,[12–14] and retinoids[4,15–18] also occurs by means of transcriptional mechanisms and also depends on several elements in the promoter. Retinoids affect gene expression through two classes of nuclear receptors, the retinoic acid receptors (RARs) (refs. 19 and 20 and references therein) and retinoid X receptors (RXRs),[21–23] each with α, β, and γ subtypes. RARs and RXRs are ligand-dependent transcription factors that are members of the superfamily of steroid-thyroid hormone receptors.[24] These two classes of receptors share only 29% homology in their ligand binding domains. RARs bind with high affinity ($K_d \sim 0.5$ nM) to both all-*trans* and 9-*cis* retinoic acid, whereas RXRs bind with a lower affinity only to 9-*cis*-retinoic acid ($K_d \sim 18$ nM).[25]

[a]This work was supported by grants from the National Institutes of Health (AR26599), The Council for Tobacco Research, and The RGK Foundation to C.E.B., and by an Arthritis Foundation postdoctoral fellowship to L.P.

[d]Corresponding author.

In addition, heterodimer formation of RARs with RXRs enhances retinoid-mediated effects.[22,25–27] Retinoids affect numerous genes, and it is probably the combination of ligand specificity and heterodimerization among the various RAR/RXR subtypes that mediates the diverse effects of retinoids on numerous cells and tissues.[22,25,26,28]

As an initial step in understanding how retinoids suppress transcription of the collagenase gene, we used a series of five chimeric constructs that contain from 182 to 1176 bp of promoter DNA linked to the chloramphenicol acetyl transferase (CAT) gene (FIG. 1). We transiently transfected these constructs into primary cultures of rabbit fibroblasts, the same cells that express the collagenase gene endogenously, and we measured the ability of retinoic acid to suppress phorbol-induced CAT expression.[4] We found that the level of repression varied among the fragments tested. When repression was seen, it was only modest, and some constructs actually showed enhanced CAT activity. These results suggested the presence of a series of retinoid-responsive elements throughout the collagenase promoter.

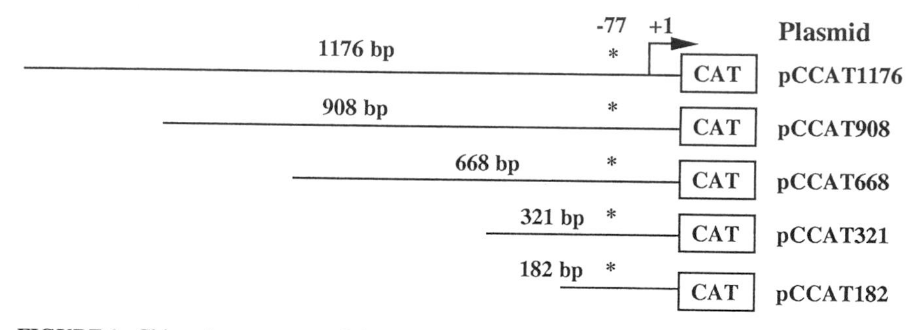

FIGURE 1. Chimeric constructs of the rabbit collagenase promoter linked to chloramphenicol acetyl transferase (CAT). Fragments of the collagenase promoter which contained up to 1176 bp of promoter DNA were linked to the Hind III site of pSVOCAT, a promoterless CAT reporter. Plasmids are designated "pCCAT" followed by a number which indicates the size (in bp) of the rabbit collagenase 5′ flanking DNA inserted. Asterisk indicates AP-I sequence (TGAGTCAC) located at −77 to −70.

Under these same conditions, however, the endogenous collagenase gene showed at least 50% suppression, consistent with results previously reported.[4] To address the possibility that the modest repression may have been due to limited availability of RARs, we cotransfected each RAR into fibroblasts along with 182 bp of promoter DNA.[15] We found that, indeed, cotransfection resulted in a level of repression that paralleled the endogenous gene.[15] Further, CAT expression was regulated in an RAR type-specific manner. RAR-γ was the most potent repressor and the only one that could suppress basal activity.[15]

In the present study we investigate the ability of two retinoids, all-*trans* and 9-*cis* RA, to suppress collagenase gene expression and to modulate levels of mRNA for RAR-α, -β, and -γ and for RXR-α. In addition, we document the presence of both RARs and RXRs in proteins that complex with the AP-1 site in the collagenase promoter, and, finally, we provide further evidence that repression of the collagenase gene by retinoic acid is mediated in an RAR type-specific manner.

MATERIALS AND METHODS

Cell Culture

HIG82 cells are a spontaneously immortalized cell line of rabbit synovial fibroblasts[15,29] and were obtained from American Tissue Culture Collection. Stock cultures were grown to confluence in Dulbecco's modified Eagle's medium (DMEM) supplemented with 10% fetal calf serum, penicillin, and streptomycin. These cells remain contact inhibited and do not grow in soft agar.[15] However, compared to primary cultures of synovial fibroblasts, they reach a higher saturation density and express higher constitutive levels of collagenase,[15,29] although they are still inducible with phorbol esters.[15]

Western Blot Analysis

Cells were grown to confluence in 35-mm 6-well plates. They were then washed 3 times with Hanks' balanced salt solution to remove traces of serum, placed in 2 mL of serum-free DMEM medium supplemented with 0.2% lactalbumin hydrolysate (DMEM/LH) and treated for 24 h with all-*trans*-RA or 9-*cis*-RA (10^{-6} or 10^{-7} M). One milliliter of medium was used for Western blot analysis as previously described.[30]

Northern Blot Analysis

RNA was isolated from HIG82 cells by the guanidine thiocyanate-CsCl method.[15] RNA (20 μg per sample) was electrophoresed on 1% formaldehyde agarose gel and transferred to GeneScreen Plus membrane. Membranes were hybridized for 20 h at 56 °C with denatured [α-^{32}P] dCTP-labeled probe (2×10^6 cpm/mL). They were washed with 0.2× SSC/0.1% SDS, twice at room temperature for 10 min and twice at 56 °C for 30 min.

cDNA Probes

Full-length cDNA of mouse RAR-γ, 900 bp cDNA fragment of RAR-β (EcoRI) (ref. 15 and references therein), 745 bp cDNA of RAR-α (EcoRI-Sac I), full-length or 500 bp of human RXR-α, 550 bp cDNA of GAPDH (HindIII-XbaI) (ATCC), and 630 bp cDNA fragment for collagenase (EcoRI-HaeIII) were used as probes in Northern analysis. Probes were labeled with [α-^{32}P]dCTP by random priming (Pharmacia, Piscataway, NJ).[5–7,15] Glyceraldehyde phosphate dehydrogenase (GAPDH) was used as a control for loading of RNA.

Plasmids

Fragments of the rabbit collagenase promoter linked to the CAT gene have been described previously and are shown in FIGURE 1.[6] The cDNAs encoding mouse RAR-α, -β, and -γ (gift of Dr. P. Chambon) have also been described (ref. 15 and references therein).

Transfection and CAT Assay

Transient transfection and CAT assays were performed as described.[4,5,7,15] Briefly, the calcium phosphate coprecipitation method was used to cotransfect equal amounts of DNA containing the chimeric CAT construct and RAR expression vector into HIG82 cells (2.5 μg DNA per 60-mm plate) with a shock of 10% glycerol for 3 min. Cells were allowed to recover overnight and were then placed in DMEM/LH with or without 10^{-6} M all-*trans*-RA or 9-*cis*-RA for 26 h, 10^{-8} M PMA for 24 h, or 10^{-6} M RA plus 10^{-8} M PMA. Cell extracts were prepared by the freeze-thaw method, and protein concentrations in the cell extracts were determined by Bradford assay (Bio-Rad, Melville, NY). Protein (5 μg per sample) was used for CAT assay. Reaction products were resolved by thin-layer chromatography (TLC) and autoradiography, and were quantitated by scintillation counting of the TLC. Relative CAT activities were normalized to the CAT activity in the absence of treatment within the same transfection group. Student's paired *t* test was used for statistical analysis.

Preparation of Nuclear Extracts and Mobility Shift Assay

Nuclear extracts were prepared from untreated cells and from cells treated for 24 h with all-*trans* or 9-*cis* RA (10^{-6} M).[6,7,15] Nuclear extracts (5 μg) were incubated with 30,000 cpm [γ-^{32}P]ATP-labeled double-stranded oligonucleotides in binding buffer (12 mM Hepes pH 7.9, 4 mM Tris-HCl pH 7.9, 12% glycerol, 60 mM KCl, 1 mM EDTA, 1 μg poly dI-dC) for 15 min on ice, and then incubated with diluted ascites (final concentration 1:30) specific for RAR-α, -β, -γ and RXRs for an additional 30 min on ice. Complexes were resolved by electrophoresis on 5% nondenaturing polyacrylamide gels. Antibodies to RARs and RXRs are generous gifts from Dr. P. Chambon. Antibodies to RARs were prepared against the F region of mouse RAR-α, -β, -γ, and recognize RAR-α, -β, and -γ, respectively.[31] Antibody to RXRs was raised against the E/D region of RXRs, and recognizes all three subtypes of RXRs.[22]

RESULTS

*Inhibition of Endogenous Collagenase Gene Expression by All-***trans** *and 9-***cis** *Retinoic Acid*

Because RARs play a role in repressing collagenase synthesis and because 9-cis RA binds to RARs as well as to RXRs,[25] we compared the suppressive effect of all-*trans* RA and 9-*cis* RA on collagenase synthesis. HIG82 cells were treated for 24 h with all-*trans* or 9-*cis* RA. Collagenase mRNA and protein were then visualized by Western and Northern blot analyses. FIGURE 2 confirms the constitutive expression of collagenase in HIG 82 cells[15,29] and demonstrates the ability of both all-*trans* and 9-*cis* RA to repress markedly collagenase protein (FIG. 2A) and collagenase mRNA (FIG. 2B). The results illustrate the linkage between collagenase mRNA and protein,[5] and also indicate that 9-*cis* RA is as effective as all-*trans* RA in repressing the endogenous gene. Inasmuch as 9-*cis* RA is a ligand for RXRs,[25] this receptor may play a role in repression.

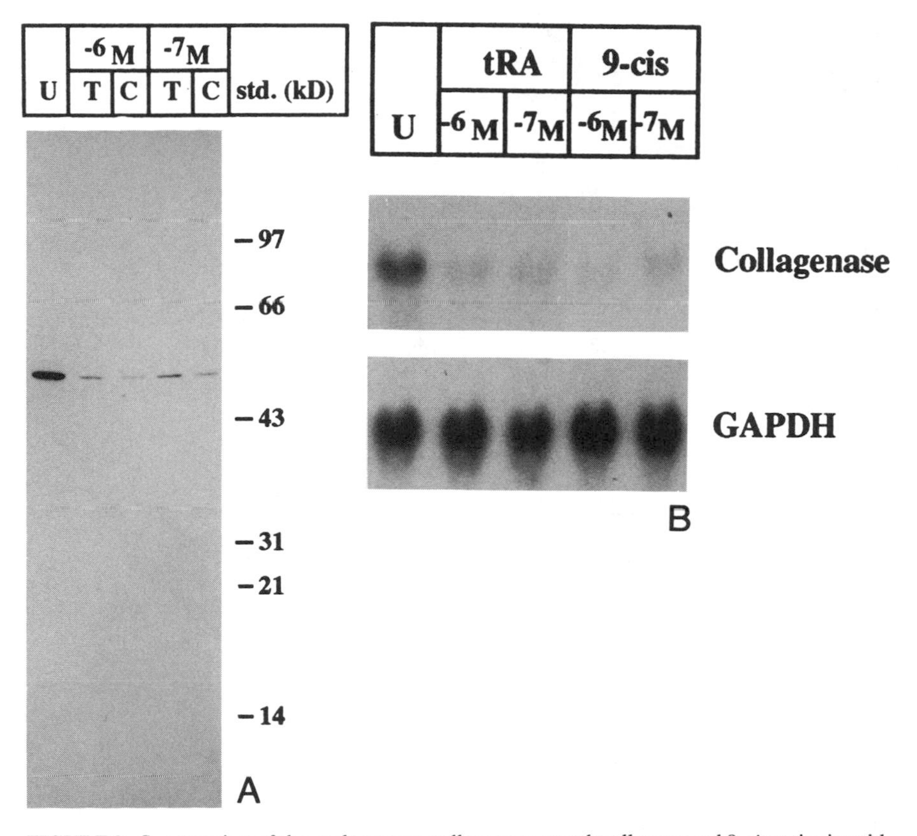

FIGURE 2. Suppression of the endogenous collagenase gene by all-*trans* and 9-*cis* retinoic acid. HIG82 cells were cultured in DMEM/LH medium in the absence (U) or presence of all-*trans* RA (T) or 9-*cis* RA (C) at the concentration indicated. After 24 h, RNA and culture medium were harvested for Western and Northern blot analyses. **(A)** Western analysis: Culture media (1 mL) was TCA precipitated and electrophoresed through a 7.5% acrylamide gel. Proteins were transferred to a nylon membrane, and procollagenase protein was detected with sheep antiserum specific to rabbit collagenase. **(B)** Northern analysis: Total RNA (20 μg) was electrophoresed through a 1% formaldehyde agarose gel. RNA was transferred to a nylon membrane and probed first with a [^{32}P]labeled cDNA for rabbit collagenase. After stripping, the membrane was then probed with a [^{32}P]labeled cDNA for GAPDH to control for variations in loading of the gel. GAPDH, glyceraldehyde phosphate dehydrogenase.

*Induction of RAR mRNAs by All-*trans *and 9-*cis *Retinoic Acid*

We next examined the ability of all-*trans* and 9-*cis* RA to modulate expression of mRNA for RAR-α, -β, and -γ and for RXR-α (FIG. 3). The figure shows a low level of constitutive expression of mRNAs for RAR-α and -γ, whereas mRNA for RAR-β is hardly detectable. However, treatment with either all-*trans* or 9-*cis* RA increases the steady-state levels of mRNAs for all three RARs. This induction was observed within 3 h and peaked between 6 and 12 h of treatment (data not shown).

In contrast, there is a higher level of constitutive expression of mRNA for RXR-α, and neither 9-*cis* RA nor all-*trans* RA affected this expression (FIG. 3). To

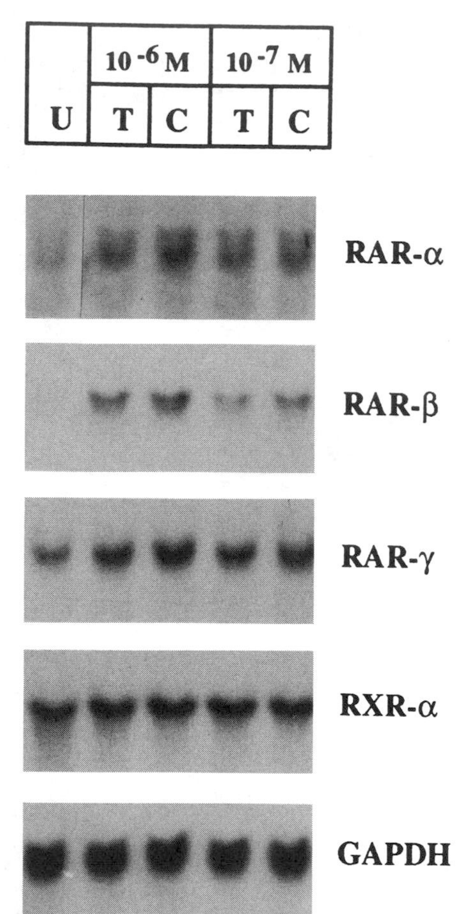

FIGURE 3. Induction of retinoic acid receptor mRNAs by all-*trans* and 9-*cis* RA. HIG82 cells were cultured in DMEM/LH medium in the absence (U) or in the presence of all-*trans* RA (T) or 9-*cis* RA (C) at the concentration indicated for 6 h. Total RNA was then prepared and 20 μg RNA per sample was subjected to Northern analysis. The membrane was hybridized sequentially with [^{32}P]labeled probes specific for RAR-α, -β, -γ, RXR-α, and GAPDH (loading control), respectively. RXR, retinoid X receptor; GAPDH, glyceraldehyde phosphate dehydrogenase.

rule out the possibility that retinol or retinoic acid might be present in the serum-containing culture medium and might account for the relatively high level of expression of RXR-α mRNA, we cultured HIG cells in serum-free medium for up to 4 days. As shown in FIGURE 4, these culture conditions did not alter the level of RXR-α mRNA. However, after 24 h in serum-free medium, constitutive expression of collagenase mRNA increased considerably. These findings indicate that constitutive expression of RXR-α, by itself, cannot repress collagenase synthesis and suggest that repression is ligand dependent. Perhaps repression occurs through RAR/RXR heterodimer formation, resulting from the induction of RARs by all-*trans* and 9-*cis* RA.

Both RAR and RXR Are Involved in Protein/DNA Complexes with the Collagenase Promoter

We used gel mobility shift assays to test the hypothesis that RAR/RXR heterodimers might be involved in suppressing transcription. Because retinoids mediate some of their effects on the collagenase gene through the AP-1 site,[15–18] we designed an oligonucleotide which spanned the sequences between −83 and −56 and which contained an AP-1 consensus at position −77 to −70.[6,7,15] Nuclear extracts from untreated cells and from cells treated with either all-*trans* or 9-*cis* RA were incubated with this oligo- in the presence or absence of antibodies specific to RARs and RXRs.

We found that nuclear proteins derived from untreated cells bound poorly to this oligonucleotide (FIG. 5). In contrast, extracts from cells treated with all-*trans* and 9-*cis* RA did bind, and binding was competed by a 100-fold molar excess of self (FIG. 5, bands 1 and 2). Of importance, addition of antibodies to RARs and RXRs in the shift reaction gave rise to a "supershifted" band (FIG. 5, band 3), suggesting that the complex contains both RARs and RXRs, perhaps as heterodimers. A single T to G base-pair mutation at position −77 of this oligo-abolished binding activity (data not shown), and no "supershifts" were seen when antibody was added to extracts from untreated cells (data not shown). Others[16,18,32] have reported that there is no direct binding of RARs to the AP-1 sequence. However, we do not yet know whether RXRs can bind to this sequence, either as homo- or heterodimers. Thus, the binding we observe may indicate that RAR/RXR heterodimers are able to bind directly to the AP-1 site. Alternatively, the RAR/RXR complexes may bind to other proteins which, in turn, interact with the DNA.

*Repression of CAT Activity by All-*trans *Retinoic Acid*

Although our earlier studies suggested the presence of several RA response elements along 1200 bp of the collagenase promoter,[4] the role of each RAR in mediating these responses was not known. More recent studies with the smallest construct of 182 bp of promoter DNA showed that each RAR participated in repression, with RAR-γ being the most potent.[15] We also found that repression required several regions of promoter DNA: the AP-1 site, a TTCA site at −109, and a 20 bp region located at −180 to −160. [15] To test the ability of larger fragments of the collagenase promoter to respond to each RAR, we cotransfected our nested series of promoter fragments (see FIG. 1) and each RAR into HIG cells, and then treated the cells with all-*trans* or 9-*cis* RA.

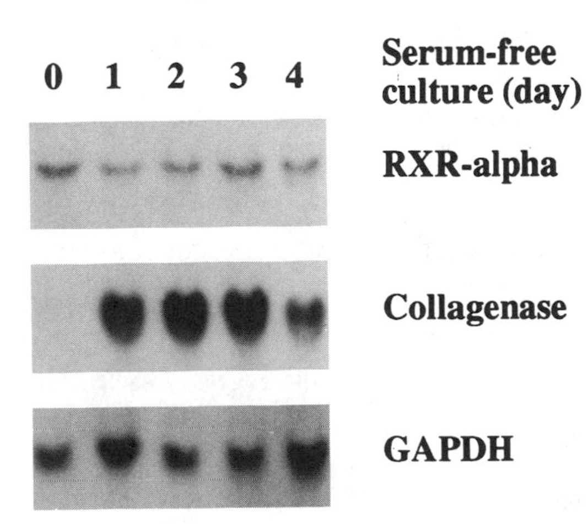

FIGURE 4. Constitutive expression of RXR-α mRNA. HIG82 cells were cultured in DMEM/LH medium up to 4 days. Total RNA was prepared at each time point and 20 μg of RNA per sample were subjected to Northern analysis. The membrane was hybridized with [^{32}P]labeled probes specific for RXR-α, collagenase, and GAPDH (loading control), respectively. RXR, retinoid X receptor; GAPDH, glyceraldehyde phosphate dehydrogenase.

Treatment	none	all-trans RA	9-cis RA
Competitor	- +	- + - - - -	- + - - - -
Ab to RAR	- -	- - α β γ -	- - α β γ -
Ab to RXR	- -	- - - - - +	- - - - - +

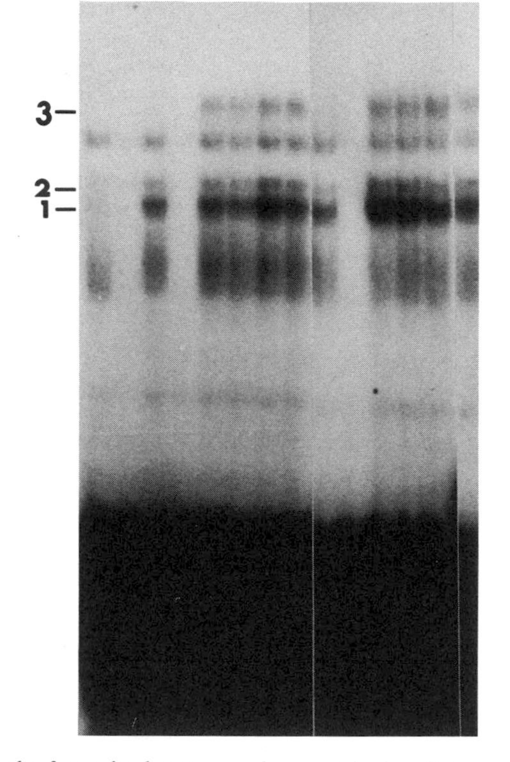

FIGURE 5. Specific complex formation between nuclear proteins in cells treated with all-*trans* and 9-*cis* retinoic acid and an oligonucleotide corresponding to $-83/-56$ of the rabbit collagenase promoter. HIG82 cells were cultured in DMEM/LH (none) or treated with 10^{-6} M all-*trans* RA, 10^{-6} M 9-*cis* RA for 24 h. Nuclear extracts were prepared and 5 μg of protein were incubated with the oligoprobe alone, or in the presence of 100X molar excess of unlabeled self-competitor on ice for 10 min. Antibodies to RARs and RXR, at 1:30 final concentration, were then added to reaction mixture for another 20-min incubation. The reaction products were analyzed by 5% nondenaturing acrylamide gel. The oligoprobe represents the sequence 5′-GAAAGCATGAGTCACACAAGCCCTCAGCT-3′ from -83 to -56 of the collagenase promoter which contains the AP-1 consensus (underlined).

TABLE 1 illustrates the effects of each RAR on constitutive CAT expression. Similar to our earlier findings when 182 bp of promoter DNA was tested with all-*trans* RA,[15] only RAR-γ repressed transcription. Here we also tested 9-*cis* RA and found that both ligands were equally effective, with approximately 35–40% repression. The level of repression was somewhat less with the larger constructs,

TABLE 1. Repression of Basal CAT Activity by Retinoic Acid Receptors upon Treatment with All-*trans* and 9-*cis* RA

Cotransfection	pSG5		RAR-α		RAR-β		RAR-γ	
pCCAT1176	1.16	1.22	0.81	0.78	0.65	0.62	0.71	0.65
pCCAT908	1.26	1.16	0.84	0.80	0.70	0.66	0.75	0.75
pCCAT668	1.44	1.36	0.96	0.90	0.87	0.69	0.63	0.51
pCCAT321	1.16	1.09	1.02	1.01	0.82	0.86	0.75	0.69
pCCAT182	1.03	1.04	0.95	0.90	0.98	1.02	0.61	0.64
Treatment	*t*RA	9-*cis*	*t*RA	9-*cis*	*t*RA	9-*cis*	*t*RA	9-*cis*

NOTE: Nested 5′-deletions of pCCAT1176 constructs were cotransfected with individual RAR expression vectors or expression vector without RAR insert (pSG5) into HIG82 cells by the calcium-phosphate coprecipitation method. Cells then were treated with 10^{-6} M all-*trans* RA or 9-*cis* RA for 24 h. Cell lysates were harvested by the freeze-thaw method, and 5 μg of protein were used for the CAT assay. Results represent the mean of 3–4 independent transfections. The CAT activity in untreated cells was given the value of 1, and the values presented are the CAT activities relative to it. CAT, chloramphenical acetyl transferase.

suggesting the possible presence of some positive regulatory elements. RAR-β could not decrease constitutive transcription with the 182 bp construct, but it could mediate a modest decrease with 320 and 670 bp of promoter, and a more substantial (30–40%) one with the two largest constructs. Finally, RAR-α was hardly effective. A 20% repression was seen only with the two largest fragments of promoter DNA. The effects of all-*trans* and 9-*cis* RA were always the same, suggesting either that RARs are binding both retinoids and/or that cells can convert all-*trans* to 9-*cis* RA to allow specific interaction of RXRs with 9-*cis* RA.

We next compared the effect of each RAR and all-*trans* or 9-*cis* RA on cells transfected with the same series of promoter constructs and induced with phorbol esters. In agreement with our previous studies in primary cultures,[4] our results showed that, in the absence of cotransfected RARs, repression was only modest and some constructs showed a slight enhancement of CAT activity (TABLE 2). Cotransfection of each RAR suppressed transcription. However, in contrast to the effects on

TABLE 2. Repression of PMA-Induced CAT Activity by All-*Trans* or 9-*cis* Retinoic Acid

Cotransfection	pSG5		RAR-α		RAR-β		RAR-γ	
pCCAT1176	1.13	0.92	0.67	0.55	0.56	0.50	0.58	0.51
pCCAT908	0.89	0.90	0.60	0.63	0.55	0.54	0.45	0.48
pCCAT668	1.11	1.12	0.72	0.60	0.50	0.49	0.53	0.51
pCCAT321	1.05	1.14	0.83	0.78	0.68	0.63	0.49	0.57
pCCAT182	1.03	1.05	0.68	0.62	0.47	0.53	0.42	0.54
Treatment	*t*RA	9-*cis*	*t*RA	9-*cis*	*t*RA	9-*cis*	*t*RA	9-*cis*

NOTE: Nested 5′-deletions of pCCAT1176 constructs were cotransfected with individual RAR expression vectors or expression vector without RAR insert (pSG5) into HIG82 cells by the calcium-phosphate coprecipitation method. Cells then were treated with 10^{-8} M PMA with or without 10^{-6} M all-*trans* RA or 9-*cis* RA for 24 h. Cell lysates were harvested by the freeze-thaw method, and 5 μg of protein were used for CAT assay. Results represent the mean of 3–4 independent transfections. Compared with untreated cells, PMA increased CAT activity 5- to 10-fold. CAT activity in PMA-treated cells is given as 1, and values seen with retinoid treatment are relative to it. PMA, phorbol myristate acetate; CAT, chloramphenicol acetyl transferase.

constitutive expression (TABLE 1) where there were distinct differences among the RARs in their ability to modulate each construct, suppression of phorbol-induced transcription by each RAR was similar for all the constructs. Again, RAR-γ was the most potent, suppressing transcription by about 50%, compared to only a 30% repression with RAR-α ($p < 0.005$ for all-*trans* RA and $p < 0.02$ for 9-*cis* RA). RAR β-mediated repression was about 45%. Data in TABLES 1 and 2 indicate that the mechanisms suppressing basal versus phorbol-induced transcription are different and underscore a distinctive role for each RAR.

DISCUSSION

In this paper, we show that RXR-α mRNA is constitutively expressed in fibroblasts and is not regulated by retinoids. In contrast, mRNAs for RAR-α, -β, and -γ are constitutively low but are inducible with either all-*trans* and 9-*cis* retinoic acid. We also demonstrate that nuclear extracts from normal fibroblasts treated with either all-*trans* or 9-*cis* RA contain proteins that specifically complex with an oligonucleotide containing the AP-1 site and that these complexes can be "super-shifted" with antibodies to the RARs and to the RXRs (FIG. 5). These results suggest that both RARs and RXRs are involved in complex formation with the DNA sequences, and that RAR/RXR heterodimers mediate RA suppression of collagenase gene expression. Because no evidence exists that either RARs or RXRs bind directly to the AP-1 site in either collagenase or stromelysin,[16–18,32] it is possible that, like the glucocorticoid receptor/hormone (GR/h) complex,[12–14,33,34] the RARs and RXRs repress transcription indirectly by protein-protein interactions rather than by binding to the DNA directly.

This model is similar to that described for the down-regulation of gene expression by glucocorticoids.[33,34] Receptors (GRs) for these hormones reside in the cytoplasm, and in the presence of the glucocorticoid hormone (h), the ligand-receptor complex enters the nucleus where it usually binds directly to specific nucleotide sequences within the promoter (called *g*lucocorticoid *r*esponse *e*lements, or GREs) and influences gene expression.[33,34] However, in some instances, including regulation of metalloproteinase gene expression, the GR/h complex interacts indirectly with the promoter.[12–14,33,34] This has been called a "composite" GRE because it binds both the GR/h complex and additional proteins, such as Fos or Jun.[33,34] In the absence of glucocorticoid, Fos and Jun heterodimerize, bind to the AP-1 site in the collagenase promoter, and activate transcription. Then, in the presence of hormone, the GR/h complex interacts with Jun and induces a conformational change in it.[34] This change interferes with the ability of the Fos:Jun heterodimer to bind to DNA, thereby suppressing transcription.[12–14,33,34] However, it not yet clear whether retinoids induce a conformational change in Fos or Jun. Retinoids suppress mRNA levels for Fos and Jun[15] and may also sequester one or both of these proteins,[18,32] thus preventing their interaction with the DNA.

The fact that both all-*trans* and 9-*cis* RA suppress transcription of the collagenase gene equally well provides support for the concept of heterodimerization. Although RXR expression is constitutive in HIG-82 cells, it cannot suppress transcription by itself (FIG. 4). Further, RXRs bind only to 9-*cis* retinoic acid with relatively low affinity,[25] whereas both all-*trans* and 9-*cis* RA interact with and induce expression of the RARs (ref. 25 and FIG. 3). Conceivably, upon exposure to either retinoid, levels of the RARs increase; this provides an opportunity for heterodimerization. The increased affinity for the ligand afforded by RARs (K_d = 0.5 nM versus 18 nM for

RXRs)[25] may allow the ligand/receptor complex to bind to other proteins and/or to DNA, thereby efficiently modulating gene expression.

It is now well accepted that the AP-1 site plays an important role in inducing and suppressing metalloproteinase gene expression.[6–10,12–18] However, increasing evidence indicates that additional sites within the promoters of the metalloproteinase genes contribute significantly to both induction and repression.[4,7,15,35–37] Our earlier data had suggested the presence of a series of positive and negative regulatory elements within the collagenase promoter that mediated RA responsiveness,[4] and data presented in TABLE 1 support this. The table illustrates clear differences in the ability of each RAR to regulate constitutive expression of the collagenase gene. This suggests either that distinctive elements along the promoter selectively bind each RAR and/or that these elements bind all the RARs, but with different affinities.

The data in TABLE 2 are somewhat more subtle. All constructs are repressible by all three RARs. Phorbol induction seems to augment the ability of RARs to repress, perhaps because the phorbol treatment up-regulates the level of proteins such as Fos and Jun[15,37] with which the RARs may interact.[15–18] In addition, for any RAR tested, the level of repression seen with all constructs is about the same; larger fragments of DNA do not give substantially more repression. This suggests that repression might be mediated by sequences contained within the smallest fragment tested, that is, 182 bp of DNA, and that all three RARs can bind to these sequences. However, slight differences exist among the RARs in their ability to repress. For example, for the two smallest constructs containing 182 bp and 320 bp of promoter, RAR-γ is slightly more potent that RAR-β. In addition, although RAR-α is the least potent, there is some variability among the constructs in their responsiveness, with the 320 bp construct being the least sensitive.

Despite the complexities observed at the subcellular level, studies with animal models of arthritis suggest that retinoids are therapeutically useful[38] and that they can prevent the connective tissue degradation caused by overproduction of collagenase in arthritis.[3,39] This suggests that inhibiting enzyme synthesis is a viable clinical option. However, these compounds can also affect the expression of many other genes[24,28] and thereby may compromise efficacy. Because we already know that repression of collagenase occurs in an RAR type-specific manner[15] and that certain retinoids have affinities for certain RARs/RXRs,[25–27] it is possible that eventually we may be able to develop a transcriptional inhibitor that is targeted specifically to the gene(s) of interest—in this case collagenase. This will require knowledge of exactly which sequences of DNA are involved in transcriptional activation and repression, what transcription factors bind to these sequences, and how specific receptors participate.

REFERENCES

1. MATRISIAN, L. M. 1990. Metalloproteinases and their inhibitors in matrix remodeling. Trends Genet. **6:** 121–125.
2. WOESSNER, J. J. 1991. Matrix metalloproteinases and their inhibitors in connective tissue remodeling. FASEB J. **5:** 2145–2154.
3. BIRKEDAL-HANSEN, H., Z. WERB, H. WELGUS & H. VAN WORT, EDS. 1992. Matrix Metalloproteinases and Inhibitors. :165–175. Gustav Fischer Verlag. New York.
4. NAGASE, H., C. E. BRINCKERHOFF, C. A. VATER & E. D. HARRIS, JR. 1983. Biosynthesis and secretion of procollagenase by rabbit synovial fibroblasts. Biochem. J. **214:** 281–288.
5. BRINCKERHOFF, C. E. & D. T. AUBLE. 1990. Regulation of collagenase gene expression in synovial fibroblasts. Ann. N.Y. Acad. Sci. **580:** 355–374.
6. AUBLE, D. T. & C. E. BRINCKERHOFF. 1991. The AP-1 sequence is necessary but not

sufficient for phorbol induction of collagenase in fibroblasts. Biochemistry **30:** 4629–4635.

7. Chamberlain, S. H., R. M. Hemmer & C. E. Brinckerhoff. 1993. A novel phorbol ester response region in the collagenase promoter binds Fos and Jun. J. Cell. Biochem. **52:** 337–351.
8. Angel, P., I. Baumann, B. Stein, H. Delius, H. J. Rahmsdorf & P. Herrlich. 1987. 12-*O*-Tetradecanoyl-phorbol-13-acetate induction of the human collagenase gene is mediated by an inducible enhancer element located in the 5′-flanking region. Mol. Cell. Biol. **7:** 2256–2266.
9. Angel, P., M. Imagawa, R. Chiu, B. Stein, R. J. Imbra, H. J. Rahmsdorf, C. Jonat, P. Herrlich & M. Karin. 1987. Phorbol ester-inducible genes contain a common cis element recognized by a TPA-modulated trans-acting factor. Cell **49:** 729–739.
10. Rauscher, F. E., P. J. Voulalas, B. J. Franza & T. Curran. 1988. Fos and Jun bind cooperatively to the AP-1 site: Reconstitution in vitro. Genes Dev. **2:** 1687–1699.
11. Kerr, L. D., D. B. Miller & L. M. Matrisian. 1990. TGF β inhibition of transin/stromelysin gene expression is mediated through a fos binding sequence. Cell **61:** 267–278.
12. Jonat, C., H. J. Rahmsdorf, K. K. Park, A. C. Cato, S. Gebel, H. Ponta & P. Herrlich. 1990. Antitumor promotion and antiinflammation: Down-modulation of AP-1 (Fos/Jun) activity by glucocorticoid hormone. Cell **62:** 1189–1204.
13. Konig, H., H. Ponta, H. J. Rahmsdorf & P. Herrlich. 1992. Interference between pathway-specific transcription factors: Glucocorticoids antagonize phorbol ester-induced AP-1 activity without altering AP-1 site occupation *in vivo.* EMBO J. **11:** 2241–2246.
14. Lucibello, F. C., E. P. Slater, K. U. Jooss, M. Beato & R. Muller. 1990. Mutual transrepression of Fos and the glucocorticoid receptor: Involvement of a functional domain in Fos which is absent in FosB. EMBO J. **9:** 2827–2834.
15. Pan, L., S. H. Chamberlain, D. T. Auble & C. E. Brinckerhoff. 1992. Differential regulation of collagenase gene expression by retinoic acid receptors-α, β and γ. Nucleic Acids Res. **20:** 3105–3111.
16. Nicolson, R. C., S. Mader, S. Nagpal, M. Leid, C. Rochette-Egly & P. Chambon. 1990. Negative regulation of the rat stromelysin gene promoter by retinoic acid is mediated by an AP-1 binding site. EMBO J. **13:** 4443–4445.
17. Lafyatis, R., S. J. Kim, P. Angel, A. B. Roberts, M. B. Sporn, M. Karin & R. L. Wilder. 1990. Interleukin-1 stimulates and *all-trans*-retinoic acid inhibits collagenase gene expression through its 5′ activator protein-1-binding site. Mol. Endocrinol. **4:** 973–980.
18. Schule, R., P. Rangarajan, N. Yang, S. Kliewer, L. J. Ransone, J. Bolado, I. M. Verma & R. M. Evans. 1991. Retinoic acid is a negative regulator of AP-1-responsive genes. Proc. Natl. Acad. Sci. USA **88:** 6092–6096.
19. Giguere, V., E. S. Ong, P. Segui & R. M. Evans. 1987. Identification of a receptor for the morphogen retinoic acid. Nature **330:** 624–629.
20. Krust, A., P. Kastner, M. Petkovich, A. Zelent & P. Chambon. 1989. A third human retinoic acid receptor, hRAR-gamma. Proc. Natl. Acad. Sci. USA **86:** 5310–5314.
21. Mangelsdorf, D. J., E. S. Ong, J. A. Dyck & R. M. Evans. 1990. Nuclear receptor that identifies a novel retinoic acid response pathway. Nature **345:** 224–229.
22. Leid, M., P. Kastner, R. Lyons, H. Nakshatri, M. Saunders, T. Zachareski, J. Y. Chen, A. Staub, J. M. Garnier, S. Mader & P. Chambon. 1992. Purification, cloning, and RXR identity of the HeLa cell factor which RAR or TR heterodimerizes to bind target sequences efficiently. Cell **68:** 377–395.
23. Mangelsdorf, D. J., U. Borgmeyer, R. A. Heyman, J. Y. Zhou, E. S. Ong, A. E. Oro, A. Kakizuka & R. M. Evans. 1992. Characterization of three RXR genes that mediate the action of 9-*cis* retinoic acid. Genes Dev. **6:** 329–344.
24. Evans, R. M. 1988. The steroid and thyroid hormone receptor superfamily. Science **240:** 889–895.
25. Allenby, G., M.-T. Bocquel, M. Saunders, S. Kazmer, J. Speck, M. Rosenberger, A. Lovey, P. Kastner, J. Grippo, P. Chambon & A. Levin. 1993. Retinoic acid receptors

and retinoid X receptors: Interactions with endogenous retinoic acids. Proc. Natl. Acad. Sci. USA **90:** 30–34.

26. LEHMANN, J. M., L. JONG, A. FANJUL, J. F. CAMERON, X. P. LU, P. HAEFNER, M. I. DAWSON & M. PFAHL. 1992. Retinoids selective for retinoid X receptor response pathways. Science **258:** 1944–1946.
27. LEVIN, A. A., L. J. STURZENBECKER, S. KAZMER, T. BOSAKOWSKI, C. HUSELTON, G. ALLENBY, J. SPECK, C. KRATZEISEN, M. ROSENBERGER, A. LOVEY & J. F. GRIPPO. 1992. 9-cis retinoic acid stereoisomer binds and activates the nuclear receptor RXRα. Nature **355:** 359–361.
28. LEID, M., P. KASTNER & P. CHAMBON. 1992. Multiplicity generates diversity in the retinoic acid signalling pathways. Trends Biochem. Sci. **17:** 427–433.
29. GEORGESCU, H. I., D. MENDLOW & C. H. EVANS. 1988. HIG-82: An established cell line from rabbit periarticular soft tissue which retains the "activatable" phenotype. In Vitro Cell. Dev. Biol. **24:** 1015–1022.
30. BRINCKERHOFF, C. E., K. SUZUKI, T. I. MITCHELL, F. ORAM, C. I. COON, R. D. PALMITER, & H. NAGASE. 1990. Rabbit procollagenase synthesized and secreted by a high yield mammalian expression vector requires stromelysin (matrix metalloproteinase 3) for maximal activation. J. Biol. Chem. **265:** 22262–22269.
31. ROCHETTE-EGLY, C., Y. LUTZ, M. SAUNDERS, M. P. GAUB & P. CHAMBON. 1991. Retinoic acid receptor g: Specific immunodetection and phosphorylation. J. Cell. Biol. **115:** 535–545.
32. YANG-YEN, H.-F., X.-K. ZHANG, G. GRAUPNER, M. ZUKERMAN, B. SAKAMOTO, M. KARIN & M. PFAHL. 1991. Antagonism between retinoic acid receptors and AP-1: Implications for tumor protection and inflammation. New Biologist **3:** 1206–1219.
33. DIAMOND, M. I., J. N. MINER, S. K. YOSHINGA & K. R. YAMAMOTO. 1990. Transcription factor interactions: Selectors of positive or negative regulation from a single DNA element. Science **249:** 1266–1272.
34. MINER, J. & K. R. YAMAMOTO. 1991. Regulatory crosstalk at composite response elements. Trends Biochem. Sci. **16:** 423–426.
35. BUTTICE, G., S. QUINONES & M. KURKINEN. 1991. The AP-1 site is required for basal expression but is not necessary for TPA-response of the human stromelysin gene. Nucleic Acids Res. **19:** 3723–3731.
36. BUTTICE, G. & M. KURKINEN. 1993. A polyoma enhancer A-binding protein-3 site and Ets-2 protein have a major role in the 12-*O*-tetradecanoylphorbol-13-acetate response of the human stromelysin gene. J. Biol. Chem. **268:** 7196–7204.
37. CONCA, W., P. E. AURON, W. M. AOUN, N. BENNETT, P. SECKINGER, H. G. WELGUS, S. R. GOLDRING, S. P. EISENBERG, J. M. DAYER & S. M. KRANE. 1991. An interleukin 1 beta point mutant demonstrates that *jun/fos* expression is not sufficient for fibroblast metalloproteinase expression. J. Biol. Chem. **266:** 16265–16268.
38. HAROUI, B., R. W. WILDER, M. B. ALLEN, M. B. SPORN, R. K. HELFGOTT & C. E. BRINCKERHOFF. 1985. Dose-dependent suppression by the synthetic retinoid 4-hydroxyphenylretinamide of streptococcal cell wall induced arthritis in rats. Int. J. Immunopharmacol. **7:** 903–916.
39. HARRIS, E. J. 1990. Rheumatoid arthritis: Pathophysiology and implications for therapy. N. Engl. J. Med. **322:** 1277–1289.

Confocal Laser Scanning Immunofluorescence Imaging for Quantitation of Stromelysin in Human Synovial Fibroblasts

V. M. BARAGI,[a,b] D. A. BROTT,[c] L. QIU,[a] M. C. CONROY,[a]
AND N. D. LALWANI[c]

Departments of Immunopathology[a] and
Pathology & Experimental Toxicology[c]
Parke-Davis Pharmaceutical Research
Warner Lambert Co.
Ann Arbor, Michigan 48105

Cartilage degradation associated with osteoarthritis is mediated, in part, by stromelysin produced by chondrocytes and synovial fibroblasts.[1] Human synovial fibroblasts (HSF) express elevated levels of stromelysin when incubated in the presence of interleukin-1 (IL-1) *in vitro.*[2] Elevated levels of stromelysin have been reported in humans with osteoarthritis as well as in animal models of arthritis.[3,4] The arthritic lesions are generally focal, and chondrocytes from these lesions express higher levels of stromelysin than those present in surrounding tissue.[4] Variability in the expression of stromelysin in these cells is also observed *in vitro* with IL-1 stimulation.[5] Procedures such as measurement of enzyme activity, ELISA- or immunocytochemistry-based protein detection have been used to quantitate stromelysin from normal and arthritic tissues. Nevertheless, these procedures do not take into account variability among cell populations. Therefore, to obtain information on spatial distribution and to quantitate stromelysin expressed by individual cells, HSF were treated with IL-1 in the presence of different concentrations of dexamethasone and analyzed by confocal laser scanning cytometry.

As shown in FIGURE 1, stromelysin protein-associated fluorescence was detected as discrete deposits in the perinuclear cytoplasmic region in the form of an envelope. However, not all the cells showed increased levels of stromelysin synthesis in response to stimulation with IL-1 *in vitro.* Approximately 34% of the cells showed a higher level of stromelysin expression in IL-1–treated HSF (FIG. 2). A dose-dependent inhibition (IC_{50} = 0.38 nM) in the expression of stromelysin was observed when HSF were stimulated with IL-1 in the presence of increasing concentrations of dexamethasone. The dose-dependent changes in stromelysin mRNA expression correlated well with the accumulation of stromelysin (data not shown). These results suggest that *de novo* biosynthesis of stromelysin by individual cells could be quantitated by fluorescence confocal imaging, and therefore this technique is useful in analyzing heterogeneity in the expression of the enzymes by cells in culture.

On the basis of fluorescence intensity, the present study suggests that increased numbers of cells are involved in the synthesis of stromelysin in HSF treated with IL-1 and that dexamethasone decreased the number of these cells in a dose-dependent

[b] Address correspondence to Vijaykumar M. Baragi, Ph.D., Department of Immunopathology, Parke-Davis Pharmaceutical Research, 2800 Plymouth Road, Ann Arbor, Michigan 48105.

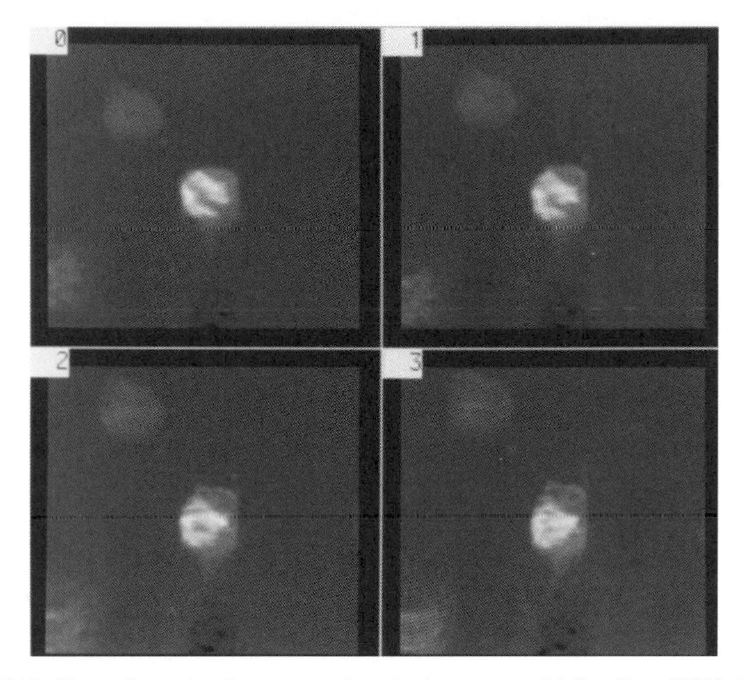

FIGURE 1. Three-dimensional reconstruction of a human synovial fibroblast (HSF) showing the distribution of stromelysin. Human synovial fibroblasts from a patient with osteoarthritis were incubated with IL-1β (40 μg/mL) in DMEM containing 10% fetal calf serum at 37 °C for 20 h followed by monensin (0.5 mM) for 5 h. Formalin-fixed preparations were stained with rabbit anti-human stromelysin and counterstained with goat anti-rabbit TRITC-labeled IgG. Fluorescence imaging was performed on an ACAS 570 Interactive Laser Cytometer, using 225-μm pinhole, 200-mW laser power along with 10% neutral density filter and 10% scanning strength. TRITC was excited at 514 nm and emission >514 nm was collected by PMT set at 40% efficiency. A 72 × 72 μm area was scanned at 0.2 μm step size and 32 samples/point. Image scans were collected at 1 μm step size in z-direction. The figure shows three-dimensional reconstructs in four rotational steps, of a stimulated and an unstimulated fibroblast at an illumination angle of 80° and opacity level of 25%. These images were generated using azimuth = 180°, elevation rotation = 40°, and azimuth rotation = 0°.

manner. Because elevated levels of stromelysin protein in fibroblasts were preceded by elevated stromelysin mRNA levels, and because monensin, an inhibitor of protein secretion, increased intracellular accumulation of the enzyme, the majority of stromelysin in IL-1–treated HSF should have resulted from *de novo* protein synthesis. These results are in agreement with earlier reports that IL-1 induces, whereas dexamethasone represses, transcription of stromelysin and collagenase in cells.[6] Interestingly, the concentration of dexamethasone effective in the present study (3.8×10^{-10} M) is similar to that reported for inhibition of phorbol ester-induced collagenase gene expression (1.5×10^{-9} M) in HeLa-tk$^-$ cells that were transiently transfected with collagenase promoter-cat constructs.[7]

In summary, our study demonstrates that confocal microscopy in combination with immunofluorescence can be used to quantitate stromelysin expression in individual cells and thus evaluate heterogeneity of cells *in vitro*.

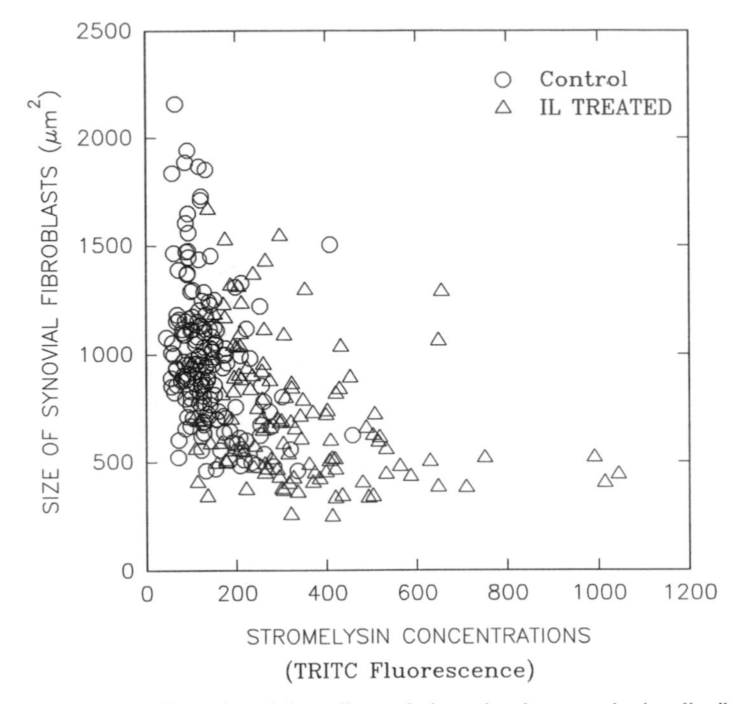

FIGURE 2. A scatter chart plot of the cell populations showing quantitative distribution of stromelysin-associated fluorescence. Fluorescence imaging was performed under similar conditions as in FIGURE 1. Selected areas 180 × 180 μm in size were scanned at 0.5 μm step size. The levels of stromelysin were quantitated as *average fluorescence* per cell. The average fluorescence per cell was obtained by dividing the total integrated fluorescence value by the number of data points in that cell. The scatter chart plot was generated by plotting the fluorescence levels of stromelysin on the abscissa relative to the size of fibroblasts (μm^2) on the ordinate. The chart shows fluorescence distribution of the control synovial fibroblasts (*circles*) and fibroblasts treated with IL-1 (*triangles*). Cells stimulated with IL-1 in the presence of dexamethasone showed a similar pattern to that of control cell populations.

REFERENCES

1. DEAN, D. D., J. MARTEL-PELLETIER, J. P. PELLETIER & J. F. WOESSNER, JR. 1987. J. Rheumatol. **14:** 43–44.
2. MACNAUL, K. L., N. CHARTRAIN, M. LARK, M. J. TOCCI & N. I. HUTCHINSON. 1990. J. Biol. Chem. **265:** 17238–17245.
3. PELLETIER, J. P. & J. MARTEL-PELLETIER. 1985. Arthritis Rheum. **28:** 1393–1401.
4. PELLETIER, J. P., J. MARTEL-PELLETIER, J. M. CLOUTIER & J. F. WOESSNER, JR. 1987. Arthritis Rheum. **30:** 541–548.
5. BRINCKERHOFF, C. E., I. M. PLUCINSKA, L. A. SHELDON & G. O'CONNOR. 1986. Biochemistry **25:** 7378–7384.
6. FRISCH, S. M. & H. E. RULEY. 1987. J. Biol. Chem. **262:** 16300–16304.
7. JONAT, C., H. J. RAHMSDORF, K. K. PARK, A. C. B. CATO, S. GEBEL, H. PONTA & P. HERRLICH. 1990. Cell **62:** 1189–1204.

A High Throughput Fluorogenic Substrate for Stromelysin (MMP-3)

D. MARK BICKETT,[a] MICHAEL D. GREEN,
CRAIG WAGNER, JEREMY T. ROTH, JUDD BERMAN,
AND GERARD M. McGEEHAN

Glaxo Research Institute
Departments of Biochemistry and Medicinal Chemistry
5 Moore Drive
Research Triangle Park, North Carolina 27709

Aberrant regulation of stromelysin, a member of the matrix metalloproteinase (MMP) family of enzymes, has been implicated in the degradation of basement membrane components in rheumatoid arthritis[1–4] and tumor metastasis.[5,6] Many synthetic peptide substrates have been developed to assay this family of enzymes. Peptide substrates and cleavage products can be separated by high performance liquid chromatography (HPLC) and detected at 214 nm[7] or dinitrophenyl groups may be incorporated, making detection possible at 372 nm.[8,9] Many of these assay systems are not well suited for high throughput screening of MMP inhibitors because of sensitivity and chemical interferences. This has led to the development of new fluorogenic substrates for several of the MMPs.[10–14] We sought to develop a fluorogenic stromelysin assay compatible with the filters available on 96-well commercial instrumentation that could be used for rapid evaluation of inhibitors. Fluorogenic peptide substrates were developed based on peptide reported by Teahan *et al.*[7] and Niedzwiecki *et al.*[15] using either *N*-methylanthranilic acid (NMA) (EX_{340}/EM_{440}) as the fluorophore with a dinitrophenyl quencher or 7-dimethylaminocoumarin-4-acetate (DMC) (EX_{368}/EM_{459}) as the fluorophore and 6-(*N*-[7-nitrobenz-2-oxa-1,3-diazol-4-yl] amino) hexanoic acid (NBD) (EX_{467}/EM_{534}) as the quencher. The evaluation of these substrates is detailed.

MATERIALS AND METHODS

Kinetic Measurements. The k_{cat}/K_m values were determined at a substrate concentration of 0.2 μM for substrate 7 and 0.75 μM for substrate 2 (TABLE 1) in a quartz cuvette containing 3 mL of assay buffer (200 mM NaCl, 50 mM Tris, 5 mM $CaCl_2$, 20 μM $ZnSO_4$, and 0.05% Brij 35, pH 7.6) at 23 °C. Fluorescence increases for substrates 7 and 2 were monitored at EX_{385}/EM_{475} and EX_{280}/EM_{346} nm, respectively, using a Perkin-Elmer LS-5B luminescence spectrometer.

Relative k_{cat}/K_m values of substrates were determined in competition assays at 37 °C. Stromelysin was adjusted to 5 nM in a 2-mL reaction volume. Reactions were initiated by the addition of substrate (10 μM). Aliquots were quenched after 15, 30, and 60 minutes with 20 mM EDTA. HPLC analysis of the dinitrophenyl (DNP) and

[a] Address correspondence to D. Mark Bickett, Glaxo Research Institute, 5 Moore Drive, Research Triangle Park, North Carolina 27709.

TABLE 1. Stromelysin Fluorogenic Substrate Development

Designation					Substrate P$_3$	P$_2$	P$_1$	P$_1'$	P$_2'$	P$_3'$				k_{cat}/K_m ($M^{-1}s^{-1}$)	Relative k_{cat}/K_m
Substance P		Arg	Pro	Lys	Pro	Gln	Gln	Phe	Phe	Gly	Leu	Met	NH_2	1.79×10^{3a}	ND
1		Arg	Pro	Lys	Pro	Gln	Gln	Phe	Trp	NH_2[a]				4.1×10^{3b}	1
2	DNP	Arg	Pro	Lys	Pro	Gln	Gln	Phe	Trp	NH_2				2.23×10^{4}	5.2
3				DNP	Pro	Gln	Gln	Phe	Lys	Arg	Lys(NMA)NH_2			ND	0.83
4		DNP	Pro	Lys	Pro	Gln	Gln	Phe	Lys(NMA)NH_2					ND	1.14
5		Arg	Pro	Lys	Pro	Leu	Ala	Nva	Trp	NH_2[b]				3.1×10^{4b}	13
6	NBD	Arg	Pro	Lys	Pro	Leu	Ala	Nva	Trp	NH_2				ND	11.1
7	NBD	Arg	Pro	Lys	Pro	Leu	Ala	Nva	Trp	Lys(DMC)NH_2				2.14×10^{4}	4.8

Abbreviations: ND = not determined; Nva = norvaline.
[a]Teahan *et al.*[7]
[b]Niedzwiecki *et al.*[15]

NBD peptide substrates was monitored at 372 and 475 nm, respectively. Tryptophan fluorescence was monitored at EX_{280}/EM_{346}. Under competitive conditions, turnover of each substrate is first order with respect to other substrates, and relative k_{cat}/K_m can be determined directly using the following equation[16]:

$$\frac{\ln(S/So)}{\ln(S'/S'o)} = \frac{kcat/Km}{k'cat/K'm}$$

RESULTS AND DISCUSSION

Initial peptide sequences for the design of a fluorogenic substrate for stromelysin (MMP-3) were chosen from known substrates for stromelysin.[7,15] Several fluorophores/quenchers were incorporated into the peptides to produce internally quenched substrates with highly fluorescent products (TABLE 1). The substrates were evaluated with recombinant stromelysin. Substrates 4 and 7 were suitable for use in a 96-well format. Both are 95% quenched in the intact peptide, and both gave the same K_i value with a carboxyalkylamino inhibitor (FIG. 1). However, the hydrolysis product of substrate 7 has a 1.6-fold greater fluorescence signal than does the product of substrate 4 (data not shown), giving greater sensitivity and improving accuracy at lower substrate turnover (10–15% hydrolysis). Substrate 7 also has a fivefold greater relative k_{cat}/K_m than does substrate 4, which significantly reduces assay times (TABLE 1).

The DNP/NMA combination used in substrate 4 has been used successfully for peptide substrates containing <8 amino acids.[14] The DNP/NMA combination also works well in these shorter substrates where quenching through intramolecular collisions effectively reduces the background fluorescence of the intact substrate. For longer substrates, where resonance energy transfer[10] becomes the required quenching mechanism, the fluorophore/quencher pair must be chosen to maximize energy transfer within the molecule by matching the emission spectra of the fluorophore with the absorbance spectra of the quencher. The emission spectrum of DMC and the absorbance spectrum of NBD give this precise spectral overlap. NBD absorbs at 467 nm ($= 22$ mM^{-1}) and the emission maximum at 534 nm is well outside the 450-nm detection wavelength of the plate reader. These characteristics make substrate 7 excellent for routine screening of stromelysin inhibitors.

SUMMARY

Stromelysin, a member of the matrix metalloproteinase family of enzymes, has been implicated in the pathogenesis of tumor metastasis and inflammatory diseases such as rheumatoid arthritis. To screen prospective inhibitors of this protease, we developed a fluorogenic substrate with excitation and emission spectra compatible with commercially available 96-well plate readers. The substrate is based on the addition of 6-[*N*-(7-nitrobenz-2-oxa-1,3-diazol-4-yl) amino] hexanoic acid (NBD) (EX_{467}/EM_{534}) and 7-dimethylaminocoumarin-4-acetate (DMC) (EX_{368}/EM_{459}) to the previously reported peptide substrate for stromelysin, Arg-Pro-Lys-Pro-Leu-Ala-Nva-Trp-NH_2. The new substrate, NBD-Arg-Pro-Lys-Pro-Leu-Ala-Nva-Trp-Lys-(DMC)-NH_2 is 95% quenched and the fluorescent product, Nva-Trp-Lys(DMC)-NH_2, is easily detected (EX_{350}/EM_{465}). In competition assays the new fluorogenic substrate has a relative k_{cat}/K_m that is one half that of the parent peptide. The

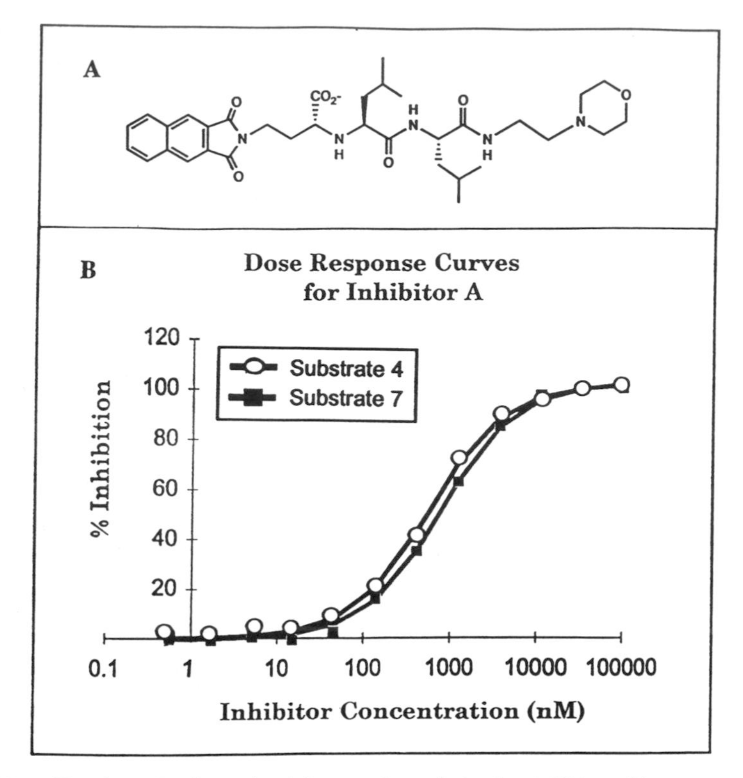

FIGURE 1. K_i values (**B**) determined for a carboxyalkylamino inhibitor (**A**) were compared using fluorogenic substrates 4 and 7. Assays were conducted in a total volume of 0.3 mL of assay buffer (200 mM NaCl, 50 mM Tris, 5 mM $CaCl_2$, 10 μM $ZnSO_4$, and 0.05% Brij 35, pH 7.6) in a black 96-well microtiter plate. Stromelysin concentrations were adjusted to 5 nM. Inhibitor A was diluted from 100,000 nM to 0.56 nM by serial threefold dilution. Assays were initiated with the addition of substrate (10 μM substrate 4 and 5 μM substrate 7). Product formation was measured at EX_{365}/EM_{450} nm after 20 min for substrate 7 and 120 min for substrate 4 using a Baxter (FCA) fluorescence analyzer. Under these conditions the K_m and K_i can be determined directly by plotting the percentage of inhibition versus the log of the inhibitor concentration.

fluorophores NBD and DMC were chosen based on the high fluorescence yield of DMC and the overlap of the emission spectrum of DMC and excitation spectrum of NBD which results in an efficient energy transfer system in the intact substrate. These characteristics make this an excellent substrate for routine determination of *in vitro* activities of stromelysin inhibitors.

REFERENCES

1. Chin, J. R., G. Murphy & Z. Werb. 1985. J. Biol. Chem. **260:** 12367–12376.
2. Okada, Y., H. Nagase & E. D. Harris, Jr. 1986. J. Biol. Chem. **261:** 14245–14255.
3. Hasty, K. H., R. A. Reife, A. H. Kang & J. M. Stuart. 1990. Arthritis Rheum. **33:** 388–397.

4. CASE, J. P., H. SANO, R. LAFYATIS, E. F. REMMERS, G. K. KUMKUMIAN & R. L. WILDER. 1989. J. Clin. Invest. **84:** 1731–1740.
5. MATRISIAN, L. M., G. T. BOWDEN, P. KRIEG, G. FURSTENBERGER, J. P. BRIAND, P. LEROY & R. BREATHNACH. 1986. Proc. Natl. Acad. Sci. USA **83:** 9413–9417.
6. OSTROWSKI, L. E., J. FINCH, P. KRIEG, L. MATRISIAN, G. PATSKAN, J. F. O'CONNELL, J. PHILLIPS, T. J. SLAGA, R. BREATHNACH & G. T. BOWDEN. 1988. Mol. Carcinog. **1:** 13–19.
7. TEAHAN, J., R. HARRISON, M. IZQUIERDO & R. STEIN. 1989. Biochemistry **28:** 8497–8501.
8. GRAY, R. D. & H. H. SANEII. 1982. Anal. Biochem. **120:** 339–346.
9. MASUI, Y., T. TAKEMOTO, S. SAKAKIBARA, H. HORI & Y. NAGAI. 1977. Biochem. Med. **17:** 215–221.
10. YARON, A., A. CARMEL & E. KATCHALSKI-KATZIR. 1979. Anal. Biochem. **95:** 228–235.
11. STACK, M. S. & R. D. GRAY. 1989. J. Biol. Chem. **264:** 4277–4289.
12. NETZEL-ARNETT, S., S. K. MALLY, H. NAGASE, H. BIRKEDAL-HANSEN & H. E. VAN WART. 1991. Anal. Biochem. **195:** 86–92.
13. KNIGHT, C. G., F. WILLENBROCK & G. MURPHY. 1992. FEBS Lett. **296:** 263–266.
14. BICKETT, D. M., M. D. GREEN, J. BERMAN, M. DEZUBE, A. S. HOWE, P. J. BROWN, J. T. ROTH & G. M. MCGEEHAN. 1993. Anal. Biochem. **212:** 58–64.
15. NIEDZWIECKI, L., J. TEAHAN, R. K. HARRISON & R. L. STEIN. 1992. Biochemistry **31:** 12618–12623.
16. ABELES, R. H., W. R. FRISELL & C. G. MACKENZIE. 1960. J. Biol. Chem. **235:** 853–856.

Characterization of Stromelysin Enzyme Activity and Its Inhibition Using an Enzyme-Linked Immunosorbent Assay for Transferrin

FELICIA R. COCHRAN[a] AND DOUGLAS B. SHERMAN

Departments of Pharmacology and Organic Chemistry
Burroughs Wellcome Co.
Research Triangle Park, North Carolina 27709

Stromelysin (MMP-3) is a member of the zinc-dependent family of endopeptidases that degrade the extracellular matrix under normal and pathologic conditions.[1] Although stromelysin has been characterized by its ability to degrade native substrates, such as cartilage proteoglycan, it is often impractical and cumbersome to employ such substrates for routine enzyme assays. In the present study, we describe a new and convenient ELISA suitable for the determination of stromelysin enzyme activity and its inhibition.

MATERIALS AND METHODS

Prostromelysin was concentrated from the media of HIG-82 cells (ATCC CRL 1832) exposed to 640 pg/mL interleukin 1β (IL-1β) plus 0.1 μM PMA for 96 h.[2] Prostromelysin was activated with 2 μg/mL trypsin for 1 h at 37 °C followed by the addition of 50 μg/mL soybean trypsin inhibitor. The concentrated media from PMA-treated WI-38 cells (ATCC CCL 75) was used as the source for TIMP.[3] Peptides were synthesized on a Millipore 9050 peptide synthesizer and were purified by RP-HPLC. For the protease assay, 96-well plates were coated with human apotransferrin and then incubated at 37 °C with test compounds diluted in 50 mM Tris-buffered saline solution, pH 8.0, containing 10 mM $CaCl_2$, 0.05% Brij 35, and 0.02% NaN_3 with or without stromelysin (2.86 μg protein/well). After 3 h, the plates were washed and blocked to prevent nonspecific binding. Goat anti-transferrin (1:5000) was then incubated 1.5 h at 37 °C followed by rabbit anti-goat IgG peroxidase (1:5000). OPD was the hydrogen donor; optical density was measured at 450 nm.

RESULTS AND DISCUSSION

Inasmuch as an ELISA for collagenase activity was reported previously,[4] it seemed feasible to develop a transferrin ELISA to measure stromelysin enzyme

[a]Address correspondence to Felicia R. Cochran, Ph.D., Department of Pharmacology, Burroughs Wellcome Co., 3030 Cornwallis Road, Research Triangle Park, North Carolina 27709.

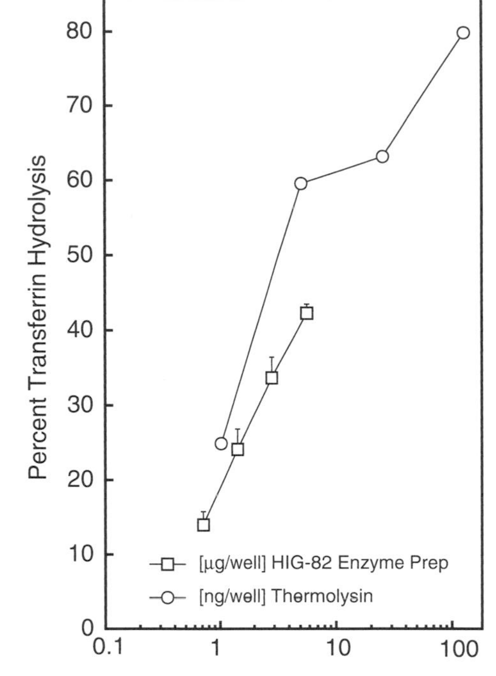

FIGURE 1. Concentration-dependent transferrin digestion by thermolysin versus stromelysin. Conditioned media from HIG-82 cells[2] or thermolysin was incubated in the presence of 350 ng/well transferrin. The amount of transferrin remaining bound to the plate after 3 h was determined by ELISA against a standard curve of 60–350 ng/well transferrin as described under MATERIALS AND METHODS.

activity. The choice of transferrin as a native substrate is based on its greater specificity for stromelysin versus interstitial collagenase or gelatinase.[5] FIGURE 1 illustrates the concentration-dependent protease activity of thermolysin versus stromelysin. As shown in TABLE 1, transferrin degradation was inhibited by chelators, TIMP, and synthetic peptides based on the highly conserved proenzyme domain (PRCGVPDV) of the matrix metalloproteinases. Replacement of the Cys residue with Ala abolished inhibitory activity.[6] In conclusion, our results suggest that the transferrin ELISA may be a useful and convenient tool for the evaluation of stromelysin enzyme activity and its inhibition.

TABLE 1. Inhibitors of Stromelysin-Mediated Degradation of Transferrin as Assessed by ELISA[a]

Compound	Concentration	% Inhibition
EDTA	10 mM	100
1,10-Phenanthroline	5 mM	100
TIMP[b]	50 µg/mL	100
Ac-RCGVPDV-NH_2	10 nM	50
Ac-RAGVPDV-NH_2	1 mM	16

[a]Concentrated conditioned media from HIG-82 cells[2] was incubated with test compounds at 37 °C in the presence of 350 ng/well transferrin. The amount of transferrin remaining bound to the plate after 3 h was determined by ELISA against a standard curve of 60–350 ng/well transferrin as described under MATERIALS AND METHODS.

[b]Concentrated conditioned media from WI-38 cells.[3]

REFERENCES

1. WOESSNER, J. F. 1991. Matrix metalloproteinases and their inhibitors in connective tissue remodeling. FASEB J. **5:** 2145–2154.
2. GEORGESCU, H. I., D. MENDELOW & C. H. EVANS. 1988. HIG-82: An established cell line from rabbit periarticular soft tissue, which retains the "activatable" phenotype. In Vitro Cell. Dev. Biol. **24:** 1015–1022.
3. STETLER-STEVENSON, W. G., P. D. BROWN, M. ONISTO, A. T. LEVY & L. A. LIOTTA. 1990. Tissue inhibitors of metalloproteinases-2 (TIMP-2) mRNA expression in tumor cells and human tumor tissues. J. Biol. Chem. **265:** 13933–13938.
4. YOSHIOKA, H., I. OYAMADA & G. USUKU. 1987. An assay of collagenase activity using enzyme-linked immunosorbent assay for mammalian collagenase. Anal. Biochem. **166:** 172–177.
5. NAGASE, H. & J. F. WOESSNER. 1993. Role of endogenous proteinases in the degradation of cartilage matrix. *In* Joint Cartilage Degradation. J. F. Woessner & D. S. Howell, Eds.: 159–185. Marcel Dekker. New York.
6. PARK, A. J., L. M. MATRISIAN, R. PEARSON, Z. YUAN & M. NAVRE. 1991. Mutational analysis of the transin (rat stromelysin) autoinhibitor region demonstrates a role for residues surrounding the "cysteine switch." J. Biol. Chem. **266:** 1584–1590.

A One-Step Sandwich Enzyme Immunoassay for Human Matrix Metalloproteinase 9 Using Monoclonal Antibodies

N. FUJIMOTO,[a] N. HOSOKAWA,[a] K. IWATA,[a] Y. OKADA,[b] AND T. HAYAKAWA[c]

[a]*Biotechnology Section*
Research Laboratories I
Fuji Chemical Industries, Ltd.
Takaoka, Toyama 933, Japan

[b]*Department of Pathology*
School of Medicine
Kanazawa University
Kanazawa, Ishikawa 920, Japan

[c]*Department of Biochemistry*
School of Dentistry
Aichi-Gakuin University
Chikusa-ku, Nagoya 464, Japan

Matrix metalloproteinase 9 (MMP-9) has been implicated in various pathophysiologic conditions such as cancer invasion and metastasis[1–3] and bone resorption.[4] To study the release and function of MMP-9 in body fluid and tissue, we established a one-step sandwich enzyme immunoassay (EIA) system for MMP-9 and applied it to measure MMP-9 in human plasma from normal subjects and patients with disease.

ProMMP-9 was purified from the conditioned medium of HT1080 cells stimulated with tumor necrosis factor-alpha as previously described.[5] Monoclonal antibodies for MMP-9 were prepared against the purified proMMP-9. A monoclonal antibody for TIMP-1 from bovine dental pulp is capable of cross-reacting with human TIMP-1.[6] A one-step sandwich EIA system for MMP-9 was developed using a combination of monoclonal antibodies (clones 73–18B3 and 73–74E11) as a solid phase and a peroxidase conjugate, respectively.

The sensitivity of this EIA for standard proMMP-9 was 2.4 pg/assay (0.24 ng/mL), and linearity was obtained between 0.24 and 250 ng/mL (2.4–2,500 pg/assay) (FIG. 1). The intra-assay coefficients of variation (CVs) of normal plasma having MMP-9 at a concentration of 35, 104, and 109 ng/mL were 1.8–3.9% (n = 10). Inter-assay CVs were 1.8–6.2% (n = 8) at plasma MMP-9 levels of 38, 47, and 114 ng/mL. Recovery of standard proMMP-9 added to human plasma having an MMP-9 concentration of 35, 45, and 100 ng/mL over a range of 12.5–200 ng/mL was 103 ± 7.6% (mean ± SD). This EIA system recognized proMMP-9, the intermediate form with M_r = 83 kDa and their complexes formed with TIMP-1.

The concentration of MMP-9 in normal human sera by our EIA was 346 ± 197 ng/mL. MMP-9 levels in human sera increased with time after venipuncture. Therefore, our one-step sandwich EIA technique was applied to evaluate MMP-9

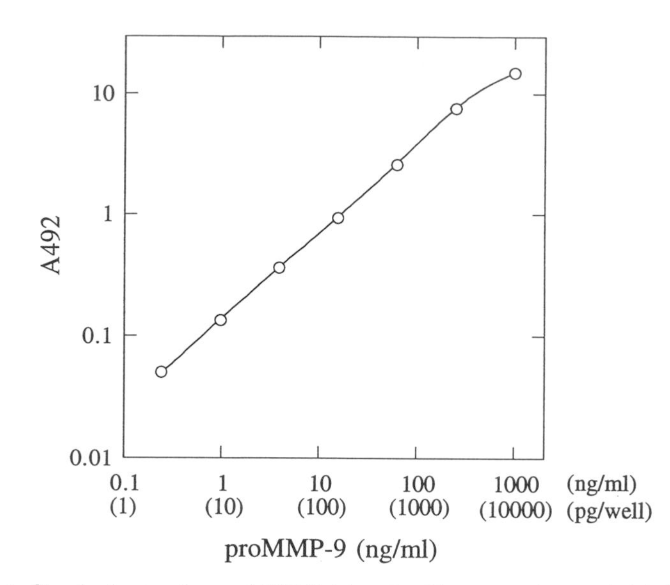

FIGURE 1. Standard curve for proMMP-9 determined by a one-step sandwich EIA system. Purified proMMP-9 was determined with a combination of two monoclonal antibodies (clone 73–18B3 as a solid phase and clone 73–74E11 as a peroxidase conjugate) using microplate wells.

levels in plasma from normal subjects and patients with disease. As shown in TABLE 1, MMP-9 levels in normal plasma were 38 ± 15 ng/mL (mean ± SD). A significantly higher level of MMP-9 was detected in plasma from patients with hepatocellular carcinoma (107 ± 185 ng/mL). No significant difference, however, was observed between normal plasma and samples from patients with renal disease including IgA nephropathy, membranous nephropathy and mesangial proliferative glomerulonephritis (85 ± 90 ng/mL), or systemic lupus erythematosus (42 ± 12 ng/mL).

Human plasma was applied to a column packed with anti-TIMP-1 IgG (clone 7-21B12)-Sepharose 4B, and after collection of the pass-through volume, the column was eluted with 0.2 M glycine-HCl buffer (pH 2.0) and fractions were collected. Both immunoreactive MMP-9 and its complex with TIMP-1 were detected only in the eluted fractions but not in the pass-through fraction, indicating that a majority of the proMMP-9 in human plasma exists as a complex with TIMP-1.

TABLE 1. MMP-9 Levels in Human Plasma from Normal Subjects and Patients with Disease

Diseases	*n*	Mean ± SD (ng/mL)
None (normal subjects)	15	38 ± 15
Renal disease	13	85 ± 90
Hepatocellular carcinoma	40	107 ± 185[a]
Systemic lupus erythematosus	10	42 ± 12

[a] $p < 0.05$.

REFERENCES

1. MOLL, U. M., G. L. YOUNGLEIB, K. B. ROSINSKI & J. P. QUIGLEY. 1990. Tumor promoter-stimulated Mr 92,000 gelatinase secreted by normal and malignant human cells: Isolation and characterization of the enzyme from HT1080 tumor cells. Cancer Res. **50:** 6162–6170.
2. BALLIN, M., D. E. GOMEZ, C. C. SINHA & U. P. THORGEIRSSON. 1988. *Ras* oncogene mediated induction of a 92 kDa metalloproteinase: Strong correlation with the malignant phenotype. Biochem. Biophys. Res. Commun. **154:** 832–838.
3. TRYGGVASON, K., M. HÖYHTYÄ & C. PYKE. 1993. Type IV collagenases in invasive tumors. Breast Cancer Res. Treat. **24:** 209–218.
4. OKADA, Y., K. NAKA, K. KAWAMURA, T. MATSUMOTO, I. NAKANISHI, N. FUJIMOTO, H. SATO & M. SEIKI. 1993. Localization of matrix metalloproteinase 9 (92 kDa gelatinase/type IV collagenase) in osteoclasts: Implication for bone resorption. Lab. Invest. Submitted.
5. OKADA, Y., Y. GONOJI, K. NAKA, K. TOMITA, I. NAKANISHI, K. IWATA, K. YAMASHITA & T. HAYAKAWA. 1992. Matrix metalloproteinase 9 (92 kDa gelatinase/type IV collagenase) from HT 1080 human fibrosarcoma cells: Purification and activation of the precursor, and enzymic properties. J. Biol. Chem. **267:** 21712–21719.
6. KODAMA, S., K. IWATA, H. IWATA, K. YAMASHITA & T. HAYAKAWA. 1990. Rapid one-step sandwich enzyme immunoassay for tissue inhibitor of metalloproteinases: An application for rheumatoid arthritis serum and plasma. J. Immunol. Methods **127:** 103–108.

Characterization of Rat Matrilysin and Its cDNA[a]

SUSAN R. ABRAMSON AND J. FREDERICK WOESSNER, JR.[b]

Departments of Medicine[b] and
Biochemistry and Molecular Biology
University of Miami School of Medicine
Miami, Florida 33101

Matrilysin is the least common denominator of the matrix metalloproteinase (MMP) family; it contains only the putative catalytic and metal binding domains and it is inhibited by TIMP-1.[1] For this reason, we believe it is an important tool in understanding the mechanisms of activity and inhibition of the whole family of enzymes. A uterine metalloproteinase (ump) was previously isolated from the involuting rat uterus.[2] A cDNA for a human enzyme, pump-1, was also isolated.[3] It is believed that ump and pump-1 are the same; both are now called matrilysin, or MMP-7.

FIGURE 1 shows a digest of the isolated α2(I) chain (using the method of Piez *et al.*[4]) by rat MMP-7. Rat MMP-7 is capable of digesting the α2 chain of collagen I at a limited number of sites, but it has almost no ability to digest the α1(I) chain (lane 7) and it has no effect on native collagen I. A pattern of discrete bands is apparent (lanes 2–4) and EDTA was able to completely inhibit digestion (lane 1). Most of the substrate is eventually converted to a fragment of about 40 kDa (lane 2). The initial fragments created indicate that one or two sites should be present on each side of the MMP-1 cutting site. Human pump-1 (kindly provided by A.J.P. Docherty from Celltech) creates a similar gelatin digestion pattern as rat ump (data not shown).

Rat MMP-1 was activated by ump using a preparation of rat MMP-1 from the involuting uterus. The addition of 50 nM ump results in 50% activation compared to 4-aminophenylmercuric acetate (APMA) activation. Rat MMP-1 is extremely sensitive to APMA, and maximal activation occurred with only 5 μM APMA. Rat MMP-1 is more easily activated than is human MMP-1, suggesting that the rat propeptide is less tightly folded over the active site and/or that the activation cutting site is more exposed. The MMP-1 preparation is not completely pure, and the possibility that a contaminating enzyme (not necessarily a metallo) in the MMP-1 is capable of mediating its activation cannot be ruled out. If this turns out to be the case, however, the "contaminant" may well be a natural activator of MMP-1.

Inhibition of MMP-7 was achieved by two synthetic inhibitors (kindly provided by Searle & Co.), SC 44463, a hydroxamate, and SC 40827, a pseudopeptide. FIGURE 2 shows inhibition curves for both; the IC_{50} for SC 44463 is about 7–8 nM, whereas that for SC 40827 is about 3 μM. Therefore, the hydroxamate is about 400 times more powerful as an inhibitor of MMP-7.

A clone containing the complete open reading frame (804 bp) with 33 and 220 bp of 5′ and 3′ UTR was isolated and sequenced. The cDNA and protein sequences are, respectively, 74% and 70% identical to human MMP-7, and the deduced zymogen is of the expected size (28 kDa). Regions of high identity of the deduced amino acid sequences include the cysteine switch and metal binding sequences.

[a] This work was supported by National Institutes of Health grant HD 06773.

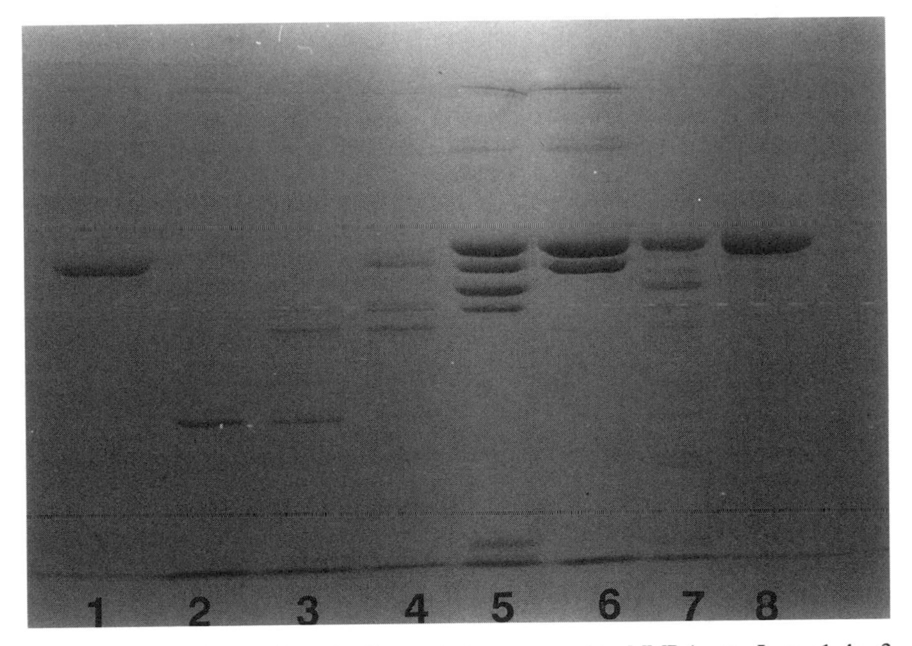

FIGURE 1. MMP-7 digestion of collagen chains compared to MMP-1 cuts. **Lanes 1–4:** α2 digestion by MMP-7 (*lane 1,* 18 h plus EDTA; *lanes 2–4,* 18, 3, and 0.5 h, respectively). **Lanes 5–6**: telopeptide-free collagen digested with MMP-1 without (*lane 5*) or with (*lane 6*) EDTA; **lanes 7–8:** α1 digestion by MMP-7 without (*lane 7*) and with (*lane 8*) EDTA, 18 h.

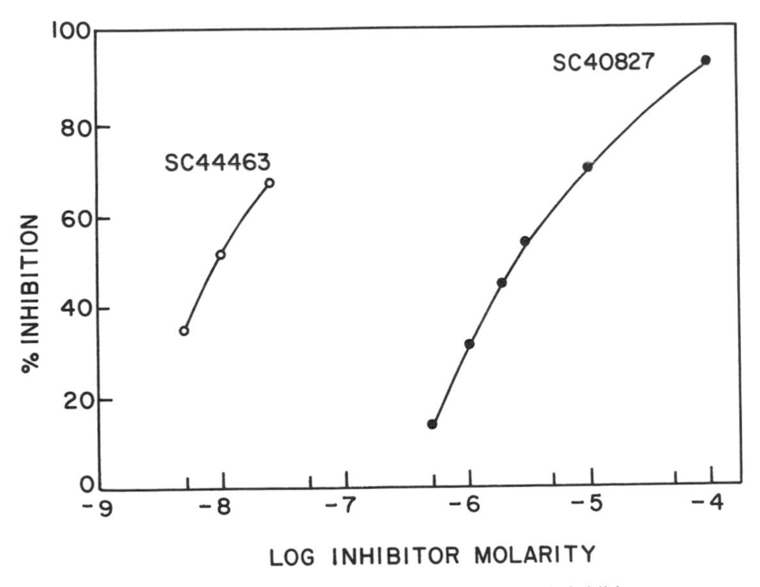

FIGURE 2. Inhibition of MMP-7 by two synthetic inhibitors.

Because the highest levels of ump are present in the uterus at 2 days postpartum, we believe that ump is involved in the loss and reorganization of the extracellular matrix that occurs during this time. We support this finding here with evidence that ump can digest gelatin and, perhaps more importantly, is capable of activating MMP-1.

REFERENCES

1. Woessner, J. F. 1991. Matrix metalloproteinases and their inhibitors in connective tissue remodeling. FASEB J. **5:** 2145–2154.
2. Woessner, J. F. & C. J. Taplin. 1988. Purification and properties of a small latent matrix metalloproteinase of the rat uterus. J. Biol. Chem. **263:** 16918–16925.
3. Quantin, B., G. Murphy & R. Breathnach. 1989. Pump-1 cDNA codes for a protein with characteristics similar to those of classical collagenase family members. Biochemistry **28:** 5327–5334.
4. Piez, K. A., E. A. Eigner & M. S. Lewis. 1963. The chromatographic separation and amino acid composition of the subunits of several collagens. Biochemistry **2:** 58–66.

The Binding of Gelatinases A and B to Type I Collagen Yields Both High and Low Affinity Sites

J. A. ALLAN,[a] A. J. P. DOCHERTY,[b] AND G. MURPHY[a]

[a] *Strangeways Research Laboratory*
Worts Causeway
Cambridge CB1 4RN, United Kingdom

[b] *Celltech Ltd.*
Slough, SL1 4EN, United Kingdom

Immunolocalization studies have shown that gelatinase A as well as collagenase and stromelysin is able to bind to matrix components in rapidly resorbing tissues.[1] The ability of latent and active forms of gelatinases A and B and the contribution made by different enzyme domains to binding to various matrix components including collagen are described. The roles of the various enzyme domains were investigated by the preparation of enzymes lacking one or more domains by both biochemical and protein engineering approaches. The COOH-terminal domain of gelatinase A (delta$_{1\text{-}414}$GLA), gelatinase A deletion mutants lacking either the COOH-terminal domain ($\Delta_{418\text{-}631}$GLA) or the fibronectin-like domain ($\Delta_{191\text{-}364}$GLA), and gelatinase B deletion mutants lacking either the COOH-terminal domain ($\Delta_{471\text{-}688}$GLB) or both this domain and the collagen-like region ($\Delta_{426\text{-}688}$GLB) were used in this study.

The sheep antibodies to human gelatinase A[2] and pig gelatinase B[3] were characterized previously and were used in two ELISAs to deduce how well pro and active gelatinase A, gelatinase B, and their different domains bind to collagen films.

Gelatinase A, gelatinase B, or their different domains were applied in 50-μL volumes in 100 mM Tris-HCl, pH 7.6, 150 mM NaCl, 0.2% casein to types I and IV collagens, gelatin, laminin, fibronectin, and heparan sulfate proteoglycan films and incubated at 17 °C for 1 h. Bound enzyme was revealed by using, where appropriate, anti-gelatinase A IgG or anti-gelatinase B IgG, both at 20 μg/mL for 2 h at 17 °C. This was followed by a donkey anti-sheep IgG conjugated to peroxidase at a 1:20,000 dilution. Both 3,3′,5,5′-tetramethyl benzidine (0.3 mM) and 0.004% H_2O_2 in 0.1 M acetate buffer (pH 6.0) were used as substrates. The reaction was stopped by the addition of 2.5 M H_2SO_4, and absorbance at 450 nm was measured.

Pro and active forms of gelatinase A bound to types I and IV collagens, gelatin, and laminin films; however, they bound very poorly to fibronectin and heparan sulfate proteoglycan. Scatchard analysis of progelatinase A binding to type I collagen yielded both high and low affinity binding sites with K_d values of 1.2×10^{-11} M and 3.6×10^{-8} M, respectively. When isolated NH_2-($\Delta_{418\text{-}631}$GLA) and COOH-terminal ($\Delta_{1\text{-}414}$GLA) domains were applied to type I collagen films, binding occurred via the NH_2-terminal domain; deletion of the fibronectin-like domain ($\Delta_{191\text{-}364}$GLA) abolished binding. Also, fibronectin competed with gelatinase A, confirming that binding occurs through this domain. Studies on gelatinase B showed that, like gelatinase A, both pro and active forms of the enzyme bound to type I and IV collagens and gelatin. Binding also occurred to laminin but not to fibronectin or heparan sulfate proteoglycan. Scatchard analysis of progelatinase B binding to type I collagen also yielded high and low affinity binding sites with K_d values of 6.1×10^{-13} M and 9.7 ×

10^{-8} M. A mutant of gelatinase B lacking the COOH-terminal domain ($\Delta_{471\text{-}688}$GLB) bound to the collagen film, as did a mutant lacking both this domain and the collagen-like domain ($\Delta_{426\text{-}688}$GLB). The binding of the latter was reduced, suggesting that the collagen-like domain may also contribute to collagen binding. Preformed complexes of pro and active gelatinase A with TIMP-2 and progelatinase B with TIMP-1 did not interfere with the binding of gelatinases A and B to collagen.

This study demonstrates that gelatinases A and B interact with collagen through the fibronectin-like domain and that this involves high and low affinity binding sites. The significance of gelatinase binding to collagen is unknown, but it may confer retention, stability and bioactivity for prolonged periods, thus facilitating its role in pericellular proteolysis.

REFERENCES

1. GAVRILOVIC, J., R. M. HEMBRY, J. J. REYNOLDS & G. MURPHY. 1989. Collagenase is expressed by rabbit VX2 tumour cells in syngeneic and xenogeneic hosts. Matrix **9:** 206–213.
2. HIPPS, D. S., R. M. HEMBRY, A. J. P. DOCHERTY, J. J. REYNOLDS & G. MURPHY. 1991. Purification and characterisation of human 72-kDa gelatinase (type IV collagenase). Use of immunolocalisation to demonstrate the non-coordinated regulation of the 72-kDa and 95-kDa gelatinases by human fibroblasts. Biol. Chem. Hoppe-Seyler **372:** 287–296.
3. MURPHY, G., R. WARD, R. M. HEMBRY, J. J. REYNOLDS, K. KUHN & K. TRYGGVASON. 1989. Characterisation of gelatinase from pig polymorphonuclear leucocytes. A metalloproteinase resembling tumour type IV collagenase. Biochem. J. **258:** 463–472.

Neutrophil Procollagenase Can Be Activated by Stromelysin-2

VERA KNÄUPER,[a] GILLIAN MURPHY,[a]
AND HARALD TSCHESCHE[b]

[a]*Cell and Molecular Biology Department*
Strangeways Research Laboratory
Worts Causeway
Cambridge, CB1 4RN, United Kingdom

[b]*Department of Biochemistry*
University of Bielefeld
Bielefeld, Germany

Human neutrophil procollagenase (MMP-8), the second member of the interstitial collagenase subfamily, plays an important role during extracellular matrix remodeling processes induced by neutrophils. Procollagenase is released from the specific granules of the cells through receptor-mediated exocytosis, followed by extracellular proenzyme activation. The active enzyme degrades the interstitial collagens (type 1–3), proteoglycans, and various serine proteinase inhibitors of the serpin superfamily. One key event in the regulation of collagenolytic activity is proenzyme activation, which can be initiated by reactive oxygen species, different endopeptidases, or mercurial compounds *in vitro.* The final active enzyme lacks the complete propeptide domain, and the different active forms have either Phe^{79}, Met^{80}, or Leu^{81} as the N-terminal amino acid.[1–3]

Stromelysin-2 (MMP-10), the second member of the stromelysin subfamily and a potent matrix metalloproteinase, has a similar substrate specificity to stromelysin-1.[4] Its mRNA is expressed by different tumors, keratinocytes, and human alveolar macrophages (H. Welgus, personal communication), which implies an important role for this enzyme in inflammation.

We investigated whether neutrophil procollagenase can be activated by stromelysin-2 using SDS-PAGE, activity assay, and N-terminal sequence determination.

Human neutrophil procollagenase was incubated with recombinant stromelysin-2 in a 19:1 molar ratio. The generation of collagenolytic activity was observed in a time-dependent manner. Using SDS-PAGE, a reduction in the molecular mass of neutrophil procollagenase was demonstrated which was suppressed by preincubation of active stromelysin-2 with equimolar amounts of TIMP-2. N-terminal sequence determination of the final active form revealed cleavage of the Gly^{78}-Phe^{79} peptide bond at the end of the propeptide domain. Phe^{79} N-terminal collagenase displayed specific collagenolytic activity that was four times greater than that of Met^{80} or Leu^{81} N-terminal collagenase, which was generated by mercurial treatment of the proenzyme. Thus, stromelysin-2, like homologous stromelysin-1, superactivates human neutrophil procollagenase, which might be of physiologic relevance during inflammatory reactions.

Critical for superactivation of both fibroblast and neutrophil collagenase is the generation of an enzyme having either Phe^{81} or Phe^{79} as the N-terminal amino acid residue. Recent crystallographic analysis of both the Phe^{79}-Gly^{242} and Met^{80}-Gly^{242} catalytic domain of neutrophil collagenase reveals that the Phe^{79} ammonium group forms a salt bridge with the carboxylate group of Asp^{232}, thereby stabilizing the N-terminal segment from Phe^{79} to Pro^{86} of the enzyme, which is disordered within the Met^{80}-Gly^{242} enzyme.[5] The higher ordered structure seems linked to the observed activity enhancement, because conformation of the active site residues shows no obvious differences between both forms. Nevertheless, it cannot be excluded that the Phe^{79} N-terminus of intact active collagenase (Phe^{79}-Gly^{447}) may induce dramatic structural changes within the active site cleft due to possible interactions with the C-terminal domain.

REFERENCES

1. Knäuper, V., S. Krämer, H. Reinke & H. Tschesche. 1990. Eur. J. Biochem. **189:** 295–300.
2. Bläser, J., V. Knäuper, A. Osthues, H. Reinke & H. Tschesche. 1991. Eur. J. Biochem. **202:** 1223–1230.
3. Knäuper, V., S. M. Wilhelm, P. K. Seperack, Y. A. DeClerck, K. E. Langley, A. Osthues & H. Tschesche. 1993. Biochem. J. **295:** 581–586.
4. Nicholson, R., G. Murphy & R. Breathnach. 1989. Biochemistry **28:** 5195–5203.
5. Reinemer, P., F. Grams, R. Huber, T. Kleine, S. Schnierer, H. Tschesche, M. Pieper & W. Bode. 1994. FEBS Lett. **338:** 227–233.

Structure of the Mouse 92-kDa Type IV Collagenase Gene

In Vitro and *in Vivo* Expression in Transient Transfection Studies and Transgenic Mice

C. MUNAUT,[a] P. REPONEN,[a] P. HUHTALA,[a]
S. KONTUSAARI,[a] J.-M. FOIDART,[b] AND K. TRYGGVASON[a]

[a]*Biocenter Oulu and*
Department of Biochemistry
University of Oulu
FIN-90570 Oulu, Finland

[b]*Laboratory of Biology*
University of Liège
B-4000 Liège, Belgium

The 92-kDa type IV collagenase (gelatinase B) belongs to a family of extracellular matrix metalloproteinases. Different members of this family play a role in turnover of extracellular matrix during embryo implantation, morphogenesis, and wound healing. Their expression has to be tightly regulated at many levels. Increased secretion is seen in pathologic conditions such as rheumatoid arthritis, inflammation, and tumor invasion. The specific role of each enzyme and details about their function at the cellular level *in vivo* are still largely unknown. Despite the fact that both type IV collagenases (72 and 92 kDa), also termed gelatinases A and B, apparently have the same substrate specificity, their temporal and spatial expression *in vivo* is totally different. We previously showed by *in situ* hybridization studies that gelatinase A is primarily expressed in stromal fibroblast-like cells during mouse development,[1] whereas the expression of gelatinase B is almost completely restricted to osteoclasts at the site of bone formation during mouse development.[2] To better understand the physiologic role of the 92-kDa type IV collagenase *in vivo,* we cloned and characterized the mouse gene. Its regulation was investigated *in vitro* by cell transfections and *in vivo* using transgenic mice.

The complete structure of the mouse gene was determined from one cosmid clone (about 34 kb) containing 3 kb and about 23 kb of the 5′ and 3′ flanking regions. The sequencing of exons and exon/intron boundaries was determined. The mouse 92-kDa type IV collagenase gene contains 13 exons which mainly correspond in size to those in the human gene. The only differences are exons 9 and 13 which, respectively, contain 54 and 15 more bases than do those in the human. On the other hand, introns 1, 2, 3, and 12 in the mouse gene were about half the size of those in the human gene, whereas intron 8 was larger (1150 bp in mouse and 700 bp in human). The initiation site for transcription was determined by primer extension analysis performed with total RNA extracted from mouse skulls. This analysis revealed a double initiation start site located 19 and 20 bp upstream of the translated sequence.

Sequencing of 3000 bp of the 5′ flanking region revealed several common promoter elements. There is a TATA box at position −29 to −25 bp. There is no CCAAT motif in the promoter region, but three GC boxes that can serve as binding sites for the transcription factor Sp1 were localized (at position −62 to −56, −451 to −445, and −595 to −589). Four AP1-like binding sites were also identified (at position −88 to −80, −316 to −308, −472 to −465, and −1080 to −1072). Several AP2 sites and CA repeats were found in the mouse promoter region. A putative transforming-growth factor β inhibitory element found in the human gene was absent in the murine promoter.

To study regulation of the expression of gelatinase B, several human and mouse cell lines were transiently transfected by the calcium phosphate precipitation proce-

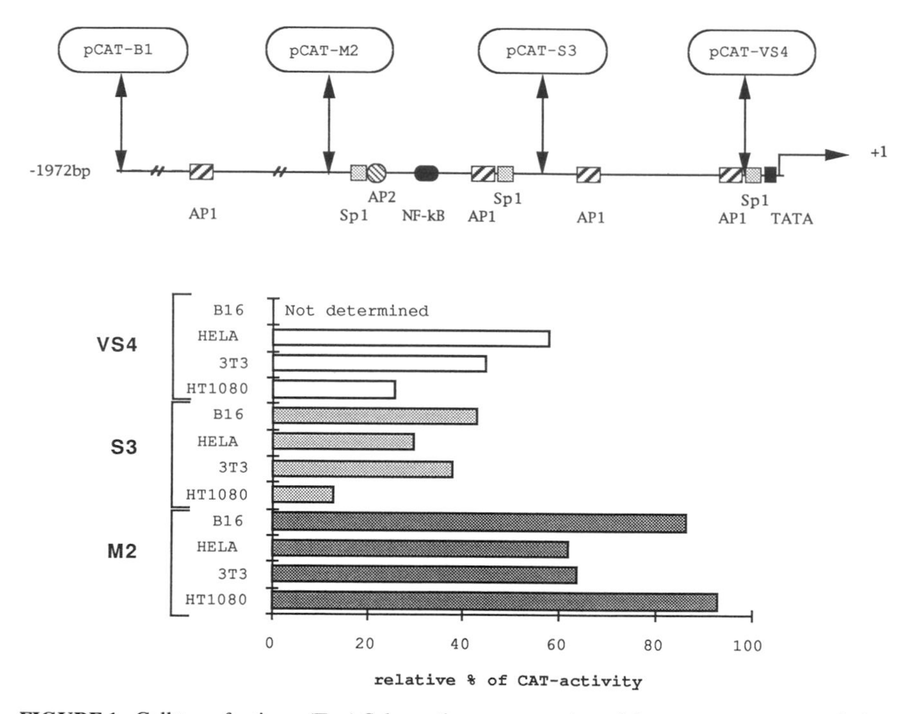

FIGURE 1. Cell transfections. **(Top)** Schematic representation of the mouse promoter gelatinase B-CAT reporter constructs. **(Bottom)** CAT assay results in HT1080, NIH-3T3, HeLa, and B16 cells transfected with the same constructs. The amounts of cell extract taken for CAT assays were normalized using β-galactosidase as a control of transfection efficiency. To normalize the assay, CAT activity of the pCAT-B1 construct was set to 100% in each cell line. Results are the mean of three to five different experiments performed in duplicate.

dure with CAT constructs containing 1970, 629, 243, and 82 bp of the 5′ flanking region. As shown in FIGURE 1, the 92-kDa type IV collagenase promoter ligated to the CAT reporter vector was active in a size-dependent manner regardless of the origin and/or nature of the cell line transfected. Even cell lines (mouse B16 melanoma and human HeLa cells) that do not express the enzyme *in vitro* gave the same results as did the expressing cell lines (mouse NIH-3T3 and human HT1080 cells). Because we were unable to obtain information about cell-specific expression and regulation of this enzyme from the *in vitro* experiments, we decided to analyze the regulation *in vivo* using the transgenic mice technology. The transgenic animals were generated by the injection of a 92-kDa type IV collagenase–*LacZ* fusion gene into the pronuclei of one-cell mouse fertilized embryos which were transferred into the oviduct of pseudopregnant mice. Transgenic animals were identified by Southern analysis, and expression of the fusion gene was analyzed in mouse embryos fixed in paraformaldehyde-glutaraldehyde and stained with X-gal. At 14.5 days the expression was mainly in the growing bones (arms, fingers, mandibles, and calvaria). This X-gal expression pattern corresponds well with that observed by *in situ* hybridization where strong expression was detected in cells in which bone resorption occurs. These transgenic mouse experiments reveal that the element(s) required for specific bone expression is located within a 2-kb 5′ flanking region of the gene.

REFERENCES

1. Reponen, P., C. Sahlberg, P. Huhtala, T. Hurskainen, I. Thesleff & K. Tryggvason. 1992. Molecular cloning of murine 72-kDa type IV collagenase and its expression during mouse development. J. Biol. Chem. **267:** 7856–7862.
2. Reponen, P., C. Sahlberg, C. Munaut, I. Thesleff & K. Tryggvason. 1994. Expression of 92-kDa type IV collagenase (gelatinase B) is confined to the osteoclasts lineage during mouse development. J. Cell. Biol. **124:** 1091–1102.

Relation between Substrate Specificity and Domain Structure of 92-kDa Type IV Collagenase[a]

TAYEBEH POURMOTABBED

Department of Biochemistry
University of Tennessee, Memphis
858 Madison Avenue
Memphis, Tennessee 38163

The 92-kDa type IV collagenase (MMP-9) is a member of the matrix metalloproteinase (MMP) family that degrades gelatin, types IV, V, and XI collagen. MMP-9, similar to the other member of the MMP family,[1–3] contains a central catalytic domain homologous to the zinc-binding domain of thermolysin.[4] This putative active site is flanked by an amino-terminal domain important in the conversion from the latent to the active form of the enzyme and by a highly variable carboxy-terminal domain with some homology to hemopexin. MMP-9 has two additional domains, one with homology to the gelatin-binding region of fibronectin and the other with a collagen-like sequence.[1,2] Thus, it is logical to assume that properties that are unique to a given metalloproteinase are determined by peptide sequences in variable regions of the proteins. Collier and co-workers[5] demonstrated that the fibronectin-like gelatin-binding domain of MMP-9 is essential for the binding of gelatin; therefore, it is tempting to speculate if the type V collagen-like domain is involved in interaction of the enzyme with the collagen substrates (types IV, V, and XI). It is also possible that degradation of gelatin and native collagen substrates may be unrelated to these domains. Here we show that the fibronectin-like gelatin-binding domain not only is important for binding of MMP-9 to gelatin substrate, but also is essential for gelatinolytic activity of the enzyme. Contrary to expectation, collagenolytic activity (types IV, V, and XI) of the enzyme is independent of type V collagen-like binding domain.

RESULTS AND DISCUSSION

To assess the significance of type V collagen-like (amino acid residues 452–512) and fibronectin-like gelatin-binding domains (amino acid residues 214–395) in the activity and substrate specificity of MMP-9, we deleted independently, nucleotides 642–885 corresponding to amino acid residues 214–395 (NG Δ 214–395) and nucleotides 1356–1536 corresponding to amino acid residues 452–512 (NG Δ 452–512) from the collagenase cDNA,[2] using the polymerase chain reaction method. The wild-type and mutant enzymes were expressed in *Escherichia coli* cells as described previously,[2] and their enzymatic activities were determined by zymography. The NG

[a]This work was supported by an Arthritis Foundation Investigator award and National Institutes of Health grant AR-41843.

Δ 452–512 protein was purified from cell lysates using gelatin-affinity chromatography. We designated types IV, V, and XI collagen as the specific substrates for NG Δ 452–512 mutant because other metalloproteinases show activity against gelatin.

Fibronectin-like gelatin-binding domain. NG Δ 214–395 mutant enzyme was inactive toward gelatin substrate. Thus, this domain in MMP-9 is important both for binding to gelatin, as demonstrated by Collier and coworkers,[5] and for gelatinolytic activity of the enzyme. Other MMPs that lack the gelatin-binding domain have gelatinolytic activity, but MMP-9 which lacks this domain is devoid of any gelatinolytic activity. This finding suggests that MMP-9 has evolved to degrade gelatin produced from cleavage and/or denaturation of different types of collagen by other proteases.

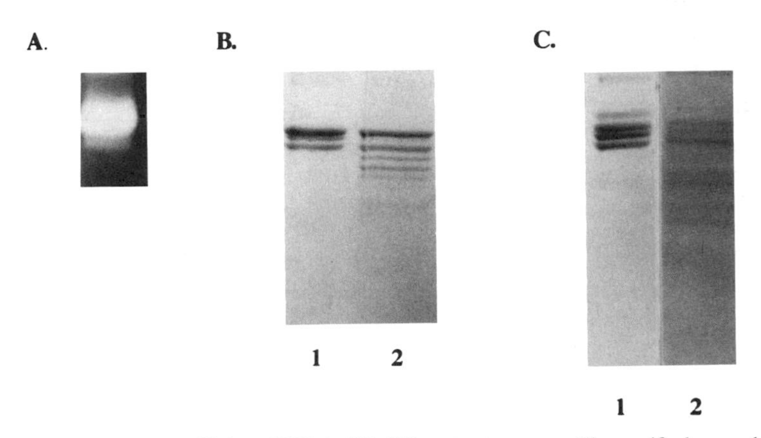

FIGURE 1. Substrate specificity of NG Δ 452–512 mutant enzyme. The purified recombinant enzyme was (**A**) separated on an 8% SDS-polyacrylamide gel containing 1.0 mg/mL gelatin. After overnight incubation with 50 mM Tris buffer, pH 7.6, containing 1.0 mM APMA and 1% Triton at 37 °C, the gel was stained with Coomassie brilliant blue, (**B**) incubated with type V collagen (2 μg) for 16 h in the (1) absence or (2) presence of 1.0 mM APMA at 32.5 °C, and (**C**) incubated with type XI collagen (5 μg) for 12 h in the (1) absence or (2) presence of 1.0 mM APMA at 36.4 °C. Reaction products were separated on SDS-PAGE and the polyacrylamide gels stained with Coomassie brilliant blue.

Type V collagen-like domain. Unlike the NG Δ 214–395 mutant, the NG Δ 452–512 mutant degrades gelatin (FIG. 1A), is latent, and upon activation with APMA, digests types V (FIG. 1B) and XI (FIG. 1C) collagen substrates. The pattern of products obtained with NG Δ 452–512 and wild-type enzymes was identical (data not shown). These data suggest that the type V collagen-like domain is not involved in substrate specificity of the enzyme. The collagen-like domain is unique to MMP-9 and consists of a 60 amino acid long proline-rich sequence.[1] Proline-rich sequences are believed to serve as a module that mediates protein-protein associations.[6] Whether the type V collagen-like domain is involved in interaction of MMP-9 with basement membrane component(s) is under investigation.

REFERENCES

1. WILHELM, S. M., I. E. COLLIER, B. L. MARMER, A. Z. EISEN, G. A. GRANT & G. I. GOLDBERG. 1989. J. Biol. Chem. **264:** 17213–17221.
2. POURMOTABBED, T., T. L. SOLOMON, K. A. HASTY & C. L. MAINARDI. 1994. Biochim. Biophys. Acta. **1204:** 97–107.
3. WOESSNER, J. F., JR. 1991. FASEB J. **5:** 2145–2154.
4. MATTHEWS, B. W., L. H. WEANER & W. R. KESTER. 1974. J. Biol. Chem. **249:** 8030–8044.
5. COLLIER, J. I. V., P. A. KRASNOV, A. Y. STRONGIN, H. BIRKEDAL-HANSEN & G. I. GOLDBERG. 1992. J. Biol. Chem. **267:** 6776–6781.
6. REN, R., B. J. MAYER, P. CICCHETTI & D. BALTIMORE. 1993. Science **259:** 1157–1161.

Structural Analysis of the Catalytic Domain of Human Fibroblast Collagenase

BRETT LOVEJOY, ANNE CLEASBY, ANNE M. HASSELL,
MICHAEL A. LUTHER, DEBRA WEIGL,
GERARD McGEEHAN, MILLARD H. LAMBERT,
AND STEVEN R. JORDAN

Glaxo Research Institute
5 Moore Drive
Research Triangle Park, North Carolina 27709

After secretion, matrix metalloproteinases (MMPs) are processed to form the mature active enzyme.[1,2] The active enzyme contains a catalytic domain and a COOH-terminal domain. The MMP catalytic domain shares the highest sequence homology. Recombinant expressed collagenase constructs containing only the catalytic domain can cleave casein, gelatin, and peptide substrates, but, unlike the full-length enzyme, they cannot cleave collagen.[3,4] With peptide substrates, however, collagenase inhibitors exhibit similar inhibition activity toward the whole enzyme as they do on the catalytic domain alone.[4] Previous models of the MMP catalytic domain were based on available structures of the bacterial endoproteinase thermolysin[5] complexed with inhibitors as well as the structure of crayfish astacin.[6] However, such models do not provide the detail necessary to design highly selective MMP inhibitors. Here we report the crystal structure of the collagenase catalytic domain in both the presence and the absence of an inhibitor.

The catalytic domain of human fibroblast collagenase bound to a carboxyalkylamine-based inhibitor (CPLX) was solved at 2.4 Å resolution by multiple isomorphous replacement and refined to a crystallographic *R* factor of 0.186.[7] In addition, two native crystal forms of the collagenase catalytic domain (CF1 and CF2) were solved using CPLX as a starting model. CF1 was determined at 1.9 Å resolution with an *R* factor of 19.8%, and CF2 was determined at 2.1 Å resolution with an *R* factor of 19.7%.[8] The catalytic domain of recombinant human fibroblast collagenase is a spherical molecule that contains a twisted five-stranded β-sheet and three α helices (helices A, B, and C) (FIG. 1).[7] The domain contains a distinct active site cleft and the catalytic zinc is at the bottom of this cleft. The catalytic zinc is ligated by His 218, His 222, and His 228. In addition to the catalytic zinc, there is a second zinc ion and a calcium ion that stabilize an extended loop between β strands III and IV. The amino termini (Leu 102 to Gly 105) of CF1 and CF2 differ from the conformation found in CPLX by bending away from the molecule and interacting with the active site of a crystallographic symmetry-related molecule.[8]

Many interactions between the inhibitor and the collagenase active site cleft in CPLX are comparable to interactions between the amino terminus and the symmetry-related active site cleft in CF1 and CF2.[8] In CPLX, the inhibitor carboxylate group ligates the catalytic zinc (FIG. 2). In CF1 and CF2, both the nitrogen and the carbonyl of Leu 102 ligate the symmetry-related catalytic zinc. The complexes found in CPLX, CF1, and CF2 are each stabilized by a network of hydrogen bonds. Although some of these bonds are unique to CPLX or CF1 and CF2, hydrogen bonds to Ala 182 CO, Leu 181 N, Tyr 240 N, and Gly 179 CO are comparable. In collagenase, Glu 219 represents the glutamate of the HEXGH motif (where X is any residue) that is conserved in metalloproteinases.[1,2] In CPLX, the main chain nitrogen of the inhibitor P1′ group, which mimics the nitrogen of the substrate scissile bond, forms a hydrogen bond with Glu 219. In CF1 and CF2, the amino terminus (Leu 102 nitrogen) forms a hydrogen bond with the symmetry-related Glu 219. Hydrophobic interactions present in CPLX, CF1, and CF2 are also comparable. For example, in

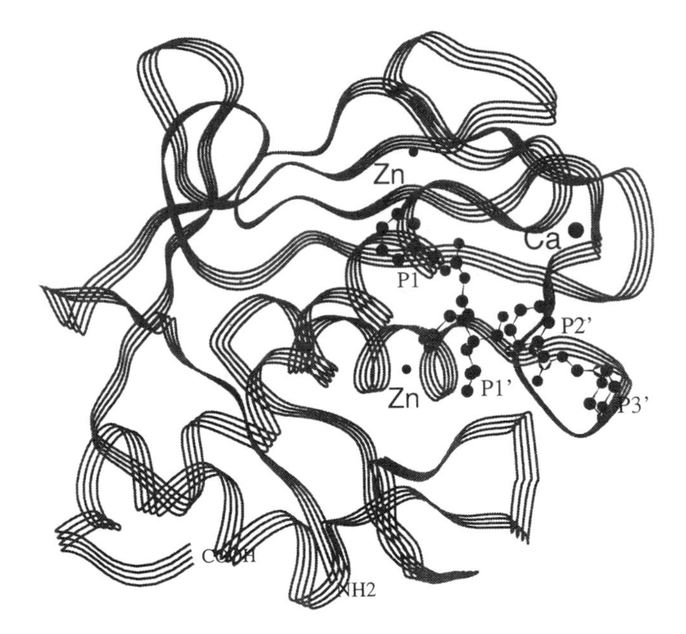

FIGURE 1A. Ribbon diagram of the catalytic domain of human fibroblast collagenase. In CPLX, an inhibitor occupies the active site cleft. A second zinc ion and a calcium ion are present in CPLX, CF1, and CF2.

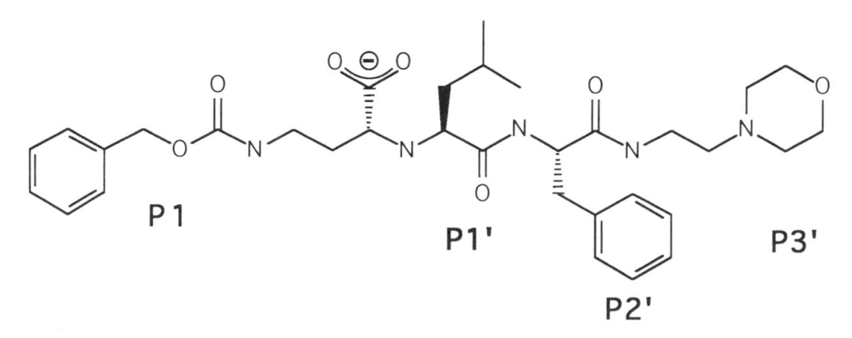

FIGURE 1B. Structure of the CPLX inhibitor. The central carboxylate group binds the active site zinc. P1′ is leucine, P2′ is phenylalanine, P3′ is a morpholino group, and P1 is a benzyloxycarbonylamino group. The inhibitor is bound to collagenase with a K_i of 135 nM.[9]

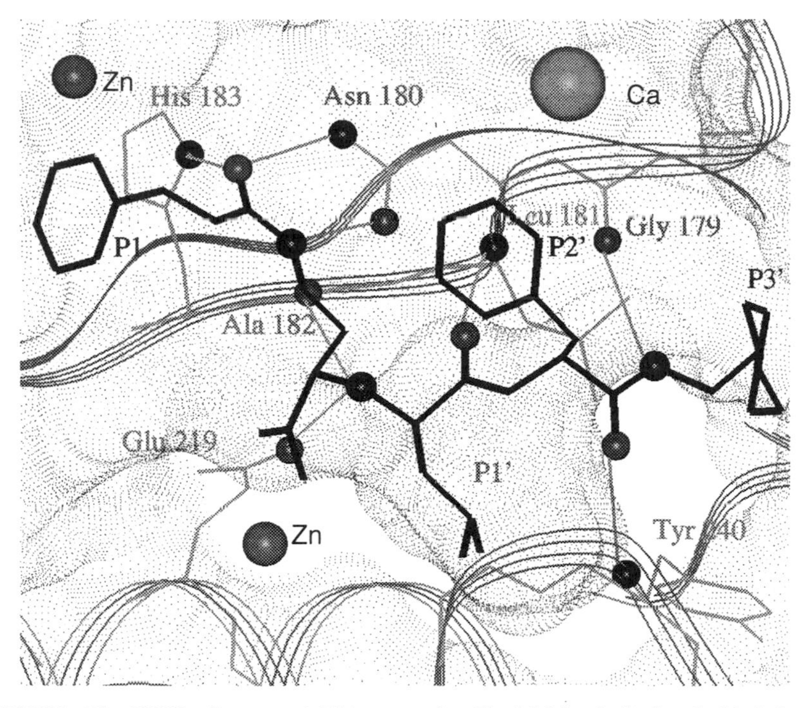

FIGURE 2. The CPLX collagenase-inhibitor complex. The inhibitor is displayed with darkened bonds and collagenase is represented by bonds, ribbons, and a Connolly surface (with a rolling sphere of 1.4 Å radius). The catalytic zinc (*lower left*), secondary zinc (*upper left*), and calcium (*right*) ions are also shown. Atoms that participate in the eight hydrogen bonds between collagenase and the inhibitor are drawn as spheres and the hydrogen bonds are represented by lines. The P1′ leucine of the CPLX inhibitor is buried in the collagenase P1′ pocket.

CPLX, the inhibitor P1′ Leu side chain occupies a hydrophobic pocket (P1′ pocket) that is adjacent to the catalytic zinc. In CF1 and CF2, Thr 103 is buried in the symmetry-related P1′ pocket.

The role of MMPs in a wide array of physiologic and pathologic processes is complex and largely undefined. The structure of collagenase complexed to an inhibitor will facilitate the detailed modeling of other MMPs. These models will provide new direction for the design of compounds that selectively inhibit individual members of the MMP family. Such inhibitors will be useful for examining the function of MMPs and may have therapeutic value.

REFERENCES

1. Birkedal-Hansen, H., W. G. I. Moore, M. K. Bodden, L. J. Windsor, B. Birkedal-Hansen, A. DeCarlo & J. A. Engler. 1993. Crit. Rev. in Oral Biol. Med. **4:** 197–250.
2. Woessner, J. F., Jr. 1991. FASEB **5:** 2145–2154.
3. Murphy, G., J. A. Allan, F. Willenbrock, M. I. Cockett, J. P. O'Connell & A. J. P. Docherty. 1992. J. Biol. Chem. **267:** 9612–9618.
4. Becherer, J. D., A. S. Howe, I. Patel, B. Wisely, H. Levine & G. M. McGeehan. 1992. J. Cell Biochem. (Suppl. 15G): 139.

5. MATTHEWS, B. W. 1988. Acc. Chem. Res. **21:** 333–340.
6. BODE, W., F. X. GOMIS-RÜTH, R. HUBER, R. ZWILLING & W. STÖCKER. 1992. Nature **358:** 164–167.
7. LOVEJOY, B., A. CLEASBY, A. M. HASSELL, K. LONGLEY, M. A. LUTHER, D. WEIGL, G. MCGEEHAN, B. MCELROY, D. DREWRY, M. H. LAMBERT & S. R. JORDAN. 1994. Science, **263:** 293–440.
8. LOVEJOY, B., A. M. HASSELL, M. A. LUTHER, D. WEIGL & S. R. JORDAN. 1994. Biochemistry. In press.
9. BROWN, F. K., P. J. BROWN, D. M. BICKETT, C. L. CHAMBERS, H. G. DAVIES, D. N. DEATON, D. DREWRY, M. FOLEY, M. GREGSON, A. B. MCELROY, G. M. MCGEEHAN, P. L. MEYERS, D. NORTON, J. M. SALOVICH, F. J. SCHOENEN & P. WARD. 1994. J.Med. Chem. **37:** 674–688.

Nutritional Supplementation with a Wheat Sprout Complex (a Metalloprotein) on Interleukin-1β–induced Bone Resorption

B. DONN,[a] L. TUROCZI, Z. BARANOWSKI,
AND M. SILESKI

Wilkes University
Wilkes-Barre, Pennsylvania 18766

Bone resorption, a characteristic accompaniment of chronic inflammatory diseases including rheumatoid arthritis and periodontal disease, is induced by interleukin-1β (IL-1β) through stimulating fibroblast production of collagenase and prostaglandins (PGE_2). Recent research has shown that superoxide dismutase levels are high in healthy periodontal tissue and low in diseased tissue, suggesting that antioxidant status may be important in controlling inflammation and resulting bone resorption.

The effects of supplementation with a superoxide dismutase active wheat sprout complex (Perioguard®, Biogenetics Food Corp., Naples, FL) on IL-1β–induced bone resorption were studied in three human subjects. Human recombinant IL-1β was mixed with fetal rat long bones prelabeled with calcium 45 in a neutral culture medium. Calcium 45 released from the bone was measured to determine bone resorption by IL-1β. Serum and diseased periodontal tissues prior to supplementation with the wheat sprout complex were removed from the three subjects and again on the contralateral side after supplementation of 3 g per day for 4 weeks.

Serum and periodontal tissues before supplementation, when combined with calcium 45 labeled fetal rat bones, demonstrated average bone resorption activity as measured by release of calcium 45 of 63 and 79%, respectively. Serum and periodontal tissues after supplementation demonstrated average bone resorption activity as measured by the release of calcium 45 of 23 and 29%, respectively.

Results indicate that this superoxide dismutase active wheat sprout complex is effective in the management of periodontal disease states because of its ability to reduce IL-1β–induced bone resorption activity.

[a] Address correspondence to Burt J. Donn, Jr., D.M.D., Banc Florida Center #101, 5811 Pelican Bay Boulevard, Naples, Florida 33963.

Effects of Proteinase Inhibitors on Aggrecan Catabolism in Chondrocyte Cultures[a]

CARL R. FLANNERY[b] AND JOHN D. SANDY

Shriners Hospital for Crippled Children (Tampa Unit) and
Department of Medical Microbiology & Immunology
University of South Florida
Tampa, Florida 33612-9499

The catabolism of aggrecan in cartilage can be achieved exclusively by chondrocyte-derived proteinases; however, the specific enzyme(s) responsible has yet to be identified. Analysis of fragments from chondrocyte-mediated catabolic systems indicates that a novel proteinase (aggrecanase), which cleaves the Glu[373]-Ala[374] bond of the interglobular domain (IGD) of aggrecan, plays a central role *in situ.* Thus, catabolic products that initiate at Ala[374] are released from articular cartilage[1] and growth plate[2] explants, from the cell layer of chondrocyte cultures,[3] and are present in human synovial fluids from patients with either osteoarthritic[4] or inflammatory joint diseases.[5] In addition, the release of aggrecan fragments generated by aggrecanase cleavage is markedly elevated in explant or cell cultures treated with retinoic acid or interleukin-1 (IL-1). Attempts to generate aggrecanase cleavage products with purified and recombinant proteinases and with subcellular fractions from chondrocytes[3] have been unsuccessful to date, so that the identity of aggrecanase remains unknown. To further characterize aggrecanase, we examined the effects of several class-specific proteinase inhibitors on retinoic acid or IL-1–induced aggrecanase activity in rat and bovine chondrocyte cultures.

METHODS

Confluent monolayers of calf articular chondrocytes or Swarm rat chondrosarcoma cells (LTC cell line from Dr. J. Kimura, Henry Ford Hospital, Detroit, MI) were cultured in Ham's F12 (calf cells) or DMEM (rat cells) containing 10% fetal calf serum and 1 μM retinoic acid (calf and rat cells) or 50 U/mL human IL-1α (calf cells) ± proteinase inhibitors. Medium was replaced daily and cultures were maintained for 3 days. The percentage of proteoglycan released was determined by dimethylmethylene blue assays of daily medium collections and papain digests of residual cell layers. A kit for assaying medium lactate concentrations was obtained from Sigma Chemical Co. (St. Louis, MO), as were the proteinase inhibitors E64d (Ep453), E64c (Ep475), D-penicillamine, phosphoramidon, actinonin, and doxycycline. Cycloheximide and AEBSF were from Calbiochem (La Jolla, CA). Recombinant human TIMP-1 (tissue inhibitor of metalloproteinases-1) was from Dr. M.

[a]This work was supported by grants from Shriners Hospitals (15960), National Institutes of Health (AR38580), and Merck & Co.

[b]Address correspondence to Carl R. Flannery, Shriners Hospital (Tampa Unit), 12502 North Pine Drive, Tampa, Florida 33612–9499.

Lark, Merck & Co. (Rahway, NJ). To demonstrate effective inhibition of cysteine proteinases by the lipophilic inhibitors Z-F-A-CHN_2 (Enzyme Systems Products, Dublin, CA), Z-Y-A-CHN_2 (provided, along with the calpain inhibitor Z-L-L-Y-CHN_2, by Dr. E. Shaw, Friedrich Miescher-Institut, Basel), and E64d, cell lysates from cultures treated with these compounds were incubated with a radioiodinated active site probe (E64 derivative) designated JPM-565[6] and subjected to SDS-PAGE and autoradiography (analyses performed by Dr. J. Munger, Brigham & Women's Hospital, Boston, MA).

RESULTS AND DISCUSSION

For the rat cultures that released ~70% or more aggrecan, retinoic acid-induced catabolism was unaffected by the addition of any of the proteinase inhibitors tested.

TABLE 1. Three-Day Culture of Rat Cells Treated with Retinoic Acid (1 μM)

Inhibitor Tested	% PG Released −/+ Inhibitor	% Inhibition of Release
Z-F-A-CHN_2 (20 μM)[a]	87.2/90.1	0
Z-Y-A-CHN_2 (100 μM)[a]	87.2/91.3	0
Z-L-L-Y-CHN_2 (100 μM)[a]	68.3/91.3	0
E64d (100 μM)	27.7/37.3	0
E64d (100 μM)	87.2/90.3	0
Penicillamine (100 μg/mL)	48.4/66.2	0
Phosphoramidon (50 μM)	35.2/39.9	0
Actinonin (10 μM)	48.4/43.1	11.0
Actinonin (50 μM)	27.7/20.4	26.4
Actinonin (100 μM)	23.4/16.0	25.3
Actinonin (100 μM)	68.3/85.2	0
Doxycycline (10 μM)	68.3/69.4	0
Doxycycline (50 μM)	68.3/81.2	0
TIMP (29 μg/mL)	78.6/84.8	0
TIMP (58 μg/mL)	78.6/83.9	0
AEBSF (1 mM)	20.8/7.5	63.9
Cycloheximide (10 μg/mL)	37.2/9.2	75.3

[a]Z = carbobenzoxy.

For rat cultures that responded to retinoic acid by releasing less than 50% aggrecan, reproducible inhibition was obtained using the metalloproteinase inhibitor actinonin (TABLE 1). Lactate production was unaffected in the presence of actinonin, demonstrating that the inhibitory effect was not due to cytotoxicity. In contrast, inhibition of proteoglycan release by 1 mM AEBSF (a water soluble, stable alternative to PMSF) was due to a cytotoxic effect (lactate production was reduced 71% by day 3 of culture). Treatment of cultures with 10 μg/mL cycloheximide resulted in a 75% inhibition of proteoglycan release (TABLE 1), demonstrating a positive correlation between aggrecanase activity and active protein synthesis. Treatment of calf chondrocytes with retinoic acid or IL-1 typically caused the loss of ~90% or ~70%, respectively, of cell layer aggrecan into the medium. The addition of inhibitors had no major effect in calf chondrocyte cultures (TABLE 2). The highest inhibition obtained using calf cells (10.3% in the presence of TIMP; see TABLE 2, bottom) was for IL-1–treated cultures which released less than 50% aggrecan in the absence of inhibitor.

Following treatment in culture ± retinoic acid or IL-1 and ± peptide diazomethanes (Z-F-A-CHN_2, Z-Y-A-CHN_2) or E64 derivatives, cell layer lysates were incubated with a radioiodinated E64 analog (JPM-565) which reacts with the active site of cysteine proteinases, then subjected to SDS-PAGE/autoradiography. Based on molecular size and abundance, a major ~31-kDa protein produced by rat cells corresponded to single chain cathepsin B. This band increased in quantity following retinoic acid treatment, but inactivation of cathepsin B, and indeed other reactive

TABLE 2. Three-Day Culture of Calf Cells Treated with Retinoic Acid (1 μM) or Interleukin-1 (50 U/mL)

Inhibitor Tested	% PG Released −/+ Inhibitor	% Inhibition of Release
Treatment with Retinoic Acid (1 μM)		
Z-F-A-CHN_2 (20 μM)	91.6/90.1	1.6
Z-Y-A-CHN_2 (100 μM)	91.6/87.0	5.0
E64d (100 μM)	91.6/92.5	0
Penicillamine (100 μg/mL)	91.6/91.7	0
Phosphoramidon (10 μM)	91.6/91.2	0.4
TIMP (20 μg/mL)[a]	38.9/41.3	0
Treatment with Interleukin-1 (50 U/mL)		
Z-F-A-CHN_2 (20 μM)	70.7/71.0	0
Z-Y-A-CHN_2 (100 μM)	70.7/68.7	2.8
E64d (100 μM)	70.7/68.7	7.0
Penicillamine (100 μg/mL)	70.7/77.5	0
Phosphoramidon (10 μM)	70.7/72.1	0
TIMP (20 μg/mL)[a]	39.8/35.7	10.3

[a]Three-day continuous cultures (no daily medium change).

proteins, by peptide diazomethanes and E64d (shown by the "blocking" of JPM-565 labeling) had no effect on aggrecanase activity (TABLE 1). Similar results were obtained for calf cells treated with retinoic acid: Z-F-A-CHN_2 blocked labeling at a concentration of 0.1 μM and E64c (Ep475) blocked at 10 μM. To demonstrate that the mechanism of retinoic acid–induced catabolism was unchanged for cultures in which cysteine proteinases had been inactivated, aggrecan present in media collected on day 2 from rat cultures treated with and without 100 μM E64d were incubated with hyaluronan and assessed for aggregability by Sepharose CL-2B chromatography. The aggrecan released in both the presence and the absence of E64d was ~30% aggregating and nonaggregating fragments eluted with a peak KAv of ~0.55, consistent with aggrecanase-mediated catabolism.

These results show that high levels of aggrecanase activity cannot be readily inhibited in isolated chondrocyte cultures using cysteine proteinase inhibitors (E64d and Z-Y-A-CHN_2) which were recently shown by others to be effective in cartilage explant cultures.[7,8] A role for metalloproteinases in aggrecan catabolism is indicated based on the partial inhibition of proteoglycan release by actinonin in cultures exhibiting a reduced response to retinoic acid; the concentrations of inhibitor tested, however, were ineffective against increased aggrecanase activity.

REFERENCES

1. Sandy, J. D., P. J. Neame, R. E. Boynton & C. R. Flannery. 1991. J. Biol. Chem. **266:** 8683–8685.
2. Plaas, A. H. K. & J. D. Sandy. 1992. Matrix **13:** 135–147.
3. Flannery, C. R. & J. D. Sandy. 1993. Orthop. Trans. **17:** 677.
4. Sandy, J. D., C. R. Flannery, P. J. Neame & L. S. Lohmander. 1992. J. Clin. Invest. **89:** 1512–1516.
5. Lohmander, L. S., P. J. Neame & J. D. Sandy. 1993. Arthritis Rheum. **36:** 1214–1222.
6. Shi, G.-P., J. S. Munger, J. P. Meara, D. H. Rich & H. A. Chapman. 1992. J. Biol. Chem. **267:** 7258–7262.
7. Buttle, D. J., J. Saklatvala, M. Tamai & A. J. Barrett. 1992. Biochem. J. **281:** 175–177.
8. Buttle, D. J. & J. Saklatvala. 1992. Biochem. J. **287:** 657–661.

Induction of Tissue Inhibitor of Metalloproteinase-2 in Injured Rat Artery[a]

REZA FOROUGH,[b] DAVID HASENSTAB,[b]
NOBUYA ZEMPO,[b] KEITH LANGLEY,[c]
YVES DE CLERCK,[d] AND ALEXANDER CLOWES[b]

[b]*Department of Surgery, RF-25,*
University of Washington
Seattle, Washington 98195

[c]*Amgen Inc.*
Thousand Oaks, California 91320

[d]*Division of Hematology/Oncology*
Childrens Hospital of Los Angeles and
University of Southern California
Los Angeles, California 90027

Arteries damaged by the passage of a balloon catheter become thicker.[1] The vascular response to injury includes smooth muscle cell (SMC) proliferation and migration from the media into the intima, and it results in the formation of a neointima. The neointima is further enlarged in mass by the accumulation of matrix molecules synthesized by the SMCs.

A shift in the balance of protease and protease inhibitors is probably necessary for vascular remodeling to occur. For example, a shift in favor of proteolysis might facilitate medial SMC movement by increasing the rate of degradation of the pericellular matrix. Recent studies from this laboratory showed an increased expression of the serine protease tissue-type plasminogen activator (t-PA) by migrating SMCs in the balloon-injured rat carotid artery. Furthermore, heparin, a known inhibitor of SMC migration and intimal thickening, inhibits t-PA expression by SMCs.[2] Matrix metalloproteinases (MMPs) are also expressed by SMCs and increased in injured vessels. The effect of these proteases is, to some extent, offset by the activity of the corresponding protease inhibitors. For example, the coordinated expression of extracellular MMPs and their inhibitors during transition of the mammary gland from a lactating phenotype to a nonlactating phenotype during involution was described in mice.[3] Matrix accumulation in the forming neointima might be influenced as much by a decline in proteolytic activity as by an increase in matrix synthesis. In this study, we measured the expression of TIMP-2, a tissue inhibitor of the MMPs, in the injured artery to define better the role of proteases and their inhibitors in vascular remodeling.

RESULTS AND DISCUSSIONS

TIMP-2 expression was measured in rat carotid artery injured by the passage of a balloon catheter (FIG. 1). FIGURE 2 shows that TIMP-2 mRNA levels began to

[a]This work was supported by National Institutes of Health grant HL 18645.

increase at 24 h after balloon injury and reached a maximum of fourfold induction at day 7. Induction of the TIMP-2 1.0-kb but not the 3.5-kb transcript coincides with the onset of SMC replication, which occurs approximately 24 h after carotid balloon injury. TIMP-1 message was not significantly modulated in the injured arterial wall (data not shown). Approximately 4 to 5 days after carotid injury, both cycling and quiescent medial SMCs start migrating from the media to the intima. FIGURE 3 shows that the activity of TIMP-2 peaks around day 5 as demonstrated by reverse zymography.

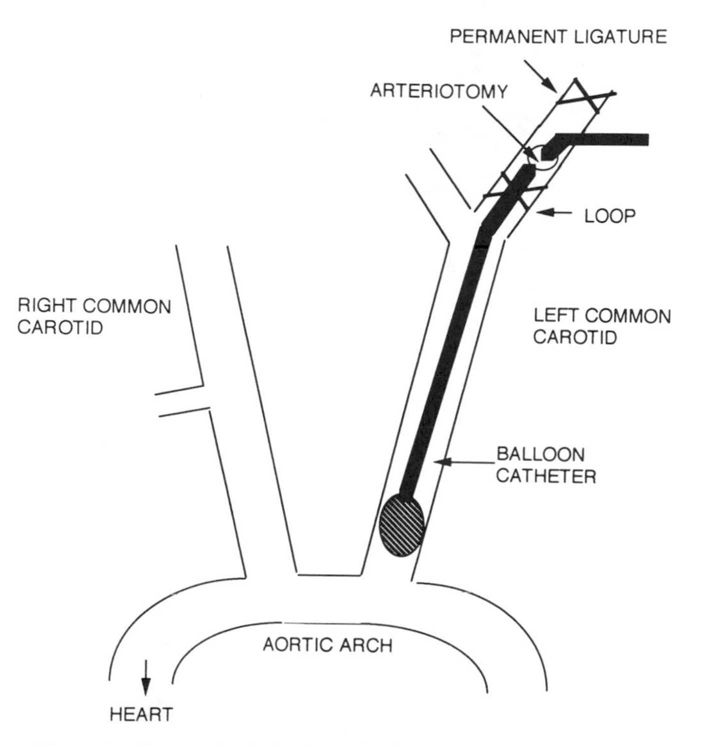

FIGURE 1. Schematic diagram depicting how the left common carotid artery of the rat is stripped of endothelium by the passage of a balloon catheter introduced through the external branch.

Both MMPs and TIMPs are produced by the cells in the vessel wall. Focal expression of MMPs by SMCs might facilitate SMC movement across tissue compartment boundaries as it does for invasive tumor cells. The vascular response to injury includes SMC proliferation, migration, and matrix turnover and may require a delicate shift of balance between MMPs and their inhibitors. The recent observation of increased 92-kD gelatinolytic activity following balloon denudation of the rat carotid artery at the time of SMC replication and migration further supports the argument for a role of MMPs in vascular remodeling.[4] An increase in TIMP-2 might be an endogenous response aimed at limiting excessive MMP activity and generalized tissue degradation while still permitting smooth muscle cell movement.

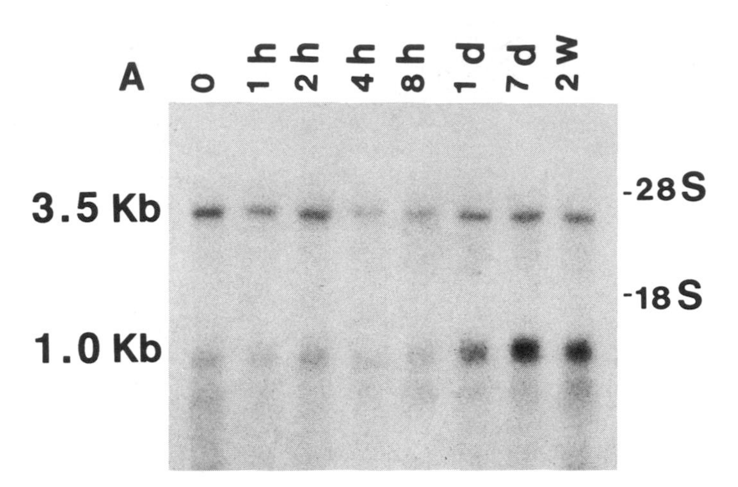

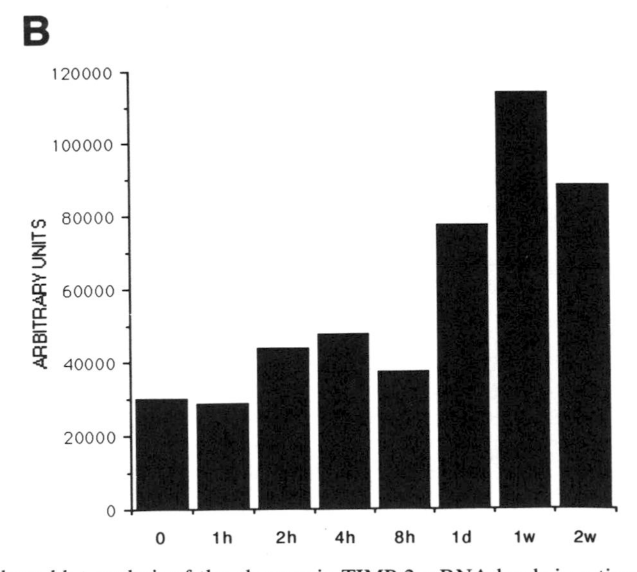

FIGURE 2. Northern blot analysis of the changes in TIMP-2 mRNA levels in a time course injury of rat carotid arteries. Total RNA was isolated from uninjured and injured carotid arteries (3 arteries per time point) on the indicated days. **(A)** Ten micrograms of each RNA sample were electrophoresed, transferred to a nylon membrane, and hybridized with a [^{32}P]-labeled human cDNA TIMP-2 probe. **(B)** The blot was subsequently probed with a cDNA for 28S ribosomal RNA to document RNA loading in each lane. Analysis was carried out on the PhosphorImager Facility at the Markey Molecular Medicine Center, University of Washington, Seattle, WA.

The balance of proteases and their inhibitors is important for endothelial function as well. TIMP-2 was shown to inhibit the proliferation of human microvascular endothelial cells stimulated with basic fibroblast growth factor (bFGF).[5] Others have used an *in vitro* endothelial wound model and measured levels of urokinase-type plasminogen activator (u-PA) and its natural inhibitor plasminogen activator inhibitor (PAI-1) to define the balance of protease-antiprotease expression during cell migration.[6] They demonstrated that the increase in u-PA and PAI-1 transcripts is colocalized to the same population of migrating endothelial cells at the edge of a wound. Furthermore, antibodies to bFGF inhibit PAI-1 mRNA induction in wounded cells. In endothelial cells treated with bFGF, generation of plasminogen activator contributes to the formation of the active form of transforming growth factor-β, which in turn opposes the mitogenic activity of bFGF by increasing PAI-1 expression

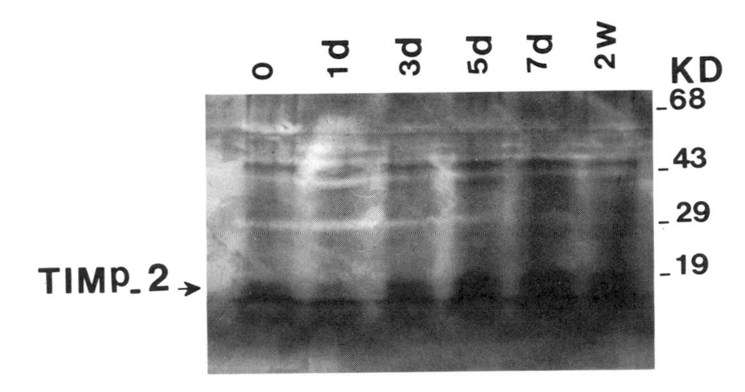

FIGURE 3. Zymogram demonstrating changes in the activity of TIMP-2 protein in injured rat carotid arteries. Five arteries per time point were extracted in 200 μL of a buffer containing 0.05 M Tris-HCl, pH 7.5, 0.01 M $CaCl_2$, 2.0 M guanidine, and 0.2% Triton X-100 using a homogenizer. Samples were centrifuged at 16,000 × *g* for 15 min, and the supernatants were dialyzed against large volumes of 0.05 M Tris-HCl, pH 7.5, and 0.2% Triton X-100 for 2 days. Then 273 μg of each sample were loaded on a 10% SDS-polyacrylamide gel containing 1 mg/mL gelatin and 4 mL of a 24-h postpartum rat uterus conditioned medium as the protease source (26% vol/vol) in the resolving gel. After gel electrophoresis, the gel was rinsed in 2.5% Triton X-100 at room temperature for 30 min and was developed for 16.5 h at 37 °C in 50 mM Tris-HCl, pH 7.8, and 10 mM $CaCl_2$. The gel was then stained with Coomassie blue.

and modulating plasminogen activator activity.[7] These observations support the conclusion that there is a critical signaling pathway that maintains the proteolytic balance in the pericellular environment of migrating cells. The future goal is to understand the molecular events leading to disruption of the balance between MMPs and their inhibitors in order to develop a pharmacologic approach to inhibiting intimal hyperplasia after vascular reconstruction.

ACKNOWLEDGMENT

We thank Holly Henson for excellent technical assistance.

REFERENCES

1. Clowes, A., M. Reidy & M. Clowes. 1983. Lab. Invest. **49:** 327–333.
2. Clowes, A., M. Clowes, T. Kirkman, C. Jackson, Y. Au & R. Kenagy. 1992. Circ. Res. **70:** 1128–1136.
3. Talhouk, R., M. Bissell & Z. Werb. 1992. J. Cell Biol. **118:** 1271–1282.
4. Bendeck, M., N. Zempo, A. Clowes & M. Reidy. 1994. Circ. Res. In press.
5. Murphy, A., E. Unsworth & W. Stetler-Stevenson. 1993. J. Cell. Physiol. **157:** 351–358.
6. Pepper, M., A. Sappino, R. Montesano, L. Orci & J. Vassalli. 1992. J. Cell. Physiol. **153:** 129–139.
7. Flaumenhaft, R., M. Abe, P. Mignatti & D. Rifkin. 1992. J. Cell Biol. **118:** 901–909.

Interaction between Progelatinase A and TIMP-2

T. HAYAKAWA,[a] N. FUJIMOTO,[b] R. V. WARD,[c]
AND K. IWATA[b]

[a]*Department of Biochemistry*
School of Dentistry
Aichi-Gakuin University
Nagoya 464, Japan

[b]*Research Institute*
Fuji Chemical Industries, Ltd.
Toyama 933, Japan

[c]*Department of Cell and Molecular Biology*
Strangeways Research Laboratory
Cambridge CB1 4RN, United Kingdom

Tissue inhibitor of metalloproteinase-2 (TIMP-2) forms a stable complex with progelatinase A (proMMP-2) and inhibits the autoactivation of proMMP-2.[1] In this study we reveal the interaction between proMMP-2 and TIMP-2 in detail by immunoreactivity analyses.

Both N-terminal (P_{1-417}) and C-terminal ($P_{415-631}$) domains of proMMP-2 were prepared as reported previously.[2,3] A monoclonal antibody (clone 43-3F9) prepared against the synthetic peptide representing a part (P_{17-35}) of the propeptide sequence of MMP-2 was used for the detection of both proMMP-2 and its N-terminal domain, and that (clone 75-7F7) prepared against native proMMP-2 was used for the detection of both proMMP-2 and its C-terminal domain.[4] Monoclonal antibodies for TIMP-2 were prepared against synthetic peptides representing sequences of TIMP-2, that is, P_{30-44} (clone 68-6H4) and $P_{178-193}$ (clone 67-4H11), which recognized the N-terminal domain and the C-terminal tail of TIMP-2, respectively.[5] A monoclonal antibody that reacts with either interstitial collagenase (MMP-1) or gelatinase B (MMP-9) was prepared against the corresponding native proenzyme.

We first confirmed that the C-terminal domain of proMMP-2 bound to the C-terminal domain of TIMP-2 in a 1:1 molar ratio through the same binding site as that of proMMP-2 as already reported previously.[2,6,7] We initially demonstrated that the N-terminal domain also bound to TIMP-2 in a 1:1 molar ratio (FIG. 1) through the same binding site as that of proMMP-2 (FIG. 2). Either N- or C-terminal domain complexed with TIMP-2, however, was easily replaced with proMMP-2, suggesting that the binding affinity of those terminal domains to TIMP-2 is not as strong as that of proMMP-2 (FIG. 2). We speculate that the binding of N-terminal domain to TIMP-2 might not be strong enough to be retained as a complex on a gelatin-Sepharose column.[2] Either MMP-1–TIMP-2 or MMP-9–TIMP-2 complex formed a tertiary complex with proMMP-2, suggesting that the proMMP-2 binding site(s) on TIMP-2 is independent of those of active MMP-1 and MMP-9. This fact is in good agreement with the result which suggested the presence of two distinct TIMP-2–binding sites on proMMP-2.[6] The binding of TIMP-2 with monoclonal antibody prepared against the tail peptide of TIMP-2 was dose-dependently suppressed in the presence of proMMP-2. This is consistent with the result reported by Willenbrock *et*

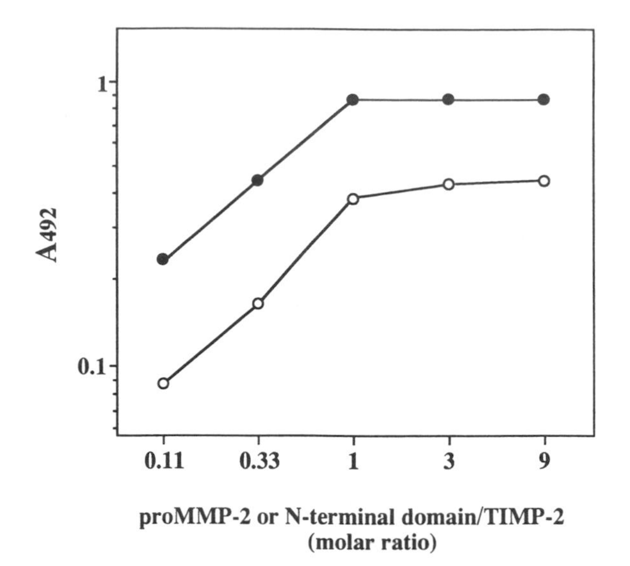

FIGURE 1. Comparison of the immunoreactivity between proMMP-2–TIMP-2 and N-terminal domain–TIMP-2 complex. A fixed amount (500 ng/mL) of TIMP-2 was mixed with either proMMP-2 (*open circles*) or N-terminal domain (*filled circles*) to give the indicated molar ratio and was then subjected to a sandwich enzyme immunoassay using anti-TIMP-2 antibody (clone 68-6H4) as a solid phase and anti-proMMP-2 antibody (clone 43-3F9) as an HRP-labeled phase.

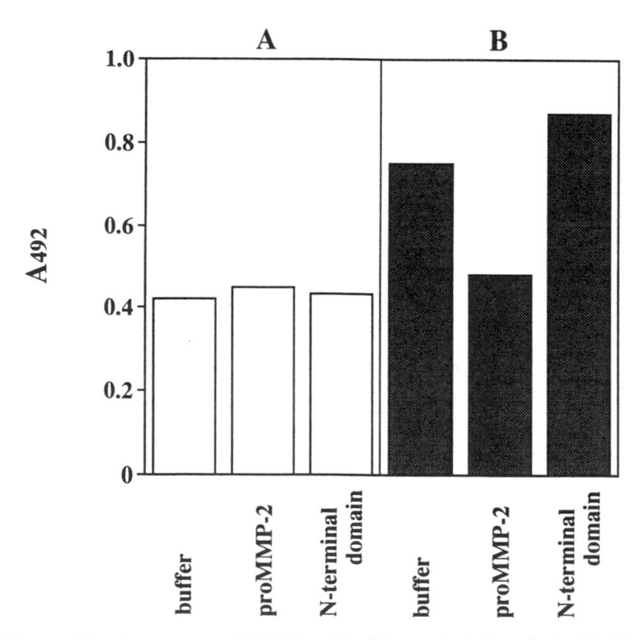

FIGURE 2. Competition between proMMP-2 and its N-terminal domain for binding to TIMP-2. First, 2 μg/well of proMMP-2 (**A**) or 1 μg/well of N-terminal domain (**B**) was transferred to three individual microtiter wells previously coated with TIMP-2 (50 ng/well). Then each individual well from both groups A and B was reacted with either incubation buffer, proMMP-2 (2 μg/well), or N-terminal domain (1 μg/well). Both proMMP-2 and N-terminal domain bound to TIMP-2 were detected by using the HRP-labeled anti-proMMP-2 antibody (clone 43-3F9).

al.[7] which suggested the interaction between proMMP-2 and the tail portion of TIMP-2.

In conclusion, proMMP-2 interacts with TIMP-2 through both N-terminal and C-terminal domains. The binding sites between proMMP-2 and TIMP-2 on either N-terminal or C-terminal domain are independent of those between active MMP and TIMP-2.

ACKNOWLEDGMENTS

We thank Drs. Gillian Murphy and Andrew J. P. Docherty for providing the N-terminal domain of progelatinase A.

REFERENCES

1. Birkedal-Hansen, H., W. G. I. Moore, M. K. Bodden, L. J. Windsor, B. Birkedal-Hansen, A. DeCarlo & J. A. Engler. 1993. Matrix metalloproteinases: A review. Crit. Rev. Oral Biol. Med. **4:** 197 250.
2. Murphy, G., F. Willenbrock, R. V. Ward, M. I. Cockett, D. Eaton & A. J. P. Docherty. 1992. The C-terminal domain of 72 kDa gelatinase A is not required for catalysis, but is essential for membrane activation and modulates interactions with tissue inhibitors of metalloproteinases. Biochem. J. **283:** 637–641.
3. Ward, R. V., R. M. Hembry, J. J. Reynolds & G. Murphy. 1991. The purification of tissue inhibitor of metalloproteinases-2 from its 72 kDa progelatinase complex. Biochem. J. **278:** 179–187.
4. Fujimoto, N., N. Mouri, K. Iwata, E. Ohuchi, Y. Okada & T. Hayakawa. 1993. A one-step sandwich enzyme immunoassay for human matrix metalloproteinase 2 (72-kDa gelatinase/type IV collagenase) using monoclonal antibodies. Clin. Chim. Acta. **221:** 91–103.
5. Fujimoto, N., J. Zhang, K. Iwata, T. Shinya, Y. Okada & T. Hayakawa. 1993. A one-step sandwich enzyme immunoassay for tissue inhibitor of metalloproteinases-2 (TIMP-2) using monoclonal antibodies. Clin. Chim. Acta **220:** 31–45.
6. Howard, E. W. & M. J. Banda. 1991. Binding of tissue inhibitor of metalloproteinases 2 to two distinct sites on human 72-kDa gelatinase. Identification of a stabilization site. J. Biol. Chem. **266:** 17972–17977.
7. Willenbrock, F., T. Creabbe, P. M. Slocombe, C. W. Sutton, A. J. P. Docherty, M. I. Cockett, M. O'Shea, K. Brocklehurst, I. R. Phillips & G. Murphy. 1993. The activity of the tissue inhibitor of metalloproteinases is regulated by C-terminal domain interactions: A kinetic analysis of the inhibition of gelatinase A. Biochemistry **32:** 4330–4337.

Treatment of Canine Osteoarthritis with Sodium Pentosan Polysulfate and Insulin-Like Growth Factor-1

RICHARD A. ROGACHEFSKY, DAVID D. DEAN,[a]
DAVID S. HOWELL,[b] AND ROY D. ALTMAN

GRECC—VA Medical Center and
Departments of Orthopaedics and Rehabilitation and Medicine
University of Miami School of Medicine
Miami, Florida 33101

[a] *Department of Orthopaedics*
University of Texas Health Science Center
San Antonio, Texas 78284

Current medical therapy of osteoarthritis (OA) is symptomatic and fails overall to address the underlying pathologic processes in cartilage. Therefore, some researchers are focusing on delaying cartilage breakdown or stimulating hyaline cartilage regeneration in the early stages of OA.[1] The terms "chondroprotection" for specific effects and "disease-modifying anti-OA drug" (DMOAD) for the overall properties have evolved. There are several classes of such compounds including heparinoids,[1] one of the latter being sodium pentosan polysulfate (Cartrophen® or SP54; PPS). Sodium pentosan polysulfate is a polysulfated polysaccharide semisynthetic product of beech hemicellulose[1] and a potent inhibitor of multiple proteases found in joint tissues.[1,2] We postulated that an additive effect might be found if administration of an anabolic agent insulin-like growth factor-1 (IGF-1) were combined with PPS in the treatment of a canine model of OA.[3]

METHODS

Thirty mongrel adult dogs were divided into five groups (n = 6 per group): (1) normal controls; (2) OA-positive controls subjected to anterior cruciate ligament transection (ACLT) and no medication; (3) an experimental group receiving ACLT and PPS treatment alone (2 mg/kg/week intramuscularly from the third to the sixth postoperative week); (4) a second experimental group receiving ACLT and IGF-1 (1 μg intraarticularly 3 times weekly) from the third to the sixth postoperative week; and (5) a group receiving both agents in the same manner as just described.[4] Following sacrifice of all animals at 6 weeks, medial femoral condylar cartilages were reproducibly sampled at lesional sites for *en bloc* gross photography, histologic sectioning, and biochemistry.

After extraction and dialysis, neutral metalloproteinase, collagenase, and TIMP were measured as reviewed in references 3 and 4.

[b] Reprint requests to: David S. Howell, M.D., Department of Medicine (D26), University of Miami School of Medicine, P.O. Box 016960, Miami, Florida 33101.

RESULTS AND DISCUSSION

Grossly all cartilage in the control OA group was eroded, ulcerated, and discolored. The group with IGF-1 alone showed slightly less erosion. The group with PPS alone showed no erosion but some pitting. Only the group with PPS + IGF-1 showed glistening white cartilage with no erosion or pitting. Histologically, the Mankin scale (an index of OA lesional severity) showed no OA changes in the normal group, whereas all of the ACLT OA controls showed chondrocyte cloning and loss of safranin-o staining, with flaking and fissures. The PPS alone group showed a lack of cloning and reduction in flaking and fissures. Only the PPS + IGF-1 group appeared almost normal.

Total and active cartilage collagenase and neutral metalloproteinases were significantly elevated in OA versus normal negative controls. P values for group comparison are given in TABLE 1. Cartilage from the OA group had less TIMP than did that of normal controls. The OA group receiving IGF-1 alone showed no improvement in either protease or TIMP levels compared to the OA controls (TABLE 1). In contrast, the OA group receiving PPS alone showed less active neutral metalloproteinase and higher levels of TIMP than did the OA controls. Only in the group with PPS + IGF-1 were both active and total neutral metalloproteinase, total collagenase, and TIMP levels closely comparable to those of normal controls.

TABLE 1. *p* Values by a Two-tailed *t* Test in an Intergroup Comparison for Total Collagenase, Total Neutral Metalloproteinase, and TIMP

Parameter	Normal	OA	OA + PPS	OA + IGF-1	OA + IGF-1/PPS	*p*
Collagenase	*	<*				<0.05
		*>			*	<0.05
Metalloproteinase	*	<*				0.02
	*			<*		0.001
	*		*			0.01
		*>			*	0.03
				*>	*	0.02
TIMP	*>	*				0.001
	*>			*		0.007
		*	<*			0.04
		*			<*	0.005
				*	<*	0.03

Abbreviations: TIMP = tissue inhibitor of metalloproteinase; IGF = insulin-like growth factor; OA = osteoarthritis; PPS = sodium pentosan polysulfate; * = paired comparisons; < > = level of one group relative to the other.

In conclusion, we demonstrated that probably through inhibition of extracellular matrix-degrading enzymes and stimulation of extracellular matrix synthesis, the dual action of PPS + IGF-1 was effective in the repair of early OA lesions in a canine model. Whether this treatment regimen can be similarly successful in more severe OA models or in human OA requires further study.

ACKNOWLEDGMENTS

We thank Victoria A. Grayson for laboratory assistance and acknowledge the secretarial assistance of Geneva Jackson.

REFERENCES

1. BURKHARDT, P. & P. GHOSH. 1987. Laboratory evaluation of antiarthritic drugs as potential chondroprotective agents. Semin. Arthritis Rheum. **17**(Suppl.): 3–34.
2. FRANCIS, D. J., M. J. FORREST, P. M. BROOKS & P. GHOSH. 1989. Retardation of articular cartilage degradation by glycosaminoglycan polysulfate, pentosan polysulphate, and DH-40J in the rat air pouch model. Arthritis Rheum. **32:** 608–616.
3. ROGACHEFSKY, R. A., D. D. DEAN, D. S. HOWELL & R. D. ALTMAN. 1993. Osteoarthritis and cartilage: Treatment of canine osteoarthritis (OA) with insulin-like growth factor-1 (IGF-1) and sodium pentosan polysulfate. Osteoarthritis Cartilage **1:** 105–114.
4. DEAN, D. D. & J. F. WOESSNER, JR. 1985. A sensitive specific assay for tissue collagenase using telopeptide-free ^{3}H-acetylated collagen. Anal. Biochem. **148:** 174–181.

Inhibition of Interleukin-1–Stimulated Collagen Degradation in Cartilage Explants

PETER G. MITCHELL,[a] LORI LOPRESTI-MORROW,[a]
SUE A. YOCUM,[a] FRANCIS J. SWEENEY,[a]
AND LAWRENCE A. REITER[b]

[a]Department of Immunology and Infectious Diseases and
[b]Department of Medicinal Chemistry
Central Research Division
Pfizer Inc.
Groton, Connecticut 06340

One of the characteristics of joint diseases such as osteoarthritis is a progressive loss of matrix constituents, leading to fibrillation and thinning of the articular cartilage. Members of the family of matrix metalloproteinases are capable of degrading all major protein constituents of the cartilage matrix, and evidence suggests that these enzymes are involved in the pathology of arthritis. *In vitro,* cytokines such as interleukin-1 (IL-1) can stimulate the breakdown of cartilage and secretion of metalloproteinases from chondrocyte monolayers.[1] In addition, metalloproteinases have been detected in the cartilage of patients with osteoarthritis.[2] Included in the matrix metalloproteinase family is collagenase, the only enzyme known to cleave the helical region of native type II collagen at physiologic temperature and pH. Because loss of the collagen matrix may be the rate-limiting step in the advance of osteoarthritis, it is expected that inhibition of collagenase would slow disease progression.

Bovine nasal cartilage explants have been used as a model system to study the effect of inhibitors on cytokine-mediated proteoglycan and collagen degradation *in vitro.*[3] The release of collagen fragments from IL-1–stimulated nasal cartilage explants requires several weeks of culture, and very little work has been done to characterize the appearance and activation of collagenase in this model. In this work we assayed the appearance of collagenase (latent and active) in the culture media and correlated this with collagen degradation. We also studied the effects of both dexamethasone and a potent collagenase inhibitor (SC 44463) on IL-1–induced collagen degradation in cartilage explants.

MATERIALS AND METHODS

Fresh bovine nasal cartilage was cut into strips, and uniform plugs (approximately 20 mg/plug) were cut out using a sterile hole punch. Most experiments were carried out in 24-well plates, each well containing three cartilage plugs and 1 mL of serum-free medium ± IL-1 (human recombinant IL-1α; 5 ng/mL). Collagen degradation was quantified by hydroxyproline determination of hydrolyzed conditioned media samples using a Waters Pico Tag system. Conditioned media were assayed for collagenase activity using [^{14}C]labeled type I collagen dried to thin films in 96-well plates. Latent collagenase was activated by the addition of trypsin before assay.

RESULTS

Interleukin-1 stimulated collagen degradation in explant cultures of bovine nasal cartilage beginning after day 12 and being complete by day 24 (FIG. 1). Collagenase activity secreted into the medium was entirely latent until day 14 of culture (FIG. 2). After day 14 there was a progressive increase in the proportion of active collagenase in the conditioned medium. Between 10 and 14 days of IL-1 treatment were required before the IL-1 could be removed and collagen degradation would still proceed. Only low levels of active collagenase were measured if IL-1 was removed on or before day 10. The hydroxamate metalloproteinase inhibitor SC 44463 inhibited IL-1–induced collagen degradation (IC_{50} < 1 μM) and could be added up to 18 days after IL-1, and almost complete inhibition of collagen degradation was still observed. Dexamethasone at concentrations down to at least 100 nM completely inhibited IL-1–induced collagen loss, and this was associated with inhibition of secreted collagenase.

CONCLUSIONS

Interleukin-1 stimulates the appearance of active collagenase in bovine nasal cartilage explant cultures, and this correlates with the time at which degraded collagen fragments are detected in the media and the time at which a collagenase enzyme inhibitor can be added and still remain effective. Extended IL-1 treatment times (10–14 days) are required to produce both degraded collagen and the appearance of active collagenase. Interleukin-1 may induce an enzyme activator or inhibit the production of a collagenase inhibitor (e.g., TIMP). This model provides a useful

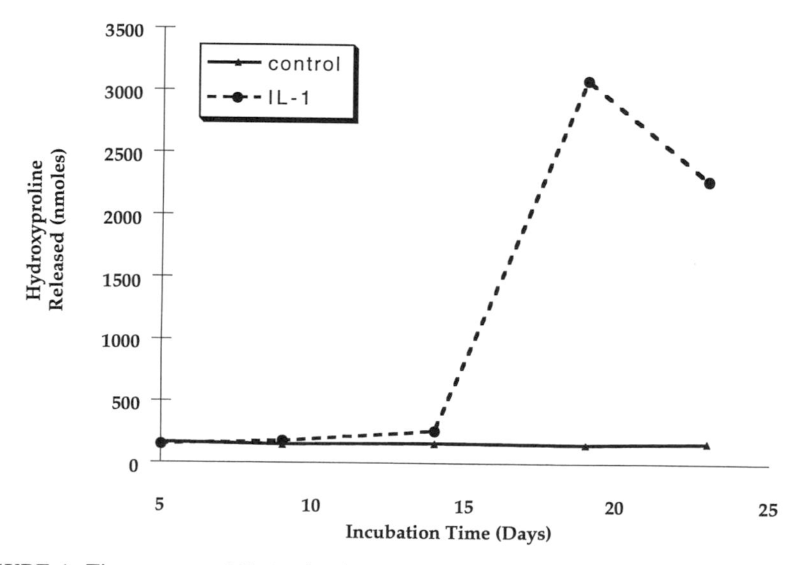

FIGURE 1. Time course of IL-1–stimulated collagen degradation in bovine nasal cartilage explants. Cartilage explants were incubated ± IL-1 for 24 days. Conditioned media were removed and fresh media added every 6 days. Collagen degradation was determined by hydroxyproline analysis of hydrolyzed conditioned media.

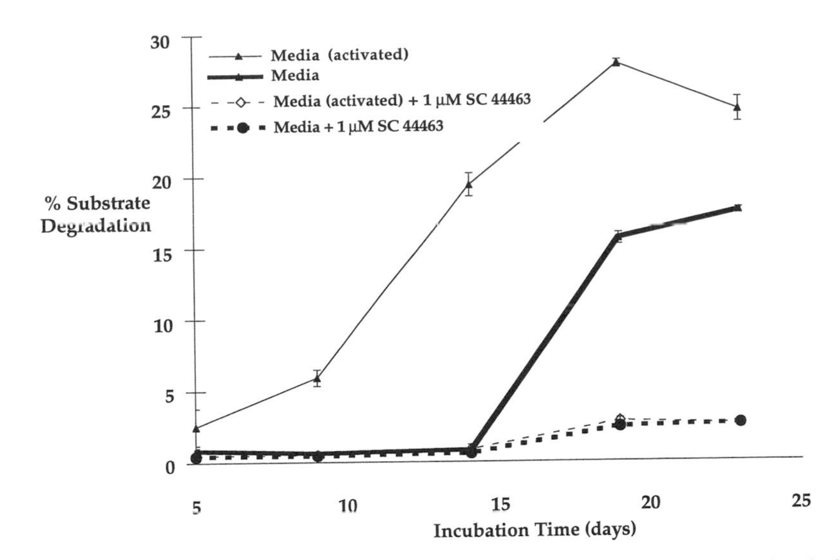

FIGURE 2. Collagenase activity in the media of bovine nasal cartilage explants stimulated with IL-1: active versus latent activity. Conditioned media from IL-1–stimulated bovine nasal cartilage explant cultures were collected at 6-day intervals. The media were assayed for collagenase activity (latent and total). The specific metalloproteinase inhibitor SC 44463 was used to inhibit collagenase activity. No activity was found in control cultures (data not shown).

in vitro determination of collagenase-induced collagen damage in cartilage. Both SC 44463, a potent collagenase enzyme inhibitor, and dexamethasone, which may inhibit collagenase transcription, were effective in protecting against IL-1–induced collagen loss in cartilage explants.

REFERENCES

1. Gowen, M., D. D. Wood, E. J. Ihrie, J. E. Meats & R. G. G. Russell. 1984. Biochim. Biophys. Acta **797:** 186–193.
2. Pelletier, J. P., J. Martel-Pelletier, D. S. Howell & J. F. Woessner. 1983. Arthritis Rheum. **26:** 63–68.
3. Nixon, J. S., K. M. K. Bottomly, M. J. Broadhurst, P. A. Brown, W. H. Johnson, G. Lawton, J. Marley, A. D. Sedgwick & S. E. Wilkinson. 1991. Int. J. Tissue React. **13:** 237–243.

Inhibition of Cartilage Metalloproteoglycanases and Stromelysin by Anionic Homopolymers

MICHAEL F. SHEFF AND ASHER I. SAPOLSKY

The Miriam Hospital and
Brown University School of Medicine
164 Summit Avenue
Providence, Rhode Island 02906

Inhibiting chondrocyte-secreted acid and neutral proteoglycanases[1,2] may permit healing of some osteoarthritic lesions. These metalloproteinases, separable by physical methods,[3] are metal dependent; therefore, they are inhibited by EDTA and *o*-phenanthroline. We also found that 1 mM $Na_4P_2O_7$ and 3 mM tetracycline produced nearly 100% inhibition. "Ideal" inhibitors would be soluble, nontoxic, nonimmunogenic, and only slowly lost by diffusion or metabolism. TIMP, the physiologic inhibitor, may be cleared too quickly.

To model possible inhibitors we prepared metalloproteinases from human articular cartilage by chromatography[1] and from the media of cultures of rabbit articular cartilage.[4] Recombinant stromelysin was a gift from Dr. Charles A. Maniglia and Miles Laboratories. Substrate (proteoglycan subunit, PGS, or A_1D_1) free of any proteases was derived from bovine nasal cartilage.[5] The viscosities of PGS solutions, both control and experimental, were measured in an Ostwald type microviscometer. Enzyme activity was expressed as a second-order function of the decline in viscosity of PGS over a 24-h period[1] at a pH of 5–7.4 in 50 mM TRIS-maleate buffer. As a buffer, 2-[N-morpholino]ethane sulfonic acid strongly altered viscosity.

We tested several classes of polymers for inhibitory activity. Albumin and casein are strongly ($>90\%$) inhibitory, but they are immunologically unsuitable as blocking agents. Studies with amino acid homopolymers (HPAAs) showed that at concentrations greater than 0.63 mg/mL, cationic HPAAs precipitated PGS. Furthermore, 50-kDa polylysine at 0.63 mg/mL had no inhibitory activity, but polyaspartate at the same concentration produced 79% inhibition.

Anionic HPAAs (polyaspartate [PASP] and polyglutamate) proved to be effective inhibitors. Most experiments were performed using PASP as the inhibitor. We obtained the following results:

1. Monomeric aspartic acid is without inhibitory activity.
2. PASP with a molecular weight of 6 kDa is partially inhibitory at pH 7, in a dose-dependent manner, over concentrations of 2.5–10 mg/mL. PASP 10 mg/mL gave 74% inhibition.
3. PASP with a molecular weight of 30 kDa also inhibits in a dose-dependent manner. Inhibition is 97+% for concentrations of 2.5 mg/mL, 5 mg/mL, and above. A concentration of 0.63 mg/mL inhibited 79%. Thus, high molecular weight PASP is approximately 16 times as inhibitory as low molecular weight PASP.
4. Polyglutamic acid (MW of 74 kDa) only produces 57% inhibition at concentrations of 5 mg/mL and 70% at 10 mg/mL. Differences in "R" group size may account for this difference.
5. Inhibition is similar at pH 7.0, 6.0, and 5.0 for all metalloproteinase enzymes.

6. Inhibition is independent of steric conformation. DL:PASP is as inhibitory as L:PASP.

7. Inhibition is partially reversible by divalent cations such as Ca^{2+} and less so by Zn^{2+}.

8. Trypsin digestion of PGS at pH 7.0 is also inhibited by PASP, but to a lesser degree, with 30-kDa molecular weight PASP inhibiting trypsin 80% at 10 mg/mL.

9. Trypsin digestion of albumin at pH 7.0 is entirely unaffected by PASP. Under the same conditions inhibition by soybean antitrypsin was nearly complete.

We conclude that the effects of PASP suggest the possibility of developing an *in vivo* agent based on a similar molecular structure and capable of blocking metalloproteinase activity in cartilage.

REFERENCES

1. Sapolsky, A. I. *et al.* 1976. J. Clin. Invest. **56:** 1030–1041.
2. Woessner, J. F., Jr. & M. F. Selzer. 1984. J. Biol. Chem. **259:** 3633–3638.
3. Sapolsky, A. I. & D. S. Howell. 1982. Arthritis Rheum. **25:** 981–988.
4. Malemud, C. J. *et al.* 1979. J. Lab. Clin. Med. **93:** 1018–1030.
5. Hascall, V. C. & S. W. Sadjera. 1969. J. Biol. Chem. **244:** 2384–2396.

Inhibition of Matrix Metalloproteinases in Rheumatoid Arthritis and the Crystallographic Binding Mode of a Peptide Inhibitor

H. TSCHESCHE,[a] J. BLÄSER,[a] T. KLEINE,[a]
S. SCHNIERER,[a] P. REINEMER,[b] W. BODE,[b]
U. MAASJOSHUSMANN, AND C. FRICKE[c]

[a] *Lehrstuhl für Biochemie*
Fakultät für Chemie
Universität Bielefeld
33615 Bielefeld, Germany

[b] *Max-Planck-Institut für Biochemie*
82152 Martinsried, Germany

[c] *Klinik für Rheumatologie*
St. Josefs-Stift
48324 Sendenhorst, Germany

In rheumatoid arthritis, invasive and proliferating tissue degrades cartilage and bone, resulting in joint destruction. Matrix metalloproteinases (MMPs) are involved in this process and have been detected by ELISA in synovial fluids.[1] A therapy used in Europe is destruction of the pannus by chemical synovectomy using OsO_4 or Varicocid (sodium salt of fatty acid from cod liver oil). This treatment is always associated with an inflammatory reaction. The synovial fluid of patients with rheumatoid arthritis demonstrated a significant increase in interstitial leukocyte collagenase (MMP-8), whereas the fibroblast enzyme (MMP-1) decreased with Varicocid (FIG. 1).

Along with immunosuppressives, steroids, and nonsteroidal anti-inflammatory drugs (NSAIDs), the most commonly employed therapeutics in the treatment of rheumatoid arthritis are the disease-modifying antirheumatic drugs (DMARDs) including antimalarials, gold compounds, and D-penicillamine. Their influence on MMPs was investigated, and kinetic measurements with fluorogenic peptide substrates against the recombinant catalytic domain of leukocyte collagenase (MMP-8)[2] showed that they exert inhibitory activity against the enzyme. The chrysotherapeutics auro and gold sodium thiomalate were the most potent inhibitors with K_i values of ≤ 5.0 and 1.15 μM, respectively, followed by Auranofin and gold thioglucose with a K_i of 30 μM. The chelator D-penicillamine inhibited with a K_i of 28 μM. The inhibition constants suggest that under therapy by intramuscular injection, plasma concentrations can be reached that can exert an inhibitory effect on the metalloproteinases, thus explaining the potential antiarthritic action of these chrysotherapeutics.

The binding mode of a low molecular weight Pro-Leu-Gly-hydroxamat inhibitor to the active site of the catalytic domain of leukocyte collagenase (MMP-8) was revealed by X-ray crystallography at a resolution of 2.0 Å. The structure of the Phe^{79}-Gly^{242} catalytic domain is shown as a ribbon model in FIGURE 2. The inhibitor is bound to a shallow active site cleft formed from an upper (N-terminal) and a lower

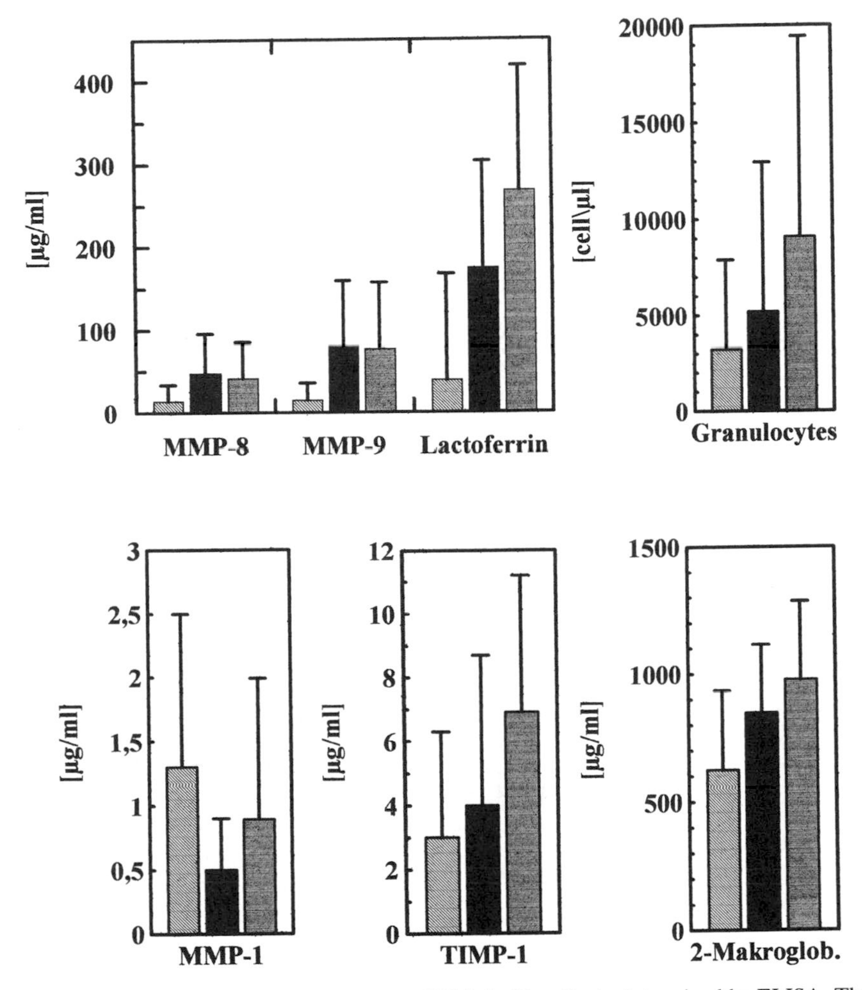

FIGURE 1. Synovial fluid content (mean ± SEM) in 38 patients determined by ELISA. The knee was punctured before (▧) and 8 h (■) and 24 h (▩) after treatment with Varicocid.

(C-terminal) subdomain of the enzyme. The upper part locates two zinc ions, the "structural" zinc on top, coordinated by three His and one Asp, and the "catalytic" zinc in the middle, coordinated by three His and the inhibitor's two hydroxamic acid oxygens (thick lines), and the two structural calcium ions.

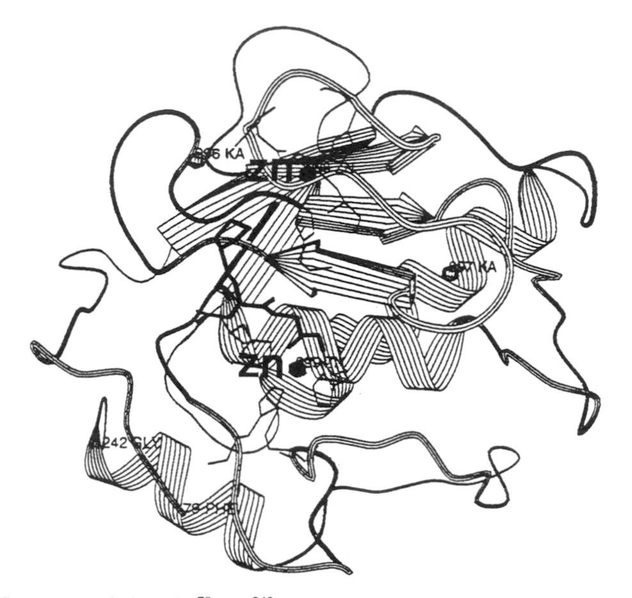

FIGURE 2. Structure of the Phe[79]-Gly[242] catalytic domain of human leukocyte collagenase (MMP-8) shown as a ribbon model with the "structural" and "catalytic" zinc ions. The Pro-Leu-Gly-NHOH inhibitor is shown as *thick lines.* The inhibitor's two hydroxamic acid oxygens coordinate the catalytic zinc ion. (See refs. 3 and 4.)

REFERENCES

1. Bergmann, U., J. Michaelis, R. Oberhoff, V. Knäuper, R. Beckmann & H. Tschesche. 1989. J. Clin. Chem. Clin. Biochem. **27:** 351–359.
2. Schnierer, S., T. Kleine, T. Gote, A. Hillemann, V. Knäuper & H. Tschesche. 1993. Biochem. Biophys. Res. Commun. **191:** 319–326.
3. Bode, W., P. Reinemer, R. Huber, T. Kleine, S. Schnierer & H. Tschesche. 1994. EMBO J. **13:** 1263–1269.
4. Reinemer, P., F. Grams, R. Huber, T. Kleine, S. Schnierer, M. Pieper, H. Tschesche & W. Bode. 1994. FEBS Lett. **338:** 227–233.

Human TIMP-1 Binds to Pro-M_r 92K GL (Gelatinase B, MMP-9) through the "Second Disulfide Knot"[a]

M. KIRBY BODDEN,[b-d] L. JACK WINDSOR,[b,e]
NANCY C. M. CATERINA,[e] GREGORY J. HARBER,[b]
BENTE BIRKEDAL-HANSEN,[f] AND
HENNING BIRKEDAL-HANSEN[b]

[b]*Department of Oral Biology and Research Center in Oral Biology*
University of Alabama School of Dentistry
University of Alabama at Birmingham
Birmingham, Alabama 35294

[c]*Department of Pathology*
University of Alabama at Birmingham
Birmingham, Alabama 35294

[e]*Department of Biochemistry and Molecular Genetics*
University of Alabama at Birmingham
Birmingham, Alabama 35294

[f]*Department of Diagnostic Sciences*
University of Alabama School of Dentistry
University of Alabama at Birmingham
Birmingham, Alabama 35294

Remodeling of the extracellular matrix is controlled in part by tissue inhibitors of metalloproteinases (TIMPs) through their ability to inhibit matrix metalloproteinases (MMP) and to block or delay the activation of MMP zymogens.[1,2] TIMP-1 and TIMP-2 form complexes with the zymogen forms of the two (M_r 72K and M_r 92K) gelatinases, TIMP-1 with pro-M_r 92K GL and TIMP-2 with pro-M_r 72K GL.[3,4] The purpose of this study was to identify domains that are important for binding of TIMP-1 to pro-M_r 92K GL.

METHODS

To determine which domains of TIMP-1 interact with pro-M_r 92K GL, we used a peptide competition approach.[5] Aliquots of pro-M_r 92K GL–TIMP-1 complex isolated by gelatin-sepharose chromatography from serum-free media conditioned by phorbol ester-stimulated HT-1080 cells were incubated with a panel of synthetic peptides (1 mM) modeled after the human TIMP-1 sequence[6,7] (FIG. 1A). Our

[a]This work was supported by U. S. Public Health Service grants DE08228, DE00283, and DE10631.

[d]Address correspondence to Dr. M. Kirby Bodden, Department of Oral Biology, University of Alabama School of Dentistry, University of Alabama at Birmingham, Birmingham, Alabama 35294.

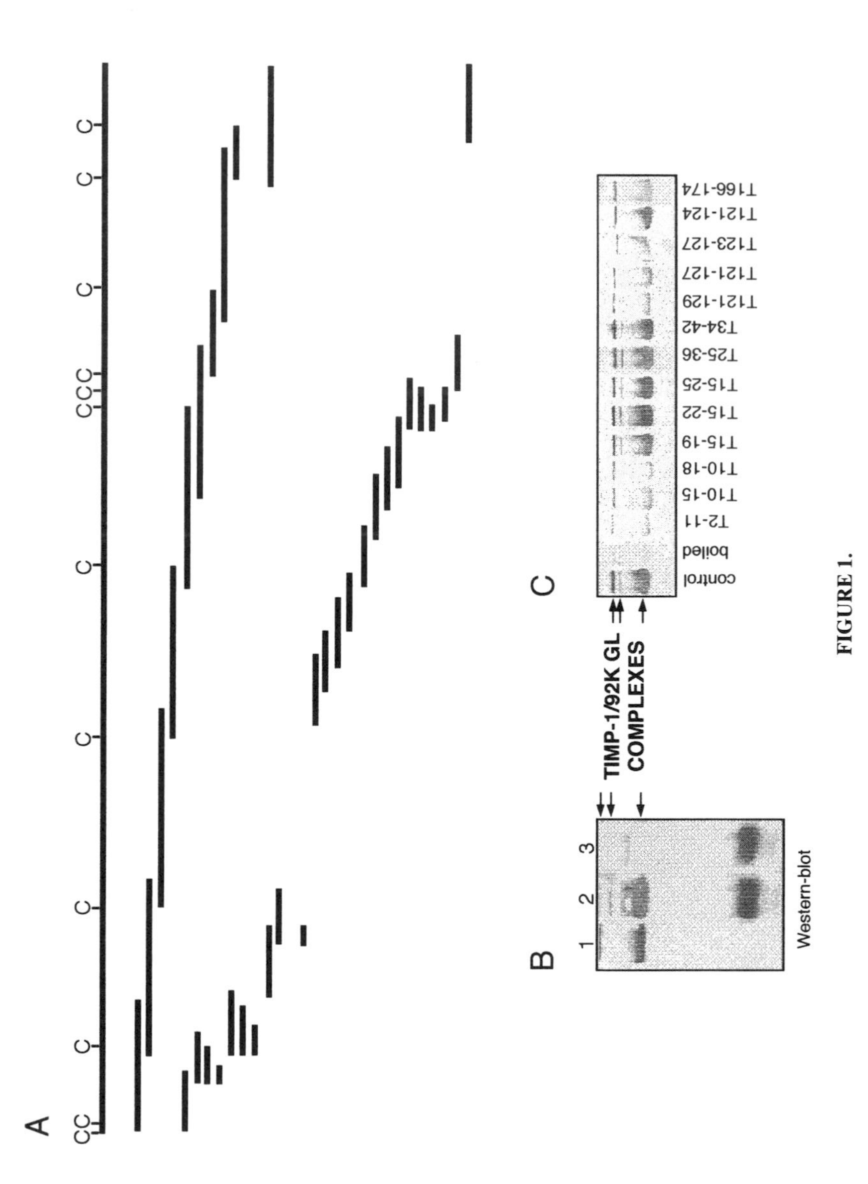
A
B
C
TIMP-1/92K GL
COMPLEXES
1
2
3
Western-blot
control
boiled
T2-11
T10-15
T10-18
T15-19
T15-22
T15-25
T25-36
T34-42
T121-129
T121-127
T123-127
T121-124
T166-174

FIGURE 1.

approach took advantage of the observation that the complex is resistant to 0.1% SDS. The samples were resolved by SDS-PAGE (0.1% SDS) using the method of DeClerck *et al.*[8] and were transferred to nitrocellulose blots. Peptides that were able to displace TIMP-1 from the complex were identified by Western analysis using antibodies to human TIMP-1 and M_r 92K GL. "Forward" and "reverse" zymography, biosynthetic radiolabeling, and immunoprecipitation were conducted as previously described.[9,10]

RESULTS

The pro-M_r 92K GL–TIMP-1 complex isolated by gelatin-sepharose chromatography contained equimolar ratios of the two components, and forward and reverse zymography showed that TIMP-1 and pro-M_r 92K GL, isolated individually and as a complex, were fully functional.[11] M_r 92K GL–TIMP-1 complexes visualized by staining with anti-TIMP-1 antibody migrated as a major $\approx M_r$ 95 kDa band and two minor (200 and 150 kDa) bands of somewhat varying intensity, the latter presumably representing complexes between pro-M_r 92K GL dimers and TIMP-1 (FIG. 1B). Once dissociated, however, the two components were no longer capable of forming complexes resistant to SDS (data not shown). This is in marked contrast to FIB-CL which, under the same conditions, forms an SDS-resistant complex with TIMP-1 after activation but not in the zymogen form.[8]

Peptides that displaced TIMP-1 from its complex with pro-M_r 92K GL abolished or markedly reduced the subsequent staining of the M_r 95 kDa and M_r 150/200 kDa bands with anti-TIMP-1 antibody on Western blots (FIG. 2A). The peptide competition studies showed that a number of peptides derived from regions of loop 1, loop 3, loop 4, and from the COOH-terminal tail displaced TIMP-1 from the complex. From the highly conserved NH_2-terminus, peptides T2–11, 10–15, and 10–18 were highly effective, as indicated by the lack of TIMP-1 staining on Western blots (FIG. 2A). Peptides T121–129 and 128–136 of loops 3 and 4, which in the native TIMP-1 molecule are attached to the NH_2-terminal sequence through disulfide bond Cys13-Cys124, also displaced TIMP-1. Shorter versions of peptide T121–129 (TVGCE-ECTV) which represents the "upper strand" of the "second disulfide knot," such as T121–127 (TVGCEEC), T123–127 (GCEEC), but not T121–124 (TVGC), also displaced TIMP-1 from the complex. Peptide T166–174 from the COOH-terminal domain also was effective.

FIGURE 1. **(A)** Schematic showing design of synthetic peptides used for competition experiments. **(B)** M_r 92K GL–TIMP-1 complexes identified by Western analysis using anti-TIMP-1 antibody. Western blots of pro-M_r 92K GL–TIMP-1 complex isolated by gelatin-sepharose chromatography: **lane 1,** staining with anti-M_r 92K GL monoclonal antibody (mAb); **lane 2,** staining with anti-TIMP-1 mAb; **lane 3,** staining with anti-TIMP-1 mAb after dissociation of complex by heating in 2% SDS. (Data from Bodden *et al.*[5]) **(C)** Peptide competition experiments with pro-M_r 92K GL-TIMP-1 complexes examined by Western analysis. Aliquots of complex were incubated at 4 °C for 60 min with 1-mM solutions of the peptides indicated, resolved by 0.1% SDS-PAGE, blotted onto nitrocellulose, and stained with anti-TIMP-1 pAb. Controls consisted of untreated M_r 92K GL–TIMP-1 complex and complex dissociated in boiling 0.1% SDS. (Data from Bodden *et al.*[5])

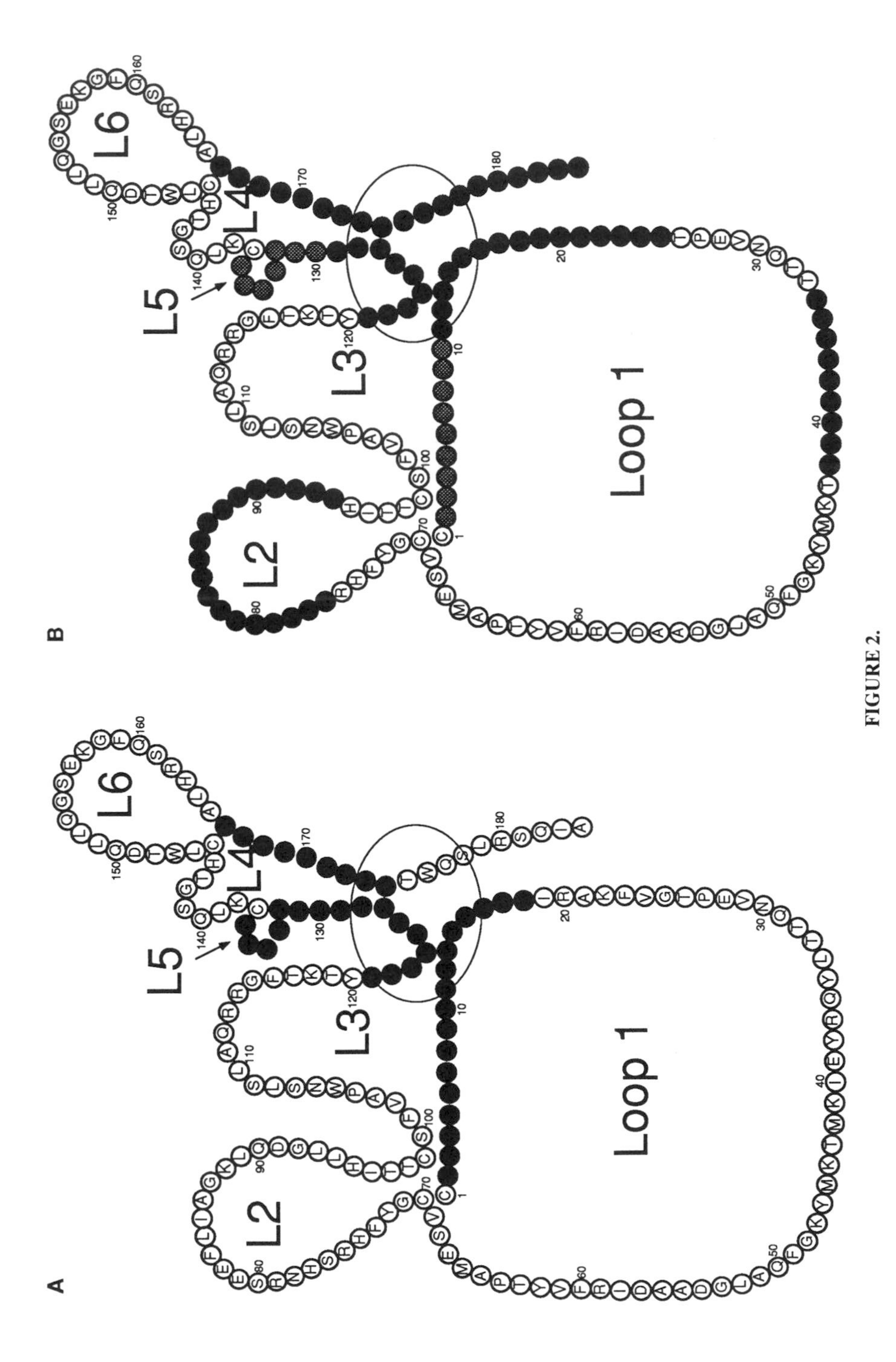

FIGURE 2.

DISCUSSION

These findings suggest that TIMP-1 binds to pro-M_r 92K GL through sequences surrounding the second disulfide knot (T2–11, T10–18, and T10–15 of loop 1 and T121–129 and 128–136 of loop 3) in addition to a short segment of the COOH-terminal domain, T166-174 (FIG. 2B). The second disulfide knot (Cys13–Cys124; Cys127–Cys174) marks the transition between the NH_2-terminal and COOH-terminal domains which are connected through the $C_{124}EEC_{127}$ sequence. Because this area is also involved in the binding of TIMP-1 to activated human FIB-CL,[12] our findings suggest that TIMP-1 uses common domains for binding to pro-M_r 92K GL and active FIB-CL.

REFERENCES

1. BIRKEDAL-HANSEN, H., W. G. I. MOORE, M. K. BODDEN, L. J. WINDSOR, B. BIRKEDAL-HANSEN, A. DECARLO & J. A. ENGLER. 1993. Crit. Rev. Oral Biol. Med. **4:** 197–250.
2. WOESSNER, J. F., JR. 1991. FASEB J. **5:** 2145–2154.
3. GOLDBERG, G. I., B. L. MARMER, G. A. GRANT, A. Z. EISEN, S. WILHELM & C. HE. 1989. Proc. Natl. Acad. Sci. USA **86:** 8207–8211.
4. WILHELM, S. M., I. E. COLLIER, B. L. MARMER, A. Z. EISEN, G. A. GRANT & G. I. GOLDBERG. 1989. J. Biol. Chem. **264:** 17213–17221.
5. BODDEN, M. K., L. J. WINDSOR, N. C. M. CATERINA, G. J. HARBER, B. BIRKEDAL-HANSEN, A. A. DECARLO, J. A. ENGLER & H. BIRKEDAL-HANSEN. 1994. Unpublished data.
6. DOCHERTY, A. J. P., A. LYONS, B. J. SMITH, E. M. WRIGHT, P. E. STEPHENS, T. J. R. HARRIS, G. MURPHY & J. J. REYNOLDS. 1985. Nature **318:** 66–69.
7. CARMICHAEL, D. F., A. SOMMER, R. C. THOMPSON, D. C. ANDERSON, C. G. SMITH, H. G. WELGUS & G. P. STRICKLIN. 1986. Proc. Natl. Acad. Sci. USA **83:** 2407–2411.
8. DECLERCK, Y. A., T. D. YEAN, H. S. LU, J. TING & K. E. LANGLEY. 1991. J. Biol. Chem. **266:** 3893–3899.
9. HERRON, G. S., M. J. BANDA, E. J. CLARK, J. GAVRILOVIC & Z. WERB. 1986. J. Biol. Chem. **261:** 2810–2813.
10. BIRKEDAL-HANSEN, H. 1987. Methods Enzymol. **144:** 140–171.
11. HERRON, G. S., M. J. BANDA, M. J. CLARK, J. GAVRILOVIC & Z. WERB. 1986. J. Biol. Chem. **261:** 2814–2818.
12. BODDEN, M. K., G. J. HARBER, B. BIRKEDAL-HANSEN, L. J. WINDSOR, N. C. M. CATERINA, J. A. ENGLER & H. BIRKEDAL-HANSEN. 1994. J. Biol. Chem. In press.

FIGURE 2. Summary of peptide competition experiments. **(A)** Peptides that displace TIMP-1 from its complex with human pro-M_r 92K GL are shown in *black.* **(B)** For comparison, peptides that compete with human TIMP-1 for binding to FIB-CL are shown (Bodden *et al.*[12]); peptides that neutralize TIMP-1 with IC_{50} 0.2–0.5 mM are shown in *black;* peptides with $IC_{50} \geq 0.5$ mM are shown in *gray;* peptides that fail to neutralize TIMP-1 at concentrations up to 1 mM are shown in *white.* (Data from Bodden *et al.*[5,12])

Inhibition of Growth of Human Tumor Cells in Nude Mice by a Metalloproteinase Inhibitor

AKIRA OKUYAMA,[a] KYOZO NAITO,
HAJIME MORISHIMA, HIROYUKI SUDA,
SUSUMU NISHIMURA, AND NOBUO TANAKA

Banyu Tsukuba Research Institute, in collaboration with
Merck Research Laboratories
Okubo 3
Tsukuba 300-33, Japan

The imbalance of matrix metalloproteinases (MMPs), especially gelatinases (MMP-2 and MMP-9) and tissue inhibitors of metalloproteinases (TIMPs), seems to be crucial in tumor metastasis and growth.[1] Low molecular weight MMP inhibitors may be a new type of anticancer therapy that inhibits tumor growth as well as tumor invasion without any cytotoxicity. Therefore, we screened MMP inhibitors from microbial cultures using MMP-9 as a target enzyme.

We isolated a new low molecular weight MMP inhibitor, BE16627B, whose structure was determined to be L-*N*-(*N*-hydroxy-2-isobutylsuccinamoyl)-seryl-L-valine (MW 375). BE16627B inhibited MMPs such as MMP-2 and MMP-9 (IC_{50}s were 0.58 and 0.85 μM, respectively) but not other proteinases.[2] BE16627B (100 μg/mL) showed no cytotoxicity against human tumor cells such as HT1080 fibrosarcoma cells and HCT116 colon carcinoma cells, its LD_{50} in mice being more than 1,000 mg/kg (i.p.). BE16627B inhibited 77.8% of gelatinolytic and 74.0% of collagenolytic activity after activation by aminophenyl mercuric acetate (APMA) in the conditioned medium of HT1080 cells overexpressing MMPs, respectively, when it was cocultured with the cells at 1 μg/mL for 3 days. Furthermore, our preliminary data suggested that BE16627B inhibited the synthesis and/or secretion of MMPs in the culture.

On the basis of these results, we investigated its antitumor effects in nude mice. BE16627B was administered in an osmotic pump subcutaneously implanted before and after the intravenous inoculation of HT1080 cells. Twenty-eight days after the inoculation of HT1080 cells all mice were sacrificed, and the number and size of nodules on lung and the lung weight were measured. BE16627B at a dose of 2 mg/mouse/day reduced to 24.3 and 46.4% the number and diameter of the control nodule of HT1080 cells, respectively, and inhibited 85.5% of the increase in lung weight due to tumor cell growth when it was administered for 3 weeks beginning 7 days after the inoculation of tumor cells, suggesting that BE16627B inhibited the growth of tumor cells in lung.

We therefore examined the effect of BE16627B on the subcutaneous growth of HT1080 cells in mice. BE16627B inhibited the increase in tumor volume dose dependently at a dose of 0.5, 1, and 2 mg/mouse/day when it was administered in osmotic pumps for 3 weeks from 7 days after the subcutaneous inoculation of HT1080 cells (FIG. 1A). At a dose of 2 mg/mouse/day, BE16627B inhibited 71.2% of

[a] Corresponding author.

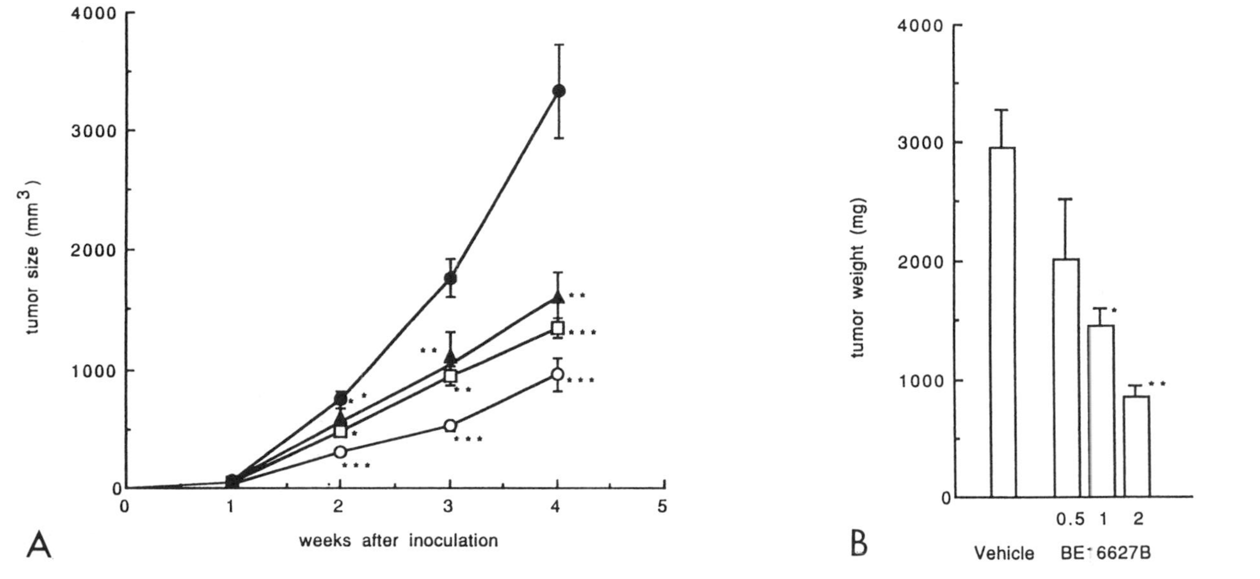

FIGURE 1. **(A)** Growth curves of HT1080 cells in nude mice treated with BE16627B. Nude mice were inoculated s.c. with HT1080 cells (1×10^6 cells) on day 0. One week after inoculation, osmotic pumps filled with BE16627B were implanted s.c. BE16627B was released from the osmotic pumps at rates of 0.5 (▲), 1 (□), and 2 mg/mouse/day (○) from 1 to 4 weeks after s.c. inoculation. Tumor size was measured every week. There were 8 or 10 mice tested in a group. Statistical significance from vehicle control (●) was $^{*}p < 0.05$, $^{**}p < 0.01$, and $^{***}p < 0.001$. **(B)** Growth inhibition of HT1080 cells by BE16627B. Subcutaneous tumors were dissected 4 weeks after inoculation and weighed. Data shown are mean ± SE. Numbers under the columns are doses of BE16627B administered (mg/mouse/day). Statistical significance from vehicle control was $^{*}p < 0.05$ and $^{**}p < 0.01$.

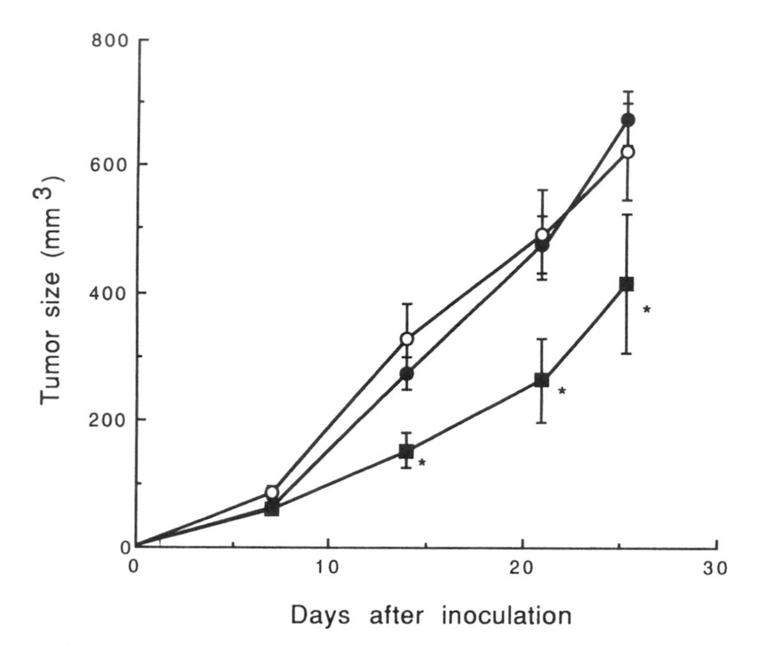

FIGURE 2. Growth curves of HCT116 cells in nude mice treated with BE16627B or adriamycin. Nude mice were inoculated s.c. with HCT116 cells (1×10^7 cells) on day 0. One week after inoculation, osmotic pumps filled with BE16627B or adriamycin were implanted s.c. BE16627B (2 mg/mouse/day [○]) or adriamycin (0.06 mg/mouse/day [■]) was released from the osmotic pumps for 7–25 days after inoculation. Tumor size was measured every week. There were 5 mice tested in a group. Statistical significance from vehicle control (●) was $^*p < 0.05$.

the tumor growth in mice (FIG. 1B). Under the same conditions as those for HT1080 cells, BE16627B did not significantly affect the tumor size and weight of HCT116 cells scarcely overexpressing MMPs (FIG. 2).[3,4] BE16627B did not show any significant effect on body weight, number of platelet and white blood cells, and spleen weight under these conditions. Our data clearly show that a small molecular weight metalloproteinase inhibitor, BE16627B, inhibited metalloproteinase-dependent human tumor growth without cytotoxicity and obvious toxicity in nude mice.

REFERENCES

1. LIOTTA, L. A., P. S. STEEG & W. G. STETLER-STEVENSON. 1991. Cell **64:** 327–336.
2. NAITO, K., S. NAKAJIMA, N. KANBAYASHI, A. OKUYAMA & M. GOTO. 1993. Agents and Actions **39:** 182–186.
3. OKUYAMA, A., K. NAITO, H. MORISHIMA, H. SUDA, S. NISHIMURA & N. TANAKA. Gordon Research Conference on Matrix Metalloproteinase, August 16–20, 1993. Plymouth, NH.
4. NAITO, K., N. KANBAYASHI, S. NAKAJIMA, T. MURAI, K. ARAKAWA, S. NISHIMURA & A. OKUYAMA. 1994. Int. J. Cancer. In press.

Preclinical Antiarthritic Activity of Matrix Metalloproteinase Inhibitors

M. J. DIMARTINO,[a,b] W. HIGH,[a] W. A. GALLOWAY,[c] AND M. J. CRIMMIN[c]

[a] *SmithKline Beecham Pharmaceuticals King of Prussia, Pennsylvania 19406-0939*

[c] *British Bio-technology Ltd. Oxford, OX4 5LY, United Kingdom*

Overexpression of matrix metalloproteinases (MMPs) (e.g., collagenase, stromelysin, and gelatinase) is believed to contribute to tissue damage occurring in rheumatoid arthritis. Indeed, MMPs have been identified in synovial fluid, in synovial tissues, and at sites of cartilage erosion.[1] Although numerous potent MMP inhibitors (MMPIs) were described over the last several years, few reports concerned their *in vivo* antiarthritic activity. Some of the most potent MMPIs that have been identified are small substrate analog peptides which contain a zinc-chelating group (e.g., hydroxamic acid). This report describes the *in vivo* antiarthritic activity of potent hydroxamic acid-containing peptide MMPIs (IC_{50} values < 50 nM) in the adjuvant arthritic (AA) rat model. Adjuvant arthritis was produced in male Lewis rats by a single injection of 0.75 mg *Mycobacterium butyricum* in light paraffin oil.[2] Hindpaw swelling occurred by day 10 and was measured plethysmographically by water displacement. The intraperitoneal (i.p.) administration of MMPIs during the development of hindpaw lesions (i.e., days 13–23) was found to inhibit progression of the swelling from day 13.

BB 94 (4-(*N*-hydroxyamino)-2*R*-isobutyl-3*S*-(thiophen-2- ylthiomethyl)-succinyl)-L-phenylalanine-*N*-methylamide) is a potent (IC_{50} < 20 nM), broad spectrum MMPI that has antitumor activity[3] and is in clinical trials for the treatment of cancer. As shown in FIGURE 1, BB 94 administered to AA rats at 5 mg/kg i.p. twice a day on days 13–17 markedly inhibited the increase in paw swelling by day 17. In contrast, BB 1722, the enantiomer of BB 94 which is ineffective as an MMPI (IC_{50} > 100 μM), did not inhibit paw swelling. The efficacy of BB 94 was still evident several days after drug administration (FIG. 1).

In dose response studies, the ED_{50} (95% confidence limits) of BB 94 for inhibiting increases in paw swelling was calculated to be 1.3 (0.3–3.6) mg/kg i.p. once a day. BB 16 (2*S*, 3*R*,-*N*-[3(*N*-hydroxycarboxyamido)-2-(2-methylpropyl)-butanoyl]-*O*-methyl-L-tyrosine-*N*-methylamide), a related MMPI (IC_{50} < 50 nM), also inhibited increases in AA rat paw swelling over a wide range of doses (i.e., 0.6–100 mg/kg i.p. twice a day). In contrast, BB 15, a diastereoisomer of BB 16, which is less potent as an MMPI (IC_{50} > 1 μM), inhibited paw swelling only at doses ≥50 mg/kg i.p. twice a day.

Results of *in vitro* and *in vivo* studies indicate that the effects of MMPI on paw swelling in the AA rat are not due to inhibition of eicosanoid production. For example, BB 16 (10 μM) did not inhibit production of PGE_2 or LTB_4 from human

[b] Address correspondence to Michael J. DiMartino, Department of Inflammation/Respiratory Pharmacology, SmithKline Beecham Pharmaceuticals, 709 Swedeland Road, King of Prussia, Pennsylvania 19406-0939.

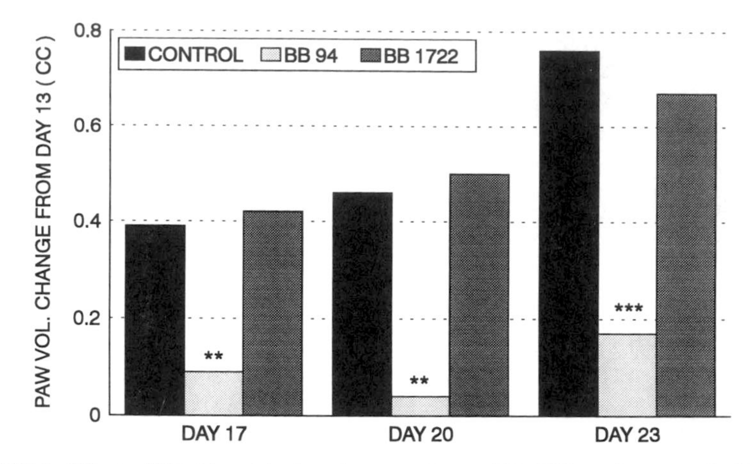

FIGURE 1. Effect of BB 94 and its inactive enantiomer, BB 1722, on increases in hindpaw swelling in adjuvant arthritic rats from day 13. BB 94 and BB 1722 were administered at a dose of 5 mg/kg i.p. twice a day from days 13–17. Mean values of hindpaw swelling from day 13 (Δ cc) were based on 8 rats per test group and 16 rats in the control group. Significant difference from control ($^{**}p < 0.01$; $^{***}p < 0.001$) was determined by Student's *t* test.

blood monocytes activated by calcium ionophore A23187. Moreover, BB 16 administered to rats did not inhibit carrageenan-induced paw edema or *ex vivo* production of LTB_4 or TXB_2 from peripheral blood cells.

The effects of MMPI on AA rat lesions were also evaluated radiographically and histologically and compared to the effects of indomethacin, a standard nonsteroidal anti-inflammatory drug. The compounds were administered daily for 9 days (days 13–17 and 20–23). On day 24 or 27, rats were sacrificed and their hind limbs fixed in buffered formalin for radiographic and histologic assessment. In control AA rats, bones of the tibiotarsal joints (distal tibia, talus, and calcaneous) exhibited radiolucent regions and marked loss of morphology and cartilage proteoglycan. As summarized in TABLE 1, BB 94 and BB 16 (5 mg/kg i.p. twice a day) produced moderate

TABLE 1. Inhibitory Effects[a] of MMPI (BB 16 and BB 94) and Indomethacin on AA Rat Lesions Assessed Radiographically and Histologically

Parameters of Arthritis or Tissue Damage	MMPI (BB 16 and BB 94) (mg/kg i.p. bid) 5	Indomethacin (mg/kg p.o. uid) 0.1	0.3	0.9
Hind paw volume	++	++	+++	++++
Radiography				
Soft tissue swelling	++	++		
Skeletal	++	0	ND	ND
Histology[b]				
Inflammation (extent)	++	0	++	+++
Inflammation (severity)	0	0	0	+++
Hard tissue (cartilage/bone)	++[c]	0	0	+++[d]

[a]Degree of inhibition: 0 = no effect; ++ = moderate; +++ = marked; ++++ = profound; ND = no data.
[b]Tibiotarsal joints.
[c]Cartilage proteoglycan content appears normal.
[d]Evidence of proteoglycan loss from cartilage.

decreases in paw swelling, bone/cartilage destruction, and the extent of inflammation. Moreover, despite histologic evidence of intense inflammation overlying the bones, the morphology of bone/cartilage remained intact and the cartilage proteoglycan, stained with alcian blue, appeared normal in the majority (i.e., 70%) of AA rats treated with the MMPI. In contrast, indomethacin administered orally at either an equivalent antiedema dose (0.1 mg/kg) or an anti-inflammatory dose (0.3 mg/kg) was not effective in preventing bone or cartilage loss. Even the large dose of indomethacin (i.e., 0.9 mg/kg), which markedly inhibited paw edema, inflammation, and skeletal changes, did not normalize the loss of cartilage or proteoglycan.

In summary, the *in vivo* administration of hydroxamic acid–containing peptide MMPI to AA rats produced potent, stereospecific, antiarthritic activity characterized by long-lasting efficacy and inhibition of bone/cartilage degradation. The results of this investigation support the hypothesis that MMPI may offer a new disease-modifying therapy for rheumatoid arthritis. Moreover, the availability of *in vivo* effective MMPI may provide the pharmacologic tools to determine the role of MMP in tissue destruction occurring in various diseases and animal models.

REFERENCES

1. BRINCKERHOFF, C. E. 1991. Arthritis Rheum. **34:** 1073–1075.
2. DIMARTINO, M., M. SLIVJAK, K. ESSER, C. WOLFF, E. SMITH III & R. GAGNON. 1993. Agents Actions **39:** C58–C60.
3. DAVIES, B., P. D. BROWN, N. EAST, M. J. CRIMMIN & F. R. BALKWILL. 1993. Cancer Res. **53:** 2087–2091.

Doxycycline Inhibition of Cartilage Matrix Degradation[a]

A. A. COLE, S. CHUBINSKAYA, K. CHLEBEK, M. W. ORTH,
L. L. LUCHENE, AND T. M. SCHMID

Department of Biochemistry
Rush Medical College
Rush-Presbyterian-St. Luke's Medical Center
Chicago, Illinois 60612

Tetracyclines decreased collagenase activity in several animal models[1–5] as well as in the synovial tissues and fluid of seven patients with rheumatoid arthritis who had received minocycline prior to joint replacement. The effects of doxycycline were tested in an *in vitro* system in which the cartilage of embryonic avian tibias was degraded during culture in the absence of serum[6,7]; during culture, cartilage matrix loss occurs in two phases with proteoglycan loss preceding collagen loss. The phase of collagen loss is accompanied by the release of active collagenase and gelatinase into the conditioned media.

To test the potential of doxycycline to inhibit degradation of cartilage during culture, tibias from 12-day chicken embryos were cultured for 30 days with 5, 20, or 40 μg/mL of doxycycline (Sigma). Control tibias were cultured under identical conditions but without doxycycline. Following the 30-day culture, conditioned media were examined to detect matrix metalloproteinase activity, collagenase, and gelatinase using gelatin zymograms.[8] The conditioned media were also analyzed for hydroxyproline content[9] and sulfated glycosaminoglycan and lactate dehydrogenase activity (Sigma Diagnostic Kit-LD No. 340-UV).[11] The tibias were processed for histologic observation after being fixed in 4% paraformaldehyde and embedded in paraffin. Sections were stained with safranin O and fast green to observe proteoglycan loss.[12]

In the presence of all three concentrations of doxycycline, cartilage was not resorbed, whereas control cartilage was completely degraded as previously described.[6] In the gelatin zymograms, collagenase activity was identified by two bands at an M_r of 57,000 and 50,000 on culture days 18–30 as previously described[7]; collagenase activity in the conditioned media of the doxycycline-treated tibias was not detectable on gelatin zymograms. Gelatinase is constitutively produced by the cultures and appears on the zymograms as 66,000 and 62,000 cleared bands. Gelatinase activity was present throughout the culture in the presence of 5 μg/mL of doxycycline. At both 20 and 40 μg/mL of doxycycline, gelatinase was undetectable by culture day 18. Collagen degradation, evaluated as hydroxyproline content in the conditioned media, was reduced at all doses of doxycycline, whereas proteoglycan degradation was inhibited in a dose-dependent manner. Lactate dehydrogenase in the conditioned media of the control cultures showed peak activity (18,000 units/tibia) on culture day 10 and returned to baseline (400 units/tibia) by culture day 14. Doxycycline reduced peak lactate dehydrogenase activity in a dose-dependent fashion. At higher concentrations of doxycycline the cartilage length was increased

[a] This work was supported by a research grant from the Arthritis Foundation, Illinois Chapter Grant (A.A.C.), and National Institutes of Health grant 2-P50-AR-39239.

during culture. The effects of doxycycline extend beyond inhibition of the activity of proteolytic enzymes. In this system, collagenase was completely inhibited, whereas gelatinase activity was reduced. Doxycycline prevented not only collagen loss from the cartilage but also proteoglycan loss. Doxycycline also reduced the cell death associated with proteoglycan loss.

REFERENCES

1. GOLUB, L. M., H. M. LEE, G. LEHRER, A. NEMIROFF, T. F. MCNAMARA, R. KAPLAN & N. S. RAMAMURTHY. 1983. Minocycline reduces gingival collagenolytic activity during diabetes. Preliminary observations and a proposed new mechanism of action. J. Periodontal Res. **18:** 516–526.
2. GOLUB, L. M., N. S. RAMAMURTHY & T. F. MCNAMARA. 1991. Tetracyclines inhibit connective tissue breakdown: New therapeutic implications for an old family of drugs. Crit. Rev. Oral Biol. Med. **2:** 297–322.
3. GREENWALD, R. A., L. M. GOLUB, B. LAVIETES, N. S. RAMAMURTHY, B. GRUBER, R. S. LASKIN & T. F. MCNAMARA. 1987. Tetracyclines inhibit human synovial collagenase in vivo and in vitro. J. Rheumatol. **14:** 28–32.
4. GREENWALD, R. A., B. G. SIMONSON, S. A. MOAK, S. W. RUSH, N. S. RAMAMURTHY, R. S. LASKIN & L. M. GOLUB. 1988. Inhibition of epiphyseal cartilage collagenase by tetracyclines in low phosphate rickets in rats. J. Orthop. Res. **6:** 695–703.
5. GREENWALD, R. A., S. A. MOAK, N. S. RAMAMURTHY & L. M. GOLUB. 1992. Tetracyclines suppress matrix metalloproteinase activity in adjuvant arthritis and in combination with flurbiprofen, ameliorate bone damage. J. Rheumatol. **19:** 927–938.
6. COLE, A. A., L. J. LUCHENE, T. F. LINSENMAYER & T. M. SCHMID. 1992. The influence of bone and marrow on cartilage hypertrophy and degradation during 30-day, serum-free culture of the embryonic chick tibia. Dev. Dynamics **193:** 277–285.
7. COLE, A. A., T. BOYD, L. LUCHENE, K. E. KUETTNER & T. M. SCHMID. 1993. Type X collagen degradation in long-term serum-free culture of the embryonic chick tibia following production of active collagenase and gelatinase. Dev. Biol. **159:** 528–534.
8. HEUSSEN, C. & E. B. DOWDLE. 1980. Electrophoretic analysis of plasminogen activators in polyacrylamide gels containing sodium dodecyl sulfate and copolymerized substrates. Anal. Biochem. **102:** 196–202.
9. SCHWARTZ, D. E., Y. CHOI, L. J. SANDELL & W. R. HANSON. 1985. Quantitative analysis of collagen, protein and DNA in fixed, paraffin-embedded and sectioned tissue. Histochem. J. **17:** 655–663.
10. CHANDRASEKHAR, S. 1987. Microdetermination of PGs and glycosaminoglycans in the presence of guanidine hydrochloride. Anal. Biochem. **161:** 103–108.
11. ROACH, H. I. 1990. Long-term organ culture of embryonic chick femora: A system for investigating bone and cartilage formation at an intermediate level of organization. J. Bone Miner. Res. **5:** 85–100.
12. ROSENBERG, L. 1971. Chemical basis for the histological use of safranin O in the study of articular cartilage. J. Bone Jt. Surg. **53A:** 69–82.

Elevation of Urinary Pyridinoline in Adjuvant Arthritic Rats and Its Inhibition by Doxycycline

V. GANU,[a] J. DOUGHTY, S. SPIRITO, AND R. GOLDBERG

Research Department
Pharmaceuticals Division
Ciba-Geigy Corp.
Summit, New Jersey 07901

Loss of articular cartilage and bone resorption is a common occurrence in chronic joint diseases such as rheumatoid arthritis and osteoarthritis. These changes can be visualized by histopathologic study and X-rays. An alternative to these methods could be the quantitation of a serum or urine marker, such as urinary pyridinoline, a breakdown product of mature type I collagen of bone and type II collagen of cartilage. Urinary pyridinoline was reported to be useful in judging the turnover of bone.[1]

Rat adjuvant arthritis is a rapidly progressive inflammatory disease which shares some components of human rheumatoid arthritis and has been used to test the efficacy of therapeutic agents.[2] In the studies presented here, we used an enzyme-linked immunosorbent assay (ELISA) for quantitating urinary pyridinoline in adjuvant arthritic rats to determine its relation to inflammation measured by paw swelling and the effect of doxycycline, a collagenase and gelatinase inhibitor,[3] on the urinary excretion of pyridinoline.

MATERIALS AND METHOD

Reagents

Nonviable desiccated *Mycobacterium tuberculosis* H37 RA from Difco Laboratories (Detroit, Michigan) was used for preparing a complete adjuvant. Doxycycline was purchased from Sigma Chemical Co. (St. Louis, MO). Male Lewis rats (150–200 g) were obtained from Charles River Research Laboratories (Kingston, NY).

Induction of Arthritis in Rats and Measurement of Inflammation and Urinary Pyridinoline

Arthritis in male Lewis rats (6–8 animals per group) was initiated by injecting 50 μL of complete adjuvant subcutaneously into the base of the tail near the hairline. Doxycycline, 20 mg/kg/day orally, was administered in cornstarch on the day of induction of the disease and continued until day 21. The normal group and diseased control group received cornstarch only. Paw volume was assessed with a water

[a]Corresponding author at Ciba-Geigy Corp., LSB 3207, 556 Morris Avenue, Summit, New Jersey 07901.

displacement technique by dipping the hind paw of the rat into water to just below the hairline. Animals terminated on day 15, 22, or 34 were anesthetized before urine was withdrawn from the bladder. All urine samples were centrifuged, and the supernatants were removed and stored at −20 °C. Urinary pyridinoline and creatinine were quantitated using kits purchased from Metra Biosystems (Palo Alto, CA).

RESULTS AND DISCUSSION

FIGURE 1 shows that as the disease progressed in the rats, the increase in paw swelling preceded the rise in urinary pyridinoline (nanomole of pyridinoline per

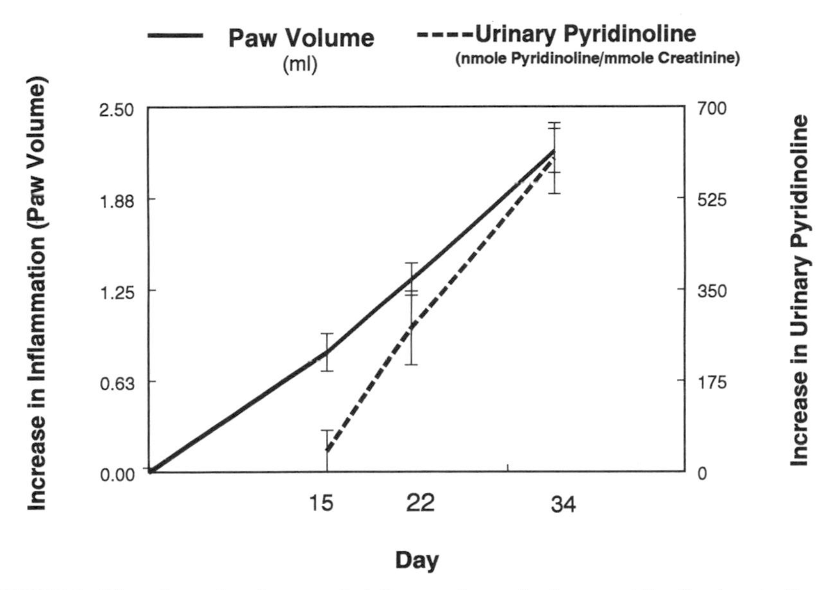

FIGURE 1. Time-dependent increase in inflammation and urinary pyridinoline in rat adjuvant arthritis. The change in paw volume is the difference between the volume (mL) of the inflamed paw on the day of measurement minus the volume on day 0. The increase in urinary pyridinoline (nanomole of pyridinoline per millimole of creatinine) is the difference between the pyridinoline content in the urine of diseased animals and that of normal animals on the day of measurement. Five to eight animals were used for all data points except for the pyridinoline value on day 14; urine from only two animals was used. Data points are mean ± SE.

millimole of creatinine). The increased excretion of urinary pyridinoline indicates an increase in the turnover of mature type I and type II collagen in adjuvant arthritis. Urinary creatinine levels in diseased animals were comparable to those in normal animals. Because the identity of enzyme(s) responsible for the degradation of collagen to produce urinary pyridinoline is unknown, we investigated the role of collagenase and gelatinase using doxycycline. When administered orally to rats at 20 mg/kg/day for 21 days, doxycycline normalized the urinary pyridinoline level without a concomitant reduction in swelling (FIG. 2).

We conclude that: (1) the increased excretion of urinary pyridinoline is not necessarily directly related to paw swelling, (2) enzymes such as collagenase and

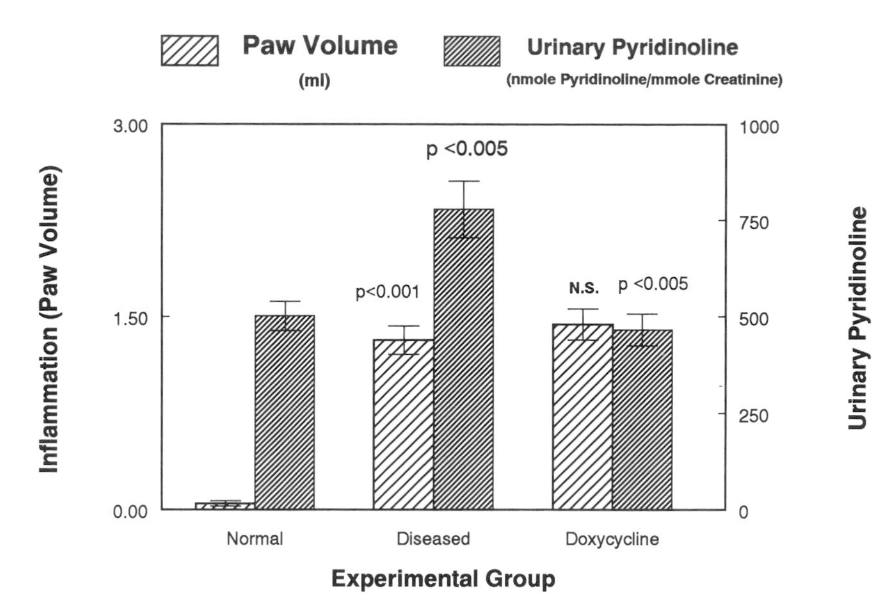

FIGURE 2. Effect of doxycycline on inflammation and urinary pyridinoline in adjuvant arthritic rats. Doxycycline was administered orally from day 0 to day 21 at 20 mg/kg. The change in paw volume is the difference between paw volume (mL) on day 22 and that on day 1. Urinary pyridinoline is the ratio of pyridinoline (nM) to creatinine (mM) in rat urine on day 22. The p value using Student's t test compares: (1) diseased to normal group and (2) doxycycline to diseased group. Values are mean ± SE.

gelatinase may be involved in the degradation of collagen in this model, and (3) doxycycline may be useful in preventing the destruction of joint architecture in rheumatoid arthritis by preventing collagen breakdown.

REFERENCES

1. BLACK, D., M. MARABANI, R. D. STURROCK & S. P. ROBINS. 1989. Urinary excretion of the hydroxypyridinium cross links of collagen in patients with rheumatoid arthritis. Ann. Rheum. Dis. **48:** 641–644.
2. BLACKMAN, A., J. H. BURNS, J. B. FARMER, H. RADZIWONIK & J. WESTWICK. 1977. An X-ray analysis of adjuvant arthritis in the rat. The effect of prednisolone and indomethacine. Agents Action **7:** 145–151.
3. YU, L. P., G. N. SMITH, K. A. HASTY & K. D. BRANDT. 1991. Doxycycline inhibits type IX collagenolytic activity of extracts from human osteoarthritic cartilage and of gelatinase. J. Rheumatol. **18:** 1450–1452.

Low Dose Doxycycline Inhibits Pyridinoline Excretion in Selected Patients with Rheumatoid Arthritis

ROBERT A. GREENWALD,[a] SUSAN A. MOAK,[a]
AND LORNE M. GOLUB[b]

[a] *Division of Rheumatology*
Long Island Jewish Medical Center
New Hyde Park, New York 11042

[b] *School of Dental Medicine*
State University of New York at Stony Brook
Stony Brook, New York 11794

Rheumatoid arthritis is characterized by pathologically excessive collagenase activity and increased urinary excretion of collagen-derived pyridinium cross-links. Doxycycline is a potent inhibitor of several matrix metalloproteinases (MMPs), and a low dose doxycycline (LDD) regimen (20 mg orally twice a day for 2–3 months) has been used successfully in periodontal disease to produce dramatic decreases of MMP activity *in vivo*.[1] Furthermore, tetracycline MMP inhibitors, when combined with NSAIDs, substantially ameliorate joint damage in adjuvant rats as observed radiologically.[2] Based on this rationale, we administered LDD to 13 patients with rheumatoid arthritis for 3 months (all of whom were on NSAIDs and a DMARD), collecting 24-h urine specimens for pyridinoline determination at baseline and every 4 weeks thereafter.

Patient data are summarized in TABLE 1. All but IN were female. The prognostic index (PI), from 0–5 for overall disease severity, and the disease activity index (DAI), 0–8 for current level of disease, were assessed by the method of Wilke and Clough.[3] The patient's global sense of well-being, hematocrit, DAI, and health activity questionnaire (HAQ) score were recorded at start and finish. (The HAQ is an 8-point questionnaire of functional status scored from 0–32.) Disease duration was 1–16 years.

Aliquots of urine were frozen and all pyridinoline measurements made simultaneously for each subject. Pyridinoline excretion was determined by ELISA (Metra Biosystems) and expressed as nanomoles of pyridinoline per millimole of creatinine. The patients segregated into three general groups by pyridinoline level, which was positively correlated with the combined clinical index of disease activity, such as PI + DAI ($r = 0.65$, $p < 0.02$). Five patients (rows 1–5) had a PI of 4 or 5, and their pyridinoline excretion ranged from 76–108 at baseline (left side of figure), with some values as high as 179 during the 3-month trial; they all had DAI values greater than 6. Low dose doxycycline had no effect on pyridinoline in this group. Four patients with a PI of less than 4 and a DAI of 0–5 had an elevated pyridinoline level at baseline (61–96), and they showed a mean fall of 34% (from 73 to 48). Four patients had a normal PYD (less than 46), and their excretion was unaffected by LDD (data not shown). No major change was noted in overall global rating, HAQ score, or hematocrit, and no toxicities were observed.

We have confirmed that pyridinoline excretion is increased in rheumatoid

TABLE 1. Characteristics of Patients with Rheumatoid Arthritis Treated with Low Dose Doxycycline

				DAI		Global		Hematocrit		HAQ		Pyridinoline			
ID	PI	Age	Yr.	Start	End	Start	End	Start	End	Start	End	Base	4 Wk.	8 Wk.	End
NT	4	59	6	8	8	Poor	Poor	32	26	16	16	108	107	153	87
LB	4	61	14	8	8	Poor	Very pr	35	36	19	16	85	60	94	73
TL	5	50	9	8	8	Very pr	Very pr	38	32	20	21	82	83	98	179
HB	4	57	13	7	7	Fair	Very pr	34	35	12	14	76	110	116	117
MS	4	65	9	4	6	Fair	Poor	32	32	8	10	76	105	98	89
SR	3	32	2	5	5	Fair	Good	34	34	16	15	96	66	91	62
LR	2	50	3	3	3	Good	Fair	35	35	8	8	64	37	35	45
FS	3	67	4	3	3	Poor	Fair	36	35	11	11	61	45	39	52
RM	1	59	3	0	0	Good	Good	34	37	10	10	73	115	67	35
CL	1	46	1	3	2	Good	Good	31	33	14	8	41	59	68	48
AB	3	65	13	2	1	Fair	Fair	37	40	8	8	46	36	54	33
SM	1	35	16	3	3	Very gd	Very gd	42	42	10	11	20	27	30	28
IN	3	59	15	3	3	Poor	Fair	40	40	13	12	42	39	37	40

Abbreviations: PI = prognostic index (0–5); Yr. = duration of rheumatoid arthritis; DAI = disease activity index (0–8); very pr = very poor; very gd = very good; HAQ = health activity questionnaire; PYD = pyridinium collagen-derived cross-link excretion (nanomoles of pyridinoline per millimoles of creatinine).

arthritis, correlating with severity and activity. If pyridinoline excretion was very high, LDD, a very low dose of doxycycline, had no effect or an inconsistent one. With moderate elevation, LDD had a salutary effect over a 3-month period (FIG. 1). In future follow-up trials, we suggest screening patients on the basis of baseline pyridinoline and then choosing a larger dose of doxycycline, perhaps 60–80 mg per day, for those in the very high range, and continuing to study LDD for those with modest elevations. In a long-term study, we hope to correlate suppression of

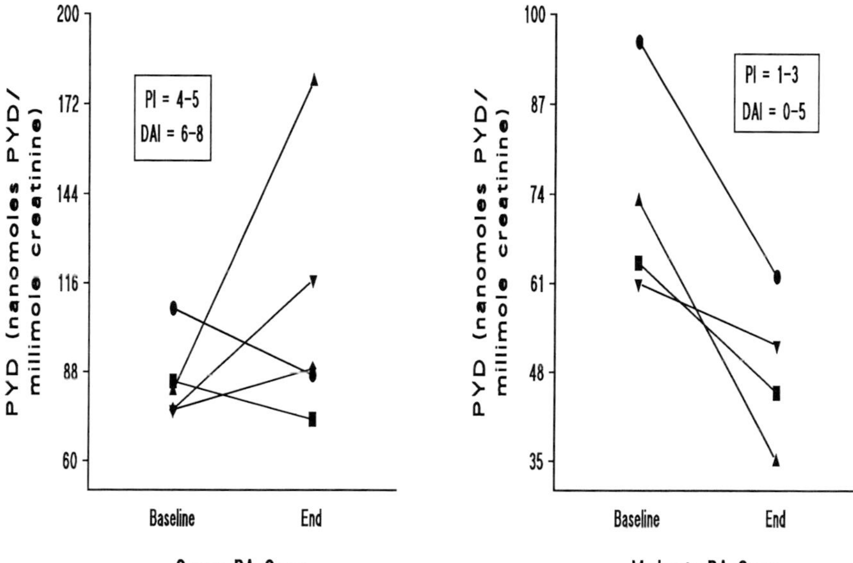

FIGURE 1. Effect of LDD (3 months of doxycycline 20 mg orally twice a day) on urinary pyridinium collagen cross-link excretion in two groups of patients with rheumatoid arthritis (RA). PI = prognostic index, range 0–5; DAI = disease activity index, range 0–8. The two patients in the severe group whose end-of-study value was less than baseline had flare-ups during the trial with interim excretion values greater than baseline. PYD = pyridinoline.

pyridinoline excretion with decreased radiologic destruction and reduced severity of functional disability.

REFERENCES

1. GOLUB, L. M., S. CIANCIO, N. S. RAMAMURTHY, M. LEUNG & T. F. MCNAMARA. 1990. Low dose doxycycline therapy: Effect on gingival and crevicular fluid collagenase activity in humans. J. Periodontal Res. **25:** 321–331.
2. GREENWALD, R. A., S. A. MOAK, N. S. RAMAMURTHY & L. M. GOLUB. 1992. Tetracyclines suppress metalloproteinase activity in adjuvant arthritis and, in combination with flurbiprofen, ameliorate bone damage. J. Rheumatol. **19:** 927–938.
3. WILKE, W. S. & J. D. CLOUGH. 1991. Therapy for rheumatoid arthritis. Semin. Arthritis Rheum. **21:** 21–34.

Minocycline in Active Rheumatoid Arthritis

A Placebo-Controlled Trial

M. KLOPPENBURG,[a] J. PH. TERWIEL,[b] C. MALLÉE,[b]
F. C. BREEDVELD,[a] AND B. A. C. DIJKMANS[a]

Departments of Rheumatology
[a]University Hospital Leiden
and
[b]Spaarne Ziekenhuis Haarlem
the Netherlands

Interest has developed in the use of the tetracyclines, especially minocycline, in connective tissue diseases. *In vitro* work described decreased matrix metalloproteinase (MMP) activity in inflamed gingival and synovial tissue after administration of the tetracyclines.[1,2] The inhibitory effect of the tetracyclines on MMP activity appears to be mediated by the chelation of calcium, because inhibition of purified collagenase by the tetracyclines was partly reversible by the addition of calcium.[1] Other characteristics of minocycline, including inhibition of neutrophil function,[3] immunomodulatory properties,[4] and inhibition of phospholipase A_2 activity,[5] provide additional support for the use of minocycline in connective tissue diseases.

Antiarthritic properties of minocycline were identified in two widely used animal models of chronic arthritis; minocycline decreased the incidence and severity of the collagen and adjuvant arthritis in rats[6] and CMT, a chemically modified tetracycline without antibacterial activity, combined with flurbiprofen, prevented bone loss in rats with adjuvant arthritis.[7]

CLINICAL STUDIES WITH MINOCYCLINE IN ARTHRITIS

The effect of the tetracyclines in reactive and rheumatoid arthritis has been studied. Clinical studies on the therapeutic effect of the tetracyclines in reactive arthritis have given conflicting results.[8–10] However, long-term treatment was beneficial in reactive arthritis when *Chlamydia trachomatis* was the triggering organism.[10] Furthermore, the very encouraging effect of minocycline in rheumatoid arthritis was established in two open studies.[11,12] Ten patients with active definite or classical rheumatoid arthritis were treated with oral minocycline (maximal daily dose 400 mg) for 16 weeks in an open study. Seven patients reported adverse effects, mostly vestibular, leading to premature discontinuation in one. Half of the efficacy variables improved significantly after 4 weeks of therapy. At the end of the study all variables were significantly changed from their pretreatment values. This open study concluded that minocycline may be beneficial in rheumatoid arthritis; however, this conclusion needs to be confirmed in controlled trials.

Recently, the results of two double-blind placebo-controlled studies to investigate the effects of minocycline in active rheumatoid arthritis were presented.[13,14] Our study compared the effects of minocycline with a placebo in a 26-week trial. Minocycline (maximal oral dose 200 mg daily) and placebo were given as adjuvant therapy in two groups of 40 patients each. Adverse reactions were more frequent in the minocycline group and consisted mainly of gastrointestinal complaints and dizziness. The patients' subjective clinical parameters improved either very slightly or not at all during treatment with minocycline. Clinical parameters as measured by the physician, that is, the Ritchie articular index and the number of swollen joints, improved more in the minocycline group than in the placebo group; however, this improvement was not impressive. Minocycline did significantly improve the objective laboratory parameters of disease activity such as c-reactive protein and the erythrocyte sedimentation rate.

FUTURE RESEARCH WITH MINOCYCLINE IN RHEUMATOID ARTHRITIS

Minocycline appears to have beneficial properties in rheumatoid arthritis especially when laboratory parameters of disease activity are considered. Further studies comparing minocycline with established disease-modifying antirheumatic drugs should be planned. Moreover, it may be worthwhile to study the antirheumatic potential of other tetracycline derivatives which produce less toxicity than does minocycline.

REFERENCES

1. GOLUB, L. M., N. S. RAMAMURTHY, T. F. MCNAMARA *et al.* 1984. Tetracyclines inhibit tissue collagenase activity. J. Periodontal Res. **19:** 651–655.
2. GREENWALD, R. A., L. M. GOLUB, B. LAVIETES, N. S. RAMAMURTHY, B. GRUBER, R. S. LASKIN & T. F. MCNAMARA. 1987. Tetracyclines inhibit human synovial collagenase *in vivo* and *in vitro*. J. Rheumatol. **14:** 28–32.
3. GABLER, W. L. & H. R. CREAMER. 1991. Suppressions of human neutrophil function by tetracyclines. J. Periodontal Res. **26:** 52–58.
4. INGHAM, E., L. TURNBULL & J. N. KEARNEY. 1991. The effects of minocycline and tetracycline on the mitotic response of human peripheral blood lymphocytes. J. Antimicrob. Chemother. **27:** 607–617.
5. PRUZANSKI, W., R. A. GREENWALD, I. P. STREET, F. LALIBERTE, E. STEFANSKI & P. VADAS. 1992. Inhibition of enzymatic activity of phospholipases A_2 by minocycline and doxycycline. Biochem. Pharmacol. **49:** 1165–1170.
6. BREEDVELD, F. C. & D. E. TRENTHAM. 1988. Suppression of collagen and adjuvant arthritis by a tetracycline. Arthritis Rheum. **31** (Suppl. 1): R3.
7. GREENWALD, R. A., S. A. MOAK, N. S. RAMAMURTHY & L. M. GOLUB. 1992. Tetracyclines suppress matrix metalloproteinase activity in adjuvant arthritis and in combination with flurbiprofen, ameliorate bone damage. J. Rheumatol. **19:** 927–938.
8. POTT, H. G., A. WITTENBORG & G. JUNGE-HÜLSING. 1988. Long-term antibiotic treatment in reactive arthritis. Lancet **i:** 245–246.
9. PANAYI, G. S. & B. CLARK. 1989. Minocycline in the treatment of patients with Reiter's syndrome. Clin. Exp. Immunol. **39:** 728–732.
10. LAUHIO, A., M. LEIRISALO-REPO, J. LÄHDEVIRTA, P. SAIKKO & H. REPO. 1991. Double-blind, placebo-controlled study of three-month treatment with lymecycline in reactive arthritis, with special reference to Chlamydia arthritis. Arthritis Rheum. **34:** 6–14.
11. BREEDVELD, F. C., B. A. C. DIJKMANS & H. MATTIE. 1990. Minocycline treatment for rheumatoid arthritis: An open dose finding study. J. Rheumatol. **17:** 43–46.
12. LANGEVITZ, P., D. ZEMER, M. BOOK & M. PRAS. 1992. Treatment of resistant rheumatoid arthritis with minocycline: An open study. J. Rheumatol. **19:** 1502–1504.
13. TILLEY, B., G. ALARCON, S. HEYSE, D. TRENTHAM, R. NEUNER, D. CLEGG, J. LELSEN, L. BUCKLEY, S. FOWLER, S. PILLEMER, M. TUTTLEMAN & H. DUNCAN. 1993. Minocycline for treatment of rheumatoid arthritis: The MIRA trial experience. Arthritis Rheum. **36**(Suppl.): S46.
14. KLOPPENBURG, M., F. C. BREEDVELD, J. PH. TERWIEL, C. MALLÉE & B. A. C. DIJKMANS. 1994. Minocycline in active rheumatoid arthritis (RA): A double blind placebo-controlled trial. Arthritis Rheum. **37:** 629–636.

Placebo-Controlled Study of the Effects of Three-Month Lymecycline Treatment on Serum Matrix Metalloproteinases in Reactive Arthritis[a]

A. LAUHIO,[b–g] Y. T. KONTTINEN,[b,g] T. SALO,[h]
H. TSCHESCHE,[i] J. LÄHDEVIRTA,[f] F. WOESSNER, JR.,[j]
L. M. GOLUB,[k] AND T. SORSA[c,e,l]

Departments of [b]Anatomy, [c]Medical Chemistry, [d]Bacteriology and Immunology, and [e]Periodontology
University of Helsinki
Helsinki, Finland

[f]Department of Internal Medicine
Aurora Hospital
Helsinki, Finland

[g]IVth Department of Internal Medicine of
Helsinki University Central Hospital
Helsinki, Finland

[h]Department of Oral Surgery and Pathology
University of Oulu
Oulu, Finland

[i]Department of Biochemistry
University of Bielefeld
Bielefeld, Germany

[j]Department of Biochemistry and Molecular Biology
Miami Medical School
Miami, Florida

[k]Department of Oral Biology and Pathology
State University of New York at Stony Brook
Stony Brook, New York 11794

The utility of antibiotic treatment in reactive arthritis has been controversial.[1] In a double-blind, placebo-controlled study by Lauhio *et al.*[1] we found that 3 months of treatment with lymecycline combined with a nonsteroidal anti-inflammatory drug (NSAID) significantly decreased the duration of reactive arthritis triggered by *Chlamydia trachomatis.* This study also found that among all the patients with

[a]This work was supported by the Academy of Finland, the Yrjö Jahnsson Foundation, the Duodecim Foundation, the Astra Foundation, the Arthritis Foundation in Finland, the Deutsche Forschungsgemeinshaft, Bonn, SFB 223, B01, the ITI Straumann Foundation, and NIDR grant R37DE-03987.

[l]Address correspondence to T. Sorsa, Department of Periodontology, Institute of Dentistry, University of Helsinki, P.O. Box 41 (Mannerheimintie 172) FIN-00014, Helsinki, Finland.

reactive arthritis (triggered by urogenital or enteric infections), tissue destruction determined by x-ray changes was significantly lower among patients treated with lymecycline than with placebo.[1] Therapeutically attainable levels of lymecycline can inhibit the oxidative activation of latent human neutrophil collagenase (matrix metalloproteinase [MMP]-8) *in vitro.*[2] Greenwald *et al.*[3] showed that the combination of a chemically modified nonantimicrobial, but anticollagenolytic, tetracycline derivative (CMT-1) with NSAID efficiently reduced collagenase activity and bone resorption in rat adjuvant arthritis. Yu *et al.*[4] recently reported that the severity of canine osteoarthritis was reduced by two months of treatment with oral doxycycline.

In this study, we treated 20 patients with acute reactive arthritis with lymecycline (300 mg twice a day) or with placebo for 3 months; all patients were also treated with

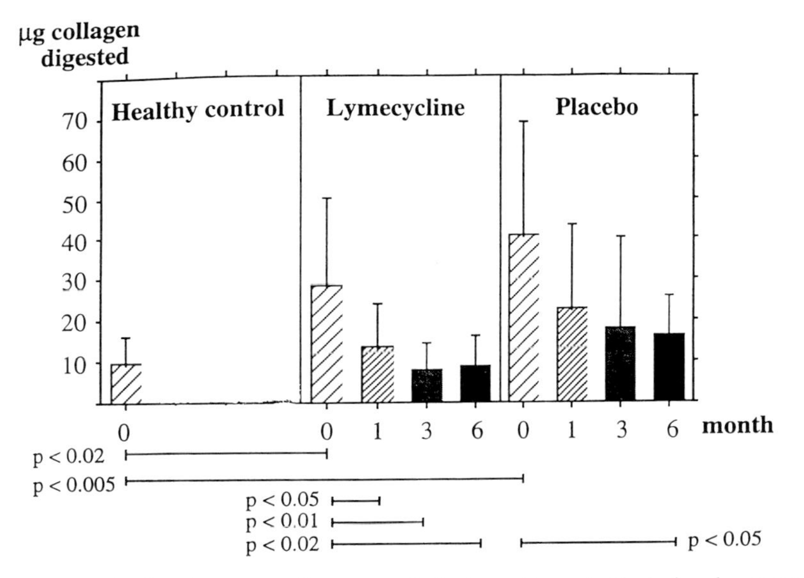

FIGURE 1. Reductions in serum collagenase activity in reactive arthritis during the course of acute reactive arthritis and 3-month tetracycline (lymecycline) treatment. Serum collagenase activities are expressed as micrograms of type I collagen fibrils digested at 37 °C. Note especially the increased serum collagenase activities at the acute phase of the disease compared to the recovered state or to healthy controls. Statistical significances are indicated.

NSAID according to Lauhio *et al.*[1] Before treatment and at 1, 3, and 6 months after recovery, serum samples were collected from the patients and also from sex- and age-matched healthy controls (n = 10). Serum collagenase activity was measured by type I collagen fibril assay and expressed as micrograms of collagen digested. Gelatinase activity was measured by functional assay. Lactoferrin and tissue inhibitor of matrix metalloproteinase-1 (TIMP-1) levels were assayed by ELISA.

We found that serum collagenase activity was increased significantly during the acute phase of the disease compared to the recovered state or to healthy controls (n = 10) in both the lymecycline group (n = 10) and the placebo group (n = 10) (FIG. 1). Increased serum collagenase activity in the acute phase of the disease decreased during the course of healing in both the lymecycline and the placebo groups, but the decrease was significantly faster in the lymecycline group (FIG. 1).

Serum gelatinase did not change during the course of the disease in the lymecycline or the placebo group. Our findings of elevated serum collagenase activity in the acute phase of reactive arthritis may reflect and/or contribute to tissue destruction during the course of arthritis.[5] The decrease in collagenase activity during the acute phase of the disease, faster in the lymecycline than in the placebo group, may be due to the healing tendency of reactive arthritis strengthened by long-term tetracycline treatment.

ACKNOWLEDGMENT

We are grateful to Mrs. Vera Süwer for her skillful technical assistance.

REFERENCES

1. Lauhio, A., M. Leirisalo-Repo, J. Lähdevirta, P. Saikku & H. Repo. 1991. Double-blind, placebo-controlled study of the three month treatment with lymecycline in reactive arthritis, with special reference to Chlamydia arthritis. Arthritis Rheum. **24:** 6–14.
2. Lauhio, A., T. Sorsa, O. Lindy, K. Suomalainen, H. Saari, L. M. Golub & Y. T. Konttinen. 1992. The anti-collagenolytic potential of lymecycline in the long-term treatment of reactive arthritis. Arthritis Rheum. **35:** 195–198.
3. Greenwald, R. A., S. A. Moak, N. S. Ramamurthy & L. M. Golub. 1992. Tetracyclines suppress matrix metalloproteinase activity in adjuvant arthritis and in combination with flurbiprofen, ameliorate bone damage. J. Rheumatol. **19:** 927–938.
4. Yu, L. P., G. N. Smith, K. D. Brandt, S. L. Meyers, R. O'Connor & D. A. Brandt. 1992. Reduction of the severity of canine osteoarthritis by prophylactic treatment with oral doxycycline. Arthritis Rheum. **35:** 1150–1159.
5. Lauhio, A., T. Sorsa, O. Lindy, K. Suomalainen, H. Saari, L. M. Golub & Y. T. Konttinen. 1993. The regulation of doxycycline/tetracycline in collagenolytic activity and tissue destruction in joint diseases. Arthritis Rheum. **36:** 1335–1336.

CMT/Tenidap Treatment Inhibits Temporomandibular Joint Destruction in Adjuvant Arthritic Rats[a]

N. RAMAMURTHY,[b] R. GREENWALD, S. MOAK,
J. SCUIBBA, A. GOREN, G. TURNER, B. RIFKIN,
AND L. GOLUB

State University of New York at Stony Brook
Stony Brook, New York 11794–8702

Long Island Jewish Medical Center
New Hyde Park, New York 11042

New York University Dental Center
New York, New York 10010

Temporomandibular joint dysfunction has been observed in patients with systemic polyarthritis (rheumatoid or psoriatic). Rats with systemic adjuvant arthritis (AA) also develop temporomandibular joint pathology. Collagenase and gelatinase are matrix metalloproteinases (MMPs) involved in the destruction of arthritic joints.[1] Published reports indicate that tetracyclines including their nonantimicrobial analogs (CMTs) can inhibit MMPs and prevent tissue destruction during various diseases.[2] Recently, Ramamurthy *et al.*[4] reported that the administration of a combination of CMT + NSAID (Tenidap [TD]) to arthritic rats inhibited bone resorption more than did either drug alone because NSAID therapy increased CMT uptake in the diseased joints possibly by reducing inflammatory stasis. In the current study, we determined the effect of CMT-1/TD combination therapy on the amelioration of temporomandibular joint pathology. Male Lewis rats (150 g of body weight) were made arthritic by injecting Freund's adjuvant at the base of the tail. The rats were then distributed into the following groups: normal, untreated AA, AA + TD (2 mg/rat), AA + CMT-1 (4 mg/rat), and AA + TD + CMT-1 (Combo). All drugs were suspended in 2% carboxymethylcellulose and administered by oral gavage once a day for 23 days. The rats became arthritic by day 13.

At the termination of the experiments (day 21), some rat heads were fixed in formalin for histologic and X-ray examination. Other specimens were frozen for later measurement of MMP levels in extracts of the temporomandibular joint. The jaws were x-rayed using mammography x-ray film. Histologic study of the temporomandibular joint was carried out on serial sections stained with hematoxylin and eosin. Both collagenase and gelatinase were determined after dissecting the intact frozen temporomandibular joint. Matrix metalloproteinases were extracted from pooled tissues[3] and assayed for collagenase (using [^{3}H-methyl] collagen as substrate) by SDS-PAGE/fluorography and for gelatinase by zymography.

X-ray analysis showed increased temporomandibular joint destruction in AA rats

[a]This study was supported by grants from the National Institute of Dental Research (R37 DE-03987 and DE-09576) and Collagenex, Inc., Wayne, Pennsylvania.

[b]Address for correspondence: Department of Oral Biology & Pathology, School of Dental Medicine, Westchester Hall, SUNY/Stony Brook, Stony Brook, New York 11794-8702.

compared to normal animals. Individually administered CMT and TD each partially reduced joint destruction, whereas Combo-treated rats showed much greater reduction ($p < 0.01$) in articular destruction than did rats treated with either drug alone. Histologically, normal rats were devoid of any articular lesions, whereas the temporomandibular joint of untreated AA rats exhibited fusion of the fibroid disk to the condylar head, with thinness of articular cartilage and mild synovitis. The Combo-treated group showed a marked decrease in fibroid ankylosis with normal thickness of articular cartilage. In contrast, both CMT and TD alone still showed some temporomandibular joint pathology (FIG. 1). Both the 72- and 92-kDa gelatinases were found in the temporomandibular joint extracts of rats in all treatment groups by gelatin zymography. However, only the 92-kDa proteinase was increased in the AA group. CMT alone produced a slight decrease in this enzyme, TD alone had no

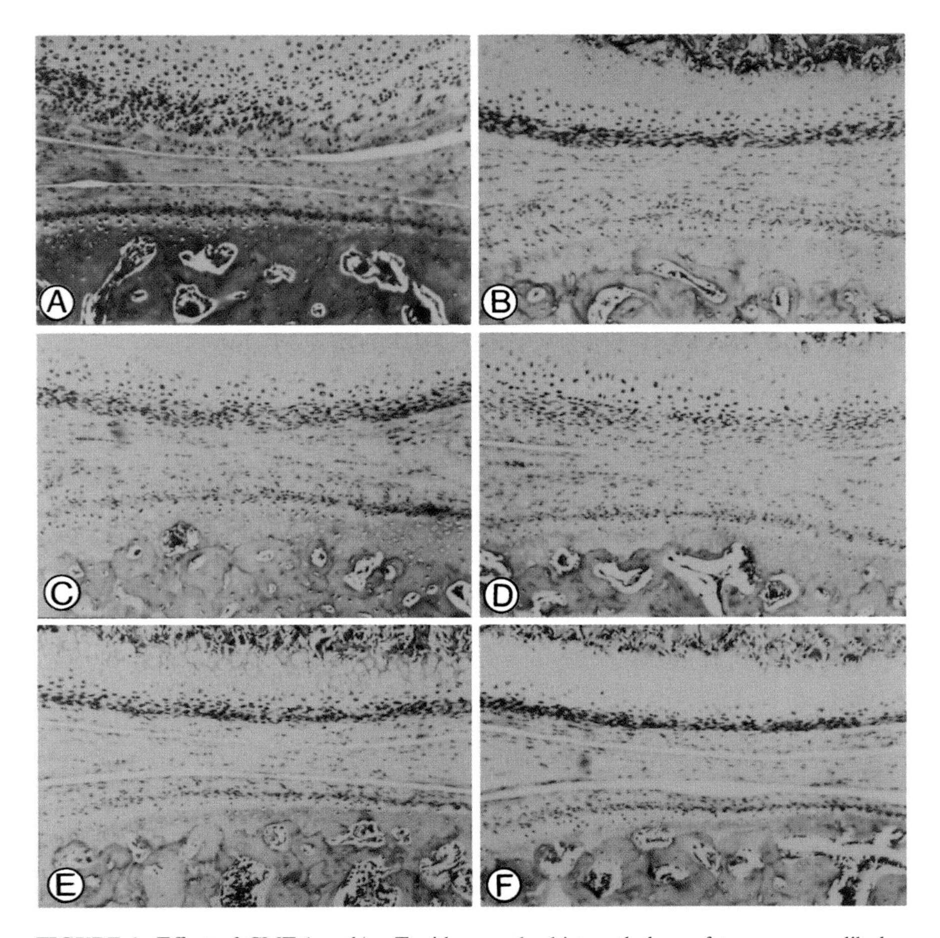

FIGURE 1. Effect of CMT-1 and/or Tenidap on the histopathology of temporomandibular joints of arthritic rats. (**A**) Normal. (**B**) Arthritic rats. Note the fibrous disk ankylosed to the condylar bone. (**C**) AA + Tenidap. This group showed no improvement. (**D**) AA + CMT-1. These temporomandibular joints showed less ankylosis than did those after Tenidap alone. (**E & F**) AA + CMT-1 + Tenidap. In Combo-treated rats, the temporomandibular joint looks normal with no ankylosis.

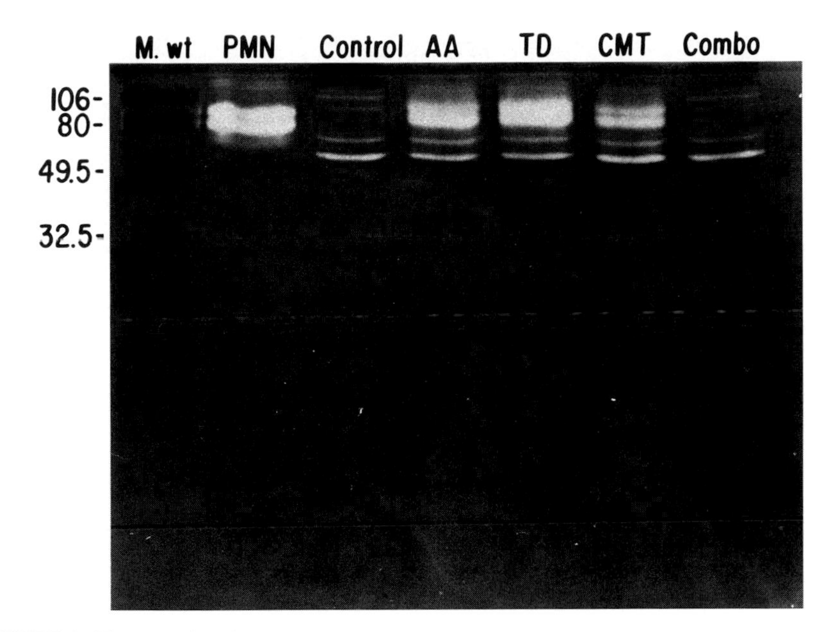

FIGURE 2. Zymography of temporomandibular joint extract; gelatinase assessed according to the method of Dean and Woessner.[3] Temporomandibular joint tissues were excised, pooled by group, weighed, and homogenized in 0.05 M Tris/HCl buffer (pH 7.5) containing 0.1 M $CaCl_2$ and 0.15 M NaCl. MMP activity was extracted by heating at 60 °C for 4 min, the extract was centrifuged, and the supernatant collected and subjected to zymography using denatured type I collagen as substrate. **Lane 1:** molecular weight protein standards; **lane 2:** polymorphonucleocyte extract showing 92-kDa gelatinase; **lane 3:** control rats showing minimal gelatinase activity mostly representing 72-kDa and little or no 92-kDa MMP; **lane 4:** untreated arthritic rats showing dramatic elevation of 92-kDa gelatinase but no increase in 72-kDa enzyme; **lane 5:** Tenidap treatment did not reduce either 72- or 92-kDa gelatinase; **lane 6:** CMT-1 treatment slightly reduced 92-kDa but not 72-kDa gelatinase; **lane 7:** Combo treatment of adjuvant arthritic rats reduced 92-kDa gelatinase to normal levels (72-kDa enzyme again unchanged).

effect, and the Combo-treated group showed the most dramatic reduction in pathologically excessive 92-kDa gelatinase (FIG. 2). In fact, the zymogram for this group was indistinguishable from that for the nonarthritic controls. Monitoring collagenase activity by assessing conversion of the radiolabeled α collagen components to α^A digestion products, using SDS-PAGE/fluorography, showed essentially the same pattern of change: elevated collagenase activity in extracts of temporomandibular joint from the untreated AA rats and maximum reduction when AA rats were treated with the Combo. In conclusion, anti-MMP therapy using the CMT/NSAID Combo decreased temporomandibular joint destruction in systemically induced experimental arthritis.

ACKNOWLEDGMENTS

The authors gratefully acknowledge the helpful suggestions of Dr. Leon Sokoloff, Professor, Department of Pathology, School of Medicine, SUNY at Stony Brook, and the excellent technical assistance of Dr. Xia Zhang with analysis by zymography.

REFERENCES

1. Greenwald, R. A., S. A. Moak, N. S. Ramamurthy & L. M. Golub. 1992. Tetracyclines suppress matrix metalloproteinase activity in adjuvant arthritis and in combination with flurbiprofen, ameliorate bone damage. J. Rheumatol. **19:** 926–938.
2. Golub, L. M., N. S. Ramamurthy, T. F. McNamara, R. A. Greenwald & B. R. Rifkin. 1991. Tetracyclines inhibit connective tissue breakdown: New therapeutic implications for an old family of drugs. Crit. Rev. Oral Biol. Med. **2:** 297–322.
3. Dean, D. D. & J. F. Woessner, Jr. 1984. Extracts of human articular cartilage contain an inhibitor of tissue metalloproteinases. Biochem. J. **218:** 277–280.
4. Ramamurthy, N., M. Leung, S. Moak, R. Greenwald & L. Golub. 1993. CMT/NSAID combination increases bone CMT uptake and inhibits bone resorption. *In* Immunosuppressive and Anti-inflammatory Drugs. Ann. N.Y. Acad. Sci. **696:** 420–421.

The *in Vivo* Effect of Doxycycline Treatment on Matrix Metalloproteinases in Reactive Arthritis[a]

A. LAUHIO,[b–g] Y. T. KONTTINEN,[b,g] T. SALO,[h]
H. TSCHESCHE,[i] D. NORDSTRÖM,[b,g] J. LÄHDEVIRTA,[f]
L. M. GOLUB,[j] AND T. SORSA[c,e,k]

Departments of [b]Anatomy, [c]Medical Chemistry, [d]Bacteriology and Immunology, and [e]Periodontology
University of Helsinki
Helsinki, Finland

[f]Department of Internal Medicine
Aurora Hospital
Helsinki, Finland

[g]IVth Department of Medicine
University Central Hospital of Helsinki
Helsinki, Finland

[h]Department of Oral Surgery and Pathology
University of Oulu
Oulu, Finland

[i]Department of Biochemistry
University of Bielefeld
Bielefeld, Germany

[j]Department of Oral Biology and Pathology
State University of New York at Stony Brook
Stony Brook, New York 11794

The tetracycline family of antibiotics exhibits anticollagenolytic activity independent of antimicrobial activity.[1] In a double-blind, placebo-controlled study, Lauhio *et al.*[2] found that in *Chlamydia*-triggered reactive arthritis, 3 months of lymecycline (tetracycline-L-methylenelysine) therapy combined with a nonsteroidal anti-inflammatory drug (NSAID) significantly decreased the duration of the disease. It was also found that among all patients with reactive arthritis (triggered by either urogenital or enteric infections), tissue destruction determined by x-ray changes was significantly less among patients treated with a 3-month regimen of lymecycline than with placebo. We also found that therapeutically attainable levels of lymecycline *in vitro* can inhibit the oxidative activation of latent human neutrophil collagenase (matrix metalloproteinase-8, MMP-8).[3]

[a]This work was supported by the Academy of Finland, the Yrjö Jahnsson Foundation, the Duodecim Foundation, the Arthritis Foundation in Finland, the Astra Foundation, the Finska Läkaresällskapets Foundation, the Finsk-Norsk-Medicinsk Stiftelse, the Deutsche Forschungsgemeinshaft, Bonn, SFB 223, B01, the ITI Straumann Foundation, and National Institute of Dental Research grant R37DE-03987.

[k]Address correspondence to: Department of Periodontology, Institute of Dentistry, University of Helsinki, P.O. Box 41 (Mannerheimintie 172), FIN-00014 Helsinki, Finland.

In the current work, we treated 10 patients with acute reactive arthritis with doxycycline (150 mg a day) and ketoprofen (100 mg 3 times a day) for 2 months. Serum and saliva samples were collected before treatment and after 2 months. Collagenase activity was measured by quantitative SDS-PAGE electrophoresis and MMP-8 levels by ELISA.[4] Gelatinase was measured by functional assay, zymography, and Western blots. Lactoferrin and tissue inhibitor of matrix metalloproteinase-1 (TIMP-1) were assayed by ELISA. Elastase, trypsin, and cathepsin-G–like activities were measured with and without reduction using a spectrophotometric assay and specific synthetic peptide (SAAANA, SAAVNA, SAAPNA, and BAPNA) substrates.

In saliva we found that both endogenously active (APMA-) and total (APMA+) collagenase activities as well as MMP-8 levels decreased significantly during treatment ($p < 0.02$, $p < 0.02$, and $p < 0.05$, respectively). The cellular source of salivary collagenases of patients with reactive arthritis was assessed by Western blotting using specific antibodies against human MMP-8 and MMP-1. MMP-8 was the major collagenase present in reactive arthritis saliva. SDS-PAGE and Western blot analysis of reactive arthritis salivary collagenase activity and MMP-8 immunoreactivities before and after 2 months of doxycycline and NSAID treatment showed significant decreases of collagenase activities, and immunoreactivities of both pro 75-kDa and active 65-kDa MMP-8 were detected. No fragmentation of either 75- or 65-kDa MMP-8 was detected after 2 months of doxycycline/NSAID treatment. Serum MMP-8 levels decreased, but not statistically significantly (mean ± SEM, 678.9 ± 185.6 vs 491.1 ± 144.8 ng MMP-8/mL). The other MMPs (gelatinase) and serine proteases as well as TIMP-1 levels measured from serum or saliva did not decrease significantly. Our findings of *in vivo* inhibition of collagenase activity in body fluid (i.e., saliva) containing inflammatory exudates during long-term doxycycline treatment may contribute to decreased tissue destruction events found recently in clinical and animal model studies in arthritides during long-term doxycycline/tetracycline therapy[2–5] and in other inflammatory diseases such as periodontal diseases.

ACKNOWLEDGMENT

We are grateful to Mrs. Vera Süwer for her skillful technical assistance.

REFERENCES

1. Golub, L. M., H. M. Lee, G. Lehrer, A. Nemiroff, T. F. McNamara, R. Kaplan & N. S. Ramamurthy. 1983. Minocycline reduces gingival collagenolytic activity during diabetes: Preliminary observations and a proposed new mechanism of action. J. Periodontal Res. **18:** 515–525.
2. Lauhio, A., M. Leirisalo-Repo, J. Lähdevirta, P. Saikku & H. Repo. 1991. Double-blind, placebo-controlled study of the three month treatment with lymecycline in reactive arthritis, with special reference to Chlamydia arthritis. Arthritis Rheum. **24:** 6–14.
3. Lauhio, A., T. Sorsa, O. Lindy, K. Suomalainen, H. Saari, L. M. Golub & Y. T. Konttinen. 1992. The anti-collagenolytic potential of lymecycline in the long-term treatment of reactive arthritis. Arthritis Rheum. **35:** 195–198.
4. Lauhio, A., Y. T. Konttinen, H. Tschesche, D. Nordstrom, T. Salo, J. Lahdevirta, L. M. Golub & T. Sorsa. 1994. Reduction of matrix metalloproteinase-8/neutrophil collagenase levels during long-term doxycycline treatment in reactive arthritis. Antimicrob. Agents Chemother. **38:** 1400–1402.
5. Lauhio, A., T. Sorsa, O. Lindy, K. Suomalainen, H. Saari, L. M. Golub & Y. T. Konttinen. 1993. The regulation of doxycycline/tetracycline in collagenolytic activity and tissue destruction in joint diseases. Arthritis Rheum. **36:** 1335–1336.

Tissue Healing with Doxycycline and Chemically Modified Tetracycline Treatments in Rats with *Porphyromonas gingivalis*–induced Periodontitis[a]

N. Y. KARIMBUX,[b] N. S. RAMAMURTHY,[c] L. M. GOLUB,[c]
AND I. NISHIMURA[b]

[b]*Departments of Prosthetic Dentistry and Periodontology*
Harvard School of Dental Medicine
Boston, Massachusetts 02115

[c]*Department of Oral Biology and Pathology*
School of Dental Medicine
State University of New York at Stony Brook
Stony Brook, New York 11794

The extracellular matrix of the periodontal ligament (PDL) is primarily composed of collagens such as types I, III,[1] and XII.[2] Whereas type I and III collagens polymerize to form a fibril during the early stages of PDL development, similar to other connective tissues, the expression of type XII collagen is found to be specific to ligamentous tissues including the PDL.[2] Therefore, it has been postulated that type XII collagen plays an integral role in PDL function. Periodontitis is the most prevalent oral disease in adults worldwide and results in the destruction of periodontal tissues by proteolytic enzymes elaborated by the host and by infecting microorganisms.[3] Specifically, matrix metalloproteinases (MMPs) and other tissue-destructive neutral proteinases, such as elastase, have been identified at the diseased sites of the periodontium.[4] Doxycycline (DOXY) and chemically modified tetracyclines (CMTs) were found to inhibit inflammatory connective tissue breakdown by blocking the action of the MMPs collagenase and gelatinase.[5] After infection and/or the inflammatory host reaction subsides, the damaged tissue undergoes a wound healing reaction which results in the formation of scar tissue through repair mechanisms or in functional tissue regeneration. The purpose of this study was to observe whether MMP inhibitors (DOXY and CMT) result in functionally regenerated periodontal tissue in an established rat periodontitis model.[6]

Twenty-four adult male pathogen suppressed Sprague-Dawley rats were inoculated with *Porphyromonas gingivalis* for a 3-day period to induce periodontitis. Ten days after the initial *P. gingivalis* inoculation, six rats per group were randomly selected for the following daily treatments by oral gavage: (1) DOXY, (2) CMT (both at 5 mg/rat), and (3) no treatment. Noninfected animals served as controls. Because type XII collagen is a tissue-specific extracellular matrix molecule of functional PDL, we focused our investigations on the expression of type XII collagen mRNA using *in situ* hybridization. Two maxillary dentoalveolar segments from each group were harvested and after 1 week of decalcification, were prepared for histologic study.

[a]This work was supported by National Institutes of Health grants DE05603, DE00351, AR36820, and R37DE03987 and Collagenex, Inc.

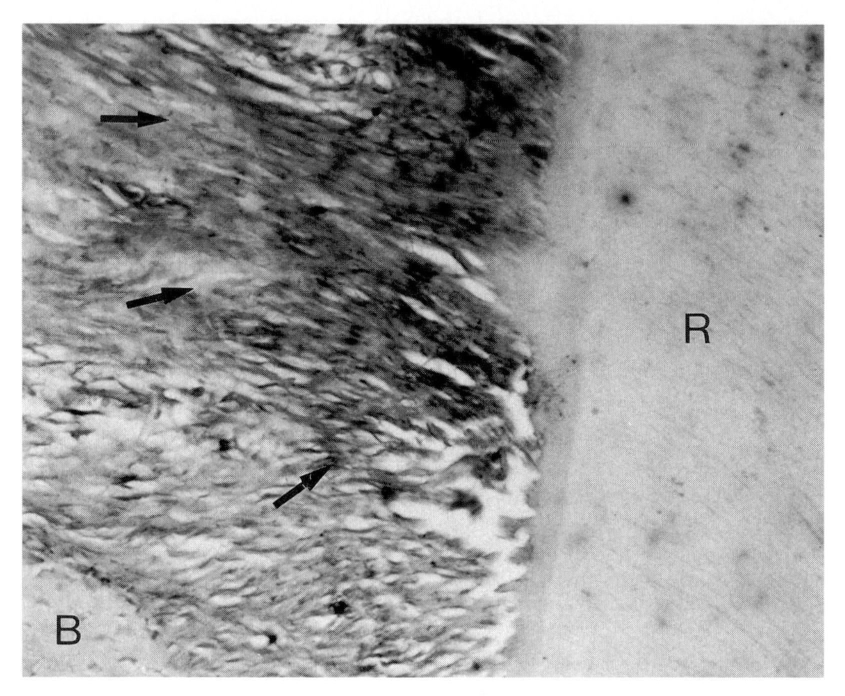

FIGURE 1. *In situ* hybridization with the αI (XII) collagen probe in CMT-treated rat specimen, distal of the first molar. The cervical region of the gingival connective tissue is labeled with the αI (XII) probe (*arrows*). R = root; B = bone; original magnification 200×.

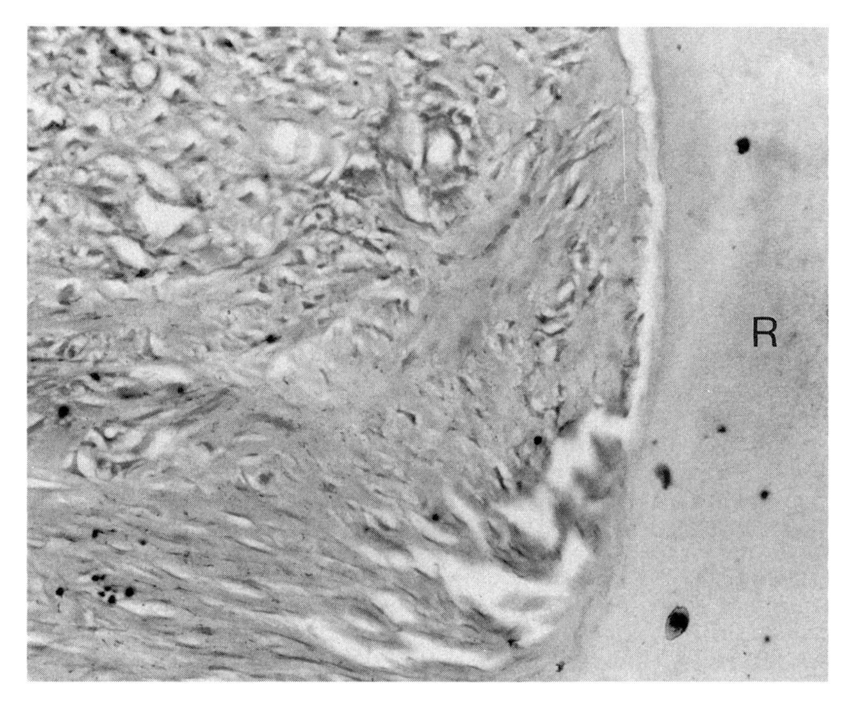

FIGURE 2. *In situ* hybridization with the αI (XII) collagen probe in the *P. gingivalis*-infected and untreated rat specimen, distal of the first molar. There is no labeling with the αI (XII) probe. R = root; original magnification 200×.

Some of the sections were prepared for nonradioactive *in situ* hybridization with a type XII cDNA probe. The groups treated with DOXY and CMT showed intense type XII collagen probe hybridization in the cervical periodontal tissue region (FIG. 1), whereas the no-treatment group lacked localized hybridization (FIG. 2). The control group showed probe hybridization throughout PDL tissue (Karimbux *et al.*, in preparation).

In conclusion, the use of DOXY and CMTs in progressing periodontitis may have therapeutic potential and result in functional PDL regeneration. After injury, DOXY and CMT therapy may up-regulate type XII collagen gene transcription by promoting connective tissue and fibroblast adhesion to the tooth surface and by directly blocking collagenolytic enzymes.

REFERENCES

1. BERKOVITZ, B. K. B. 1990. The structure of the periodontal ligament: An update. Eur. J. Orthod. **12:** 51–76.
2. KARIMBUX, N. Y., N. D. ROSENBLUM & I. NISHIMURA. 1992. Site-specific expression of collagen I and XII mRNAs in the periodontal ligament at two developmental stages. J. Dent. Res. **71:** 1355–1362.
3. GENCO, R. J., H. GOLDMAN & W. D. COHEN. Eds. 1985. Contemporary Periodontics. : 42–52. C. V. Mosby. Baltimore, MD.
4. CERGNEAUS, M., E. ANDERSEN & G. CIMASONI. 1982. *In vitro* breakdown of gingival tissue by elastase from human polymorhonuclcar leukocytes. An electron microscopic study. J. Periodontal Res. **17:** 169–182.
5. GOLUB, L. M., N. S. RAMAMURTHY, T. MCNAMARA, R. A. GREENWALD & B. RIFKIN. 1991. Tetracyclines inhibit connective tissue breakdown: New therapeutic implications for an old family of drugs. Crit. Rev. Oral Biol. Med. **2:** 297–322.
6. CHANG, K. M., N. S. RAMAMURTHY, T. MCNAMARA, R. T. EVANS, B. KLAUSEN, P. A. MURRAY & L. M. GOLUB. 1994. Tetracyclines inhibit *Porphyromonas gingivalis*-induced alveolar bone loss in rats by a non-antimicrobial mechanism. J. Periodontal Res. In press.

Procollagenase Is Reduced to Inactive Fragments upon Activation in the Presence of Doxycycline[a]

GERALD N. SMITH, JR.,[b] KENNETH D. BRANDT,[b] AND KAREN A. HASTY[c]

[b]*Arthritis Center*
Indiana University School of Medicine
Indianapolis, Indiana 46202

[c]*University of Tennessee Memphis and*
Veterans Administration Medical Center
Memphis, Tennessee 38104

Tetracyclines have been shown to inhibit matrix destruction and metalloproteinase activity in a number of pathologic conditions.[1] We showed that oral doxycycline administration slows the progression of cartilage destruction in osteoarthritis induced in dogs by anterior cruciate ligament transection.[2,3] Articular cartilage collagenase was strikingly reduced in treated dogs, and inhibition of total collagenase was more pronounced than was inhibition of active enzyme. This observation and the variability in reported levels of enzyme inhibition by the tetracyclines led us to examine the effect of doxycycline on proenzyme activation. For this purpose, we activated recombinant human neutrophil procollagenase *in vitro* in the presence or absence of doxycycline and assayed activity and the molecular weight changes that accompany activation.

Procollagenase was activated with trypsin or APMA, and collagenase activity was assayed on either collagen fibers[4] or a small peptolide substrate.[5] When the enzyme was activated before the addition of doxycycline, approximately 50% inhibition of enzyme activity was noted in the presence of 30 μM doxycycline with either substrate or mode of activation.

By contrast, when latent enzyme was activated in the presence of doxycycline, 30 μM doxycycline inhibited almost all of the activity, and 50% inhibition was obtained with 5–12 μM doxycycline depending on the conditions of activation (TABLE 1). When trypsin was used to activate latent enzyme, the extent of inhibition was related to the rate of activation. At higher trypsin concentrations, when activation was rapid, inhibition was less than that at lower trypsin concentrations, when activation required extended incubation. For the data shown in TABLE 1, IC_{50} at the lower trypsin concentration (6 μM) was half that obtained at the higher trypsin concentration. Similar results were obtained with APMA activation.

Upon examination by Western blotting of the molecular weight changes that accompany activation, the basis for the foregoing observation became apparent (FIG. 1). In the absence of doxycycline, most of the proenzyme (55 kDa) was converted to active enzyme (46 kDa with trypsin, 45 kDa with APMA). In the presence of doxycycline, however, little active enzyme accumulated. Instead, trypsin cleaved the

[a]This work was supported by the National Institutes of Health (grants AI22603, AR39166, and AR20582), the Arthritis Foundation, and Procter & Gamble.

TABLE 1. Activity of Trypsin-Activated Human Recombinant Neutrophil Procollagenase against the Thiopeptolide Substrate in the Presence and Absence of Doxycycyline[a]

Doxycycline (μM)	Trypsin-Activated Collagenase[b]: Units of Activity (% Inhibition)	Procollagenase + Trypsin (rapid)[c]: Units of Activity (% Inhibition)	Procollagenase + Trypsin (slow)[d]: Units of Activity (% Inhibition)
0	17 ± 0.6	22 ± 2.8	10 ± 1.5
5	15 ± 0.8 (10)	15 ± 0.7 (32)	6 ± 1.0 (38)
15	15 ± 1.2 (15)	10 ± 0.7 (55)	1 ± 1.3 (88)
30	8 ± 0.4 (55)	3 ± 1.2 (86)	−1 ± 0.6 (109)

[a]Activity (mean ± standard deviation, $n = 3$) is expressed as micromoles of substrate hydrolyzed per minute per microgram of procollagenase. All mean activity values from experiments with doxycycline are different from control activity values within that group ($p < 0.05$).

[b]For trypsin activation, 6.7 μg proenzyme were incubated with 2 μg trypsin at 37 °C for 30 min. Each assay contained 0.32 μg proenzyme.

[c]For rapid activation, 0.32 μg proenzyme were incubated with 2 μg trypsin at 37 °C for 30 min.

[d]For slow activation, 0.32 μg proenzyme were incubated with 0.2 μg trypsin for 4 h.

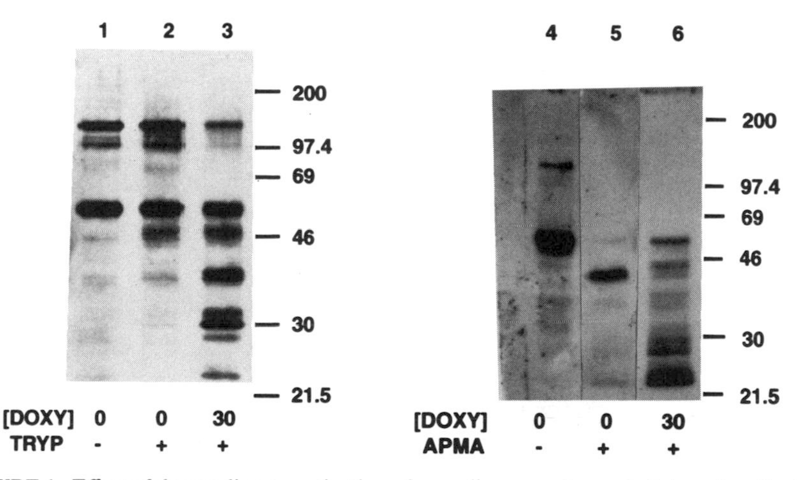

FIGURE 1. Effect of doxycycline on activation of procollagenase. **Lanes 1–3** show the effect of doxycycline on activation by trypsin, while **lanes 4–6** show the effect of doxycycline on activation with APMA. **Lanes 1 and 4** contain proenzyme (55 kDa) incubated in the absence of trypsin or APMA. **Lane 2** contains procollagenase (0.375 μg) incubated for 7 h with 0.4 μg trypsin in buffer alone, while **lane 3** contains the proenzyme, trypsin, and 30 μM doxycycline. This low concentration of trypsin converted some but not all of the proenzyme to active enzyme (46 kDa). In the presence of doxycycline, incubation with trypsin produced large amounts of 40- and 30-kDa fragments. Similarly, **lane 5** contains proenzyme incubated for 4 h with 0.2 M APMA and **lane 6** contains proenzyme incubated for 4 h with both APMA and doxycycline. With APMA the active enzyme was smaller (45 kDa), and in the presence of doxycycline 26- and 28-kDa fragments were generated.

enzyme to prominent 30- and 40-kDa fragments and additional, smaller pieces. Similar results were obtained after activation with 0.2 or 1 mM aminophenylmercuric acetate except that the prominent products were 26- and 28-kDa fragments. Active enzyme incubated in the presence of doxycycline was not degraded under the same conditions.

These data suggest that doxycycline alters the conformation of procollagenase (or of partially activated procollagenase) and renders it more susceptible to proteolysis. Active enzyme is resistant to this effect of doxycycline. Thus, an irreversible loss of enzyme protein occurs during activation in the presence of doxycycline, and degradation is enhanced when proenzyme conversion to active enzyme is delayed.

REFERENCES

1. GOLUB, L. M., N. S. RAMAMURTHY, T. F. MCNAMARA, R. A. GREENWALD & B. R. RIFKIN. 1991. Crit. Rev. Oral Biol. Med. **2:** 297–321.
2. YU, L. C., JR., G. N. SMITH, JR., K. D. BRANDT, S. L. MYERS, B. O'CONNOR & D. A. BRANDT. 1992. Arthritis Rheum. **35:** 1150–1159.
3. YU, L. C., JR., G. N. SMITH, JR., K. D. BRANDT, B. L. O'CONNOR & S. L. MYERS. 1993. Trans. Orthop. Res. Soc. **18:** 724.
4. NAGAI, Y., C. M. LAPIERRE & J. GROSS. 1966. Tadpole collagenase: Preparation and purification. Biochemistry **5:** 3123–3130.
5. WEINGARTEN, H. & J. FEDER. 1985. Spectrophotometric assay for vertebrate collagenase. Anal. Biochem. **147:** 437–440.

Effects of Matrix Metalloproteinases on Cartilage Biophysical Properties *in Vitro* and *in Vivo*[a]

L. J. BONASSAR,[b] C. G. PAGUIO,[b] E. H. FRANK,[b]
K. A. JEFFRIES,[b] V. L. MOORE,[c] M. W. LARK,[c]
C. G. CALDWELL,[c] W. K. HAGMANN,[c]
AND A. J. GRODZINSKY[b]

[b]*Continuum Electromechanics Group*
Department of Electrical Engineering and Computer Science
Massachusetts Institute of Technology
Cambridge, Massachusetts 02139

[c]*Immunology and Inflammation Research*
Merck Research Laboratories
Rahway, New Jersey 07065

Matrix metalloproteinases (MMPs) have been linked to the degradation of the cartilage matrix in normal and pathologic tissue.[1–3] Cartilage explants subjected to recombinant human stromelysin (rhSLN)[4] or MMP-activating APMA[5] showed marked loss in proteoglycan. The objectives of this study were to (1) quantify the changes in cartilage physical properties due to APMA-activated endogenous MMP, (2) characterize the effect of selected MMP inhibitors on these physical changes, and (3) quantify the effect of intraarticular injection of rhSLN on cartilage material properties and modulation of these effects by the systemic administration of an MMP inhibitor.

METHODS

In Vitro *Model*

Cartilage disks 3 mm in diameter and 1 mm thick were harvested from the femoropatellar groove of 1–2-week-old calves. Groups of four plugs were maintained in culture in 1 mL DMEM, and matrix degradation was induced by the addition of 1 mM APMA to the culture media. To inhibit matrix degradation, plugs were incubated with tissue inhibitor of metalloproteinases (TIMP) or synthetic MMP inhibitor *N*-[1(*R*)-carboxy-ethyl]-α-(*S*)-(2-phenyl-ethyl)glycine-(L)-leucine, *N*-phenylamide[6] in concentrations from 40 nM to 4 μM simultaneous to treatment with APMA. At selected times up to 72 h, groups of plugs were removed from culture, weighed, twice equilibrated in 0.01 M NaCl for 1 h to augment swelling,[7] then weighed a second time. Data plotted are $n = 4$ of mean ± SEM.

[a]This work was supported by grants from Merck Research Laboratories and the National Institutes of Health (AR33236).

Physical Properties

Cartilage plugs were tested in uniaxially confined compression using a Dynastat mechanical spectrometer to measure the dynamic stiffness and streaming potential in the 0.005 to 1 Hz range, and equilibrium modulus, hydraulic permeability, and electrokinetic coupling coefficient, using established methods.[8] Media and papain-digested plugs were analyzed for GAG by DMB dye binding as a measure of proteoglycan loss.

In Vivo *Model*

New Zealand white female rabbits 6–8 weeks old received intravenous injections of saline solution or buffer containing 25 mg/kg synthetic inhibitor, and 15 min later received intraarticular injections of 100 μg of trypsin-activated rhSLN in buffer into one hind stifle joint, with the contralateral control joint receiving only buffer. After 1 h the animals were sacrificed and joints were lavaged. Stifle joints were dissected and 3-mm cartilage/bone cylinders were cut from the femoropatellar groove. Immediately before testing, cartilage was separated from the bone, and each plug was tested in confined compression as just described, then weighed, digested with papain, and analyzed for GAG content.

RESULTS

In Vitro *Model*

FIGURE 1 shows the effects of MMP activation by APMA and MMP inhibition by TIMP on tissue proteoglycan content and swelling. Samples treated with 1 mM APMA were ~90% depleted of GAG (FIG. 1A) and swelled by ~20% (FIG. 1B) in 0.01 M NaCl after 72 h. Treatment with 1 mM APMA and graded levels of TIMP resulted in a dose-dependent inhibition GAG loss, with negligible loss at the highest TIMP dose (4 μM) and considerable loss (~70%) at the lowest TIMP dose (40 nM) after 72 h. All treatment doses of TIMP resulted in swelling in 0.01 M NaCl that was indistinguishable from that of controls (~5%) after 72 h. Treatment with 1 mM APMA and synthetic inhibitor resulted in GAG loss and swelling indistinguishable from that of controls at 72 h for all concentrations tested (40 nM to 4 μM) (data not shown). By 3 days in culture, the dynamic stiffness, streaming potential, and equilibrium modulus decreased to <30% and electrokinetic coupling coefficient to ~50% of the value at $t = 0$ (data not shown), while control disks did not change significantly (all data normalized values at $t = 0$).

In Vivo *Model*

FIGURE 2 shows the effect of intraarticular injection of rhSLN and intravenous injection of the synthetic inhibitor on the physical properties of cartilage from lapine stifle joints. With values from 2–3 disks averaged for each joint, the streaming potential at 1 Hz and electrokinetic coupling coefficient of disks decreased significantly ($p < 0.002$ and $p < 0.04$) for joints treated with 100 μg/mL SLN in animals that did not receive intravenous inhibitor. Changes in tissue GAG content and physical properties were not significant ($p > 0.05$) in these animals. In animals receiving intravenous injection of 25 mg/kg synthetic inhibitor, GAG content and all physical properties, including streaming potential and electrokinetic coupling coefficient, were indistinguishable from those of controls.

DISCUSSION

Given that endogenous levels of MMP in normal cartilage are estimated to be on the order of 20 nM,[9] it is noteworthy that concentrations of exogenous TIMP up to

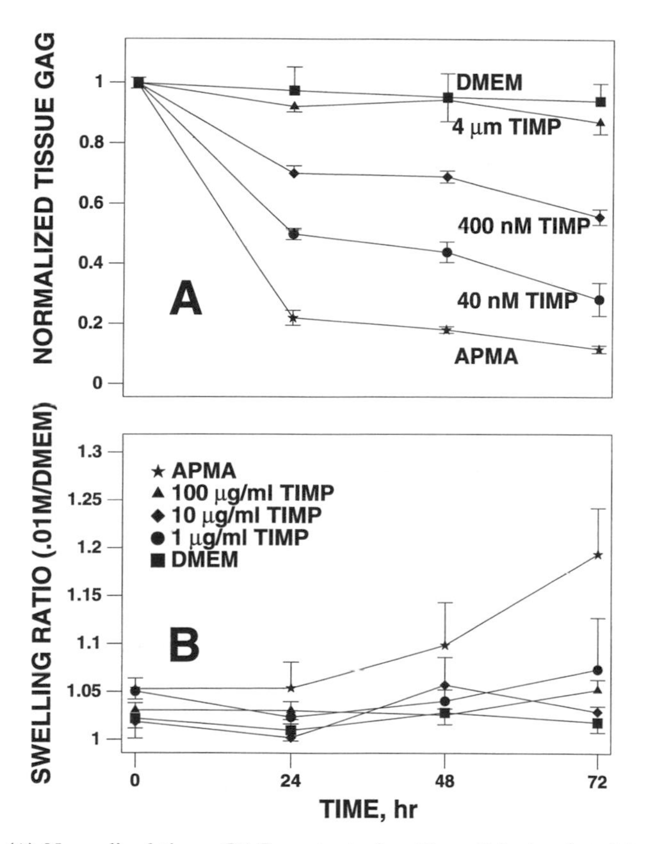

FIGURE 1. (**A**) Normalized tissue GAG content of cartilage disks incubated in DMEM or media containing 1 mM APMA, 1 mM APMA and 4 μM TIMP, 1 mM APMA and 400 nM TIMP, or 1 mM APMA and 40 nM TIMP. (**B**) Swelling ratio (wet weight in 0.01 M NaCl normalized to wet weight in DMEM) for control disks and disk treated with APMA and graded levels of TIMP. Data plotted are $n = 4$; mean ± SEM.

400 nM were unable to completely inhibit GAG loss (FIG. 1A). Given that APMA is a small molecule (~500 Da) and is able to activate MMP throughout the tissue whereas TIMP is much larger (~28 kDa), it seems likely that lower concentrations of TIMP were unable to penetrate the tissue quickly enough to inhibit all the activated

MMP. This is consistent with the observation that the smaller synthetic inhibitor (~500 Da) was able to significantly inhibit GAG loss even at concentrations of 40 nM. In contrast, all doses of TIMP and synthetic inhibitor were able to prevent the matrix damage which results in tissue swelling (FIG. 1B).

Changes in the properties of rabbit knee cartilage exposed to rhSLN *in vivo* without intravenous inhibitor resemble changes in calf cartilage explants after limited exposure to rhSLN *in vitro*.[4] After 1 h of rhSLN exposure, there was little

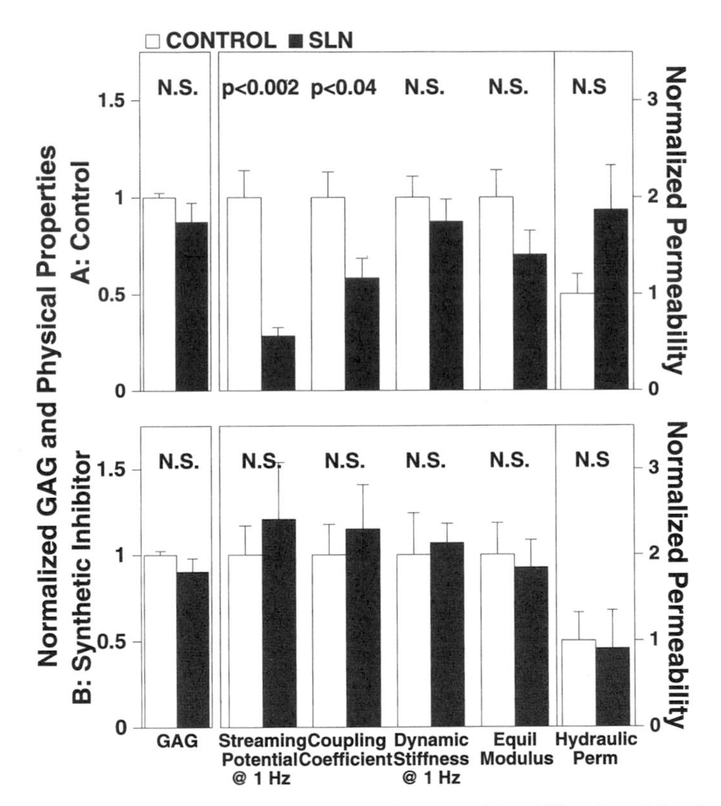

FIGURE 2. (**A**) Normalized GAG content, streaming potential at 1 Hz, electrokinetic coupling coefficient, dynamic stiffness at 1 Hz, equilibrium modulus, and hydraulic permeability of cartilage disks from lapine stifle joints injected with either 100 μg rhSLN or buffer in contralateral control. Rabbits received intravenous saline solution 1 h before intraarticular rhSLN/buffer injection. Data are $n = 10$; mean ± SEM. (**B**) Normalized GAG and physical properties of cartilage disks from lapine stifle joints receiving intraarticular rhSLN or buffer 1 h after intravenous injection of Merck metalloproteinase synthetic inhibitor. Data are $n = 4$; mean ± SEM.

GAG loss and no associated change in mechanical properties, but significant decreases were noted in high frequency streaming potential and electrokinetic coupling coefficient (FIG. 2A), consistent with proteoglycan degradation limited to the tissue *surface*.[10] Intravenous treatment with synthetic inhibitor before rhSLN exposure prevented these changes in high frequency streaming potential and electrokinetic coupling coefficient, implying that degradation at the tissue surface was prevented as well.

REFERENCES

1. OKADA, Y., M. SHINMEI, O. TANAKA, K. NAKA, A. KIMURA, I. NAKAHASHI, M. T. BAYLISS, K. IWATA & H. NAGASE. 1992. J. Clin. Invest. **66:** 680–690.
2. REYNOLDS, J. J. & R. H. HEMBRY. 1993. Matrix **1:** 375–379.
3. PELLETIER, J. P., J. MARTEL-PELLETIER, J. A. DIBATTISTA & P. J. ROUGHLEY. 1993. Am. J. Pathol. **142:** 95–101.
4. BONASSAR, L. J., E. H. FRANK, C. P. PAGUIO, J. C. MURRAY, V. L. MOORE, M. W. LARK & A. J. GRODZINSKY. 1993. Trans. ORS **18:** 192.
5. GOLDBERG, R. L., S. SPIRITO & G. DIPASQUALE. 1993. Trans. ORS **18:** 741.
6. CHAPMAN, K. T., I. E. KOPKA, P. L. DURETTE, C. K. ESSER, T. J. LANZA, M. IZQUIERDO-MARTIN, L. NIEDZWIECKI, B. CHANG, R. K. HARRISON, D. W. KUO, T-Y. LIN, R. L. STEIN & W. K. HAGMANN. 1993. J. Med. Chem. **36:** 4293–4301.
7. MAROUDAS, A. 1976. Nature **260:** 808–812.
8. FRANK, E. H., A. J. GRODZINSKY, T. J. KOOB & D. R. EYRE. 1987. J. Orthop. Res. **5:** 497–508.
9. LOHMANDER, L. S., L. A. HOERRNER & M. W. LARK. 1993. Arthritis Rheum. **36:** 181–189.
10. FRANK, E. H. & A. J. GRODZINSKY. 1987. J. Biomech. **20:** 629–639.

Gelatinases and Endogenous Inhibitors in the Preovulatory Rat Ovary[a]

T. A. BUTLER AND J. F. WOESSNER, JR.

Department of Biochemistry & Molecular Biology
University of Miami School of Medicine
Miami, Florida 33101

The role of metalloproteinases in follicular rupture is well documented.[1,2] Although the levels of matrix metalloproteinase-1 (MMP-1) have been determined, considerably less attention has been focused, until recently, on the gelatinases (MMP-2 and MMP-9). Curry *et al.*[3] found latent enzymes in heat extracts and to a lesser extent in Triton extracts of human chorionic gonadotropin (hCG)-primed ovaries.

We find high levels of latent MMP-9 and MMP-2 in both heat and Triton extracts. Twenty-seven-day-old, immature CD Charles River female rats were primed with 20 IU pregnant mare serum gonadotropin to induce maturation. Forty-eight hours later they were injected with 10 IU hCG to cause ovulation. Ovaries were harvested at 0, 4, and 8 h post-hCG injection and extracted twice, as previously described[4]—first with Triton X100 and then with heating.

Both extracts were reduced, alkylated to deactivate alpha-2 macroglobulin and free tissue inhibitor of metalloproteinase (TIMP), and then treated with methylamine (0.2 M, 30 min, 22 °C) to remove alpha-1 inhibitor 3. Extracts were dialyzed overnight, clarified, and incubated against ^{3}H-type I rat gelatin (20 μg) with and without aminophenyl mercuric acetate (APMA) (0.5 mM) and phenylmethylsulfonyl fluoride (PMSF) (1 mM) for 3 h at 37 °C. At the completion of incubation, the undigested gelatin was precipitated and the supernatants were counted for activity.

Total metalloproteinase activity increased with time in the ovary of the hCG-primed rat. Total (active and latent) activity was divided almost evenly between heat and Triton extracts. Zymography (FIG. 1) reveals an even more apparent buildup of active enzyme not measurable by gelatin assay, including a lower molecular weight form which appears to be very active against gelatin in zymographic gels. Presumably this "active" enzyme is already complexed with endogenous inhibitors in the ovary.

Our laboratory has looked previously at the buildup of natural inhibitors in the primed rat ovary.[5] In the current work, individual ovarian follicles were considered in the search for endogenous inhibitors. Follicles were carefully dissected from rats primed as just described. Individual follicles were homogenized and pelleted as also described. Aliquots of supernatant were assayed with uterine metalloproteinase (MMP-7) in an azocoll assay.[6] Inhibition was determined by measuring the decrease in activity of MMP-7 on the substrate. Inhibitory activity increased as much as five times in isolated follicles over time for the first 6 hours post-hCG injection (FIG. 2). From these data we conclude that the buildup of proteolytic enzymes is much greater in the preovulatory follicle than was previously documented. A parallel increase in endogenous inhibitor probably acts to keep proteolytic activity in check, controlling the proteolysis by active enzymes.

[a] This work was supported by National Institutes of Health grant HD-06773.

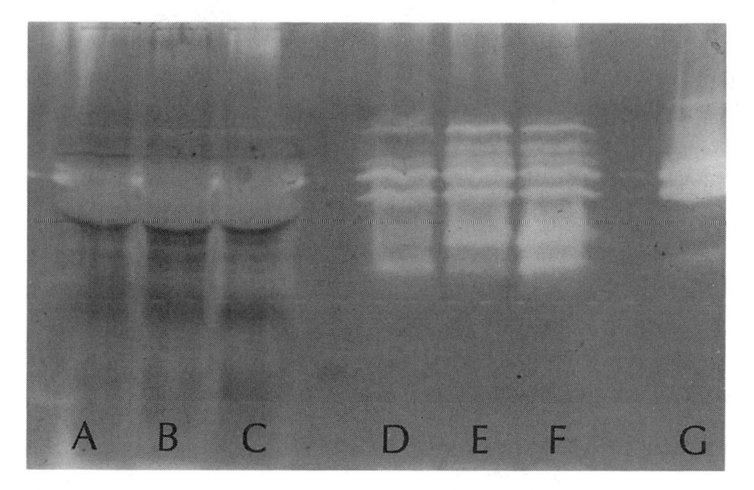

FIGURE 1. Crude extracts of rat ovarian tissue on a 10% polyacrylamide gel containing 0.2 mg/mL gelatin. **Lanes A, B, and C:** Triton X100 extracts of 0, 4, and 8 h post-hCG injection, respectively. **Lanes D, E, and F:** Heat extracts of 0, 4, and 8 h post-hCG injection, respectively. **Lane G:** MMP-2 marker.

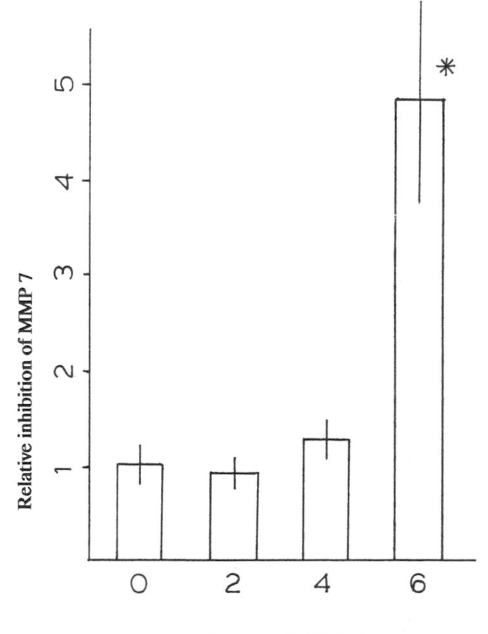

FIGURE 2. Relative levels of endogenous inhibitors in ovarian tissue from hCG-primed rats over time. Inhibition of MMP-7 with whole-follicle homogenates. *Asterisk* indicates significant difference from control; $n \geq 7$ for each time point.

REFERENCES

1. ESPEY, L. L. 1978. The Vertebrate Ovary.: 503–532. Plenum Publishers. New York.
2. BUTLER, T. A., C. ZHU, R. A. MUELLER, G. C. FULLER, W. J. LEMAIRE & J. F. WOESSNER, JR. 1991. Biol. Reprod. **44:** 1183–1188.
3. CURRY, T. E., J. S. MANN, M. H. HUANG & S. C. KEEBLE. 1992. Biol. Reprod. **46:** 256–264.
4. CURRY, T. E., D. D. DEAN, J. F. WOESSNER, JR. & W. J. LEMAIRE. 1985. Biol. Reprod. **33:** 981–991.
5. WOESSNER, J. F., JR. & C. J. TAPLIN. 1988. J. Biol. Chem. **263:** 16918–16925.
6. ZHU, C. & J. F. WOESSNER, JR. 1991. Biol. Reprod. **45:** 334–342.

Orthovanadate Inhibits Interleukin-1 and Phorbol Ester Induced Collagenase Production by Chondrocytes[a]

JULIE A. CONQUER, DANIEL T. GRIMA,
AND TONY F. CRUZ[b]

Connective Tissue Research Group
Samuel Lunenfeld Research Institute and
Department of Pathology
Mount Sinai Hospital and
University of Toronto
Toronto, Canada

Interleukin-1 (IL-1) and phorbol esters (PMA) induce collagenase production by chondrocytes and other cell types.[1] In addition, both IL-1 and PMA induce expression of AP-1 transcription factors c-fos and c-jun which bind to the AP-1 binding site or 12-*O*-tetraderanoylphorbol-13-acetate (TPA)-responsive element in the promoter region of the collagenase gene.[2-4] Increased expression of c-fos and c-jun was necessary, although not sufficient, for the induction of collagenase gene expression by these agents.[5,6]

Orthovanadate, a potent inhibitor of phosphotyrosine phosphatases, inhibits constitutive collagenase production by chondrocytes in a concentration-dependent manner.[7] The mechanism(s) by which orthovanadate inhibited collagenase production by chondrocytes is not known. In the present study, we investigated whether orthovanadate inhibited IL-1 and PMA-induced collagenase production and whether orthovanadate suppression of c-fos and c-jun expression is responsible for this inhibition.

METHODS

Bovine chondrocytes were isolated and plated as previously described. Collagenase activity was assayed using an ELISA method.[1] RNA was extracted from chondrocytes, and Northern blotting was performed as described previously[8] using cDNA probes to human collagenase and mouse c-fos and c-jun.

RESULTS AND DISCUSSION

As demonstrated in FIGURE 1, IL-1 and PMA stimulated collagenase production by chondrocytes. Orthovanadate (50 μM) inhibited IL-1 and PMA-induced collage-

[a]This work was supported by the Medical Research Council of Canada and Angiogenesis Technologies Inc.

[b]Address correspondence to Dr. Tony Cruz, Samuel Lunenfeld Research Institute, Mount Sinai Hospital, 600 University Avenue, Toronto M5G 1X5, Canada.

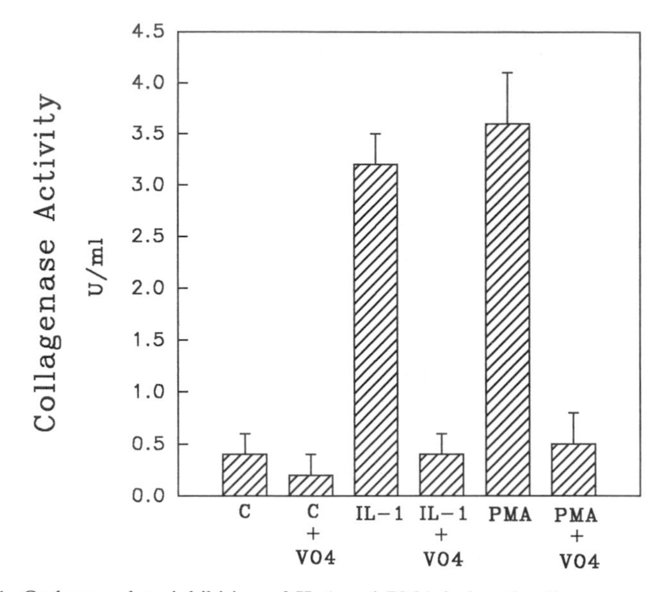

FIGURE 1. Orthovanadate inhibition of IL-1 and PMA-induced collagenase production by chondrocytes. Chondrocytes were incubated with 50 μM orthovanadate in Ham's F12 medium containing 5% FBS for 3 h and then 10 ng/mL of IL-1 or 100 ng/mL of PMA was added and incubated for a further 24 h. The medium was removed and assayed for collagenase activity.

nase production. Furthermore, IL-1 and PMA-induced increases in collagenase mRNA levels in chondrocytes were completely inhibited by orthovanadate, indicating that this inhibition is at the level of transcription (results not shown). Expression of other genes, such as aggrecan, was not affected, suggesting that orthovanadate

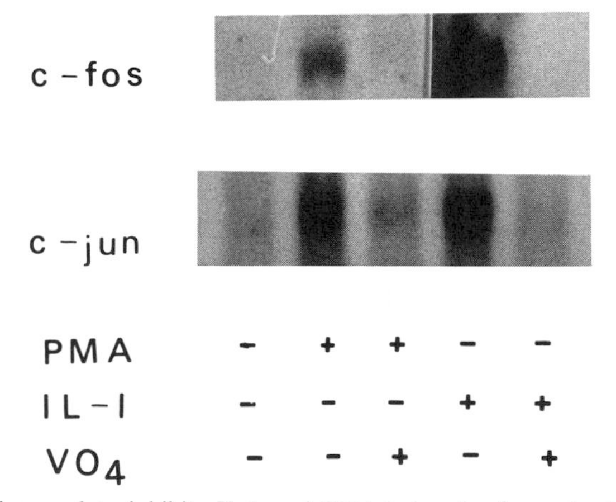

FIGURE 2. Orthovanadate inhibits IL-1 and PMA-induced c-fos and c-jun expression by chondrocytes. Chondrocyte cultures were treated in the presence or absence of orthovanadate (100 μM) for 2 h before the addition of IL-1 (20 ng/mL) or PMA (40 ng/mL) for a further 1 h. The total RNA was extracted and c-fos and c-jun mRNA analyzed by Northern blot analysis.

inhibition of collagenase production and expression is not due to general inhibition of transcription.

Inasmuch as c-fos and c-jun are required for collagenase expression, we investigated the effect of orthovanadate on the expression of these two genes. Low constitutive levels of c-fos and c-jun mRNA were present in unstimulated chondrocytes. IL-1 and PMA stimulated the expression of c-fos and c-jun maximally by 1 h and by 6 h levels had returned to baseline (results not shown). Orthovanadate completely inhibited IL-1 and PMA-induced c-fos and c-jun expression in chondrocytes (FIG. 2). These data raise the possibility that orthovanadate inhibition of collagenase expression may occur by suppressing fos and jun expression in chondrocytes. Although the exact mechanism by which orthovanadate inhibits c-fos and c-jun expression remains to be elucidated, orthovanadate may have the potential to inhibit cellular functions such as proliferation and matrix metalloproteinase production, which require expression of these two transcription factors.

REFERENCES

1. KANDEL, R. A., K. P. H. PRITZKER, G. B. MILLS & T. F. CRUZ. 1990. Biochim. Biophys. Acta **1053:** 130–134.
2. CONCA, W., P. B. KAPLAN & S. M. KRANE. 1989. J. Clin. Invest. **83:** 1753–1758.
3. AUBLE, D. T. & C. E. BRINCKERHOFF. 1991. Biochemistry **30:** 4629–4635.
4. ANGEL, P. & M. KARIN. 1991. Biochim. Biophys. Acta **1072:** 129–157.
5. ANGEL, P., M. IMAGAWA, R. CHIU, B. STEIN, R. J. IMBRA, H. J. RAHMSDORF, C. JONAT, P. HERRLICH & M. KARIN. 1987. Cell **49:** 729–739.
6. SCHONTHAL, A., P. HERRLICH, J. RAHMSDORF & H. PENTO. 1988. Cell **54:** 325–334.
7. CRUZ, T. F., G. MILLS, K. P. H. PRITZKER & R. A. KANDEL. 1990. Biochem. J. **269:** 717–721.
8. CRUZ, T. F., R. A. KANDEL & I. R. BROWN. 1991. Biochem. J. **277:** 327–330.

Gelatinase A Expression in Human Malignant Gliomas

P. C. COSTELLO,[a] R. F. DEL MAESTRO,[b]
AND W. G. STETLER-STEVENSON[c]

[a]*Department of Neurosurgery*
Georgetown University
Washington, DC 20007

[b]*Department of Neurosurgery*
University of Western Ontario
London, Ontario, Canada

[c]*Department of Pathology*
National Institutes of Health
Bethesda, Maryland 20902

Human malignant glial tumors are highly vascular compared to normal brain. An *in vitro* model of brain microvasculature was used to determine the proteolytic contribution of human cerebral endothelium and glial cells to basement membrane degradation. The specific enzymes involved in type IV collagen-degrading activity in human glial tumors are unknown. Human type IV collagen-degrading enzymes of the metalloproteinase family, gelatinase A (72-kDa collagenase IV or MMP-2) and gelatinase B (92-kDa type IV gelatinase or MMP-9), and interstitial collagenase (MMP-1) were assessed in human gliomas.[1] Isolation of human cerebral microvascular endothelial cells (HCME) *in vitro* provides a model for the characterization of collagenase enzymes at the levels of gene expression, protein production, and activity.[2–4]

MATERIALS AND METHODS

Human cerebral tissue specimens were obtained at the time of surgical resection for treatment of either malignant glioma (tumor) (n = 6) or intractable seizure disorder (nontumor) (n = 4). A 1–2-g sample was immediately frozen using isopentane at −40 °C for 2 min and stored at −80 °C until used in the Northern and *in situ* hybridization studies. A 1-g portion of three nontumor specimens was used to isolate HCME monolayers in culture as previously described. (U251 and HTB14 human glioblastoma multiforme-derived cell lines were a generous gift from Dr. G. Cairncross.) Total cellular RNA was extracted from brain tissue samples, HCME, HT1080, U251, and C6 astrocytoma cell monolayers using a previously described method with modifications. For Northern blot analysis, 10 μg of total cellular RNA was electrophoretically separated in a 1.2% agarose gel containing 2.2 M formaldehyde for 3 h. BMV viral RNA was used for RNA size determination (Pharmacia). (Plasmids containing sequences for human interstitial collagenase, PX15 (sense) and PX7 (antisense), were a generous gift from Dr. H. J. Rhamsdorf. Plasmids PH3a-puc18 and p92mo containing the sequences for gelatinase A and B were provided by W. G. Stetler-Stevenson. The plasmid containing the sequence for glyceraldehyde-3-phosphate dehydrogenase was a generous gift from D. T. Denhardt.) The ^{32}P-radiolabeled interstitial collagenase, gelatinase A, gelatinase B, and glyceraldehyde-3-phosphate dehydrogenase probes were generated using random primer extension. HCME cells were plated at an initial density of 10^4 cells/glass slide culture chambers and cultured for 3 and 5 days. Tissue from three malignant astrocytomas, three glioblastoma multiforme, and four nontumor specimens, were cryosectioned at -18 °C. The slides were then subjected to *in situ* hybridization according to the procedure of Whitfield *et al.*[5] Collagenase IV degrading activity released by the endothelium, U251, C6, and HT1080 cell lines was assayed as previously described. Briefly, degradation of ^{3}H-labeled collagen IV molecules was measured in units of activity per milligram of protein in the conditioned media samples. Zymography was carried out as previously described with modifications.[3] A 20-μL aliquot of HCME-conditioned media was pretreated with 0.1 mM EDTA for 30 min at room temperature to inhibit metalloproteinases. Conditioned media from U251 and HT1080 cell lines cultured under serum-free conditions after 4 days in culture were collected. HCME cells were cultured in M199 + 20% FBS for 5 days. Western blot analysis was performed as previously described. The polyclonal antibody to the C-terminus of gelatinase A, Ab KLH-31, was used at a concentration of 1/1,500 and incubated with the blot for 2 h for detection of gelatinase A protein released from these cells lines *in vitro.* Paraffin-embedded sections from patient surgical specimens displaying high collagenase IV activity (157 U/mg protein) and detectable mRNA for gelatinase A were exposed to the KLH-31 antibody at a dilution of 1/250. KLH-31 reactivity was detected using goat anti-rabbit immunoglobulin G and a Vectastain ABC kit (Vector Laboratories).

RESULTS AND SUMMARY

Type IV collagen-degrading enzymes were investigated at the levels of mRNA, protein, and activity production in human brain tumors, cerebral microvascular endothelium, and glial cells. Gelatinase A mRNA was detected using Northern hybridization analysis at higher levels in human malignant glioma tissue samples than in nontumor controls. Gelatinase A mRNA was detected using Northern hybridization analysis in HCME, C6 astrocytoma, and HTB14 cells *in vitro.* The production of gelatinase A mRNA by HCME *in vitro* was detected using *in situ* hybridization and was highest at subconfluency. Type IV collagenolytic activity was released by the C6 astrocytoma, U251 cells, and HT1080 fibrosarcoma cells. C6 released significantly higher levels of collagenase IV activity during log phase growth. HCME released collagenase IV activity that was highest during subconfluent growth. HCME activity was characterized as a single band of approximately 62 kDa in size by gelatin zymography. The single band of gelatinolytic activity released from the HCME was identified as a 62-kDa gelatinase A using Western blot analysis. A relatively light gelatinase A band at 72 kDa was also detected, indicating that zymogen is also released from HCME. Gelatinase A antibody stained both tumor cells and endothelium in tumor tissue sections. Gelatinase A was immunohistochemically localized to glial tumor cells and microvasculature of surgical specimens.

CONCLUSIONS

Gelatinase A mRNA, protein, and collagenase IV activity was detected at higher levels in malignant gliomas than in nonneoplastic tissue. Gelatinase A was detected in migrating and dividing cerebral endothelium, C6 glioma, and HTB14 cells *in vitro*. Inhibition of gelatinase A activity may be effective in reducing angiogenesis and tumor growth in malignant gliomas.

REFERENCES

1. BROWN, P. D., A. T. LEVY, I. M. K. MARGULIES, L. A. LIOTTA & W. G. STETLER-STEVENSON. 1990. Independent expression and cellular processing of mol. wt. 72,000 type IV collagenase and interstitial collagenase in human tumorigenic cell lines. Cancer Res. **50:** 6184–6191.
2. COSTELLO, P. & R. DEL MAESTRO. 1990. Human cerebral endothelium: Isolation and characterization of cells derived from microvessels of non-neoplastic and malignant glial tissue. J. Neuro-oncol. **8:** 231–243.
3. STETLER-STEVENSON, W. G. 1990. Type IV collagenases in tumor invasion and metastasis. Cancer Metastasis Rev. **9:** 289–303.
4. VAITHILINGAM, I. S., E. C. STROUDE, W. MCDONALD & R. F. DEL MAESTRO. 1991. General protease and collagenase (IV) activity in C6 astrocytoma cells, C6 spheroids and implanted C6 spheroids. J. Neuro-oncol. **10:** 203–212.
5. WHITFIELD, H. J., JR., L. S. BRADY, M. A. SMITH, E. MAMALAKI, R. J. FOX & M. HERKENHAM. 1990. Optimization of cRNA probe *in situ* hybridization methodology for localization of glucocorticoid receptor mRNA in rat brain: A detailed protocol. Cell. Mol. Neurobiol. **10:** 145–155.2

Gingival Crevicular Fluid and Salivary Matrix Metalloproteinases of Heavy Smokers as Indicators of Periodontal Health[a]

Y. DING,[b] K. LIEDE,[b] S. LEPPÄ,[b] T. INGMAN,[b,c]
R. SEPPER,[c] Y. T. KONTTINEN,[c] AND T. SORSA[b,d]

Departments of Periodontology[b] and Anatomy[c]
University of Helsinki
Helsinki, Finland

Cigarette smoking is one of the major risk factors in the pathogenesis of periodontal disease.[1] Matrix metalloproteinases (MMPs)[2] are key mediators of periodontal tissue destruction.[3–5] We studied, by functional and immunologic means, the levels and cellular sources of MMPs and their noncollagenous substrate α_1-antitrypsin,[2] a serum protein and a major endogenous neutrophil elastase inhibitor, in heavy smokers' gingival crevicular fluid (GCF) and saliva. In addition, these biochemical parameters were related to the periodontal health of heavy smokers.

Gingival crevicular fluid, stimulated saliva, and resting saliva were collected from smokers with periodontitis and periodontally healthy sites using methods described elsewhere.[3–5] Gingival crevicular fluid was not collected from nonsmokers in this study. Type I gelatin zymography, SDS-polyacrylamide gel electrophoresis (SDS-PAGE) assay of collagenase, Western blot, and functional assays of serine protease activities of elastase (SAAVNA-peptide), cathepsin G (SAAPNA-peptide), and trypsin-like (BAPNA-peptide) were used in this study. Collagenase assay by SDS-PAGE showed low total and endogenously active levels of collagenase in either healthy or periodontally diseased smokers' GCF (FIG. 1A, lanes 6–9). Zymography analysis showed the presence of 92-kDa MMP-9, 72-kDa MMP-2, and low molecular weight forms of gelatinases in GCF of heavy smokers (FIG. 1B, lanes 5–10). Similar findings were observed in salivary samples of heavy smokers (data not shown). Low molecular weight forms of gelatinases were especially found in GCF and saliva of heavy smokers with periodontitis (FIG. 1B, lane 10). Using specific anti–MMP-9 antiserum 92-kDa proMMP-9, 82-kDa active MMP-9 and low molecular weight fragmented forms of MMP-9 were identified in GCF and saliva of smokers with periodontitis by Western blotting. Western blot analysis using anti-MMP-8 and MMP-2 antibodies showed the presence of 75-kDa and 65-kDa MMP-8 as well as 72-kDa and 68-kDa MMP-2 (FIG. 1A-C). Western blot analysis using anti-α_1-antitrypsin showed the presence of 75-kDa α_1 AT form, probably elastase-α_1-antitrypsin complex, 54-kDa native, 45-kDa, and some other low molecular weight fragmented α_1AT forms in GCF and saliva (FIG. 1C, lanes 4–8). Functional assays of serine proteases revealed that low activities of elastase, cathepsin G, and trypsin-like

[a]This work was supported by the Academy of Finland, the ITI Straumann Foundation, the Arthritis Foundation in Finland, and the Yrjö Jahnsson Foundation.

[d]Address correspondence to: Department of Periodontology, Institute of Dentistry, University of Helsinki, P.O. Box 41, (Mannerheimintie 172), FIN 00014 Helsinki, Finland.

proteases were present in smokers' GCF from either periodontitis or healthy sites. Our results demonstrate that surprisingly low activities of MMP-2, MMP-8, and MMP-9 are present in these oral inflammatory exudates of smokers, but all of the MMPs were found to exist at least in part in their endogenously/proteolytically activated forms. The GCF and saliva of smokers with periodontitis in particular were found to contain fragmented forms of MMP-9. Polymorphonuclear leukocyte-derived MMP-9 and MMP-8 seem to be the major MMPs in GCF and saliva of smokers with periodontitis. However, slight amounts of endogenously activated MMP-2 apparently also reflect gingival stromal cell (fibroblasts, etc.) activation. α_1-Antitrypsin was proteolytically fragmented, possibly due at least partially to the action of endogenously active MMPs. Because low activities of proteases were detected in heavy smokers' (with periodontitis) GCF and saliva (data not shown), care should be taken in interpreting GCF/salivary protease activities (i.e., saliva/mouth rinse analysis systems) as direct indicators of smokers' periodontal health.[5]

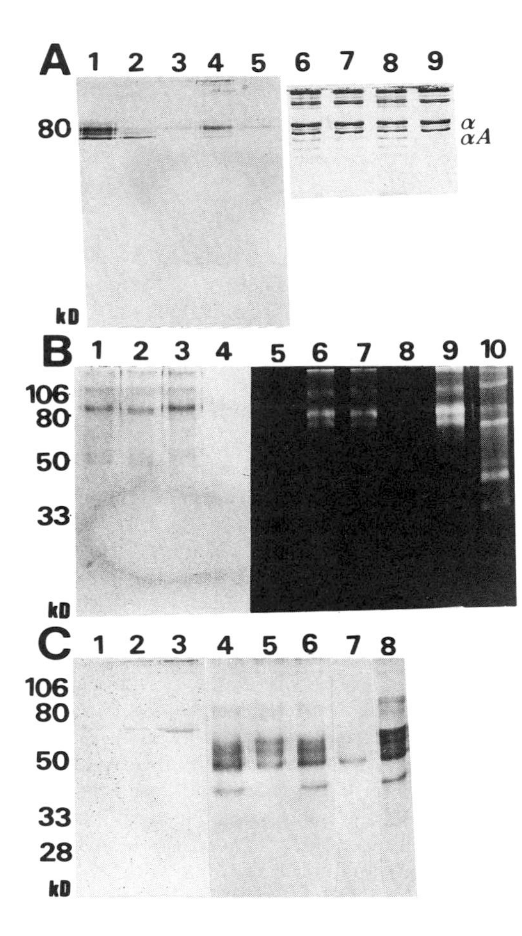

FIGURE 1. Western blot, SDS-PAGE, and zymography analysis of GCF from teeth of periodontally healthy and diseased heavy smokers. (**A**) Western blot analysis by anti–MMP-8. **Lanes 1–4**: diseased GCF; **lane 5**: healthy GCF. SDS-PAGE analysis of type I collagen degradation. **Lanes 6 and 8**: GCF incubated with type I collagen and 1 mM phenylmercuric chloride; **lanes 7 and 9**: as lanes 6 and 8 but no phenylmercuric chloride. α indicates intact type I collagen chains and α*A* indicates ¾ degradation products resulting from collagenase action. (**B**) Western blot analysis by anti–MMP-9. **Lanes 1–3**: diseased GCF samples; **lane 4**: healthy GCF; **lane 5**: type I gelatin zymography analysis of healthy GCF; **lanes 6 and 7**: gingivitis GCF; **lane 8**: healthy GCF; **lanes 9 and 10**: periodontitis GCF. (**C**) Western blot analysis by anti–MMP-2 (*lanes 1–3*) and anti-α_1 AT (*lanes 4–8*). **Lanes 1–3**: healthy, gingivitis, and periodontitis GCF; **lanes 4, 6, and 8**: periodontitis GCF; **lanes 5 and 7**: healthy GCF. Molecular weight markers are indicated.

REFERENCES

1. HABER, J., J. WATTLES, M. CROWLEY, R. MANDELL, K. JOSHIPURA & R. L. KENT. 1993. J. Periodontol. **64:** 16–23.
2. WEISS, S. J. 1989. N. Engl. J. Med. **320:** 365–376.
3. SORSA, T., T. INGMAN, K. SUOMALAINEN, M. HAAPASALO, Y. T. KONTTINEN, O. LINDY, H. SAARI & V. J. UITTO. 1992. Infect. Immun. **60:** 4491–4495.
4. UITTO, V. J., K. SUOMALAINEN & T. SORSA. 1990. J. Periodontal Res. **25:** 135–142.
5. INGMAN, T., T. SORSA, Y. T. KONTTINEN, K. LIEDE, H. SAARI, O. LINDY & K. SUOMALAINEN 1993. Oral Microbiol. Immunol. **8:** 298–305.

Differential Regulation of Matrix Metalloproteinase-2 Activation in Human Breast Cancer Cell Lines

SOURINDRA N. MAITI,[a] MING YU,[a] JORGE BUENO, REZA H. TIRGARI, FERNANDO L. PALAO-MARCO, HELENA PULYAEVA,[a] AND ERIK W. THOMPSON[a,b]

Vincent T. Lombardi Cancer Research Center
and
[a]Department of Cell Biology
Georgetown University Medical Center
3800 Reservoir Road, N.W.
Washington, D.C. 20007

Matrix metalloproteinase-2 (MMP-2), like all MMPs, is secreted in the latent zymogen form (proenzyme) and requires activation into a mature form before it can degrade substrate. Recent studies demonstrated cell-associated (cellular) activation of MMP-2 in certain cells.[1–8] Cellular activation appears to be 12-*O*-tetradecanoylphorbol-13-acetate (TPA) regulatable in HT1080 fibrosarcoma and A2058 melanoma cells[1,6,8] and concanavalin A (Con A) inducible in fibroblasts.[2,3,7] We found that a similar cell membrane-associated MMP-2 activator can be induced when fibroblasts or invasive (fibroblastoid) human breast cancer cells are cultured on a three-dimensional gel of pepsin-extracted type I collagen (Vitrogen, Collagen Corp, CA), but not on a variety of other flat or three-dimensional substrates.[4,5,10] As reported by others for Con A-[2,3,7] or TPA-[1,6,8] induced activation, collagen induction requires protein synthesis, appears to be localized to the cell surface, can be inhibited by metalloproteinase inhibitors, and requires an intact MMP-2 hemopexin-like domain.[4,5,9] In this study, we further compared a range of different human breast cancer cell lines for the MMP-2 activational responses to three agents: collagen, Con A, and TPA.

Cell lines, culture methods, and zymographic analyses used were previously described,[4,5] with 100,000 cells in 12-well culture plates. SFM/MMP-2 (75% serum-free medium [SFM, see refs. 4 and 5] and 25% MMP-2 [SFM conditioned for 72 h by MMP-2-transfected MCF-7 cells])[11] was used as the source of MMP-2. Con A (50 μg/mL) and TPA (10^{-7} M) were added when changing to SFM/MMP-2. Seventy-two-hour SFM/MMP-2 culture supernatants were analyzed by zymography for activation of the added MMP-2.

Five vimentin-negative (VIM^-) poorly invasive (MCF-7, T47D, SK-Br-3, MDA-MB-453, and BT-20) and five VIM^+ highly invasive cell lines ($MCF\text{-}7_{ADR}$, MDA-MB-231, Hs578T, MDA-MB-435, and MDA-MB-436) were used.[10] Vitrogen (Vg)-induced MMP-2 activation was not evident for 48 h and then increased in intensity by 72 h, whereas TPA and Con A induced MMP-2 activation in 24 h (FIG. 1). We

[b]Address correspondence to Dr. Erik W. Thompson, Room S178, Vincent T. Lombardi Cancer Research Center, Georgetown University Medical Center, 3800 Reservoir Road, N.W. Washington, D.C. 20007.

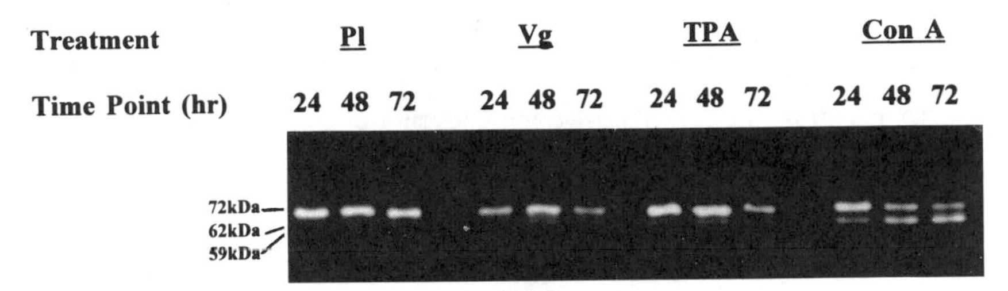

FIGURE 1. Time course for activation of exogenous MMP-2 by MDA-MB-231 cells cultured on vitrogen (Vg) or cultured on plastic and treated with either TPA (10^{-7} M) or Con A (50 μg/mL). Plastic (P1) culture cells are shown for control purposes.

compared induction of MMP-2 activation by all three agents at 72 h (FIG. 2). Low levels of MMP-2 activation were seen on plastic for most of the cell lines probably because of the use of recombinant MMP-2 which is essentially free of tissue inhibitor of metalloproteinase (TIMP-2) (unpublished observation). This MMP-2 may be more susceptible to autocatalytic activation than is the serum-derived MMP-2

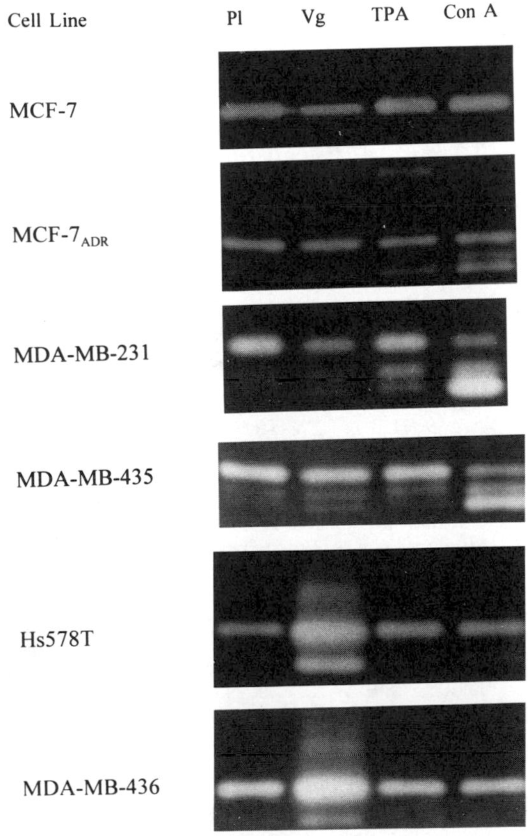

Cell Line Pl Vg TPA Con A

MCF-7

$MCF\text{-}7_{ADR}$

MDA-MB-231

MDA-MB-435

Hs578T

MDA-MB-436

FIGURE 2. Comparative effects of Con A, TPA, and vitrogen (Vg) on induction of MMP-2 activation by either the poorly invasive, vimentin-negative (MCF-7) or highly invasive, vimentin-positive ($MCF\text{-}7_{ADR}$, MDA-MB-231, MDA-MB-435, Hs578T, and MDA-MB-436) human breast cancer cell lines. Plastic (P1) cultures cells are shown for control purposes, because some autodegradation of exogenous MMP-2 did occur over the incubation period.

and/or endogenously produced MMP-2/TIMP-2 complex used in previous studies.[4,5] All VIM$^-$ human breast cancer cell lines were confirmed as unresponsive to Vg for MMP-2 activation and did not respond to either TPA or Con A (data shown for MCF-7). In contrast, all VIM$^+$ invasive cell lines showed different levels of MMP-2 activation in response to Vg, findings consistent with our previous studies.[5] Con A induced high levels of MMP-2 activation in MDA-MB-231 and MDA-MB-435, moderate levels in MCF-7_{ADR} cells responded to TPA for MMP-2 activation, with very weak or no activation in the other VIM$^+$ human breast cancer and cell lines (FIGS. 1 and 2). This did not reflect the overall responsitivity to TPA, because secretion of TPA-responsive MMP-9 (92-kDa gelatinase B) was induced strongly in MCF-7, MCF-7_{ADR}, and MDA-MB-231 cells (FIG. 2) and to a lesser extent in MDA-MB-435, Hs578T, and MDA-MB-436 cells.

Our results demonstrated differential responsiveness to these regimens and suggest that regulation of MMP-2 activation may be mediated by different cellular pathways. Collagen I and Con A both induce MMP-2 activation in fibroblasts.[2,4] Inasmuch as both agents interact with the cell surface, we hypothesized that they may induce MMP-2 activation through interaction with a common cell surface molecule(s). The inability of VIM$^+$ Hs578T and MDA-MB-436 to activate MMP-9 in response to Con A, however, brings this potential association into question. The lack of a Con A response in these two cell lines, however, may be due to increased production of molecules that inhibit the autocatalytic conversion of MMP-2 (e.g., TIMPs), possibly masking our ability to detect the presence of the putative MMP-2 activator on the cell surface. The TPA-induced pathway apparently is distinct from that induced by collagen I and/or Con A and independent of the effects of TPA on MMP-9 expression. The observation that all VIM$^-$ cell lines were completely noninducible for MMP-2 activation strengthens our previous association between MMP-2 activation potential and VIM$^+$-associated invasiveness/metastasis in human breast cancer cells.

REFERENCES

1. BROWN, P. D., A. T. LEVY, I. M. MARGULIES, L. A. LIOTTA & W. G. STETLER-STEVENSON. 1990. Cancer Res. **50:** 6184–6191.
2. OVERALL, C. M. & J. SODEK. J. Biol. Chem. 1990. **265:** 21141–21151.
3. WARD, R. V., S. J. ATKINSON, P. M. SLOCOMBE, A. J. P. DOCHERTY, J. J. REYNOLDS & G. MURPHY. 1991. Biochim. Biophys. Acta **1079:** 242–246.
4. AZZAM, H. S. & E. W. THOMPSON. 1992. Cancer Res. **52:** 4540–4544.
5. AZZAM, H. S., G. A. ARAND, M. E. LIPPMAN & E. W. THOMPSON. 1993. J. Natl. Cancer Inst. **85:** 1758–1764.
6. BROWN, P. D., D. E. KLEINER, E. J. UNSWORTH & W. G. STETLER-STEVENSON. 1993. Kidney Int. **43:** 163–170.
7. MURPHY, G., F. WILLENBROCK, R. V. WARD, M. I. COCKETT, D. EATON & A. J. DOCHERTY. 1992. Biochem. J. **283:** 637–641.
8. STRONGIN, A. Y., B. L. MARMER, G. A. GRANT & G. I. GOLDBERG. 1993. J. Biol. Chem. **268:** 14033–14039.
9. AZZAM, H. S., S.-N. BAE, G. ARAND, J. YOON, F. X. KERN, W. G. STETLER-STEVENSON & E. W. THOMPSON. 1994. Proc. Am. Assoc. Cancer Res. **35** (abstr.).
10. THOMPSON, E. W., S. PAIK, N. BRÜNNER, C. L. SOMMERS, G. ZUGMAIER, R. CLARKE, T. B. SHIMA, J. TORRI, S. DONAHUE, M. E. LIPPMAN, G. R. MARTIN & R. B. DICKSON. 1992. J. Cell Physiol. **150:** 534–544.
11. BAE, S.-N., G. S. ARAND, W. G. STETLER-STEVENSEN, M. E. LIPPMAN, F. X. KERN & E. W. THOMPSON. 1993. Proc. Am. Assoc. Cancer Res. **34** (abstr. 485): 82.

Matrix Metalloproteinases-1, -3, and -8 in Adult Periodontitis *in Situ*

An Immunohistochemical Study[a]

T. INGMAN,[b,c] T. SORSA,[b,c] J. MICHAELIS,[d]
AND Y. T. KONTTINEN[b]

Departments of [b]Anatomy and [c]Periodontology
University of Helsinki
Helsinki, Finland

[d]Department of Pathology
Christchurch Hospital
Christchurch
New Zealand

Matrix metalloproteinases (MMPs) take part in the destruction of the components of the extracellular matrix (ECM) and in remodeling in such physiologic and pathologic conditions as inflammatory periodontal diseases. Among MMPs, MMP-1 and MMP-8 share the unique ability to cleave types I, II, and III collagen. These fibrillar collagens form about 60% of the content of gingival connective tissue in healthy periodontium. Thus, synthesis and activation of MMP-1 and MMP-8 are important steps in the pathologic ECM destruction associated with inflammatory periodontal diseases.[1,2]

Several investigators have utilized methods based on MMP-specific functional characteristics (e.g., substrate specificity, activation studies, and neutralizing antibodies) to show that MMP-8 is the main collagenase in gingival crevicular fluid, saliva, and inflamed gingiva in adult periodontitis.[3,4] However, Takeyama *et al.*[5] recently found weak immunoreactivity for MMP-1 in fibroblasts, macrophages, and epithelial cells in inflamed periodontal tissues. Our objective in this study was to provide direct evidence of MMPs in inflamed gingiva in adult periodontitis. Therefore, we used a series of specific polyclonal rabbit antisera and the avidin-biotin-peroxidase staining method[6] to visualize immunohistochemically the presence of MMP-1, MMP-3, and MMP-8 in marginal gingival tissue and gingival granulation tissue of patients with adult periodontitis. For ethical reasons, inflamed tissue specimens were collected only after scaling and root planning before flap surgery. Control gingival tissue specimens were collected from subjects with clinically healthy gingiva during impacted third molar extraction operations.

MMP-8–positive polymorphonuclear leukocytes, but not MMP-1– or MMP-3–positive cells, were detected in marginal gingival connective tissue of patients with adult periodontitis (FIG. 1). In gingival granulation tissue, MMP-1– and MMP-3–positive fibroblasts and macrophages and MMP-8–positive polymorphonuclear leukocytes were detected (FIG. 2). In healthy gingival tissue, staining for these three MMPs was negative.

[a]This work was supported by the Academy of Finland, the 350-year foundation of the University of Helsinki, the Oskar Öflund Foundation, and the Association of Finnish Women Dentists.

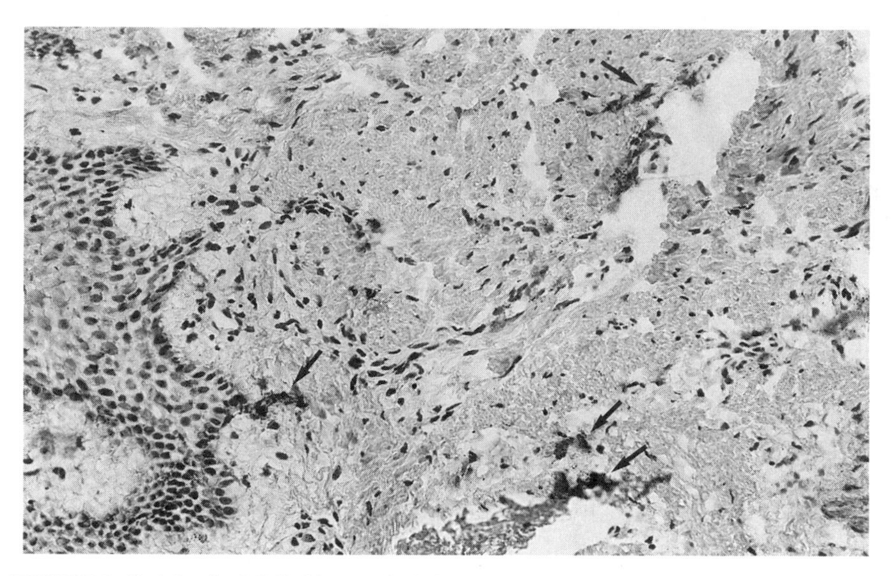

FIGURE 1. Staining for MMP-8 in marginal gingival tissue of a patient with adult periodontitis (*arrows*). MMP-8–positive neutrophils are present in connective tissue, but staining is negative in the epithelium. (Rabbit polyclonal anti-MMP-8 antiserum; ABC staining; hematoxylin counterstaining. Original magnification: 250×; reduced to 65%.)

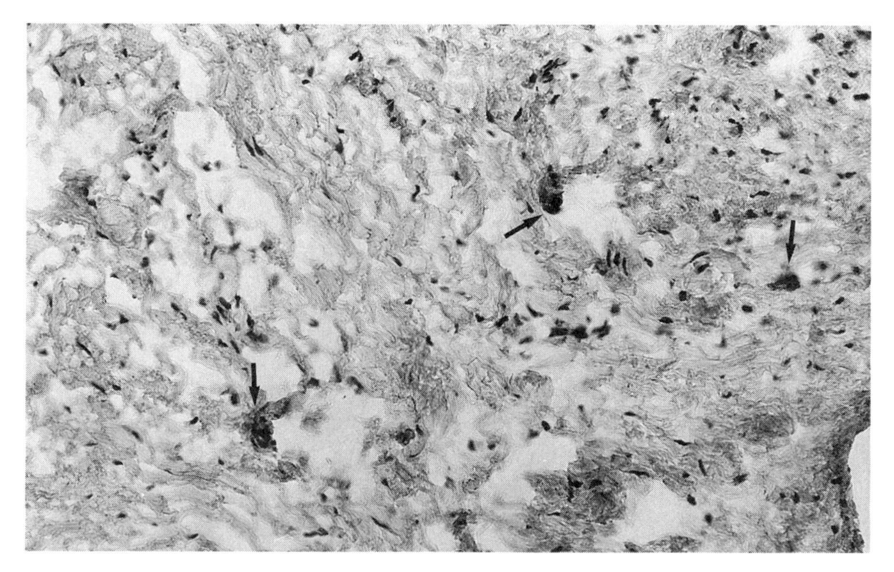

FIGURE 2. MMP-8–positive neutrophils in gingival granulation tissue of a patient with adult periodontitis (*arrows*). (Rabbit anti-MMP-8 antiserum; ABC staining; hematoxylin counterstaining. Original magnification: 250×, reduced to 65%.)

These results clearly demonstrate that MMP-8, but not MMP-1, is the predominant collagenase in inflamed marginal gingival tissue in adult periodontitis *in situ*. However, our evidence also shows that MMP-1 and MMP-3 can be synthesized (and presumably activated) during adult periodontitis. It is likely that MMP-1 and MMP-3 may play a role in more chronic inflammatory processes, whereas MMP-8 contributes more to acute injury.

REFERENCES

1. Birkedal-Hansen, H., W. G. I. Moore, M. K. Bodden, L. J. Windsor, B. Birkedal-Hansen, A. DeCarlo & J. A. Engler. 1993. Matrix metalloproteinases: A review. Crit. Rev. Oral Biol. Med. **4:** 197–250.
2. Lindhe, J. 1985. Textbook of Clinical Periodontology. W.B. Saunders Co. Munksgaard.
3. Sorsa, T., V.-J. Uitto, K. Suomalainen, M. Vauhkonen & S. Lindy. 1988. Comparison of interstitial collagenases from human gingiva, sulcular fluid and polymorphonuclear leucocytes. J. Periodontal Res. **23:** 386–393.
4. Ingman, T., T. Sorsa, K. Suomalainen, S. Halinen, O. Lindy, A. Lauhio, H. Saari, Y. T. Konttinen & L. M. Golub. 1993. Tetracycline inhibition and the cellular sources of collagenase in different human periodontal diseases. J. Periodontol. **64:** 82–88.
5. Takeyama, H., B. Birkedal-Hansen, W. G. I. Moore, M. K. Bodden & H. Birkedal-Hansen. 1993. Collagenase, Mr 72 K gelatinase and TIMP-1 in inflamed human gingiva. J. Dent. Res. **72** (abstr. 437): 158.
6. Hsu, S. M., L. Raine & H. Fanger. 1981. Use of avidin-biotin-peroxidase complex (ABC) in immunoperoxidase techniques: A comparison between ABC and unlabelled antibody (PAP) procedures. J. Histochem. Cytochem. **29:** 577–580.

A Possible Role for MMP-2 and MMP-9 in the Migration of Primate Arterial Smooth Muscle Cells through Native Matrix

RICHARD D. KENAGY AND ALEXANDER W. CLOWES

Department of Surgery RF-25
University of Washington School of Medicine
Seattle, Washington 98195

Migration of vascular cells is a major factor in vessel development and atherosclerosis.[1,2] Although the regulation of this process is not well understood, it is known that basic fibroblast growth factor (bFGF)[3] and platelet-derived growth factor (PDGF)[4] stimulate smooth muscle cell (SMC) migration after arterial injury. The role of proteinases such as matrix metalloproteinase (MMP)-2, MMP-9, and urokinase (uPA) in the migration of cancer cells has been clearly established,[5] but this is not true for SMCs. We studied the role of proteinases in SMC migration using baboon aortic explants. After removal of the endothelial layer, the inner media was stripped off and made into 1-mm^2 explants.[6] One milliliter of Dulbecco's modified Eagles medium with transferrin 5 μg/mL, insulin 6 μg/mL, and ovalbumin 1 mg/mL with or without test substances was added to 15 explants per 25 cm^2 flask. Medium was changed at 7 days. Explants were called positive if at least one SMC was present on the plastic.

Smooth muscle cells first appeared after 3 days, and at 7 days 56 ± 3% were positive (n = 32). Maximal migration (93 ± 4%, n = 10) was achieved by 14 days. Explants initially expressed MMP-2 and uPA (FIG. 1A and B). By day 1 both activities were increased and were seen in the medium. MMP-9 was barely, but consistently, detectable at 3 days and clearly visible by 6 days. Tissue plasminogen activator was variably detected in the medium as a high molecular weight complex. Collagenase and stromelysin were not detected by zymography or Western analysis (data not presented). Migration was inhibited by antibodies to bFGF and was stimulated by bFGF. Antibodies to the α and β subunits of the PDGF receptor inhibited and PDGF-BB stimulated migration (FIG. 2A). Plasminogen also increased migration nearly twofold. Migration of SMCs was inhibited by the general proteinase inhibitor α_2-macroglobulin, the metalloproteinase inhibitor 1,10-phenanthroline, and the MMP inhibitor BB-94 (FIG. 2B) at concentrations that did not alter the DNA content of explants (data not presented). By zymography, 1,10-phenanthroline and BB-94 also decreased MMP-9 and a 60-kDa gelatinase that may be activated MMP-2 (data not presented).

These data suggest that MMPs are needed for optimal migration of SMCs through extracellular matrix and confirm the report of Southgate *et al.*[7] Whether this is true *in vivo* and whether bFGF and PDGF stimulate the activity of MMPs *in vivo* will be determined in ongoing investigations.

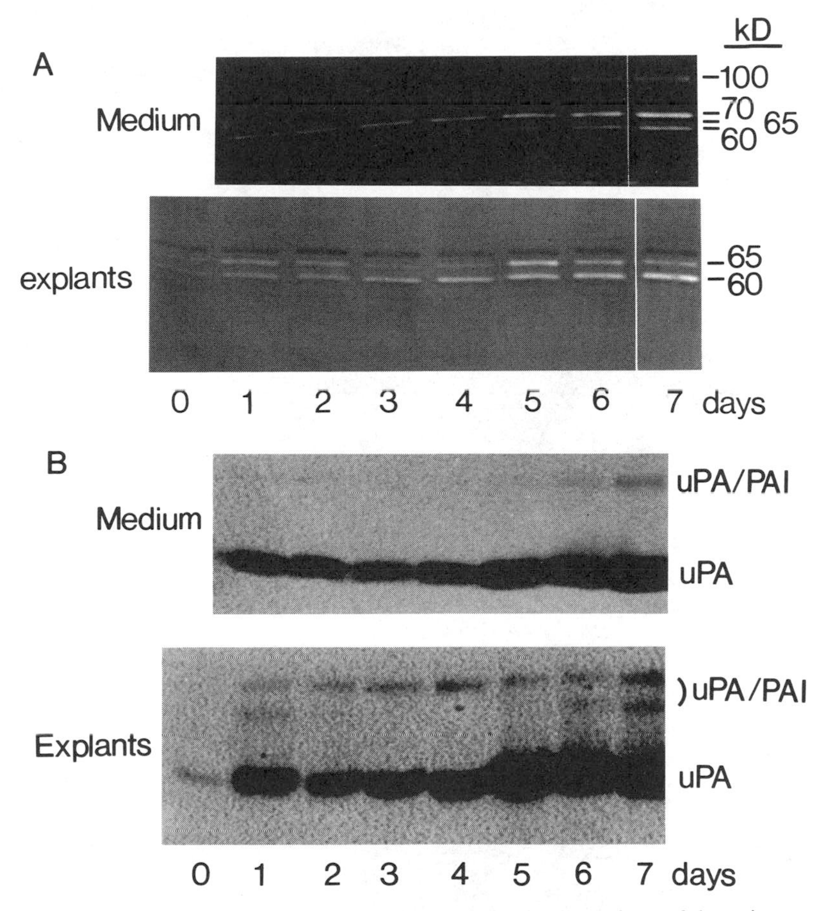

FIGURE 1. Time course of changes in proteinase production by explants of thoracic aortas of normal 3-year-old male baboons. These are gelatin (**A**) and casein (**B**, dark field photography) zymograms of both medium and extracts of explants. Explants were extracted in 200 μL of 2 M guanidine HCl, 0.01 M $CaCl_2$, 0.2% Triton X-100, and 0.05 M Tris, pH 7.5, and dialyzed against the same buffer without guanidine. Zymography for plasminogen activators was performed using casein agar gels containing plasminogen.[8] Gelatin zymography was performed by including 1 mg gelatin per milliliter in 10% acrylamide gels for electrophoresis. After electrophoresis, gels were washed in 2.5% Triton X-100, incubated 18 h in 50 mM Tris, 10 mM $CaCl_2$, pH 7.5 at 37 °C, and stained with Coumassie blue R.

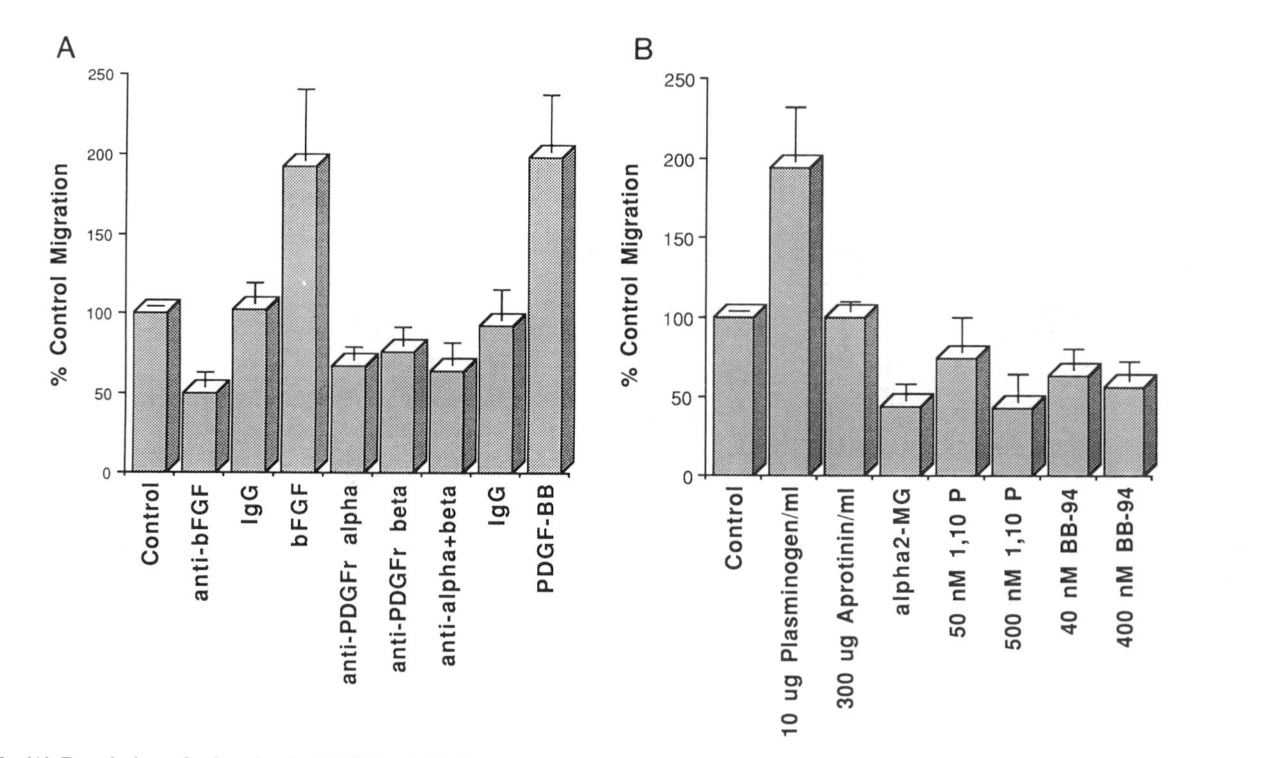

FIGURE 2. (**A**) Regulation of migration by bFGF and PDGF-BB. Explants were exposed to either 50 ng/mL bFGF (n = 7), 50 ng/mL PDGF-BB (n = 8), 100 μg/mL anti-bFGF (n = 7), 100 μg/mL rabbit IgG (n = 7), 50 μg/mL anti-α chain of the PDGF receptor (n = 13), 50 μg/mL anti-β chain of the PDGF receptor (n = 12), 25 μg/mL each of the anti-α and anti-β chain antibodies (n = 12), or 50 μg/mL anticarboxylase as a control (n = 9). bFGF and antibodies to bFGF were from R and D Systems, Inc. (Minneapolis, MN). PDGF-BB and PDGF receptor antibodies were from C. Hart of Zymogenetics, Inc. (Seattle, WA). Data are the mean ± SEM of migration at 7 days of explants from the indicated number of animals. (**B**) Regulation of migration by proteinases. Explants were exposed to either 10 μg/mL Glu-plasminogen (American Diagnostica, Greenwich, CT; n = 12), 300 μg/mL recombinant aprotinin (Zymogenetics, Inc.; n = 3), 200 μg/mL α_2-macroglobulin (ART, Inc., Athens, GA; n = 11), 50 or 500 nM 1,10-phenanthroline (n = 3 and 5, respectively), or 40 or 400 nM BB-94 (British Bio-technology, Ltd, Oxford, England; n = 6 and 4, respectively). Data are the mean ± SEM of migration at 7 days.

REFERENCES

1. SCHWARTZ, S. M., R. L. HEIMARK & M. W. MAJESKY. 1990. Physiol. Rev. **70:** 1177–1210.
2. ROSS, R. 1993. Nature **362:** 801–809.
3. JACKSON, C. L. & M. A. REIDY. 1993. Am. J. Pathol. **143:** 1024–1031.
4. JACKSON, C. L., E. W. RAINES, R. ROSS & M. A. REIDY. 1993. Arterioscler. Thromb. **13:** 1218–1226.
5. LIOTTA, L. A. & W. G. STETLER-STEVENSON. 1991. Cancer Res. **51**(Suppl.): 5054s–5059s.
6. MCMURRAY, H. F., D. P. PARROTT & D. E. BOWYER. 1991. Atherosclerosis **86:** 227–237.
7. SOUTHGATE, K. M., M. DAVIES, R. F. G. BOOTH & A. C. NEWBY. 1992. Biochem. J. **288:** 93–99.
8. VASSALLI, J.-D., J.-M. DAYER, A. WOHLWEND & D. BELIN. 1984. J. Exp. Med. **159:** 1653–1668.

Regulation of 92-kDa Gelatinase B Activity in the Extracellular Matrix by Tissue Kallikrein

SUZANNE MENASHI,[a] RAFAEL FRIDMAN,[b]
SYLVANE DESRIVIERES,[a] HE LU,[a]
YVES LEGRAND,[a] AND CLAUDINE SORIA[a]

[a]Unité 353 INSERM
Hôpital Saint Louis
1, avenue Claude Vellefaux
75010 Paris, France

[b]Department of Pathology
Wayne State University
Detroit, Michigan 48201

Gelatinases are implicated in extracellular matrix remodeling which accompanies several pathophysiologic processes such as embryogenesis, angiogenesis, and metastasis. Regulation of gelatinases to inhibit uncontrolled extracellular matrix degradation has been studied extensively. Several strategies can be used including inhibition of gelatinase biosynthesis, inhibition of its activity, or inhibition of activation. For the latter, it is necessary to understand the mechanism of *in vivo* activation of these proteases. We recently showed that tissue kallikrein can activate 92-kDa gelatinase B in endothelial cell conditioned medium.[1] Inasmuch as endothelial cell-conditioned medium may contain other proteinases and inhibitors, we have now tested the activation of purified recombinant gelatinase B by tissue kallikrein. Plasmin is usually considered responsible for the activation of most metalloproteinases; therefore, we have compared collagenase activation by kallikrein with that obtained by plasmin.

Purified human recombinant gelatinase B was incubated at 37 °C for 45 min with tissue kallikrein (0.01–1 μg/mL) or plasmin (50–200 μg/mL). After incubation, 5 mM pefablock was added to inhibit the activating serine proteases. Samples were analyzed directly by SDS-PAGE on 10% gels containing 1 mg/mL gelatin[2] and by the ^{3}H-gelatin degradation assay.[3]

Results of zymography (FIG. 1) indicate that kallikrein activates gelatinase at doses of 0.05–0.5 μg/mL, as observed by the appearance of a cleavage product of 86 kDa and another, at higher kallikrein concentrations, of approximately 60 kDa. The extent of activation was a function of kallikrein concentration and the time of activation. By contrast, under the same conditions, much greater doses of plasmin (1,000 times more than that of kallikrein) were required to cause activation, and even then, only partial activation was observed.

To demonstrate that the lower molecular weight fragment generated with tissue kallikrein corresponds to an active product, we tested the effect of kallikrein by the ^{3}H-gelatin degradation assay. The results (FIG. 2), in agreement with the zymographic results, show that kallikrein activates gelatinase in a dose-dependent manner, demonstrating that the cleavage observed by zymography leads to the formation of an active product. Under the conditions used, after activation of gelatinase by 0.5 μg/mL kallikrein for 45 min, total degradation of the labeled substrate was observed

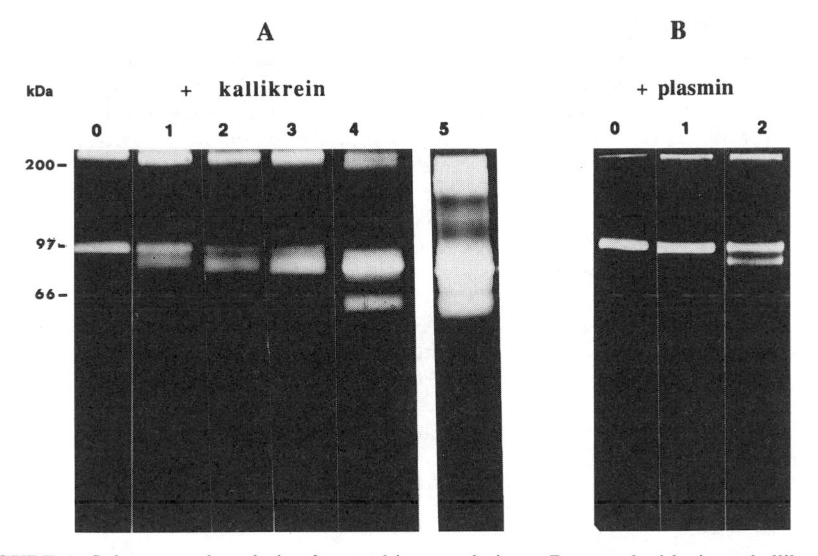

FIGURE 1. Substrate gel analysis of recombinant gelatinase B treated with tissue kallikrein and plasmin. (**A**) Proenzyme 15 ng was incubated at 37 °C alone as control (*lane 0*) or with 0.01, 0.05, 0.1, and 0.5 μg/mL kallikrein (*lanes 1, 2, 3, and 4,* respectively) for 45 min or with 0.5 μg/mL kallikrein for 90 min (*lane 5*). (**B**) Proenzyme 15 ng was incubated at 37 °C alone (*lane 0*) or with 50 and 200 μg/mL plasmin (*lanes 1 and 2*) for 45 min. Pefablock 5 mM was then added to inhibit kallikrein and plasmin and the samples were analyzed by gelatin-gel electrophoresis.

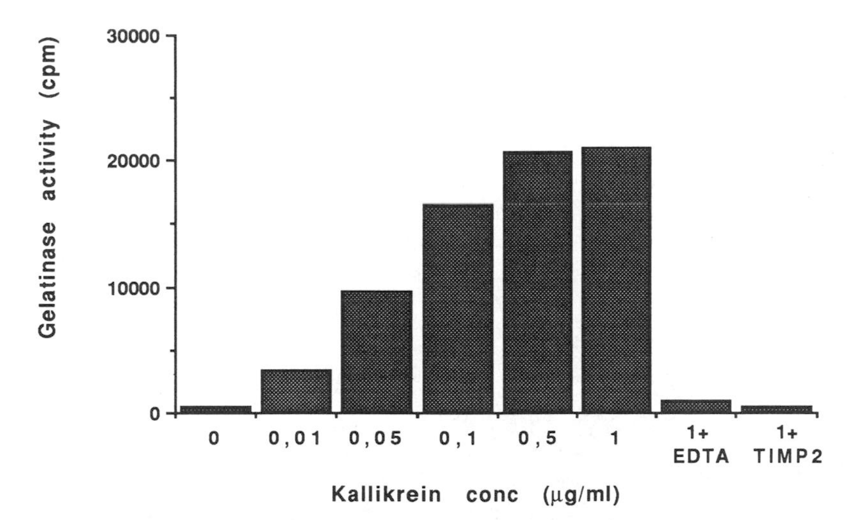

FIGURE 2. ^{3}H-gelatin degradation by gelatinase B activated by tissue kallikrein. Proenzyme 15 ng was incubated in a 100-μL volume with kallikrein concentration ranging between 0 and 1 μg/mL at 37 °C for 45 min. Pefablock was then added at a 5-mM final concentration to inhibit kallikrein before the addition of 100 μg ^{3}H-gelatin substrate (200 cpm/μg), and the reaction was allowed to proceed for 2 h. EDTA 5 mM or TIMP-2 1 μg was added to samples treated by 1 μg/mL kallikrein as controls.

after a 2-h assay. Controls in the presence of EDTA or TIMP-2 show total inhibition of kallikrein-induced gelatinase activity. The 72-kDa gelatinase A was studied at the same time, but no activation was observed with either kallikrein or plasmin.

These findings confirm our previous results with crude gelatinases expressed in porcine aortic endothelial cell-conditioned medium[1] and demonstrate that activation of gelatinase B by kallikrein is a direct effect. This activation by tissue kallikrein may be an important mechanism of gelatinase B activation, as we have already shown by immunolocalization that tissue kallikrein is abundant in the extracellular matrix, particularly the subendothelium.[1] By contrast to plasmin, which binds to the endothelial cell surface,[4] kallikrein binding sites could not be detected on endothelial cells (results not shown); therefore, we suggest a possible activation cascade in which plasmin would activate prokallikrein[5] in the pericellular space, which would in turn amplify gelatinase B activation in the matrix.

REFERENCES

1. Desrivieres, S., H. Lu, N. Peyri, C. Soria, Y. Legrand & S. Menashi. 1993. J. Cell. Physiol. **157:** 587–593.
2. Herron, G. S., M. J. Banda, E. J. Clark, J. Gavrilovic & Z. Werb. 1986. J. Biol. Chem. **261:** 2814–2818.
3. Vissers, M. C. M. & C. C. Winterbourn. 1988. Biochem. J. **249:** 327–331.
4. Hajjar, K. A., P. C. Harpel, E. A. Jaffe & R. L. Nachman. 1986. J. Biol. Chem. **261:** 11656–11662.
5. Vogt, W. 1964. J. Physiol. **170:** 153–166.

Monoclonal Antibodies to Human MMP-9

NOEMI RAMOS-DESIMONE AND DEBORAH L. FRENCH

Department of Pathology
State University of New York at Stony Brook
Stony Brook, New York 11794–8691

The degradation of extracellular matrix (ECM) components in normal and pathologic processes has been linked to the presence of a family of hydrolytic enzymes known as matrix metalloproteinases (MMPs).[1] Two members of this family, the 72-kDa (MMP-2) and the 92-kDa (MMP-9) enzyme, are potent gelatinases. MMP-2 and MMP-9 are highly expressed in invasive tumors, suggesting a role in the degradation of ECM and basement membrane during tumor cell invasion and metastasis.[1] Studies with natural inhibitors, such as the tissue inhibitor of metalloproteinases (TIMP-1 and TIMP-2), have demonstrated the importance of MMPs in cancer in that they inhibit tumor cell invasion *in vitro* and reduce metastasis in *in vivo* animal models.

Monoclonal antibodies (mAbs) are useful reagents in assessing the role(s) of specific MMPs in tumor cell invasion and metastasis. Antihuman MMP-1 and MMP-2 mAbs neutralized enzymatic activity in solution using radiolabeled collagens type I and IV, respectively.[1] The physiologic importance of matrix degradation by MMP-2 during tumor cell invasion *in vitro* was demonstrated using anti–MMP-2 mAbs.[2] Similar studies with anti–MMP-9 mAbs have not been reported. Monoclonal antibodies generated against amino- and carboxy-terminal peptides of MMP-9 have been described, but their ability to inhibit enzyme activity has not been tested.[3]

To study the role of MMP-9 in tumor cell invasion we generated mAbs against partially purified preparations of MMP-2 and MMP-9 from human fibrosarcoma cell line HT1080.[4] Three IgG_1 mAbs designated 6-6B, 7-11C, and 8-3H were identified and an IgG_{2b}-producing switch variant of 6-6B was isolated. The mAbs are specific for MMP-9 and do not react with MMP-2 by ELISA.[4] All four mAbs are reactive by immunoblot to MMP-9 and not to MMP-2, as shown by French *et al.* (this volume). In addition to the immunoreactivity analyses, the 6-6B mAb neutralizes the *in vitro* activation of MMP-9.[4] During *in vitro* activation of proMMP-9, products with apparent molecular weights of 82–83,000 and 67–75,000 are generated.[3,5,6] The formation of the lowest molecular weight species has been correlated with an increase in catalytic activity[3,6] and may be generated following cleavage of a carboxy-terminal peptide.[3] To determine if 6-6B and 8-3H mAbs react with this low molecular form of MMP-9, immunoprecipitation analyses were performed. With a highly purified preparation of MMP-9, conversion of the enzyme to the lower molecular species after activation with APMA was assessed by silver-stained SDS-PAGE and gelatin zymography (FIG. 1). The starting material contained predominantly the 92-kDa enzyme and a small amount of the 83-kDa form of the enzyme. Over a 6-h period with APMA, most of the enzyme converted to the 83-kDa form and a small amount of the 67-kDa form was visible by silver staining. With gelatin zymography, the lowest molecular species was apparent by 3 h and had increased by 6-h. Immunoreactivity of mAbs with various molecular weight species of MMP-9 was assessed by immunoprecipitation followed by gelatin zymography (FIG. 2). The same

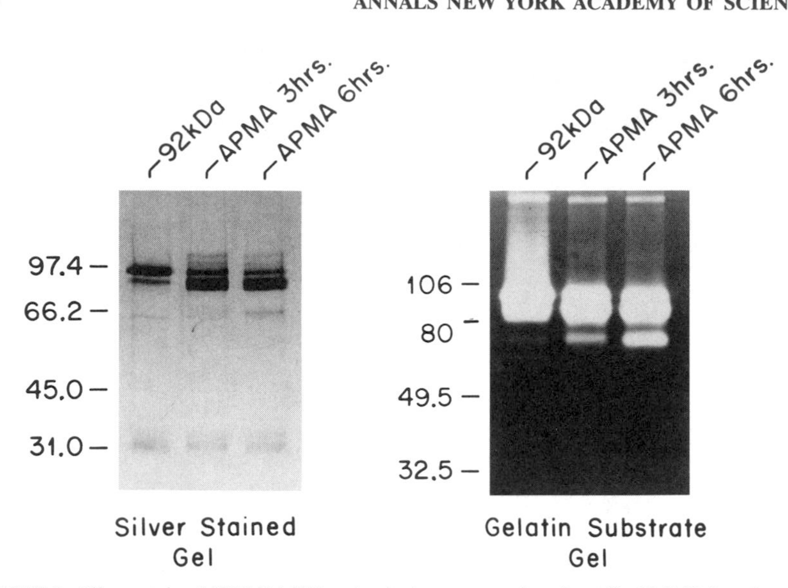

FIGURE 1. Silver-stained SDS-PAGE and gelatin zymography of purified MMP-9 activated *in vitro* with APMA. **Left:** Samples of purified MMP-9 were preincubated with 2 mM APMA in TNC/Brij buffer (50 mM Tris, pH 7.6, 0.2M NaCl, 10 mM $CaCl_2$, 0.05% Brij-35, and 0.02% NaN_3) at 37 °C for 3 and 6 h on 7.5% SDS-polyacrylamide gel and silver stained. **Right:** Aliquots of activated enzyme were electrophoresed on a 7.5% SDS-polyacrylamide gel containing gelatin (0.19 mg/mL), washed in 2.5% Triton X-100, and incubated overnight at 37 °C in TNC/Brij buffer. Gels were stained with Coomassie brilliant blue, and clear zones of lysis indicate the presence of enzyme.

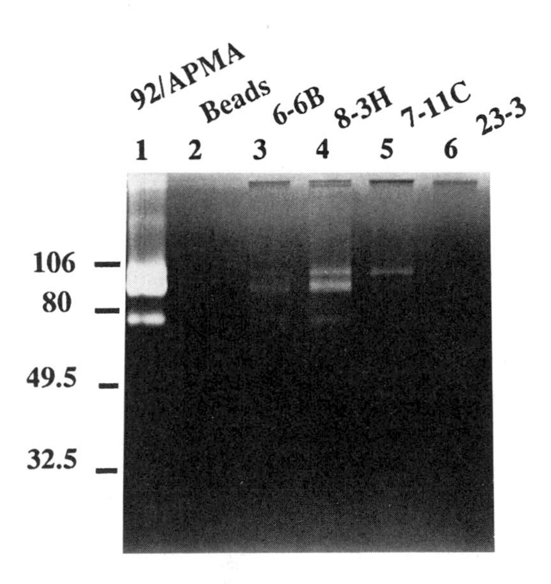

FIGURE 2. Reactivity of 6-6B, 7-11C, and 8-3H monoclonal antibodies (mAbs) with APMA-activated MMP-9 by immunoprecipitation. Purified MMP-9 (400 ng) was activated with APMA for 18 h at 37 °C. Aliquots were incubated with the mAbs (5 μg/tube) in the presence of protein-G Sepharose beads for 2 h at 25 °C. Beads were washed three times with TBS/Brij buffer, spun down, and resuspended in SDS-containing sample buffer. Samples were boiled for 10 min, and small aliquots of the supernatant were analyzed by gelatin zymography.

preparation of MMP-9 was incubated with APMA overnight, and mAbs were added to the enzyme mixture followed by protein-G sepharose beads. The 6-6B and 8-3H mAbs immunoprecipitated the 92-kDa, 82-kDa, and 67-kDa forms of the enzyme and the 7-11C mAb reacted predominantly with the 92-kDa form. Background reactivity of MMP-9 with Sepharose beads and an isotype control mAb was negligible, demonstrating the specificity of the reaction. In summary, we generated antihuman MMP-9 mAbs that can be used to quantitate and assess the role of MMP-9 in normal and pathologic processes.

REFERENCES

1. BIRKEDAL-HANSEN, H., W. G. MOORE, M. K. BODDEN, L. J. WINDSOR, B. BIRKEDAL-HANSEN, A. DECARLO & J. A. ENGLER. 1993. Matrix metalloproteinases: A review. Crit. Rev. Oral Biol. Med. **4:** 197–250.
2. HÖYHTYÄ, M., E. HUJANEN, T. TURPEENNIEMI-HUJANEN, U. THORGEIRSSON, L. A. LIOTTA & K. TRYGGVASON. 1990. Modulation of type-IV collagenase activity and invasive behavior of metastatic human melanoma (A2058) cells *in vitro* by monoclonal antibodies to type-IV collagenase. Int. J. Cancer **46:** 282–286.
3. OKADA, Y., Y. GONOJI, K. NAKA, K. TOMITA, I. NAKANISHI, K. IWATA, K. YAMASHITA & T. HAYAKAWA. 1992. Matrix metalloproteinase 9 (92-kDa gelatinase/type IV collagenase) from HT1080 human fibrosarcoma cells. J. Biol. Chem. **267:** 21712–21719.
4. RAMOS-DESIMONE, N., U. M. MOLL, J. P. QUIGLEY & D. L. FRENCH. 1993. Inhibition of matrix metalloproteinase 9 activation by a specific monoclonal antibody. Hybridoma **12:** 349–363.
5. MOLL, U. M., G. L. YOUNGLEIB, K. B. ROSINSKI & J. P. QUIGLEY. 1990. Tumor promoter-stimulated M_r 92,000 gelatinase secreted by normal and malignant human cells: Isolation and characterization of the enzyme from HT1080 tumor cells. Cancer Res. **50:** 6162–6170.
6. MORODOMI, T., Y. OGATA, Y. SASAGURI, M. MORIMATSU & H. NAGASE. 1992. Purification and characterization of matrix metalloproteinase 9 from U937 monocytic leukemia and HT1080 fibrosarcoma cells. Biochem. J. **285:** 603–611.

High Expression of 92-kDa Type IV Collagenase (Gelatinase) in the Osteoclast Lineage during Mouse Development

P. REPONEN,[a] C. SAHLBERG,[b] C. MUNAUT,[a]
I. THESLEFF,[b] AND K. TRYGGVASON[a]

[a]*Biocenter and Department of Biochemistry*
University of Oulu
FIN-90570 Oulu, Finland

[b]*Department of Pedodontics and Orthodontics*
University of Helsinki
FIN-00300 Helsinki, Finland

Type IV collagenases (gelatinases) belong to a family of mammalian extracellular metalloproteinases that are the products of related genes. The metalloproteinases share several structural and functional properties, and all of them exhibit the capacity to degrade one or more of the molecules that constitute the extracellular matrix. The 92-kDa type IV collagenase is involved in physiologic tissue remodeling by extraembryonic trophoblasts during embryonic implantation, and increased activity can be detected in pathologic states, such as tumor invasion and inflammation, where it may be secreted by macrophages.

cDNA clones for murine 92-kDa type IV collagenase (gelatinase B) were generated for the determination of its primary structure and for analysis of temporal and spatial expression *in vivo*.[1] *In situ* hybridization analyses of embryonic and postnatal mouse tissues revealed intense signals in cells of the osteoclast cell lineage. Clear expression above background was not observed in any epithelial cells of skin, lung, intestine, or kidney, or in subepithelial mesenchymal cells. Therefore, this enzyme is not likely to play a major role in the normal turnover of basement membranes. Furthermore, clear signals were not observed in tissues usually rich in macrophages; only scattered cells of bone marrow, possibly representing the monocyte/macrophage/osteoclast lineage, were positive. Expression of the gene was first observed at an early stage of cartilage and tooth development at E12 and E13, where signals were seen transiently in surrounding mesenchymal cells. At later developmental stages and postnatally strong expression was seen in large cells at the surface of bones. These cells presumably were osteoclasts as their location correlated with that of tartrate resistant acid phosphatase (TRAP)-positive cells. TRAP-positive cells. Signals above background were not observed in other tissues studied. The expression pattern of 92-kDa type IV collagenase differs completely from that of the 72-kDa enzyme, the expression of which was shown to be widely distributed in mesenchymal tissues during development of all organs.[2] The results suggest that during normal development of embryonic organs the 92-kDa type IV collagenase does not have a

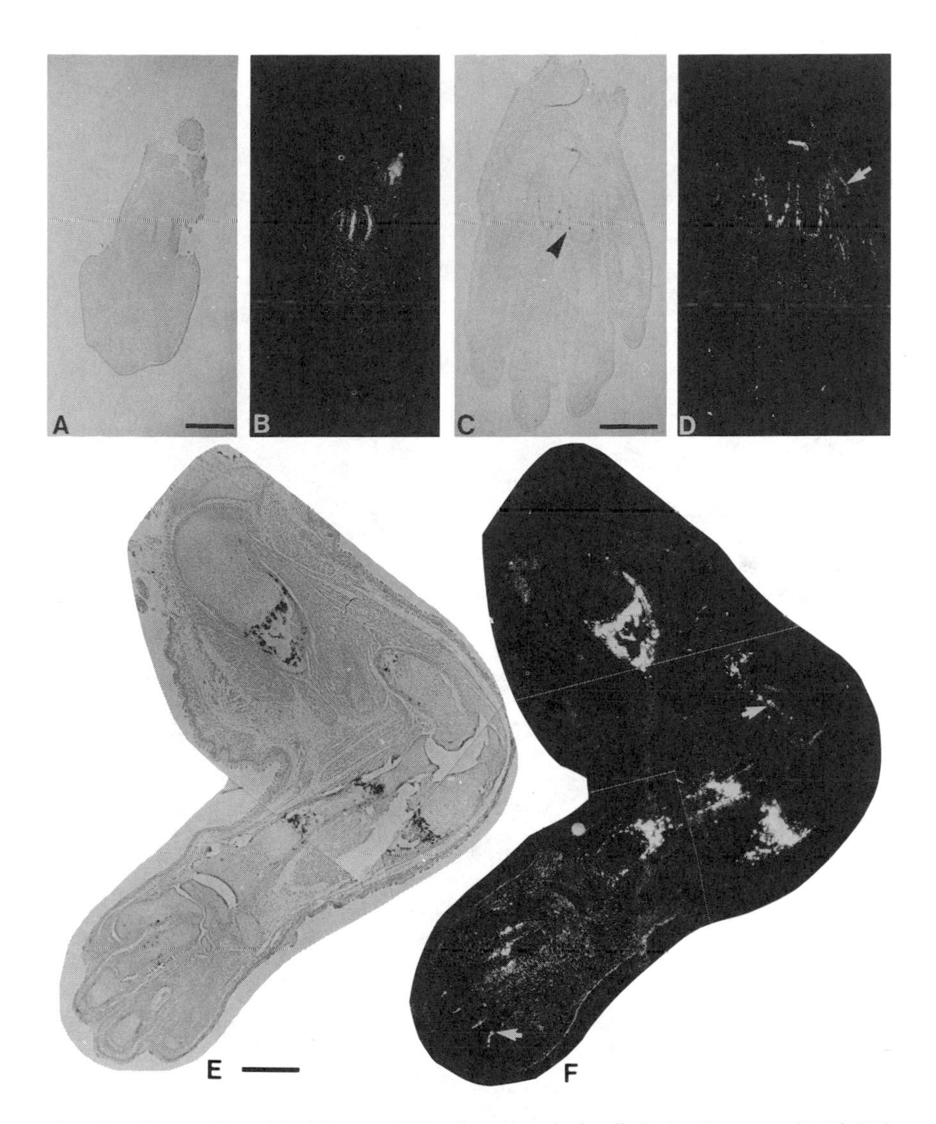

FIGURE 1. Expression of 92-kDa type IV collagenase during limb development. (**A, B**) E12 forelimb. Mesenchymal cells around the radius and ulna show expression of 92-kDa type IV collagenase. (**C, D**) In the E15 forelimb, expression is detectable around the cartilage (*white arrow*). Single cells in the perichondrium of the metacarpal cartilage as well as at the tips of the digits intensely express 92-kDa type IV collagenase. In the E16 forelimb (**E, F**), expression of 92-kDa type IV collagenase is very intense in the ossifying regions of the long bones. Also the perichondrium and periosteum show some expression (*arrows*). **A,C,E:** bright field; **B,D,F:** dark field. Bar = 400 μm.

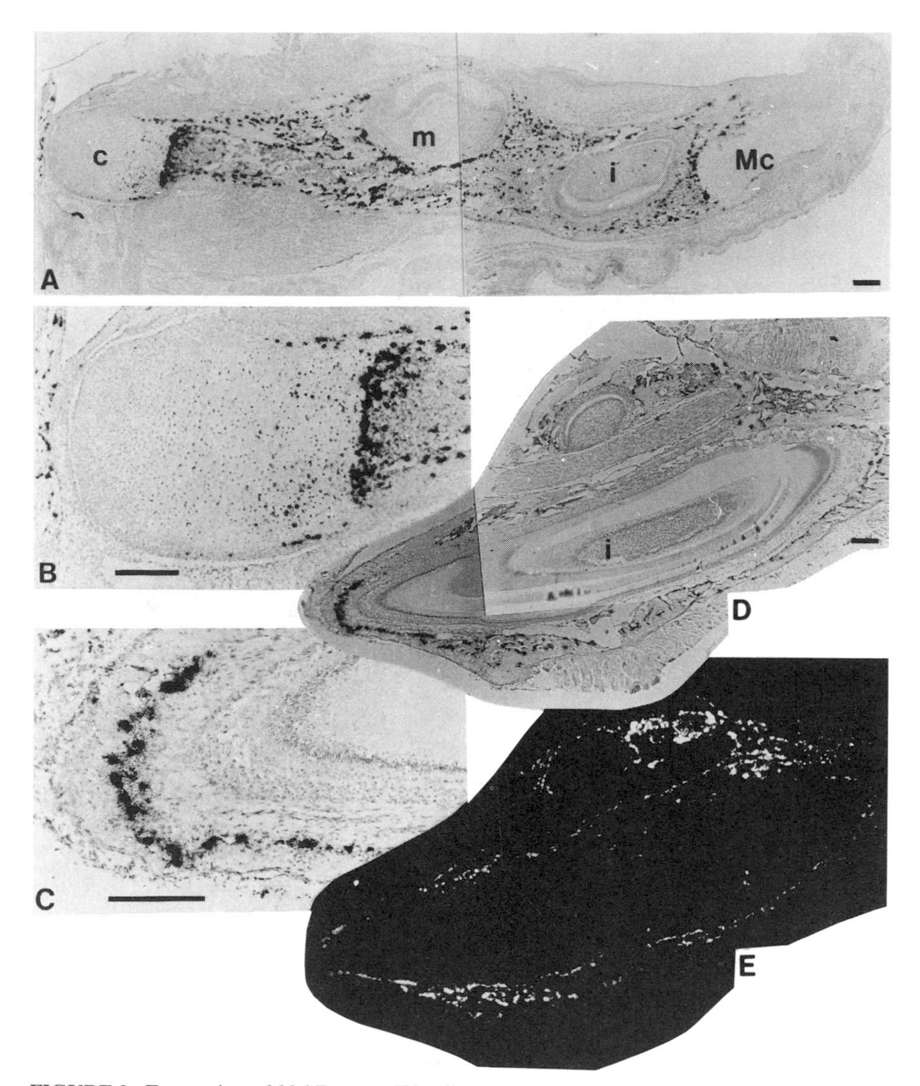

FIGURE 2. Expression of 92-kDa type IV collagenase during fetal and postnatal development of the jaw. (**A**) In the E18 mandible, 92-kDa type IV collagenase gene expression is intense throughout the developing bone of the mandible. The developing teeth are completely negative. (**B**) Higher magnification of the condylar process. (**D, E**) Eleven-day postnatal mandible with incisor tooth. Expression in the bone is most intense at the resorptive surface around the growing tooth germ. (**C**) Magnification of the tip of the incisor. The area of bone through which the incisor is about to erupt shows especially intense 92-kDa type IV collagenase expression. c = condylar process; i = incisor; m = molar; Mc = Meckel's cartilage. **A,B,C,D:** bright field; **E:** dark field of **D.** Bar = 200 μm.

major role in basement membrane degradation, but rather is utilized for the turnover of bone matrix, possibly as a gelatinase required for the removal of denatured collagen fragments (gelatin) generated by other proteases (FIGS. 1 and 2).

REFERENCES

1. REPONEN, P., C. SAHLBERG, C. MUNAUT, I. THESLEFF & K. TRYGGVASON. 1994. J. Cell Biol. **124**:1091–1102.
2. REPONEN, P., C. SAHLBERG, P. HUHTALA, T. HURSKAINEN, I. THESLEFF & K. TRYGGVASON. 1992. J. Biol. Chem. **267**: 7856–7862.

Matrix Metalloproteinases, Their Endogenous Inhibitors, and Microbial Activators in Gingival Crevicular Fluid and Saliva of HIV(+)-Subjects[a]

T. SALO,[b] T. SORSA,[c,d] A. LAUHIO,[e] Y. T. KONTTINEN,[f]
A. AINAMO,[g] L. KJELDSEN,[h] N. BORREGAARD,[i]
H. RANTA,[j] AND J. LÄHDEVIRTA[e]

[b]*Department of Oral Surgery & Pathology*
University of Oulu
Oulu, Finland

Departments of Anatomy,[f] *Periodontology,*[c]
Prosthetic Dentistry,[g] *and Forensic Medicine*[j]
University of Helsinki
Helsinki, Finland

[e]*Aurora Hospital*
Helsinki, Finland

[h]*Granulocyte Research Laboratory*
Rigshospitalet
Copenhagen, Denmark

Human immunodeficiency virus (HIV) infection has been associated with gingivitis and periodontitis in the oral cavity. In fact, two unique forms of periodontal disease have been described in HIV-infected patients. The first, HIV gingivitis, is a distinct form of gingival inflammation characterized by erythema and edema in the attached gingiva with a potential to extend beyond the mucogingival junction into the alveolar mucosa. The second, HIV periodontitis, is a particularly virulent form of periodontitis characterized by severe gingival inflammation, spontaneous gingival bleeding, and extremely rapid alveolar bone destruction.[1] Recently, HIV infection was found to stimulate T-cell invasiveness and synthesis of 92-kDa type IV collagenase/gelatinase (matrix metalloproteinase [MMP]-9).[2] Gingivitis and periodontitis are reflected in increased levels of proteolytic enzymes in oral inflammatory exudates, that is, gingival crevicular fluid (GCF) and saliva.[3] Among these proteinases, members of the MMP family are considered key initiators of extracellular matrix degradation associated with periodontal and other oral diseases.[3] In this report, we characterized by functional and immunologic means the matrix metalloproteinases and their inhibitors present in oral inflammatory exudates (GCF and saliva) of HIV(+)-subjects. We employed Western blotting to detect the presence of *Treponema*

[a]This work was supported by the Academy of Finland, Duodecim Foundation, Arthritis Foundation in Finland, Astra Foundation, the ITI-Straumann Foundation, and Yrjö Jahnsson Foundation.

[d]Address correspondence to: Department of Periodontology, Institute of Dentistry, University of Helsinki, P.O. Box 41 (Mannerheimintie 172), FIN-00014 Helsinki, Finland.

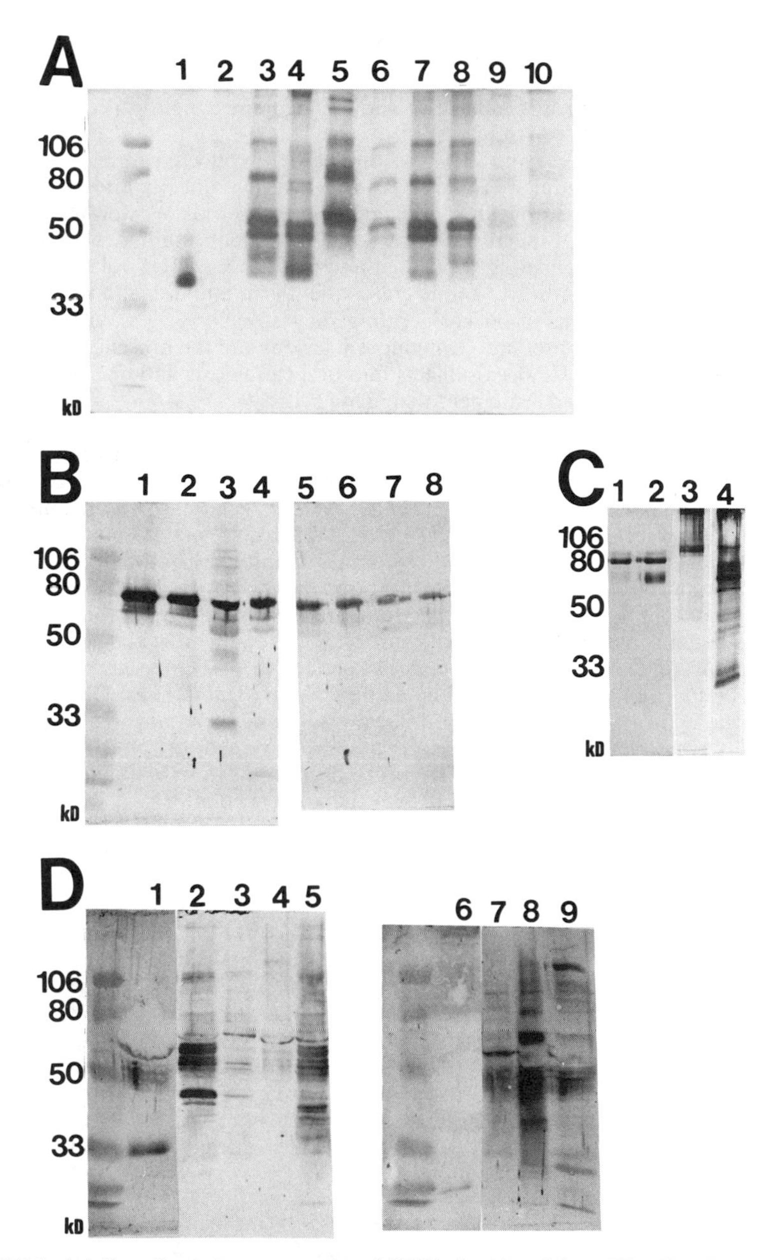

FIGURE 1. (**A**) Type I gelatin zymography of HIV(+)-subjects' (n = 10) saliva; **lanes 1–10.** Fluorescently labeled type I gelatin was used as substrate. (**B**) Western blot of HIV(+)-patients' saliva using antineutrophil collagenase (MMP-8) antibody; **lanes 1–4.** Western-blot analysis of HIV(+)-patients' saliva using anti-*Treponema denticola* protease-antibody; **lanes 5–8.** (**C**) Western blot analysis of activation of neutrophil procollagenase (proMMP-8, *lane 1*) and gelatinase (proMMP-9, *lane 3*) by *T. denticola* protease; activated/fragmented polymorphonuclear leukocyte MMPs can be seen in *lanes 2 and 4,* respectively. (**D**) Western blot analysis of TIMP-1 (*lanes 1–5*) and TIMP-2 (*lanes 6–9*) in HIV(+)-patients' saliva. Molecular weight markers are indicated.

denticola proteinase (which is capable of activating latent human procollagenases[4]) in GCF and saliva of HIV(+)-subjects.

Gingival crevicular fluid and salivary samples were collected from 10 HIV(+)-subjects, patients with adult periodontitis, and healthy controls and assayed for collagenase and gelatinase by functional activity assays and Western blotting as previously described.[3,4] Of the 10 HIV-antibody(+)-patients studied, 5 were classified as having lymphoadenopathy syndrome and 5 as having AIDS-related complex at the time of GCF and salivary sample collection. Periodontal health was assessed by standard measurements of pocket depth, radiographic bone loss, visible plaque index, and bleeding after probing. Oral mucosal lesions and the presence of *Candida* were also noted; 2 of 10 HIV(+)-subjects had oral candidosis and one patient with AIDS-related complex had acute necrotizing gingivitis.

Regardless of the clinically assessed periodontal status, each of the HIV(+)-subjects' saliva contained gelatinase activity that could not be further activated by 1 mM aminophenylmercuric acetate (APMA), indicating endogenous activation of gelatinase; type I gelatin zymography revealed multiple gelatinases, with a predominance of low molecular weight species (FIG. 1A). Salivary gelatinase activity of the HIV(+)-subjects was not activated but rather was inhibited by 1 mM dithiothreitol, a reductant known to efficiently activate proteases released by oral bacteria.[5] The low molecular weight gelatinases could be detected by Western blotting using an antihuman neutrophil gelatinase specific antibody. Latent 75-kDa and active 65-kDa human neutrophil collagenase could also be detected in GCF and saliva of HIV(+)-subjects by specific antibody against human neutrophil MMP-8 (FIG. 1B, lanes 1–4). Each of the HIV(+)-subjects' GCF and saliva was found to contain an 80–90-kDa *T. denticola* chymotrypsin-like proteinase as assessed by Western blotting (FIG. 1B, lanes 5–8). *In vitro,* the *T. denticola* proteinase converted both 75-kDa PMN proMMP-8 and 92-kDa PMN proMMP-9 to their active counterparts with apparent molecular weights of 65 and 82 kDa respectively (FIG. 1C, lanes 1–4). In addition, 92-kDa MMP-9 was further fragmented to lower molecular weight gelatinase/MMP-9 species. Also, high molecular weight forms (20–100 kDa) of tissue inhibitors of matrix metalloproteinases (TIMP-1 and TIMP-2) were detected in saliva of HIV(+)-subjects by Western blotting (FIG. 1D).

In HIV-related gingivitis and periodontitis, polymorphonuclear leukocyte-derived MMPs may well be involved in periodontal tissue destruction, because these MMPs were found to exist in endogenously/proteolytically activated forms in GCF and saliva of HIV(+)-patients. In addition, these oral inflammatory exudates (GCF and saliva) contained *T. denticola* proteinase capable of converting the polymorphonuclear leukocyte proMMPs to their active counterparts *in vitro.* Thus, periodontal inflammation associated with HIV infection may involve a cascade of activation of polymorphonuclear leukocyte proMMPs, triggered by *T. denticola* proteinase.

REFERENCES

1. WINKLER, J. R. & P. A. MURRAY. 1987. Calif. Dent. Assoc. J. **15:** 20–27.
2. WEEKS, B. S., M. E. KLOTMAN, E. HOLLOWAY, W. G. STETLER-STEVENSON, H. K. KLEIMAN & P. E. KLOTMAN. 1993. AIDS Res. Hum. Retrovir. **9:** 513–518.
3. SORSA, T., K. SUOMALAINEN & V.-J. UITTO. 1990. Arch. Oral. Biol. **36S:** 195–196.
4. SORSA, T., T. INGMAN, K. SUOMALAINEN, M. HAAPASALO, Y. T. KONTTINEN, O. LINDY, H. SAARI & V.-J. UITTO. 1992. Infect. Immun. **60:** 4491–4495.
5. SORSA, T., V.-J. UITTO, K. SUOMALAINEN, H. TURTO & S. LINDY. 1987. J. Periodontal Res. **22:** 375–380.

In Vitro Regulation of Stromelysin-1 in Human Endometrial Stromal Cells

FREDERICK SCHATZ, CSABA PAPP, ERNO TOTH-PAL,
STEFAN AIGNER, VIRGINIA HAUSKNECHT,
GRACIELA KRIKUN, RONALD E. GORDON,
RICHARD BERKOWITZ, AND CHARLES J. LOCKWOOD

Department of Obstetrics, Gynecology and Reproductive Science
The Mount Sinai School of Medicine
One Gustave L. Levy Place
New York, New York 10029

Progesterone acts with estradiol to initiate the conversion of stromal cells to decidual cells around the spiral arterioles of the midluteal phase human endometrium, with decidual cells subsequently spreading throughout the late luteal phase and early pregnant endometrium. Thus, decidual cells are positioned to interact with trophoblast cells as they invade the endometrium and its vasculature.[1] Stromelysin-1 (matrix metalloproteinase-3, MMP-3) is well suited to mediate extracellular matrix (ECM) degradation during decidual cell–trophoblast cell interactions because it uses a broad range of ECM substrates and is capable of activating other metalloproteinase zymogens.[2] The existence of a link between MMP-3 expression and decidualization was investigated in stromal cell monolayers from cycling human endometrium. Progestins regulate the expression of several decidualization markers in these cultures. By contrast, estradiol, while ineffective alone, augments progestin-mediated effects (reviewed in ref. 3), presumably by elevating progesterone receptor levels.

RESULTS AND DISCUSSION

Immunoblot analysis for the presence of MMP-3 was performed on the conditioned medium of stromal cell monolayers obtained from consecutive 3-day incubations with 0.1% ethanol (vehicle control), estradiol (E), medroxyprogesterone acetate (P), and E + P. (See ref. 3 for details of cell isolation and culture.) Medium from control incubations contained a single band of about 50,000 MW. FIGURE 1 shows the effects of added steroids on the MMP-3 band in five experiments. Consistent with the differential effects of ovarian steroids on various endpoints of decidualization,[3] the cells were refractory to E, whereas P produced a time-dependent reduction in levels of secreted MMP-3. Moreover, despite the lack of response to E alone, the inhibitory effects of P were enhanced in cultures incubated with E + P.

On Northern analysis cultured stromal cells contained mRNA of similar size to the reported MMP-3 message (approximately 2.1 kb). FIGURE 2 shows that exogenous steroids altered steady-state mRNA levels that paralleled their effects on secreted MMP-3 (i.e., E was ineffective alone, whereas E + P markedly reduced MMP-3 mRNA levels). Adding E together with concentrations of P that mimic

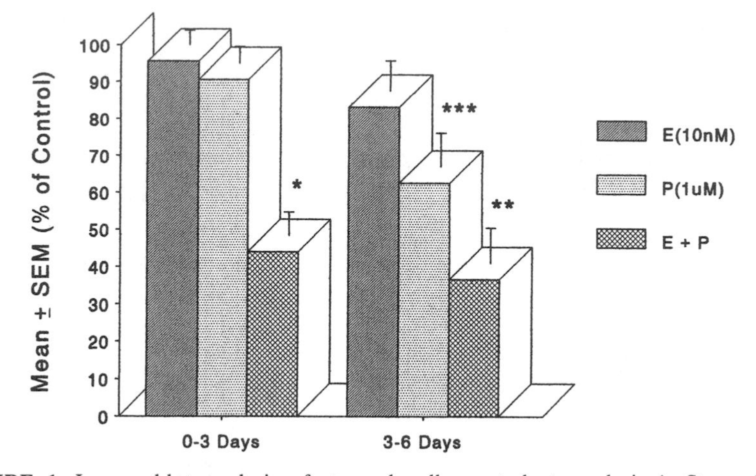

FIGURE 1. Immunoblot analysis of stromal cell-secreted stromelysin-1. Stromal cell-conditioned medium was subjected to 7% PAGE and electrotransferred onto nitrocellulose. After blocking nonspecific sites, blots were incubated with a specific rabbit polyclonal anti–MMP-3 antibody (Dr. M. Lark, Merck Research Laboratories, Rahway, NJ), then exposed to [^{125}I]-protein A. After autoradiography, the relative effects of the various treatments were assessed by densitometry of the MMP-3 zone for five separate experiments. Statistical significance was determined by one-way ANOVA. *Control vs E + P, $p < 0.005$; **control vs E + P, $p < 0.003$; ***control vs P, $p < 0.05$.

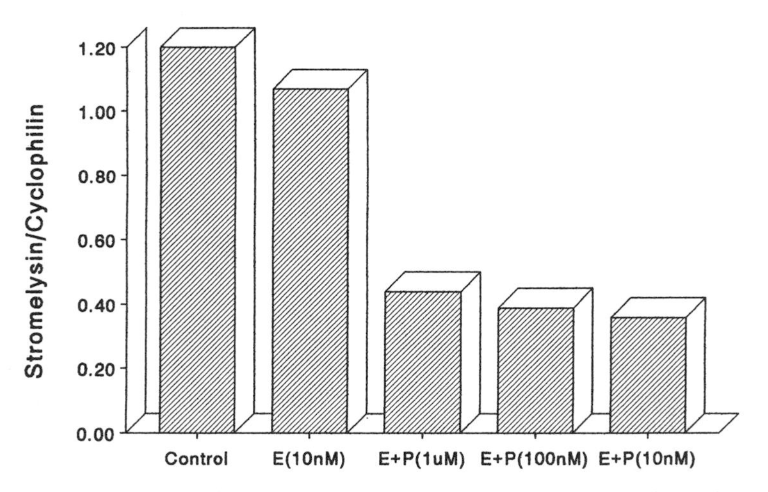

FIGURE 2. Northern analysis of stromelysin-1 mRNA levels in human endometrial stromal cells. After 3 days of experimental incubation, total RNA was extracted from stromal cell cultures by guanidinium thiocyanate/chloroform, and equal amounts along with molecular weight RNA standards were separated on a 1% agarose:2.2 M formaldehyde gel. Following transfer to a Zeta Probe nylon membrane, levels of stromelysin mRNA were detected with a probe (Dr. N. Hutchinson, Merck Research Laboratories) labeled with ^{32}P-dCTP by random priming. After hybridization and autoradiography, signals were evaluated by densitometry. To standardize RNA loads, membranes were stripped and reprobed with a cDNA for the cyclophillin housekeeping gene. *Y axis* shows the relative intensity of MMP-3 mRNA normalized to the cyclophillin load for each sample.

physiologic progesterone levels (10 nM of luteal phase–1 μM of pregnancy) reduced MMP-3 mRNA levels by about two-thirds.

Decidualization involves transformation of the interstitial-type ECM surrounding stromal cells to the ECM surrounding decidual cells, which is rich in basal laminar components. This ECM is thought to limit trophoblastic invasion of the endometrium.[1] The reduction in MMP-3 expression that accompanies *in vitro* decidualization is expected to promote this ECM transformation in view of the wide range of ECM substrates utilized by MMP-3 and its capability of activating MMP zymogens.[2] Moreover, inasmuch as activation of the 92-kDa gelatinase/type IV collagenase, a crucial mediator of trophoblast invasiveness, is MMP-3 dependent,[4] reduced decidual MMP-3 production could help to limit trophoblast invasion.

REFERENCES

1. BELL, S. C. 1990. Decidualization and relevance to menstruation. *In* Contraception and Mechanisms of Endometrial Bleeding. C. d'Arcangues, I. S. Fraser & J. R. Newton, Eds.: 187–212. Cambridge University Press. Cambridge.
2. WOESSNER, J. F. 1991. Matrix metalloproteinases and their inhibitors in connective tissue remodeling. FASEB J. **5:** 2145–2154.
3. SCHATZ, F. & C. L. LOCKWOOD. 1993. Progestin regulation of plasminogen activator inhibitor type 1 in primary cultures of endometrial stromal and decidual cells. J. Clin. Endocrinol. & Metab. **77:** 621–625.
4. OGATA, Y., J. J. ENGHILD & H. NAGASE. 1992. Matrix metalloproteinase 3 (stromelysin) activates the precursor for the human matrix metalloproteinase 9. J. Biol. Chem. **267:** 3581–3584.

Pharmacologic Influence on the Activity of Stromelysin from Bovine Articular Cartilage

J. STEINMEYER, S. DAUFELDT, AND D. A. KALBHEN

Department of Pharmacology and Toxicology
University of Bonn
Reuterstrasse 2b
53113 Bonn, Germany

The destruction of articular cartilage during osteoarthritis and rheumatoid arthritis is characterized by degradation and loss of collagen and proteoglycans. The metalloproteinase stromelysin (matrix metalloproteinase-3) is an important enzyme in proteolytic destruction of articular cartilage during chronic rheumatic diseases. Proteolytic degradation of cartilage is assumed to occur only if an imbalance exists between the proteinases and their natural inhibitors. For instance, in normal cartilage there is a sufficient quantity of the tissue inhibitor of metalloproteinases (TIMP) to counteract the metalloproteinases present in the tissue, whereas in osteoarthritis the TIMP levels do not increase to the same extent as do the proteinases. Therefore, it is suggested that in osteoarthritic cartilage the excess of proteinases over TIMP may contribute to cartilage breakdown during osteoarthritis. Compensating for the deficit in the natural occurring proteinase inhibitor TIMP to stop or slow down the proteolytic destruction of articular cartilage is of therapeutic significance. This may be accomplished by drugs that can turn off the activity of stromelysin and/or collagenase, can inhibit the synthesis and/or release of these catabolic enzymes, or can stimulate the synthesis of the endogenous inhibitor TIMP. Such drugs are expected to retard cartilage breakdown and thus may be a promising new therapy.

Our work has focused on the inhibitory potential of commercially available drugs on the neutral, Zn^{2+}-dependent proteoglycanolytic activity found in the media of interleukin-1 (IL-1)–treated articular cartilage explants, which was taken as a measure of the activity of stromelysin present. The activity of stromelysin was determined by the ^{3}H-proteoglycan bead assay of Nagase and Woessner[1] as modified by Azzo and Woessner.[2] To prepare an enzyme solution containing stromelysin, articular cartilage from the metacarpophalangeal joints of 18–24-month-old steers was removed in full thickness under sterile conditions and incubated for 4 days with DMEM containing 50 U IL-1α. After 4 days in culture, media were collected and combined. Stromelysin was activated by the addition of 1.0 mM 4-aminophenylmercuric acetate (APMA) for 6 h at 37 °C followed by dialysis of the media to remove APMA. The resulting APMA-free medium was used as an enzyme solution containing activated stromelysin. For stromelysin assay, enzyme solution, buffer containing 300 mM 3,3-dimethylglutaric acid, pH 7.0, 0.4 M NaCl, 20 mM $CaCL_2$, 0.1% Brij-35®, 0.04% NaN_3, solution with or without drug, proteinase inhibitor cocktail containing 2 mM PMSF, 10 mM *N*-ethylmaleimide, and 100 μM 6-aminocaproic acid were mixed, incubated for 30 min, and then transferred to a vial containing 2 mg polyacrylamide gel beads in which ^{3}H-proteoglycan monomers were entrapped. This mixture was incubated for 18 h at 37 °C. The activity of stromelysin was calculated from the release of tritium per milligram of proteoglycan-containing beads. The following 43 drugs were tested at a concentration of 10^{-4} M (or lower) for their inhibitory potential on the activity of stromelysin: (1) Disease-modifying drugs: auranofin, aurothiopolypeptide, chloroquine, and D-penicillamine; (2) possible chondroprotective drugs: glycosaminoglycan polysulfate, sodium pentosan polysulfate, L-cysteine, ademethionine, lyophilized bone marrow and cartilage extract (Rumalon®), glycosaminoglycan-peptide complex (DAK 16), and glucosamine sulfate; (3) nonsteroidal anti-inflammatory drugs: tiaprofenic acid, acetylsalicylic acid, phenylbutazone, diclofenac-Na, indomethacin, ketoprofen, naproxen, flufenaminic acid, and piroxicam; (4) glucocorticoids: dexamethasone; (5) ACE inhibitors: captopril, cilazapril, and lisinopril; (6) antibiotics: minocycline, doxycycline, tetracycline, and ciprofloxacine; (7) diverse drugs: methotrexate, eglin C, theophylline, nifedipine, aescine, nystatin, clonidine-HCl, propranolol-HCl, neostigmin bromide, tranylcypromine, etidron acid, prazosin-HCl, allopurinol, epimestrol, and omeprazol-Na. Only lyophilized Rumalon® (46 mg/mL assay: 83%; 4.6 mg/mL assay: 19%; 0.46 mg/mL assay: 2%), DAK 16 (10^{-4} M: 46%; 10^{-5} M: 12%; 10^{-6} M: 0%), minocycline (10^{-4} M: 26%; 10^{-5} M: 14%; 10^{-6} M: 0%), doxycycline (10^{-4} M: 14%; 10^{-5} M: 10%; 10^{-6} M: 0%), and tetracycline (10^{-4} M: 31%; 10^{-5} M: 14%; 10^{-6} M: 0%) dose dependently inhibited proteoglycanolytic activity. None of the other drugs tested at a concentration of 10^{-4} M displayed any significant inhibition.

We observed marked inhibition of stromelysin, tested as a neutral Zn^{2+}-dependent proteoglycanase, with only a few drugs tested at relatively high concentrations. To our knowledge only limited information about the *in vivo* concentrations of these drugs within cartilage is available. Taking into consideration the already known serum and/or synovial fluid levels of these drugs and assuming that they are not accumulating within cartilage, we conclude that the *in vitro* inhibition of bovine articular cartilage proteoglycanolytic activity may not be of therapeutic relevance.

REFERENCES

1. NAGASE, H. & J. F. WOESSNER, JR. 1980. An improved assay for proteases and polysaccharidases employing a cartilage proteoglycan substrate entrapped in polyacrylamide particles. Anal. Biochem. **107:** 385–392.
2. AZZO W. & J. F. WOESSNER, JR. 1986. Purification and characterization of an acid metalloproteinase from human articular cartilage. J. Biol. Chem. **261:** 5434–5441.

Localization and Characterization of Matrix Metalloproteinase-9 in Experimental Rat Adjuvant Arthritis

MONICA STEIN-PICARELLA, DIANE AHRENS,
CAROL MASE, HARRY GOLDEN,
AND MICHAEL J. NIEDBALA[a]

Institute of Bone and Joint Disorders
Miles Research Center, Miles Inc.
West Haven, Connecticut 06516

Rat adjuvant arthritis (AA), an experimental model for chronic inflammation associated with human rheumatoid arthritis (RA), has been used extensively in the discovery of potential antirheumatic agents by the pharmaceutical industry. Although no "ideal" animal model mimics human RA at this time, similarities between human disease and AA include (1) sensitivity of this animal model to antiarthritic agents, (2) edema and swelling of multiple joints, (3) synovitis which leads to progressive and destructive joint disease, and (4) an intense mononuclear immune response.[1,2]

Involvement of matrix metalloproteinases (MMPs) in the pathogenesis of human arthritis was previously reported.[3] Matrix metalloproteinases are a family of Zn^{2+}-containing neutral endopeptidases expressed by epithelial, mesenchymal, and hematopoietic cells.[4] This family of enzymes can be divided into three classes, collagenases, stromelysins, and gelatinases, each of which displays distinct substrate specificity and cellular expression. Gelatinase B (matrix metalloproteinase-9, MMP-9) has been implicated in the cellular turnover and invasion of both hard and soft tissue extracellular matrices by a variety of cells.[5–7]

In this report, we investigated the immunolocalization and characterization of MMP-9 during progression of the inflammatory reaction in experimental rat AA. Mouse monoclonal antisera which selectively recognize MMP-9 were used to study the distribution of MMP-9 in the pathogenesis of rat AA. A comparison of normal and AA joint tissue was used to assess the synovium, cartilage, and bone at days 2, 5, 9, 12, and 16 for MMP-9 antigen expression. Normal rat articular cartilage of the tibiotarsal joint displayed focal areas of MMP-9 immunoreactive chondrocytes. These cells were primarily located at the margin of the bone and may represent the need for lateral expansion in this area during growth. The remainder of the articular cartilage contained small clusters of MMP-9 immunoreactive cells at various depths within the cartilage.

At day 2 following adjuvant administration, infiltrating synovial neutrophils increased in number and stained positively for MMP-9. At day 9, synovial macrophages and fibroblasts also displayed MMP-9 antigen, and this expression was increased at days 9, 12, and 16. Cartilage expression of MMP-9 in the tibiotarsal joint of AA animals was limited to the area of joint capsule attachment at day 2 and

[a] Address correspondence to Michael J. Niedbala, Ph.D., Institute of Bone and Joint Disorders, Miles Inc., 400 Morgan Lane, West Haven, Connecticut 06516.

progressively increased at later times ($\geq$ day 12) to involve the upper half of the weight-bearing cartilage surface. Bone changes in MMP-9 expression were first detected at day 5 and included cambial layer staining of the periosteum. Positive staining of infiltrating osteoclasts was observed at days 9, 12, and 16 in AA. Biochemical analysis of rat paw homogenates demonstrated predominant 92-kDa gelatinolytic activity (specifically inhibited by EDTA) which was evident at days 2 through 16 in AA based on gelatin zymography. In contrast, normal paw homogenates displayed only 72-kDa gelatinase and its activated/processed 62-kDa form. Plasma MMP-9 activity was elevated in AA rats at day 9 and progressively increased through day 16.

These findings indicate the presence of MMP-9 in rat AA and suggest its association with disease progression, implicating this enzyme in the abnormal connective tissue turnover and joint degeneration which is a characteristic feature of this chronic inflammatory model for arthritis.

REFERENCES

1. Pearson, C. M. 1963. Experimental joint disease: Observations on adjuvant induced arthritis. J. Chronic Dis. **16:** 863–874.
2. Billingham, M. E. J. 1983. Models of arthritis and the search for anti-arthritic drugs. Pharmacol. Ther. **21:** 389–428.
3. Murphy, G. & M. Hembry. 1992. Proteinases in rheumatoid arthritis. J. Rheumatol. **19:** 61–64.
4. Woessner, J. F. 1991. Matrix metalloproteinases and their inhibitors in connective tissue remodelling. FASEB J. **5:** 2145–2154.
5. Hibbs, M. S., K. A. Hasty, A. H. Kang & C. L. Mainardi. 1984. Biochemical and immunologic characterization of the secreted forms of human neutrophil gelatinase. J. Biol. Chem. **260:** 2493–2500.
6. Unemori, E. N., M. S. Hibbs & E. P. Amento. 1991. Constitutive expression of a 92-kD gelatinase (type V) collagenase by rheumatoid synovial fibroblasts and its induction in normal human fibroblasts by inflammatory cytokines. J. Clin. Invest. **88:** 1656–1662.
7. Welgus, H., E. J. Campbell, J. D. Cury, A. Z. Eisen, R. M. Senior, S. M. Wilhelm & G. I. Goldberg. 1990. Neutral metalloproteinases produced by human mononuclear phagocytes: Enzyme profile, regulation, and expression during cellular development. J. Clin. Invest. **86:** 1496–1502.
8. Mohtai, M., R. L. Smith, D. J. Schurman, Y. Tsuji, F. Torti, N. I. Hutchinson, W. G. Stetler-Stevenson & G. I. Goldberg. 1993. Expression of 92-kD type IV collagenase/gelatinase (gelatinase B) in osteoarthritic cartilage and its induction in normal human articular cartilage by interleukin 1. J. Clin Invest. **92:** 179–185.

Gelatinases/Type IV Collagenases in Jaw Cyst Expansion[a]

O. TERONEN,[b] T. SALO, Y. T. KONTTINEN, B. RIFKIN, A. VERNILLO, N. S. RAMAMURTHY, L. KJELDSEN, N. BORREGAARD, C. P. SOMMERHOFF, AND T. SORSA

Departments of Oral and Maxillofacial Surgery, Periodontology, and Anatomy
University of Helsinki
Helsinki, Finland

Department of Oral Surgery
University of Oulu
Oulu, Finland

Department of Oral Medicine
New York University School of Dentistry
New York, New York 10010

Granulocyte Research Laboratory
Rijkshospitalet
Copenhagen, Denmark

Department of Oral Biology and Pathology
State University of New York at Stony Brook
Stony Brook, New York 11794

Department of Clinical Biochemistry
Ludwig-Maximillians University
Munich, Germany

Jaw cysts are destructive lesions of the facial skeleton that occur frequently. The expansile growth of cysts has been extensively studied, but the molecular mechanism(s) responsible for cyst expansion is still unclear. Interstitial collagenase can initiate the destruction of extracellular collagenous matrix (ECM), and collagenolytic activity has been demonstrated in radicular cyst and keratocyst homogenates.[1–4] Tissue degradation also involves other members of the matrix metalloproteinase (MMP) family, namely, the gelatinases/type IV collagenases. In this study we characterize the molecular forms of gelatinases/type IV collagenases present in radicular cysts utilizing zymographic analysis and Western blotting. In addition, we studied the effects of different compounds with regard to regulation—activation and inhibition—on jaw cyst gelatinases as well as 92-kDa polymorphonuclear leukocyte

[a] This work was supported by grants from the Academy of Finland, Finnish Dental Society, ITI Straumann Foundation, Pharmacal Research Foundation, Arthritis Foundation in Finland, and Yrjö Jahnsson Foundation.

[b] Address correspondence to Dr. Olli Teronen, Department of Oral and Maxillofacial Surgery, Institute of Dentistry, P. O. Box 41, SF-00014 University of Helsinki, Helsinki, Finland.

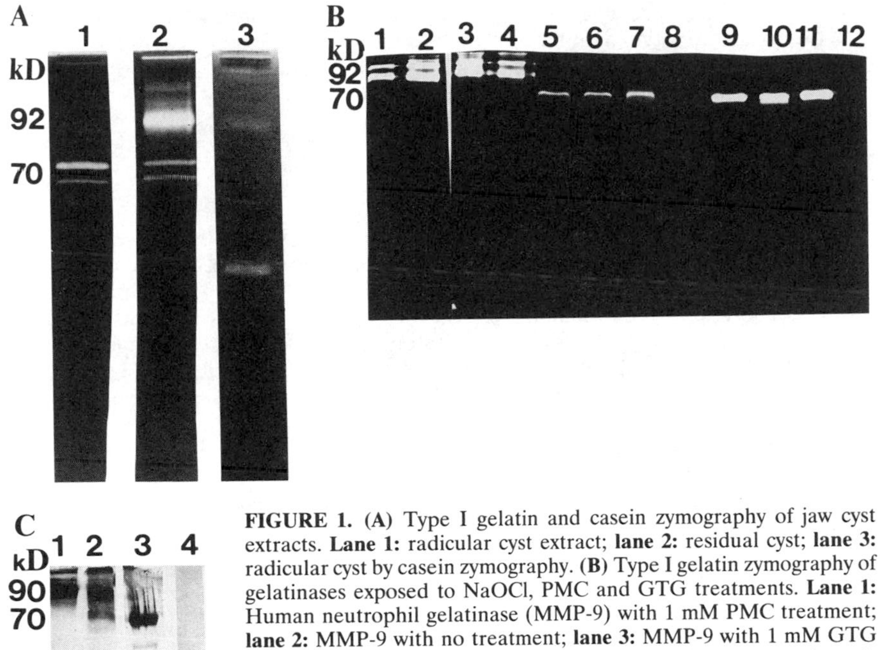

FIGURE 1. (**A**) Type I gelatin and casein zymography of jaw cyst extracts. **Lane 1:** radicular cyst extract; **lane 2:** residual cyst; **lane 3:** radicular cyst by casein zymography. (**B**) Type I gelatin zymography of gelatinases exposed to NaOCl, PMC and GTG treatments. **Lane 1:** Human neutrophil gelatinase (MMP-9) with 1 mM PMC treatment; **lane 2:** MMP-9 with no treatment; **lane 3:** MMP-9 with 1 mM GTG treatment; **lane 4:** MMP-9 with 1 mM NaOCl treatment; **lane 5:** UMR-rat osteoblast with no treatment; **lane 6:** UMR-rat osteoblast with 1 mM PMC treatment; **lane 7:** UMR-rat osteoblast with 1 mM GTG treatment; **lane 8:** UMR-rat osteoblast with 1 mM NaOCl treatment; **lane 9:** fibroblast supernatant with no treatment; **lane 10:** fibroblast supernatant with 1 mM PMC treatment; **lane 11:** fibroblast supernatant with 1 mM GTG treatment; **lane 12:** fibroblast supernatant with 1 mM NaOCl treatment. (**C**) Western-blot of neutrophil and jaw cyst extracts. **Lane 1:** PMN extract analyzed by anti-human MMP-9; **lane 2:** radicular cyst extract analyzed by anti-human MMP-9; **lane 3:** radicular cyst extract analyzed by anti-human MMP-2; **lane 4:** radicular cyst extract analyzed by anti-human tryptase. Mobilities of molecular weight markers are indicated.

type MMP-9 and 72-kDa neutral gelatinase from cultured UMR-rat osteoblasts and human gingival fibroblasts.

MATERIAL AND METHODS

Specimens used in this experiment were obtained during surgical removal of 14 jaw cysts from patients treated at the Department of Oral and Maxillofacial Surgery, University of Helsinki, Finland, later diagnosed as radicular, follicular, residual cysts and keratocysts. Both cyst wall tissue and fluid were collected and stored frozen until analyzed. Neutral salt extracts of tissue specimens were prepared as described[3,4] for protease analysis.

Functional activities of gelatinases in cyst extracts and fluids were measured with gelatin and casein zymographies as well as by using radioactive type I gelatin as substrate. Immunologic characterization of MMP-2 and MMP-9 was performed by Western blotting using specific antibodies.

RESULTS

Zymographic analysis revealed both 92-kDa gelatinase (MMP-9) and 72-kDa gelatinase (MMP-2) in cyst wall tissue extracts, the latter in small amounts (FIG. 1A). MMP-2 was also detected in 65–68-kDa molecular form, which represents the proteolytically activated form of the enzyme. The presence of MMP-9 and MMP-2 in wall tissue extracts was confirmed by Western blot analysis utilizing specific anti-human MMP-2 and MMP-9 antibodies (FIG. 1C). The antibodies labeled the MMP-9s from both cultured oral epithelial cells and polymorphonuclear neutrophils (PMNs) used as controls, leaving us uncertain as to the specific cellular origins of MMP-9 in these cysts. In addition, mast cell tryptase was detected in jaw cyst wall tissue extracts by Western blotting. Phenylmercuric chloride (PMC) and gold thioglucose (GTG) activated all types of gelatinases studied. Sodium hypochlorite (NaOCl) activated MMP-9 (FIG. 1B), whereas 100–200 μM concentrations of NaOCl clearly inhibited MMP-2. No inhibition of MMP-9 was detected even with larger doses of NaOCl.

CONCLUSIONS

Jaw cyst enlargement apparently involves MMP-2 and MMP-9 gelatinases/type IV collagenases to complete the ECM degradation initiated by interstitial collagenases[3,4] evidently from multiple local and circulating cellular sources. Mast cell tryptase may act as an activator of proMMP-9 but apparently does not activate proMMP-2 during jaw cyst enlargement. In response to cysteine-switch reagents PMC, GTG, and NaOCl, MMP-9 was clearly activated by PMC, slightly by NaOCl and GTG, whereas MMP-2 was activated by PMC, not affected by GTG, and inhibited by NaOCl. MMPs from PMN sources can be activated and are also quite resistant to inactivation by oxidative agents, allowing them to be catalytically functional in molecular microenvironments generated by PMNs during inflammatory processes. Furthermore, MMP-2 appears to be rather specifically regulated, that is, inhibited by oxidant NaOCl. These findings may contribute, at least partially, to the molecular mechanisms of jaw cyst expansion.

REFERENCES

1. HARRIS, M., M. V. JENKINS, A. BENNETT & M. R. WILLS. 1973. Nature **245:** 213–214.
2. DONOFF, R. B., E. HARPER & W. C. CURALNICK. 1972. J. Oral Surg. **30:** 879–884.
3. SORSA, T., P. YLIPAAVALNIEMI, K. SUOMALAINEN, M. VAUHKONEN & S. LINDY. 1988. Med. Sci. Res. **16:** 1189–1190.
4. YLIPAAVALNIEMI, P. 1978. Proc. Finn. Dent. Soc. **74** (Suppl. 1).

Electrospray Mass Spectrometry Study of Metal Ions in Matrilysin

Evidence That Two Zincs and Two Calciums Are Required for Inhibitor Binding

ZHENGYU YUAN,[a,b] RONG FENG,[c]
ARLINDO CASTELHANO,[a]
AND ROLAND BILLEDEAU[a]

[a]*Syntex Research, Canada*
2100 Syntex Court
Mississauga, Ontario, Canada

[c]*National Research Council of Canada*
Biotechnology Research Institute
Montreal, Quebec, Canada

It is generally accepted that matrix metalloproteinases (MMPs) need Zn^{2+} and Ca^{2+} for catalytic activity and structural stability. Recently, a few studies focused on the structure and catalytic roles of these divalent cations in MMPs.[1-3] In this study, we used electrospray mass spectrometry (ES-MS) to investigate the stoichiometry of the metals bound to matrilysin, a member of the MMP family, and their roles in the enzyme function.

EXPERIMENT

Matrilysin was expressed and purified from mammalian cells as previously described.[4] The activated enzyme was subject to dialysis against 5 mM ammonium acetate (pH 7.0). All ES-MS analyses were performed on a quadruple mass spectrometer (API III LC/MS/MS system). Ammonium acetate solutions were adjusted to different pHs by acetic acid or ammonium hydroxide. The enzyme was diluted to 50 μM by ammonium acetate buffer at the desired pH and, when necessary, mixed with the inhibitor before ES-MS analysis.

RESULTS AND DISCUSSION

The molecular weight of matrilysin in 10% AcOH, pH 2.2, is 18,720 ± 3 daltons (Da) as measured by ES-MS (calculated value 18,720). This number changes to 18,924 ± 3 Da in 20 mM ammonium acetate, pH 7.0, as shown in FIGURE 1. The difference is equivalent to the binding of 2 zinc and 2 calcium ions to the enzyme, that is, En-Zn_2-Ca_2. Note the minor peak in the insert in FIGURE 1, which has a molecular weight of 39 Da less than that of the major peak. This minor peak may be due to the

[b]Current Address: Affymax Research Institute, 4001 Miranda Avenue, Palo Alto, California 94304.

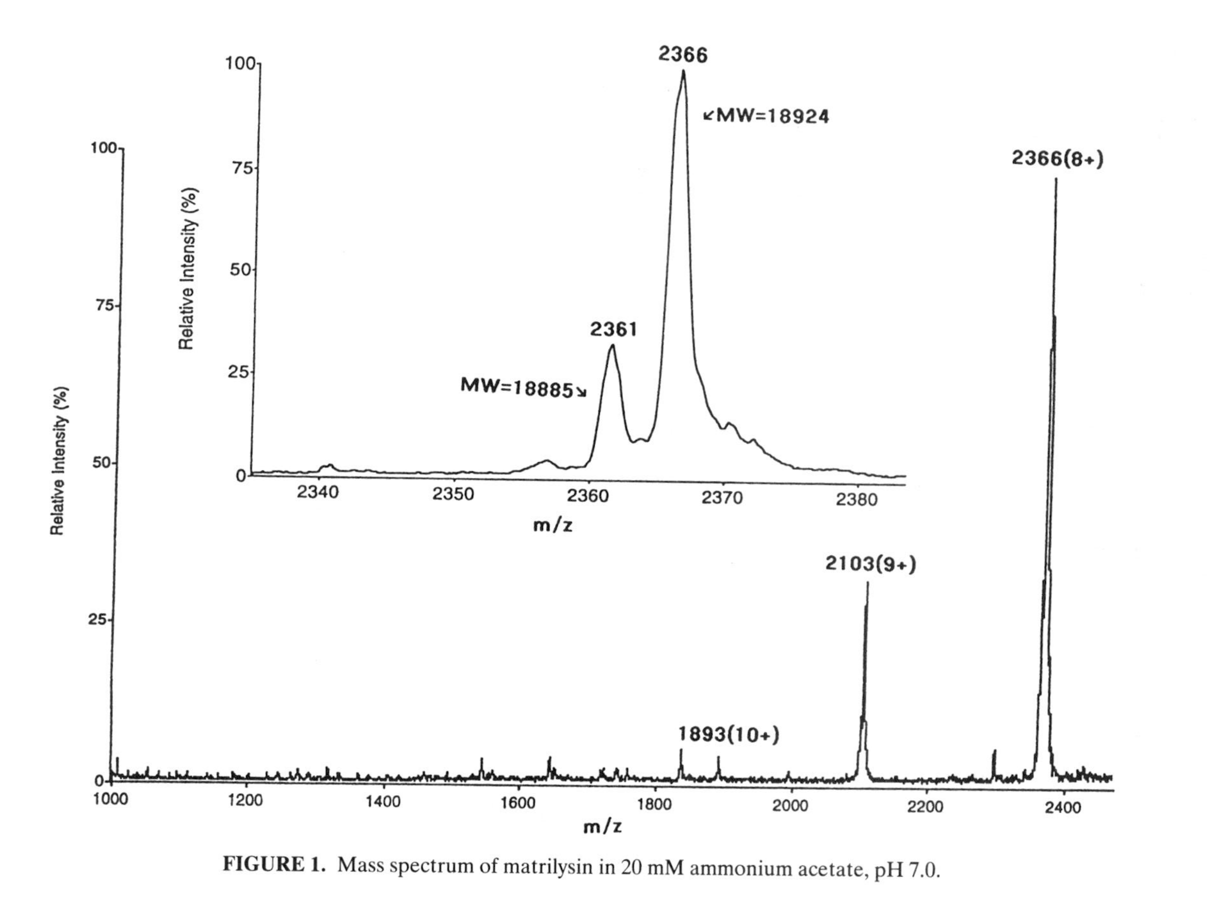

FIGURE 1. Mass spectrum of matrilysin in 20 mM ammonium acetate, pH 7.0.

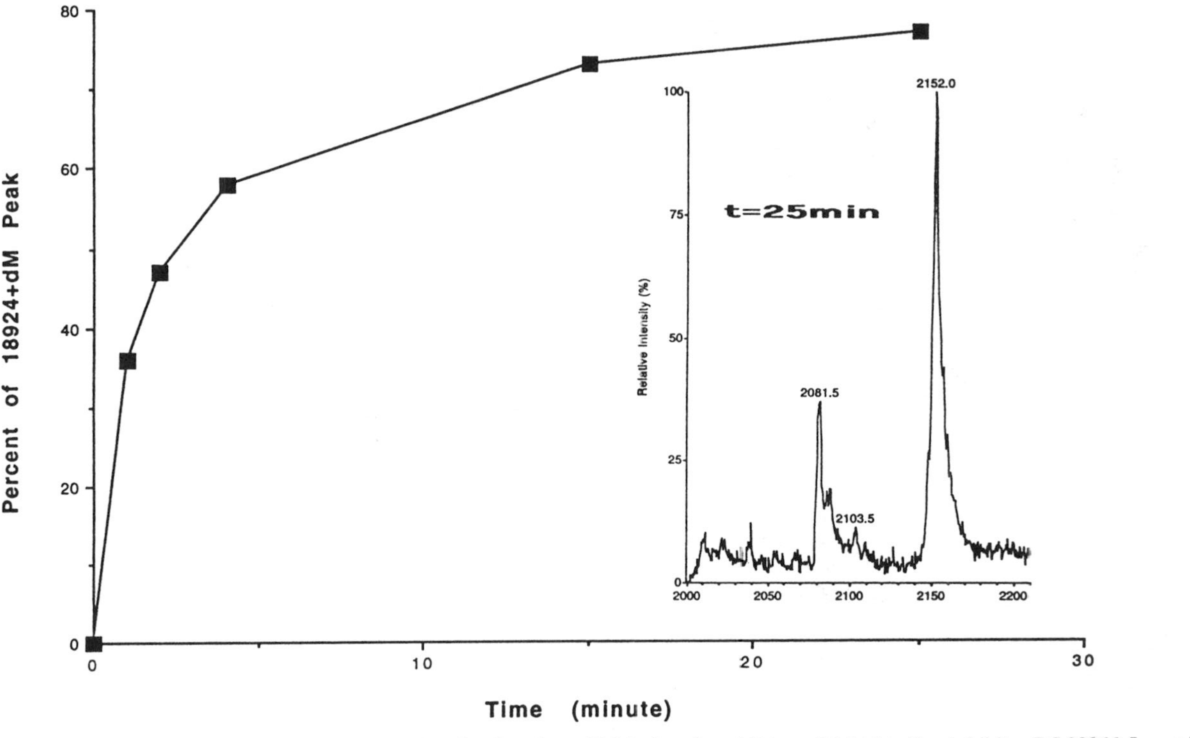

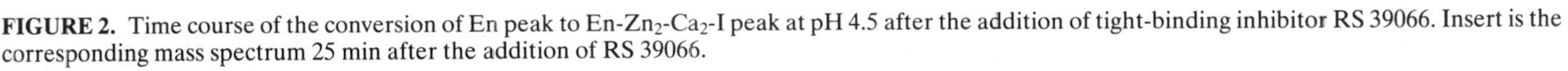

FIGURE 2. Time course of the conversion of En peak to En-Zn_2-Ca_2-I peak at pH 4.5 after the addition of tight-binding inhibitor RS 39066. Insert is the corresponding mass spectrum 25 min after the addition of RS 39066.

dissociation of 1 calcium ion from the En-Zn_2-Ca_2 complex. On this basis, we propose that matrilysin at neutral pH exists as a complex of En-Zn_2-Ca_2. Our results agree with previous observations.[1,2] The fact that these enzyme-bound calcium ions can be detected in our ES-MS analysis at neutral pH suggests that matrilysin has high affinity for calcium ions. In addition, we found that in the absence of calcium (but without EDTA or EGTA chelators) in a thiopeptide assay, extensively dialyzed matrilysin has the same activity as it does in the presence of the calcium.

Attempts were also made to detect the enzyme inhibitor complex. RS 39066 is a potent slow-binding inhibitor of matrilysin with a molecular weight of dM Da. At neutral pH, we detected a major peak of 18,924 + dM Da, which suggests that this peak represents the tight-binding enzyme-inhibitor complex (with all 4 metal ions). We also detected a minor peak which corresponds to RS 39066 bound to the enzyme complex with 2 zinc and 1 calcium ions. The binding of RS 39066 is noncovalent inasmuch as the 18,924 + dM Da peak could be converted to the 18,924 Da peak as we increased the "internal temperature" of the particles entering the ES-MS.

As we decreased the pH from 7.0 to 4.5, most of the metals were lost from the enzyme. However, we could still detect En-Zn and En-Zn_2-Ca_2 complexes. When RS 39066 was introduced to enzyme solution at pH 4.5, the 18,720-Da peak gradually changed to the 18,924 + dM peak with time. This suggests that RS 39066 quickly bound to the enzyme fraction which has metals to form the En-Zn_2-Ca_2-I complex (I is tight-binding inhibitor RS 39066). This En-Zn_2-Ca_2-I peak increased with time and reached equilibrium after 25 min, as shown in FIGURE 2.

In our ES-MS analysis, we detected at least four enzyme-metal complexes: (a) En, (b) En-Zn, (c) En-Zn_2-Ca, and (d) En-Zn_2-Ca_2. Our results suggest that inhibitor RS 39066 will only bind to complex c and d, which indicates that En-Zn_2-Ca is the minimum structure for enzyme to have the correct conformation to bind its inhibitors. At pH 7.0, equilibrium favors complex c and d. At pH 4.5, equilibrium favors a and b, although there is still a small fraction of c and d. Because RS 39066 can only form a tight complex with c and d, the addition of RS 39066 slowly shifts the equilibrium to the corresponding inhibitor complexes with a and b.

REFERENCES

1. SALOWE, S. P., A. I. MARCY, G. C. CUCA, C. K. SMITH, I. E. KOPKA, W. K. HAGMANN & J. D. HERMES. 1992. Biochemistry **31:** 4535–4540.
2. HOUSELY, T. J., A. P. BAUMANN, I. D. BRAUN, G. DAVIS, P. K. SEPERACK & S. M. WILHELM. 1993. J. Biol. Chem. **268:** 4481–4487.
3. GOMIS-RUTH, F., L. F. KRESS & W. BODE. 1993. EMBO J. **12:** 4151–4157.
4. YUAN, Z., R. UPPINGTON, J. ANDREW, P. HOY & M. NAVRE. 1994. Submitted.
5. OKADA, Y., H. NAGASE & E. D. HARRIS, JR. 1986. J. Biol. Chem. **261:** 14245–14255.

Doxycycline May Inhibit Postmenopausal Bone Damage

Preliminary Observations

F. SCARPELLINI, L. SCARPELLINI, S. ANDREASSI, AND E. V. COSMI

II Department of Obstetrics/Gynecology
University "La Sapienza"
Rome, Italy

Postmenopausal calcium bone loss is one of the most common problems of public health. In recent years various therapeutic strategies have been proposed and tested to counteract this event. However, until now these treatments were principally based on stimulation of new bone synthesis and increased osteoblastic activity. Doxycycline, a semisynthetic tetracycline, has been found to block metalloproteinases, especially collagenase.[1] Moreover, tetracyclines, including doxycycline, have been found to inhibit osteoclast-mediated bone resorption in culture.[2] Accordingly, we carried out a preliminary study to test our hypothesis that doxycycline, by inhibiting bone cell matrix metalloproteinases and bone resorption, could be beneficial in early postmenopausal women.

MATERIAL AND METHODS

Thirty-seven early postmenopausal women (minimum 1 to a maximum of 4 years after cessation of menstruation), with calcium bone loss diagnosed by evaluation of fibronectin, osteocalcin (bone-GLA protein), estrone/estradiol ratio, alkaline phosphatase in plasma, urinary excretion of hydroxyproline, and Nordin index, underwent clinical assessment of the doxycycline treatment. Sixteen of them were treated with doxycycline, 100 mg per day for three months, whereas the others were managed with a placebo (vitamin A + vitamin E, Rovigon Roche) for the same period.

RESULTS

In the doxycycline-treated group of early postmenopausal women, the plasma values for osteocalcin and fibronectin were consistently and significantly ($p < 0.01$) decreased, whereas serum alkaline phosphatase (6 cases), hydroxyprolinuria (6 cases), and Nordin index (5 cases) were less consistently reduced. The untreated group showed a persistence of high levels of osteocalcin, fibronectin, Nordin index, and alkaline phosphatase. The estrone/estradiol (E_1/E_2) ratio and the levels of both steroids were unaffected by the treatment. These findings, expressed as a mean ± SD, are summarized in TABLE 1.

TABLE 1. Comparison between Doxycycline-treated and Placebo Group for Calcium Bone Loss Parameters

Measurements	Doxycycline-treated Group (n = 16)	Placebo Group (n = 21)	Statistical Significance
Osteocalcin, ng/mL	4.7 ± 0.376	8.9 ± 0.756	$p < 0.01$
Fibronectin, mg/mL	326 ± 29.4	489 ± 32.42	$p < 0.01$
Alkaline phosphatase, ng/mL	126 ± 11.87	177 ± 32.42	$p < 0.01$
Hydroxyprolinuria, mg/24 h	34.8 ± 7.1	55.9 ± 8.5	$p < 0.01$
Estradiol, pg/mL	38 ± 3.2	36 ± 4.5	N.S.
Estrone, pg/mL	56 ± 10.2	49 ± 12.4	N.S.
Estrone/estradiol ratio	1.29 ± 0.29	1.42 ± 0.34	N.S.

DISCUSSION

The findings suggest a possible action of doxycycline on bone mineral metabolism. The modifications after treatment involve only mineral-dependent metabolites and not hormonal parameters. This particular behavior may be explained by a reduction and, in some cases, an inhibition of the matrix metalloproteinases that are involved in postmenopausal bone remodeling. If the results are confirmed by bone mineral studies, this preliminary study suggests that doxycycline treatment may be helpful in ameliorating calcium bone loss, particularly during the early postmenopausal stage.

REFERENCES

1. Greenwald, R. A., S. A. Moak, N. S. Ramamurthy & L. M. Golub.1992. Tetracyclines suppress matrix metalloproteinase activity in adjuvant arthritis and in combination with flurbiprofen ameliorate bone damage. J. Rheumatol. **19(6):** 927.
2. Vernillo, A. T., N. S. Ramamurthy, L. M. Golub & B. R. Rifkin.1994. The non-antimicrobial properties of tetracycline for the treatment of periodontal disease. Curr. Opin. Periodontol. 111–118.

Effectiveness of a Cephalosporin, Cefotetan, in Inhibition of Uterine Matrix Metalloproteinases by a Non-Antibacterial Mechanism

F. SCARPELLINI, L. SCARPELLINI, M. SBRACIA, AND C. CURTO

II Department of Obstetrics / Gynecology
University "La Sapienza"
Rome, Italy

During pregnancy, especially in the final weeks, several biochemical modifications occur in the connective tissue of the human uterine cervix.[1,2] This process is controlled by matrix metalloproteinases (MMPs) which have a metal atom in the molecule and the property of cleaving proteins. Elastase and collagenase are the most important proteins in cervical ripening during labor. The possible clinical use of cefotetan, a third generation cephalosporin, in the prevention of preterm delivery was suggested by the recent observation of an *in vitro* inhibitory effect of elastase activity by cephalosporanic/penicillanic (β-lactamic)-formed drugs. This study describes the findings of a clinical trial of cefotetan in the treatment of preterm labor.

MATERIAL AND METHODS

Sixty-three women between 21 and 34 gestational weeks with PROM were evaluated. Twenty-seven were treated with oral ritodrine and intramuscular cefotetan at daily doses, respectively, of 30 mg and 2 gr, 15 were treated only with cefotetan at the same dosage per day, and 11 received no treatment. Each patient underwent plasmatic titration of CA-125, α_1-antitrypsin, fibronectin, urinary determination of hydroxyproline and mucopolysaccharides, ultrasonographic assessment especially of the cervical region twice a week and cardiotocography daily, as markers of the onset and staging of labor. Infection as a cause of preterm labor was excluded by bacterial culture of amniotic fluid.

TABLE 1. Clinical Evolution of Three Studied Groups

	Ritodrine + Cefotetan	Cefotetan	No Treatment
Patients (*n*)	27	15	11
Delivery before 34 wk	3	2	6
Delivery before 37 wk	5	7	4
Delivery at term	19	6	1

TABLE 2. Biochemical Labor Markers in Three Observed Groups (mean values)

	Ritodrine + Cefotetan	Cefotetan	No Treatment
CA-125 (U/mL)	21	29	51
α_1-Antitrypsin (ng/dL)	293	311	486
Fibronectin (ng/mL)	376	407	543
Hydroxyprolinuria (mg/24 h)	28	32	48
Mucopolysacchariduria (mg/24 h)	18	16	28

RESULTS

The clinical evolution of the three groups is shown in the TABLE 1. The untreated group showed substantially different mean levels of biochemical markers of labor from those of the other groups, as shown in TABLE 2.

DISCUSSION

During labor, levels of connective tissue metabolites are increased by enzymatic degradation by matrix metalloproteinases. This is probably the last step in a chain of events determining labor and cervical ripening. Cefotetan was as effective as ritodrine in inhibiting the progress of labor. The unsuspected tocolytic property of this antibiotic is probably due to inhibition of MMPs. β-Lactamic antibiotics have been shown to inhibit this family of enzymes.[3] More studies on the biochemical mechanisms of labor are needed to confirm our findings that MMPs are inhibited by this treatment.

REFERENCES

1. JUNQUEIRA, L. C. U., M. ZUGAIB, G. S. MONTES, O. M. S. TOLEDO, R. M. KRISTZAN & K. M. SHIGIHARA. 1980. Morphologic and histochemical evidence for the occurrence of collagenolysis and for the role of neutrophilic polymorphonuclear leukocytes during cervical dilatation. Am. J. Obstet. Gynecol. **138:** 273.
2. SCARPELLINI, L., R. PAESANO, M. P. INNACOLI & M. CASTELLANO. 1984. Aspetti attuali ed esperienze cliniche sulla capacitazione cervicale al travaglio. Pat. Clin. Ost. Gin. **12:** 217.
3. KNIGHT, W. B., B. G. GREEN, R. M. CHABIN *et al.* 1992. Specificity, stability, and potency of monocyclic β-lactam inhibitors of human leukocyte elastase. Biochemistry **31:** 8160.

Subject Index

Index of Contributors